BLOOD–BRAIN BARRIER IN DRUG DISCOVERY

BLOOD–BRAIN BARRIER IN DRUG DISCOVERY

Optimizing Brain Exposure of CNS Drugs and Minimizing Brain Side Effects for Peripheral Drugs

Edited by

LI DI
Pharmacokinetics, Dynamics and Metabolism, Pfizer, Inc.
Groton, Connecticut, USA

EDWARD H. KERNS
Laytonsville, MD, USA

WILEY

Published by John Wiley & Sons, Inc., Hoboken, New Jersey
Published simultaneously in Canada

Library of Congress Cataloging-in-Publication Data:

Blood-brain barrier in drug discovery : optimizing brain exposure of CNS drugs and minimizing brain side effects for peripheral drugs / edited by Li Di, Edward H. Kerns.
p. ; cm.
Includes bibliographical references and index.
ISBN 978-1-118-78835-6 (hardback)
I. Di, Li, editor. II. Kerns, Edward Harvel, editor.
[DNLM: 1. Blood-Brain Barrier–metabolism. 2. Central Nervous System Agents–pharmacokinetics. 3. Blood-Brain Barrier–drug effects. 4. Central Nervous System Diseases–drug therapy. 5. Drug Discovery–methods. 6. Drug-Related Side Effects and Adverse Reactions–prevention & control. WL 200]
RM315
615.7′8–dc23

2014040935

Cover image: iStock © (pills) Martin Wimmer / (brain) Henrik5000 / (background) derrrek

Printed in the United States of America

10 9 8 7 6 5 4 3 2 1

CONTENTS

CONTRIBUTORS

N. Joan Abbott, Institute of Pharmaceutical Science, King's College London, London, UK

Alex Avdeef, *in-ADME* Research, New York, USA

Sharan Bagal, Worldwide Medicinal Chemistry, Pfizer Neusentis, Great Abington, Cambridge, UK

Benny Bang-Andersen, H. Lundbeck A/S, Department of Medicinal Chemistry, Valby, Denmark

Christoffer Bundgaard, H. Lundbeck A/S, Department of Discovery DMPK, Valby, Denmark

Peter Bungay, Pfizer Neusentis, Great Abington, Cambridge, UK

Cheng Chang, Pharmacokinetics, Dynamics and Metabolism, Pfizer Inc., Groton, CT, USA

Cuiping Chen, Depomed Inc., Newark, CA, USA

Hongming Chen, Computational Chemistry, Chemistry Innovation Center, AstraZeneca R&D, Mölndal, Sweden

Liangfu Chen, GlaxoSmithKline R&D, Platform Technology and Science, Pennsylvania, USA

Ziqiang Cheng, Drug Metabolism and Pharmacokinetics, AstraZeneca Innovation Center China, Shanghai, China

Andrew Crowe, School of Pharmacy and CHIRI-Biosciences, Curtin University, Perth, WA, Australia

Mária A. Deli, Institute of Biophysics, Biological Research Centre, Hungarian Academy of Sciences, Szeged, Hungary

Li Di, Pharmacokinetics, Dynamics and Metabolism, Pfizer Inc., Groton, CT, USA

Ola Engkvist, Computational Chemistry, Chemistry Innovation Center, AstraZeneca R&D, Mölndal, Sweden

Jim Ferry, Eisai Neuroscience Product Creation Unit, Woodcliff Lake, NJ, USA

Johan Gabrielsson, Department of Biomedical Sciences and Veterinary Public Health, Division of Pharmacology and Toxicology, Swedish University of Agricultural Sciences, Uppsala, Sweden

Huile Gao, Key Laboratory of Drug Targeting and Drug Delivery Systems, West China School of Pharmacy, Sichuan University, Chengdu, China

Sookhee N. Ha, Department of Structural Chemistry, Merck Research Laboratories, Kenilworth, NJ, USA

Takahisa Hanada, Global Biopharmacology, Neuroscience & General Medicine Product Creation System, Eisai Co Ltd., Tsukuba, Ibaraki, Japan; and Center for Tsukuba Advanced Research Alliance, Graduate School of Life and Environmental Sciences, University of Tsukuba, Tsukuba, Ibaraki, Japan

Shigeki Hibi, Global Biopharmacology, Neuroscience & General Medicine Product Creation System, Eisai Co Ltd., Tsukuba, Ibaraki, Japan

Stephan Hjorth, Department of Bioscience, AstraZeneca R&D, Pepparedsleden, Mölndal, Sweden

Jerome H. Hochman, Department of PPDM, Merck Research Laboratories, West Point, PA, USA

Kevin J. Hodgetts, Laboratory for Drug Discovery in Neurodegeneration, Harvard NeuroDiscovery Center, Cambridge, MA, USA; and Brigham and Women's Hospital, Cambridge, MA, USA

Lisbeth Illum, IDentity, The Park, Nottingham, UK

Phil Jeffrey, Rare Disease Research Unit, Pfizer Ltd, Cambridge, UK

Xinguo Jiang, Key Laboratory of Smart Drug Delivery (Fudan University), Ministry of Education; and Department of Pharmaceutics Sciences, School of Pharmacy, Fudan University, Shanghai, China

Edward H. Kerns, Laytonsville, MD, USA

William Kielbasa, Eli Lilly and Company, Indianapolis, IN, USA

Elizabeth C.M. de Lange, Division of Pharmacology, Leiden Academic Center of Drug Research, Gorlaeus Laboratories, Leiden University, Leiden, The Netherlands

Antonio Laurenza, Eisai Neuroscience Product Creation Unit, Woodcliff Lake, NJ, USA

Qian Liu, Platform Technology and Science, GlaxoSmithKline R&D, Shanghai, China

Xingrong Liu, Genentech Inc., South San Francisco, CA, USA

Winfried Neuhaus, Department of Pharmaceutical Chemistry, University of Vienna, Vienna, Austria

Jean-Marie Nicolas, UCB Pharma S.A., Chemin du Foriest, Braine-l'Alleud, Belgium

William M. Pardridge, ArmaGen Technologies, Inc., Calabasas, CA, USA

Alan L. Pehrson, Lundbeck Research USA, Department of Neuroscience, Paramus, NJ, USA

Lambertus A. Peletier, Mathematical Institute, Leiden University, Leiden, the Netherlands

Andy Pike, Pfizer Neusentis, Great Abington, Cambridge, UK

Zoran Rankovic, Eli Lilly and Company, Indianapolis, IN, USA

Andreas Reichel, DMPK, Global Drug Discovery, Bayer Healthcare, Berlin, Germany

Jasminder Sahi, GlaxoSmithKline R&D, Platform Technology and Science, Shanghai, China

Connie Sánchez, Lundbeck Research USA, Department of Neuroscience, Paramus, NJ, USA

Andrew Satlin, Eisai Neuroscience Product Creation Unit, Woodcliff Lake, NJ, USA

Robert P. Sheridan, Department of Structural Chemistry, Merck Research Laboratories, Rahway, NJ, USA

Eric V. Shusta, Department of Chemical and Biological Engineering, University of Wisconsin–Madison, Madison, WI, USA

Robert E. Stratford Jr., Xavier University of Louisiana, New Orleans, LA, USA

Scott Summerfield, GlaxoSmithKline R&D, Platform Technology and Science, David Jack Centre for R&D, Hertfordshire, UK

Anabella Villalobos, Pfizer Worldwide Research and Development, Neuroscience Medicinal Chemistry, Cambridge, MA, USA

Hannah K. Wilson, Department of Chemical and Biological Engineering, University of Wisconsin–Madison, Madison, WI, USA

Susanne Winiwarter, Computational ADME/Safety, Drug Safety and Metabolism, AstraZeneca R&D, Mölndal, Sweden

Haichen Yang, Eisai Neuroscience Product Creation Unit, Woodcliff Lake, NJ, USA

Lei Zhang, Pfizer Worldwide Research and Development, Neuroscience Medicinal Chemistry, Cambridge, MA, USA

PREFACE

Innovation of treatments for human disease is an engaging endeavor that inspires the intellects, skills, risks and ethics of medical caregivers, researchers, business, and government to prolong and improve the quality of human life. In particular, central nervous system (CNS) disorders are a major area of undertreated medical need. Many researchers aspire to discover successful new treatments for these devastating CNS diseases. Pharmaceuticals are a major contributor to CNS and peripheral disease treatment. They emerge from a long and complex process of drug discovery and development. The science and process of drug discovery and development is an ever-evolving and challenging venture that is attaining steady improvements in understanding the underlying mechanisms affecting pharmaceuticals and innovating new approaches that lead to enhanced quality of patient care.

Past improvements in knowledge regarding the blood–brain barrier (BBB) have contributed to the development of CNS drugs. In recent years, the depth and breadth of knowledge of the BBB and drug interactions in the brain have accelerated. They are yielding innovative drugs that improve on current CNS disease treatments and, excitingly, treat previously intractable CNS diseases. In the broader pharmaceutical field, BBB knowledge is reducing unwanted CNS side effects of drugs that treat peripheral diseases. Improvements in both CNS and peripheral drugs, based on this knowledge, are highly beneficial for patients.

The chapters in this book were written by researchers that are actively involved in increasing the understanding of the BBB and drug interactions in the brain and applying this to more quickly discover and develop better drugs. All of us who work toward better human disease treatment can really appreciate the contributions of these authors in sharing their knowledge and insights. Furthermore, they offer highly valuable guidance for researchers for successful drug discovery and development.

Early chapters provide an overview of the unique pharmacokinetics of brain exposure. The fundamentals of drug binding to CNS tissue and plasma are described, with emphasis on the primary role of free drug concentration in determining *in vivo* efficacy. Free drug concepts have recently been widely accepted in the field and are crucial for researchers to understand and apply in practice.

The extraordinary mechanisms affecting drug permeation through the BBB are discussed in four chapters (Chapters 5–8). These describe the BBB tight junctions that limit access of some drugs to the brain, the constraints on BBB transcellular passive diffusion, the efflux transport that reduces brain exposure, and the uptake transporters that offer intriguing opportunities for enhancing brain penetration.

Furthermore, ground-breaking research is discussed, which uses BBB receptors to enable uptake of biological molecule constructs. These biologics would otherwise not be able to cross the BBB and have advantages for treatment of certain CNS diseases not treatable by small molecule drugs.

Predictions and measured data are important in discovering new drugs. These indicate the behavior of drug candidates at various barriers that limit CNS exposure. A series of chapters discuss state-of-the-art approaches for *in silico* prediction, *in vitro* data measurement of specific barriers, and *in vivo* methods for measuring the free drug concentration and imaging compound locations in the CNS tissues. Recent advancements in physiologically based pharmacokinetic and pharmacokinetic/pharmacodynamic (PK/PD) tools are effective new approaches to predict PK and efficacy in preclinical and clinical space.

Medicinal chemists and project leaders will benefit from the CNS drug design strategies that use current insights on the BBB and CNS barriers to achieve enhanced CNS drug exposure. Case studies examine how integration of the data and design strategies advanced successful new drugs to the market. Nanotechnology and nasal CNS drug delivery techniques are also discussed as other alternative approaches to enhance brain access.

The editors greatly thank the individual chapter authors for kindly sharing their knowledge, strategies, and experience. It was a great pleasure to collaborate with them on development of this book. We admire the outstanding and heart-felt work they contributed so that all drug researchers could benefit and achieve increased success in development of the drugs of the future. We wish for you success in achieving your goals for new disease treatments for the benefit of the patients.

June, 2014

Li Di
Edward H. Kerns

1

INTRODUCTION AND OVERVIEW

LI DI[1] AND EDWARD H. KERNS[2]
[1]*Pharmacokinetics, Dynamics and Metabolism, Pfizer Inc., Groton, CT, USA*
[2]*Laytonsville, MD, USA*

Brain exposure can affect drug development success for all diseases. For neuroscience therapeutics, a leading area of pharmaceutical research, development, and product portfolios in pharmaceutical companies and research institutions, insufficient brain exposure leaves many central nervous system (CNS) diseases untreated or without optimum drugs, despite the vast resources applied to the problem. Researchers working to treat CNS diseases were stymied by the blood–brain barrier (BBB), but, in recent years, experience led to improved drug exposure at brain targets. Conversely, researchers working on peripheral diseases encountered CNS side effects owing to brain exposure at unintended CNS targets, but they are increasingly successful at reducing brain exposure. These advances on brain exposure came as pharmaceutical science uncovered the intricacies of drug molecule interactions at the BBB and within brain tissue. Newly discovered interactions provide an opportunity to overcome previous project disappointments, understand previously unexplained observations, and enable new tools for successful drug development.

This book comprises the contributions of experts regarding the complex interactions encountered by drug molecules that affect brain exposure and their successful solution in drug discovery, development, and clinical studies, including the following:

- Complexities of brain physiology and anatomy
- Designing CNS drug candidates to reduce transporter BBB efflux or increase BBB uptake

Blood–Brain Barrier in Drug Discovery: Optimizing Brain Exposure of CNS Drugs and Minimizing Brain Side Effects for Peripheral Drugs, First Edition. Edited by Li Di and Edward H. Kerns.

- Designing peripheral drugs to increase BBB efflux
- Focus on brain free drug concentration for efficacy
- Constructing novel biologics to deliver therapeutic molecules to the CNS
- Building pharmacokinetic–pharmacodynamic (PK/PD) and physiologically-based pharmacokinetic (PBPK) models for CNS therapy
- Projecting *in vivo* CNS exposure
- Nanotechnology and nasal dosing for CNS delivery
- In silico, *in vitro*, and *in vivo* methods of predicting and measuring CNS barriers, exposure, and free drug concentration
- Imaging for CNS therapy
- Case studies of successful recent drug product advances in brain delivery enhancement or reduction

RESTRICTED BRAIN EXPOSURE REDUCES CNS DRUG EFFICACY

A primary cause of the disappointment in developing CNS disease treatments is that the brain is a difficult organ for drug therapy. In past years, a high percentage of promising CNS drug candidates have failed. A major cause of this failure is the restricted access of many drug candidates circulating in the blood to penetrate into the brain owing to the BBB. Chapters 2 and 4 discuss the physiology of the BBB and differences among species and disease states. For most organs, drug molecules freely move between the blood and tissue via open junctions between capillary cells, but the BBB presents greater restrictions via tight junctions that reduce drug molecule access to brain tissue. Thus, molecules that do not have facile passive transcellular diffusion (e.g., acids, biologics) are restricted. In addition, efflux transporters (e.g., Pgp, BCRP), actively pump the molecules of some compounds out of the brain. These barriers to BBB permeation and the general characteristics of compounds that are efflux substrates are detailed in Chapters 5 and 6. These barriers effectively reduce the concentration, and therefore the efficacy, of some potentially therapeutic drug molecules to brain cells.

Another component of brain exposure restriction is binding of drug molecules to blood and brain tissue components. This restricts the free drug concentration that is available to bind to the therapeutic target protein molecules. In past years, the concentration of drug molecules that are available to bind to the brain target was assumed to be the total concentration measured in the brain tissue. However, in recent years, there has been a major shift in acceptance and application of the Free Drug Hypothesis, which states that only the unbound drug molecules are available to bind to the target to produce efficacy. Binding varies with the structure and physicochemical properties of each compound. This recognition has solved many previously unexplained failures in translation from *in vitro* activity to *in vivo* efficacy. The primary role of *free drug concentration* in determining *in vivo* efficacy is now being widely applied to CNS research and is reviewed in Chapter 3.

PERMITTED CNS ACCESS INCREASES SIDE EFFECTS OF PERIPHERAL DRUGS

Many drug candidates for peripheral therapeutic targets have minimal restrictions in penetrating the BBB and affecting brain targets. For example, they may have high passive diffusion through the BBB endothelial cells and not be efflux substrates. These drugs penetrate into the brain and may interact with CNS targets to cause difficult side effects for patients. Such effects lead to research project cancelation, regulatory rejection, drug product use restrictions, reduction of patient administration compliance, and long-term toxicities. For these reasons, drug researchers and developers need to investigate whether a new drug candidate causes unfavorable CNS effects *in vivo*. Chapters 20 and 21 explain this issue for peripheral drugs and how it may be overcome.

A NEW GENERATION OF CNS EXPOSURE TOOLS

As the interactions affecting brain exposure are elucidated, in silico, *in vitro*, and *in vivo* methods for these interactions are developed. In addition, these interactions are included in methods for *in vivo* projection. Such methods allow drug researchers to screen for potential problems, measure specific interactions (e.g., Pgp efflux), and quantitate how they affect drug tissue concentrations *in vivo*. These tools provide reliable information for lead selection and optimization to benefit drug research projects throughout their progress. Chapter 9 discusses the development and state of the art of in silico BBB predictions. BBB permeability is often predicted using *in vitro* artificial and cell membrane assays (Chapters 10 and 11). Another component of brain exposure assessment is *in vitro* assays for brain binding, as discussed in Chapter 12. This information is typically used in combination with *in vivo* brain exposure studies (Chapter 13) to determine the free drug concentration in brain tissue. Direct measurement of free drug concentration using microdialysis is reviewed in Chapter 16. Another important advance in the field of brain exposure is the replacement of the Log BB and B/P parameters by the more valuable $K_{p,uu}$, the free drug distribution coefficient between brain and plasma, as discussed in Chapters 2, 3, 4, and 18. There is an increasing sophistication in PBPK modeling for the BBB (Chapter 14) and PK/PD model building (Chapter 15) for CNS drug candidates, which improve interpretations of biological efficacy, as well as projections for higher animals and human clinical studies. Recent advances in imaging techniques for CNS discovery research are discussed in Chapter 17.

Drug design advancements for brain exposure enhancement have taken advantage of the growing knowledge of drug interactions at the BBB. Small molecule design to optimize exposure (Chapter 18) and case study examples (Chapter 19) report successful strategies in CNS drug discovery. Concepts for the enhancement of brain exposure by designing drugs as substrates for BBB uptake transporters are advancing and are reviewed in Chapter 7. Biological drugs (antibodies, proteins) of higher-molecular weight generally do not pass the BBB. However, recent success in delivering biologics was achieved by making constructs that contain the biological drug and a group that binds to a BBB receptor that promotes transport across the BBB. Chapter 8 explains the exciting progress in this promising field.

Researchers working on peripheral drugs will benefit from insights into the design suggestions for minimizing brain exposure in Chapter 20 and the successful case studies on nonsedative antihistamines in Chapter 21 that have efficacy at peripheral targets but are restricted by the BBB from producing effects at the same or closely related receptors in the brain.

For more compounds that are very recalcitrant to CNS exposure, researchers are developing new concepts for CNS delivery. Technologies using nanotechnology have the possibility to enhance delivery across the BBB (Chapter 24). Concepts and evidence for CNS drug delivery using the nasal route is also reviewed in Chapter 25.

CASE STUDIES OF DRUG DEVELOPMENT SUCCESSES

We often find guidance from successful case study examples. Thus, colleagues have kindly provided successful CNS exposure case studies for fycompa, an AMPA receptor antagonist (Chapter 22) and for vortioxetine, a serotonin modulator and simulator (Chapter 23). The inspiration and enlightenment of these experienced examples provide encouragement and direction for our research projects.

CONCLUSION

This book was prepared with the purpose of benefiting drug researchers in the following areas:

- Fundamental knowledge about the BBB and drug binding
- Implications of these restrictions for brain pharmacokinetics (PK) and pharmacodynamics (PD)
- Drug structure design elements that overcome BBB barriers for CNS drugs
- Drug design principles that enhance these brain exposure barriers for peripheral drugs
- Methods for assessing compound restrictions at the BBB
- Case studies from CNS drug discovery

Valuable perspectives on the future of BBB research and CNS drug development are provided by distinguished BBB researcher Dr. Joan Abbott in Chapter 26. The sharing of valuable insights about brain exposure by the chapter authors in this volume is intended to advance the fundamental knowledge about the BBB and brain exposure throughout all areas of drug development. Practical real-world information and examples are emphasized for the purpose of developing therapies for underserved diseases and of developing improved drugs.

PART 1

PHARMACOKINETICS OF BRAIN EXPOSURE

2

PHARMACOKINETICS OF CNS PENETRATION

ANDREAS REICHEL

DMPK, Global Drug Discovery, Bayer Healthcare, Berlin, Germany

INTRODUCTION

Drugs are likely to exert their pharmacological effects only if they have a proper chance to engage with their molecular targets at the site of action in the body. This is true for all drug targets, including those that reside within the central nervous system (CNS). A prerequisite for drug target engagement, that is, binding of a drug to its molecular target protein, is the exposure of the target at concentrations in excess of the pharmacological potency of the compound for a sufficient period of time. Adequate CNS exposure of a drug at the site of its pharmacological target is, therefore, paramount for a drug to be able to elicit CNS activity [1]. The concept of active target site exposure has now become a central tenet for the pharmacokinetic (PK) optimization in drug discovery projects focusing on optimizing unbound rather than total drug concentrations [2–4]. Without sufficient exposure of the drug target, the likelihood is very low that a drug will be able to express target-mediated pharmacology and, ultimately, the desired effects on the course of the disease.

Both target exposure and target engagement have been identified as two out of three "pillars of success" of drug discovery programs during a retrospective analysis of about 40 clinical Phase II programs running at Pfizer between 2005 and 2009 [5]. The third pillar of success is the demonstration of the relevance of the expression of the pharmacology for the intended therapeutic intervention (Fig. 2.1). This holds particularly true for CNS drug discovery and development, which are suffering from

Blood–Brain Barrier in Drug Discovery: Optimizing Brain Exposure of CNS Drugs and Minimizing Brain Side Effects for Peripheral Drugs, First Edition. Edited by Li Di and Edward H. Kerns.

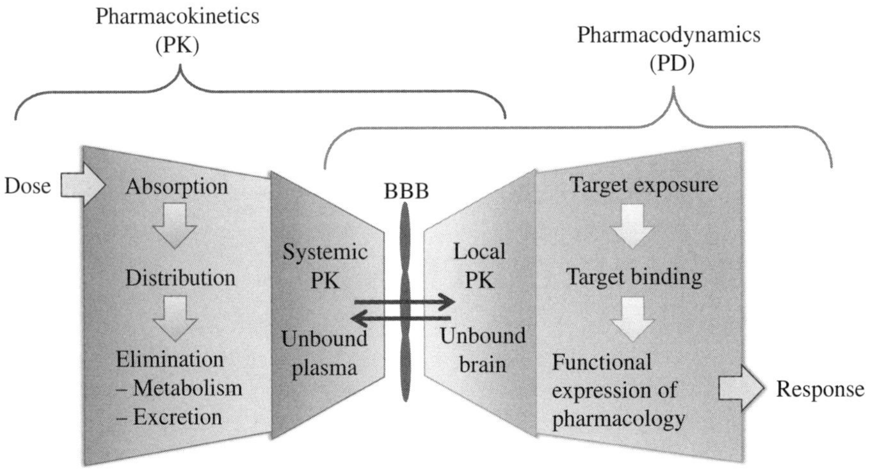

FIGURE 2.1 Schematic presentation of the key processes and the link between pharmacokinetics and pharmacodynamics of CNS drugs, which ultimately translate a drug dose into a drug response.

dauntingly low clinical trial success rate and the lack of a clear understanding of underlying reasons for the failures [6, 7]. Very often it remains unclear whether a failure was due to the pharmacological target hypothesis being wrong or the target exposure being insufficient to exert the desired pharmacological effects. Some authors [8–10] suggest that not only will the development of new CNS medications benefit from a better understanding of target exposure and engagement, but applying these principals may also help redefine dose and dosing regimens of already existing "old" medications, for example, many classical antipsychotics which have never undergone rational dose-finding studies. The authors suggest applying positron emission tomography (PET) occupancy studies in patients as a basis to readjust the currently recommended doses in order to make their use more efficient and safer compared to the traditional doses and, hopefully, also improve their often poor response rates within the patient population. PET studies are ideal as they allow addressing several key questions directly in patients [11]: Does the drug reach the target site? Does the drug interact with the desired target? Is the concentration of the drug at the target site sufficient to elicit an effect? What is the temporal nature of such an interaction? What is the relationship between the target site concentration and the administered dose and plasma concentration?

Although PET studies, which can also be carried out in animals, are able to answer many of these key questions directly, PET technology is not applicable in most CNS drug discovery projects due to the absence of suitable PET tracers for novel targets. Therefore, and due to the inherent difficulty to directly measure the active site concentrations at the CNS target, alternative methodologies and surrogate approaches that are compatible with modern-day drug discovery and development have been developed [12–14].

This chapter summarises the key processes that control the drug concentrations at the site of the CNS target, in particular the pharmacokinetics of CNS penetration and distribution.

CNS PENETRATION

Unlike most other organs in the mammalian body, the brain is separated from the blood circulation by the existence of physiological barriers. In order to get access to the brain tissue, a drug needs to be able to cross these barriers.

Barriers within the Brain

There are two important barriers between the CNS and the blood circulation: the blood–brain barrier (BBB) and the blood–cerebrospinal fluid (CSF) barrier (BCSFB), which are introduced here only very briefly. Although the BBB is highly complex and formed by multiple cell types (Fig. 2.3, left), the gatekeeper function is essentially a result of the endothelial cells lining the brain capillaries as they are very tightly sealed together by an intricate network of tight junctions [15]. Since these effectively prevent paracellular transport between the cells, movement of any material can only occur through the endothelial cells, thereby allowing the brain to control all traffic including that of ions, solutes, nutrients, hormones, larger molecules, or even cells (e.g., immune cells).

Besides the BBB, which separates the blood circulation from the brain's parenchyma, there is also a barrier between the blood circulation and the CSF. This barrier, the BCSFB, which is located at the level of the choroid plexus, differs from the BBB in that its barrier function originates from the tight epithelium lining, the choroid plexus of the ventricles of the brain, which are supplied by leaky capillaries [16].

From a PK point of view, the following anatomical and physiological parameters of the BBB are of relevance: brain capillary length and volume in humans are about 650 km and 1 ml, respectively, with the area of the luminal capillary surface approximately 12 m^2, which is equivalent to 100–240 cm^2/g brain depending on the brain region [15, 17]. The thickness of the BBB is between 200 and 500 nm. The luminal diameter of brain capillaries is about 4 μm in rats and 7 μm in humans, with a mean distance between two capillaries of about 40 μm and the transit time of blood of about 5 s. The capillary volume of 11 μl/g brain is very low, that is, less than 1% of the brain. In contrast, the compartment of the brain interstitial fluid (ISF) amounts to about 20% of the brain parenchyma [18, 19]. In rats, ISF flows with a bulk flow rate of approximately 0.15–0.29 μl/min/g toward the CSF [20]. The volume of CSF is approximately 250 μl in rats and 160 ml in humans, with the rate of CSF secretion being approximately 2.1 and 350 μl/min, respectively [21, 22].

Because the ratio of the surface areas between the BBB and the BCSFB is in the range of 5000:1, and the density of the capillaries within the brain parenchyma is so high that virtually every neuron can be supplied by its own capillary, the BBB

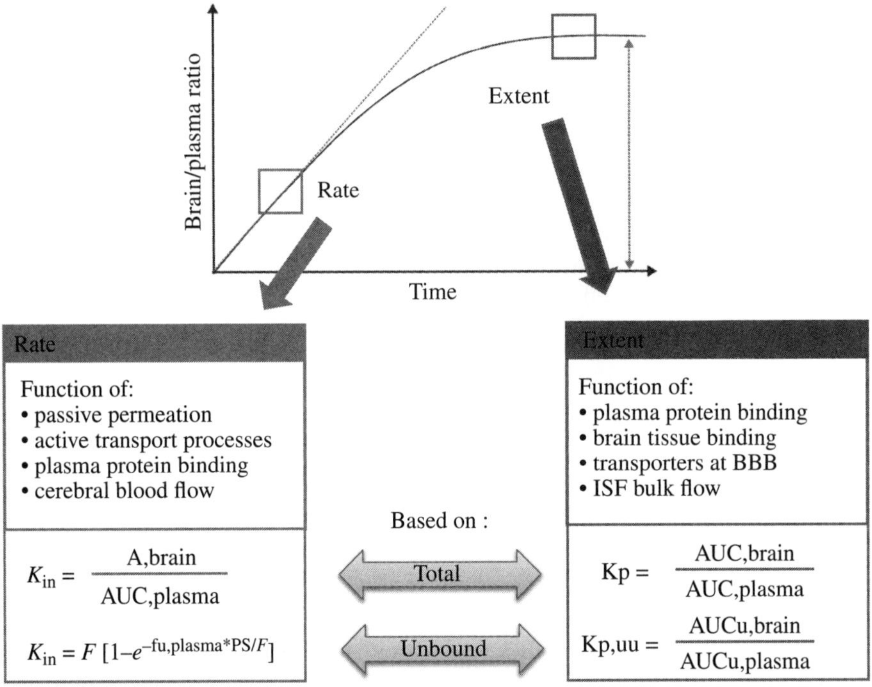

FIGURE 2.2 Schematic illustration of rate and extent as the two independent properties of CNS penetration. Shown also are PK parameters that are used to describe rate and extent of CNS penetration and the factors which are the most important determinants.

is generally considered to play the major role in the transfer of CNS drugs to the brain [17, 19, 23].

Understanding the BBB is therefore an essential element for optimizing CNS penetration in drug discovery. The BBB impacts both the rate and the extent of CNS penetration (Fig. 2.2), which is discussed in more detail in the following sections.

Rate of CNS Penetration

The rate of CNS penetration relates to the speed at which a compound enters the CNS, independent of how much drug will enter the brain or to the degree of CNS penetration. The rate of CNS penetration is controlled by two factors: the cerebral blood flow (CBF), which controls the amount of drug delivered to the brain, and the permeability of the compound across the BBB. According to the classical principles of PK, either of these two factors can become the rate-limiting step in the process of tissue penetration [24].

Cerebral Blood Flow (CBF, F) In rats, blood flow through brain capillaries is about 0.5–2 ml/min/g brain, which varies between brain regions, neuronal activity, and CNS diseases [18, 25]. In humans, CBF is slower than in rats, with values of

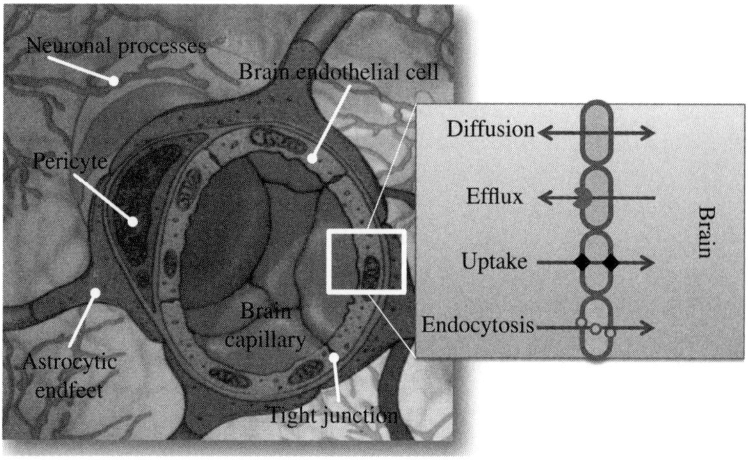

FIGURE 2.3 Schematic illustration of the blood–brain barrier and typical cell types constituting it (left) and principal pathways available to drugs, in order to gain access to the brain parenchyma: passive diffusion, which may be restricted by active efflux, carrier-mediated uptake via transporters expressed on the brain endothelium, and endocytosis, which may be mediated by specific receptors on the luminal endothelial cell surface or less specifically triggered by adsorption to the endothelial cell membrane.

0.15 and 0.6 ml/min/g brain for white and gray matter, respectively [26]. The CBF delivers the maximum amount of drug the brain is exposed to and, thus, constitutes the upper limit of the rate of brain penetration *in vivo*. For drugs whose rate of CNS penetration is perfusion-limited, changes in CBF will thus affect their CNS penetration, for example, under the influence of anesthetics, which often decrease CBF [27].

Permeability (P) The permeability relates to the speed of crossing the BBB of a drug and depends on the membrane properties of the BBB and the physicochemical properties of the crossing compound. There are several mechanisms by which a compound can pass through the BBB (Fig. 2.3): passive transcellular diffusion, which may be limited by efflux pumps, facilitated by carrier-mediated uptake, and adsorptive or receptor-mediated transcytosis, which is more relevant for large molecules [28]. Paracellular diffusion, which is an important mechanism for drug penetration of peripheral tissues, is virtually nonexistent in the CNS, due to the complex network of tight junctions between the brain endothelial cells.

The BBB permeability can be examined *in vivo* and *in vitro*. Since permeability (*P*) and surface area (*S*) cannot be easily distinguished *in vivo*, the permeability surface (PS) area product is most often given readout [29, 30]. The PS product is equivalent to the net influx clearance (CL_{in}) and both are measured in units of flow: µl/min/g brain [31]. PS products may span a range of about 10,000-fold [32, 33] and cannot easily be compared across studies or with the CBF, as such, since the results depend on the exact conditions of the *in situ* brain perfusion method used, in particular, the rate and the duration of the perfusion and the composition of the perfusion fluid (e.g., the amount of plasma protein). The *in situ* brain perfusion technique requires high technical skills, is very labor-intensive, and, hence, not suitable for routine drug discovery screening. An alternative *in vivo* method to determine the rate of CNS penetration is to determine the amount of compound in the brain after oral or systemic administration as K_{in} value, which relates the amount of compound in the brain (homogenate) at time *t* (A,brain(*t*)) to the plasma exposure up to this time point (AUC,plasma(0 – *t*)). To be more exact, A,brain(*t*) should be corrected for the amount of drug remaining in the cerebral vasculature at the end of the experiment [25, 29]. The experimental setup to generate K_{in} data follows that of regular *in vivo* PK studies, making K_{in} a more popular *in vivo* estimate than the PS product.

$$K_{in} = \frac{\text{A,brain}(t)}{\text{AUC,plasma}(0-t)} \tag{2.1}$$

The Renkin–Crone equation relates K_{in} and PS based on the basic principles of capillary flow, with *F* being the flow in the system in question, that is, either CBF or the rate of perfusion in the experiment [34]. K_{in} may be correct for the unbound fraction in plasma or the perfusate (see Fig. 2.2) [25]. The classical equation without protein-binding correction is

$$K_{in} = F^*(1 - e^{PS/F}) \tag{2.2}$$

The capillary flow model underlying this equation ensures that the rate of CNS penetration cannot be higher than the CBF, even if the intrinsic permeability is very high. Hence, the upper limit of K_{in} is the CBF (for high-permeability compounds) and the lower limit is PS (for low-permeability compounds). It has been estimated that for a drug to be permeability-limited the PS product has to be 10% of the CBF or less, resulting in a tissue extraction of less than 20% compared to blood. PS products in the range of, or greater than, the CBF make the tissue penetration of the drug perfusion-limited. The available data on PS products suggest that CNS drugs typically do not belong in the category of permeability-limited compounds [22]. This may well be a result of the availability of high-throughput *in vitro* permeability models in drug discovery and the successful use of in silico tools to predict PS products based on the physicochemical properties of the drug. Very good results have been made with the following equation, which predicts the passive PS product expressed as log PS [35, 36]:

$$\log \text{PS} = 0.123 * \log D - 0.00656 * \text{TPSA} + 0.0588 * \text{Vbase} - 1.76 \tag{2.3}$$

where log *D* is the partition coefficient in octanol/water at pH 7.4, TPSA is the topological van der Waals polar surface area, and Vbase is the van der Waals surface area of the basic atoms.

Today, *in vitro* assays have become the method of choice to assess permeability as they reflect both the passive diffusion and the transporter-mediated component (in particular, efflux). Typically, the rate of transport is measured across a tight monolayer of cells, which resemble most of the critical barrier properties of the BBB [23, 28, 37]. The permeability is expressed as apparent permeability coefficient (P_{app}):

$$P_{app} = \frac{d\mathrm{Cr}}{dt} * \frac{\mathrm{Vr}}{A * C0} \qquad (2.4)$$

where *d*Cr/*dt* is the slope of the cumulative concentration in the receiver compartment versus time, Vr is the volume of the receiver compartment, *A* is the surface area of the monolayer, and *C*0 is the initial concentration in the donor compartment. The assay is often run in both directions in order to assess the susceptibility of the test compound toward drug efflux using the efflux ratio (ER):

$$\mathrm{ER} = \frac{P_{app}(B\_A)}{P_{app}(A\_B)} \qquad (2.5)$$

While there is still no one *in vitro* BBB model available that resembles all key aspects of the BBB, MDCK-MDR1 cells have become the most widely used cell line to determine the *in vitro* permeability in CNS drug discovery [28, 38–42]. Typically, P_{app} values at approximately, or greater than, 100 nm/s are taken as evidence for high permeability, with ER values ideally around 1, or below 2–3.

While in the past there has been too much emphasis on optimizing permeability, that is, the rate of CNS penetration, it is now being accepted that, in particular for chronic treatment of CNS disorders, the extent is the more important parameter to be examined. Although low permeability may delay the time to equilibrium, it will not affect the level of the drug equilibrium between blood and brain.

Extent of CNS Penetration

While rate is an important parameter to describe CNS penetration, it does not determine the extent (i.e., degree) to which a compound will enter brain tissue (Fig. 2.2). This has sometimes been confused in the past, leading to overemphasis of *in vitro* permeability assays in drug discovery programs, whose actual purpose was to increase the extent of brain penetration. It was the seminal review by Hammarlund-Udenaes that made very clear the distinction between rate and extent of CNS penetration [18]. Another source of confusion was the misleading assessment of CNS penetration based on the ratio of total brain/plasma concentrations [13, 43].

Total Brain/Plasma Ratio (Kp) Traditionally, a typical *in vivo* study to assess brain penetration involved the measurement of brain and plasma sample concentrations at 3–4 time points after ip, sc, iv, or oral administration to rodents. At selected time points, plasma samples were drawn and brain tissue was removed and subsequently homogenized for quantification by liquid chromatography–mass spectrometry (LCMS) analysis [14, 44, 45]. The method was amenable to cassette dosing, thereby allowing significant reduction of the number of animals to be used [46]. The extent of brain penetration was expressed as follows:

$$\mathrm{Kp} = \frac{\mathrm{AUCtot,brain}}{\mathrm{AUCtot,plasma}} \tag{2.6}$$

Whenever a project team went on to improve brain penetration by increasing Kp, however, they ran into the problem that compounds with higher Kps, despite an often improved potency, did not result in better efficacy in animal models [43, 47]. On examining the brain's unbound concentrations it was found that increasing total concentrations does not necessarily lead to higher unbound concentrations, which most closely relate to the active site concentrations [1, 14, 48–51]. Since the brain has a relatively high lipid content [52], increasing Kp was often a simple consequence of increasing the lipophilicity of the drug, thereby steering project teams into a drug lipidization trap [10, 13, 43, 49]. Computational methods aimed at predicting Kp are therefore of highly questionable value as guiding tools for CNS drug discovery.

The key caveat of Kp is its composite nature of three independent factors: nonspecific drug binding to brain tissue, nonspecific drug binding to plasma proteins, and specific drug transport across the BBB [18]. For CNS drugs which often are very lipophilic, Kp is dominated by nonspecific drug binding, masking the transport properties of the drug.

Unbound Brain/Plasma Ratio (Kp,uu) Correction for nonspecific binding both to brain tissue and plasma proteins [43, 53, 54] is therefore an essential element which led to the concept of Kp,uu [18]. Kp,uu is not confounded by nonspecific binding of the drug and is thus a parameter that directly reflects the transport equilibrium across the BBB. Kp,uu is calculated from Kp, the fraction unbound in plasma (fu,plasma) and the fraction unbound in brain tissue (fu,brain):

$$\mathrm{Kp,uu} = \mathrm{Kp} * \frac{\mathrm{fu,brain}}{\mathrm{fu,plasma}} \tag{2.7}$$

Hence, Kp,uu can therefore also be expressed as

$$\mathrm{Kp,uu} = \frac{\mathrm{AUCtot,brain}}{\mathrm{AUCtot,plasma}} * \frac{\mathrm{fu,brain}}{\mathrm{fu,plasma}} \tag{2.8}$$

Both fu,plasma and fu,brain can be measured readily *in vitro* by equilibrium dialysis [53, 55]. See also Chapter 12 for more details on the method. Since the ratio of fu,plasma and fu,tissue can be regarded as an *in vitro* estimate of Kp [24], the following equation can be used as an alternative:

$$\mathrm{Kp,uu} = \frac{\mathrm{Kp}(\textit{in vivo})}{\mathrm{Kp}(\textit{in vitro})} \tag{2.9}$$

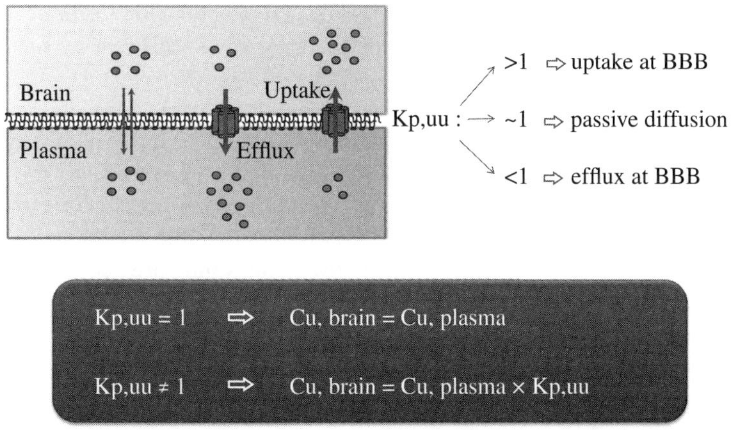

FIGURE 2.4 Kp,uu—the unbound brain to unbound plasma concentration ratio as the true measure of the extent of brain penetration, which is purely reflecting the transport properties of a drug across the BBB.

where Kp(*in vitro*) solely describes the ratio of the nonspecific binding of the compound to plasma proteins relative to the binding to brain tissue constituents:

$$\mathrm{Kp}(\textit{in vitro}) = \frac{\mathrm{fu,plasma}}{\mathrm{fu,brain}} \tag{2.10}$$

whereas Kp(*in vivo*), in addition to nonspecific drug binding, also carries information with regard to the BBB transport properties of the drug.

While Kp may span a more than 3000-fold range from values of less than 0.01 to above 30 [56, 57], the operating range of Kp,uu is smaller: typically between below 0.01 and 5 at the most [18, 51, 57]. The smaller range of Kp,uu compared to Kp illustrates how strong the impact on the brain-to-plasma distribution of nonspecific drug binding can be relative to BBB transport.

Since the parameter Kp,uu is devoid of nonspecific drug binding, it reflects the distributional drug properties as a result of transport across the BBB. A Kp,uu near 1 suggests passive diffusion, while a Kp,uu different from 1 suggests either active efflux back into blood ($\mathrm{Kp,uu} < 1$) or active uptake into brain ($\mathrm{Kp,uu} > 1$), which, however, is much more rare than the former (Fig. 2.4). Mechanistically, the following parameters determine Kp,uu:

$$\mathrm{Kp,uu} = \frac{\mathrm{CL}_{\mathrm{passive}} + \mathrm{CL}_{\mathrm{uptake}}}{\mathrm{CL}_{\mathrm{passive}} + \mathrm{CL}_{\mathrm{efflux}} + \mathrm{CL}_{\mathrm{bulkflow}} + \mathrm{CL}_{\mathrm{metabolism}}} \tag{2.11}$$

where $\mathrm{CL}_{\mathrm{passive}}$ is the diffusional clearance of the drug across the BBB (i.e., passive PS, see earlier), $\mathrm{CL}_{\mathrm{uptake}}$ is the active uptake transporter clearance, $\mathrm{CL}_{\mathrm{efflux}}$ is the efflux transporter clearance, $CL_{bulkflow}$ is the clearance due to bulk flow of brain ISF into CSF, and $CL_{metabolism}$ is the elimination of the drug from the brain by metabolism within the CNS [12].

According to Equation 2.11, Kp,uu can become smaller than unity due to (i) dominance of drug efflux relative to $CL_{passive}$ and/or CL_{uptake}, or (ii) lower passive diffusion into brain compared to the bulk flow of ISF into CSF. An example of the latter is mannitol, which shows a very low $CL_{passive}$ (PS <1 µl/min/g brain) relative to the bulk flow resulting in a Kp,uu of only 0.01 [12]. The same is true for the low Kp,uu of 0.09 of atenolol [58]. More often, however, low Kp,uu values are due to active efflux across the BBB, for example, loperamide (Kp,uu = 0.014), quinidine (Kp,uu = 0.07), or imatinib (Kp,uu = 0.18) [45].

Kp,uu values in excess of unity are much less frequent, as there seem to be few drugs which are actively taken up into brain from the blood circulation, for example, oxycodone (Kp,uu = 3.1) and diphenhydramine (Kp,uu = 5.5) [59, 60].

It follows from Equation 2.8 that Kp,uu can also be regarded as

$$\text{Kp,uu} = \frac{\text{AUCu,brain}}{\text{AUCu,plasma}} \tag{2.12}$$

or at steady-state:

$$\text{Kp,uu} = \frac{\text{Cu,brain}}{\text{Cu,plasma}} \tag{2.13}$$

These equations illustrate the power of Kp,uu as compared to Kp: while Kp is confounded by nonspecific binding and, hence, is difficult to interpret, Kp,uu purely reflects the transport properties of a drug across the BBB and, hence, can be used directly as a link between the unbound concentrations in brain and those in plasma.

$$\text{Cu,brain} = \text{Cu,plasma} * \text{Kp,uu} \tag{2.14}$$

For drugs whose pharmacological target is accessible from the brain's ISF compartment, the unbound brain concentration seems to be the most relevant PK compartment [48, 50, 61]. Kp,uu therefore replaces Kp as a more useful and more meaningful measure of the extent of CNS penetration in drug discovery and development [9, 30, 58, 62–65].

NEUROPK

With the general acceptance of the free drug hypothesis (see Chapter 3) the pivotal role of the unbound drug concentration at the site of the pharmacological drug target is now well established. This chapter looks at the pharmacokinetics in the CNS versus plasma and the factors which control the dynamics of the concentration–time profile, that is, NeuroPK.

Basic PK Compartments and PK Processes

Although the CNS is among the most complicated organs of the mammalian body, in terms of its anatomy, physiology, and pathophysiology, the pharmacokinetics of a drug in the brain can be described based on just a few key compartments (Fig. 2.5).

Principal NeuroPK Compartments The following are the principal NeuroPK compartments:

1. Brain vasculature with the cerebral blood supply
2. Brain parenchyma brain ISF
3. Brain parenchyma intracellular fluid (ICF)
4. Ventricles containing CSF

While the transfer of drugs between blood and ISF and blood and CSF is restricted by the BBB and BCSFB, respectively, there is no such tight barrier between ISF and CSF, that is, the ependymal cell layer which is lining the inner surface of the ventricles does not restrict the movement between these two intrabrain compartments.

For drugs whose pharmacological target resides within the brain ISF (e.g., receptor proteins, transporter proteins, ion channels), the brain ISF can be considered to be the

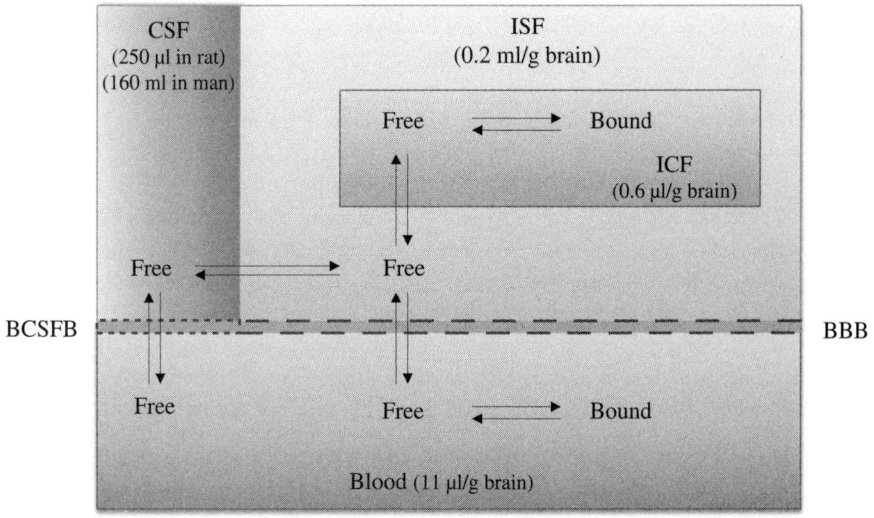

FIGURE 2.5 Schematic representation of the four key compartments within the CNS commonly used to describe the NeuroPK of compounds, including typical physiological volumes: CSF, cerebrospinal fluid; ICF, intracellular fluid; ISF, interstitial fluid. Compound flows are restricted between blood and ISF by the BBB (blood–brain barrier) and between blood and CSF by the BCSFB (blood–CSF barrier), which can be crossed both passively or with the involvement of carrier-mediated processes and/or active drug efflux (see text).

most relevant surrogate PK compartment for the pharmacological effect. For drugs which bind to pharmacological target proteins within brain cells (e.g., intracellular enzymes), the ICF is the more relevant effect compartment.

Principal NeuroPK Processes The following are the most important processes that describe the PK of a compound in the CNS:

1. Absorption, distribution, metabolism, excretion (ADME) processes which determine the systemic concentration–time profile in the blood circulation (as the source of input of the drug to the CNS)
2. Transport processes which determine the BBB transfer of the drug between blood and brain (as the measure of the rate and extent of brain penetration)
3. Intrabrain distribution of the drug (as the key element to determine the active site concentration of the drug in the effect compartment)
4. Elimination of the drug from the CNS (often with ISF bulk flow into the CSF or back to blood across BBB)

PK Measurements, PK Parameters, and Key Equations

This paragraph focuses on those PK parameters and equations which are most critical for the understanding and optimization of CNS penetration and distribution of compounds in a drug discovery setting. A more complete picture, with regard to the full range of assays and approaches used to optimize the ADME properties of compounds in drug discovery, can be found elsewhere [9, 66–68].

Systemic Exposure and Plasma Concentration–Time Profile The unbound plasma concentration–time profile is typically obtained from iv, ip, sc, or po dosing of the test compound to animals and the collection of blood samples over a sufficient period of time. Blood samples are used to prepare plasma, which is subjected to liquid chromatography/mass spectroscopy (LC–MS)/MS analysis to measure the plasma concentration of the test compound.

After oral administration, and considering one-compartmental PK behavior of the compound, the time course of the plasma concentration–time profile (C,plasma(t)) depends on the oral bioavailability (F), the dose administered (D), and the volume of distribution (V) of the compound, as well as on the rate constants of absorption and elimination, ka and ke, respectively.

$$C,\text{plasma}(t) = \frac{F * D}{V}(e^{-\text{ke}*t} - e^{-\text{ka}*t}) \tag{2.15}$$

The fraction unbound in plasma (fu,plasma) is used to convert total plasma concentrations into unbound plasma concentrations:

$$\text{Cu,plasma} = \text{Cplasma(total)} * \text{fu,plasma} \tag{2.16}$$

In the case of unequal distribution of the compound between blood plasma and blood cells the blood/plasma ratio (BPR) should be used to calculate unbound blood concentrations:

$$\text{Cu,blood} = \text{Cu,plasma} * \text{BPR} \tag{2.17}$$

For compounds with BPR close to 1, unbound plasma concentrations and unbound blood concentrations are the same and can be used interchangeably. The unbound concentration in blood represents the maximum level of the drug to which the brain is exposed, that is, the upper limit of the amount of drug the brain can extract from the circulation.

Unbound Brain Exposure and Brain Extracellular Fluid (ECF) Concentration–Time Profile Besides repeated blood sampling, as described earlier, brain tissue is also collected from the animals at different time points. After brain homogenization and protein precipitation, the samples are subjected to LC–MS/MS analysis for quantitation. The fraction unbound in brain homogenate (fu,brain) is used to convert total brain concentrations into unbound brain concentrations:

$$\text{Cu,brain} = \text{Cbrain(total)} * \text{fu,brain} \tag{2.18}$$

In addition, from the previous data, the following parameters can be calculated: Kp(*in vivo*) based on Equation 2.6, Kp(*in vitro*) based on Equation 2.10, and Kp,uu based on Equations 2.8 and 2.9.

Parallel Plasma and Brain Concentration–Time Profiles For most compounds with low to medium lipophilicity (log P = 0–3) and good *in vitro* permeability (P_{app} > 60 nm/s; MDCK-MDR1 cells), total plasma, and total brain concentrations run in parallel over time, with a typical example shown in Figure 2.6. In the case of a Kp,uu near unity, that is, if Kp(*in vivo*) equals Kp(*in vitro*), the unbound brain concentration–time profile matches the unbound plasma concentration–time profile (Fig. 2.4). For such compounds the unbound plasma concentration–time profile can thus be taken as a reliable PK compartment to describe the concentration in the effect compartment of PK/pharmacodynamic (PD) relationships.

Nonparallel Plasma and Brain Concentration–Time Profiles A more complicated situation arises for compounds which bind more extensively to brain tissue than to plasma proteins, which have a slow rate of CNS penetration, or both. Such compounds typically show a markedly nonparallel concentration–time profile between total plasma and total brain, and hence also between unbound plasma and brain (Fig. 2.7). Extensive binding to brain tissue leads to a long time required to achieve equilibrium between systemic and brain levels [69]:

$$T_{1/2(\text{equilibrium})} = \frac{\ln 2 * \text{V,brain}}{\text{PS} * \text{fu,brain}} \tag{2.19}$$

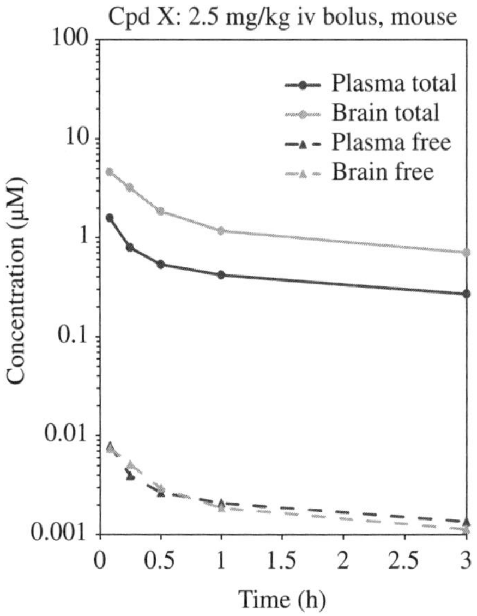

	AUC (total)	fu (%)	AUC (free)
Brain	4.3	0.16	0.0068
Plasma	1.4	0.51	0.0071
Ratio	3.1	3.2	0.96
	Kp (*in vivo*)	Kp (*in vitro*)	Kp,uu

- Total plasma and brain conc–time profiles run in parallel
- Kp(*in vivo*) = Kp(*in vitro*), hence Kp,uu = 1

⇒ Cu,brain = Cu,plasma

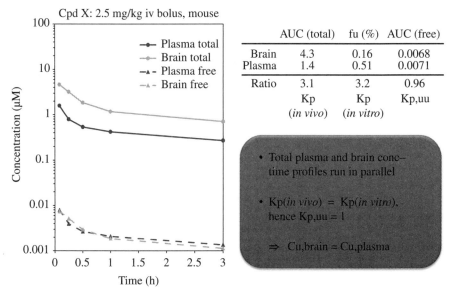

FIGURE 2.6 Concentration–time profile of compound X, dosed iv at 2.5 mg/kg into mice with plasma and brain sampling over a time period of 3 h, both for the total and the unbound concentrations in plasma and brain. Shown also are the AUC in plasma and brain for total and unbound, and the fu values for brain and plasma, with the resulting NeuroPK parameters Kp (*in vivo*), Kp (*in vitro*) and Kp,uu.

where V,brain represents physiological brain volume, PS is the permeability surface area product as a measure of the rate of penetration into the CNS, and fu,brain is the fraction unbound in brain tissue as a measure of the extent of brain tissue binding. The latter parameter appears to have the strongest impact on the time to equilibrium, because strong tissue binding increases the apparent tissue volume to be filled and, hence, the time needed to do so. In contrast, compounds which are substrates for transporters at the BBB achieve equilibrium faster than passively distributing compounds, and thus efflux at the BBB does not delay equilibrium between plasma and brain [31].

Duloxetine, for instance, shows a 16-fold lower fraction unbound in brain tissue compared to plasma resulting in a flat concentration–time profile in brain compared to a steep profile in plasma with the corresponding elimination half-lives in plasma and brain of 16 and 92 h, respectively [70].

For compounds where the concentration–time profiles in plasma and brain do not run in parallel, unbound brain concentrations cannot directly be taken from unbound plasma concentrations and Kp,uu. Therefore, the unbound brain concentrations of such compounds carry much more uncertainty compared to compounds with a parallel concentration–time profile. Whenever possible, preference should therefore

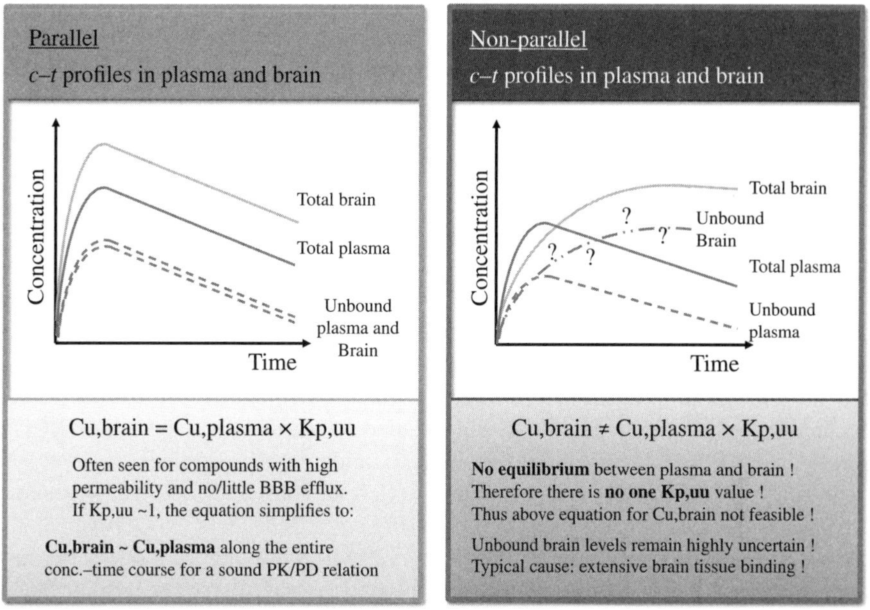

FIGURE 2.7 Differences in the predictivity of parallel versus nonparallel plasma and brain concentration–time profiles with regard to the estimation of unbound brain concentrations as relevant effect compartment for the active site concentrations at the drug target. *c–t*, concentration–time profile. Modified from Ref. [8]. © 2014 Springer. With kind permission of Springer Science+Business Media.

be given to the selection of the latter type of compounds as they make PK and PK/PD predictions easier and more reliable, in particular with regard to translation from animal to human (Fig. 2.7).

Some authors have successfully applied physiologically based PK (PBPK) modeling approaches to describe and predict concentration–time profiles which can accommodate time delays [69–71]. Although this approach requires more input data and a sound knowledge of the compound to make reliable assumptions, PBPK modeling is a very sophisticated method to understand and predict the PK of compounds in the CNS. This is because it allows more insight into the interplay between passive permeability and active transport processes at the BBB, plasma protein binding, and brain tissue distribution, as well as their combined impact on drug concentrations in the CNS (see also Chapter 14).

CNS Distribution

Besides systemic PK and BBB transport, the distribution of compounds within the CNS is another important determinant of the processes which control unbound drug concentrations at the site of the target within the brain, that is, NeuroPK.

In principle, the following methods are available to address this question: two *in vivo* techniques (brain microdialysis as a direct measurement of brain ISF concentrations and collection of the CSF), and two *in vitro* techniques (binding to brain homogenate and uptake by freshly prepared brain slices).

Both *in vivo* techniques suffer from strong limitations: brain microdialysis is unsuitable for routine use in drug discovery if the test compounds are very lipophilic and extensively stick to the material of the dialysis probe [51, 62]. CSF levels do not always correlate well with ISF levels, in particular if transporters control brain penetration [9, 63, 72, 73]. In two systematic studies of 39 compounds by Fridén et al. [57] and 25 compounds by Kodaira et al. [73], both groups demonstrated that good correspondence is to be expected for compounds that show a high permeability and little or no drug efflux. For those compounds, CSF levels can be a reliable surrogate for the concentration in the effect compartment of the brain. Deviations, however, occur for compounds which show a relevant net transport by Pgp and/or BCRP across the BBB (e.g., verapamil, loperamide, quinidine, or cimetidine).

In spite of CSF samples being routinely collected from *in vivo* NeuroPK studies, the *in vitro* methods have become the method of choice. Indeed, both *in vitro* methods are much easier to apply and can be used in a higher-throughput mode. They are cost-effective and compatible with assay needs in a drug discovery setting. For a more detailed description of the methods see also Chapters 11-13 and 16.

Brain Homogenate Technique—fu,brain The brain homogenate binding technique was first introduced by Kalvass and Maurer [53] and has since been enjoying wide acceptance by many drug discovery DMPK groups. This is because the same equipment can be used for both plasma protein binding and brain homogenate binding. The brain homogenate method can even be run in a cassette format [74] and, since brain composition and, hence, fu,brain is species-independent [14, 75], it is sufficient to measure this parameter in only one species, typically the PD species (e.g., rat, mouse) [64]. The key parameter obtained from the method is fu,brain, which is calculated by the following equation:

$$\text{fu,brain} = \frac{1/D}{[(1/\text{fu,dh}) - 1] + (1/D)} \tag{2.20}$$

where D is the dilution factor of the brain homogenate and fu,dh the fraction unbound in diluted brain homogenate.

As stated earlier, determining fu,brain allows calculation of unbound brain concentrations from total brain concentrations by Equation 2.18. Because the composition of plasma and brain differs significantly, with plasma being more rich in proteins and brain being more rich in lipids, fu,plasma and fu,brain do not correlate and thus cannot be used interchangeably [4, 13, 51, 63, 76].

Brain Slice Technique—Vu,brain As homogenizing brain tissue will destroy all intratissue compartments, the brain tissue binding method cannot provide information on compound levels in specific subcellular effect compartments, for example, cytosol and subcellular organelles. This may be particularly critical if the drug target resides within the cells of the CNS. Becker and Liu [77] and Fridén et al. [55] therefore developed an alternative *in vitro* method to determine the brain free fraction by using a slice technique, which, in contrast to the brain homogenization method, maintains the cellular structure of the brain tissue. Consequently, this method allows (i) measurement of differences between ISF and ICF concentrations and (ii) determination of the unbound volume of distribution of a compound in brain (Vu,brain).

$$\text{Vu, brain} = \frac{A(\text{brain slice})}{C(\text{buffer})} \tag{2.21}$$

where C(buffer) is the concentration of the compound in the incubation buffer and A(brain slice) the amount to compound in the brain slice at the end of the incubation [78]. The unbound distribution volume of a compound in brain is interpreted in relation to physiological volumes of ISF (0.2 ml/g brain) and ISF+ICF=0.8 ml/g brain. Thus, it is suggested that compounds with Vu,brain values of around 0.2 ml/g brain distribute mainly into the brain ECF, while compounds with Vu,brain values of around 0.8 ml/g brain distribute equally into brain ISF and brain ICF. For many compounds, however, Vu,brain values of much larger than 0.8 ml/g brain are obtained, which is highly indicative of strong tissue binding, active transport into the brain cells, lysosomal trapping, sequestration into organelles, or a combination thereof. Vu,brain therefore allows calculation of Kp,uu,cell as a measure of the ICF/ISF concentration ratio. Kp,uu,cell can be obtained from the following equation:

$$\text{Kp, uu, cell} = \text{fu, brain} * \text{Vu, brain} \tag{2.22}$$

In principle, Kp,uu,cell gives access to brain intracellular unbound concentrations (Cu,cell), which, of course, would be highly desirable for intracellular CNS drug targets:

$$\text{Cu, cell} = \text{Cu, plasma} * \text{Kp, uu} * \text{Kp, uu, cell} \tag{2.23}$$

However, these calculations should be used with caution: Fridén et al. [79] have shown that the largest differences between the two methods are observed for basic compounds and are due to lysosomal trapping. Since lysosomal trapping is a physicochemical consequence of pH partitioning, it can be corrected for by applying the Henderson–Hasselbalch equation using the pK_a values of the drug and the pH values for plasma, cytosol, and lysosomes [78, 79]. Thus, the brain homogenate technique combined with correction for pH partitioning provides equivalent information for both methods, that is, fu,brain is approximately equal to 1/Vu,brain within a twofold range [51, 79].

It also has to be kept in mind that intracellular targets may only in some cases reside within lysosomes and more often are located to other compartments of brain cells, for example, cytosol or other organelles. The unbound drug concentrations in the cytosol or the subcellular compartments of individual brain cells will depend on the micro-pH and the expression and functional activity of the transport processes.

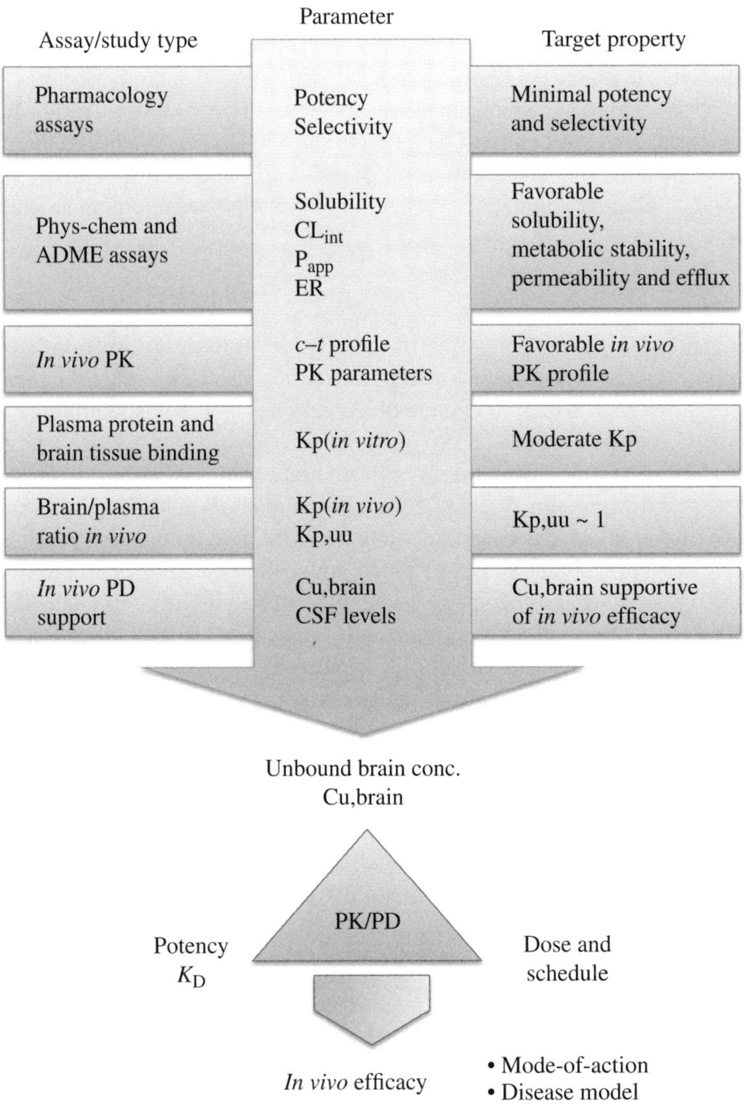

FIGURE 2.8 Generic workflow of *in vitro* assays and *in vivo* studies in the compound optimization phase of a drug discovery project with emphasis on those assays and parameters which have a direct influence on the NeuroPK. The left panel of the illustration shows the assay and study types to be employed and the middle panel shows the parameters obtained with target values shown on the left. The focus of the workflow is on optimizing unbound brain concentrations as the basis for target exposure, PK/PD, and *in vivo* efficacy. *c–t*, concentration–time profile; CL_{int}, intrinsic metabolic clearance; ER, efflux ratio; K_D, potency of compound P_{app}, apparent permeability; PD, pharmacodynamics.

The latter is highly specific for the cell membrane of different cell types [80] and the membranes of the intracellular organelles [81, 82]. Since the brain slice technique will only provide an average Kp,uu,cell for all cell types present in the material used, it may have limited value when it comes to a particular target cell type. Indeed it is known that the different cell types vary in their transporter expression and function [80], and they differ between species [83] and also between healthy and diseased brain [84, 85]. Therefore, the advantage of the brain slice technique may also be its caveat, and much remains to be learned about the application of this technique in order to unravel its true power for specific questions. For instance, it may be of particular interest to see whether addressing novel Alzheimer's disease drug targets, which are associated with lysosomes [86], will particularly benefit from the brain slice technique.

Integrated Method Spectrum to Assess and Optimize NeuroPK

There is now a complete set of methods available which allows characterization of the CNS penetration and distribution, that is, the NeuroPK, of compounds both in drug discovery and in drug development.

Figure 2.8 shows a cascade of NeuroPK *in vitro* assays and *in vivo* studies illustrating their place and use during routine compound optimization cycles in the drug discovery phase of research projects. Since the flowchart shown is generic, the screening tree of a given project may be further adjusted to accommodate the specific issues of a compound class and the demands, based on the actual target and intended disease. Absorption, distribution, metabolism, excretion, and toxicity (ADMET) assays and *in vivo* studies, which are not directly related to NeuroPK, for example, solubility, metabolic stability, intestinal permeability, efflux, CYP inhibition, CYP induction, hERG, and others, are not subjects of this chapter and can be found in greater detail elsewhere [9, 66, 68]. The flowchart illustrates that the NeuroPK assays are inherently interwoven with other PK and PD assays, in order to allow compound optimization cycles to generate compounds with a good balance of all the properties that make up a successful clinical candidate. The flowchart shown puts particular emphasis on optimizing the unbound brain concentrations as a direct link to PK/PD and *in vivo* efficacy. The approach shown will allow (i) optimization of compounds with regard to the properties which favour CNS target exposure, (ii) guidance of the dosing and dosing schedule for *in vivo* pharmacology studies, and thereby (iii) enhancement of the understanding of the target disease hypothesis of the project.

STRATEGIES TO INCREASE/AVOID CNS PENETRATION/EXPOSURE TO TARGETS WITHIN THE CNS

In principle, the strategies to increase and to avoid CNS penetration of compounds are based on the same key equations which determine the unbound brain concentrations as the relevant CNS effect compartment (Fig. 2.9).

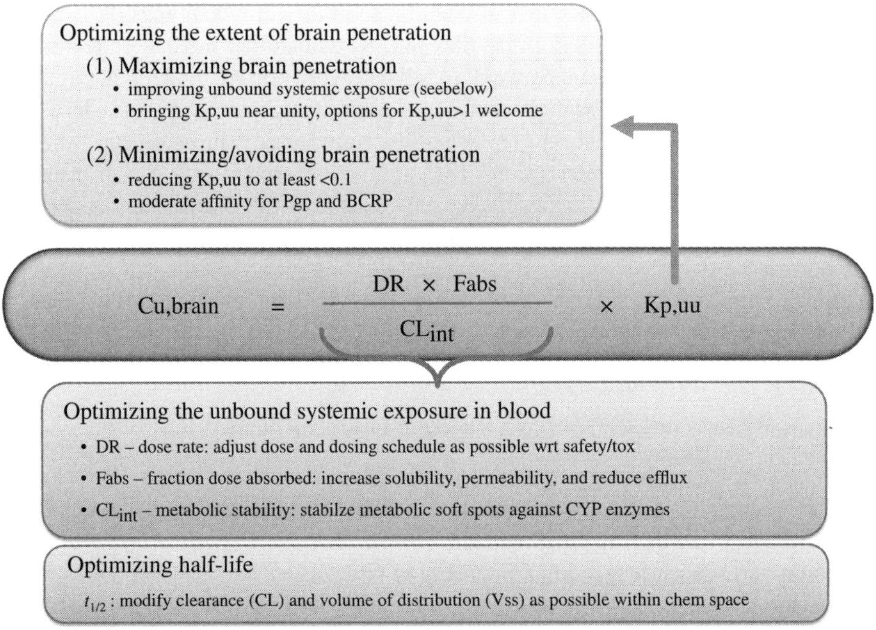

FIGURE 2.9 Graphic illustration of the PK parameters that lead to the maximization and the minimization of unbound drug concentrations within the CNS, depending on the needs of the therapeutic intervention of a given project. Note that maximizing unbound CNS exposure may need addressing of all PK parameters shown in the equation, whereas minimizing CNS exposure relies only on Kp,uu.

Strategy to Increase CNS Penetration

The most powerful ways to increase the unbound drug concentrations in the CNS are (i) to increase the unbound plasma exposure and/or (ii) to bring Kp,uu to unity (Fig. 2.9).

Increasing Unbound Plasma Concentrations It can be achieved by optimizing the following *in vitro* ADME parameters: increasing the intestinal permeability and removing any drug efflux using the Caco-2 assay, increasing the aqueous solubility of the compound to avoid any solubility- or dissolution-limited absorption, and reducing the intrinsic metabolic clearance of the compound. All of these approaches will contribute to maximizing the unbound concentration in the blood circulation as the maximum of compound available to the brain [9].

Getting Kp,uu to Unity Since a Kp,uu < 1 will reduce unbound brain levels relative to unbound plasma levels, raising a low Kp,uu value toward unity will be beneficial for brain exposure. Since low Kp,uu values are associated with efflux, the team should aim for removing recognition by active efflux pumps, such as Pgp and BCRP, using *in vitro* cell lines such as Caco-2 cells expressing both transporters and overexpressing cell lines (e.g., MCDK-MDR1 and MDCK-BCRP).

The data from the Caco-2 assay, which is a routine assay for optimizing the oral absorption properties of lead optimization (LO) compounds, should be used concurrently. A high ER in the Caco-2 assay can be taken as a surrogate of low Kp,uu [53]. It has been shown by several groups that the *in vitro* ER can be used for a first approximation of Kp,uu [14, 36, 87–89]:

$$\text{Kp,uu} \sim \frac{1}{\text{ER}} \tag{2.24}$$

This approximation is particularly attractive as it does not require any *in vivo* study and can be readily obtained from routine high-throughput permeability assays. However, because this empirical relationship is based on overexpressing cells of nonbrain origin, it should only be used to rank order compounds during compound optimization cycles and not as a proper Kp,uu value.

Equation 2.14 could also be exploited with regard to increasing Kp,uu above 1 in order to generate compounds which selectively enter the CNS via BBB-specific carrier systems. However, there are very few drugs which are capable of utilizing transport systems at the BBB and which accordingly have a Kp,uu > 1, that is, oxycodone and diphenhydramine with Kp,uu values of 3.0 and 5.5, respectively [31]. Both drugs are thought to use the postulated pyrilamine transporter, which seems to be functionally preserved across species and may thus be a suitable BBB transporter for CNS drug delivery [60]. Other BBB transporters, such as amino acid and nucleoside transporters, may also have potential for CNS drug delivery, but have not yet attracted wider interest [90, 91].

Strategy to Avoid CNS Penetration

Several examples of drug classes exist where distribution into the CNS correlates with CNS side effects due to interference with centrally located receptors, including opioid receptor agonists, H1-receptor antagonists, and antimuscarinic agents.

Loperamide is a peripherally acting opioid receptor agonist used for the management of chronic diarrhea. It lacks central opioid effects, for example, respiratory depression, since it does not sufficiently reach CNS tissue and, hence, central opioid receptors, due to significant Pgp efflux at the level of the BBB. Similarly, central side effects of antimuscarinic agents used to treat overactive bladder are lower for drugs which are Pgp substrates (e.g., 5-hydroxymethyl tolterodine, darifenacin, trospium) as compared to other muscarinic agents which are not recognized by Pgp and are more prone to central side effects, for example, oxybutynin, tolterodine, and solifenacin [92].

The power of Pgp efflux at the BBB in effectively limiting central side effects has first been postulated for second-generation H1 receptor antagonists, which lack central side effects, such as dizziness and somnolence, compared to their first-generation counterparts [92]. It has since been shown that subjecting compounds to drug efflux at the BBB opens the way for a therapeutic window with regard to central side effects of peripherally acting drugs whose target also resides within the CNS [9, 94–96].

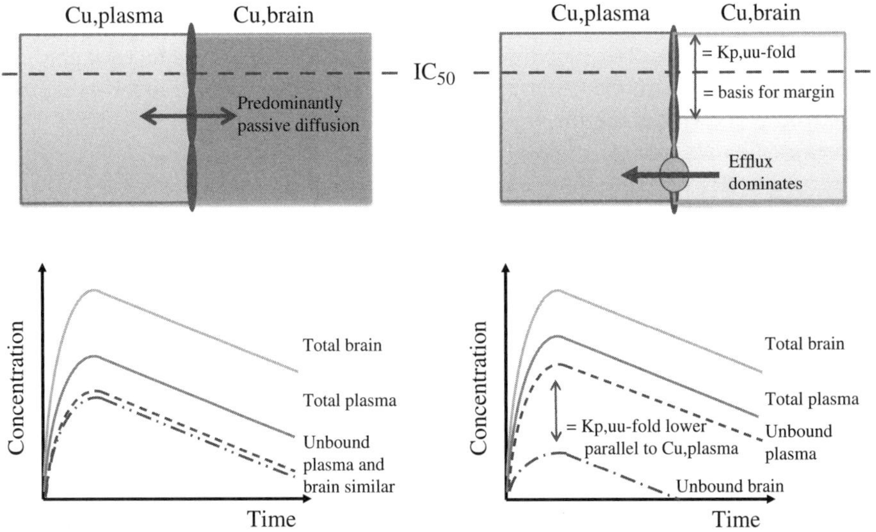

FIGURE 2.10 Graphic illustration of the effect of drug efflux at the BBB as the basis for the peripheral restriction of the drug activity via a very low Kp,uu, reducing the unbound brain concentration below efficacy, whereas unbound systemic concentrations in plasma remain sufficient to expose peripheral target sites, as indicated by the pharmacological potency (IC_{50}) of the drug.

In contrast to drugs which are optimized to act centrally and where Kp,uu is ideally near or above unity, peripherally acting drugs can be designed to avoid central side effects by significantly reducing Kp,uu below unity (Fig. 2.10). For instance, the peripherally restricted antimuscarinic agents 5-hydroxymethyl tolterodine, darifenacin, and trospium have Kp,uu values between 0.01 and 0.04, while those with central side effects show Kp,uu values between 0.23 and 3.3 [92].

It can thus be concluded that the target value for Kp,uu for compounds devoid of central activity should be <0.1. This may be best achieved by designing the compounds to be medium substrates for both Pgp and BCRP, as both efflux pumps show functional cooperation at the BBB, making their combined impact on brain penetration very strong [97], while minimizing the risk for poor intestinal absorption [92, 96, 98]. To achieve this purpose, MDR1-MDCK cells, BCRP-MDCK cells, and Caco-2 cells may be used in concert to optimize compounds for a balance of sufficient intestinal permeability versus poor BBB permeability [9].

Friden et al. [57] have shown that the addition of two hydrogen-binding acceptors (HBAs) to a compound on average results in a twofold reduction in Kp,uu and, hence, its unbound brain exposure. It is possible that the concomitant increase in hydrophilicity attenuates the intrinsic passive permeability, thereby extending the membrane residence time and, hence, the efficiency for drug efflux pumps to expel the compounds from the brain endothelial cell membrane.

TRANSLATING NEUROPK FROM ANIMALS TO HUMANS

Being fully in line with the free drug hypothesis, unbound drug concentrations in the brain are considered to be the most relevant effect compartment to drive the CNS pharmacological effect of drugs. Unbound brain concentrations can, therefore, be successfully used to link the pharmacokinetics with the pharmacodynamics of a drug. Predicting unbound concentrations in humans from animal and *in vitro* data is, therefore, an ultimate goal of drug discovery and development projects and a key piece of information to project human dosing schedules, which are likely to exert therapeutic effects in the patients.

PK/PD—Translating Exposure to Efficacy

The unbound drug concentrations obtained using C,brain from *in vivo* studies and fu,brain from brain homogenate have widely been considered to be a suitable surrogate of the unbound concentration in brain ISF [57, 62]. They have also successfully been demonstrated to allow a link between target exposure, target receptor occupancy, and efficacy [99, 100].

An estimate of the degree of occupancy of a target receptor can be obtained from the following equation:

$$\mathrm{RO} = \frac{\mathrm{Cu,brain}}{\mathrm{Cu,brain} + K_\mathrm{D}} * 100\% \qquad (2.25)$$

where the receptor occupancy (RO) represents the percentage of receptors occupied in relation to the unbound concentration at the target receptor (Cu,brain) and the potency of the drug (K_D). For an antagonist, a greater than 75% RO, which can be achieved by the unbound concentrations exceeding K_D by at least threefold, is often sufficient to elicit pharmacological effects in the target cell population.

The level of occupancy needed for efficacy should be explored carefully in animal models, in order to identify the concentration–time profile and, hence, the dose size and dosing schedule, which enables efficacy from the point of view of the target exposure and occupancy [101, 102].

The receptor occupancy of a drug in relation to the dose administered, the concentration–time profiles in plasma and brain, and the efficacy can be tested ex vivo [49, 102], *in vivo* [100], and using PET technology [11, 103]. Confirming the assumed receptor occupancy *in vivo* seems to be important to increase the validity of PK/PD relationships, as has been demonstrated, for instance, with the dual serotonin and norepinephrine transport inhibitors venlafaxine and milnacipran, for which *in vitro* binding was not sufficiently predictive of *in vivo* receptor occupancy [104]. It has also to be kept in mind that a high level of receptor occupancy may not be sufficient to elicit the desired pharmacological effects, as has been seen for NK-1 receptor antagonists in pain [105].

Another highly valuable benefit of *in vivo* PET studies is that they allow linking receptor occupancy with plasma concentration–time profiles and, hence, validating the relationship between Cu,plasma and Cu,brain, that is, the validity of the Kp,uu estimate. PET studies also allow examining the time course of the concentration–time profiles and potential temporal disconnects between the concentrations in plasma and brain, in relation to target occupancy.

Translating PK and PK/PD from Animals to Humans

Once a solid dose-exposure–response relationship has been established in animal disease models, which allows linking plasma concentration–time profiles with brain target exposure and receptor binding, the question arises as to how this PK/PD relationship will translate to other species, including nonrodent species used for safety testing (in order to determine the therapeutic window) and humans (in order to predict dose regimens which will be efficacious in patients).

Translating PK and PK/PD from animals to humans has to occur on several levels: (i) potency, to accommodate species differences in the target receptor inhibition, (ii) PK, to accommodate for dose-exposure differences between species, and (iii) on the level of the PK/PD relationship itself.

Potency estimates for the human receptor may be obtained *in vitro* from cells expressing the human target. Predicting the human PK from *in vivo* and *in vitro* data is a more complex undertaking, where potential species differences in ADME processes that control the plasma and brain concentration–time profile have to be taken into account. There are well-established approaches in place to predict the PK profile in humans [106]. The predicted plasma concentration–time profiles have to be translated into unbound concentration–time profiles in brain tissue, in order to estimate the target receptor exposure in humans as a basis for the PK/PD in CNS patients.

While species differences, with regard to the unbound fraction in plasma, can be significant, there do not seem to be relevant species differences in the unbound fraction in brain tissue [14, 75], allowing to use fu,brain estimates from one species for other species as well as for humans [9, 64].

For compounds with a Kp,uu near unity, which have a high permeability and whose plasma and brain concentration–time profiles in animals run in parallel, the predicted unbound plasma concentration–time profile in humans can be taken directly as a reliable surrogate for the effect compartment in the brain, that is, it represents the unbound concentration–time profile in the brain. For such compounds the predictions of the human PK profile in the CNS can be made with a very high reliability and confidence.

The situation is more complicated for compounds (i) whose Kp,uu differs significantly from unity and/or (ii) whose plasma and brain concentration–time profiles do not run in parallel [9]. The first type of compounds are subject to significant active transport across the BBB causing their Kp,uu to differ from unity. Since the transporters expressed at the level of the BBB may show significant species differences in their expression and function [107], Kp,uu values which markedly deviate from unity cannot simply be used across species, even though species differences in substrate properties may be minimal between humans and mice for Pgp [108] and BCRP [109], and some uptake transporters [60]; in addition, there are other transporters for which stronger species differences have been implied [110]. For compounds that are strong substrates for efflux transporters, Doran et al. [65] have shown that CSF levels are more predictive of the unbound brain concentrations in the respective species. The utility of CSF levels in translational neuroscience has been summarized extensively by de Lange [111].

The second type of compounds tends to bind extensively to brain tissue, causing the plasma and brain concentration–time profiles to go out of phase [9], prohibiting the application of the equation shown in Figure 2.4, that is, to translate the plasma PK into NeuroPK simply by correcting for Kp,uu. For such compounds, physiologically based approaches may have to be applied to predict NeuroPK from animals to humans [70, 112–114].

The use of translational PK/PD modeling for CNS drugs ultimately allows prediction of the response in the human patient based on target exposure and the corresponding expression of pharmacology in animal models of the human disease. There are a growing number of examples of how translational PK/PD can facilitate CNS drug discovery and development [102, 115, 116], thereby bringing to life the scheme shown in Figure 2.1.

SUMMARY AND OUTLOOK

After the recent change in paradigm of how to assess the CNS penetration and distribution of drugs [1, 47, 51, 58], a target exposure–driven approach to CNS drug discovery and development is evolving now [9, 13]. The approach gives full credit to unbound brain concentration as the most relevant surrogate compartment for the pharmacological activities of CNS drugs, with Kp,uu replacing the old and often misleading brain/plasma ratio, Kp, as the measure of the extent of brain penetration. In contrast to Kp, the unbound brain to unbound plasma ratio, Kp,uu, has the advantage of allowing a direct link between the plasma concentrations to the brain target exposure and further to the pharmacological effects elicited upon target binding of the drug. This paradigm, therefore, provides a powerful framework for the full integration of the receptor theory and PK/PD into the establishment of dose-exposure–response relationships and their translation from animals to humans [101, 115, 117, 118].

The current concept illustrates that CNS penetration and distribution is multifactorial; hence, to capture the most important processes several key *in vitro* assays and *in vivo* studies need to be performed in concert and evaluated using an integrated framework, the so-called NeuroPK. There are proposals to capture several aspects of brain penetration in combined *in vitro* assay formats [119].

For drug discovery scientists, knowledge of the factors controlling the relationship between drug concentration and pharmacological activity is key for successful lead optimization of CNS drug discovery compounds and the selection of successful drug candidates for preclinical and clinical drug development [2, 47]. Recently, significant progress has been made in the establishment of physiologically based PK models of CNS penetration and distribution, which take into account the kinetics of the drug concentration–time profiles [112–114, 120] as the basis for the understanding of the dynamics of the pharmacological effects over time [115, 118, 121]. The understanding of these dynamics may pave the way to move beyond the "one target–one drug–one disease" paradigm of current CNS drug discovery toward a more pathway-oriented therapeutic intervention, which is hoped to be more compatible with the complex pathology of CNS disorders, such as neurodegenerative disease, where several pathophysiological pathways play a role [122]. A well-tuned interference with these pathways in a temporal fashion may be a promising, if demanding, approach for the treatment of chronic neurodegenerative disease.

When assessing the brain penetration and CNS distribution of compounds in drug discovery it has also to be kept in mind that the data generated typically originate from healthy brains (e.g., of inbred animals) which may not reflect the function of the BBB and, indeed, the NeuroPK under the condition of diseased brain tissue in a diverse human patient population. Indeed, there is strong evidence that the BBB function changes in response to age, diet, lifestyle and stress, diseases (central and peripheral diseases), and drug treatment [22, 123–125]. Furthermore, our understanding of the role of brain metabolism [110, 126] in CNS drug disposition and the fluid and drug flow through the complex network of the brain's interstitial space is still insufficient [20, 21]. The very recent discovery of the glymphatic system of the brain may shed more light onto the movement of solutes and drugs within the brain interstitial space and their clearance from the CNS [127].

At present, NeuroPK generally looks at the brain as a whole rather than appreciating its heterogeneity and regional differences, which may well impact the exposure in a specific brain area of interest [111, 120, 121]. While our knowledge of the transport processes at the level of the BBB is growing rapidly [107, 110, 128], much needs to be learned about transport processes at the aging BBB and the BBB under disease conditions, as any such changes would have an impact on Kp,uu.

Although confidence may be high for the prediction of relevant drug concentrations in the brain ISF, that is, for extracellular drug targets, a high level of uncertainty remains with regard to the exposure of intracellular targets, since brain intracellular concentrations—in particular in the target brain cells—are likely to be driven by processes to which we currently have no or only insufficient access. The generation of relevant data for intracellular concentrations of target brain cells will be an important area of future research, including the expression, regulation, and function of transporter proteins in the target brain cell population.

The characterization of primary NeuroPK parameters, as described in this chapter, allows the understanding and prediction of the processes which govern the CNS penetration and distribution. The described NeuroPK concept forms a sound basis for the delivery of several key tasks of both drug discovery and development DMPK: guidance of the optimization of the CNS properties of drug discovery compounds and translation of the PK/PD from animals to humans, in order to aid the design of efficacious dose regimen. The power of the approach lies in its target exposure–driven paradigm, which is able to remove much of the uncertainty from which numerous previous CNS drug discovery and development programs have suffered across the pharmaceutical industry.

The presented NeuroPK concept also allows more stringent examination of the mechanism(s) of action of novel targets and the type of target exposure needed to modulate the target, in order to have the desired impact on the course of the disease. The concept may also be able to support projections aiming at the interference with whole cellular pathways, rather than discrete drug targets, to validate biomarkers of response, and to improve the design of clinical trials in humans and patients, thereby connecting CNS drug discovery programs with model-based drug development [129].

REFERENCES

[1] Hammarlund-Udenaes M (2009) Active-site concentrations of chemicals—are they a better predictor of effect than plasma/organ/tissue concentrations? Basic Clin Pharmacol Toxicol 106(3):215–220.

[2] Gabrielsson J, Dolgos H, Gillberg PG, Bredberg U, Benthem B, Duker G (2009) Early integration of pharmacokinetic and dynamic reasoning is essential for optimal development of lead compounds: strategic considerations. Drug Discov Today 14(7–8):358–372.

[3] Smith DA, Di L, Kerns EH (2010) The effect of plasma protein binding on in vivo efficacy: misconceptions in drug discovery. Nat Rev Drug Discov 9(12):929–939.

[4] Liu X, Chen C, Hop CE (2011) Do we need to optimize plasma protein and tissue binding in drug discovery? Curr Top Med Chem 11(4):450–466.

[5] Morgan P, Van Der Graaf PH, Arrowsmith J, Feltner DE, Drummond KS, Wegner CD, Street SD (2012) Can the flow of medicines be improved? Fundamental pharmacokinetic and pharmacological principles toward improving Phase II survival. Drug Discov Today 17(9–10):419–424.

[6] Abbott A (2011) Novartis to shut brain research facility. Drug giant redirects psychiatric efforts to genetics. Nature 480(7376):161–162.

[7] Tufts CSDD Impact Reports (2012) Pace of CNS drug development and FDA approvals lags other drug classes, March/April 2012, Vol. 14, No. 2.

[8] Palmer AM, Alavijeh MS (2012) Translational CNS medicines research. Drug Discov Today 17(19–20):1068–1078.

[9] Reichel A (2014) Integrated Approach to Optimizing CNS Penetration in Drug Discovery: From the Old to the New Paradigm and Assessment of Drug–Transporter Interactions. In: *Drug Delivery to the Brain, Physiological Concepts, Methodologies and Approaches.* Hammarlund-Udenaes M, de Lange ECM, Thorne RG (eds.), AAPS Advances in the Pharmaceutical Sciences Series 10, Springer Press, New York, p. 731.

[10] Gründer G, Hiemke C, Paulzen M, Veselinovic T, Vernaleken I (2011) Therapeutic plasma concentrations of antidepressants and antipsychotics: lessons from PET imaging. Pharmacopsychiatry 44(6):236–248.

[11] Syvänen S, Gunn RN (2014) Principles of PET and Its Role in Understanding Drug Delivery to the Brain. In: *Drug Delivery to the Brain, Physiological Concepts, Methodologies and Approaches.* Hammarlund-Udenaes M, de Lange ECM, Thorne RG (eds.), AAPS Advances in the Pharmaceutical Sciences Series 10, Springer Press, New York, p. 731.

[12] Liu X, Chen C (2005) Strategies to optimize brain penetration in drug discovery. Curr Opin Drug Discov Dev 8(4):505–512.

[13] Reichel A (2009) Addressing central nervous system (CNS) penetration in drug discovery: basics and implications of the evolving new concept. Chem Biodivers 6(11):2030–2049.

[14] Summerfield SG, Lucas AJ, Porter RA, Jeffrey P, Gunn RN, Read KR, Stevens AJ, Metcalf AC, Osuna MC, Kilford PJ, Passchier J, Ruffo AD (2008) Toward an improved prediction of human in vivo brain penetration. Xenobiotica 38(12):1518–1535.

[15] Abbott NJ, Rönnbäck L, Hansson E (2006) Astrocyte–endothelial interactions at the blood-brain barrier. Nat Rev Neurosci 7(1):41–53.

[16] Segal MB (2000) The choroid plexuses and the barriers between the blood and the cerebrospinal fluid. Cell Mol Neurobiol 20(2):183–196.

[17] Begley DJ, Brightman MW (2003) Structural and functional aspects of the blood-brain barrier. Prog Drug Res 61:39–78.

[18] Hammarlund-Udenaes M, Fridén M, Syvänen S, Gupta A (2008) On the rate and extent of drug delivery to the brain. Pharm Res 25(8):1737–1750.

[19] Wong AD, Ye M, Levy AF, Rothstein JD, Bergles DE, Searson PC (2013) The blood-brain barrier: an engineering perspective. Front Neuroeng 6:1–22.

[20] Nicholson C, Kamali-Zare P, Tao L (2011) Brain extracellular space as a diffusion barrier. Comput Vis Sci 14(7):309–325.

[21] Abbott NJ (2004) Evidence for bulk flow of brain interstitial fluid: significance for physiology and pathology. Neurochem Int 45(4):545–552.

[22] Deo AK, Theil FP, Nicolas JM (2013) Confounding parameters in preclinical assessment of blood-brain barrier permeation: an overview with emphasis on species differences and effect of disease states. Mol Pharm 10(5):1581–1595.

[23] Abbott NJ, Dolman DE, Patabendige AK (2008) Assays to predict drug permeation across the blood-brain barrier, and distribution to brain. Curr Drug Metab 9(9):901–910.

[24] Rowland M, Tozer T (2011) *Clinical Pharmacokinetics and Pharmacodynamics: Concepts and Applications*. Lippincott, Williams & Wilkins, Baltimore, MD/Philadelphia, PA, p. 839.

[25] Smith QR, Samala R (2014) In Situ and In Vivo Animal Models. In: *Drug Delivery to the Brain, Physiological Concepts, Methodologies and Approaches*. Hammarlund-Udenaes M, de Lange ECM, Thorne RG (eds.), AAPS Advances in the Pharmaceutical Sciences Series 10, Springer Press, New York, p. 731.

[26] Fung EK, Carson RE (2013) Cerebral blood flow with [15O] water PET studies using an image-derived input function and MR-defined carotid centerlines. Phys Med Biol 58(6):1903–1923.

[27] Harada N, Ohba H, Fukumoto D, Kakiuchi T, Tsukada H (2004) Potential of [(18)F]beta-CFT-FE (2beta-carbomethoxy-3beta-(4-fluorophenyl)-8-(2-[(18)F]fluoroethyl)nortropane) as a dopamine transporter ligand: a PET study in the conscious monkey brain. Synapse 54(1):37–45.

[28] Palmer AM, Alavijeh MS (2013) Overview of experimental models of the blood-brain barrier in CNS drug discovery. Curr Protoc Pharmacol 62:7.15.1–30.

[29] Smith QR (2003) A review of blood-brain barrier transport techniques. Methods Mol Med 89:193–208.

[30] Zhao R, Kalvass JC, Pollack GM (2009) Assessment of blood–brain barrier permeability using the in situ mouse brain perfusion technique. Pharm Res 26(7):1657–1664.

[31] Hammarlund-Udenaes (2014) Pharmacokinetic Concepts in Brain Drug Delivery. In: *Drug Delivery to the Brain, Physiological Concepts, Methodologies and Approaches.* Hammarlund-Udenaes M, de Lange ECM, Thorne RG (eds.), AAPS Advances in the Pharmaceutical Sciences Series 10, Springer Press, New York, p. 731.

[32] Gratton JA, Abraham MH, Bradbury MW, Chadha HS (1997) Molecular factors influencing drug transfer across the blood-brain barrier. J Pharm Pharmacol. 49(12):1211–1216.

[33] Dagenais C, Avdeef A, Tsinman O, Dudley A, Beliveau R (2009) P-glycoprotein deficient mouse in situ blood-brain barrier permeability and its prediction using an in combo PAMPA model. Eur J Pharm Sci 38(2):121–137.

[34] Crone C (1964) Permeability of capillaries in various organs as determined by use of the indicator diffusion method. Acta Physiol Scand 58: 292–305.

[35] Liu X, Tu M, Kelly RS, Chen C, Smith BJ (2004) Development of a computational approach to predict blood-brain barrier permeability. Drug Metab Dispos 2004 Jan;32(1):132–139.

[36] Kodaira H, Kusuhara H, Fuse E, Ushiki J, Sugiyama Y (2014) Quantitative investigation of the brain-to-cerebrospinal fluid unbound drug concentration ratio under steady-state conditions in rats using a pharmacokinetic model and scaling factors for active efflux transporters. Drug Metab Dispos 42(6):983–989.

[37] Reichel A, Begley DJ, Abbott NJ (2003) An overview of in vitro techniques for blood–brain barrier studies. Methods Mol Med 89:307–324.

[38] Garberg P, Ball M, Borg N, Cecchelli R, Fenart L, Hurst RD, Lindmark T, Mabondzo A, Nilsson JE, Raub TJ, Stanimirovic D, Terasaki T, Oberg JO, Osterberg T (2005) In vitro models for the blood-brain barrier. Toxicol In Vitro 19:299–334.

[39] Cecchelli R, Berezowski V, Lundquist S, Culot M, Renftel M, Dehouck M-P, Fenart L (2007) Modelling of the blood–brain barrier in drug discovery and development. Nature Rev Drug Discov. 6:650–661.

[40] Bicker J, Alves G, Fortuna A, Falcão A (2014) Blood-brain barrier models and their relevance for a successful development of CNS drug delivery systems: a review. Eur J Pharm Biopharm 87(3):409–432.

[41] Wilhelm I, Krizbai IA (2014) In vitro models of the blood-brain barrier for the study of drug delivery to the brain. Mol Pharm 1(7):1949–1963.

[42] Abbott NJ, Dolman DEM, Yusof SR, Reichel A (2014) In Vitro Models of CNS Barriers. In: *Drug Delivery to the Brain, Physiological Concepts, Methodologies and Approaches.* Hammarlund-Udenaes M, de Lange ECM, Thorne RG (eds.), AAPS Advances in the Pharmaceutical Sciences Series 10, Springer Press, New York, p. 731.

[43] Summerfield SG, Jeffrey P (2006) In vitro prediction of brain penetration—a case for free thinking? Expert Opin Drug Discov 1(6):595–607.

[44] Reichel A (2006) The role of blood–brain barrier studies in the pharmaceutical industry. Curr Drug Metab 7(2):183–203.

[45] Liu X, Cheong J, Ding X, Deshmukh G (2014) Use of cassette dosing approach to examine the effects of P-glycoprotein on the brain and cerebral spinal fluid concentrations in wild-type and P-glycoprotein knockout rats. Drug Metab Dispos 42(4):482–491.

[46] Liu X, Ding X, Deshmukh G, Liederer BM, Hop CE (2012) Use of the cassette-dosing approach to assess brain penetration in drug discovery. Drug Metab Dispos 40(5):963–969.

[47] Jeffrey P, Summerfield S. (2010) Assessment of the blood-brain barrier in CNS drug discovery. Neurobiol Dis 37(1):33–37.

[48] Kalvass JC, Olson ER, Cassidy MP, Selley DE, Pollack GM (2007) Pharmacokinetics and pharmacodynamics of seven opioids in P-glycoprotein-competent mice: assessment of unbound brain EC50,u and correlation of in vitro, preclinical, and clinical data. J Pharmacol Exp Ther 323(1):346–355.

[49] Summerfield SG, Read K, Begley DJ, Obradovic T, Hidalgo IJ, Coggon S, Lewis AV, Porter RA, Jeffrey P. (2007) Central nervous system drug disposition: the relationship between in situ brain permeability and brain free fraction. J Pharmacol Exp Ther 322(1):205–213.

[50] Watson J, Wright S, Lucas A, Clarke KL, Viggers J, Cheetham S, Jeffrey P, Porter R, Read KD (2009) Receptor occupancy and brain free fraction. Drug Metab Dispos 37(4):753–760.

[51] Loryan I, Sinha V, Mackie C, Van Peer A, Drinkenburg W, Vermeulen A, Morrison D, Monshouwer M, Heald D, Hammarlund-Udenaes M, (2014) Mechanistic understanding of brain drug disposition to optimize the selection of potential neurotherapeutics in drug discovery. Pharm Res 31(8):2203–2219.

[52] Di L, Kerns EH, Carter GT (2008) Strategies to assess blood–brain barrier penetration. Exp Opin Drug Disc 3(6):677–687.

[53] Kalvass JC, Maurer TS (2002) Influence of nonspecific brain and plasma binding on CNS exposure: implications for rational drug discovery. Biopharm Drug Dispos 23(8):327–338.

[54] Maurer TS, Debartolo DB, Tess DA, Scott DO (2005) Relationship between exposure and nonspecific binding of thirty-three central nervous system drugs in mice. Drug Metab Dispos 33(1):175–181.

[55] Fridén M, Gupta A, Antonsson M, Bredberg U, Hammarlund-Udenaes M (2007) In vitro methods for estimating unbound drug concentrations in the brain interstitial and intracellular fluids. Drug Metab Dispos 35(9):1711–1719.

[56] Doran A, Obach RS, Smith BJ, Hosea NA, Becker S, Callegari E, Chen C, Chen X, Choo E, Cianfrogna J, Cox LM, Gibbs JP, Gibbs MA, Hatch H, Hop CE, Kasman IN, Laperle J, Liu J, Liu X, Logman M, Maclin D, Nedza FM, Nelson F, Olson E, Rahematpura S, Raunig D, Rogers S, Schmidt K, Spracklin DK, Szewc M, Troutman M, Tseng E, Tu M, Van Deusen JW, Venkatakrishnan K, Walens G, Wang EQ, Wong D, Yasgar AS, Zhang C (2005) The impact of P-glycoprotein on the disposition of drugs targeted for indications of the central nervous system: evaluation using the MDR1A/1B knockout mouse model. Drug Metab Dispos 33(1):165–174.

[57] Fridén M, Winiwarter S, Jerndal G, Bengtsson O, Wan H, Bredberg U, Hammarlund-Udenaes M, Antonsson M (2009) Structure–brain exposure relationships in rat and human using a novel data set of unbound drug concentrations in brain interstitial and cerebrospinal fluids. J Med Chem 52(20):6233–6243.

[58] Di L, Rong H, Feng B (2013) Demystifying brain penetration in central nervous system drug discovery. J Med Chem 56(1):2–12.

[59] Sadiq MW, Borgs A, Okura T, Shimomura K, Kato S, Deguchi Y, Jansson B, Björkman S, Terasaki T, Hammarlund-Udenaes M (2011) Diphenhydramine active uptake at the blood-brain barrier and its interaction with oxycodone in vitro and in vivo. J Pharm Sci 100(9):3912–3923.

[60] Shaffer CL, Osgood SM, Mancuso JY, Doran AC (2014) Diphenhydramine has similar interspecies net active influx at the blood-brain barrier. J Pharm Sci 103(5):1557–1562.

[61] Hammarlund-Udenaes M (2010) Active-site concentrations of chemicals—are they a better predictor of effect than plasma/organ/tissue concentrations? Basic Clin Pharmacol Toxicol 106(3):215–220.

[62] Hammarlund-Udenaes M, Bredberg U, Fridén M (2009) Methodologies to assess brain drug delivery in lead optimization. Top Med Chem 9(2):148–162.

[63] Liu X, Van Natta K, Yeo H, Vilenski O, Weller PE, Worboys PD, Monshouwer M (2009) Unbound drug concentration in brain homogenate and cerebral spinal fluid at steady state as a surrogate for unbound concentration in brain interstitial fluid. Drug Metab Dispos 37(4):787–793.

[64] Read KD, Braggio S (2010) Assessing brain free fraction in early drug discovery. Exp Opin Drug Metab Toxicol 6(3):337–344.

[65] Doran AC, Osgood SM, Mancuso JY, Shaffer CL (2012) An evaluation of using rat-derived single-dose neuropharmacokinetic parameters to project accurately large animal unbound brain drug concentrations. Drug Metab Dispos 40(11):2162–2173.

[66] Kerns E, Di L (2008) Drug-Like Properties: Concepts, Structure Design and Methods: From ADME to Toxicity Optimization. Academic Press, Amsterdam, p. 552.

[67] Summerfield S, Jeffrey P. (2009) Discovery DMPK: changing paradigms in the eighties, nineties and noughties. Exp Opin Drug Discov 4(3):207–218.

[68] Zhang D, Surapaneni S (2012) ADME-Enabling Technologies in Drug Design and Development. John Wiley & Sons, NJ, p. 622.

[69] Liu X, Smith BJ, Chen C, Callegari E, Becker SL, Chen X, Cianfrogna J, Doran AC, Doran SD, Gibbs JP, Hosea N, Liu J, Nelson FR, Szewc MA, Van Deusen J. (2005) Use of a physiologically based pharmacokinetic model to study the time to reach brain equilibrium: an experimental analysis of the role of blood-brain barrier permeability, plasma protein binding, and brain tissue binding. J Pharmacol Exp Ther 313(3):1254–1262.

[70] Kielbasa W, Stratford RE Jr. (2012) Exploratory translational modeling approach in drug development to predict human brain pharmacokinetics and pharmacologically relevant clinical doses. Drug Metab Dispos 40(5):877–883.

[71] Ball K, Bouzom F, Scherrmann JM, Walther B, Declèves X (2012) Development of a physiologically based pharmacokinetic model for the rat central nervous system and determination of an in vitro–in vivo scaling methodology for the blood-brain barrier permeability of two transporter substrates, morphine and oxycodone. J Pharm Sci 101(11):4277–4292.

[72] de Lange EC, Danhof M (2002) Considerations in the use of cerebrospinal fluid pharmacokinetics to predict brain target concentrations in the clinical setting: implications of the barriers between blood and brain. Clin Pharmacokinet 41(10):691–703.

[73] Kodaira H, Kusuhara H, Fujita T, Ushiki J, Fuse E, Sugiyama Y (2011) Quantitative evaluation of the impact of active efflux by p-glycoprotein and breast cancer resistance protein at the blood–brain barrier on the predictability of the unbound concentrations of drugs in the brain using cerebrospinal fluid concentration as a surrogate. J Pharmacol Exp Ther 339(3):935–944.

[74] Wan H, Rehngren M, Giordanetto F, Bergström F, Tunek A (2007) High-throughput screening of drug–brain tissue binding and in silico prediction for assessment of central nervous system drug delivery. J Med Chem 50(19):4606–4615.

[75] Di L, Umland JP, Chang G, Huang Y, Lin Z, Scott DO, Troutman MD, Liston TE (2011) Species independence in brain tissue binding using brain homogenates. Drug Metab Dispos 39(7):1270–1277.

[76] Liu X, Wright M, Hop CE (2014) Rational use of plasma protein and tissue binding data in drug design. J Med Chem. [Epub ahead of print].

[77] Becker S, Liu X (2006) Evaluation of the utility of brain slice methods to study brain penetration. Drug Metab Dispos 34(5):855–861.

[78] Loryan I, Hammarlund-Udenaes M (2014) Drug Discovery Methods for Studying Brain Drug Delivery and Distribution. In: *Drug Delivery to the Brain, Physiological Concepts, Methodologies and Approaches.* Hammarlund-Udenaes M, de Lange ECM, Thorne RG (eds.), AAPS Advances in the Pharmaceutical Sciences Series 10, Springer Press, New York, p. 731.

[79] Fridén M, Bergström F, Wan H, Rehngren M, Ahlin G, Hammarlund-Udenaes M, Bredberg U (2011) Measurement of unbound drug exposure in brain: modeling of pH partitioning explains diverging results between the brain slice and brain homogenate methods. Drug Metab Dispos 39(3):353–362.

[80] Sanchez-Covarrubias L, Slosky LM, Thompson BJ, Davis TP, Ronaldson PT (2014) Transporters at CNS barrier sites: obstacles or opportunities for drug delivery? Curr Pharm Des 20(10):1422–1449.

[81] Eskelinen EL, Tanaka Y, Saftig P (2003) At the acidic edge: emerging functions for lysosomal membrane proteins. Trends Cell Biol 13(3):137–145.

[82] Saftig P, Klumperman J (2009) Lysosome biogenesis and lysosomal membrane proteins: trafficking meets function. Nat Rev Mol Cell Biol 10(9):623–635.

[83] Ohtsuki S, Hirayama M, Ito S, Uchida Y, Tachikawa M, Terasaki T (2014) Quantitative targeted proteomics for understanding the blood-brain barrier: towards pharmacoproteomics. Expert Rev Proteom 11(3):303–313.

[84] Zlokovic BV (2008) The blood-brain barrier in health and chronic neurodegenerative disorders. Neuron 57(2):178–201.

[85] Miller DS, Cannon RE (2014) Signaling pathways that regulate basal ABC transporter activity at the blood-brain barrier. Curr Pharm Des 20(10):1463–1471.

[86] Caglayan S1, Takagi-Niidome S, Liao F, Carlo AS, Schmidt V, Burgert T, Kitago Y, Füchtbauer EM, Füchtbauer A, Holtzman DM, Takagi J, Willnow TE (2014) Lysosomal sorting of amyloid-β by the SORLA receptor is impaired by a familial Alzheimer's disease mutation. Sci Transl Med 12;6(223):1–10.

[87] Kalvass JC, Maurer TS, Pollack GM (2007) Use of plasma and brain unbound fractions to assess the extent of brain distribution of 34 drugs: comparison of unbound concentration ratios to in vivo p-glycoprotein efflux ratios. Drug Metab Dispos 35(4):660–666.

[88] Tang C, Kuo Y, Pudvah NT, Ellis JD, Michener MS, Egbertson M, Graham SL, Cook JJ, Hochman JH, Prueksaritanont T (2009) Effect of P-glycoprotein-mediated efflux on cerebrospinal fluid concentrations in rhesus monkeys. Biochem Pharmacol 78(6):642–647.

[89] Caruso A, Alvarez-Sánchez R, Hillebrecht A, Poirier A, Schuler F, Lavé T, Funk C, Belli S (2013) PK/PD assessment in CNS drug discovery: prediction of CSF concentration in rodents for P-glycoprotein substrates and application to in vivo potency estimation. Biochem Pharmacol 85(11):1684–1699.

[90] Reichel A, Begley DJ, Abbott NJ (2000) Carrier-mediated delivery of metabotrophic glutamate receptor ligands to the central nervous system: structural tolerance and potential of the l-system amino acid transporter at the blood–brain barrier. J Cereb Blood Flow Metab 20(1):168–174.

[91] Reichel A, Abbott NJ, Begley DJ (2002) Evaluation of the RBE4 cell line to explore carrier-mediated drug delivery to the CNS via the L-system amino acid transporter at the blood-brain barrier. J Drug Target 10(4):277–283.

[92] Callegari E, Malhotra B, Bungay PJ, Webster R, Fenner KS, Kempshall S, LaPerle JL, Michel MC, Kay GG (2011) A comprehensive non-clinical evaluation of the CNS penetration potential of antimuscarinic agents for the treatment of overactive bladder. Br J Clin Pharmacol 72(2):235–246.

[93] Chishty M, Reichel A, Siva J, Abbott NJ, Begley DJ (2001) Affinity for the P-glycoprotein efflux pump at the blood-brain barrier may explain the lack of CNS side-effects of modern antihistamines. J Drug Target 9(3):223–228.

[94] Chen C, Hanson E, Watson JW, Lee JS (2003) P-glycoprotein limits the brain penetration of nonsedating but not sedating H1-antagonists. Drug Metab Dispos 31(3):312–318.

[95] Obradovic T, Dobson GG, Shingaki T, Kungu T, Hidalgo IJ (2007) Assessment of the first and second generation antihistamines brain penetration and role of P-glycoprotein. Pharm Res 24(2):318–327.

[96] Wager TT, Liras JL, Mente S, Trapa P (2012) Strategies to minimize CNS toxicity: in vitro high-throughput assays and computational modeling. Exp Opin Drug Metab Toxicol 8(5):531–542.

[97] Kusuhara H, Sugiyama Y (2009) In vitro–in vivo extrapolation of transporter-mediated clearance in the liver and kidney. Drug Metab Pharmacokinet 24(1):37–52.

[98] Cole S, Bagal S, El-Kattan A, Fenner K, Hay T, Kempshall S, Lunn G, Varma M, Stupple P, Speed W (2012) Full efficacy with no CNS side-effects: unachievable panacea or reality? DMPK considerations in design of drugs with limited brain penetration. Xenobiotica 42(1):11–27.

[99] Kalvass JC, Olson ER, Cassidy MP, Selley DE, Pollack GM (2007) Pharmacokinetics and pharmacodynamics of seven opioids in P-glycoprotein-competent mice: assessment of unbound brain EC50,u and correlation of in vitro, preclinical, and clinical data. J Pharmacol Exp Ther 323(1):346–355.

[100] Bundgaard C, Sveigaard C, Brennum LT, Stensbøl TB (2012) Associating in vitro target binding and in vivo CNS occupancy of serotonin reuptake inhibitors in rats: the role of free drug concentrations. Xenobiotica 42(3):256–265.

[101] Amore BM, Gibbs JP, Emery MG (2010) Application of in vivo animal models to characterize the pharmacokinetic and pharmacodynamic properties of drug candidates in discovery settings. Comb Chem High Throughput Screen 13(2):207–218.

[102] Bourdet DL, Tsuruda PR, Obedencio GP, Smith JA (2012) Prediction of human serotonin and norepinephrine transporter occupancy of duloxetine by pharmacokinetic/pharmacodynamic modeling in the rat. J Pharmacol Exp Ther 341(1):137–145.

[103] Gunn RN, Summerfield SG, Salinas CA, Read KD, Guo Q, Searle GE, Parker CA, Jeffrey P, Laruelle M (2012) Combining PET biodistribution and equilibrium dialysis assays to assess the free brain concentration and BBB transport of CNS drugs. J Cereb Blood Flow Metab 32(5):874–883.

[104] Takano A, Halldin C, Farde L (2013) SERT and NET occupancy by venlafaxine and milnacipran in nonhuman primates: a PET study. Psychopharmacology (Berl) 226(1):147–153.

[105] Borsook D, Upadhyay J, Klimas M, Schwarz AJ, Coimbra A, Baumgartner R, George E, Potter WZ, Large T, Bleakman D, Evelhoch J, Iyengar S, Becerra L, Hargreaves RJ (2012) Decision-making using fMRI in clinical drug development: revisiting NK-1 receptor antagonists for pain. Drug Discov Today 17(17–18):964–973.

[106] Poulin P, Jones HM, Jones RD, Yates JW, Gibson CR, Chien JY, Ring BJ, Adkison KK, He H, Vuppugalla R, Marathe P, Fischer V, Dutta S, Sinha VK, Björnsson T, Lavé T, Ku MS

(2011) PhRMA CPCDC initiative on predictive models of human pharmacokinetics. J Pharm Sci 100(10): 4050–4073.

[107] Uchida Y, Tachikawa M, Obuchi W, Hoshi Y, Tomioka Y, Ohtsuki S, Terasaki T (2013) A study protocol for quantitative targeted absolute proteomics (QTAP) by LC–MS/MS: application for inter-strain differences in protein expression levels of transporters, receptors, claudin-5, and marker proteins at the blood-brain barrier in ddY, FVB, and C57BL/6J mice. Fluids Barriers CNS 10(1):21.

[108] Feng B, Mills JB, Davidson RE, Mireles RJ, Janiszewski JS, Troutman MD, de Morais SM (2008) In vitro P-glycoprotein assays to predict the in vivo interactions of P-glycoprotein with drugs in the central nervous system. Drug Metab Dispos 36(2):268–275.

[109] Zhou L, Schmidt K, Nelson FR, Zelesky V, Troutman MD, Feng B (2009) The effect of breast cancer resistance protein and P-glycoprotein on the brain penetration of flavopiridol, imatinib mesylate (Gleevec), prazosin, and 2-methoxy-3-(4-(2-(5-methyl-2-phenyloxazol-4-yl)ethoxy)phenyl) propanoic acid (PF-407288) in mice. Drug Metab Dispos 37(5):946–955.

[110] Shawahna R, Decleves X, Scherrmann JM (2013) Hurdles with using in vitro models to predict human blood-brain barrier drug permeability: a special focus on transporters and metabolizing enzymes. Curr Drug Metab 14(1):120–136.

[111] de Lange EC. (2013) Utility of CSF in translational neuroscience. J Pharmacokinet Pharmacodyn 40(3):315–326.

[112] Ball K, Bouzom F, Scherrmann JM, Walther B, Declèves X (2013) Physiologically based pharmacokinetic modelling of drug penetration across the blood-brain barrier—towards a mechanistic IVIVE-based approach. AAPS J 15(4):913–932.

[113] Ball K, Bouzom F, Scherrmann JM, Walther B, Declèves Xww (2014) A Physiologically Based Modeling Strategy during Preclinical CNS Drug Development. Mol Pharm. 11(3):836–848.

[114] Badhan RK, Chenel M, Penny JI (2014) Development of a physiologically-based pharmacokinetic model of the rat central nervous system. Pharmaceutics 6(1):97–136.

[115] Bueters T, Ploeger BA, Visser SA (2013) The virtue of translational PKPD modeling in drug discovery: selecting the right clinical candidate while sparing animal lives. Drug Discov Today 18(17–18):853–862.

[116] Melhem M (2013) Translation of central nervous system occupancy from animal models: application of pharmacokinetic/pharmacodynamic modeling. J Pharmacol Exp Ther 347(1):2–6.

[117] Ploeger BA, van der Graaf PH, Danhof M (2009) Incorporating receptor theory in mechanism-based pharmacokinetic–pharmacodynamic (PK–PD) modeling. Drug Metab Pharmacokinet 2009;24(1):3–15.

[118] Westerhout J, Danhof M, De Lange EC (2011) Preclinical prediction of human brain target site concentrations: considerations in extrapolating to the clinical setting. J Pharm Sci 100(9):3577–3593.

[119] Mangas-Sanjuan V, González-Álvarez I, González-Álvarez M, Casabó VG, Bermejo M (2013) Innovative in vitro method to predict rate and extent of drug delivery to the brain across the blood-brain barrier. Mol Pharm 10(10):3822–3831.

[120] Westerhout J, Ploeger B, Smeets J, Danhof M, de Lange EC (2012) Physiologically based pharmacokinetic modeling to investigate regional brain distribution kinetics in rats. AAPS J 14(3):543–553.

[121] de Lange EC (2013) The mastermind approach to CNS drug therapy: translational prediction of human brain distribution, target site kinetics, and therapeutic effects. Fluids Barriers CNS 10(1):12.

[122] Enna SJ, Williams M (2009) Challenges in the search for drugs to treat central nervous system disorders. J Pharmacol Exp Ther 329(2):404–411.

[123] Neuwelt E, Abbott NJ, Abrey L, Banks WA, Blakley B, Davis T, Engelhardt B, Grammas P, Nedergaard M, Nutt J, Pardridge W, Rosenberg GA, Smith Q, Drewes LR (2008) Strategies to advance translational research into brain barriers. Lancet Neurol 7(1):84–96.

[124] Dominguez A, Alvarez A., Hilario E, Suarez-Merino B, Goni-de-Cerio F (2013) Central nervous system diseases and the role of the blood-brain barrier in their treatment. Neurosci Discov 1(3):1–11.

[125] Agarwal S, Manchanda P, Vogelbaum MA, Ohlfest JR, Elmquist WF (2013) Function of the blood-brain barrier and restriction of drug delivery to invasive glioma cells: findings in an orthotopic rat xenograft model of glioma. Drug Metab Dispos 41(1):33–39.

[126] Meyer RP, Gehlhaus M, Knoth R, Volk B (2007) Expression and function of cytochrome p450 in brain drug metabolism. Curr Drug Metab 8(4):297–306.

[127] Iliff JJ, Lee H, Yu M, Feng T, Logan J, Nedergaard M, Benveniste H (2013) Brain-wide pathway for waste clearance captured by contrast-enhanced MRI. J Clin Invest 123(3):1299–1309.

[128] Chaves C, Shawahna R, Jacob A, Scherrmann JM, Declèves X (2014) Human ABC transporters at blood–CNS interfaces as determinants of CNS drug penetration. Curr Pharm Des 20(10):1450–1462.

[129] Grasela TH, Slusser R (2010) Improving productivity with model-based drug development: an enterprise perspective. Clin Pharmacol Ther 88(2):263–268.

3

FREE DRUG HYPOTHESIS FOR CNS DRUG CANDIDATES

XINGRONG LIU[1] AND CUIPING CHEN[2]
[1]*Genentech Inc., South San Francisco, CA, USA*
[2]*Depomed Inc., Newark, CA, USA*

INTRODUCTION

For most organs, no barriers exist between blood and the organ to restrict the diffusion of small molecular–weight drugs between the blood and interstitial space of the organ, resulting in identical unbound or free drug concentrations in the blood and the organ at steady state or equilibrium. The brain and a few other organs are exceptions to this rule. The brain is separated from the systemic circulation by two barriers: the blood–brain barrier (BBB) and the blood–cerebrospinal fluid barrier (BCSFB). The BBB is composed of cerebral endothelial cells of the blood capillaries that differ from those in the rest of the body by the presence of extensive tight junctions, absence of fenestrations, and sparse pinocytotic vesicular transport. The BCSFB is formed by a continuous layer of epithelial cells that line the choroid plexus in the ventricles. The BBB and BCSFB exhibit very low paracellular permeability and express multiple drug transporters, such as efflux drug transporters P-glycoprotein (P-gp) and breast cancer resistance protein (BCRP). These characteristics restrict the diffusion of hydrophilic compounds and regulate the diffusion of lipophilic efflux transport substrates between the blood and the brain [1]. The drug in the brain can be divided into intravascular space (within the blood capillaries), interstitial space, and intracellular space. For most central nervous system (CNS) compounds, it is generally assumed that the interstitial unbound drug concentration is the same as the intracellular unbound concentration. The unbound concentration in the interstitial space is defined

Blood–Brain Barrier in Drug Discovery: Optimizing Brain Exposure of CNS Drugs and Minimizing Brain Side Effects for Peripheral Drugs, First Edition. Edited by Li Di and Edward H. Kerns.

as the unbound brain concentration. At steady state, the unbound concentration in the brain is not necessarily equal to the unbound concentration in the blood. If a compound is a substrate of an efflux transporter, such as P-gp or BCRP, then the unbound brain concentration can be lower than the unbound blood concentration. In this chapter we will discuss the free drug hypothesis with specific emphasis on those drugs or compounds that are targeted for the CNS. We will summarize recent literature data that support the free drug hypothesis for CNS drugs and the authors' opinions on how to design CNS drugs to maximize their CNS activities.

Modern pharmacology is based upon the principle, referred to as the free drug hypothesis, that drug effects are determined by the interactions of unbound drug molecules at the site of action and drug receptors [2]. Drug receptors can be broadly defined as macromolecules, either on the surface of cells, such as G-protein coupled receptors (GPCRs), kinase receptors, ion channels, and transporters, or inside cells, such as enzymes, proteins on calcium storage organelles, and nuclear receptors. This principle also serves as the foundation for pharmacokinetics, where it is assumed that only unbound drug molecules can diffuse across a lipid membrane, such as a cell plasma membrane, following their concentration gradient. At equilibrium the concentrations of the unbound and unionized drug are the same on both sides of the biological membrane in the absence of active drug transporters on the lipid membrane. In the presence of active drug transporters, the drug concentrations at equilibrium are governed by the passive permeability of the drug and the characteristics of the interaction between the drug and the active drug transporters such as affinity, maximal transport capacity, and transport direction. In a recent review, Smith et al. [3] examined the *in vitro* activities including equilibrium binding constant K_b, equilibrium inhibition constant K_I, half-maximal inhibitory concentration IC_{50}, or minimum inhibitory concentration (MIC) and their *in vivo* observed average efficacious unbound plasma drug concentration for 16 drugs in humans (Table 3.1). They found that the *in vivo* efficacious unbound plasma concentration is generally within one- to threefold of their *in vitro* activities. This data set strongly supports the fact that it is the unbound drug concentration and not the total drug concentration that determines *in vivo* activity.

Due to difficulties in accurately measuring unbound brain concentration, there were limited reports in the literature comparing the relationship between unbound versus total brain concentration and *in vivo* activity. Typically, the concentration for the drug receptors in the brain is generally very low and a majority of drug molecules in the brain bind to lipids and/or partition into organelles, such as lysosomes, and only a very small fraction of molecules exist in unbound form. The traditional approach for determining unbound brain concentration is brain microdialysis. Brain microdialysis is a direct approach to determine unbound brain concentration using a semipermeable dialysis probe and represents the gold standard for determination of unbound brain concentration. A detailed discussion of this method can be found in Chapter 18 in this book. However, the requirement for extensive resources, low throughput, and the challenges of nonspecific binding to the apparatus limit the use of this methodology in the drug discovery setting and do not render rapid throughput to examine a large set of compounds to test the free drug hypothesis for CNS targets.

TABLE 3.1 *In vitro* potency and *in vivo* free drug concentrations at the mean efficacious dose[a]

Drug target	Compound	*In vitro* measure	*In vitro* concentration (nM)	Average free drug concentration (nM)
Ca^{2+} channel	Nifedipine	IC_{50}	4	6
Ca^{2+} channel	Amlodipine	IC_{50}	2	1
5-HT transporter	Sertraline	K_i	7	4
K^+ channel	Dofetilide	EC_{15}	7	3
M3 muscarinic receptor	Darifenacin	K_b	4	10
M3 muscarinic receptor	Zamifenacin	K_b	10	20
M3 muscarinic receptor	UK-112,166	A_2	1	3
β-adrenergic receptor	Propranolol	K_i	4.5	3
β-adrenergic receptor	Alprenolol	K_i	8	18
α_{1A}-adrenergic receptor	Tamsulosin	K_i	0.04	0.03–0.16
α_{1A}-adrenergic receptor	Terazosin	K_i	1	1–9
A_{2A} adenosine receptor	2-chloroadenosine	K_i	80	202–225
PDE5 inhibitor	Sildenafil	K_i	4	10
Thromboxin receptor antagonist	UK-147,535	A_2	0.1	0.5
CYP51	Fluconazole	MIC	2600	Exceeds MIC for 8 h; $C_{ave,u}$ over this period is 4000
CYP51	Ketoconazole	MIC	20	Exceeds MIC for 8 h; $C_{ave,u}$ over this period is 200

[a]Reproduced from Smith et al. [3]. Used with permission from Nature Publishing Group.

5-HT, 5-hydroxytryptamine (serotonin); A_2, a value obtained from a Schild plot analysis, which calculates antagonist potency in a functional assay; $C_{av,u}$, the average free drug concentration; IC_{50}, half-maximal inhibitory concentration; K_b, the equilibrium binding constant, a measure of drug potency (affinity) in a receptor binding assay; K_i, equilibrium inhibition constant, a measure of the potency of enzyme inhibition; Mic; minimum inhibitory concentration; PDE5, phosphodiesterase 5.

Particularly, poor recovery for lipophilic compounds has severely limited its wide usage in drug discovery because currently many compounds designed for CNS targets are lipophilic and exhibit high protein binding. Alternatively, unbound plasma concentration and CSF drug concentrations have been used as surrogates for the unbound brain concentrations. This surrogate approach is valid only when a drug is not a substrate for the drug transporters at the BBB [4, 5]. Often times it is difficult to rule out confidently whether a novel compound is a substrate of drug transporters at the BBB.

In the last few years, these difficulties have been overcome, at least partially, by the advancement and application of several surrogate *in vitro* methods, such as the brain homogenate and brain slice methods for the estimation of unbound fraction or unbound concentration in the brain tissue [6–20]. The advancement of these methods has enabled several studies in the last few years to extensively examine the free drug hypothesis for a large number of CNS compounds/drugs. We will first briefly review these methods and then discuss their applications to several *in vivo* studies.

USE OF BRAIN HOMOGENATE AND BRAIN SLICE METHODS TO DETERMINE UNBOUND BRAIN CONCENTRATION

For the brain homogenate method, one part of blank brain tissue is typically mixed with two parts of buffer and then the brain homogenate is prepared using a tissue homogenizer. The unbound fraction of a drug in the brain homogenate is determined using a conventional equilibrium dialysis apparatus, where the brain homogenate is placed on one side of a semipermeable membrane and a buffer on the other side, identical to the equilibrium dialysis method of the plasma protein binding assay [11, 13]. Then the unbound fraction determined in the brain homogenate is extrapolated to calculate the unbound fraction in the intact brain tissue using Equation 3.1, which incorporates a correction factor for the dilution effects of the brain homogenate method:

$$f_{\text{u,brain}} = \frac{1}{1 + \left(\dfrac{1}{f_{\text{u,homogenate}}} - 1 \right) * D} \tag{3.1}$$

where $f_{\text{u,brain}}$ represents the unbound fraction in brain tissue and $f_{\text{u,homogenate}}$ the unbound fraction in brain homogenate. D is the dilution factor for the brain homogenate. For example, if the brain homogenate is prepared by mixing one part of brain with two parts of buffer, then the dilution factor is 3. Once the unbound brain fraction is estimated, the unbound brain concentration ($C_{\text{u,brain}}$) can be calculated from the product of the unbound brain fraction and the observed *in vivo* total brain concentration.

A good correlation was observed for model compounds between the *in vitro* brain homogenate method and *in vivo* brain microdialysis [9]. In this work, the accuracy of using the brain homogenate method to estimate $C_{\text{u,brain}}$ and the brain

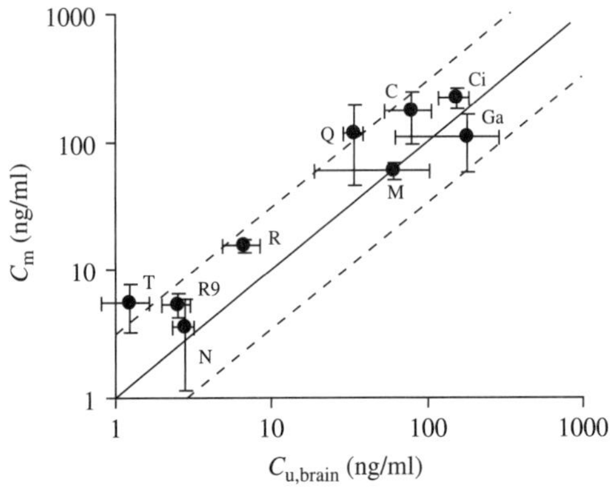

FIGURE 3.1 The relationship of brain interstitial unbound drug concentration measured by brain microdialysis (C_m) and brain unbound drug concentration measured by brain homogenate method ($C_{u,brain}$) at steady state in rats (mean ± SD, n = 3–6). The $C_{u,brain}$ values were calculated from the total brain concentration and brain unbound fraction using Equation . All of the concentrations represent the values at 6 h post start of an intravenous bolus and followed with a constant intravenous infusion. The solid and dashed lines represent unity and threefold boundaries, respectively. C, Carbamazepine; Ci, Citalopram; G, Ganciclovir; M, Metoclopramide; N, *N*-desmethylclozapine; Q, Quinidine; R, Risperidone; R9, 9-OH-Risperidone; and T, Thiopental. Reproduced from Liu et al. [9]. © 2009 American Society of Pharmacology and Experimental Therapeutics (ASPET). With kind permission of ASPET.

interstitial fluid concentration determined by brain microdialysis (C_m) were examined. Nine compounds—carbamazepine, citalopram, ganciclovir, metoclopramide, *N*-desmethylclozapine, quinidine, risperidone, 9-hydroxyrisperidone, and thiopental—were selected and each was administered as an intravenous bolus followed by a constant intravenous infusion for 6h in rats. For 8 of the 9 compounds, the $C_{u,brain}$ values were within threefold of their C_m values; the $C_{u,brain}$ of thiopental was within fourfold of its C_m (Fig. 3.1). Similar results were also reported by Friden et al. [15]. These authors used unbound volume of distribution in the brain, $V_{u,brain}$, which is equivalent to the reciprocal of the unbound fraction in the brain tissue in their study. They determined $V_{u,brain}$ using the brain homogenate method for 15 compounds—alovudine, R-apomorphine, S-apomorphine, R-cetirizine, S-cetirizine, codeine, CP-122721, diazepam, gabapentin, morphine, morphine-3-glucuronide, morphine-6-glucuronide, norfloxacin, oxycodone, and thiopental—and observed that $V_{u,brain}$, estimated using brain homogenate, was within threefold of the observed $V_{u,brain}$ from brain microdialysis for 10 of the 15 compounds and within fivefold for the other five compounds (Fig. 3.2a). These results suggest that the brain homogenate method is a practical approach for estimating the unbound fraction in the brain tissue.

The limitations of the brain homogenate method include the loss of cellular structure during the preparation of brain homogenate and release of the proteins normally restricted to the intracellular space. To address these concerns, the brain slice method was adapted to determine the unbound drug fraction in brain tissue [10, 15]. Brain

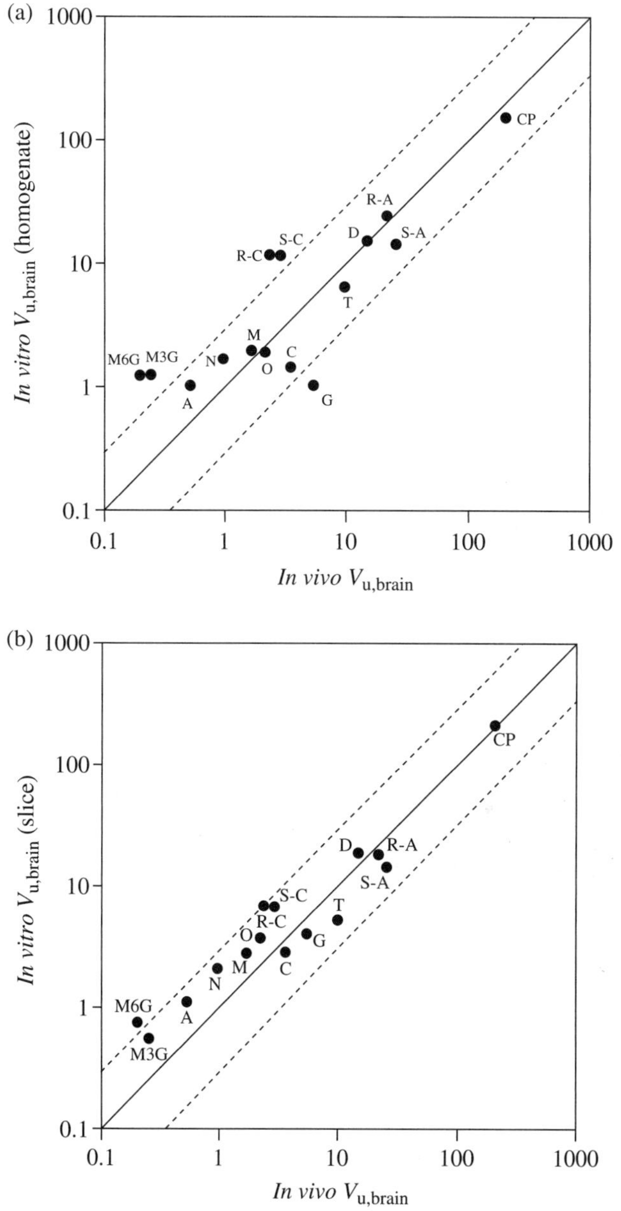

FIGURE 3.2 Relationship between *in vivo* $V_{u,brain}$ values and (a) *in vitro* brain homogenate values and (b) *in vitro* slice values. The solid line represents perfect agreement. The dashed lines represent a threefold over- or underestimation compared with *in vivo* $V_{u,brain}$ values. A, alovudine; R-A, R-apomorphine; S-A, S-apomorphine; R-C, R-cetirizine, S-C, S-cetirizine; C, codeine; CP, 1.41CP-122721; D, diazepam; G, gabapentin; M, morphine; M3G, morphine-3-glucuronide; M6G, morphine-6-glucuronide; N, norfloxacin; O, oxycodone; T, thiopental. Reproduced from Friden et al. [15]. © 2007 American Society of Pharmacology and Experimental Therapeutics (ASPET). With kind permission of ASPET.

slices have been used to study neural physiology for almost half a century [21]. Cellular structure is maintained in brain slices but the BBB is not functional, since a compound can directly penetrate into the brain slices from the incubation medium [22]. Use of brain slices to estimate the unbound fraction of drug in brain tissue has been reported in the literature [23]. In that study, 1000 μm brain slices were incubated for 1 h and the buffer-to-slice drug ratio was assumed to be equal to its unbound fraction in the brain tissue ($f_{u,brain}$). Brain slices have also been used to study the partition coefficient between the slice and the incubation buffer [24, 25]. Based on the same concept, Kakee et al. [26] used brain slices to estimate the apparent distribution volume in the brain.

Our laboratory demonstrated that the brain slice technique could be adapted to determine the unbound brain fraction in a drug discovery setting [10]. Eight compounds—caffeine, CP-141938, fluoxetine, *N*[3-(4′-fluorophenyl)-3-(4′-phenylphenoxy)-propyl] sarcosine (NFPS), propranolol, quinidine, theobromine, and theophylline—were selected to evaluate the utility of the brain slice method in this aspect. The selection of these model compounds was based on their physicochemical properties, BBB permeability, *in vivo* K_p values, and P-gp transporter activity. Two brain slice methods, indirect and direct, have been assessed and developed. In the indirect method, mouse or rat brain slices were incubated in plasma containing a testing compound and the slice-to-plasma concentration ratio (C_{slice}/C_{plasma}) was determined. In the direct method, the brain slices were incubated with a buffer containing a testing compound and the buffer-to-slice concentration ratio, that is, unbound fraction in brain tissue ($f_{u,slice}$), was determined. The $f_{u,slice}$ is equivalent to the unbound fraction in brain tissue, $f_{u,brain}$, discussed earlier.

Figure 3.3a exhibits the relationship between rat *in vivo* brain-to-plasma ratios (K_p values) and C_{slice}/C_{plasma} ratios for the eight model compounds, based on data derived from the direct method. The C_{slice}/C_{plasma} values for six non-P-gp substrates (caffeine, fluoxetine, NFPS, propranolol, theobromine, and theophylline) were within threefold of the observed *in vivo* K_p. The C_{slice}/C_{plasma} values of two P-gp substrates (CP-141938 and quinidine) were nine- and fivefold greater than their *in vivo* K_p, respectively.

Figure 3.3b shows the relationship between rat K_p and $f_{u,plasma}/f_{u,slice}$ values of the same eight model compounds based on data derived from the indirect method. The $f_{u,plasma}/f_{u,slice}$ values for five of the six non-P-gp substrates were within threefold of the observed K_p. Only NFPS showed a fourfold difference between its $f_{u,plasma}/f_{u,slice}$ and K_p. The $f_{u,plasma}/f_{u,slice}$ for two P-gp substrates, CP-141938 and quinidine, were 16- and 12-fold, respectively, greater than the K_p. Therefore, both the direct and indirect brain slice methods were able to predict *in vivo* K_p for non-P-gp substrates and overpredicted *in vivo* K_p for P-gp substrates. This overprediction for the P-gp substrates is expected because the brain slice method measures the brain-to-plasma ratio due to brain tissue binding and plasma protein binding without drug transport functions at the BBB as in the *in vivo* situation.

In order to compare the brain slice and brain homogenate methods in the estimation of unbound fraction in brain, the relationship between K_p and unbound fraction in plasma divided by the unbound fraction determined from brain

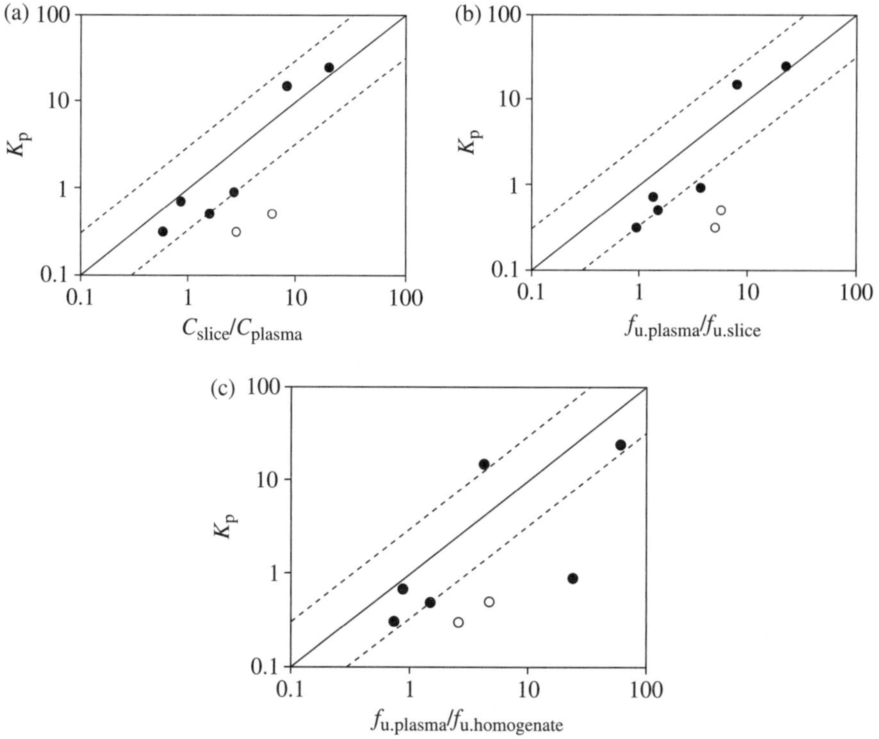

FIGURE 3.3 Relationship between *in vivo* K_p and (a) C_{slice}/C_{plasma}, (b) $f_{u,plasma}/f_{u,slice}$ ratio, and (c) $f_{u,plasma}/f_{u,homogenate}$ ratio in rats. Solid and open symbols represent non-P-gp substrates (caffeine, fluoxetine, NFPS, propranolol, theobromine, and theophylline) and P-gp substrates (CP-141938 and quinidine), respectively. Solid and dashed lines represent unity and threefold boundaries, respectively. Reproduced from Becker and Liu [10]. © 2006 American Society of Pharmacology and Experimental Therapeutics (ASPET). With kind permission of ASPET.

homogenate ($f_{u,plasma}/f_{u,homogenate}$) for the eight model compounds was examined (Fig. 3.3c). The $f_{u,plasma}/f_{u,homogenate}$ for four of the six non-P-gp substrates (caffeine, fluoxetine, theobromine, and theophylline) were within threefold of *in vivo* K_p. However, for the two non-P-gp substrates, NFPS and propranolol, $f_{u,plasma}/f_{u,homogenate}$ overpredicted 27-fold and underpredicted fourfold the *in vivo* K_p, respectively. For the two P-gp substrates, CP-141938 and quinidine, $f_{u,plasma}/f_{u,homogenate}$ overpredicted their K_p by eightfold and ninefold, respectively. These results indicate that the brain slice method is comparable to or better than the brain homogenate method in predicting K_p.

Friden at al. [15] confirmed this observation in their work in which they compared the brain homogenate and brain slice methods in the prediction of unbound volume of distribution in the brain (Fig. 3.2a). The slice method was within threefold of the *in vivo* results for all but one compound. Although successful in 10 of 15 cases, the brain homogenate method failed to estimate the $V_{u,brain}$ of drugs that reside predominantly in the interstitial space or compounds that are accumulated intracellularly [26].

Compared to the brain slice method, the brain homogenate method is easier to set up in any laboratories that are capable of the determination of plasma protein binding; thus, it has been widely used by many laboratories to estimate the unbound brain fraction [6–15]. In the situation where the protein binding is too high to be accurately measured or drug transport across the cell plasma membrane of neurons is anticipated, the brain slice method needs to be used to determine the unbound brain fraction and possibly to estimate intracellular unbound drug concentration [15].

Once we know the unbound fraction in brain tissue, the unbound brain concentration can be calculated from the total brain concentration, which is measured from *in vivo* studies, using Equation 3.2:

$$C_{\text{u,brain}} = f_{\text{u,brain}} \cdot C_{\text{brain}} \tag{3.2}$$

where $C_{\text{u,brain}}$, C_{brain}, and $f_{\text{u,brain}}$ represent, respectively, the unbound brain concentration, total brain concentration, and unbound fraction in the brain tissue. This approach is used to estimate the unbound brain concentration in the following pharmacokinetic/pharmacodynamic (PKPD) studies.

IN VIVO PHARMACOKINETIC/PHARMACODYNAMIC OBSERVATIONS TO SUPPORT FREE DRUG HYPOTHESIS FOR CNS DRUGS

Several PKPD studies have been published in the last few years to specifically examine the free drug hypothesis for CNS drugs. Liu et al. [27] examined the relationship between *ex vivo* receptor occupancy and total and unbound plasma and brain concentrations for 18 serotonin (SERT) and dopamine (DAT) dual reuptake transporter inhibitors (Fig. 3.4). Receptor occupancy was used as a biomarker for the CNS effects for these inhibitors. As each compound has a different affinity toward the targets, the concentrations were normalized by the corresponding affinity (K_I). According to the free drug hypothesis, in which the unbound brain concentration represents the drug concentration interacting with the CNS targets, one can predict that the unbound brain concentration that results in 50% occupancy for the receptor (OC_{50}) should be similar to the *in vitro* K_I (i.e., concentration/K_I ratio equals unity) if the impact of the endogenous ligands on the occupancy of the drug is negligible. In contrast, total brain OC_{50} will be greater than K_I, because most of the drug molecules are nonspecifically bound to the cellular components, such as phospholipids, and are not available to interact with the drug target. The experimental observations are consistent with these predictions. The average observed OC_{50} was 44- to 53-fold greater than their K_I, expressed as total plasma concentration, and was 408- and 410-fold greater than their K_I, expressed as total brain concentration (Fig. 3.4, Table 3.2). In contrast, the average observed IC_{50} was within 4.6-fold of their K_I, expressed as unbound plasma concentration, and 3.3- to 4.1-fold of the unbound brain concentration (Fig. 3.4, Table 3.2). This view is further supported by the correlation between plasma and brain OC_{50} and *in vitro* affinity K_I (Fig. 3.5). Clearly the correlation is better with the unbound concentration in the plasma or in the brain

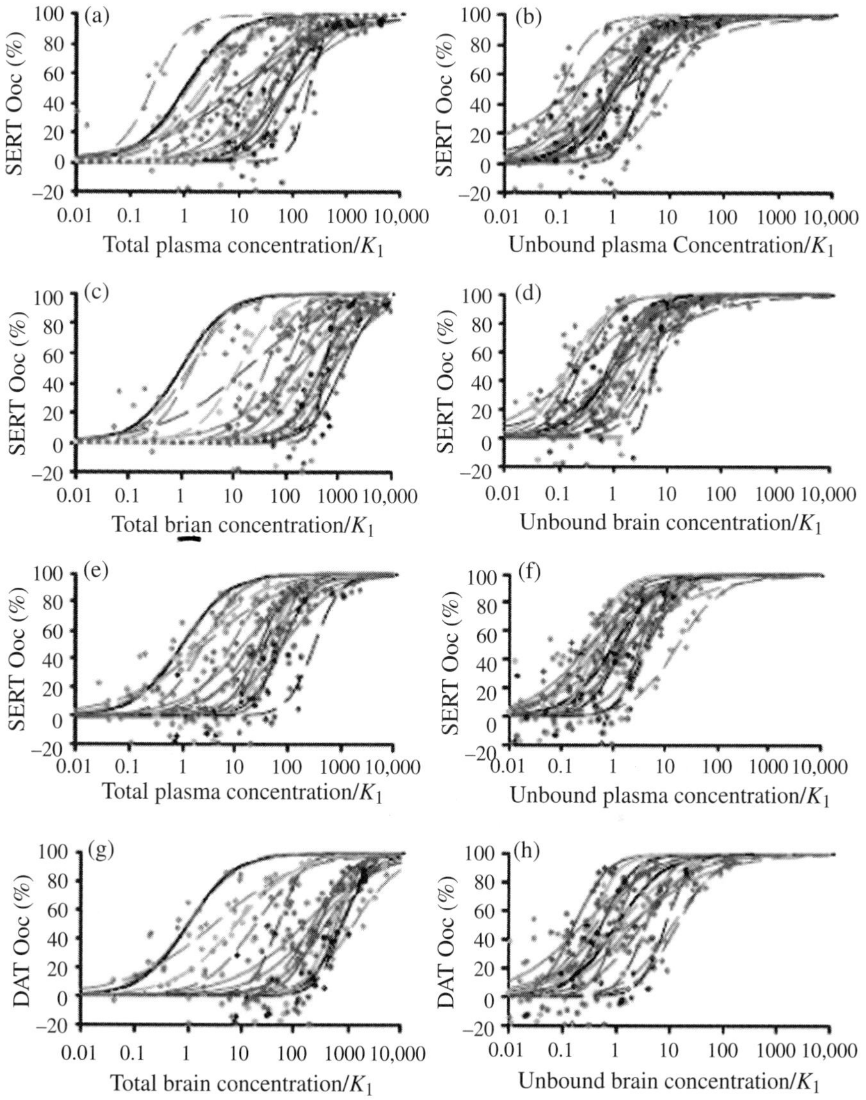

FIGURE 3.4 Relationship between *in vivo* observed (a–d) SERT and (e–h) DAT receptor occupancy and the ratio of concentration/K_I of 18 compounds after intraperitoneal administration in rats. The occupancy, and plasma and brain concentration were determined from the *in vivo* study with 3–6 dose levels for each compound. Unbound plasma and brain concentrations were calculated from the total concentration and corresponding unbound fractions. K_I for each compound was determined from *in vitro* expressed human SERT or DAT cell membranes. The solid black lines represent the theoretical simulation using Equation 3.5 assuming *in vivo* $IC_{50} = K_I$. Each type of symbol represents one compound and each symbol represents observed datum from one rat for the compound. Other lines represent the best fit of Hill's equation to the observed data for each compound. Reproduced from Liu et al. [27]. © 2009 American Society of Pharmacology and Experimental Therapeutics (ASPET). With kind permission of ASPET.

TABLE 3.2 Fold of difference of OC_{50} and K_I of the 18 compounds[a]

Compound	Plasma SERT OC_{50}	Unbound plasma SERT OC_{50}	Brain SERT OC_{50}	Unbound brain SERT OC_{50}	Plasma DAT OC_{50}	Unbound plasma DAT OC_{50}	Brain DAT OC_{50}	Unbound brain DAT OC_{50}
NS2359	45	9.0	689	6.0	86	17	1328	12
Cocaine	(4.1)	(9.9)	1.6	(5.1)	1.0	(2.5)	7.0	(1.2)
Bupropion	NA	NA	NA	NA	2.0	(3.9)	19	3.0
Duloxetine	3.5	(7.2)	612	2.8	NA	NA	NA	NA
Citalopram	2.4	(1.5)	41	2.5	NA	NA	NA	NA
Compound 1	80	3.8	833	1.3	66	3.2	625	1.0
Compound 2	44	1.0	464	(3.8)	29	(1.5)	303	(5.9)
Compound 3	69	1.2	1166	(1.3)	82	1.5	797	(1.9)
Compound 4	26	(3.8)	238	(6.9)	45	(2.2)	518	(3.2)
Compound 5	1.8	(23)	739	1.2	3.0	(17)	712	1.1
Compound 6	84	1.4	823	1.0	34	(1.8)	417	(2.0)
Compound 7	148	3.6	487	3.7	86	2.1	275	2.1
Compound 8	16	(4.5)	139	1.3	NA	NA	NA	NA
Compound 9	3.8	(1.1)	96	5.9	18	4.1	224	14
Compound 10	8.8	1.1	13	(1.1)	39	4.7	50	3.5
Compound 11	213	2.8	583	5.8	308	4.0	867	8.7
Compound 12	2.2	(1.4)	12	1.7	2.0	(1.6)	8	1.1
Compound 13	1.2	(1.3)	2	(4.4)	3.0	1.7	3	1.7
Mean	*44*	*4.6*	*408*	*3.3*	*53*	*4.6*	*410*	*4.1*
SD	*60*	*5.5*	*368*	*2.1*	*77*	*5.1*	*395*	*4.0*

[a]Reproduced from Liu et al. [27]. © 2009 American Society of Pharmacology and Experimental Therapeutics (ASPET). With kind permission of ASPET. Numbers in parentheses represent fold of OC_{50} less than K_I. N/A, data were not available.

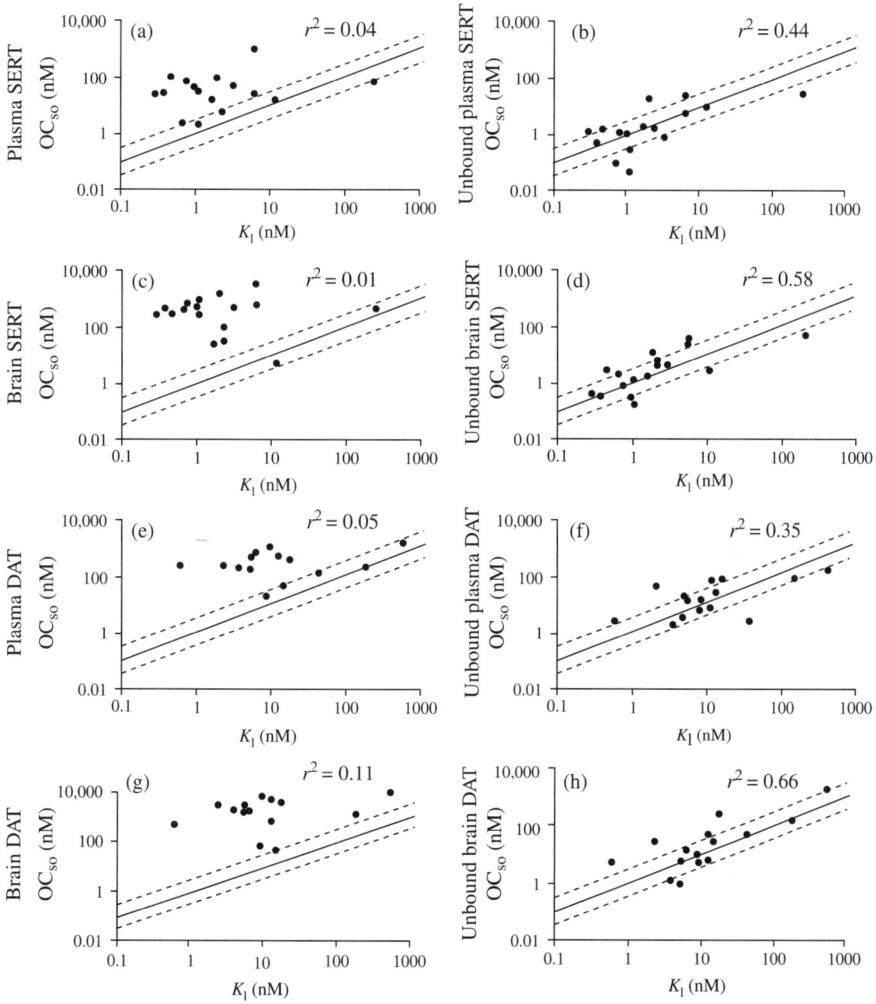

FIGURE 3.5 Relationship between *in vivo* OC_{50} and human *in vitro* K_I for (a–d) SERT and (e–h) DAT after intraperitoneal administration in rats. The OC_{50} was calculated from the data in Figure 3.4. The solid and broken lines represent unity and threefold boundaries, respectively. Reproduced from Liu et al. [27]. © 2009 American Society of Pharmacology and Experimental Therapeutics (ASPET). With kind permission of ASPET.

and the correlation was further improved for unbound brain OC_{50} compared to unbound plasma OC_{50}.

In the study discussed earlier, the unbound brain and unbound plasma OC_{50} values were similar. This was because most compounds in that study are not P-gp substrates. For P-gp substrates, the unbound plasma concentration is expected to be higher than the unbound brain concentration [28]. For example, Kalvass et al. [29] examined the

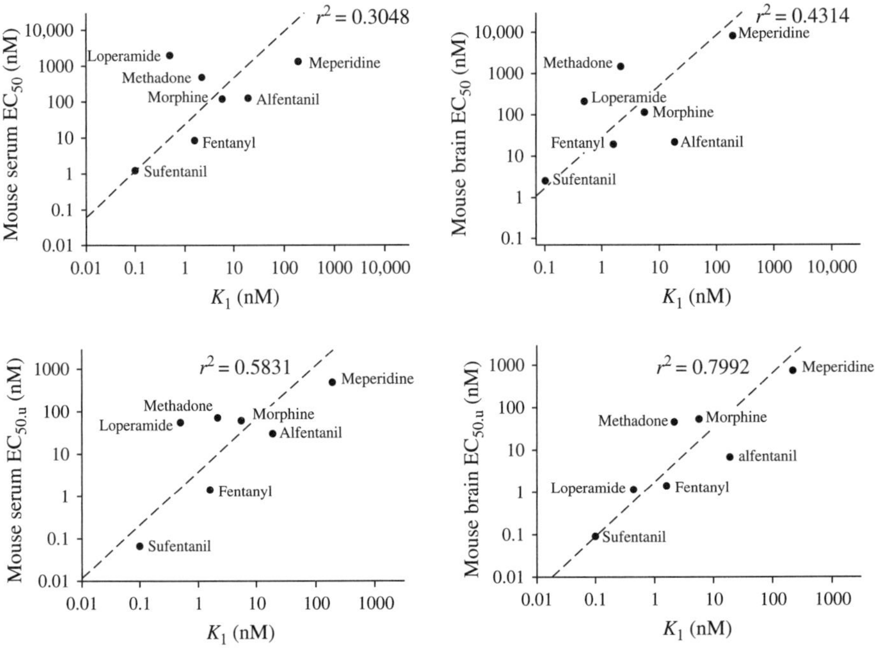

FIGURE 3.6 Correlation analysis for various *in vivo* measures relative to *in vitro* potency. The dashed line represents the line from log–log orthogonal regression analysis. Reproduced from Kalvass et al. [29]. © 2007 American Society of Pharmacology and Experimental Therapeutics (ASPET). With kind permission of ASPET.

correlation between *in vivo* EC_{50} for antinociceptive activity in the mouse and *in vitro* binding affinity (K_I) for seven μ-opioid agonists (alfentanil, fentanyl, loperamide, methadone, meperidine, morphine, and sufentanil). These compounds were selected as model CNS drugs because they elicit a readily measurable central effect (antinociception) and their clinical PKPD is well understood. The correlation between the mouse total and unbound serum and brain EC_{50} and K_I values is shown in Figure 3.6. The total serum and total brain EC_{50} values were weakly related to *in vitro* K_I (Fig. 3.6a and Fig. 3.6b, $r^2 < 0.5$). A modest improvement was observed with unbound serum $EC_{50,u}$ (Fig. 3.6c, $r^2 = 0.583$). However, the strongest correlation was observed between unbound brain $EC_{50,u}$ and K_I (Fig. 3.6d, $r^2 = 0.799$). Therefore, the unbound plasma or unbound brain EC_{50} was better correlated with the pharmacological antinociceptive activity than total plasma or total brain EC_{50}. Moreover, unbound brain EC_{50} was better correlated with the activity than the unbound plasma EC_{50}. This was due to the fact that some of the compounds in this study are P-gp substrates. For example, loperamide is a known P-gp substrate, which, at equilibrium, has a much higher unbound plasma concentration than unbound brain concentration. Consequently, its unbound plasma EC_{50} was more than 50-fold of their K_I while its unbound brain EC_{50} was similar to its K_I.

Watson et al. [17] reported similar correlations between receptor occupancy of dopamine D_2 receptor and *in vitro* binding affinity (K_I) for six marketed antipsychotic drugs. The main objective of the study was to investigate whether unbound brain concentration ($C_{u,brain}$) is a better predictor of dopamine D_2 receptor occupancy than total brain concentration (C_{brain}) or unbound blood concentration ($C_{u,blood}$). The ex vivo D_2 receptor occupancy and concentration–time profiles in blood and brain of six marketed antipsychotic drugs were determined after oral administration in rats at a range of dose levels. The relationships between the receptor occupancy and K_I normalized the total brain concentration, unbound brain concentration, and unbound blood concentrations are shown in Figure 3.7. These results showed that $C_{u,brain}$ of the antipsychotic agents is a good predictor of D_2 receptor occupancy in rats. Furthermore, $C_{u,brain}$ seemed to provide a better prediction of D_2 receptor occupancy than $C_{u,blood}$ for those compounds whose mechanism of entry into brain tissue is likely influenced by factors other than simple passive diffusion.

These results demonstrate that CNS activity is better correlated with unbound brain concentration than unbound plasma concentration, especially for those compounds whose mechanism of entry into brain tissue is influenced by drug transporters at the BBB. These results also suggest that unbound brain concentrations estimated from the brain homogenate method represents a practical approach to determine the biophase concentrations for CNS drugs. From the discussion hitherto, it is clear that higher unbound brain concentration leads to greater CNS activity for CNS-targeted drugs. In the following section, we will discuss the strategies to achieve higher unbound brain concentration and will also clarify several misconceptions and practices that should be avoided in CNS drug design.

STRATEGY TO INCREASE UNBOUND BRAIN CONCENTRATION IN CNS DRUG DESIGN

In order to understand the strategies to increase unbound brain concentration, we need to first discuss the factors determining the unbound brain concentration so we may modulate these factors in CNS drug design to achieve high unbound brain concentration. A key parameter that determines the efficiency for brain drug delivery is the ratio of unbound brain concentration to unbound plasma concentration called $K_{p,uu}$ [30].

$$K_{p,uu} = \frac{C_{u,brain}}{C_{u,blood}} \tag{3.3}$$

If $K_{p,uu}$ is 1 for a drug, the unbound brain concentration can reach 100% of the unbound plasma concentration at equilibrium, indicating that the BBB does not hinder the delivery of the drug from the blood into the brain. In contrast, if $K_{p,uu}$ is 0.1 for a drug, the unbound brain concentration can only rise up to 10% of the unbound plasma concentration at equilibrium, suggesting that the BBB severely impairs the

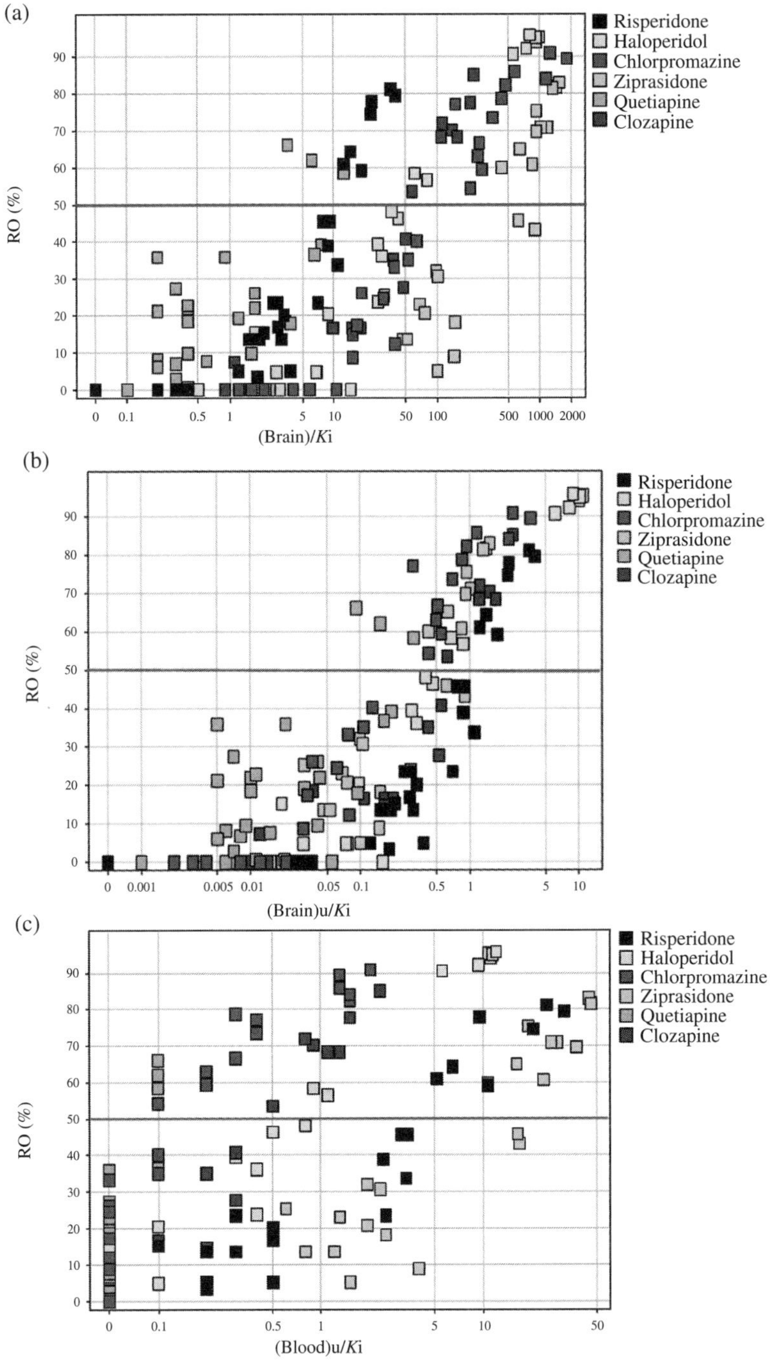

FIGURE 3.7 Relationship between receptor occupancy and (a) total brain concentration, (b) unbound brain concentration, and (c) unbound blood concentration. The concentration on the *x*-axis is normalized for the *in vitro* affinity for rat striatal D_2 receptors. Solid line represents 50% receptor occupancy. Reproduced from Watson et al. [17]. © 2009 American Society of Pharmacology and Experimental Therapeutics (ASPET). With kind permission of ASPET.

delivery of the drug into the brain. At equilibrium, the $K_{p,uu}$ can be described by the following equation [31, 32]:

$$K_{p,uu} = \frac{Cl_{diffusion} + Cl_{uptake}}{Cl_{diffusion} + Cl_{efflux} + Cl_{bulk} + Cl_{metabolism}} \quad (3.4)$$

$Cl_{diffusion}$ is the passive diffusion clearance, which equals the permeability-surface (PS) area product of the BBB. Cl_{uptake} and Cl_{efflux} represent uptake and efflux transport clearance across the BBB, respectively. Cl_{bulk} and $Cl_{metabolism}$ represent bulk flow, which is the fluid in the brain tissue moving into the CSF and blood, and metabolic clearance in the brain tissue, respectively. From Equation 3.4, it is evident that high uptake clearance and low efflux clearance lead to high $K_{p,uu}$. It is desirable to design a compound as a substrate of brain uptake transporters to enhance Cl_{uptake}. For example, large neutral amino acid transporter 1 transports L-DOPA and gabapentin across the BBB [33, 34]. Although L-DOPA has been available for more than 30 years, the same success in increasing brain penetration of other drugs has rarely been replicated except for its close-in analogs. Effective *in vitro* approaches have yet to be developed to harness endogenous transporters at the BBB for CNS drug candidates during early drug screening.

Thus far, evidence has suggested that it would be more feasible to design compounds with high passive permeability but without significant efflux transport (low Cl_{efflux}) than to design compounds as uptake transporter substrates. In general, the most effective and practical approach to enhance brain delivery for most CNS drug discovery programs is to design a CNS compound with no efflux activity. In the cases where the structure motif of a CNS compound required to interact with its target is the same as the motif that happens to be recognized by an efflux drug transporter at the BBB, the strategy to increase $K_{p,uu}$ may include designing a compound with a high passive permeability to negate the impact of the efflux transporter. One possible caveat of this approach is increased lipophilicity and, thus, increased metabolic instability. Sometimes the efflux issue for a compound becomes insurmountable and the scaffold may have to be abandoned and a new scaffold identified to avoid the efflux liability. Key considerations in a CNS drug discovery program are to understand the impact of the efflux on the overall profile, such as dose and safety margin, for a series of compounds and to assess if the efflux liability can be dialed out by structure modification without loss of the pharmacological activity. If the projected human effective dose is clinically feasible and the safety margin is acceptable for the intended therapeutic indications, then being an efflux substrate per se is not a "show stopper" for advancing to clinical development. Good examples in this category include risperidone and 9-hydroxyl risperidone. Both compounds are P-gp substrates, but both have been successfully developed for clinical use for the treatment of schizophrenia [35, 36].

Cl_{bulk} can play an important role in decreasing $K_{p,uu}$. It has been estimated that bulk flow clearance spans the range of 0.2–0.3 μl/min/g [37]. Take the example of mannitol, a compound of low permeability with a PS value of less than 1 μl/min/g. Bulk flow becomes significant compared to its permeability, resulting in a low $K_{p,uu}$ (0.01). For a compound with moderate to high permeability, Cl_{bulk} is insignificant. This is illustrated by caffeine, a compound of moderate to high permeability, with a PS value of 13 μl/min/g. In this case, bulk flow clearance is much lower than the permeability and has an insignificant effect on $K_{p,uu}$ (1.0) [38].

Additionally, brain metabolism, $Cl_{metabolism}$, could also play a significant role in reducing $K_{p,uu}$. Metabolizing enzymes such as monoamine oxidase (MAO), flavin-containing monooxygenase (FMO), cytochrome P450, and glucuronsyltransferases have been identified in brain endothelial cells and brain tissue [39–42]. Hence, the stability of a compound in brain tissue needs to be examined in early drug discovery.

Moreover, it is clear from Equation 3.3 that high unbound plasma concentration is conducive to high unbound brain concentration. For a typical orally administered drug, the key to increasing unbound plasma concentration is to design a compound with high absorption and low intrinsic metabolic clearance. Reduction of plasma protein binding will not necessarily increase the unbound plasma concentration. A more detailed discussion on the strategy to increase unbound plasma concentration can be found in recent reviews [3, 43]. The readers should be aware of and avoid several misconceptions regarding the approach to increase unbound brain concentration in CNS drug design, which are discussed in the following text.

REDUCTION OF BRAIN TISSUE BINDING CANNOT INCREASE UNBOUND BRAIN CONCENTRATION

One may postulate, based on Equation 3.2, that an increase of unbound fraction in the brain can lead to an increase of unbound brain concentration. This may have led some investigators to wonder whether reducing brain tissue binding can be used as a CNS drug design strategy to increase unbound brain concentration. This reasoning is correct only *in vitro* where the total brain concentration is determined by the total amount of drug that is introduced into the *in vitro* system and the volume in which the drug distributes. However, this reasoning is not correct *in vivo* where the unbound brain concentration is determined only by the product of the unbound plasma concentration and $K_{p,uu}$ as indicated by Equation 3.3 and is not determined by either the unbound fraction or total brain concentration. *In vivo*, Equation 3.2 needs to be rearranged as Equation 3.5 to represent the causal relationship among the total brain concentration, unbound brain concentration, and unbound brain fraction, where the total brain concentration is the dependent variable and the unbound brain concentration and unbound brain fraction are the independent variables:

$$C_{brain} = \frac{C_{u,brain}}{f_{u,brain}} \tag{3.5}$$

Therefore, *in vivo*, reduction of brain tissue binding (i.e., increase of $f_{u,brain}$) will not lead to the increase of the unbound brain concentration. Instead, it will lead to a reduced total brain concentration. Therefore, brain binding should not be considered a critical parameter in CNS drug design. Many successful CNS drugs have very high brain tissue binding. Maurer et al. [13] determined the unbound fractions in mouse

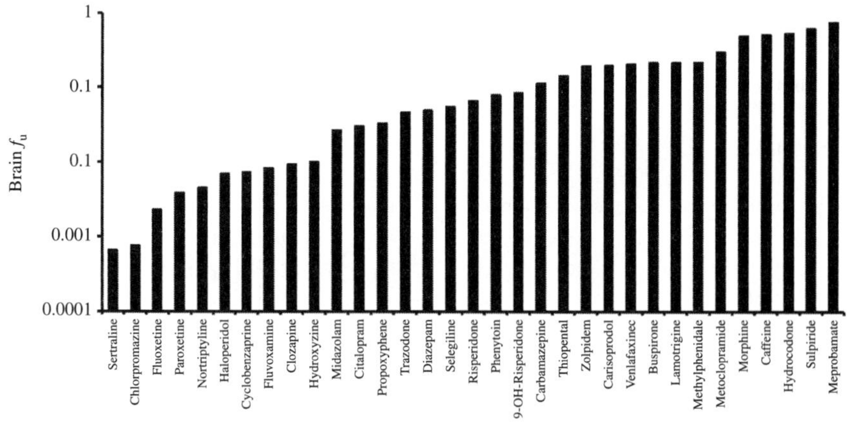

FIGURE 3.8 Unbound brain fraction of 32 frequently prescribed CNS drugs. The data were generated in mouse brain and reported by Maurer et al. [13]. Reproduced from Liu et al. [43]. ©2011 Bentham Science. With kind permission of Bentham Science.

brain for the 32 most prescribed CNS drugs (Fig. 3.8). In this data set, the lowest unbound brain fraction is 0.00066 for sertraline and the highest unbound brain fraction is 0.76 for meprobamate. Both are very successful drugs in the clinic, while their unbound brain fractions differ by 1152-fold! Of the 32 compounds 15 had unbound brain fractions of less than 0.05. A similar observation was made from another study with a data set of 41 proprietary compounds and more than 50 marketed CNS drugs [6, 7]. These results demonstrate that high brain tissue binding per se is not a liability for a CNS drug.

K_p IS A MISLEADING PARAMETER FOR THE ASSESSMENT OF BRAIN PENETRATION

K_p, the ratio of total brain concentration over total blood concentration, has been used extensively in drug design to optimize brain penetration, because it is directly derived from the experimentally measured total brain and total blood concentrations [44]. The limitations of using this parameter to characterize brain penetration have been discussed extensively in the literature [31, 32, 45–47]. A more appropriate parameter to describe the brain penetration is $K_{p,uu}$, the ratio of unbound brain concentration over unbound plasma concentration at steady state [30]. The relationship between K_p and $K_{p,uu}$ can be described by Equation 3.6:

$$K_p = K_{p,uu} \frac{f_{u,blood}}{f_{u,brain}} \tag{3.6}$$

Equation 3.6 shows that a low K_p can be due to low $K_{p,uu}$, low $f_{u,blood}$ (i.e., high plasma protein binding), or high $f_{u,brain}$ (i.e., low brain binding). This means that K_p is

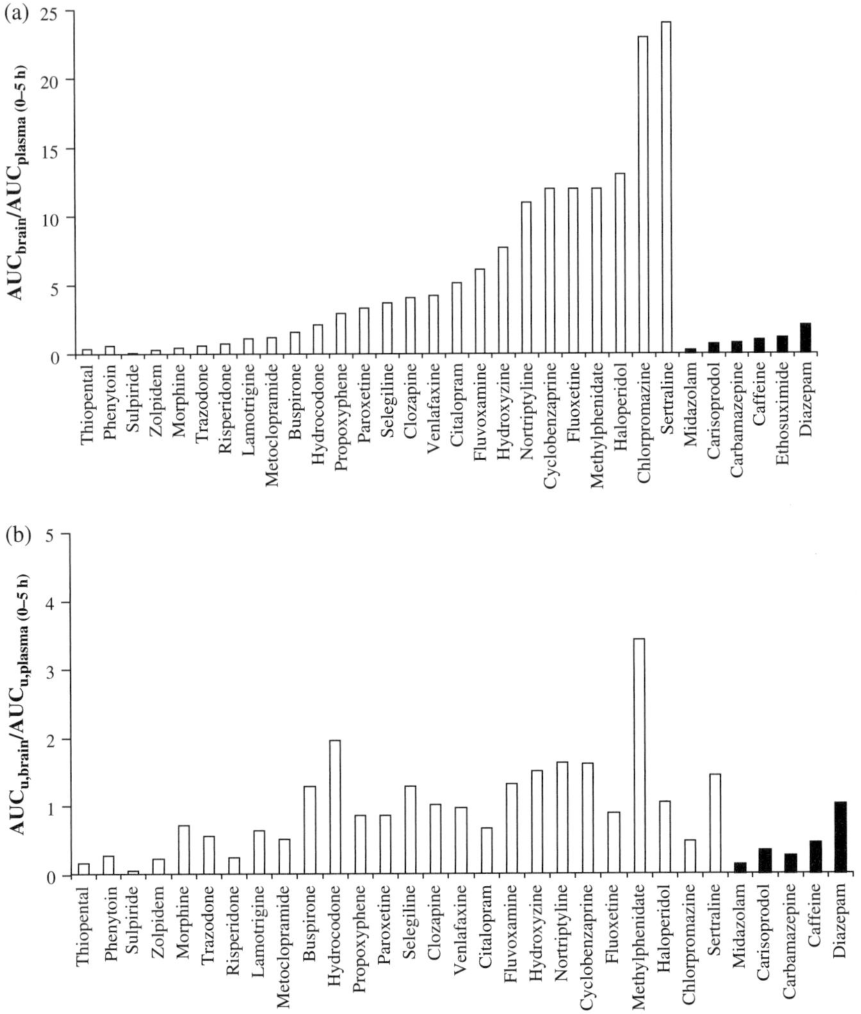

FIGURE 3.9 AUC ratios of total brain versus plasma concentrations (C_{brain}/C_{plasma}, (a) and unbound brain versus plasma concentrations ($C_{u,brain}/C_{u,plasma}$, (b) of 32 CNS compounds. The open, slashed, and solid bars represent acid, basic, and neutral compounds, respectively. The data are from Doran et al. [48] and Maurer et al. [13]. Reproduced from Liu et al. [31]. © 2008 American Society of Pharmacology and Experimental Therapeutics (ASPET). With kind permission of ASPET.

influenced by multiple factors and not always as a true reflection of brain penetration efficiency as $K_{p,uu}$. Doran et al. [48] reported a mouse data set of K_p values for the 32 most prescribed CNS drugs. There is a 240-fold difference for K_p among the 32 drugs but only a 34-fold difference for $K_{p,uu}$, indicating that protein binding in plasma and brain tissue can confound the interpretation of brain penetration when using K_p as the parameter to evaluate the brain penetration for a CNS compound (Fig. 3.9) [31]. The

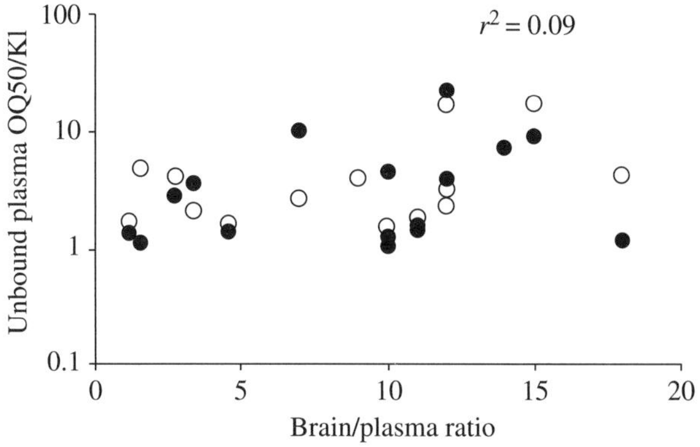

FIGURE 3.10 The relationship between *in vivo* unbound plasma OC_{50}/K_I and brain/plasma ratio (K_p) for SERT (closed circles) and DAT (open circles). The unbound plasma OC_{50} was calculated from the data in Figure 3.4 and is presented in Table 3.2. The K_p was the observed *in vivo* data at 90 min following IP dose. Reproduced from Liu et al. [27]. © 2009 American Society of Pharmacology and Experimental Therapeutics (ASPET). With kind permission of ASPET.

following example by Gupta et al. [49] better illustrates this point. They determined that K_p for S- and R-cetirizine was 0.22 and 0.04, respectively. These K_p values apparently suggest that S-cetirizine penetrates brain tissue much better than R-cetirizine. However, the $K_{p,uu}$ values were 0.17 and 0.14 for S- and R-cetirizine, respectively, indicating that there was no stereoselective brain penetration. Further investigation revealed that plasma protein binding for these enantiomers was different. The plasma unbound fractions for S- and R-cetirizine were 0.5 and 0.15, respectively. The stereoselective K_p was caused by differential binding to plasma proteins rather than an efflux transporter at the BBB. Thus, it is very important to differentiate whether a low K_p is due to efflux transporters at the BBB, high binding in the plasma, or low binding in the brain.

Liu et al. [27] observed that there was no correlation between the unbound plasma OC_{50}/K_I, which is a plasma biomarker for the *in vivo* activity and the K_p (Fig. 3.10). In other words, higher K_p values were not associated with greater CNS activities. Taken together, results so far have demonstrated that K_p is not a good indicator for brain drug penetration efficiency. A high K_p may represent a relatively high total brain drug concentration, but it does not mean a high unbound brain concentration. Therefore, K_p should not be used; instead, $K_{p,uu}$ should be used as a parameter to select CNS drugs in drug discovery.

SUMMARY

For CNS drugs, the interstitial unbound drug concentration determines the pharmacological activity for the extracellular targets and intracellular unbound drug concentration determines the pharmacological activity for the intracellular target. For most CNS compounds, it can be assumed that the interstitial unbound drug concentrations are same as the intracellular unbound concentrations and these concentrations are generally defined as unbound brain concentration. The recent advancement in the brain homogenate and brain slice methods allows us to study the relationship between CNS responses, such as receptor occupancy, or CNS pharmacologic effects, such as analgesia, and unbound brain concentration for a large number of compounds. The results from these studies support the free drug hypothesis for CNS drugs, that is, it is the unbound brain concentration not the total brain concentration that determines the CNS effects.

Since unbound brain concentration drives CNS effects, the primary objective to optimize brain penetration for CNS drugs should be to achieve high unbound brain concentrations. To that end, the strategy may include the design of CNS drugs with high affinity and potency to the pharmacological targets, high unbound plasma concentration, and high efficiency for brain penetration. High efficiency for brain penetration, quantified as high unbound brain to unbound plasma concentration ratio ($K_{p,uu}$), can be achieved by designing a CNS compound with high permeability, high uptake transport activity, low efflux transport activity, and low brain metabolism.

REFERENCES

[1] Davson, H.; Segal, M. B. *Physiology of the CSF and Blood-Brain Barriers.* CRC Press, Inc.: Boca Raton, **1995**.

[2] Brunton, L.; Lazo, J.; Parker, K. *Goodman & Gilman's the Pharmacological Basis of Therapeutics.* 11th ed.; McGraw-Hill: New York, **2006**.

[3] Smith, D. A.; Di, L.; Kerns, E. H. The effect of plasma protein binding on in vivo efficacy: misconceptions in drug discovery. *Nat Rev Drug Discov* **2010**, 9, 929–39.

[4] Liu, X.; Cheong, J.; Ding, X.; Deshmukh, G. Use of cassette dosing approach to examine the effects of p-glycoprotein on the brain and cerebrospinal fluid concentrations in wild-type and p-glycoprotein knockout rats. *Drug Metab Dispos* **2014**, 42, 482–91.

[5] Kalvass, J. C.; Polli, J. W.; Bourdet, D. L.; Feng, B.; Huang, S. M.; Liu, X.; Smith, Q. R.; Zhang, L. K.; Zamek-Gliszczynski, M. J. Why clinical modulation of efflux transport at the human blood-brain barrier is unlikely: the ITC evidence-based position. *Clin Pharmacol Ther* **2013**, 94, 80–94.

[6] Summerfield, S. G.; Read, K.; Begley, D. J.; Obradovic, T.; Hidalgo, I. J.; Coggon, S.; Lewis, A. V.; Porter, R. A.; Jeffrey, P. Central nervous system drug disposition: the relationship between in situ brain permeability and brain free fraction. *J Pharmacol Exp Ther* **2007**, 322, 205–13.

[7] Summerfield, S. G.; Stevens, A. J.; Cutler, L.; del Carmen Osuna, M.; Hammond, B.; Tang, S. P.; Hersey, A.; Spalding, D. J.; Jeffrey, P. Improving the in vitro prediction of in vivo central nervous system penetration: integrating permeability, P-glycoprotein efflux, and free fractions in blood and brain. *J Pharmacol Exp Ther* **2006**, 316, 1282–90.

[8] Summerfield, S. G.; Lucas, A. J.; Porter, R. A.; Jeffrey, P.; Gunn, R. N.; Read, K. R.; Stevens, A. J.; Metcalf, A. C.; Osuna, M. C.; Kilford, P. J.; Passchier, J.; Ruffo, A. D. Toward an improved prediction of human in vivo brain penetration. *Xenobiotica* **2008**, 38, 1518–35.

[9] Liu, X.; Van Natta, K.; Yeo, H.; Vilenski, O.; Weller, P. E.; Worboys, P. D.; Monshouwer, M. Unbound drug concentration in brain homogenate and cerebral spinal fluid at steady state as a surrogate for unbound concentration in brain interstitial fluid. *Drug Metab Dispos* **2009**, 37, 787–93.

[10] Becker, S.; Liu, X. Evaluation of the utility of brain slice methods to study brain penetration. *Drug Metab Dispos* **2006**, 34, 855–61.

[11] Kalvass, J. C.; Maurer, T. S. Influence of nonspecific brain and plasma binding on CNS exposure: implications for rational drug discovery. *Biopharm Drug Dispos* **2002**, 23, 327–38.

[12] Kalvass, J. C.; Maurer, T. S.; Pollack, G. M. Use of plasma and brain unbound fractions to assess the extent of brain distribution of 34 drugs: comparison of unbound concentration ratios to in vivo p-glycoprotein efflux ratios. *Drug Metab Dispos* **2007**, 35, 660–6.

[13] Maurer, T. S.; Debartolo, D. B.; Tess, D. A.; Scott, D. O. Relationship between exposure and nonspecific binding of thirty-three central nervous system drugs in mice. *Drug Metab Dispos* **2005**, 33, 175–81.

[14] Liu, X.; Smith, B. J.; Chen, C.; Callegari, E.; Becker, S. L.; Chen, X.; Cianfrogna, J.; Doran, A. C.; Doran, S. D.; Gibbs, J. P.; Hosea, N.; Liu, J.; Nelson, F. R.; Szewc, M. A.; Van Deusen, J. Evaluation of cerebrospinal fluid concentration and plasma free concentration as a surrogate measurement for brain free concentration. *Drug Metab Dispos* **2006**, 34, 1443–7.

[15] Friden, M.; Gupta, A.; Antonsson, M.; Bredberg, U.; Hammarlund-Udenaes, M. In vitro methods for estimating unbound drug concentrations in the brain interstitial and intracellular fluids. *Drug Metab Dispos* **2007**, 35, 1711–9.

[16] Kodaira, H.; Kusuhara, H.; Fujita, T.; Ushiki, J.; Fuse, E.; Sugiyama, Y. Quantitative evaluation of the impact of active efflux by p-glycoprotein and breast cancer resistance protein at the blood-brain barrier on the predictability of the unbound concentrations of drugs in the brain using cerebrospinal fluid concentration as a surrogate. *J Pharmacol Exp Ther* **2011**, 339, 935–44.

[17] Watson, J.; Wright, S.; Lucas, A.; Clarke, K. L.; Viggers, J.; Cheetham, S.; Jeffrey, P.; Porter, R.; Read, K. D. Receptor occupancy and brain free fraction. *Drug Metab Dispos* **2009**, 37, 753–60.

[18] Hammarlund-Udenaes, M.; Bredberg, U.; Friden, M. Methodologies to assess brain drug delivery in lead optimization. *Curr Top Med Chem* **2009**, 9, 148–62.

[19] Friden, M.; Winiwarter, S.; Jerndal, G.; Bengtsson, O.; Wan, H.; Bredberg, U.; Hammarlund-Udenaes, M.; Antonsson, M. Structure–brain exposure relationships in rat and human using a novel data set of unbound drug concentrations in brain interstitial and cerebrospinal fluids. *J Med Chem* **2009**, 52, 6233–43.

[20] Friden, M.; Ducrozet, F.; Middleton, B.; Antonsson, M.; Bredberg, U.; Hammarlund-Udenaes, M. Development of a high-throughput brain slice method for studying drug distribution in the central nervous system. *Drug Metab Dispos* **2009**, 37, 1226–33.

[21] Collingridge, G. L. The brain slice preparation: a tribute to the pioneer Henry McIlwain. *J Neurosci Methods* **1995**, 59, 5–9.

[22] Newman, G. C.; Hospod, F. E.; Wu, P. Thick brain slices model the ischemic penumbra. *J Cereb Blood Flow Metab* **1988**, 8, 586–97.

[23] Van Peer, A. P.; Belpaire, F. M.; Bogaert, M. G. Binding of drugs in serum, blood cells and tissues of rabbits with experimental acute renal failure. *Pharmacology* **1981**, 22, 146–52.

[24] Gredell, J. A.; Turnquist, P. A.; MacIver, M. B.; Pearce, R. A. Determination of diffusion and partition coefficients of propofol in rat brain tissue: implications for studies of drug action in vitro. *Br J Anaesth* **2004**, 93, 810–17.

[25] Ooie, T.; Terasaki, T.; Suzuki, H.; Sugiyama, Y. Quantitative brain microdialysis study on the mechanism of quinolones distribution in the central nervous system. *Drug Metab Dispos* **1997**, 25, 784–9.

[26] Kakee, A.; Terasaki, T.; Sugiyama, Y. Brain efflux index as a novel method of analyzing efflux transport at the blood-brain barrier. *J Pharmacol Exp Ther* **1996**, 277, 1550–9.

[27] Liu, X.; Vilenski, O.; Kwan, J.; Apparsundaram, S.; Weikert, R. Unbound brain concentration determines receptor occupancy: a correlation of drug concentration and brain serotonin and dopamine reuptake transporter occupancy for eighteen compounds in rats. *Drug Metab Dispos* **2009**, 37, 1548–56.

[28] Chen, C.; Liu, X.; Smith, B. J. Utility of Mdr1-gene deficient mice in assessing the impact of P-glycoprotein on pharmacokinetics and pharmacodynamics in drug discovery and development. *Curr Drug Metab* **2003**, 4, 272–91.

[29] Kalvass, J. C.; Olson, E. R.; Cassidy, M. P.; Selley, D. E.; Pollack, G. M. Pharmacokinetics and pharmacodynamics of seven opioids in P-glycoprotein-competent mice: assessment of unbound brain EC50,u and correlation of in vitro, preclinical, and clinical data. *J Pharmacol Exp Ther* **2007**, 323, 346–55.

[30] Syvanen, S.; Xie, R.; Sahin, S.; Hammarlund-Udenaes, M. Pharmacokinetic consequences of active drug efflux at the blood-brain barrier. *Pharm Res* **2006**, 23, 705–17.

[31] Liu, X.; Chen, C.; Smith, B. J. Progress in brain penetration evaluation in drug discovery and development. *J Pharmacol Exp Ther* **2008**, 325, 349–56.

[32] Liu, X.; Chen, C. Strategies to optimize brain penetration in drug discovery. *Curr Opin Drug Discov Dev* **2005**, 8, 505–12.

[33] Uchino, H.; Kanai, Y.; Kim, D. K.; Wempe, M. F.; Chairoungdua, A.; Morimoto, E.; Anders, M. W.; Endou, H. Transport of amino acid-related compounds mediated by L-type amino acid transporter 1 (LAT1): insights into the mechanisms of substrate recognition. *Mol Pharmacol* **2002**, 61, 729–37.

[34] Dave, M. H.; Schulz, N.; Zecevic, M.; Wagner, C. A.; Verrey, F. Expression of heteromeric amino acid transporters along the murine intestine. *J Physiol* **2004**, 558, 597–610.

[35] Invega Package Insert. www.invega.com/assets/prescribing-information (accessed on December 30, **2013**).

[36] Risperdal Package Insert. http://www.janssenpharmaceuticalsinc.com/assets/risperdal.pdf (accessed on December 30, **2013**).

[37] Cserr, H. F.; Patlak, C. S. Secretion and bulk flow of interstitial fluid. In: Bradbury, M.W.B. (Ed.), Physiology and Pharmacology of the Blood–Brain Barrier. Springer, Heidelberg, pp. 245–261; **1992**.

[38] Hansen, D. K.; Scott, D. O.; Otis, K. W.; Lunte, S. M. Comparison of in vitro BBMEC permeability and in vivo CNS uptake by microdialysis sampling. *J Pharm Biomed Anal* **2002**, 27, 945–58.

[39] El Bacha, R. S.; Minn, A. Drug metabolizing enzymes in cerebrovascular endothelial cells afford a metabolic protection to the brain. *Cell Mol Biol (Noisy-le-grand)* **1999**, 45, 15–23.

[40] Gervasini, G.; Carrillo, J. A.; Benitez, J. Potential role of cerebral cytochrome P450 in clinical pharmacokinetics: modulation by endogenous compounds. *Clin Pharmacokinet* **2004**, 43, 693–706.

[41] Fang, J. Metabolism of clozapine by rat brain: the role of flavin-containing monooxygenase (FMO) and cytochrome P450 enzymes. *Eur J Drug Metab Pharmacokinet* **2000**, 25, 109–14.

[42] Strazielle, N.; Khuth, S., T.; Ghersi, E., Jean, F. Detoxification systems, passive and specific transport for drugs at the blood–CSF barrier in normal and pathological situations. *Adv Drug Deliv Rev* **2004**, 56, 1717–40.

[43] Liu, X.; Chen, C.; Hop, C. E. Do we need to optimize plasma protein and tissue binding in drug discovery? *Curr Top Med Chem* **2011**, 11, 450–66.

[44] Hitchcock, S. A.; Pennington, L. D. Structure–brain exposure relationships. *J Med Chem* **2006**, 49, 7559–83.

[45] Hammarlund-Udenaes, M.; Paalzow, L. K.; de Lange, E. C. Drug equilibration across the blood-brain barrier—pharmacokinetic considerations based on the microdialysis method. *Pharm Res* **1997**, 14, 128–34.

[46] Hammarlund-Udenaes, M.; Friden, M.; Syvanen, S.; Gupta, A. On the rate and extent of drug delivery to the brain. *Pharm Res* **2008**, 25, 1737–50.

[47] Di, L.; Rong, H.; Feng, B. Demystifying brain penetration in central nervous system drug discovery. Miniperspective. *J Med Chem* **2013**, 56, 2–12.

[48] Doran, A.; Obach, R. S.; Smith, B. J.; Hosea, N. A.; Becker, S.; Callegari, E.; Chen, C.; Chen, X.; Choo, E.; Cianfrogna, J.; Cox, L. M.; Gibbs, J. P.; Gibbs, M. A.; Hatch, H.; Hop, C. E.; Kasman, I. N.; Laperle, J.; Liu, J.; Liu, X.; Logman, M.; Maclin, D.; Nedza, F. M.; Nelson, F.; Olson, E.; Rahematpura, S.; Raunig, D.; Rogers, S.; Schmidt, K.; Spracklin, D. K.; Szewc, M.; Troutman, M.; Tseng, E.; Tu, M.; Van Deusen, J. W.; Venkatakrishnan, K.; Walens, G.; Wang, E. Q.; Wong, D.; Yasgar, A. S.; Zhang, C. The impact of P-glycoprotein on the disposition of drugs targeted for indications of the central nervous system: evaluation using the MDR1A/1B knockout mouse model. *Drug Metab Dispos* **2005**, 33, 165–74.

[49] Gupta, A.; Hammarlund-Udenaes, M.; Chatelain, P.; Massingham, R.; Jonsson, E. N. Stereoselective pharmacokinetics of cetirizine in the guinea pig: role of protein binding. *Biopharm Drug Dispos* **2006**, 27, 291–7.

4

SPECIES DIFFERENCES AND IMPACT OF DISEASE STATE ON BBB

Jean-Marie Nicolas

UCB Pharma S.A., Chemin du Foriest, Braine-l'Alleud, Belgium

THE NEED FOR A BETTER UNDERSTANDING OF CNS DRUG DISPOSITION

Central nervous system (CNS) diseases are considered by the World Health Organization as one of the greatest threats facing our aging society [1]. By 2020, the United States will have more than 20% of its population older than 65 and the treatment of CNS diseases will cost trillions of dollars [2]. As an example, over the next three decades, the incidence of Alzheimer's disease (AD) is predicted to increase by 100% in the United States and Europe, and by more than 300% in Latin America, China, and India [3]. The unresolved CNS diseases go beyond AD and also include Parkinson's disease (PD), brain cancer, multiple sclerosis, epilepsy, schizophrenia, migraine, and stroke. Thus, the pharmaceutical industry is now under considerable pressure to deliver new effective CNS drugs.

The urgent need for better CNS medicines contrasts with the very limited number of novel therapies reaching the market. As an example, only five drugs are currently approved for the treatment of AD (tacrine, donepezil, rivastigmine, galantamine, and memantine), with no new agents approved since 2004, and with numerous programs discontinued because of disappointing efficacy in Phase 3 trials (e.g., bapineuzumab, AZD-103, AN1792, latrepirdine, semagacestat, solanezumab, tarenflurbil) [4]. It is considered that only 7% of the CNS programs entering clinical development will reach regulatory approval, compared with 15% for the average across all therapeutic

Blood–Brain Barrier in Drug Discovery: Optimizing Brain Exposure of CNS Drugs and Minimizing Brain Side Effects for Peripheral Drugs, First Edition. Edited by Li Di and Edward H. Kerns.

areas. In addition, the development of a CNS drug is a very lengthy process, taking on average 13 years compared to 6 years for cardiovascular drugs [3]. The high attrition rate of CNS drugs, their expensive development, and pricing pressure from generics has led several major companies to severely downsize their capabilities in neuroscience.

More than cutting back on research effort, the high demand for more effective CNS drugs requires a paradigm shift toward faster and more successful drug discovery and development approaches. To achieve this, it is important to identify the factors contributing to the high attrition rate of CNS drug candidates. (i) First, the notorious complexity of the human brain and our limited understanding of CNS diseases is a major hurdle in new drug development. (ii) Second, the progression of many new interesting CNS targets has been hampered by safety issues. (iii) Then clinical trials in CNS diseases are particularly difficult to design and to run as a result of slow recruitment rates, poor diagnostic techniques, and placebo effect. (iv) Another challenge is the inability of some drugs to cross the blood–brain barrier (BBB) and provide relevant concentration at the target site [5, 6]. The NMDA receptor glycine site antagonist Gavestinel has been developed by Glaxo Welcome as a neuroprotective agent for stroke. The drug progressed to clinical trials before being dropped for lack of efficacy in a randomized double-blind placebo-controlled study [7]. Retrospective analysis showed that Gavestinel did not cross the human BBB in pharmacologically relevant concentrations [8]. (v) Finally, the limited translatability of animal models for efficacy is a subject of considerable discussion [9]. Again, the Gavestinel case study is particularly illustrative. The compound was progressed because of encouraging neuroprotection activity in a rat model for stroke (rat permanent MCA occlusion). Concerns have been raised that the model might be associated with artifactual BBB breakdown, leading to a large overestimation of the brain penetration of Gavestinel [8].

Efforts must be directed toward building more knowledge on the mechanisms underlying CNS diseases and CNS drug disposition. Such knowledge will ultimately help develop better predictive tools to extrapolate animal data to diseased patients. This chapter focuses on two factors possibly contributing to the poor translatability of the animal models of CNS disorders, namely the species variability in CNS drug disposition and the effect of disease states.

THE KEY PARAMETERS DESCRIBING DRUG DELIVERY INTO THE BRAIN

The BBB is one of the most important and most studied blood–CNS interfaces. The BBB is composed of endothelial cells connected by tight junctions. These endothelial cells are surrounded by a basement membrane, pericytes, smooth muscles cells, and astrocyte end-feet, all these elements forming the neurovascular unit. It regulates brain homeostasis through multiple efflux and uptake transporters, metabolic enzymes, low pinocytotic activity, and low paracellular permeability [10, 11]. The BBB prevents the entry of potentially harmful compounds into the brain but may also reduce the permeability of therapeutic drugs, especially hydrophilic agents or substrates of efflux transporters [10]. Among the other blood–CNS interfaces, the blood–cerebrospinal fluid barrier (BCSFB) might also play a role in drug transport as its surface appears to be larger than originally thought [12]. Over the last two decades, tremendous efforts have been made to better understand BBB function. Two key parameters best describing CNS drug penetration have been identified [13, 14]: the extent of equilibration of the drug into the brain (reflected by the brain-to-plasma concentration ratio at steady state; "how much goes to the brain?") and the time required to reach equilibrium ("how long does it take?") [14].

The extent of equilibrium distribution between the brain and the plasma is described by the partition coefficient K_p using AUC or steady-state concentrations C_{ss} [15]:

$$K_p = C_{ss\ brain} / C_{ss\ plasma} = AUC_{brain} / AUC_{plasma}$$

K_p refers to total concentrations and is approximated by the well-known LogBB (i.e., log total brain-to-plasma concentration ratio as single time-point measurement). Both K_p and LogBB are potentially misleading since they do not distinguish the unbound and bound drug [16, 17]. One should instead consider the free drug concentration since free drug concentrations are more reliably linked to pharmacodynamic effects and directly indicate the presence of active transport (when the ratio differs from the unity) [18]. Free drug concentrations in brain interstitial fluid and in plasma can be incorporated into the given equation to obtain $K_{p,uu}$. According to the three-compartment model for CNS drug distribution [15], $K_{p,uu}$ depends on multiple mechanisms that include the passive diffusion permeability of the drug through the BBB (permeability surface area product, *PS*), any active uptake (Cl_{uptake}) or efflux (Cl_{efflux}) processes, the potential elimination of the drug from the brain by metabolic clearance ($Cl_{metabolism}$), and the clearance due to brain interstitial fluid bulk flow (Cl_{bulk}). At steady state, $K_{p,uu}$ can be described by the following equation [15]:

$$K_{p,free} = \frac{PS + Cl_{uptake}}{PS + Cl_{efflux} + Cl_{bulk} + Cl_{metabolism}}$$

$$K_{p,uu} = \frac{PS + Cl_{uptake}}{PS + Cl_{efflux} + Cl_{bulk} + Cl_{metabolism}}$$

The rate of drug delivery to the brain can be described by the unidirectional influx constant from the blood to the brain (K_{in}) and the permeability surface area product *PS*. These two parameters have the same dimension (ml/min/g brain) and are interconnected through the Renkin–Crone equation where CBF is the cerebral blood flow:

$$K_{in} = CBF \times \left(1 - e^{\frac{-PS}{CBF}}\right)$$

Drugs will have their rate of entry into the brain limited by either their permeability (permeability-limited; $PS < CBF$) or by the CBF (flow-limited; $PS > CBF$). The few brain *PS* values reported so far suggest that CNS drugs with low *PS* are infrequent in number [15, 19]. In addition, permeability-limited drugs will just require more time to equilibrate in the brain [20], which is not an issue for CNS disorders requiring chronic drug treatment to achieve steady-state concentrations. In such a scenario, low brain permeability has even been described as a possible advantage since it would stabilize CNS concentrations during fluctuations in drug plasma levels [21]. Low *PS* is an issue only for CNS drugs where rapid onset of action is a prerequisite (e.g., anesthesia, analgesia, status epilepsy, sleep disorders, and stroke). Under constant plasma concentrations (e.g., loading dose given intravenously followed by continuous infusion), the time to reach the brain equilibrium ($t_{1/2\text{eq}}$) is given by the following equation [20]. The assumptions inherent in this equation include passive diffusion without active transport, no influence of CSF bulk flow, and no metabolic clearance from the brain tissue. V_{brain} is the volume of the brain (ml/kg), $f_{\text{u brain}}$ is the fraction unbound in the brain tissue, and *PS* is expressed as ml/h/kg:

$$t_{\frac{1}{2}\text{eq}} = \frac{V_{\text{brain}} \times \ln 2}{PS \times f_{\text{u brain}}}$$

From this relationship, it becomes apparent that rapid equilibrium is achieved with drugs characterized by a high permeability and a high fraction unbound in the brain tissue.

Considerable advances have been made in the development of quantitative methods to measure drug exposure within the CNS [22–25], such as *in situ* brain perfusion, intracerebral microdialysis, CSF sampling, and tissue distribution (coupled with *in vitro* binding to brain homogenates). Given the invasive nature of these techniques, investigations in humans were for a long time restricted to CSF sampling. Positron emission tomography (PET) is now developing as a new noninvasive methodology applicable to animals and humans [26]. Finally, there is an increasing use of physiologically based pharmacokinetic (PBPK) modeling to integrate the pharmacokinetics data and extrapolate drug levels between species, brain regions, and dosing regimens.

SPECIES DIFFERENCES IN DRUG DELIVERY INTO THE BRAIN

Reports are emerging where drug delivery to the brain is different between species, as illustrated by AZD3839, a selective inhibitor of the β-secretase enzyme developed to treat Alzheimer's disease. Free drug concentrations were measured in plasma, CSF, and brain samples of C57BL/6 mouse, Dunkin–Hartley guinea pig and cynomolgus monkey after a single dose [27]. The free drug concentration in the brain was consistently higher in mouse than in guinea-pig (free drug brain/plasma ratios of 0.7 and 0.3, respectively) irrespective of the post-dosing time. CSF measurements overestimated free brain concentrations (2.5-fold in guinea pig) and showed time dependencies in monkey, making data interpretation difficult in the latter species. No underlying mechanisms were reported. Understanding such preclinical data and predicting the human situation require a careful look at the BBB

TABLE 4.1 Cerebral blood flow across species

Species	Brain weight (g/kg) [29]	Cerebral blood flow (L/h/kg)[a]	(ml/min/100 g brain)	(% cardiac output)[a]	References
Mouse	18	1.07	99	4.5	[30]
		0.52–2.14	48–198	2.2–8.9	[31]
		0.80–1.17	74–108	3.3–4.9	[32]
		0.92	85	3.8	[28]
		1.14	105	4.8	[33]
Rat	7	0.48	110	2.7	[34]
		0.12	28	0.7	[35]
		0.15	35	0.8	[36]
		0.27–0.96	63–224	1.5–5.3	[37]
		0.31	72	1.7	[29]
		0.72	166	4.0	[38]
		0.35–0.72	81–164	1.9–4.0	[33]
		0.86	197	4.7	[39]
		0.54	124	3.0	[40]
Dog	8	0.27	56	3.8	[29]
		0.20	42	2.8	[41]
		0.48	100	6.7	[42]
Monkey	18	0.86	80	6.6	[29]
		0.53	49	4.0	[40]
Human	20	0.60	50	12.5	[29]
		0.58	48	12.1	[34]
		0.64	53	13.3	[43]
		0.53	44	11.0	[44]
		0.75	62	15.6	[45]
		0.66	55	13.8	[39]
		0.49	41	10.3	[40]
		0.56–0.79	47–66	11.8–16.5	[21]

[a] Computed using brain weight and cardiac output values reported by Davies and Morris [29].

properties that could potentially differ between animal species. All the parameters driving the extent and the rate of drug delivery into the brain should be examined (e.g., CBF, *PS*, transporter activities, metabolism, and $f_{\mathrm{u\ brain}}$), as well as any potentially confounding parameters in drug measurement (e.g., CSF turnover and sampling site).

As mentioned earlier, CBF is the major determinant in the brain delivery of highly permeable flow-limited drugs. Putative species differences in CBF have been frequently claimed [16, 28]. In rodents, 2–4% of the cardiac output distributes to the brain, as compared with 11–16% in humans (Table 4.1). When expressed per gram brain tissue, CBF is about threefold higher in rodents than in humans. This finding fits with the allometric relationship reported by Karbowski [40], with $\log_{10}$ CBF (ml/min/100 g) = [−0.17 × $\log_{10}$ brain volume (cm^3)] + 2. The same author reported that CBF and neuron density scale with brain volume in the same way, which results in a blood flow per neuron that is invariant among mammals (i.e., 1.4×10^{-8} ml/min). CBF varies among brain areas, up to 18-fold in the rat [46]. This regional difference in the flow might explain regional drug distribution and should be further explored to help refine existing PBPK models. Of note, the reported CBF values can be largely variable (up to eightfold in rat), illustrating the difficulty to define precise reference values to be used in modeling approaches.

Permeability-limited drugs will have their transport rate affected by the drug itself (i.e., their permeability *P*) and by the surface area of the brain capillaries (*S*), both characteristics being combined in a single parameter, *PS*. The capillary surface area values reported in rats are consistently in the 100–150 cm^2/g brain range [39, 47–50]. A more detailed study showed that the capillary surface area varies between brain regions, the highest value being obtained in the cortical gray matter [51]. Human data are broadly similar to rat data, with 100–200 cm^2/g brain [18, 39, 52–54], depending on the region considered. According to Karbowski [40], the fraction of the capillary volume is invariant with respect to brain volume, which would indicate a capillary surface area (expressed per gram or ml brain) conserved across species. The non-P-gp substrate diazepam has a brain *PS* of 2.6 ml/min/g in human patients [21], indistinguishable from the value of 3.0 ml/min/g reported in rat [55]. This finding confirms similar *S* values (once expressed per gram tissue) between rodents and humans, assuming *P* is maintained through species (see later). A set of 21 diverse compounds were tested for their brain capillary permeability *PS* in mouse and rat, using *in situ* brain perfusion [56]. The measured *PS* values were similar between the two species with only one noticeable exception, vincristine, presumably because of bias due to active transport. In another study, felbamate was found to have a broadly comparable brain *PS* in mouse, rat, and rabbit (~0.20, 0.09, and 0.17 ml/min/g, respectively) [33]. The brain pharmacokinetics of selected opioids has been measured in mouse and compared to clinical data [57]. The $t_{1/2\ eq}$ values of the non-P-gp substrates [58] alfentanyl, sulfentanyl, and fentanyl were found remarkably analogous in mouse and humans. Considering the equation describing $t_{1/2\ eq}$, this finding would further confirm that *PS* values (ml/min/g) are conserved between the two species. Taken collectively, all the data available so far suggest that rodent values of *PS* should be appropriate for human PBPK modeling.

Cerebral blood volume has been claimed to differ between species [59], ranging from 6–14 µl/g in rat [60] to 23–60 µl/g in dog [61, 62]. Humans would be intermediate with values ranging from 3 to 34 µl/g depending on the technique used [14]. Another report described 30, 10, and 40 µl/g in rat, dog, and humans, respectively [28]. These data contradict Karbowski's paper reporting that the fraction of the capillary volume does not vary between species, with an average value of 16 µl/cm^3. Cerebral blood volume is used to correct drug concentration measures in brain tissue for the vascular component in order to improve the accuracy of the K_p determinations [59].

The CSF production rate is 1.03–3.00 µl/min/g brain in rodent [63, 64] compared with 0.44 in dog [65], 0.30–0.36 in monkey [66], and 0.29 in humans [64]. The higher production rate in rodent parallels a faster turnover half-life (40–100 min in rat versus 170 min in humans [67]). This species difference is of importance as CSF turnover could affect drug diffusion across the various brain compartments and, ultimately, the drug concentration in the CSF [64]. It has been suggested that the high flow of CSF in rodents could cause hydrophilic compounds to be taken along with the CSF instead of diffusing into the brain target tissue [23]. Finally, protein concentration in CSF is reported to be approximately 5–10 times higher in rat than in humans [68] and to vary along the flow path, making the CSF sampling site very important. Typically, CSF samples in human are obtained from distant sites (i.e., lumbar or thoracic) compared to rodents (cisterna magna). Overall these findings complicate any species comparison based on CSF drug concentrations. A set of 43 structurally diverse drugs were compared for their free drug concentrations in rat brain ($K_{p,\ uu}$), as measured by CSF and tissue distribution [67]. Rat CSF data were also compared to those reported in humans. There was an overall bias toward the human CSF values being on average threefold higher than in rat. The author argued that the higher CSF turnover in rat and the differences in CSF sampling sites might possibility account for the observed bias.

The lipid composition of the brain tissue could also potentially impact drug disposition into the CNS by affecting nonspecific binding and passive diffusion. Species differences in some brain lipids have been suggested [69–71]. Recently, porcine brain lipids have been demonstrated to differ from those measured in human endothelial cells [72], which could question the validity of the porcine cells as an *in vitro* BBB model [73]. This contradicts the outcome of a large *in vitro* study that measured the brain tissue binding of 47 diverse compounds in Wistar Han rat, Sprague–Dawley rat, CD-1 mouse, Hartley guinea pig, beagle dog, cynomolgus monkey, and humans. Excellent correlations were found across species (correlation coefficient ≥ 0.93) [74]. A similar study on a smaller set of compounds also demonstrated that brain tissue binding is well conserved across species, including pig [75]. It can be thus reasonably assumed that rodent values of $f_{u\ brain}$ are appropriate for human PBPK modeling.

Plasma protein binding determines the unbound fraction of the drug that can permeate through the BBB [76]. There are numerous illustrations of marked species differences in plasma protein binding [77] with potential impact on CNS drug distribution. A series of compounds were found active in an efficacy model in guinea pig, but not in the rat counterpart [78, 79]. Additional investigations demonstrated low free drug exposure in rat brain when compared to guinea pig, as a consequence of a four- to sixfold higher plasma protein binding. The $5HT_{1A}$ antagonist NAD-299 has shown 10-fold higher target occupancy in cynomolgus monkey than in human subjects. This difference has been attributed to the 10-fold lower plasma protein binding in monkey plasma compared to humans [80]. Nowadays, species differences in protein binding are incorporated into the PBPK modeling of CNS drugs [39, 81]. It has been advocated that the kinetics of the binding to the plasma proteins should be considered (as opposed to the percent bound) to better account for the dynamic nature of the drug distribution processes [23]. If the dissociation from the plasma proteins is slow when compared to the capillary transit, it will slow down the drug penetration through the BBB membrane [21].

Numerous uptake and efflux transporters are expressed at the BBB and BCSFB level. It is estimated that 10–15% of all the proteins in the neurovascular unit are transporters [82], with two major drug transporter families, the ATP-binding cassette

TABLE 4.2 Protein expression level of BBB transporters across species

	Protein expression (fmol/μg protein)				
Transporter	Mouse [86]	Rat [83]	Cynomolgus [85]	Marmoset [83]	Human [84]
MDR1	15.5	19.1	4.71	6.48	6.06
MRP4	1.59	1.53	0.29	0.32	0.20
BCRP	4.41	4.95	14.2	16.5	8.14
OATP-A	2.11	—	0.72	—	ULQ
OATP-F	2.41	—	—	—	ULQ
OAT3	1.97	1.23	ULQ	ULQ	ULQ
ABCA8	ULQ	ULQ	—	—	1.21

ULQ, under limit of quantitation.

(ABC) protein family and the solute carrier (SLC) family. Liquid chromatography–tandem mass spectrometry-based quantitative-targeted absolute proteomics (QTAP) has recently emerged as a powerful tool to compare the expression level of the various transporters in mouse, rat, monkey, and human brain capillaries [83–85]. The anion transporters MRP4, OATP-A, OATP-F, and OAT3 are more abundant in mouse than their counterparts in humans (8.0-, >3.3-, >11.6-, and >5.6-fold difference, respectively) (Table 4.2). P-glycoprotein (MDR1) is also more abundant in mouse than in humans, albeit to a lower extent (2.6-fold) [84]. Conversely, BCRP and ABCA8 transporters are more expressed in humans than their counterparts in mouse (1.8- and >1.6-fold difference). Other differences were reported in transporters for endogenous compounds or receptors (e.g., 4F2hC, LAT1, MCT1). The same group reported that the expressions of the major drug transporters were similar between rodent species (rat versus mouse), and between strain (Sprague–Dawley versus Wistar rat, marmoset versus cynomolgus monkey) [83]. Importantly, the differences between monkey and humans were less than twofold for the measured transporters. The authors proposed that marmoset might be a more translatable and a convenient species for BBB permeability studies.

Quantitative real-time PCR determination of mRNA expression levels of ABC transporters demonstrated species differences [87], in line with the protein expression data. MRP4 and P-gp showed higher expression in rodents than in humans (5.0- and 2.2-fold, respectively). On the other hand, BCRP mRNA has an expression 5.1-fold higher in humans than in rodents.

In addition to the observed cross-species differences in transporter protein and gene expression, differences in functional activities *in vitro* have also been observed. Booth-Genthe et al. [88] investigated a total of 179 compounds for their ability to be transported by rat and human P-gp expressed in LLC-PK1 cells. Eighteen compounds were found to be P-gp substrates in rat but not in humans. On the contrary, 5 compounds were substrates of human, but not rat P-gp. Katoh et al. [89] also used transfected LLC-PK1 to study the transport of diltiazem, cyclosporin, and dexamethasone by mouse, rat, dog, monkey, and human P-gp, and provided detailed kinetic parameters. Affinity for P-gp, clearance, and efflux ratios varied across species, with various patterns depending on the substrate. As an example, the

P-gp-mediated clearance of diltiazem varied from 0.1 in dog to 0.7 in humans and monkey and 2.1 ml/mg/h in rat. Takeuchi et al. [90] performed a similar species comparison, looking at 12 different drug substrates. The observed species differences (as measured by corrected efflux ratio) were strongly dependent on the substrate. Mouse, rat, dog, monkey, and humans showed an efflux ratio for vinblastine of 10, 6, 9, 4, and 4, respectively. Ratios for verapamil were 3, 7, 8, 8, and 6, respectively, in these species. Ohe et al. measured the transport of CNS drug candidates by human and mouse P-gp [91]. Seventeen compounds (out of the 20 tested) showed more efficient efflux transport in mouse than in human P-gp, with up to sixfold difference observed. Another study compared several antiepileptic drugs for their ability to be transported by human and mouse P-gp [92]. Phenytoin and levetiracetam were shown to be transported by mouse P-gp, but not by the human counterpart. Carbamazepine was not transported by any P-gp and the positive control cyclosporin was transported by P-gp in both species, with a higher rate observed in mouse when compared to humans [92].

PET data are providing growing indications of interspecies variation in BBB transporter activity *in vivo*. A PET study showed that the brain uptake of the P-gp probe substrates [^{11}C]-GR205171, [^{18}F]-altanserin, [^{11}C](*R*)-RWAY, and [^{11}C]-*N*-desmethyl-loperamide was significantly higher in humans as compared to rodent (8.6-, 4.5-, 2.5-, and up to 15-fold difference, respectively) [93–96]. The reference P-gp substrate ^{11}C-verapamil provided contradictory data with brain uptake in humans similar [97] or even lower than in rat [98]. This later report postulated that the extensive metabolism of the tracer might have interfered with the readout. ^{11}C-verapamil is rapidly transformed into lipophilic radiolabeled metabolites [99], some actively transported, and has a complex disposition that varies with the dose [100], all confounding data interpretation. Of interest, irrespective of the probe used, the increase in their brain uptake following co-administration with P-gp inhibitors (e.g., cyclosporine A and tariquidar) is higher in rat compared to humans [98, 101, 102]. This finding adds to the accumulating evidence of the higher P-gp activity in rodents. Monkey appears to better mimic the human situation [103, 104]. So far, no human PET study investigating the other major ABC and SLC transporters has been reported [105]. As illustrated with BCRP, identifying a specific probe substrate for those transporters remains a major challenge [106].

Doran et al. [107] measured the CSF concentrations of three drugs in rat, dog, and nonhuman primate. The $K_{p,uu}$ values of the two non-P-gp substrates PF-478574 and CE-157119 were similar across species (within twofold difference). This contrasted with the P-gp substrate risperidone that showed a higher $K_{p,uu}$ in dog and nonhuman primate compared to rat (ca. fivefold difference). The previously mentioned screen comparing the rat and human CSF levels of 43 compounds [67] identified compounds with a $K_{p,uu}$ much higher in humans than in rat. The most striking dissimilarities were obtained for baclofen (24-fold), moxalactam (20-fold), atenolol (15-fold), verapamil (10-fold), and methotrexate (9-fold). The authors envisaged several experimental factors that might have contributed to the observed findings, with the highest bias due to CSF sampling (already discussed earlier) and the effect of the disease state of the human subjects. One might add species difference in BBB transporter activity as another potential factor. Active uptake was reported for baclofen, the involved transporter not yet being identified [108]. Interestingly, moxalactam is a substrate of PEPT2 [109], a transporter expressed at the apical membrane of the choroid plexus that effluxes substrates from the CSF [110]. Atenolol is a substrate of OATP1A2 [111], a BBB transporter responsible for the active uptake of a number of drugs [112]. Finally, methotrexate was recently reported to be effluxed by OAT3 [113], a BBB transporter with higher expression level in rodent than in humans [84].

Most modeling approaches to describe drug kinetics into the CNS assume no metabolic elimination from the brain tissue [20, 39, 114]. Although more prominent in the liver, drug metabolism can also take place in extrahepatic tissues including the brain. Various drug metabolizing enzymes were identified in the CNS, where they can influence endogenous processes as well as drug efficacy and safety [115–117]. It took years to recognize that brain CYP is able to play a substantial role in local drug disposition despite having an average expression level 100-fold lower than in the liver [118]. The major brain CYP isoforms differ from those in the liver and are concentrated in specific brain areas reaching high expression at a cellular level [116]. Detailed data across species about the expression, activity, and regulation of drug metabolizing enzymes in the brain are missing. Interspecies differences are likely, as already reported for the liver and the intestine [119]. A very recent study reported differences in CYP transcripts between mouse and rat [120].

BBB PROPERTIES IN DISEASE STATES

Species differences in BBB properties are not the only hurdle affecting the translatability of the animal models for CNS disorders. Most models use healthy animals, while there is growing evidence that the BBB is disrupted in several acute and chronic neurological disorders [23, 121, 122]. These include epilepsy, Alzheimer's disease, Parkinson's disease, multiple sclerosis, dementia [123], cerebral ischemia [124], traumatic brain injury, stroke, neuromyelitis optica, human immunodeficiency virus (HIV) encephalopathy, glioblastoma, bacterial meningitis [125], and pain [126] (selected conditions listed in Table 4.3). The BBB also appears to be impaired in some non-CNS diseases such as rheumatoid arthritis [156], diabetes [157], liver failure [158], eclampsia [159], hypertension [160], and atherosclerosis [161].

The mechanisms underlying BBB disruption are numerous and frequently involve the loosening of the tight junctions, the down-regulation of the tight junction proteins, and the degradation of the capillary basement membrane. BBB breakdown and increased capillary permeability allow the passage of inflammatory cells into the brain tissues and the extravasation of serum proteins (e.g., albumin, immunoglobulins, and clotting factors) or other neurotoxic plasma constituents (e.g., amyloid proteins). Combined with the activation of astrocytes and microglia, all these changes lead to edema, accumulation of toxic substances in the brain interstitial fluid, oxidative stress, impaired homeostasis, neuroinflammation, remodeling of vasculature, and altered synaptic plasticity. It has been recently hypothesized that BBB

TABLE 4.3 Effect of various physiopathological conditions on BBB properties

Conditions	Changes in BBB properties
Alzheimer	Abnormal tight junctions and increased permeability [127]
	BBB leakage as indicated by increased CSF albumin index [127]
	Regional CBF decreased through ECE2 activation [128]
	Decreased P-gp expression [129–131]
	Changes in LRP-1 and RAGE leading to amyloid deposits [132]
	P-gp expression and senile plaques negatively correlated [130]
	Up-regulation of ABCG2 [129, 133]
Parkinson	Decrease in tight junction proteins
	Decreased mRNA coding for P-gp in postmortem samples [134]
	Susceptibility linked to P-gp genotype [135]
	Increased MRP2 expression [136]
Epilepsy	Decrease in tight junction proteins, tight junction opening [137]
	Albumin, IgG, and white blood cell extravasation [138–140]
	Regional CBF decreased in temporal lobe epilepsy [141]
	Increased P-gp, MRP1, MRP2, and BCRP expression [178, 142–144]
	Link between ABCC2 polymorphism and pharmacoresistance [145]
Multiple sclerosis	Increased CBF and *PS* as measured by MRI [146]
	Decrease in tight junction proteins [147]
	Leakage of the BBB and inflammatory cell infiltration [148]
	Decreased P-gp, MRP-1, and MRP-2 expression [149]
Schizophrenia	Albumin and IgG extravasation [150, 151]
Depression	Albumin extravasation [152]
Aging	Regional CBF decreased [153]
	Albumin extravasation [154]
	Decreased P-gp expression [155]

disruption might often act as the initiating trigger of many neurological disorders, as opposed to just being a consequence of the disease [162].

The BBB changes seen in neurological diseases are not limited to the physical breakdown of the barrier, but also include altered expression of transporter proteins. So far, most of the attention has been focused on P-gp, with *in vivo* functional activity measured by PET studies using [^{11}C]-verapamil (Table 4.4). P-gp activity has been reported to be decreased in Alzheimer's disease [163] and in multiple sclerosis [175]. It is thought that such P-gp underactivity at the endothelial level contributes to neuronal damage by allowing the accumulation of protein aggregates (e.g., Alzheimer's disease, Creutzfeldt–Jakob disease [176]). Epilepsy has the opposite effect on P-gp with increased expression in brain endothelial cells, especially in refractory patients [177, 178]. It has been hypothesized that the seizure-induced release of glutamate is able to up-regulate P-gp through *N*-methyl-D-aspartate receptor and cyclooxygenase-2 pathways [179]. Parkinson's disease patients showed a more complex pattern, with increased P-gp activity in the early stages of the disease, followed by decreased activity in late stages [165]. Dysfunctional P-gp might contribute to the onset of PD by allowing neurotoxins to enter the brain [166, 180, 181]. The previous opposite

TABLE 4.4 P-gp activity as assessed by PET in various physiopathological conditions[a]

Conditions	Effect on P-gp activity	
Alzheimer	↓	23% increase in [^{11}C]-verapamil binding potential in AD [163]
Parkinson	—	No change in [^{11}C]-verapamil cerebral volume of distribution in early stage PD [164]
	↑	Lower [^{11}C]-verapamil uptake in early PD [165]
	↓	Higher [^{11}C]-verapamil uptake in late-stage PD [165]
	↓	18% increase in [^{11}C]-verapamil brain uptake [166]
Epilepsy	↑	16% decrease in [^{11}C]-verapamil transport rate constant in pharmacoresistant patients compared to seizure-free subjects [167]
Schizophrenia	↑	30% decrease in [^{11}C]-verapamil brain uptake [168]
Depression	↑	31% decrease in [^{11}C]-verapamil uptake in prefrontal cortex [169]
Aging	↓	30% increase in [^{11}C]-verapamil distribution volume in elderly [170]
	↓	61% increase in [^{11}C]-verapamil distribution volume in elderly [171]
	↓	18–38% increase in [^{11}C]-verapamil distribution volume in elderly, mainly driven by male subjects [172]
	↓	18–38% increase in [^{11}C]-verapamil distribution volume in elderly [173]

[a]Adapted from Syvanen and Eriksson [174].

effects on P-gp activity originate from the intricate and multiple pathways involved in its regulation [182]. Short-term exposure to endothelin-1, VEGF, LPS, or pro-inflammatory cytokines produces a rapid but reversible decrease in P-gp activity, without affecting its expression. On the other hand, P-gp expression and activity can increase through the activation of transcription factors, such as PXR, CAR, NF-kB, or AP-1.

Many of the previously discussed BBB changes already develop during normal aging. A large meta-analysis of 31 studies with a total of 1953 individuals showed that BBB permeability increased with normal aging [154]. PET studies indicated that P-gp activity decreased with age [170–172]. These findings may be an important mechanism in the higher sensitivity of the elderly to develop CNS disorders or to show CNS side effects [183]. Many other nonpathological conditions are recognized to change BBB permeability, such as physical exercise [184] and pregnancy [158].

The BBB leakiness accompanying some diseases could allow otherwise nonpermeable drugs to reach the brain tissue and exert their pharmacological activity. Because of their hydrophilic nature and their ionization at neutral pH, β-lactam antibiotics hardly penetrate the brain under normal conditions [185]. However, during cerebral infection, such as bacterial meningitis, the BBB becomes more permeable, with opening of the tight junctions, increased pinocytosis, and changes in transporter expression. These modifications translate into higher brain exposure to β-lactam antibiotics. Three hours after IV infusion, penicillin concentration in CSF increases from 3% of free plasma concentration in control rabbits to 11% in rabbits with intracisternal inoculation of hemophilus influenza [186]. The β-lactam antibiotic moxalactam has a brain $K_{p,uu}$, as measured in the CSF, 25-fold higher in patients with bacterial meningitis compared to healthy human volunteers [67]. All angiotensin receptor blockers (ARBs) contain an acid moiety, are fully ionized at physiological pH, and are highly bound to plasma protein [187]. These properties together with extremely low brain K_p values in rat (e.g., 0.04 for candesartan) suggest that ARBs cannot cross the BBB. This contrasts with the growing evidence that treatment of hypertensive patients with ARBs is efficacious in cognitive disorders with less Alzheimer-related pathology on autopsy evaluations [188–190]. This effect is unique and is not shared by other antihypertensive medications. The underlying mechanisms are not yet fully elucidated. However, it could be tentatively assumed that the increased BBB permeability in hypertension [160, 191] could favor the entry of ARBs into the brain, allowing pharmacologically active concentrations at the CNS target site.

The modulation of BBB transporter activities in disease states is expected to impact on the brain disposition of actively transported drugs. Whether the effects are large enough to be clinically significant remains to be explored. Genetic polymorphism and co-administration with transporter inhibitors are other factors contributing to variations in BBB transporter activities. MDR1 polymorphism has been linked with the side effects and/or the therapeutic response of a number of CNS drugs acting as P-gp substrates such as escitalopram [192], citalopram [193], venlafaxine [194], methylphenidate [195], and cabergoline [196]. It remains unclear whether the observed changes were due to dysfunctional P-gp at the BBB level or impaired systemic plasma levels. For some other P-gp clinical substrates, such as amitryptiline [197] and loperamide [198], no association could be found between MDR1 polymorphism and central effects. A PET study in healthy volunteers showed that MDR1 polymorphism has a limited impact on the brain disposition of [^{11}C]-verapamil [199]. Similarly, in a recent review, the International Transporter Consortium (ITC) group concluded that changes in CNS drug disposition due to transporter inhibition remain low in amplitude and are unlikely to have any clinical significance [200]. The same authors argued that a 50% change in BBB P-gp, as a result of drug interaction or species differences, should have a modest impact on CNS drug distribution (maximum twofold).

Obviously, there is still considerable uncertainty and consequently different views on the influence of transporter modulation on drug disposition and pharmacological response. This also applies to transporter modulation induced by diseases. More *in vivo* studies in animal models with disrupted BBB (e.g., SAMP-8 senescence accelerated mouse [201]), more clinical studies comparing healthy and diseased human subjects (e.g., PET), and/or more epidemiological investigations are obviously required to draw more definitive conclusions.

CONCLUSIONS

The increasing prevalence of CNS disorders as the population becomes older and the significant difficulties in successfully developing new therapeutic agents require reconsidering the current drug discovery approaches. Animal models of efficacy need to be critically revisited to contribute more effectively to translational medicine and drug discovery. This chapter reviewed two factors that might complicate the interpretation of CNS drug distribution data obtained in animals: the species difference in BBB functionality and the effect of disease states.

A comprehensive list of BBB physiological parameters across species is still missing. Based on the available literature, it appears that some parameters measured in animals can be directly used for human prediction (e.g., f_u brain, *PS* expressed per gram tissue) whereas others (e.g., CBF, CSF bulk flow, transporter contribution) require careful scaling and adjustments. Rodent BBB does differ from human BBB in many aspects and extreme caution is recommended when interpreting the data. To add to the complexity, disease states are now recognized to potentially affect the BBB function. All these findings should be taken seriously as most of the brain exposure data are collected in healthy rodents to predict the situation in diseased patients.

A change in a single BBB parameter might overall have a rather limited influence on CNS drug disposition, as suggested for brain P-gp activity [200]. On the other hand, there are circumstances where various confounding parameters combined can lead to a large impact on the drug delivery into the brain. The study reported by Friden et al. [67] identified drugs with up to a 24-fold difference in the CSF concentrations between rat and humans. At least five experimental factors are likely to have contributed to this massive difference: CSF turnover that varies between species, bias due to timing in CSF sampling, bias due to CSF sampling site, transporter activity that differs between species, and effect of disease state (as most of the human data were collected in diseased patients, not volunteers). Efforts should be made to incorporate all these variables into more predictive PBPK models. In addition, more imaging techniques (PET, single-photon emission computed tomography (SPECT)) should be applied in preclinical assays and in early investigational clinical assays [17]. Ultimately, those efforts will allow better prediction of the CNS response in the patient population and, thus, better mitigation of the risk for failures in late-stage development.

REFERENCES

[1] World Health Organization (WHO). *Neurological disorders: public health challenges.* WHO: Geneva, **2006**; p. xi, 218 pp.

[2] Ghose, A. K.; Herbertz, T.; Hudkins, R. L.; Dorsey, B. D.; Mallamo, J. P. Knowledge-based, central nervous system (CNS) lead selection and lead optimization for CNS drug discovery. *ACS chemical neuroscience* **2012**, 3, 50–68.

[3] Pangalos, M. N.; Schechter, L. E.; Hurko, O., Drug development for CNS disorders: strategies for balancing risk and reducing attrition. *Nature reviews. Drug discovery* **2007**, 6, 521–32.

[4] Cummings, J. L.; Banks, S. J.; Gary, R. K.; Kinney, J. W.; Lombardo, J. M.; Walsh, R. R.; Zhong, K., Alzheimer's disease drug development: translational neuroscience strategies. *CNS spectrums* **2013**, 18, 128–38.

[5] Pardridge, W. M. Drug transport across the blood-brain barrier. *Journal of cerebral blood flow & metabolism* **2012**, 32, 1959–72.

[6] Pardridge, W. M. The blood-brain barrier: bottleneck in brain drug development. *NeuroRx* **2005**, 2, 3–14.

[7] Lees, K. R.; Asplund, K.; Carolei, A.; Davis, S. M.; Diener, H. C.; Kaste, M.; Orgogozo, J. M.; Whitehead, J. Glycine antagonist (gavestinel) in neuroprotection (GAIN International) in patients with acute stroke: a randomised controlled trial. GAIN International Investigators. *Lancet* **2000**, 355, 1949–54.

[8] Dawson, D. A.; Wadsworth, G.; Palmer, A. M. A comparative assessment of the efficacy and side-effect liability of neuroprotective compounds in experimental stroke. *Brain research* **2001**, 892, 344–50.

[9] Markou, A.; Chiamulera, C.; Geyer, M. A.; Tricklebank, M.; Steckler, T. Removing obstacles in neuroscience drug discovery: the future path for animal models. *Neuropsychopharmacology* **2009**, 34, 74–89.

[10] Pardridge, W. M. Drug and gene targeting to the brain with molecular Trojan horses. *Nature reviews. Drug discovery* **2002**, 1, 131–9.

[11] Ballabh, P.; Braun, A.; Nedergaard, M. The blood-brain barrier: an overview: structure, regulation, and clinical implications. *Neurobiology of disease* **2004**, 16, 1–13.

[12] Lin, J. H. CSF as a surrogate for assessing CNS exposure: an industrial perspective. *Current drug metabolism* **2008**, 9, 46–59.

[13] Reichel, A. The role of blood-brain barrier studies in the pharmaceutical industry. *Current drug metabolism* **2006**, 7, 183–203.

[14] Hammarlund-Udenaes, M.; Friden, M.; Syvanen, S.; Gupta, A. On the rate and extent of drug delivery to the brain. *Pharmaceutical research* **2008**, 25, 1737–50.

[15] Liu, X.; Chen, C.; Smith, B. J. Progress in brain penetration evaluation in drug discovery and development. *The journal of pharmacology and experimental therapeutics* **2008**, 325, 349–56.

[16] Deo, A. K.; Theil, F. P.; Nicolas, J. M. Confounding parameters in preclinical assessment of blood-brain barrier permeation: an overview with emphasis on species differences and effect of disease states. *Molecular pharmaceutics* **2013**, 10, 1581–95.

[17] Di, L.; Rong, H.; Feng, B. Demystifying brain penetration in central nervous system drug discovery. Miniperspective. *Journal of medicinal chemistry* **2013**, 56, 2–12.

[18] Reichel, A. Addressing central nervous system (CNS) penetration in drug discovery: basics and implications of the evolving new concept. *Chemistry & biodiversity* **2009**, 6, 2030–49.

[19] Fenstermacher, J. D. The blood-brain barrier is not a "barrier" for many drugs. *NIDA Research Monographs* **1992**, 120, 108–20.

[20] Liu, X.; Smith, B. J.; Chen, C.; Callegari, E.; Becker, S. L.; Chen, X.; Cianfrogna, J.; Doran, A. C.; Doran, S. D.; Gibbs, J. P.; Hosea, N.; Liu, J.; Nelson, F. R.; Szewc, M. A.; Van Deusen, J. Use of a physiologically based pharmacokinetic model to study the time to reach brain equilibrium: an experimental analysis of the role of blood-brain barrier permeability, plasma protein binding, and brain tissue binding. *The journal of pharmacology and experimental therapeutics* **2005**, 313, 1254–62.

[21] Paulson, O. B.; Gyory, A.; Hertz, M. M., Blood-brain barrier transfer and cerebral uptake of antiepileptic drugs. *Clinical pharmacology & therapeutics* **1982**, 32, 466–77.

[22] Jeffrey, P.; Summerfield, S. Assessment of the blood-brain barrier in CNS drug discovery. *Neurobiology of disease* **2010**, 37, 33–7.

[23] Westerhout, J.; Danhof, M.; De Lange, E. C. Preclinical prediction of human brain target site concentrations: considerations in extrapolating to the clinical setting. *Journal of pharmaceutical sciences* **2011**, 100, 3577–93.

[24] Fu, B. M. Experimental methods and transport models for drug delivery across the blood-brain barrier. *Current pharmaceutical biotechnology* **2012**, 13, 1346–59.

[25] Soni, V.; Jain, A.; Khare, P.; Gulbake, A.; Jain, S. K. Potential approaches for drug delivery to the brain: past, present, and future. *Critical reviews in therapeutic drug carrier systems* **2010**, 27, 187–236.

[26] Pike, V. W. PET radiotracers: crossing the blood-brain barrier and surviving metabolism. *Trends in pharmacological sciences* **2009**, 30, 431–40.

[27] Jeppsson, F.; Eketjall, S.; Janson, J.; Karlstrom, S.; Gustavsson, S.; Olsson, L. L.; Radesater, A. C.; Ploeger, B.; Cebers, G.; Kolmodin, K.; Swahn, B. M.; von Berg, S.; Bueters, T.; Falting, J. Discovery of AZD3839, a potent and selective BACE1 inhibitor clinical candidate for the treatment of Alzheimer disease. *The journal of biological chemistry* **2012**, 287, 41245–57.

[28] Brown, R. P.; Delp, M. D.; Lindstedt, S. L.; Rhomberg, L. R.; Beliles, R. P. Physiological parameter values for physiologically based pharmacokinetic models. *Toxicology and industrial health* **1997**, 13, 407–84.

[29] Davies, B.; Morris, T. Physiological parameters in laboratory animals and humans. *Pharmaceutical research* **1993**, 10, 1093–5.

[30] Lei, H.; Pilloud, Y.; Magill, A. W.; Gruetter, R. Continuous arterial spin labeling of mouse cerebral blood flow using an actively-detuned two-coil system at 9.4 T. *Conference proceedings of the Annual International Conference of the IEEE Engineering in Medicine and Biology Society; 2011 30 Aug-03 Sep; Boston, MA: IEEE Engineering in Medicine and Biology Society*. Institute of Electrical and Electronics Engineers (IEEE) Publ.; **2012**, pp. 6993–6.

[31] Jay, T. M.; Lucignani, G.; Crane, A. M.; Jehle, J.; Sokoloff, L. Measurement of local cerebral blood flow with [14C]iodoantipyrine in the mouse. *Journal of cerebral blood flow & metabolism* **1988**, 8, 121–9.

[32] Demchenko, I. T.; Boso, A. E.; Natoli, M. J.; Doar, P. O.; O'Neill, T. J.; Bennett, P. B.; Piantadosi, C. A. Measurement of cerebral blood flow in rats and mice by hydrogen clearance during hyperbaric oxygen exposure. *Undersea and hyperbaric medicine* **1998**, 25, 147–52.

[33] Cornford, E. M.; Young, D.; Paxton, J. W.; Sofia, R. D. Blood-brain barrier penetration of felbamate. *Epilepsia* **1992**, 33, 944–54.

[34] Bodenheimer, M. M.; Wackers, F. J.; Schwartz, R. G.; Brown, M. Prognostic significance of a fixed thallium defect one to six months after onset of acute myocardial infarction or unstable angina. Multicenter Myocardial Ischemia Research Group. *American journal of cardiology* **1994**, 74, 1196–200.

[35] Furuya, Y.; Ikehira, H.; Obata, T.; Koga, M.; Yoshida, K. The measurement of blood flow parameters with deuterium stable isotope MR imaging. *Annals of nuclear medicine* **1997**, 11, 281–4.

[36] Weber, B.; Spath, N.; Wyss, M.; Wild, D.; Burger, C.; Stanley, R.; Buck, A. Quantitative cerebral blood flow measurements in the rat using a beta-probe and H2 15O. *Journal of cerebral blood flow & metabolism* **2003**, 23, 1455–60.

[37] Otori, T.; Katsumata, T.; Muramatsu, H.; Kashiwagi, F.; Katayama, Y.; Terashi, A. Long-term measurement of cerebral blood flow and metabolism in a rat chronic hypoperfusion model. *Clinical and experimental pharmacology & physiology* **2003**, 30, 266–72.

[38] Moffat, B. A.; Chenevert, T. L.; Hall, D. E.; Rehemtulla, A.; Ross, B. D. Continuous arterial spin labeling using a train of adiabatic inversion pulses. *Journal of magnetic resonance imaging* **2005**, 21, 290–6.

[39] Ball, K.; Bouzom, F.; Scherrmann, J. M.; Walther, B.; Decleves, X. Development of a physiologically based pharmacokinetic model for the rat central nervous system and determination of an in vitro–in vivo scaling methodology for the blood-brain barrier permeability of two transporter substrates, morphine and oxycodone. *Journal of pharmaceutical sciences* **2012**, 101, 4277–92.

[40] Karbowski, J. Scaling of brain metabolism and blood flow in relation to capillary and neural scaling. *PloS one* **2011**, 6, e26709.

[41] Kassell, N. F.; Baumann, K. W.; Hitchon, P. W.; Gerk, M. K.; Hill, T. R.; Sokoll, M. D. The effects of high dose mannitol on cerebral blood flow in dogs with normal intracranial pressure. *Stroke* **1982**, 13, 59–61.

[42] Werner, C.; Kochs, E.; Hoffman, W. E.; Blanc, I. F.; Schulte am Esch, *J. Cerebral blood flow and cerebral blood flow velocity during angiotensin-induced arterial hypertension in dogs. Canadian journal of anaesthesia* **1993**, 40, 755–60.

[43] Ito, H.; Kanno, I.; Ibaraki, M.; Hatazawa, J.; Miura, S. Changes in human cerebral blood flow and cerebral blood volume during hypercapnia and hypocapnia measured by positron emission tomography. *Journal of cerebral blood flow & metabolism* **2003**, 23, 665–70.

[44] Ito, H.; Kanno, I.; Fukuda, H. Human cerebral circulation: positron emission tomography studies. *Annals of nuclear medicine* **2005**, 19, 65–74.

[45] Meltzer, C. C.; Cantwell, M. N.; Greer, P. J.; Ben-Eliezer, D.; Smith, G.; Frank, G.; Kaye, W. H.; Houck, P. R.; Price, J. C. Does cerebral blood flow decline in healthy aging? A PET study with partial-volume correction. *Journal of nuclear medicine* **2000**, 41, 1842–8.

[46] Fenstermacher, J.; Nakata, H.; Tajima, A.; Lin, S. Z.; Otsuka, T.; Acuff, V.; Wei, L.; Bereczki, D. Functional variations in parenchymal microvascular systems within the brain. *Magnetic resonance in medicine* **1991**, 19, 217–20.

[47] Kumar, G.; Smith, Q. R.; Hokari, M.; Parepally, J.; Duncan, M. W. Brain uptake, pharmacokinetics, and tissue distribution in the rat of neurotoxic N-butylbenzenesulfonamide. *Toxicological Sciences* **2007**, 97, 253–64.

[48] Terasaki, T.; Ohtsuki, S.; Hori, S.; Takanaga, H.; Nakashima, E.; Hosoya, K. New approaches to in vitro models of blood-brain barrier drug transport. *Drug discovery today* **2003**, 8, 944–54.

[49] Chen, W.; Yang, J. Z.; Andersen, R.; Nielsen, L. H.; Borchardt, R. T. Evaluation of the permeation characteristics of a model opioid peptide, H-Tyr-D-Ala-Gly-Phe-D-Leu-OH (DADLE), and its cyclic prodrugs across the blood-brain barrier using an in situ perfused rat brain model. *The journal of pharmacology and experimental therapeutics* **2002**, 303, 849–57.

[50] Summerfield, S. G.; Read, K.; Begley, D. J.; Obradovic, T.; Hidalgo, I. J.; Coggon, S.; Lewis, A. V.; Porter, R. A.; Jeffrey, P. Central nervous system drug disposition: the relationship between in situ brain permeability and brain free fraction. *The journal of pharmacology and experimental therapeutics* **2007**, 322, 205–13.

[51] Gross, P. M.; Sposito, N. M.; Pettersen, S. E.; Fenstermacher, J. D. Differences in function and structure of the capillary endothelium in gray matter, white matter and a circumventricular organ of rat brain. *Blood vessels* **1986**, 23, 261–70.

[52] Cucullo, L.; Hossain, M.; Rapp, E.; Manders, T.; Marchi, N.; Janigro, D. Development of a humanized in vitro blood-brain barrier model to screen for brain penetration of anti-epileptic drugs. *Epilepsia* **2007**, 48, 505–16.

[53] Crone, C. The permeability of capillaries in various organs as determined by use of the 'indicator diffusion' method. *Acta physiologica Scandinavica* **1963**, 58, 292–305.

[54] Abbott, N. J.; Patabendige, A. A.; Dolman, D. E.; Yusof, S. R.; Begley, D. J. Structure and function of the blood-brain barrier. *Neurobiology of disease* **2010**, 37, 13–25.

[55] Abraham, M. H. The permeation of neutral molecules, ions, and ionic species through membranes: brain permeation as an example. *Journal of pharmaceutical sciences* **2011**, 100, 1690–701.

[56] Murakami, H.; Takanaga, H.; Matsuo, H.; Ohtani, H.; Sawada, Y. Comparison of blood-brain barrier permeability in mice and rats using in situ brain perfusion technique. *American journal of physiology. Heart and circulatory physiology* **2000**, 279, H1022–8.

[57] Kalvass, J. C.; Olson, E. R.; Cassidy, M. P.; Selley, D. E.; Pollack, G. M. Pharmacokinetics and pharmacodynamics of seven opioids in P-glycoprotein-competent mice: assessment of unbound brain EC50,u and correlation of in vitro, preclinical, and clinical data. *The journal of pharmacology and experimental therapeutics* **2007**, 323, 346–55.

[58] Wandel, C.; Kim, R.; Wood, M.; Wood, A. Interaction of morphine, fentanyl, sufentanil, alfentanil, and loperamide with the efflux drug transporter P-glycoprotein. *Anesthesiology* **2002**, 96, 913–20.

[59] Groothuis, D. R.; Levy, R. M. The entry of antiviral and antiretroviral drugs into the central nervous system. *Journal of neurovirology* **1997**, 3, 387–400.

[60] Cremer, J. E.; Seville, M. P. Regional brain blood flow, blood volume, and haematocrit values in the adult rat. *Journal of cerebral blood flow & metabolism* **1983**, 3, 254–6.

[61] Archer, D. P.; Labrecque, P.; Tyler, J. L.; Meyer, E.; Trop, D. Cerebral blood volume is increased in dogs during administration of nitrous oxide or isoflurane. *Anesthesiology* **1987**, 67, 642–8.

[62] Tudorica, A.; Fang Li, H.; Hospod, F.; Delucia-Deranja, E.; Huang, W.; Patlak, C. S.; Newman, G. C. Cerebral blood volume measurements by rapid contrast infusion and T2*-weighted echo planar MRI. *Magnetic resonance in medicine* **2002**, 47, 1145–57.

[63] Oshio, K.; Watanabe, H.; Song, Y.; Verkman, A. S.; Manley, G. T. Reduced cerebrospinal fluid production and intracranial pressure in mice lacking choroid plexus water channel Aquaporin-1. *FASEB journal* **2005**, 19, 76–8.

[64] de Lange, E. C. The mastermind approach to CNS drug therapy: translational prediction of human brain distribution, target site kinetics, and therapeutic effects. *Fluids and barriers of the CNS* **2013**, 10, 12.

[65] Artru, A. A. Reduction of cerebrospinal fluid pressure by hypocapnia: changes in cerebral blood volume, cerebrospinal fluid volume, and brain tissue water and electrolytes. *Journal of cerebral blood flow & metabolism* **1987**, 7, 471–9.

[66] Bergman, I.; Burckart, G. J.; Pohl, C. R.; Venkataramanan, R.; Barmada, M. A.; Griffin, J. A.; Cheung, N. K. Pharmacokinetics of IgG and IgM anti-ganglioside antibodies in rats and monkeys after intrathecal administration. *The journal of pharmacology and experimental therapeutics* **1998**, 284, 111–5.

[67] Friden, M.; Winiwarter, S.; Jerndal, G.; Bengtsson, O.; Wan, H.; Bredberg, U.; Hammarlund-Udenaes, M.; Antonsson, M. Structure–brain exposure relationships in rat and human using a novel data set of unbound drug concentrations in brain interstitial and cerebrospinal fluids. *Journal of medicinal chemistry* **2009**, 52, 6233–43.

[68] Maurer, M. H. Proteomics of brain extracellular fluid (ECF) and cerebrospinal fluid (CSF). *Mass spectrometry reviews* **2010**, 29, 17–28.

[69] Poulin, P.; Theil, F. P. Prediction of pharmacokinetics prior to in vivo studies. 1. Mechanism-based prediction of volume of distribution. *Journal of pharmaceutical sciences* **2002**, 91, 129–56.

[70] Rodgers, T.; Jones, H. M.; Rowland, M. Tissue lipids and drug distribution: dog versus rat. *Journal of pharmaceutical sciences* **2012**, 101, 4615–26.

[71] Rodgers, T.; Leahy, D.; Rowland, M. Physiologically based pharmacokinetic modeling 1: predicting the tissue distribution of moderate-to-strong bases. *Journal of pharmaceutical sciences* **2005**, 94, 1259–76.

[72] Campbell, S. D.; Regina, K. J.; Kharasch, E. D. Significance of lipid composition in a blood-brain barrier-mimetic PAMPA assay. *Journal of biomolecular screening* **2014**, 19, 437–44.

[73] Zhang, Y.; Li, C. S.; Ye, Y.; Johnson, K.; Poe, J.; Johnson, S.; Bobrowski, W.; Garrido, R.; Madhu, C. Porcine brain microvessel endothelial cells as an in vitro model to predict in vivo blood-brain barrier permeability. *Drug metabolism and disposition* **2006**, 34, 1935–43.

[74] Di, L.; Umland, J. P.; Chang, G.; Huang, Y.; Lin, Z.; Scott, D. O.; Troutman, M. D.; Liston, T. E. Species independence in brain tissue binding using brain homogenates. *Drug metabolism and disposition* **2011**, 39, 1270–7.

[75] Read, K. D.; Braggio, S. Assessing brain free fraction in early drug discovery. *Expert opinion on drug metabolism & toxicology* **2010**, 6, 337–44.

[76] Tanaka, H.; Mizojiri, K. Drug-protein binding and blood-brain barrier permeability. *The journal of pharmacology and experimental therapeutics* **1999**, 288, 912–8.

[77] Bohnert, T.; Gan, L. S. Plasma protein binding: from discovery to development. *Journal of pharmaceutical sciences* **2013**, 102, 2953–94.

[78] Summerfield, S. G.; Stevens, A. J.; Cutler, L.; del Carmen Osuna, M.; Hammond, B.; Tang, S. P.; Hersey, A.; Spalding, D. J.; Jeffrey, P. Improving the in vitro prediction of in vivo central nervous system penetration: integrating permeability, P-glycoprotein efflux, and free fractions in blood and brain. *The journal of pharmacology and experimental therapeutics* **2006**, 316, 1282–90.

[79] Jeffrey, P.; Summerfield, S. G. Challenges for blood-brain barrier (BBB) screening. *Xenobiotica* **2007**, 37, 1135–51.

[80] Melhem, M. Translation of central nervous system occupancy from animal models: application of pharmacokinetic/pharmacodynamic modeling. *The journal of pharmacology and experimental therapeutics* **2013**, 347, 2–6.

[81] Kalvass, J. C.; Maurer, T. S.; Pollack, G. M. Use of plasma and brain unbound fractions to assess the extent of brain distribution of 34 drugs: comparison of unbound concentration ratios to in vivo p-glycoprotein efflux ratios. *Drug metabolism and disposition* **2007**, 35, 660–6.

[82] Neuwelt, E. A.; Bauer, B.; Fahlke, C.; Fricker, G.; Iadecola, C.; Janigro, D.; Leybaert, L.; Molnar, Z.; O'Donnell, M. E.; Povlishock, J. T.; Saunders, N. R.; Sharp, F.;

Stanimirovic, D.; Watts, R. J.; Drewes, L. R. Engaging neuroscience to advance translational research in brain barrier biology. *Nature reviews. Neuroscience* **2011**, 12, 169–82.

[83] Hoshi, Y.; Uchida, Y.; Tachikawa, M.; Inoue, T.; Ohtsuki, S.; Terasaki, T. Quantitative atlas of blood-brain barrier transporters, receptors, and tight junction proteins in rats and common marmoset. *Journal of pharmaceutical sciences* **2013**, 102, 3343–55.

[84] Uchida, Y.; Ohtsuki, S.; Katsukura, Y.; Ikeda, C.; Suzuki, T.; Kamiie, J.; Terasaki, T. Quantitative targeted absolute proteomics of human blood-brain barrier transporters and receptors. *Journal of neurochemistry* **2011**, 117, 333–45.

[85] Ito, K.; Uchida, Y.; Ohtsuki, S.; Aizawa, S.; Kawakami, H.; Katsukura, Y.; Kamiie, J.; Terasaki, T. Quantitative membrane protein expression at the blood-brain barrier of adult and younger cynomolgus monkeys. *Journal of pharmaceutical sciences* **2011**, 100, 3939–50.

[86] Agarwal, S.; Uchida, Y.; Mittapalli, R. K.; Sane, R.; Terasaki, T.; Elmquist, W. F. Quantitative proteomics of transporter expression in brain capillary endothelial cells isolated from P-glycoprotein (P-gp), breast cancer resistance protein (Bcrp), and P-gp/Bcrp knockout mice. *Drug metabolism and disposition* **2012**, 40, 1164–9.

[87] Warren, M. S.; Zerangue, N.; Woodford, K.; Roberts, L. M.; Tate, E. H.; Feng, B.; Li, C.; Feuerstein, T. J.; Gibbs, J.; Smith, B.; de Morais, S. M.; Dower, W. J.; Koller, K. J. Comparative gene expression profiles of ABC transporters in brain microvessel endothelial cells and brain in five species including human. *Pharmacological research* **2009**, 59, 404–13.

[88] Booth-Genthe, C. L.; Louie, S. W.; Carlini, E. J.; Li, B.; Leake, B. F.; Eisenhandler, R.; Hochman, J. H.; Mei, Q.; Kim, R. B.; Rushmore, T. H.; Yamazaki, M. Development and characterization of LLC-PK1 cells containing Sprague-Dawley rat Abcb1a (Mdr1a): comparison of rat P-glycoprotein transport to human and mouse. *Journal of pharmacological and toxicological methods* **2006**, 54, 78–89.

[89] Katoh, M.; Suzuyama, N.; Takeuchi, T.; Yoshitomi, S.; Asahi, S.; Yokoi, T. Kinetic analyses for species differences in P-glycoprotein-mediated drug transport. *Journal of pharmaceutical sciences* **2006**, 95, 2673–83.

[90] Takeuchi, T.; Yoshitomi, S.; Higuchi, T.; Ikemoto, K.; Niwa, S.; Ebihara, T.; Katoh, M.; Yokoi, T.; Asahi, S. Establishment and characterization of the transformants stably-expressing MDR1 derived from various animal species in LLC-PK1. *Pharmaceutical research* **2006**, 23, 1460–72.

[91] Ohe, T.; Sato, M.; Tanaka, S.; Fujino, N.; Hata, M.; Shibata, Y.; Kanatani, A.; Fukami, T.; Yamazaki, M.; Chiba, M.; Ishii, Y., Effect of P-glycoprotein-mediated efflux on cerebrospinal fluid/plasma concentration ratio. *Drug metabolism and disposition* **2003**, 31, 1251–4.

[92] Baltes, S.; Gastens, A. M.; Fedrowitz, M.; Potschka, H.; Kaever, V.; Loscher, W. Differences in the transport of the antiepileptic drugs phenytoin, levetiracetam and carbamazepine by human and mouse P-glycoprotein. *Neuropharmacology* **2007**, 52, 333–46.

[93] Syvanen, S.; Lindhe, O.; Palner, M.; Kornum, B. R.; Rahman, O.; Langstrom, B.; Knudsen, G. M.; Hammarlund-Udenaes, M. Species differences in blood-brain barrier transport of three positron emission tomography radioligands with emphasis on P-glycoprotein transport. *Drug metabolism and disposition* **2009**, 37, 635–43.

[94] Pike, V. W.; McCarron, J. A.; Lammertsma, A. A.; Osman, S.; Hume, S. P.; Sargent, P. A.; Bench, C. J.; Cliffe, I. A.; Fletcher, A.; Grasby, P. M. Exquisite delineation of 5-HT1A receptors in human brain with PET and [carbonyl-11 C]WAY-100635. *European journal of pharmacology* **1996**, 301, R5–7.

[95] Zoghbi, S. S.; Liow, J. S.; Yasuno, F.; Hong, J.; Tuan, E.; Lazarova, N.; Gladding, R. L.; Pike, V. W.; Innis, R. B. 11C-loperamide and its N-desmethyl radiometabolite are avid substrates for brain permeability-glycoprotein efflux. *Journal of nuclear medicine* **2008**, 49, 649–56.

[96] Seneca, N.; Zoghbi, S. S.; Liow, J. S.; Kreisl, W.; Herscovitch, P.; Jenko, K.; Gladding, R. L.; Taku, A.; Pike, V. W.; Innis, R. B. Human brain imaging and radiation dosimetry of 11C-N-desmethyl-loperamide, a PET radiotracer to measure the function of P-glycoprotein. *Journal of nuclear medicine* **2009**, 50, 807–13.

[97] Hsiao, P.; Sasongko, L.; Link, J. M.; Mankoff, D. A.; Muzi, M.; Collier, A. C.; Unadkat, J. D. Verapamil P-glycoprotein transport across the rat blood-brain barrier: cyclosporine, a concentration inhibition analysis, and comparison with human data. *The journal of pharmacology and experimental therapeutics* **2006**, 317, 704–10.

[98] Bauer, M.; Zeitlinger, M.; Karch, R.; Matzneller, P.; Stanek, J.; Jager, W.; Bohmdorfer, M.; Wadsak, W.; Mitterhauser, M.; Bankstahl, J. P.; Loscher, W.; Koepp, M.; Kuntner, C.; Muller, M.; Langer, O. Pgp-mediated interaction between (R)-[11C]verapamil and tariquidar at the human blood-brain barrier: a comparison with rat data. *Clinical pharmacology and therapeutics* **2012**, 91, 227–33.

[99] Pauli-Magnus, C.; von Richter, O.; Burk, O.; Ziegler, A.; Mettang, T.; Eichelbaum, M.; Fromm, M. F. Characterization of the major metabolites of verapamil as substrates and inhibitors of P-glycoprotein. *The journal of pharmacology and experimental therapeutics* **2000**, 293, 376–82.

[100] Syvanen, S.; Hooker, A.; Rahman, O.; Wilking, H.; Blomquist, G.; Langstrom, B.; Bergstrom, M.; Hammarlund-Udenaes, M. Pharmacokinetics of P-glycoprotein inhibition in the rat blood-brain barrier. *Journal of pharmaceutical sciences* **2008**, 97, 5386–400.

[101] Liow, J. S.; Lu, S.; McCarron, J. A.; Hong, J.; Musachio, J. L.; Pike, V. W.; Innis, R. B.; Zoghbi, S. S. Effect of a P-glycoprotein inhibitor, Cyclosporin A, on the disposition in rodent brain and blood of the 5-HT1A receptor radioligand, [11C](R)-(–)-RWAY. *Synapse* **2007**, 61, 96–105.

[102] Farwell, M. D.; Chong, D. J.; Iida, Y.; Bae, S. A.; Easwaramoorthy, B.; Ichise, M. Imaging P-glycoprotein function in rats using [(11)C]-N-desmethyl-loperamide. *Annals of nuclear medicine* **2013**, 27, 618–24.

[103] Yasuno, F.; Zoghbi, S. S.; McCarron, J. A.; Hong, J.; Ichise, M.; Brown, A. K.; Gladding, R. L.; Bacher, J. D.; Pike, V. W.; Innis, R. B. Quantification of serotonin 5-HT1A receptors in monkey brain with [11C](R)-(–)-RWAY. *Synapse* **2006**, 60, 510–20.

[104] Liow, J. S.; Kreisl, W.; Zoghbi, S. S.; Lazarova, N.; Seneca, N.; Gladding, R. L.; Taku, A.; Herscovitch, P.; Pike, V. W.; Innis, R. B. P-glycoprotein function at the blood-brain barrier imaged using 11C-N-desmethyl-loperamide in monkeys. *Journal of nuclear medicine* **2009**, 50, 108–15.

[105] Mairinger, S.; Erker, T.; Muller, M.; Langer, O. PET and SPECT radiotracers to assess function and expression of ABC transporters in vivo. *Current drug metabolism* **2011**, 12, 774–92.

[106] Wanek, T.; Kuntner, C.; Bankstahl, J. P.; Mairinger, S.; Bankstahl, M.; Stanek, J.; Sauberer, M.; Filip, T.; Erker, T.; Muller, M.; Loscher, W.; Langer, O. A novel PET protocol for visualization of breast cancer resistance protein function at the blood-brain barrier. *Journal of cerebral blood flow & metabolism* **2012**, 32, 2002–11.

[107] Doran, A. C.; Osgood, S. M.; Mancuso, J. Y.; Shaffer, C. L. An evaluation of using rat-derived single-dose neuropharmacokinetic parameters to project accurately large

animal unbound brain drug concentrations. *Drug metabolism and disposition* **2012**, 40, 2162–73.

[108] van Bree, J. B.; Audus, K. L.; Borchardt, R. T. Carrier-mediated transport of baclofen across monolayers of bovine brain endothelial cells in primary culture. *Pharmaceutical research* **1988**, 5, 369–71.

[109] Luckner, P.; Brandsch, M. Interaction of 31 beta-lactam antibiotics with the H+/peptide symporter PEPT2: analysis of affinity constants and comparison with PEPT1. *European journal of pharmaceutics and biopharmaceutics* **2005**, 59, 17–24.

[110] Kamal, M. A.; Keep, R. F.; Smith, D. E. Role and relevance of PEPT2 in drug disposition, dynamics, and toxicity. *Drug metabolism and pharmacokinetics* **2008**, 23, 236–42.

[111] Bailey, D. G. Fruit juice inhibition of uptake transport: a new type of food-drug interaction. *British journal of clinical pharmacology* **2010**, 70, 645–55.

[112] Lee, W.; Glaeser, H.; Smith, L. H.; Roberts, R. L.; Moeckel, G. W.; Gervasini, G.; Leake, B. F.; Kim, R. B. Polymorphisms in human organic anion-transporting polypeptide 1A2 (OATP1A2): implications for altered drug disposition and central nervous system drug entry. *The journal of biological chemistry* **2005**, 280, 9610–7.

[113] Li, L.; Agarwal, S.; Elmquist, W. F. Brain efflux index to investigate the influence of active efflux on brain distribution of pemetrexed and methotrexate. *Drug metabolism and disposition* **2013**, 41, 659–67.

[114] Kielbasa, W.; Stratford, R. E., Jr. Exploratory translational modeling approach in drug development to predict human brain pharmacokinetics and pharmacologically relevant clinical doses. *Drug metabolism and disposition* **2012**, 40, 877–83.

[115] Shawahna, R.; Decleves, X.; Scherrmann, J. M. Hurdles with using in vitro models to predict human blood-brain barrier drug permeability: a special focus on transporters and metabolizing enzymes. *Current drug metabolism* **2013**, 14, 120–36.

[116] Miksys, S.; Tyndale, R. F. Cytochrome P450-mediated drug metabolism in the brain. *Journal of psychiatry & neuroscience* **2013**, 38, 152–63.

[117] Ferguson, C. S.; Tyndale, R. F. Cytochrome P450 enzymes in the brain: emerging evidence of biological significance. *Trends in pharmacological sciences* **2011**, 32, 708–714.

[118] Warner, M.; Kohler, C.; Hansson, T.; Gustafsson, J. A. Regional distribution of cytochrome P-450 in the rat brain: spectral quantitation and contribution of P-450b,e, and P-450c,d. *Journal of neurochemistry* **1988**, 50, 1057–65.

[119] Nishimuta, H.; Nakagawa, T.; Nomura, N.; Yabuki, M. Species differences in hepatic and intestinal metabolic activities for 43 human cytochrome P450 substrates between humans and rats or dogs. *Xenobiotica* **2013**, 43, 948–55.

[120] Stamou, M.; Wu, X.; Kania-Korwel, I.; Lehmler, H. J.; Lein, P. J. Cytochrome P450 mRNA expression in the rodent brain: species-, sex- and region-dependent differences. *Drug metabolism and disposition* **2013**, 42, 239–44.

[121] Zlokovic, B. V. The blood-brain barrier in health and chronic neurodegenerative disorders. *Neuron* **2008**, 57, 178–201.

[122] Vangilder, R. L.; Rosen, C. L.; Barr, T. L.; Huber, J. D. Targeting the neurovascular unit for treatment of neurological disorders. *Pharmacology & Therapeutics* **2011**, 130, 239–47.

[123] Popescu, B. O.; Toescu, E. C.; Popescu, L. M.; Bajenaru, O.; Muresanu, D. F.; Schultzberg, M.; Bogdanovic, N. Blood-brain barrier alterations in ageing and dementia. *Journal of the neurological sciences* **2009**, 283, 99–106.

[124] del Zoppo, G. J.; Poeck, K.; Pessin, M. S.; Wolpert, S. M.; Furlan, A. J.; Ferbert, A.; Alberts, M. J.; Zivin, J. A.; Wechsler, L.; Busse, O.; et al. Recombinant tissue plasminogen activator in acute thrombotic and embolic stroke. *Annals of neurology* **1992**, 32, 78–86.

[125] Paul, R.; Lorenzl, S.; Koedel, U.; Sporer, B.; Vogel, U.; Frosch, M.; Pfister, H. W. Matrix metalloproteinases contribute to the blood-brain barrier disruption during bacterial meningitis. *Annals of neurology* **1998**, 44, 592–600.

[126] Wolka, A. M.; Huber, J. D.; Davis, T. P. Pain and the blood-brain barrier: obstacles to drug delivery. *Advanced Drug Delivery Reviews* **2003**, 55, 987–1006.

[127] Bowman, G. L.; Quinn, J. F. Alzheimer's disease and the blood-brain barrier: past, present and future. *Aging health* **2008**, 4, 47–55.

[128] Palmer, J. C.; Baig, S.; Kehoe, P. G.; Love, S. Endothelin-converting enzyme-2 is increased in Alzheimer's disease and up-regulated by Abeta. *American journal of pathology* **2009**, 175, 262–270.

[129] Dutheil, F.; Jacob, A.; Dauchy, S.; Beaune, P.; Scherrmann, J. M.; Decleves, X.; Loriot, M. A. ABC transporters and cytochromes P450 in the human central nervous system: influence on brain pharmacokinetics and contribution to neurodegenerative disorders. *Expert opinion on drug metabolism & toxicology* **2010**, 6, 1161–1174.

[130] Vogelgesang, S.; Cascorbi, I.; Schroeder, E.; Pahnke, J.; Kroemer, H. K.; Siegmund, W.; Kunert-Keil, C.; Walker, L. C.; Warzok, R. W. Deposition of Alzheimer's beta-amyloid is inversely correlated with P-glycoprotein expression in the brains of elderly nondemented humans. *Pharmacogenetics* **2002**, 12, 535–41.

[131] Hartz, A. M.; Miller, D. S.; Bauer, B. Restoring blood-brain barrier P-glycoprotein reduces brain amyloid-beta in a mouse model of Alzheimer's disease. *Molecular pharmacology* **2010**, 77, 715–23.

[132] Donahue, J. E.; Flaherty, S. L.; Johanson, C. E.; Duncan, J. A., 3rd; Silverberg, G. D.; Miller, M. C.; Tavares, R.; Yang, W.; Wu, Q.; Sabo, E.; Hovanesian, V.; Stopa, E. G. RAGE, LRP-1, and amyloid-beta protein in Alzheimer's disease. *Acta neuropathologica* **2006**, 112, 405–15.

[133] Xiong, H.; Callaghan, D.; Jones, A.; Bai, J.; Rasquinha, I.; Smith, C.; Pei, K.; Walker, D.; Lue, L. F.; Stanimirovic, D.; Zhang, W. ABCG2 is upregulated in Alzheimer's brain with cerebral amyloid angiopathy and may act as a gatekeeper at the blood-brain barrier for Abeta(1–40) peptides. *The journal of neuroscience* **2009**, 29, 5463–75.

[134] Westerlund, M.; Belin, A. C.; Olson, L.; Galter, D. Expression of multi-drug resistance 1 mRNA in human and rodent tissues: reduced levels in Parkinson patients. *Cell and tissue research* **2008**, 334, 179–185.

[135] Vautier, S.; Fernandez, C. ABCB1: the role in Parkinson's disease and pharmacokinetics of antiparkinsonian drugs. *Expert opinion on drug metabolism and toxicology* **2009**, 5, 1349–1358.

[136] Kim, W. S.; Halliday, G. M. Changes in sphingomyelin level affect alpha-synuclein and ABCA5 expression. *Journal of Parkinson's disease* **2012**, 2, 41–6.

[137] Bednarczyk, J.; Lukasiuk, K. Tight junctions in neurological diseases. *Acta Neurobiologiae Experimentalis (Wars.)* **2011**, 71, 393–408.

[138] Friedman, A.; Heinemann, U. Role of blood-brain barrier dysfunction in epileptogenesis. In: Noebels, J.L., Avoli, M., Rogawski, M.A., Olsen, R.W., Delgado-Escueta, A.V.,

editors. Jasper's Basic Mechanisms of the Epilepsies. 4th edition (internet). Bethesda: National Center for Biotechnology Information (US); **2012**.

[139] van Vliet, E. A.; da Costa, A. S.; Redeker, S.; van, S. R.; Aronica, E.; Gorter, J. A. Blood-brain barrier leakage may lead to progression of temporal lobe epilepsy. *Brain* **2007**, 130, 521–34.

[140] Marchi, N.; Teng, Q.; Ghosh, C.; Fan, Q.; Nguyen, M. T.; Desai, N. K.; Bawa, H.; Rasmussen, P.; Masaryk, T. K.; Janigro, D. Blood-brain barrier damage, but not parenchymal white blood cells, is a hallmark of seizure activity. *Brain research* **2010**, 1353, 176–86.

[141] Appel, S.; Duke, E. S.; Martinez, A. R.; Khan, O. I.; Dustin, I. M.; Reeves-Tyer, P.; Berl, M. B.; Sato, S.; Gaillard, W. D.; Theodore, W. H. Cerebral blood flow and fMRI BOLD auditory language activation in temporal lobe epilepsy. *Epilepsia* **2012**, 53, 631–638.

[142] Lazarowski, A.; Czornyj, L.; Lubienieki, F.; Girardi, E.; Vazquez, S.; D'Giano, C. ABC transporters during epilepsy and mechanisms underlying multidrug resistance in refractory epilepsy. *Epilepsia* **2007**, 48 Suppl 5, 140–9.

[143] Potschka, H. Modulating P-glycoprotein regulation: future perspectives for pharmacoresistant epilepsies? *Epilepsia* **2010**, 51, 1333–47.

[144] van Vliet, E. A.; Redeker, S.; Aronica, E.; Edelbroek, P. M.; Gorter, J. A. Expression of multidrug transporters MRP1, MRP2, and BCRP shortly after status epilepticus, during the latent period, and in chronic epileptic rats. *Epilepsia* **2005**, 46, 1569–80.

[145] Ufer, M.; von Stulpnagel, C.; Muhle, H.; Haenisch, S.; Remmler, C.; Majed, A.; Plischke, H.; Stephani, U.; Kluger, G.; Cascorbi, I. Impact of ABCC2 genotype on antiepileptic drug response in Caucasian patients with childhood epilepsy. *Pharmacogenetics and genomics* **2011**, 21, 624–30.

[146] Ingrisch, M.; Sourbron, S.; Morhard, D.; Ertl-Wagner, B.; Kumpfel, T.; Hohlfeld, R.; Reiser, M.; Glaser, C. Quantification of perfusion and permeability in multiple sclerosis: dynamic contrast-enhanced MRI in 3D at 3T. *Investigative radiology* **2012**, 47, 252–8.

[147] Weiss, N.; Miller, F.; Cazaubon, S.; Couraud, P. O. The blood-brain barrier in brain homeostasis and neurological diseases. *Biochimica et Biophysica Acta* **2009**, 1788, 842–857.

[148] de Vries, H. E.; Kooij, G.; Frenkel, D.; Georgopoulos, S.; Monsonego, A.; Janigro, D. Inflammatory events at blood-brain barrier in neuroinflammatory and neurodegenerative disorders: implications for clinical disease. *Epilepsia* **2012**, 53, Suppl 6, 45–52.

[149] Kooij, G.; Mizee, M. R.; van Horssen, J.; Reijerkerk, A.; Witte, M. E.; Drexhage, J. A.; van der Pol, S. M.; van Het Hof, B.; Scheffer, G.; Scheper, R.; Dijkstra, C. D.; van der Valk, P.; de Vries, H. E. Adenosine triphosphate-binding cassette transporters mediate chemokine (C–C motif) ligand 2 secretion from reactive astrocytes: relevance to multiple sclerosis pathogenesis. *Brain* **2011**, 134, 555–70.

[150] Muller, N.; Ackenheil, M. Immunoglobulin and albumin content of cerebrospinal fluid in schizophrenic patients: relationship to negative symptomatology. *Schizophrenia research* **1995**, 14, 223–8.

[151] Schwarz, M. J.; Ackenheil, M.; Riedel, M.; Muller, N. Blood-cerebrospinal fluid barrier impairment as indicator for an immune process in schizophrenia. *Neuroscience letters* **1998**, 253, 201–3.

[152] Gudmundsson, P.; Skoog, I.; Waern, M.; Blennow, K.; Palsson, S.; Rosengren, L.; Gustafson, D. The relationship between cerebrospinal fluid biomarkers and depression in elderly women. *The American journal of geriatric psychiatry* **2007**, 15, 832–8.

[153] Aanerud, J.; Borghammer, P.; Chakravarty, M. M.; Vang, K.; Rodell, A. B.; Jonsdottir, K. Y.; Moller, A.; Ashkanian, M.; Vafaee, M. S.; Iversen, P.; Johannsen, P.; Gjedde, A. Brain energy metabolism and blood flow differences in healthy aging. *Journal of cerebral blood flow & metabolism* **2012**, 32, 1177–87.

[154] Farrall, A. J.; Wardlaw, J. M. Blood-brain barrier: ageing and microvascular disease—systematic review and meta-analysis. *Neurobiology of aging* **2009**, 30, 337–352.

[155] Silverberg, G. D.; Messier, A. A.; Miller, M. C.; Machan, J. T.; Majmudar, S. S.; Stopa, E. G.; Donahue, J. E.; Johanson, C. E. Amyloid efflux transporter expression at the blood-brain barrier declines in normal aging. *Journal of neuropathology and experimental neurology* **2010**, 69, 1034–43.

[156] Nishioku, T.; Furusho, K.; Tomita, A.; Ohishi, H.; Dohgu, S.; Shuto, H.; Yamauchi, A.; Kataoka, Y. Potential role for S100A4 in the disruption of the blood-brain barrier in collagen-induced arthritic mice, an animal model of rheumatoid arthritis. *Neuroscience* **2011**, 189, 286–92.

[157] Starr, J. M.; Wardlaw, J.; Ferguson, K.; MacLullich, A.; Deary, I. J.; Marshall, I. Increased blood-brain barrier permeability in type II diabetes demonstrated by gadolinium magnetic resonance imaging. *Journal of neurology, neurosurgery, and psychiatry* **2003**, 74, 70–6.

[158] Nguyen, J. H. Blood-brain barrier in acute liver failure. *Neurochemistry International* **2012**, 60, 676–83.

[159] Cipolla, M. J.; Sweet, J. G.; Chan, S. L. Cerebral vascular adaptation to pregnancy and its role in the neurological complications of eclampsia. *Journal of Applied Physiology* **2011**, 110, 329–39.

[160] Tang, J. P.; Rakhit, A.; Douglas, F. L.; Melethil, S. Effect of chronic hypertension on the blood-brain barrier permeability of libenzapril. *Pharmaceutical research* **1992**, 9, 236–43.

[161] Ong, W. Y.; Halliwell, B. Iron, atherosclerosis, and neurodegeneration: a key role for cholesterol in promoting iron-dependent oxidative damage? *Annals of the New York Academy of Sciences* **2004**, 1012, 51–64.

[162] Stanimirovic, D. B.; Friedman, A. Pathophysiology of the neurovascular unit: disease cause or consequence? *Journal of cerebral blood flow & metabolism* **2012**, 32, 1207–21.

[163] van Assema, D. M.; Lubberink, M.; Bauer, M.; van der Flier, W. M.; Schuit, R. C.; Windhorst, A. D.; Comans, E. F.; Hoetjes, N. J.; Tolboom, N.; Langer, O.; Muller, M.; Scheltens, P.; Lammertsma, A. A.; van Berckel, B. N. Blood-brain barrier P-glycoprotein function in Alzheimer's disease. *Brain* **2012**, 135, 181–9.

[164] Bartels, A. L.; van Berckel, B. N.; Lubberink, M.; Luurtsema, G.; Lammertsma, A. A.; Leenders, K. L. Blood-brain barrier P-glycoprotein function is not impaired in early Parkinson's disease. *Parkinsonism & related disorders* **2008**, 14, 505–8.

[165] Bartels, A. L.; Willemsen, A. T.; Kortekaas, R.; de Jong, B. M.; de, V. R.; de, K. O.; van Oostrom, J. C.; Portman, A.; Leenders, K. L. Decreased blood-brain barrier P-glycoprotein function in the progression of Parkinson's disease, PSP and MSA. *Journal of Neural Transmission* **2008**, 115, 1001–1009.

[166] Kortekaas, R.; Leenders, K. L.; van Oostrom, J. C.; Vaalburg, W.; Bart, J.; Willemsen, A. T.; Hendrikse, N. H. Blood-brain barrier dysfunction in parkinsonian midbrain in vivo. *Annals of neurology* **2005**, 57, 176–9.

[167] Feldmann, M.; Asselin, M. C.; Liu, J.; Wang, S.; McMahon, A.; Anton-Rodriguez, J.; Walker, M.; Symms, M.; Brown, G.; Hinz, R.; Matthews, J.; Bauer, M.; Langer, O.; Thom, M.; Jones, T.; Vollmar, C.; Duncan, J. S.; Sisodiya, S. M.; Koepp, M. J. P-glycoprotein expression and function in patients with temporal lobe epilepsy: a case–control study. *Lancet neurology* **2013**, 12, 777–85.

[168] de Klerk, O. L.; Willemsen, A. T.; Bosker, F. J.; Bartels, A. L.; Hendrikse, N. H.; den Boer, J. A.; Dierckx, R. A. Regional increase in P-glycoprotein function in the blood-brain barrier of patients with chronic schizophrenia: a PET study with [(11)C]verapamil as a probe for P-glycoprotein function. *Psychiatry research* **2010**, 183, 151–6.

[169] de Klerk, O. L.; Willemsen, A. T.; Roosink, M.; Bartels, A. L.; Hendrikse, N. H.; Bosker, F. J.; den Boer, J. A. Locally increased P-glycoprotein function in major depression: a PET study with [11C]verapamil as a probe for P-glycoprotein function in the blood-brain barrier. *The international journal of neuropsychopharmacology* **2009**, 12, 895–904.

[170] Bauer, M.; Karch, R.; Neumann, F.; Abrahim, A.; Wagner, C. C.; Kletter, K.; Muller, M.; Zeitlinger, M.; Langer, O. Age dependency of cerebral P-gp function measured with (R)-[11C]verapamil and PET. *European journal of clinical pharmacology* **2009**, 65, 941–6.

[171] Bartels, A. L.; Kortekaas, R.; Bart, J.; Willemsen, A. T.; de Klerk, O. L.; de Vries, J. J.; van Oostrom, J. C.; Leenders, K. L. Blood-brain barrier P-glycoprotein function decreases in specific brain regions with aging: a possible role in progressive neurodegeneration. *Neurobiology of aging* **2009**, 30, 1818–24.

[172] van Assema, D. M.; Lubberink, M.; Boellaard, R.; Schuit, R. C.; Windhorst, A. D.; Scheltens, P.; Lammertsma, A. A.; van Berckel, B. N. P-glycoprotein function at the blood-brain barrier: effects of age and gender. *Molecular imaging and biology* **2012**, 14, 771–6.

[173] Toornvliet, R.; van Berckel, B. N.; Luurtsema, G.; Lubberink, M.; Geldof, A. A.; Bosch, T. M.; Oerlemans, R.; Lammertsma, A. A.; Franssen, E. J. Effect of age on functional P-glycoprotein in the blood-brain barrier measured by use of (R)-[(11)C]verapamil and positron emission tomography. *Clinical pharmacology and therapeutics* **2006**, 79, 540–8.

[174] Syvanen, S.; Eriksson, J. Advances in PET imaging of P-glycoprotein function at the blood-brain barrier. *ACS chemical neuroscience* **2013**, 4, 225–37.

[175] Rapposelli, S.; Digiacomo, M.; Balsamo, A. P-gp transporter and its role in neurodegenerative diseases. *Current topics in medicinal chemistry* **2009**, 9, 209–17.

[176] Vogelgesang, S.; Glatzel, M.; Walker, L. C.; Kroemer, H. K.; Aguzzi, A.; Warzok, R. W. Cerebrovascular P-glycoprotein expression is decreased in Creutzfeldt-Jakob disease. *Acta neuropathologica* **2006**, 111, 436–43.

[177] Tishler, D. M.; Weinberg, K. I.; Hinton, D. R.; Barbaro, N.; Annett, G. M.; Raffel, C. MDR1 gene expression in brain of patients with medically intractable epilepsy. *Epilepsia* **1995**, 36, 1–6.

[178] Loscher, W.; Potschka, H. Role of multidrug transporters in pharmacoresistance to antiepileptic drugs. *The journal of pharmacology and experimental therapeutics* **2002**, 301, 7–14.

[179] Hartz, A. M.; Notenboom, S.; Bauer, B. Signaling to P-glycoprotein-A new therapeutic target to treat drug-resistant epilepsy? *Drug News & Perspectives* **2009**, 22, 393–7.

[180] Lee, G.; Bendayan, R. Functional expression and localization of P-glycoprotein in the central nervous system: relevance to the pathogenesis and treatment of neurological disorders. *Pharmaceutical research* **2004**, 21, 1313–30.

[181] Drozdzik, M.; Bialecka, M.; Mysliwiec, K.; Honczarenko, K.; Stankiewicz, J.; Sych, Z. Polymorphism in the P-glycoprotein drug transporter MDR1 gene: a possible link between environmental and genetic factors in Parkinson's disease. *Pharmacogenetics* **2003**, 13, 259–63.

[182] Miller, D. S. Regulation of P-glycoprotein and other ABC drug transporters at the blood-brain barrier. *Trends in pharmacological sciences* **2010**, 31, 246–54.

[183] Trifiro, G.; Spina, E. Age-related changes in pharmacodynamics: focus on drugs acting on central nervous and cardiovascular systems. *Current drug metabolism* **2011**, 12, 611–20.

[184] Watson, P.; Shirreffs, S. M.; Maughan, R. J. Blood-brain barrier integrity may be threatened by exercise in a warm environment. *American journal of physiology. Regulatory, integrative and comparative physiology* **2005**, 288, R1689–94.

[185] Spellerberg, B.; Prasad, S.; Cabellos, C.; Burroughs, M.; Cahill, P.; Tuomanen, E. Penetration of the blood-brain barrier: enhancement of drug delivery and imaging by bacterial glycopeptides. *The Journal of experimental medicine* **1995**, 182, 1037–43.

[186] Spector, R.; Lorenzo, A. V. Inhibition of penicillin transport from the cerebrospinal fluid after intracisternal inoculation of bacteria. *Journal of clinical investigation* **1974**, 54, 316–25.

[187] Morsing, P.; Adler, G.; Brandt-Eliasson, U.; Karp, L.; Ohlson, K.; Renberg, L.; Sjoquist, P. O.; Abrahamsson, T. Mechanistic differences of various AT1-receptor blockers in isolated vessels of different origin. *Hypertension* **1999**, 33, 1406–13.

[188] Hajjar, I.; Brown, L.; Mack, W. J.; Chui, H. Impact of angiotensin receptor blockers on alzheimer disease neuropathology in a large brain autopsy series. *Archives of neurology* **2012**, 1–7.

[189] Pelisch, N.; Hosomi, N.; Ueno, M.; Masugata, H.; Murao, K.; Hitomi, H.; Nakano, D.; Kobori, H.; Nishiyama, A.; Kohno, M. Systemic candesartan reduces brain angiotensin II via downregulation of brain renin-angiotensin system. *Hypertension research* **2010**, 33, 161–4.

[190] Corbett, A.; Pickett, J.; Burns, A.; Corcoran, J.; Dunnett, S. B.; Edison, P.; Hagan, J. J.; Holmes, C.; Jones, E.; Katona, C.; Kearns, I.; Kehoe, P.; Mudher, A.; Passmore, A.; Shepherd, N.; Walsh, F.; Ballard, C. Drug repositioning for Alzheimer's disease. *Nature reviews. Drug discovery* **2012**, 11, 833–46.

[191] Fredriksson, K.; Kalimo, H.; Westergren, I.; Kahrstrom, J.; Johansson, B. B. Blood-brain barrier leakage and brain edema in stroke-prone spontaneously hypertensive rats. Effect of chronic sympathectomy and low protein/high salt diet. *Acta neuropathologica* **1987**, 74, 259–68.

[192] Lin, K. M.; Chiu, Y. F.; Tsai, I. J.; Chen, C. H.; Shen, W. W.; Liu, S. C.; Lu, S. C.; Liu, C. Y.; Hsiao, M. C.; Tang, H. S.; Liu, S. I.; Chang, L. H.; Wu, C. S.; Tsou, H. H.; Tsai, M. H.; Chen, C. Y.; Wang, S. M.; Kuo, H. W.; Hsu, Y. T.; Liu, Y. L. ABCB1 gene polymorphisms are associated with the severity of major depressive disorder and its response to escitalopram treatment. *Pharmacogenetics and genomics* **2011**, 21, 163–70.

[193] Nikisch, G.; Eap, C. B.; Baumann, P. Citalopram enantiomers in plasma and cerebrospinal fluid of ABCB1 genotyped depressive patients and clinical response: a pilot study. *Pharmacological research* **2008**, 58, 344–7.

[194] Karlsson, L.; Green, H.; Zackrisson, A. L.; Bengtsson, F.; Jakobsen Falk, I.; Carlsson, B.; Ahlner, J.; Kugelberg, F. C. ABCB1 gene polymorphisms are associated with fatal intoxications involving venlafaxine but not citalopram. *International journal of legal medicine* **2013**, 127, 579–86.

[195] Kim, S. W.; Lee, J. H.; Lee, S. H.; Hong, H. J.; Lee, M. G.; Yook, K. H. ABCB1 c.2677G>T variation is associated with adverse reactions of OROS-methylphenidate in children and adolescents with ADHD. *Journal of clinical psychopharmacology* **2013**, 33, 491–8.

[196] Athanasoulia, A. P.; Sievers, C.; Ising, M.; Brockhaus, A. C.; Yassouridis, A.; Stalla, G. K.; Uhr, M. Polymorphisms of the drug transporter gene ABCB1 predict side effects of treatment with cabergoline in patients with PRL adenomas. *European journal of endocrinology* **2012**, 167, 327–35.

[197] Laika, B.; Leucht, S.; Steimer, W., ABCB1 (P-glycoprotein/MDR1) gene G2677T/a sequence variation (polymorphism): lack of association with side effects and therapeutic response in depressed inpatients treated with amitriptyline. *Clinical chemistry* **2006**, 52, 893–5.

[198] Pauli-Magnus, C.; Feiner, J.; Brett, C.; Lin, E.; Kroetz, D. L. No effect of MDR1 C3435T variant on loperamide disposition and central nervous system effects. *Clinical pharmacology and therapeutics* **2003**, 74, 487–98.

[199] Takano, A.; Kusuhara, H.; Suhara, T.; Ieiri, I.; Morimoto, T.; Lee, Y. J.; Maeda, J.; Ikoma, Y.; Ito, H.; Suzuki, K.; Sugiyama, Y. Evaluation of in vivo P-glycoprotein function at the blood-brain barrier among MDR1 gene polymorphisms by using 11C-verapamil. *Journal of nuclear medicine* **2006**, 47, 1427–33.

[200] Kalvass, J. C.; Polli, J. W.; Bourdet, D. L.; Feng, B.; Huang, S. M.; Liu, X.; Smith, Q. R.; Zhang, L. K.; Zamek-Gliszczynski, M. J.; International Transporter C. Why clinical modulation of efflux transport at the human blood-brain barrier is unlikely: the ITC evidence-based position. *Clinical pharmacology and therapeutics* **2013**, 94, 80–94.

[201] Uchida, S.; Yamada, S.; Deguchi, Y.; Yamamoto, M.; Kimura, R. In vivo specific binding characteristics and pharmacokinetics of a 1,4-dihydropyridine calcium channel antagonist in the senescent mouse brain. *Pharmaceutical research* **2000**, 17, 844–50.

PART 2

MECHANISMS OF DRUGS ACROSS THE BLOOD–BRAIN BARRIER

5

PASSIVE DIFFUSION PERMEABILITY OF THE BBB—EXAMPLES AND SAR

SCOTT SUMMERFIELD,[1] PHIL JEFFREY,[2] JASMINDER SAHI,[3] AND LIANGFU CHEN[4]

[1]*GlaxoSmithKline R&D, Platform Technology and Science, David Jack Centre for R&D, Hertfordshire, UK*
[2]*Rare Disease Research Unit, Pfizer Ltd, Cambridge, UK*
[3]*GlaxoSmithKline R&D, Platform Technology and Science, Shanghai, China*
[4]*GlaxoSmithKline R&D, Platform Technology and Science, Pennsylvania, USA*

INTRODUCTION

The effectiveness of a drug to modulate a disease state relies on the interplay between the pharmacokinetics (PK, as the forcing function) and the pharmacodynamics (PD, as the beneficial effect or toxicity). Efficacy requires three elements to be established: (i) the molecule reaches the biophase at the requisite concentration and over the requisite time course, (ii) sufficient target engagement is established, and (iii) that this leads to a beneficial biochemical and clinical response [1]. Since few therapeutic agents can be delivered directly to the biophase, most drug molecules are required to traverse and distribute through many cellular barriers between the site of administration and the site of action. In the case of drugs targeting the central nervous system (CNS) the blood–brain barrier (BBB) is one of these intervening cellular interfaces.

A consequence of the tight junction morphology of the BBB is that passage of low molecular-weight agents in the general range of CNS drugs (<MW 500 amu) occurs primarily via transcellular mechanisms, which in turn dictates that xenobiotics have the appropriate physicochemical properties to either diffuse through lipid layers or

Blood–Brain Barrier in Drug Discovery: Optimizing Brain Exposure of CNS Drugs and Minimizing Brain Side Effects for Peripheral Drugs, First Edition. Edited by Li Di and Edward H. Kerns.

be recognized by an endogenous transporter system. Although other transcellular transport mechanisms operate, these are associated with the transfer of larger molecular–weight entities, such as proteins [2, 3].

Passive diffusion across the BBB is driven by the concentration gradient of unbound, unionized drug between the blood and the brain extracellular fluid (ECF). The rate of diffusion (dn/dt) of solute molecules (n) is governed by three factors: (i) permeability (P) of the xenobiotic across the BBB, (ii) the surface area (S) of the brain capillary endothelium, and (iii) the unbound concentration gradient established between the two sides of the cellular barrier (dC/dx). As described by Fick's first law,

$$\left(\frac{dn}{dt}\right) = P \cdot S \cdot \left(\frac{dC}{dx}\right)$$

This process is not energy-dependent, but relies on the random movement of solute molecules through the cellular barrier with an overall net flux toward the region of lower solute concentration. Passive diffusion is a linear function of the unbound concentration gradient; thus it is rarely saturable, not chirally selective, and not the subject of inhibition, as is observed for carrier-mediated transport processes. The pH of the aqueous phases and the molecule's pK_a are also important, as they affect the unionized fraction of drug on either side of the membrane. From a thermodynamic perspective, the Gibbs free energy (ΔG) for passive transfer may be described as follows:

$$\Delta G = RT \ln\left(\frac{C_{\text{in}}}{C_{\text{out}}}\right)$$

Here C_{in} and C_{out} denote the initial unbound concentrations inside and outside the cell (or barrier), respectively. $\Delta G°$ is zero, because no covalent bonds are broken in the overall diffusion process [4]. Passive transfer would occur spontaneously when ΔG is negative, for example, when $C_{\text{out}} > C_{\text{in}}$.

Under normal physiological conditions, the blood and brain ECF is maintained at pH 7.4 and so pH differences do not contribute to the unbound concentration gradient across the BBB [6]. However, pH changes may alter the fraction of unbound, unionized drugs available for passive diffusion across membranes. The ratio of the ionized to unionized forms is determined by the pK_a of the drug molecule together with the pH of the fluid compartment in which it resides. For example, phenobarbital overdose has been treated by administration of sodium bicarbonate in order to increase plasma pH and decrease CNS toxicity by reducing the fraction of unbound, unionized drug in the blood that is available to cross the BBB [5]. Conversely, administration of carbonic anhydrase inhibitors, for example, acetazolamide, can decrease plasma pH, potentially causing the unbound, unionized fraction of weakly acidic drugs to increase and concentrate more in the CNS, raising the potential of neurotoxicity [6].

Carrier-mediated transport processes follow Michaelis–Menten kinetics and are, therefore, saturable at higher substrate concentrations. Figure 5.1a compares the

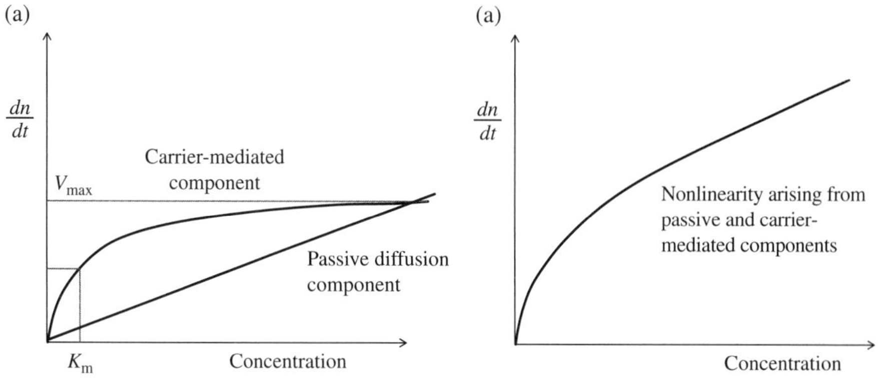

FIGURE 5.1 The rate of diffusion across the blood–brain barrier, shown as a function of the available concentration (a) individual curves for passive diffusion and carrier-mediated processes, and (b) composite profile where passive and carrier-mediated processes are occurring in concert.

transport behavior for a drug crossing the BBB by means of passive diffusion and another undergoing carrier-mediated transfer. If a drug is subjected to both processes, the resulting concentration dependency is shown in Figure 5.1b. Following saturation of the carrier-mediated process, passive diffusion will dominate and the linear dependence to concentration will be established. An example of this is nicotine [7], which appears to undergo facilitated BBB uptake by means of a proton antiporter at lower concentrations *in vivo*. The authors measured *in vivo* BBB permeability with the mouse *in situ* brain perfusion model and the composite elements of passive diffusion and carrier-mediated transport were deconvoluted by means of nonlinear regression analysis. Passive diffusion was shown to contribute approximately 20% of the BBB uptake of nicotine in the range of unbound concentrations associated with pharmacological activity in the mouse. Similarly, Parepally et al. [8] investigated the BBB transport of several nonsteroidal anti-inflammatory drugs with the rat *in situ* brain perfusion model and identified a concentration dependency for ibuprofen, which suggested a contribution from carrier-mediated uptake.

The presentation of endobiotics and xenobiotics to the BBB via the blood supply is highly efficient, as a consequence of the brain's high demand for energy and oxygen. In humans, approximately 15% of cardiac output is delivered to the brain, which as an organ accounts for only 2% of body weight [9]. The potential for rapid diffusion and distribution of solute molecules is also facilitated by the high surface area (S) of the BBB (~100 cm^2/g) and the relatively short distance between brain capillaries (~50–100 μm) [10]. Relating back to Equation 5.1, the passive permeability term (P) is determined by both the molecular properties of the drug molecule and also the chemical composition of the BBB. Akin to many cellular barriers in the body, the human endothelial brain cell membranes are rich in lipids, particularly phosphatidylethanolamine, phosphatidylcholine, and sphingomyelin [11, 12]. Relative to the gastrointestinal tract, the BBB is characterized by a higher proportion of negatively charged lipids (%wt/wt) and a lower overall proportion of triglycerides, cholesterol, and cholesterol esters [13]. These compositional differences across cell barriers in different tissues would be expected to alter the magnitude of the associated permeability as the lipophilicity, charge, and rigidity would be altered [14].

Plasma protein binding is another major determinant of drug diffusion from the plasma across the BBB. Albumin and globulins constitute approximately 55 and 38% of the total plasma protein, respectively. Human serum albumin and α-1-acid glycoprotein are the major serum proteins that bind to drugs [15]. Albumin is basic, and preferentially binds acidic and neutral compounds, while basic drugs bind preferentially to the acidic α-1-acid glycoprotein. Drug binding to plasma proteins enables circulation through the body while limiting distribution to tissues, as only the unbound portions cross tight barriers such as the BBB, essentially because the *dc/dx* terms in Fick's first law is reduced.

POINTS TO CONSIDER FOR *IN VITRO–IN VIVO* CORRELATIONS

Many *in vitro* techniques have been applied to assessing the permeability of drugs across the BBB. These results are often benchmarked against two rather distinct sets of *in vivo* data; K_p partition coefficients (e.g., brain/blood or brain/plasma total concentration ratios) or permeability-surface product (PS) values derived from *in situ* rat brain perfusion experiments. From these *in vitro–in vivo* correlations, several *in silico* predictive models have been constructed [16]. Although K_p data are widely available, it is important to note that the parameter measures the extent of brain distribution and reflects the balance of all processes affecting distribution at steady state [10]. These include passive diffusion, carrier-mediated transport, metabolism, and also the relative extents of tissue binding to blood (or plasma) and brain constituents. As Figure 5.2 shows, permeability is associated with the rate of brain distribution

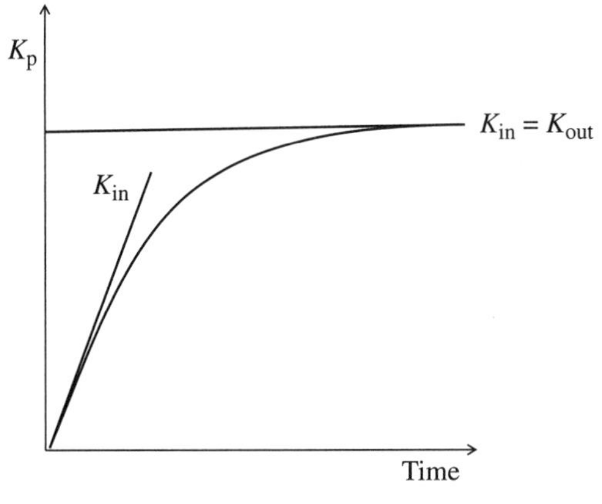

FIGURE 5.2 The *in vivo* brain/blood (or plasma) ratio, K_p, increases over time toward its steady-state value. The initial rate of uptake is characterized by the magnitude of K_{in} across the BBB.

(across the BBB) rather than the extent; hence, caution must be taken when correlating *in vitro* permeability measurements with K_p (or the *in silico* equivalent log BB, logarithm of the total brain to plasma concentration ratio), since tissue binding and other effects may be masking the true permeability component within the composite K_p parameter.

Measurements of brain PS are employed less routinely in drug discovery and development, as the *in situ* brain perfusion model requires specific surgical preparation [17] and cannot be performed readily alongside a routine PK or PK/PD screen. However, the model is available in certain specialist contract research organizations [18] and is also favored by academic BBB researchers. The technique involves drug delivery to the brain of an anesthetized animal via perfusion into the carotid artery. The perfusion time is kept very short (<1 min) so that BBB permeability is the main driver for uptake and brain tissue distribution acts as an infinite sink over that time period. The uptake clearance across the BBB, or K_{in}, can be derived as the brain/perfusate concentration ratio divided by the perfusion time, and, if the cerebral perfusate flow is known, then the drug's PS value can be determined using the Crone–Renkin equation [19]. The consistency of the data derived from *in situ* perfusion studies is excellent across literature reports, which provides a significant benefit when helping to build viable *in vitro* and in silico permeability models. The main reason for this high data consistency is that the model is generally run with several controls to demonstrate data quality. These controls include compounds to measure the vascular volume (e.g., atenolol), perfusion flow (e.g., diazepam), and consistency in the permeability measurement with reference markers (e.g., antipyrine). Since PS is derived from an *in vivo* model, the measurements may include a carrier-mediated component alongside the passive component. In this model saturation of efflux transporters such as P-glycoprotein (P-gp) may occur if the unbound drug concentration in the perfusate is sufficiently high [18, 20]. The dependency of PS on the unbound perfusate concentration has also been studied to demonstrate the presence of carrier-mediated uptake alongside passive diffusion [7, 8] by altering the concentration of perfusate (often a protein-free solution) in order to saturate carrier-mediated uptake. Given the PS parameter is related more closely to BBB permeability than $K_{p,}$ it is considered a more appropriate *in vivo* measurement to compare with *in vitro* permeability models.

IN VITRO MODELS AND MEASUREMENTS

Simple partition models, such as octanol–water (e.g., log P_{oct}), have been compared to *in situ* brain perfusion reference data and show that many molecules reside close to the line of identity in the log P_{oct} range −4 to 1 [19]. There are several outliers on either side of the line of identity which reflect carrier-mediated efflux (e.g., vincristine, azido-deoxythymidine) and facilitated uptake (e.g., glutamate, D-glucose, choline). This model is a very simple approximation, as octanol is a single surrogate for the BBB composition and the lipid bilayer nature is also lacking. Indeed, as the lipophilicity of the agents being tested increases into the range common to marketed CNS agents (e.g., log P_{oct} of 2–6), the correlation diminishes substantially [21].

Artificial membrane models build on this partitioning approach by introducing a thin lipid membrane bilayer separating aqueous compartments [22]. Essentially, the analyte of interest is spiked into one aqueous compartment (e.g., donor compartment) and allowed to diffuse through the membrane into the acceptor compartment. The concentration in the acceptor well (at a given time and allowing for adsorptive losses, etc.) is indicative of the permeability of the analyte through the membrane. Evolution of this model improved the stability of the lipid membrane by incorporating porous filters over which the lipid film could form [13, 23]. A modern variant, referred to as the parallel artificial membrane permeability assay (PAMPA), provides the best overall correlations to BBB permeability [21, 24]. Di et al. reported the first BBB-specific version of PAMPA in 2003 [24], whereby porcine brain lipid extracts were employed to create the artificial lipid membrane and this was a marked step forward for CNS drug research. The major lipid constituents were phosphatidylethanolamine, phosphatidylserine, phosphatidylinositol, and phosphatidylcholine and could be sourced from a stable commercial supply. Although the initial comparisons were made to K_p (and log BB measurements), subsequent analyses have looked toward *in vitro–in vivo* correlation with *in situ* PS values [21, 25] and have reported excellent correlations for a large set of drugs covering a wide lipophilicity range (log P_{oct} −3 to 6). On comparison with reported *in situ* PS values, the overall *in vitro–in vivo* correlations (e.g., R^2 values) were 0.77 for the entire compound test set (including weak bases, weak acids, zwitterions, and neutral molecules) and 0.97 for the weak bases. In recent examples, the experimental design has also included the presence of a surfactant mixture in the acceptor compartment to account for the lipophilic sink effect of brain tissue binding [13, 18].

Often PAMPA experiments are performed over a range of donor compartment pH values in order to fully characterize the membrane contributions to the permeability measurement, thereby eliminating the effects of the unstirred water layer [13, 23]. Throughput can be increased by performing at a single pH (e.g., 7.4 for BBB studies). Optimal incubation times and shaking conditions may also be different for hydrophilic and lipophilic compounds [21], so these factors should be well understood when generating *in vitro* permeability data by means of PAMPA-BBB.

CELL-BASED PERMEABILITY ASSAYS

Over the years, cell-based models have been reported for a wide range of cell types derived from both brain and nonbrain sources, primary or immortalized cells, single and co-cultured preparations, and grown on a range of supports (see references in [26]). Transporters have been transfected into cell lines and/or are inhibited by chemical reagents in order to study the differential effects on permeability, such as for P-gp. This variety of “flavors” of cell-based assays means that there is a large variation in the measured permeability values, not only between different cell lines [27] but also for the same cell line run in different laboratories [28]. Because of this, the apparent permeability data are often benchmarked across an internal set of CNS-active compounds or a marketed CNS drug set [26]. Furthermore, cross-literature data comparisons, even within the same cell line, must be used with caution. There is little doubt that the magnitude of the permeability values is also influenced by the nature of the cell membrane composition, for example, Caco-2 or MDCKII, which are not derived from brain cells, and are unlike primary bovine brain microvessel endothelial cells (BBMEC) [26].

Nonetheless, a central benefit of using cell-based systems for permeability measurements is the modulation of key transporters that affect brain distribution *in vivo*. MDR1-MDCKII cells are stably transfected with the MDR1 gene encoding human P-gp. The measurement of BBB permeability in the presence and absence of a P-gp inhibitor (e.g., GF120918, cyclosporine A) provides a relatively simple means to understand the interplay between the passive diffusion and carrier-mediated components of the drug molecule under test. This is not feasible in the artificial membrane models because functional transporter proteins cannot be expressed or supplemented.

IN VIVO MODELS AND MEASUREMENTS

One of the key advances in helping to understand the contributions of passive and active processes at the level of the BBB has been the development of brain tissue binding assays via equilibrium dialysis of brain homogenates [29, 30] or incubations of brain tissue slices [31, 32]. The latter offers the advantage of maintaining cellular barriers and intracellular pH gradients that allow for better overall predictivity [32, 33]. Knowing the brain tissue binding allows total brain concentrations to be converted to unbound brain concentrations, for example, the drug concentration in ECF, which is the relevant fraction of drug able to traverse the brain cell membrane or interact with the brain therapeutic target. In the space of a few years, the measurements of unbound brain concentrations have been performed on thousands of compounds, whereas, prior to this, brain microdialysis was the only viable approach. In turn, this higher-throughput approach of measuring unbound brain concentrations has enabled CNS drug discovery scientists to shift their focus away from K_p (e.g., the brain/blood or brain/plasma ratio based on total concentrations) as a tool to drive the optimization of CNS candidates and move toward $K_{p,uu}$, which is the ratio of the unbound concentrations in brain and blood (or plasma). The partition coefficient, K_p, encompasses plasma protein binding and brain nonspecific binding, so it represents a weak link to BBB transport process and pharmacological activity [10]. On the other hand, $K_{p,uu}$ represents the unbound concentration gradient across the BBB at steady state and can, therefore, be used as a more direct means to assess whether the BBB transporters are impacting distribution into brain ECF. If passive diffusion is the dominant process for drug passage across the BBB, then, at steady state, the unbound concentrations in blood (or plasma) and brain would be equal and, therefore, $K_{p,uu}$ would be unity. If an efflux transporter, such as P-gp, were acting at the BBB to shunt unbound drug back toward the blood stream, then a concentration gradient would be maintained and $K_{p,uu}$ would be less than 1, such as for loperamide, paclitaxel, saquinavir, and verapamil [34, 35]. Impairment of brain penetration may also occur for very poor membrane-permeable drugs, such as atenolol (rat brain $K_{p,uu}=0.026$; [34]), which is unable to traverse the BBB to any significant extent (hence its use as a vascular marker for *in situ* perfusion studies). The magnitude of $K_{p,uu}$ could exceed 1 if there is a facilitating process alongside passive diffusion, for example, an uptake transporter [33, 36]. $K_{p,uu}$ provides a simple means for assessing the involvement of passive diffusion (close to unity) and active transport (much less or greater than unity). It is important to note that the relative contributions of passive and active processes cannot be elucidated directly from $K_{p,uu}$ alone as it is a single point composite determination of all BBB transport processes. However, any concentration dependency noted in $K_{p,uu}$ could help to deconvolute the relative contribution of active and passive transport as with *in situ* brain perfusion [7, 8].

The partition terms K_p and $K_{p,uu}$ are related through the relationship described in Equation 5.3, where $K_{p,intrinsic}$ represents the ratio of tissue binding in blood (or plasma) and brain [16]:

$$K_p = K_{p,intrinsic} \cdot K_{p,uu}$$

For compounds that passively diffuse across the BBB (e.g., where $K_{p,uu}$ is equal to unity), K_p would correlate with the ratio of the relative tissue binding between blood (or plasma) and brain. Brain K_p measurements in humans using positron emission tomography (PET) ligands (e.g., biodistribution studies) and marketed CNS drugs (from human postmortem studies) have shown good agreement (generally within threefold) versus *in vitro* predictions of $K_{p,intrinsic}$ [37]. Similarly brain K_p measurements from pig PET biodistribution studies have shown a good correlation with the corresponding *in vitro* $K_{p,intrinsic}$ values [38]. Indeed, the *in vivo* K_p of loperamide in the pig is seen to shift to the line of identity when administered in conjunction with the P-gp inhibitor cyclosporine, suggesting that following inhibition of this transporter, the passive diffusion component for loperamide dominates.

Many marketed CNS drugs are weak bases, but this does not mean that acids are not able to diffuse passively across the BBB. Parepally et al. [8] showed that ibuprofen, flurbiprofen, and indomethacin were able to cross the rodent BBB rapidly in the *in situ* brain perfusion model from protein-free perfusate. Indeed, the BBB permeability of all three acids reduced markedly on introducing protein to the perfusate solution. Since the blood (or plasma) protein binding of acids is generally very high, it is the low unbound concentration that plays a major role in their poor brain penetration, rather than an intrinsic property of the acid class. The role of passive diffusion for the brain penetration of acids is further exemplified in Table 5.1 and Figure 5.3. The *in vivo* K_p values correlate well with the *in vitro* derived $K_{p,intrinsic}$ measurements, indicating that $K_{p,uu}$ is close to unity for these molecules.

Recent literature has questioned whether carrier-mediated transport is the prevalent means of drug uptake into tissues [39–41] or whether passive diffusion is a key mechanism too [42, 43]. In the case of the CNS, there are instances where marketed drugs are substrates for efflux transporters, but their efficacious brain exposure is still met. Table 5.2 shows the influence of P-gp on the brain exposure of several CNS drugs in a knockout mouse model, alongside K_p, $K_{p,uu}$, and PS. Although P-gp does

TABLE 5.1 *In vivo* brain K_p (brain/blood) and *in vitro* $K_{p,intrinsic}$ values (fu_{blood}/fu_{brain}) for a range of marketed and proprietary acids[a]

Compound	Rat blood[b] (% unbound)	Rat brain[b] (% unbound)	Rat $K_{p,intrinsic}$ (*in vitro*)	Rat K_p[c] (*in vivo*)	Polar surface Area	*c*Log *P*	cpK_a
GSK-A	0.00876	0.00635	1.38	0.93	65.5	7.9	3.09
GSK-B	0.207	0.225	0.92	0.53	73.3	6.3	3.09
GSK-C	0.031	0.027	1.14	0.52	68.2	7.3	3.18
GSK-D	0.139	0.152	0.92	0.39	73.3	6.5	3.22
GSK-E	0.078	0.236	0.33	0.2	56.6	7.1	0.96
GSK-F	0.047	0.089	0.53	0.18	70.5	7.1	3.09
GSK-G	0.041	0.165	0.25	0.17	47.7	7.6	3.66
GSK-H	0.021	0.047	0.46	0.05	97.9	6.8	1.25
GSK-I	0.019	0.371	0.05	0.05	47.1	7.3	2.73
GSK-J	0.112	2.609	0.04	0.05	90.8	4.1	4.20
GSK-K	0.320	1.880	0.17	0.05	74.3	4.9	4.30
GSK-L	0.206	6.680	0.03	0.05	49.1	4.7	4.53
Ibuprofen	1.60	29.64	0.05	0.03	40.5	3.7	4.08
Naproxen	1.80	54.16	0.03	0.02	50.7	2.8	4.06
Diclofenac	1.18	5.47	0.22	0.02	39.9	4.7	4.48
Flurbiprofen	0.63	12.89	0.05	0.01	40.5	3.8	3.03
Ketorolac	5.82	48.53	0.12	0.01	61.8	1.6	3.63
Indomethacin	0.57	5.92	0.10	0.01	68.3	4.2	4.26

[a] Reproduced from Ref. [30]. With kind permission of ASPET.
[b] Generated by means of equilibrium dialysis [30].
[c] Steady-state *K*p parameters generated from intravenous PK studies ($n = 3$ rats per compounds).

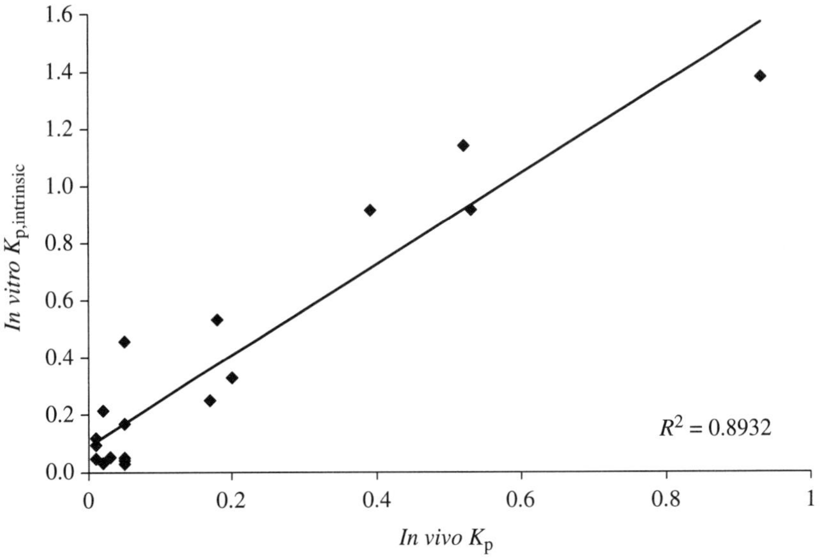

FIGURE 5.3 Plot showing the distribution of acidic drugs into the rat brain. The *x*-axis describes the *in vivo* K_p values while the *y*-axis describes the *in vitro* measurement, $K_{p,intrinsic}$, which estimates brain distribution under passive diffusion conditions.

TABLE 5.2 ***In vivo*** K_p **marketed CNS drug compared in mouse (wild type and P-gp knockout) versus the corresponding permeability-surface product values in the rats. Although P-gp reduces the magnitude of** K_p**, the *in situ* BBB permeability is sufficiently high to maintain significant brain distribution (presumably via passive diffusion)**

	Rat PS[a]	K_p (mouse)		Efflux
	(ml/s/g, brain)	(FVB/N)	mdr1a/b(–/–)	Ratio
Midazolam[b]	0.0710	0.2[c]	0.2[c]	1
Atomoxetine	0.0423	13.7	23.9	1.7
Maprotiline	0.0445	12.4	26.8	2.2
Amoxapine	0.0689	1.4	3.2	2.4
Citalopram	0.0104	5.7	17.9	3.1
Metaclopramide	0.0021	0.8	5.5	6.9
Thiothixene	0.0747	2.3	24.6	10.9
Risperidone	0.0157	0.4	5.5	12.7
Mesorizadine	0.0155	1.0	16.9	17.8

[a] Derived from Ref. [18] by means of the Crone–Renkin equation and the K_{in} sertaline as flow (F).
[b] Reference compound showing flow-limited diffusion into the rodent brain.
[c] K_p values from Ref. [55].

have an appreciable influence on the ultimate extent of brain penetration for these molecules, their intrinsic passive permeability is sufficient to drive an adequate unbound drug concentration in brain ECF. Other data supporting an important role for passive diffusion (alongside carrier-mediated effect) come from comparisons between wild-type and genetic knockout models for a range of BBB transporters such as ENT-1 [44], OATP1A4 [45], and others [46–50]. The change in K_p is generally less than twofold. There does remain the possibility that other transporter processes are balancing both the uptake and efflux. For example, *in vitro* studies indicate that zolmitriptan is a substrate of P-gp, but also for the uptake transporter OATP1A2, which is expressed at high concentrations at the human BBB. *In vivo*, it is only possible to dissect out the individual contribution of these processes if a concentration dependency is observed in parameters such as PS, K_p, or $K_{p,uu}$. Taken with other information, there is strong evidence to confer that passive diffusion is at play [42].

IN SILICO MODELS

Many *in silico* brain distribution models are related to K_p (e.g., log BB models) and often lipophilicity (log P or log $D_{7.4}$) is a key component in the model's characteristics [16]. As mentioned earlier, brain K_p is a partition term often measured at or close to steady-state conditions. Binding to brain tissue shows a significant correlation to log P [30] presumably as a result of the high lipid content within this organ [16]. Hence, factoring out the tissue binding effects is necessary for understanding what physicochemical properties of drug candidates truly impact passive permeability across the BBB. This can be addressed somewhat more readily by constructing *in silico* permeability models that are based on *in situ* brain perfusion data or $K_{p,uu}$ partition measures. Fridén et al. [34] showed that lipophilicity is not a strong determinant for the magnitude $K_{p,uu}$; rather the number of hydrogen bond acceptors is the dominant term, where unbound brain exposure is in general inversely related to the number of hydrogen bond acceptor of the drug. Similarly, Chen et al. [51] compared and contrasted the key descriptors in silico for K_p and $K_{p,uu}$ for a common set of molecules. Again reduced hydrogen-bonding potential figured prominently for increasing $K_{p,uu}$ along with topology descriptors that estimate molecular linearity. Conversely, for K_p, the key model determinants were log P and polar surface area. It is both notable and intriguing that lipophilicity plays no role despite these molecules all needing to traverse the BBB *in vivo* and the ensuing lipid membranes.

PAMPA-BBB data provide a useful means to investigate the molecular properties that influence passive diffusion since the artificial membranes are devoid of transport systems. Several in silico modeling examples have been reported based on linear free energy relationship (LFER) approaches [21, 52, 53]. Tsinman et al. [21] suggested a series of compound progression criteria for enhancing passive diffusion, including high dispersion forces for acids (introducing more polarisable n- and pi-electrons) and low dispersion forces for bases (solute dipolarity/polarizability). More recently, Abraham [54] described a general equation for permeation rate through the BBB, based on the modeling of PS data. This analysis introduces additional terms related to ion–solvent interactions for cations and anions alongside hydrogen bonding and the McGowan volume.

SUMMARY

Passive permeability plays an important role in the transfer of drug molecules across the BBB, as demonstrated by a range of correlations with artificial membranes, *in situ* brain perfusion techniques, and $K_{p,intrinsic}$ and $K_{p,uu}$ data. Sufficient literature exists for the generation of refined *in silico* predictive models of passive diffusion based on PAMPA-BBB and PS measurements. Such approaches can be adapted and readily exploited by medicinal chemistry in a drug discovery setting, establishing structure–activity relationships and correlating with drug-like properties and pharmacological activity. Acids, bases, and zwitterions are all able to traverse the BBB. There is a substantial background of evidence to support the importance of passive diffusion and there is an emerging body of evidence highlighting the importance of drug transporters from both an uptake and efflux perspective, particularly P-gp-mediated drug efflux (Table 5.2).

However, in order to further understand the importance of drug transporters and evaluate the possibility of exploiting their potential, several key questions still remain. There is still a compelling requirement for an agreed and accepted unified *in vitro* model that would enable a reliable and meaningful comparison and integration of basic permeability data from different laboratories and chemical series. Ideally, such a model would focus on mimicking the human BBB; however, potential species differences in transporter expression and function may require further study, especially when considering clinical therapeutic safety margins relative to preclinical safety data. The importance and understanding of BBB passive diffusion and drug transporters in the clinical setting should also be considered and, if necessary, factored into the screening strategy sooner rather than later. Finally, it is important to remember that while drug permeability per se is essential in order to access the site of action, optimizing and ranking compounds on the basis of "the faster the delivery and the greater the total amount delivered, the better" is no longer acceptable. The permeability rate and the extent of drug permeation required must always be contextualized with respect to the desired target engagement and hence the PD response.

REFERENCES

[1] Morgan. P., Van Der Graaf, P.H., Arrowsmith, J., Feltner, D.E., Drummond, K.S., Wegner, C.D., Street, S.D. (2012). Can the flow of medicines be improved? Fundamental pharmacokinetic and pharmacological principles toward improving Phase II survival. *Drug Discov. Today, 17(9–10)*, 419–424.

[2] Abbott, N.J., Rönnbäck, L., Hansson, E. (2006). Astrocyte–endothelial interactions at the blood-brain barrier. *Nat. Rev. Neurosci., 7(1)*, 41–53.

[3] Chen, Y., Liu, L. (2012). Modern methods for delivery of drugs across the blood-brain barrier. *Adv. Drug Deliv. Rev., 64(7)*, 640–665.

[4] Lodish, H., Berk, A., Kaiser, C.A., Krieger, M., Bretscher, A., Ploegh, H., Amon, A., Scott, M.P., Molecular Cell Biology, WH Freeman, New York, 2013, pp. 23–58.

[5] Herrington, A.M., Clifton, G.D. (1995). Toxicology and management of acute drug ingestions in adults. *Pharmacotherapy, 15(2)*, 182–200.

[6] Chesler, M. (2003). Regulation and modulation of pH in the brain. *Physiol. Rev., 83(4)*, 1183–1221.

[7] Cisternino, S., Chapy, H., André, P., Smirnova, M., Debray, M., Scherrmann, J.M. (2013). Coexistence of passive and proton antiporter-mediated processes in nicotine transport at the mouse blood-brain barrier. *AAPS J, 15(2)*, 299–307.

[8] Parepally, J.M., Mandula, H., Smith, Q.R. (2006). Brain uptake of nonsteroidal anti-inflammatory drugs: ibuprofen, flurbiprofen, and indomethacin. *Pharm. Res., 23(5)*, 873–881.

[9] Davies, B., Morris, T. (1993). Physiological parameters in laboratory animals and humans. *Pharm. Res., 10*, 1093–1095.

[10] Hammarlund-Udenaes, M., Fridén, M., Syvänen, S., Gupta, A. (2008). On the rate and extent of drug delivery to the brain. *Pharm. Res., 25*, 1737–1750.

[11] Siakotos, A.N., Rouser, G. (1969). Isolation of highly purified human and bovine brain endothelial cells and nuclei and their phospholipid composition. *Lipids, 4*, 234–239.

[12] Tewes, B.J., Galla, H.J. (2001). Lipid polarity in brain capillary endothelial cells. *Endothelium, 8*, 207–220.

[13] Avdeef, A. *Absorption and Drug Development: Solubility, Permeability, and Charge State.* John Wiley & Sons, Inc. New York, 2003, pp. 116–244.

[14] Campbell, S.D., Regina, K.J., Kharasch, E.D. (2013). Significance of lipid composition in a blood-brain barrier-mimetic PAMPA assay. *J. Biomol. Screen., 19(3)*, 437–444.

[15] Otagiri, M. (2009). Study on binding of drug to serum protein. *Yakugaku Zasshi, 129(4)*, 413–425.

[16] Summerfield, S.G., Dong, K.C. (2013). In vitro, in vivo and in silico models of drug distribution into the brain. *J. Pharmacokinet. Pharmacodyn., 40(3)*, 301–314.

[17] Takasato, Y., Rapoport, S.I., Smith, Q.R. (1984). An in situ brain perfusion technique to study cerebrovascular transport in the rat. *Am. J. Physiol., 247*, H484–H493.

[18] Summerfield, S.G., Read, K., Begley, D.J., Obradovic, T., Hidalgo, I.J., Coggon, S., Lewis, A.V., Porter, R.A., Jeffrey, P. (2007). Central nervous system drug disposition: the relationship between in situ brain permeability and brain free fraction. *J. Pharmacol. Exp. Ther. 322(1)*, 205–213 [Erratum in *J. Pharmacol. Exp. Ther., 330(3)* (2009), 971–972].

[19] Smith, Q.R. (2003). A review of blood-brain barrier transport techniques. *Methods Mol. Med., 89*, 193–208.

[20] Lin, J.H. (2004). How significant is the role of P-glycoprotein in drug absorption and brain uptake? *Drugs Today, 40(1)*, 5–22.

[21] Tsinman, O., Tsinman, K., Sun, N., Avdeef, A. (2011). Physicochemical selectivity of the BBB microenvironment governing passive diffusion-matching with a porcine brain lipid extract artificial membrane permeability model. *Pharm. Res., 28*, 337–363.

[22] Kansy, M., Senner, K., Gubernator, J. (1998) Physicochemical high throughput screening: parallel artificial membrane permeation assay in the description of passive absorption processes. *J. Med. Chem., 41(7)*, 1007–1010.

[23] Avdeef, A., Tsinman, O. (2006). PAMPA. A drug absorption in vitro model 13. Chemical selectivity due to membrane hydrogen bonding: in combo comparisons of HDM-, DOPC-, and DS-PAMPA models. *Eur. J. Pharm. Sci., 28(1–2)*, 43–50.

[24] Di, L., Kerns, E.H., Fan, K., McConnell, O.J., Carter, G.T. (2003). High throughput artificial membrane permeability assay for blood-brain barrier. *Eur. J. Med. Chem., 38(3)*, 223–232.

[25] Di, L., Kerns, E.H., Bezar, I.F., Petusky, S.L., Huang, Y. (2009). Comparison of blood-brain barrier permeability assays: in situ brain perfusion, MDR1-MDCKII and PAMPA-BBB. *J. Pharm. Sci., 98(6)*, 1980–1991.

[26] Mahar Doan, K.M., Humphreys, J.E., Webster, L.O., Wring, S.A., Shampine, L.J., Serabjit-Singh, C.J., Adkison, K.K., Polli, J.W. (2002). Passive permeability and P-glycoprotein-mediated efflux differentiate central nervous system (CNS) and non-CNS marketed drugs. *J. Pharmacol. Exp. Ther., 303(3)*, 1029–1037.

[27] Hakkarainen, J.J., Jalkanen, A.J., Kääriäinen, T.M., Keski-Rahkonen, P., Venäläinen, T., Hokkanen, J., Mönkkönen, J., Suhonen, M., Forsberg, M.M. (2010). Comparison of in vitro cell models in predicting in vivo brain entry of drugs. *Int. J. Pharm., 402(1–2)*, 27–36.

[28] Jeffrey, P., Summerfield, S.G. (2007). Challenges for blood-brain barrier (BBB) screening. *Xenobiotica, 37(10–11)*, 1135–1151.

[29] Kalvass, J.C., Maurer, T.S. (2002). Influence of nonspecific brain and plasma binding on CNS exposure: implications for rational drug discovery, *Biopharm. Drug Dispos., 23(8)*, 327–338.

[30] Summerfield, S.G., Stevens, A.J., Cutler, L., del Osuna Carmen M., Hammond, B., Tang, S.P., Hersey, A., Spalding, D.J., Jeffrey, P. (2006) Improving the in vitro prediction of in vivo central nervous system penetration: integrating permeability, P-glycoprotein efflux, and free fractions in blood and brain. *J. Pharmacol Exp. Ther., 316(3)*, 1282–1290.

[31] Fridén, M., Ducrozet, F., Middleton, B., Antonsson, M., Bredberg, U., Hammarlund-Udenaes, M. (2009). Development of a high throughput brain slice method for studying drug distribution in the central nervous system. *Drug Metab. Dispos., 37(6)*, 1226–1233.

[32] Loryan, I., Fridén, M., Hammarlund-Udenaes, M. (2013 Jan 21). The brain slice method for studying drug distribution in the CNS. *Fluids Barriers CNS, 10(1)*, 6.

[33] Fridén, M., Bergstro, M.F., Wan, H., Rehngren, M., Ahlin, G., Hammarlund-Udenaes, M., Bredberg, U. (2011). Measurement of unbound drug exposure in brain: modelling of pH partitioning explains diverging results between the brain slice and brain homogenate methods. *Drug Metab. Dispos., 39(3)*, 353–362.

[34] Fridén, M., Winiwarter, S., Jerndal, G., Bengtsson, O., Wan, H., Bredberg, U., Hammarlund-Udenaes, M., Antonsson, M. (2009). Structure-brain exposure relationships in rat and human using a novel data set of unbound drug concentrations in brain interstitial and cerebrospinal fluids. *J. Med. Chem., 52(20)*, 6233–6243.

[35] Kalvass, J.C., Maurer, T.S., Pollack, G.M. (2007). Use of plasma and brain unbound fractions to assess the extent of brain distribution of 34 drugs: comparison of unbound concentration ratios to in vivo P-glycoprotein efflux ratios. *Drug Metab. Dispos., 35(4)*, 660–666.

[36] Okura, T., Hattori, A., Takano, Y., Sato, T., Hammarlund-Udenaes, M., Terasaki, T., Deguchi, Y. (2008). Involvement of the pyrilamine transporter, a putative organic cation transporter, in blood-brain barrier transport of oxycodone. *Drug Metab. Dispos., 36(10)*, 2005–2013.

[37] Summerfield, S.G., Lucas, A.J., Porter, R.A., Jeffrey, P., Gunn, R.N., Read, K.R., Stevens, A.J., Metcalf, A.C., Osuna, M.C., Kilford, P.J., Passchier, J., Ruffo, A.D. (2008). Toward an improved prediction of human in vivo brain penetration. *Xenobiotica, 38(12)*, 1518–1535.

[38] Gunn, R.N., Summerfield, S.G., Salinas, C.A., Read, K.D., Guo, Q., Searle, G.E., Parker, C.A., Jeffrey, P., Laruelle, M. (2012). Combining PET biodistribution and equilibrium

dialysis assays to assess the free brain concentration and BBB transport of CNS drugs. *J. Cereb. Blood Flow Metab., 32(5)*, 874–883.

[39] Dobson, P.D., Kell, D.B. (2008). Carrier-mediated cellular uptake of pharmaceutical drugs: an exception or the rule? *Nat Rev. Drug Discov., 7(3)*, 205–220.

[40] Kell, D.B., Dobson, P.D., Oliver, S.G. (2011). Pharmaceutical drug transport: the issues and the implications that it is essentially carrier-mediated only. *Drug Discov. Today, 16(15–16)*, 704–714.

[41] Kell, D.B., Dobson, P.D., Bilsland, E., Oliver, S.G. (2013). The promiscuous binding of pharmaceutical drugs and their transporter-mediated uptake into cells: what we (need to) know and how we can do so. *Drug Discov. Today, 18(5–6)*, 218–239.

[42] Sugano, K., Kansy, M., Artursson, P., Avdeef, A., Bendels, S., Di, L., Ecker, G.F., Faller, B., Fischer, H., Gerebtzoff, G., Lennernaes, H., Senner, F. (2010). Coexistence of passive and carrier-mediated processes in drug transport. *Nat Rev. Drug Discov., 9(8)*, 597–614.

[43] Di, L., Artursson, P., Avdeef, A., Ecker, G.F., Faller, B., Fischer, H., Houston, J.B., Kansy, M., Kerns, E.H., Krämer, S.D., Lennernäs, H., Sugano, K. (2012). Evidence-based approach to assess passive diffusion and carrier-mediated drug transport. *Drug Discov. Today, 17(15–16)*, 905–912.

[44] Paproski, R.J., Wuest, M., Jans, H.S., Graham, K., Gati, W.P., McQuarrie, S., McEwan, A., Mercer, J., Young, J.D., Cass, C.E. (2010). Biodistribution and uptake of 3′-deoxy-3′-fluorothymidine in ENT1-knockout mice and in an ENT1-knockdown tumor model. *J. Nucl. Med., 51(9)*, 1447–1455.

[45] Ose, A., Kusuhara, H., Endo, C., Tohyama, K., Miyajima, M., Kitamura, S., Sugiyama, Y. (2010). Functional characterization of mouse organic anion transporting peptide 1a4 in the uptake and efflux of drugs across the blood-brain barrier. *Drug Metab. Dispos., 38(1)*, 168–176.

[46] Pan, W., Kastin, A.J. (2002). TNF alpha transport across the blood-brain barrier is abolished in receptor knockout mice. *Exp. Neurol., 174(2)*, 193–200.

[47] André, P., Debray, M., Scherrmann, J.M., Cisternino, S. (2009). Clonidine transport at the mouse blood-brain barrier by a new H+antiporter that interacts with addictive drugs. *J. Cereb. Blood Flow Metab., 29(7)*, 1293–1304.

[48] Ose, A., Ito, M., Kusuhara, H., Yamatsugu, K., Kanai, M., Shibasaki, M., Hosokawa, M., Schuetz, J.D., Sugiyama, Y. (2009). Limited brain distribution of [3R,4R,5S]-4-acetamido-5-amino-3-(1-ethylpropoxy)-1-cyclohexene-1-carboxylate phosphate (Ro 64–0802), a pharmacologically active form of oseltamivir, by active efflux across the blood-brain barrier mediated by organic anion transporter 3 (Oat3/Slc22a8) and multidrug resistance-associated protein 4 (Mrp4/Abcc4). *Drug Metab. Dispos., 37(2)*, 315–321.

[49] Miyajima, M., Kusuhara, H., Fujishima, M., Adachi, Y., Sugiyama, Y. (2011). Organic anion transporter 3 mediates the efflux transport of an amphipathic organic anion, dehydroepiandrosterone sulfate, across the blood-brain barrier in mice. *Drug Metab. Dispos., 39(5)*, 814–819.

[50] Smith, D.E., Hu, Y., Shen, H., Nagaraja, T.N., Fenstermacher, J.D., Keep, R.F. (2011). Distribution of glycylsarcosine and cefadroxil among cerebrospinal fluid, choroid plexus, and brain parenchyma after intracerebroventricular injection is markedly different between wild-type and Pept2 null mice. *J. Cereb. Blood Flow Metab., 31(1)*, 250–261.

[51] Chen, H., Winiwarter, S., Fridén, M., Antonsson, M., Engkvist, O. (2011). In silico prediction of unbound brain-to-plasma concentration ratio using machine learning algorithms. *J. Mol. Graph. Model., 29(8)*, 985–995.

[52] Dagenais, C., Avdeef, A., Tsinman, O., Dudley, A., Beliveau, R. (2009). P-glycoprotein deficient mouse in situ blood-brain barrier permeability and its prediction using an in combo PAMPA model. *Eur. J. Pharm. Sci., 38(2)*, 121–137.

[53] Abraham, M.H. (2004). The factors that influence permeation across the blood-brain barrier. *Eur. J. Med. Chem., 39(3)*, 235–240.

[54] Abraham, M.H. (2011). The permeation of neutral molecules, ions, and ionic species through membranes: brain permeation as an example. *J Pharm. Sci., 100(5)*, 1690–1701.

[55] Doran, A., Obach, R.S., Smith, B.J., Hosea, N.A., Becker, S., Callegari, E., Chen, C., Chen, X., Choo, E., Cianfrogna, J., Cox, L.M., Gibbs, J.P., Gibbs, M.A., Hatch, H., Hop, C.E., Kasman, I.N., Laperle, J., Liu, J., Liu, X., Logman, M., Maclin, D., Nedza, F.M., Nelson, F., Olson, E., Rahematpura, S., Raunig, D., Rogers, S., Schmidt, K., Spracklin, D.K., Szewc, M., Troutman, M., Tseng, E., Tu, M., Van Deusen, J.W., Venkatakrishnan, K., Walens, G., Wang, E.Q., Wong, D., Yasgar, A.S., Zhang, C. (2005). The impact of P-glycoprotein on the disposition of drugs targeted for indications of the central nervous system: evaluation using the MDR1A/1B knockout mouse model. *Drug Metab. Dispos., 33(1)*, 165–174.

6

ESTABLISHMENT OF P-GLYCOPROTEIN STRUCTURE–TRANSPORT RELATIONSHIPS TO OPTIMIZE CNS EXPOSURE IN DRUG DISCOVERY

JEROME H. HOCHMAN,[1] SOOKHEE N. HA,[2] AND ROBERT P. SHERIDAN[3]

[1] *Department of PPDM, Merck Research Laboratories, West Point, PA, USA*
[2] *Department of Structural Chemistry, Merck Research Laboratories, Kenilworth, NJ, USA*
[3] *Department of Structural Chemistry, Merck Research Laboratories, Rahway, NJ, USA*

TRANSPORT MODELS SURROGATED FOR THE BLOOD–BRAIN BARRIER

The development of central nervous system (CNS) drugs presents formidable challenges to drug discovery from both biological and chemical perspectives [1]. CNS drugs have low probability of success relative to other therapeutic areas in part due to the complexities of achieving therapeutic drug levels in the CNS while maintaining potency and other drug-like properties. The blood–brain barrier (BBB) is the primary constraint to achieving therapeutic drug exposure in the brain. It is comprised of a dense network of microvessels which are uniquely adapted to pose a selective barrier between the blood and the CNS compartments [2, 3]. These microvessels are nonfenestrated and neighboring cells are sealed by tight junctions, which prevents free diffusion of hydrophilic molecules between the two compartments. Consequently CNS penetration is primarily limited to transcellular transport in which membrane-permeable compounds transit the lumenal and ablumenal endothelial cell membranes. This route

Blood–Brain Barrier in Drug Discovery: Optimizing Brain Exposure of CNS Drugs and Minimizing Brain Side Effects for Peripheral Drugs, First Edition. Edited by Li Di and Edward H. Kerns.

of entry to the CNS is strongly attenuated by efflux transporters on the lumenal membrane of the microvessels. P-glycoprotein (P-gp) is the most prominent efflux transporter in the BBB, regulating CNS concentrations of a wide variety of structurally diverse compounds. Polli and coworkers [4] compared P-gp efflux and passive transport characteristics for CNS and non-CNS marketed drugs, showing that for the most part CNS drugs have higher permeability and very low P-gp transport compared to drugs acting on systemic compartments. Consequently, development of CNS-active drugs restricts chemical space to constraints compatible with low P-gp efflux and high membrane permeability while retaining features for potent target engagement, low toxicity, and favorable pharmacokinetic and physical chemical properties.

The physiology of the BBB is uniquely adapted to maintain selective chemical communication between the systemic circulation and the CNS [5]. The density of microvessels in the brain is very high with intercapillary distance of 40 μm (roughly two cell widths between microvessels) [2]. Consequently the brain is highly perfused, and steady-state unbound concentrations of drugs in the blood and CNS compartment are achieved rapidly for most drugs. Under these conditions the distribution of drug in both compartments can be represented as in Figure 6.1. In the absence of active transport (efflux or active uptake), passive diffusion drives equilibration of unbound drug between the blood and CNS. P-gp-mediated transport from the CNS to the blood generates higher unbound drug concentrations in the blood such that passive diffusion from the blood to the central compartment equals the flux in the opposite direction (passive transport from the central compartment to the blood plus P-gp-mediated drug flux from the CNS). Consequently, the unbound concentrations of drug in the blood relative to unbound concentrations in the CNS are the product of both passive and active transport [6]. To generate meaningful structure–activity relationship (SAR), the source of data should closely reflect P-gp involvement in CNS distribution and consequently should reflect the active efflux by P-gp

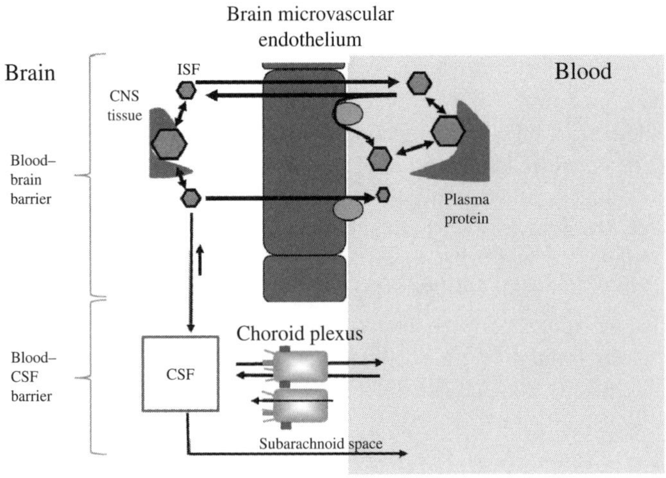

FIGURE 6.1 Drug distribution across the blood–brain and blood–CSF barriers.

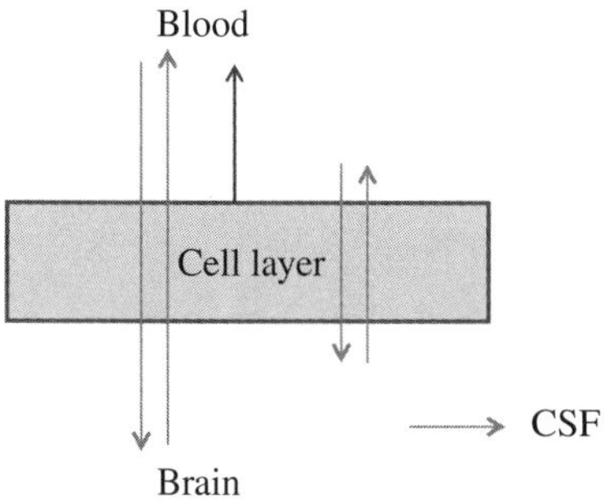

In vivo[a,b]:

$$\frac{br_u}{pl_u} = \frac{k_{\text{passive}}}{k_{\text{passive}} + k_{\text{p-gp}} + k_{\text{ISF}\rightarrow\text{CSF}}}$$

In vitro[a]:

$$\frac{A \rightarrow B}{B \rightarrow A} = \frac{k_{\text{passive}} - k_{\text{p-gp}}}{k_{passive} + k_{\text{p-gp}}}$$

[a] $k_{p\text{-}gp}$ is Vm/Km at $C \ll Km$

[b] $k_{ISF \rightarrow CSF}$ is relatively small for mos CSF compounds

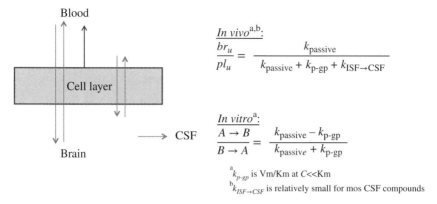

FIGURE 6.2 Model depicting kinetic properties of transport across P-gp-expressing epithelial cell models.

relative to passive permeability. Although several types of *in vitro* assays for P-gp can be applied to identify P-gp substrates and characterize P-gp kinetic parameters, the bidirectional transport assay is the most relevant to evaluate CNS distribution and is the most widely used model in practice. The typical bidirectional transport assay is illustrated in Figure 6.2. Cells are grown on permeable filter supports and drug transport in the basolateral (B) to apical (A) direction and the A to B direction is measured. The ratio of B to A relative to A to B transport in P-gp-expressing cells is used as a measure of P-gp transport. In the absence of P-gp efflux the ratio is 1, while P-gp-facilitated transport in the B to A direction and P-gp-attenuated transport in the A to B directions result in a ratio greater than 1 for P-gp substrates. In practice compounds with transport ratios greater than 2 to 3 are considered to be significant P-gp substrates. It is important to note that similar to distribution across the BBB, transport ratios from bidirectional transport studies are not a pure measure of P-gp activity alone, but rather reflect the flux mediated by P-gp relative to the passive membrane permeability of a drug (i.e., when two drugs have the same P-gp kinetic parameters but different passive permeability, the drug with higher passive permeability will have a lower B to A/A to B ratio). In this regard, transport ratios determined in the bidirectional transport models may have direct relevance to the impact of P-gp on CNS exposure. Indeed, transport ratios in this type of bidirectional transport assay have been shown to directly correlate with the impact of P-gp *in vivo*. Studies with P-gp-deficient and P-gp-competent mice evaluating P-gp impact on brain/plasma ratios showed a strong correlation between P-gp's impact on *in vivo* and *in vitro* transport ratios in mouse mdr-1a expressing LLC-PK1 cells [7]. More recently, studies (Fig. 6.3) have shown that approximations of CNS concentrations can be directly extrapolated from unbound plasma concentrations corrected for *in vitro* P-gp transport ratios for rodents and monkeys [8–11]. The implication of this correlation is that SAR established from bidirectional transport data should translate from *in vitro* to *in vivo*. Moreover, incremental improvements in *in vitro* P-gp transport ratios can be built upon to improve CNS exposure.

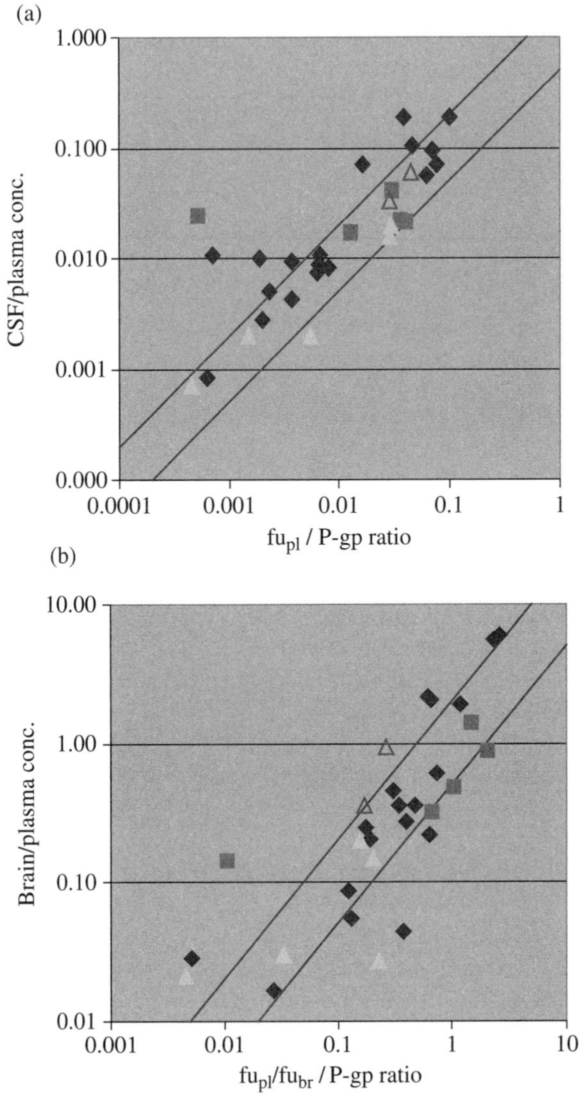

FIGURE 6.3 Correlation of *in vitro* P-gp transport on distribution of unbound drug in blood and (a) CSF or (b) unbound brain concentrations.

P-gp TRANSPORT MECHANISMS AND DRUG PROPERTIES

P-gp transport is different from many substrate–protein interactions, such as those seen in enzymes. First, P-gp recognizes and transports very disparate chemical structures. Second, in contrast to conventional enzyme–substrate interactions that occur in the aqueous phase, P-gp substrates diffuse into the cytosolic leaflet of the plasma membrane, are extracted from hydrophobic membrane location, and then transported through an aqueous channel into the aqueous extracellular compartment. Evidence for extraction of drugs from the inner leaflet of the plasma membrane was first generated using fluorescent P-gp substrates. Fluorescence resonance energy transfer studies on P-gp containing inside–out plasma membrane vesicles demonstrated that the fluorescent P-gp substrate Hoechst 33342 is extracted from the inner leaflet of the bilayer [12]. Since Hoechst 33342 fluorescence is low in an aqueous environment, the kinetics of changes in Hoechst 33342 fluorescence further indicated that Hoechst 33342 was released directly into an aqueous compartment followed by a slow passive reassociation with the outer leaflet of the bilayer. Taken together, these results indicate that P-gp can extract compounds from the cytoplasmic leaflet of the bilayer and transport the compounds directly into the extracellular aqueous medium.

Further resolution of the mechanism of P-gp substrate interactions comes from high resolution crystal structures of P-gp and closely related proteins. Three Angstrom resolution of the prokaryotic P-gp ortholog Sav1866 [13], and 3.8 Å resolution crystal structures of mouse apo-P-gp and drug- bound P-gp proteins further [14] support substrate recognition occurring within the membrane. The structure revealed a 6000 $Å^3$ internal cavity lined by 12 transmembrane α-helical stretches through which drug is transported to extracellular fluid. This channel is accessible through two portals in the cytosolic leaflet of the plasma membrane allowing direct extraction of drug from the membrane for transport. Co-crystals of the P-gp with bound stereo-isomers QZ59-RRR and QZ59-SSS revealed distinct drug binding sites for each isomer consistent with H (Hoechst) and R (rhodamine) binding sites previously proposed based on inhibition studies [14]. These sites contain primarily hydrophobic and aromatic amino acids. Within the internal channel 46 amino acid residues are associated with the translocation pathway with electrostatic interactions including cation–π, CH–π, and π–π recognition involved in protein–drug interactions. Translocation of drugs through the channel is associated with large structural changes in P-gp linked to hydrolysis of ATP [14] catalyzed by the two nucleotide-binding domains. Conformational changes are transmitted from the nucleotide-binding domains to the transmembrane helices as detected by biophysical and biochemical approaches [15–18], and modeled using targeted molecular dynamics [19]. Molecular dynamic simulations suggest substantial rearrangement of the transmembrane domains resulting in an open conformation in which the drug could disassociate to the extracellular space.

Based on this discussion it is clear that the bidirectional transport ratios being used to establish SAR are the composite of multiple processes and not solely the product of substrate–protein interactions. Establishment of SAR for P-gp efflux entails identification of properties associated with highly complex series of events. Since the measure for P-gp efflux (transport ratios) are determined by passive permeability, membrane partitioning and drug–P-gp interactions, the factors defining drug transport will reflect chemical properties that influence each of these factors. Consequently individual physical and chemical descriptors in isolation may not have consistent effects on P-gp transport across different compounds. The influence of these descriptors may have distinct effects on membrane partitioning, passive permeability, and P-gp interactions. Moreover, descriptors defining bidirectional transport ratios for drugs may be distinct from those that solely define docking with the P-gp.

ESTABLISHMENT OF SAR IN DISCOVERY PROGRAMS

For CNS programs knowing the SAR for P-gp transport can help maximize CNS exposure and minimize systemic exposure. SAR can be generated for a narrow set of analogs from a specific discovery program, or it can be generated for a large number of diverse compounds. In the first case, specific SAR can be resolved within the narrow chemical space determined by the pharmacological target which may not be resolved in more diverse chemical space. The impact of specific substituents on P-gp transport can be evaluated by comparison of analogous compounds in a structural template with different functional groups at defined positions. An example of this approach is illustrated in Figure 6.4. For this structural series, groups which tend to confer greater P-gp efflux tend to increase hydrogen bond acceptor and donor count, while modifications which lower P-gp efflux tend to lower pK_a of basic groups and present steric hindrance around hydrogen bond donors and acceptors or lower hydrogen bond potency. Combining this template specific information with SAR guidance from broader quantitative structure–activity relationship (QSAR) studies enables us to prospectively inform further chemical synthesis for a more efficient route to designing out P-gp transport, while maintaining other essential drug properties.

In Silico Models for P-gp Efflux: QSAR

Consideration of simple physical parameters gives some insight for P-gp activity. For example, P-gp substrates are more likely to have high polar surface areas and/or more hydrogen bond donors. Any single physical parameter is very weakly predictive (e.g., $R^2 \sim 0.2$). However, more complicated models using detailed chemical information are likely to make more accurate predictions. To this end, we generated a number of QSAR models based on P-gp activity. Having the models (i) allows a

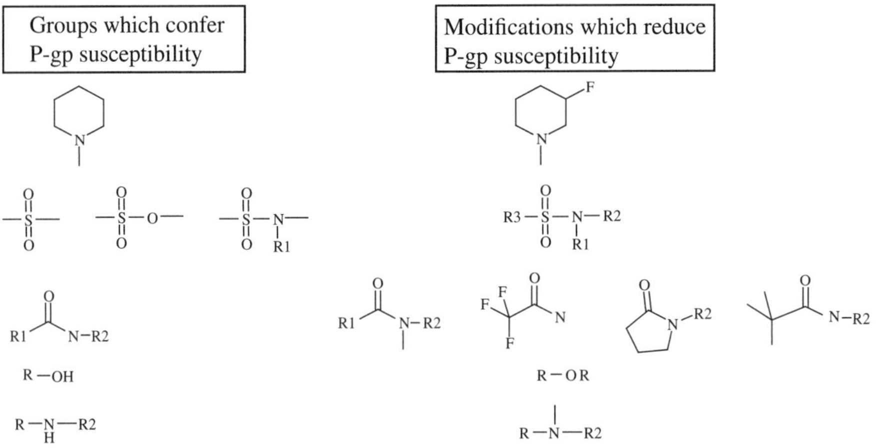

FIGURE 6.4 Chemical functional groups conferring high and low susceptibility to P-gp efflux.

better understanding of the mechanism of P-gp efflux and identifying structural and chemical characteristics of molecules conferring substrate properties, (ii) permits virtual screening of compounds to assess the impact of novel structural changes, and (iii) gives informed prospective guidance on structural modifications to minimize P-gp efflux. To generate a QSAR model, we need three things: (i) a set of molecules where each molecule has an activity, since we want to be able to predict (the "activities"), (ii) a way of describing the chemical structures of the molecules (the "descriptors"), and (iii) a statistical method of relating the activities to the descriptors (the QSAR method). Activities and descriptors together form the "training set" for the model. Typically a model is built from the entire training set and predictions for new compounds are made against that model. In house, we have data for more than 10,000 diverse compounds. We find that the most useful models use the log of the BA/AB ratio as the "activity." A variety of QSAR methods and descriptors were evaluated. Based on the cross-validation, the "best" models apply the random forest QSAR method [20, 21] with the following descriptors: AP [22], DP [23], and MOE_2D descriptors (Molecular Operating Environment (MOE), Version 2010, release 10, Chemical Computing Group, Montreal, Canada, 2009 (www.chemcomp.com)). Correlations between the cross-validated predicted ratios and observed ratios are shown in Figure 6.5. While the MOE_2D descriptors are computable physical

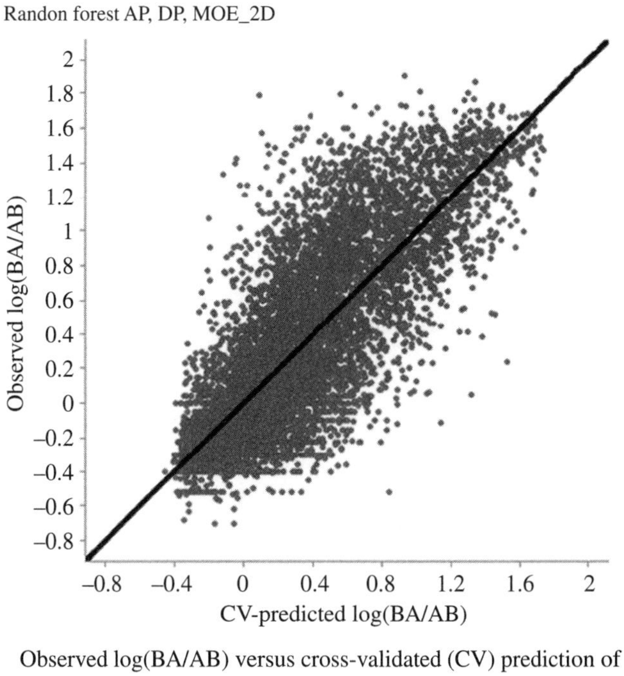

FIGURE 6.5 Observed log(BA/AB) versus cross-validated (CV) prediction of log(BA/AB) for human P-gp at 1 μM. This uses random forest as a QSAR method and AP, DP, and MOE_2D as the descriptor combination.

(a)

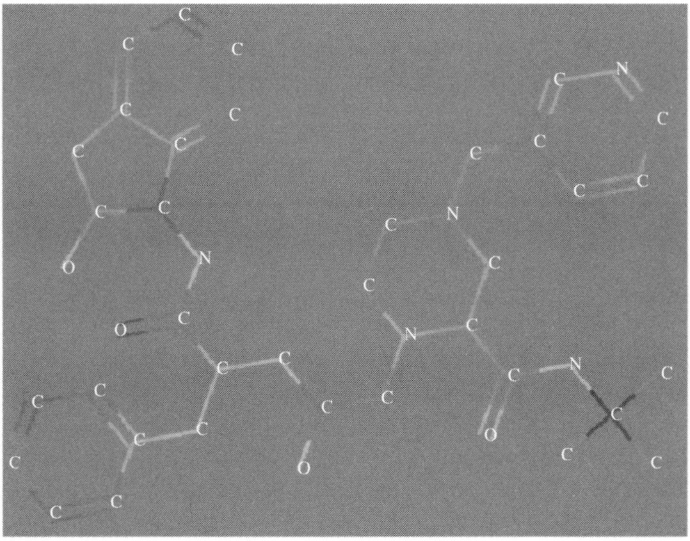

(b)

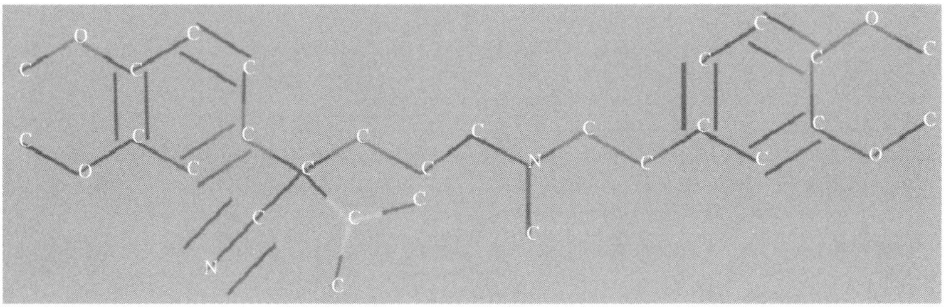

FIGURE 6.6 (a) Verapamil and (b) indinavir substrates for P-gp colored to represent the contribution of atoms to log (BA/AB) activation. The convention is that redder atoms are associated with descriptors that increase activity and blue atoms with descriptors that decrease activity. For color detail, please see color plate section.

properties, the AP and DP descriptors contain information about specific chemical groups. Some atom pair descriptors (AP, DP) also have strong contributions to determining BA/AB ratios separate from their impact on the global physical chemical properties of molecules. The contribution of descriptors can be highlighted in individual molecules. This is done by removing each atom in turn and examining how the prediction of the model changes. Figure 6.6 shows two molecules with high BA/AB ratios with the atoms colored according to the contributions of their descriptors to log(BA/AB). Redder colors represent those atoms with descriptors associated with higher log(BA/AB). With this type of display one would look to modify the atoms that contribute most positively to activity.

In Silico Models for P-gp Efflux: Molecular Docking–Based Models

Recent high-resolution crystal structures for mouse P-gp and bacterial P-gp orthologs have made molecular docking studies possible. While QSAR studies can illuminate features of compounds associated with bidirectional transport studies, molecular docking studies can resolve specific modification to alter drug–P-gp interactions. To understand P-gp susceptibility of a series of GPCR compounds, docking studies were carried out on a homology model of human P-gp. The human P-gp model was generated based on the crystal structure of mouse P-gp [14]. The overall homology between the two proteins is 87% and even higher in the transmembrane domain. Due to substrate promiscuity of P-gp and the large volume cavity in the transmembrane domain, docking studies are very challenging in the sense that it is hard to pick out the true site of interaction of substrates with P-gp. However, incorporating SAR and mutagenesis information can help. Table 6.1 shows the SAR relationship of three related compounds **1**, **2**, and **3**. Docking of these three compounds to the cavity of the transmembrane domain of the human P-gp model generated multiple docking poses. All the poses were examined with a docking score, common poses of these three compounds, and SAR (BA/AB ratio), and the pose that explains the SAR best of all was selected as shown in Figure 6.7. This pose was used to optimize this series of compounds further. Figure 6.7 illustrates the interaction between Compound **1** and P-gp while SAR is tabulated in Table 6.1. In Figure 6.7, the

TABLE 6.1 Influence of halogen placement on p-gp and efflux

Compounds	BA/AB	Structure
1	8.6	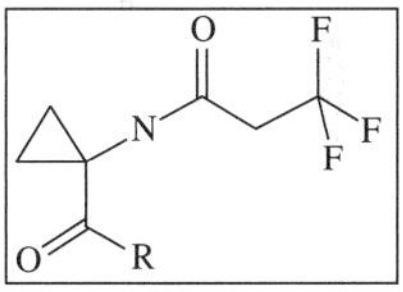
2	2.2	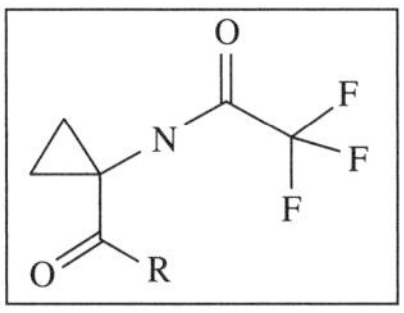
3	2.8	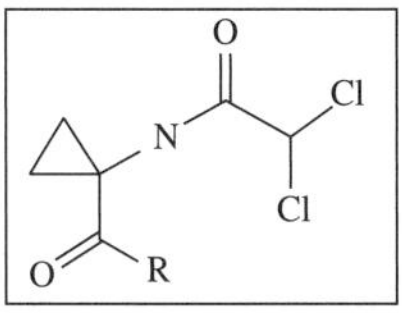

CF_3 or CCl_2 adjacent to the amide in Compounds **2** and **3** attenuate the hydrogen bond potency relative to Compound **1** in which a methylene spacer is present. Removal of the methylene spacer also weakens hydrophobic interactions with Phe994 in P-gp due to longer distance.

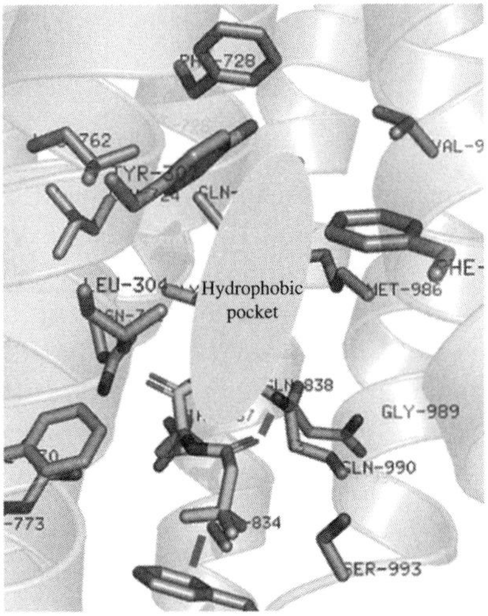

Compound 1 in P-gp model

FIGURE 6.7 Docking pose of Compound **1**. Amide group interacts with Gln990 and CF_3 with Phe994. For color detail, please see color plate section.

TABLE 6.2 Partial charges of the atoms calculated from quantum mechanics

Charge	Compound **1**	Compound **2**
O	–0.68	–0.56
C	1.01	0.61

carbonyl oxygen of the amide group of Compound **1** is making a hydrogen bond with the side chain of Gln990 while the CF_3 group is in hydrophobic contact with the phenyl ring of Phe994. Direct attachment of CF_3 and CCl_2 to the amide groups of Compound **2** and Compound **3**, respectively, by removing a carbon between them, results in attenuation of the hydrogen bond with Gln990 as well as the hydrophobic interaction with Phe994. This affirms a decrease in partial charges of the carbonyl oxygen of Compound **2** in Table 6.2. Due to an electron-withdrawing group in the neighborhood, the ability of carbonyl oxygen to make a hydrogen bond decreases.

CONCLUSIONS

Since initial studies in mdr-1a and mdr-1a and b deficient mice, P-gp has been widely recognized as a major obstacle to the discovery of CNS-active drugs. Recently, SAR-based approaches have emerged to more efficiently minimize P-gp liabilities to enhance CNS drug exposure. Establishment of these strategies was in part due to the development of *in vitro* transport models reflecting P-gp's impact on CNS exposure, better understanding of P-gp's role in regulating CNS exposure, establishment of large data sets on diverse compounds for QSAR analysis, and more sophisticated understanding of P-gp structure and transport mechanism. Global SAR, as well as SAR confined to the target restricted chemical space, can prospectively inform compound optimization. Currently, empirically defined SAR for P-gp-mediated directional transport can inform on components that influence the overall transport process (passive permeability, membrane partitioning, and P-gp transport). Refinements to molecular docking and molecular dynamic simulations should further enhance these approaches by focusing on features of drugs specifically involved in P-gp interactions. Appropriate application of these approaches should accelerate identification of compounds with good CNS penetration, ultimately resulting in improved probability of success.

REFERENCES

[1] *Trends in risks associated with new drug development: success rates for investigational drugs*. DiMasi JA, Feldman L, Seckler A, Wilson A. 2010, Clin Pharmacol Ther, Vol. 87, pp. 272–7.

[2] *The blood-brain barrier: an engineering perspective*. Wong AD, Ye M, Levy AF, Rothstein JD, Bergles DE, Searson PC. 2013, Front Neuroeng, Vol. 30, pp. 1–22.

[3] *Development, maintenance and disruption of the blood-brain barrier*. Obermeier B,Daneman R, Ransohoff RM. 2013, Nat Med, Vol. 19, pp. 1584–96.

[4] *Passive permeability and P-glycoprotein-mediated efflux differentiate central nervous system (CNS) and non-CNS marketed drugs*. Mahar Doan KM, Humphreys JE, Webster LO, Wring SA, Shampine LJ, Serabjit-Singh CJ, Adkison KK, Polli JW. 2002, J Pharmacol Exp Ther, Vol. 303, pp. 1029–37.

[5] *On the rate and extent of drug delivery to the brain*. Hammarlund-Udenaes M, Fridén M, Syvänen S, Gupta A. Pharm Res, 2008, Vol. 25, Issue 8, pp. 1737–50.

[6] *Use of plasma and brain unbound fractions to assess the extent of brain distribution of 34 drugs: comparison of unbound concentration ratios to in vivo p-glycoprotein efflux ratios*. Kalvass JC, Maurer TS, Pollack GM. 2007, Drug Metab Dispos, Vol. 24, pp. 265–76.

[7] *In vitro substrate identification studies for p-glycoprotein-mediated transport: species difference and predictability of in vivo results*. Yamazaki M, Neway WE, Ohe T, Chen I, Rowe JF, Hochman JH, Chiba M, Lin JH. 2007, J Pharmacol Exp Ther, Vol. 296, pp. 723–35.

[8] *Effect of P-glycoprotein-mediated efflux on cerebrospinal fluid concentrations in rhesus monkeys*. Tang C, Kuo Y, Pudvah NT, Ellis JD, Michener MS, Egbertson M, Graham SL, Cook JJ, Hochman JH, Prueksaritanont T. 2009, Biochem Pharmacol, Vol. 78, pp. 642–7.

[9] *Effect of P-glycoprotein-mediated efflux on cerebrospinal fluid/plasma concentration ratio*. Ohe T, Sato M, Tanaka S, Fujino N, Hata M, Shibata Y, Kanatani A, Fukami T, Yamazaki M, Chiba M, Ishii Y. 2003, Drug Metab Dispos, Vol. 31, pp. 1251–4.

[10] *PK/PD assessment in CNS drug discovery: prediction of CSF concentration in rodents for P-glycoprotein substrates and application to in vivo potency estimation*. Caruso A,

Alvarez-Sánchez R, Hillebrecht A, Poirier A, Schuler F, Lavé T, Funk C, Belli S. 2013, Biochem Pharmacol, Vol. 85, pp. 1684–99.

[11] *Utility of unbound plasma drug levels and P-glycoprotein transport data in prediction of central nervous system exposure.* He H, Lyons KA, Shen X, Yao Z, Bleasby K, Chan G, Hafey M, Li X, Xu S, Salituro GM, Cohen LH, Tang W. 2009, Xenobiotica, Vol. 39, pp. 687–93.

[12] *P-glycoprotein-mediated Hoechst 33342 transport out of the lipid bilayer.* Shapiro AB, Ling V. 1997, Eur J Biochem, Vol. 250, pp. 115–29.

[13] *Structure of a bacterial multidrug ABC transporter.* Dawson RJ, Locher KP. 2006, Nature, Vol. 443, pp. 180–5.

[14] *Structure of P-glycoprotein reveals a molecular basis for poly-specific drug binding.* Aller SG, Yu J, Ward A, Weng Y, Chittaboina S, Zhuo R, Harrell PM, Trinh YT, Zhang Q, Urbatsch IL, Chang G. 2009, Science, Vol. 323, pp. 1718–22.

[15] *Ligand mediated tertiary structure changes of reconstituted P-glycoprotein. A tryptophan fluorescence quenching analysis.* Sonveaux N, Vigano C, Shapiro AB, Ling V, Ruysschaert JM. 1999, J Biol Chem, Vol. 274, pp. 17649–54.

[16] *Conformational changes of P-glycoprotein by nucleotide binding.* Wang G, Pincheira R. Zhang JT. 1997, Biochem J, Vol. 328, pp. 897–904.

[17] *Dissection of drug-binding-induced confromational changes in P-glycoprotein.* Wang G, Pincheira R, Zhang JT. 1998, Eur J Biochem, Vol. 255, pp. 383–90.

[18] *Three dimensional structures of the mammalian multidrug resistance transporter P-glycoprotein demonstrate major conformational changes in the transmembrane domains upon nucleotide binding.* Rosenberg MF, Kamis AB, Callaghan R, Higgins CF, Ford RC. 2003, J Biol Chem, Vol. 278, pp. 8294–9.

[19] *The flexibility of P-glycoprotein for its poly-specific drug binding from molecular dynamics simulations.* Liu M, Hou T, Feng Z, Li Y. 2013, J Biomol Struct Dyn, Vol. 31, pp. 612–29.

[20] *Random forest: a classification and regression tool for compound classification and QSAR modeling.* Svetnik V, Liaw A, Tong C, Culberson JC, Sheridan RP, Feuston BP. 2003, J Chem Inf Comp Sci, Vol. 43, pp. 1947–58.

[21] *Random forests.* Breiman L. 2001, Mach Learn, Vols. 45, p. 5–45.

[22] *Atom pairs as molecular features in structure–activity studies: definition and application.* Carhart RE, Smith DH, Ventkataraghavan R. 1985, J Chem Inf Comp Sci, Vol. 25, pp. 64–73.

[23] *Chemical similarity using physiochemical property descriptors.* Kearsley SK, Sallamack S, Fluder EM, Andose JD, Mosley RT, Sheridan RP. 1996, J Chem Inf Comp Sci, Vol. 36, pp. 118–27.

7

UPTAKE TRANSPORT AT THE BBB—EXAMPLES AND SAR

ZIQIANG CHENG[1] AND QIAN LIU[2]

[1] *Drug Metabolism and Pharmacokinetics, AstraZeneca Innovation Center China, Shanghai, China*

[2] *Platform Technology and Science, GlaxoSmithKline R&D, Shanghai, China*

INTRODUCTION

The blood–brain barrier (BBB) is essential to protect the central nervous system (CNS) from potentially harmful agents in the peripheral circulation; however, it also prevents potential therapeutics from reaching the site of action. It is estimated that 98% of all small molecules do not cross the BBB [1], which presents great challenges to CNS drug discovery. Efflux transport is a major determinant of drug disposition to the CNS. Several adenosine triphophate (ATP)-dependent efflux pumps from the ATP-binding cassette (ABC) superfamily (P-glycoprotein (P-gp), BCRP, MRP4, and MRP5) have been localized at the luminal side of human brain capillary endothelial cells [2] and P-gp and BCRP have been shown to play an important role in limiting entry of various drugs into the CNS [3, 4]. Efflux transporter and passive permeability assays are usually placed early in the CNS drug discovery screening cascade. Prioritized compounds from the *in vitro* experiments are further assessed in rodent brain penetration studies. Compounds with high passive permeability that are not efflux transporter substrates and have unbound brain to unbound blood concentration ratio ($K_{puu,\ brain}$) >0.4 in animals show good CNS penetration in humans [5]. Besides the impact of passive permeability and efflux transport, physicochemical properties affect the possibility of a compound to cross the BBB. PSA < 70 Å, HBD 0–1, cLog*P* 2–4, cLog*D* (pH 7.4) 2–4, and molecular weight < 450 are preferred for a CNS drug [6]. In the real life of drug discovery, medicinal chemists have to balance

Blood–Brain Barrier in Drug Discovery: Optimizing Brain Exposure of CNS Drugs and Minimizing Brain Side Effects for Peripheral Drugs, First Edition. Edited by Li Di and Edward H. Kerns.

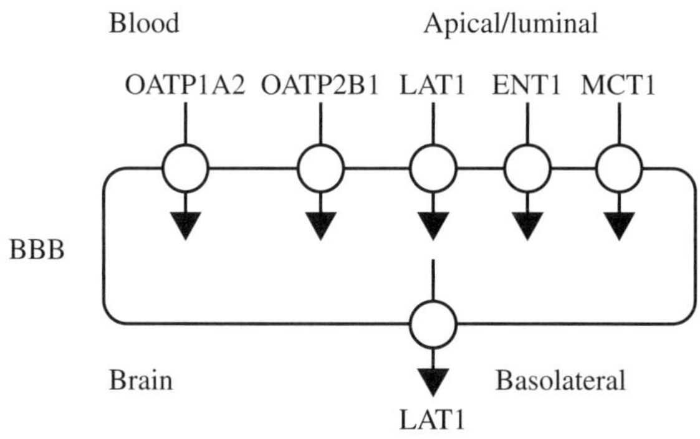

FIGURE 7.1 Uptake transporters at the human BBB.

many parameters. Compounds with preferred physicochemical properties may not have potency at the therapeutic target and/or selectivity from off-target effects. Is there an alternative approach to deliver drugs to the brain for compounds with sub-optimal physicochemical properties for brain penetration? Uptake transporters may provide a solution for CNS drug delivery across the BBB (Fig. 7.1). The role of uptake transporters for drug disposition in liver, gastrointestinal (GI) tract, and kidney has been well illustrated; however, their role at the BBB is not well understood. The International Transporter Consortium has revealed that uptake transporters OATP1A2 and OATP2B1 are on the luminal membrane of the BBB [2]. In addition, large neutral amino acid transporter 1 (LAT1), equilibrative nucleoside transporter 1 (ENT1), and monocarboxylate transporter 1 (MCT1) are also expressed on the human BBB [7]. In this chapter, the following aspects of BBB uptake transporters will be described: evidence of apical or luminal expression on the BBB, substrate specificity, structure–activity relationship (SAR), feasibility, and limitations to apply the uptake transporters for CNS drug discovery and strategies to address the limitations.

OATP1A2

OATP1A2 is also known as human OATP-A or OATP1 and its gene (*SLCO1A2*) belongs to the *SLCO* family (previously called SLC 21 family). *SLCO1A2* is located on chromosome 12p12 and the encoded 670 amino acid OATP1A2 glycoprotein has 12 transmembrane domains. OATP1A2 has the highest mRNA expression in the brain and is also observed in the liver, intestine, kidney, lung, and testis [8]. OATP1A2 protein expression was confirmed in the BBB, at the apical membrane of distal nephrons, at the apical membrane of enterocytes, and at the apical membrane of cholangiocytes [9, 10]. Bronger et al. [11] revealed that OATP1A2 protein is expressed in the luminal membrane of endothelial cells forming the human BBB, but not in astrocytes and neurons [9, 10], by immunohistochemical staining. OATP1A2 has been shown to transport a broad spectrum of substrates, including both endogenous compounds and clinically relevant pharmaceuticals [12]. As with the other human OATP transporters, OATP1A2 transports more amphipathic substrates, including bile salts, thyroid hormones, steroid conjugates, organic dyes, and anionic oligopeptides as well as several pharmaceuticals and xenobiotics. To reveal structural features frequently observed in OATP1A2 substrates, which, thus, likely contributing to OATP1A2 binding, we analyzed 26 reported OATP1A2 substrates (Table 7.1). Out of the 26 OATP1A2 substrates [13,14], 22 contained carboxylic or sulfonic acid, or functional groups which can be converted to the corresponding anion *in vivo*. The anionic center is obviously a featured functional group for the substrates of the organic anion transporter family. In addition to compounds with a single anion, compounds possessing two anionic centers are also observed in OATP1A2 substrates such as BSP and methotrexate. Other types of ionic moieties appearing in OATP1A2 substrates are zwitterions (Deltorphin II, DPDPE, fexofenadine, and levofloxacin) and cations (*N*-methylquinine, imatinib, erythromycin, rocuronium, and saquinavir). Of the 26 compounds, 10 (bamet-UD2, bamet-R2, chlorambucil taurocholate, DHEAS, estradiol-17b-glucuronide, glycocholate, ouabain, rocuronium, TCA, and TUDCA) are steroids. Among the 5 instances of the 10 steroidal compounds with reported K_m (Bamet-R2, Bamet-UD2, DHEAS, TCA, and TUDCA), the positive relationship between molecular lipophilicity (cLog*P*) and OATP1A2 binding affinity ($-\log K_m$) can be readily recognized. The result is consistent with that derived from a total of 16 OATP1A2 substrates with reported K_m (DPDPE, bamet-R2, bamet-UD2, BSP, DHEAS, deltorphin II, fexofenadine, levofloxacin, methotrexate, *N*-methylquinine, ouabain, pitavastatin, rosuvastatin, saquinavir, TCA, and TUDCA), where the square of the correlation coefficient between $-\log K_m$ and cLog*P* is 0.46. The distance between the anion and the geometry center of the hydrophobic steroid scaffold should play a role in OATP1A2 binding affinity, but the relationship between the distance and K_m could not be established due to the small number of compound samples with K_m values reported.

Uptake of Triptans by OATP1A2

Cheng et al. [15] have established a robust BacMam2-OATP1A2 transduced HEK293 system. Thirty-six CNS marketed drugs have been evaluated in this cell culture system. Hydrophilic triptans, 5-HT$_{1B/1D}$ receptor agonists for the treatment of migraine attacks, were identified as efficient OATP1A2 substrates. The triptans have relatively low molecular weights (MW, 243–382), are hydrophilic (cLog$D_{7.4}$ < 0.5), and are positively charged at physiological pH (pK_b < 5). These findings contrast with our intuition that OATPs favor negatively charged and large molecules as their efficient substrates. The K_m values were greater than 1 μM for the triptans. Unbound plasma concentrations of triptans are typically less than 300 nM in patients [16]. The results suggest that the OATP1A2 uptake transporter function is unlikely saturated at therapeutic concentrations for the triptans. As some solute carrier (SLC) transporters, such as MATE1 and MATE2K, mediate efflux of substrates from cells in the kidney and the liver [17], a SLC transporter localized at the apical side of BBB, such as OATP1A2, does not necessarily imply that it must be involved in drug uptake into the

TABLE 7.1 OATP1A2 Substrates Excluding Triptans [13, 14]

Compound	Structure	K_m (μM)	cLog*P*	Classification
[D-penicillamine 2,5] enkephalin (DPDPE)	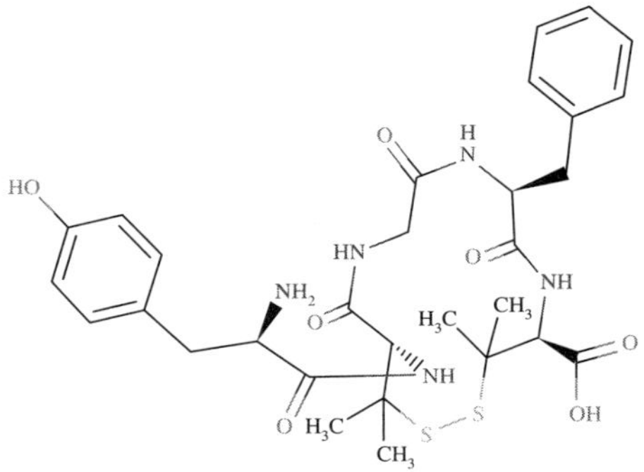	202	0.97	Zwitterion
Bamet-R2	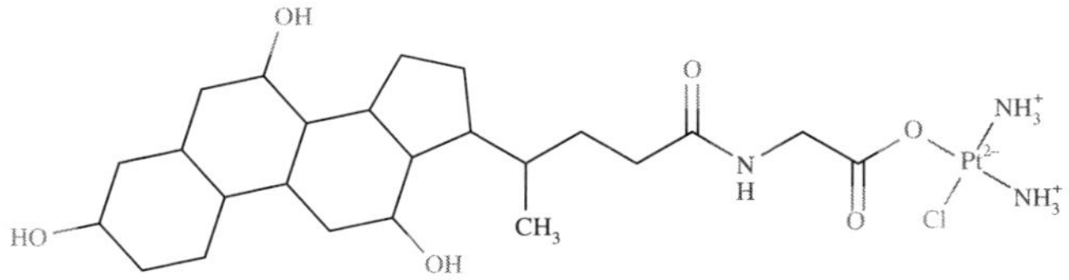	24	1.71	Anion
Bamet-UD2	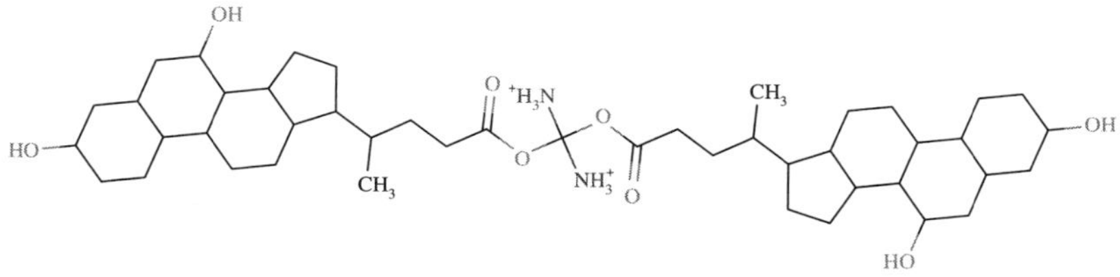	14	4.18	Anion

BQ-123	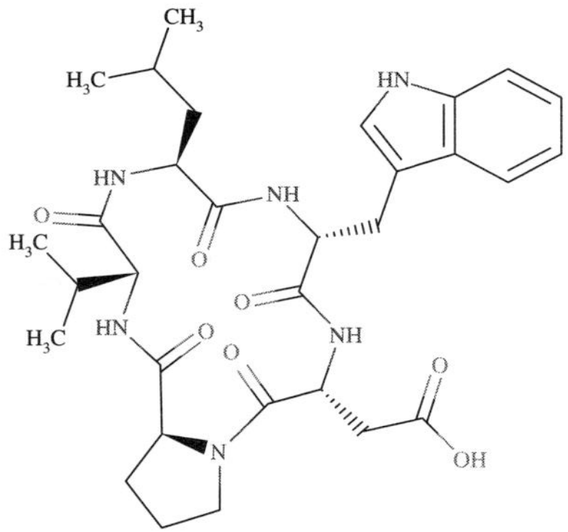			Anion
Bromosulfo-phthalein (BSP)	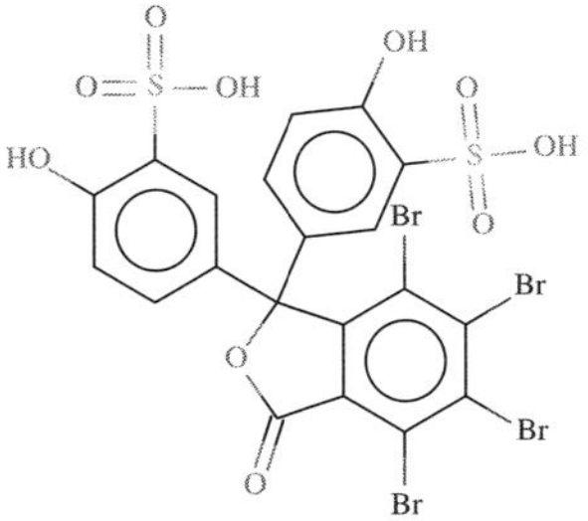	20	0.8	Multi-anion
Chlorambucil taurocholate	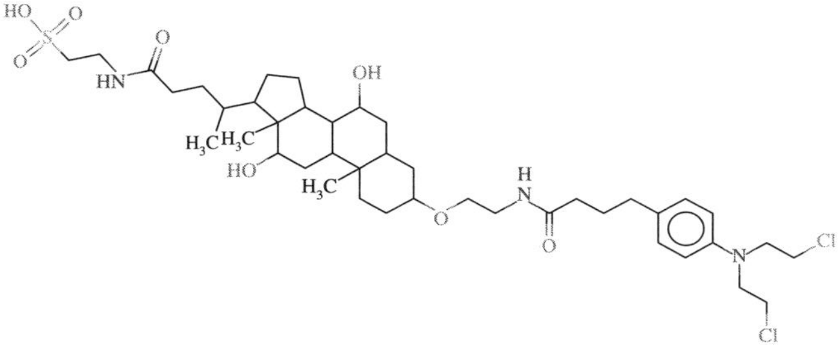			Anion

(*Continued*)

TABLE 7.1 (*Continued*)

Compound	Structure	K_m (µM)	cLog*P*	Classification
Dehydroepiandroserone-3-sulfate (DHEAS)	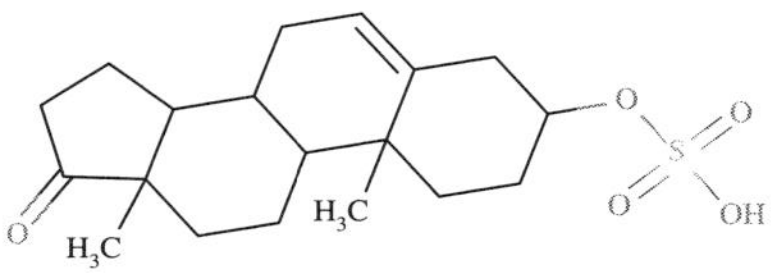	6.6	3.01	Anion
Deltorphin II	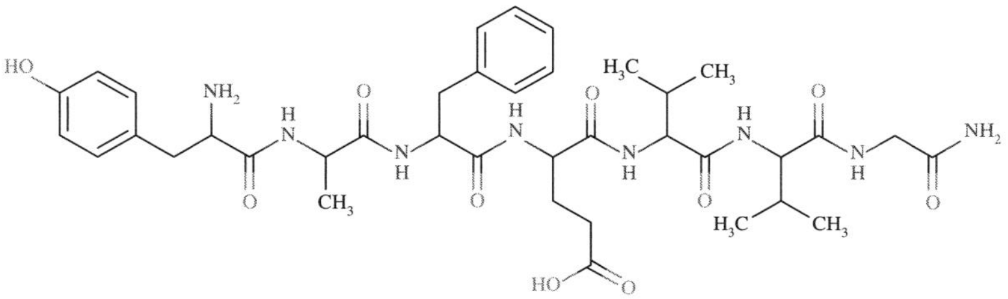	330	–2.55	Zwitterion
D-Penicillamine	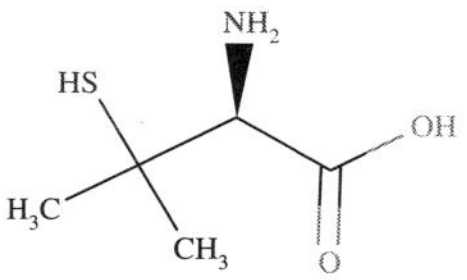			Zwitterion
Estradiol-17b-glucuronide	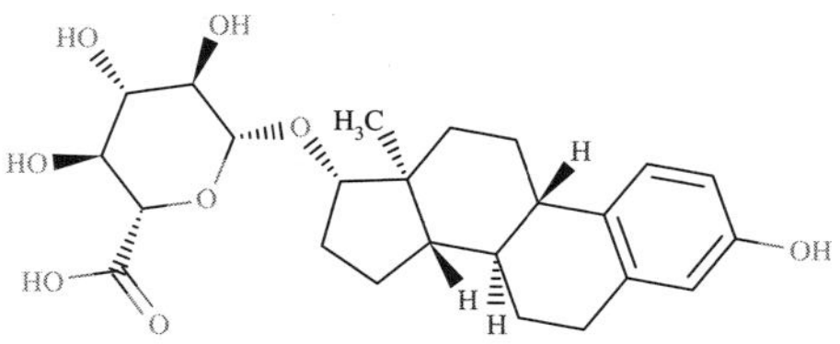			Anion

(a)

(b)

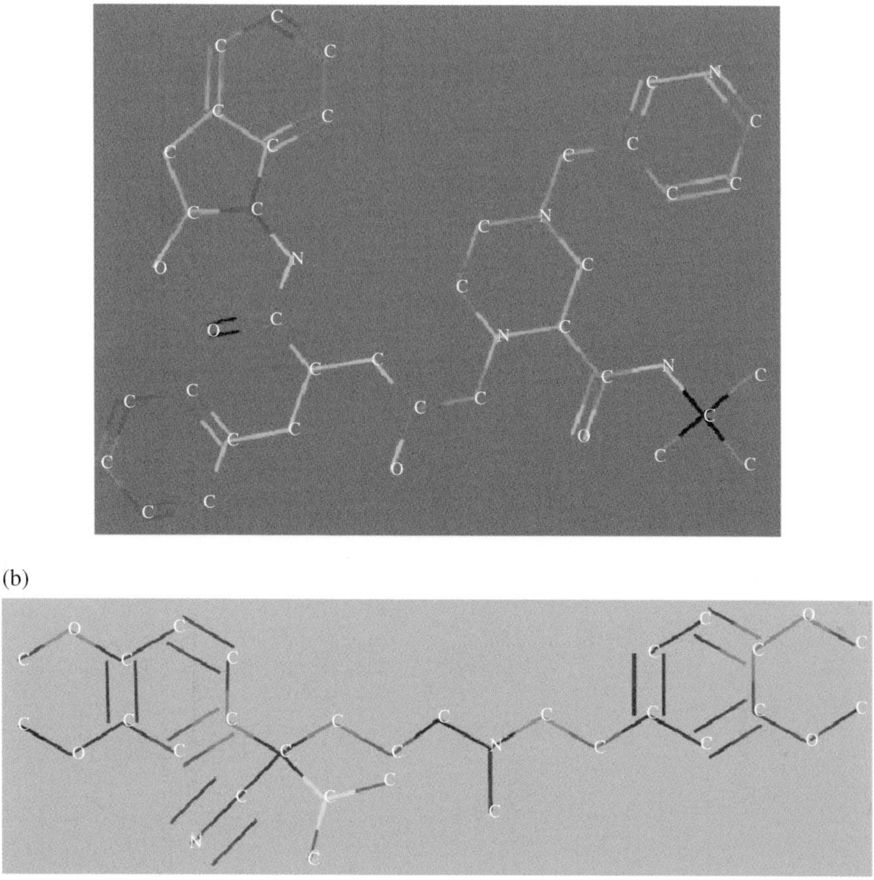

FIGURE 6.6 (a) Verapamil and (b) indinavir substrates for P-gp colored to represent the contribution of atoms to log (BA/AB) activation. The convention is that redder atoms are associated with descriptors that increase activity and blue atoms with descriptors that decrease activity.

Blood–Brain Barrier in Drug Discovery: Optimizing Brain Exposure of CNS Drugs and Minimizing Brain Side Effects for Peripheral Drugs, First Edition. Edited by Li Di and Edward H. Kerns.

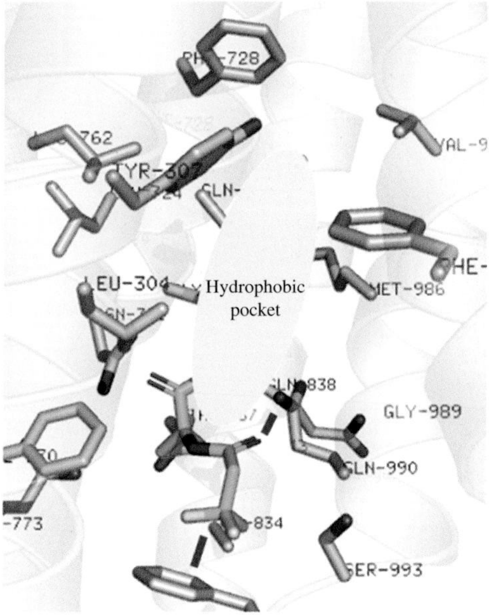

Compound 1 in P-gp model

FIGURE 6.7 Docking pose of Compound **1**. Amide group interacts with Gln990 and CF_3 with Phe994.

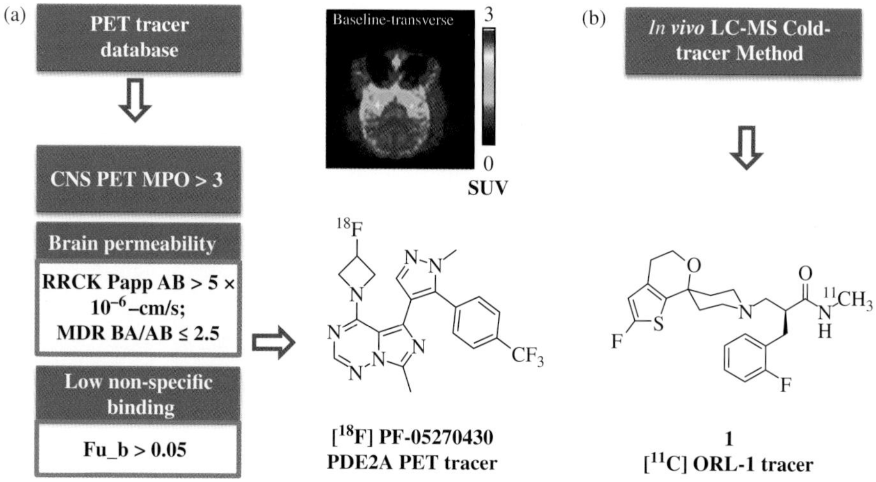

FIGURE 17.1 CNS PET MPO and *in vivo* LC-MS cold tracer method.

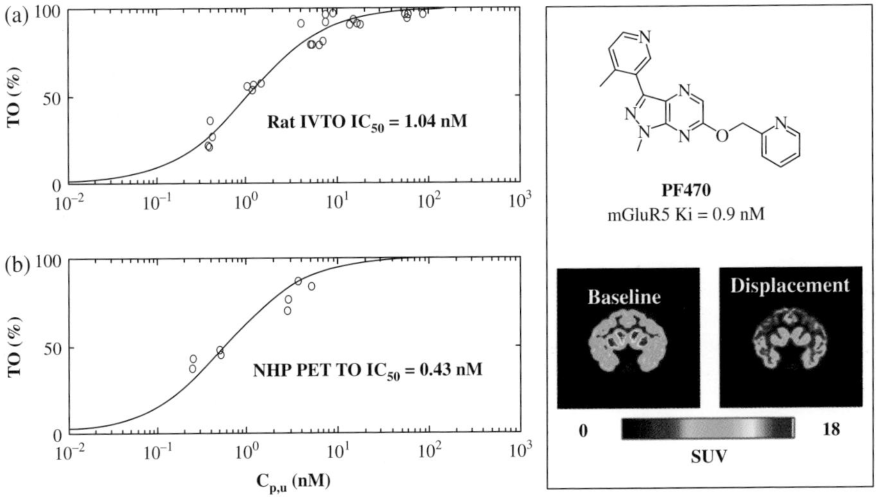

FIGURE 17.4 (a) Rat IVTO and PET TO of **PF470**; (b) the chemical structure of **PF470** and the coronal views of [^{18}F]FPEB-dependent PET images from an NHP-receiving vehicle (baseline) and **PF470** corresponding to 76% TO.

Erythromycin	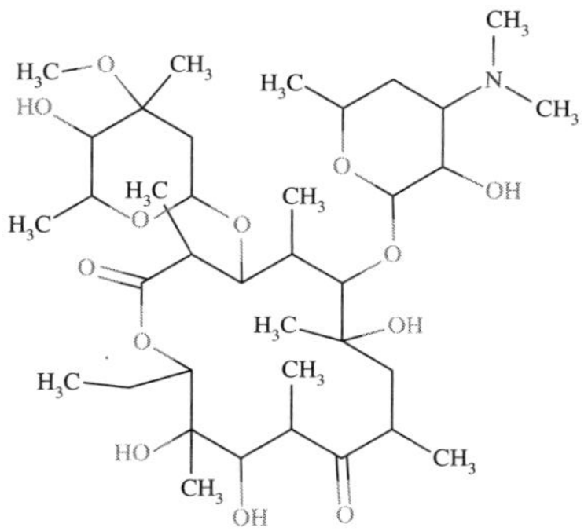			Cation
Fexofenadine	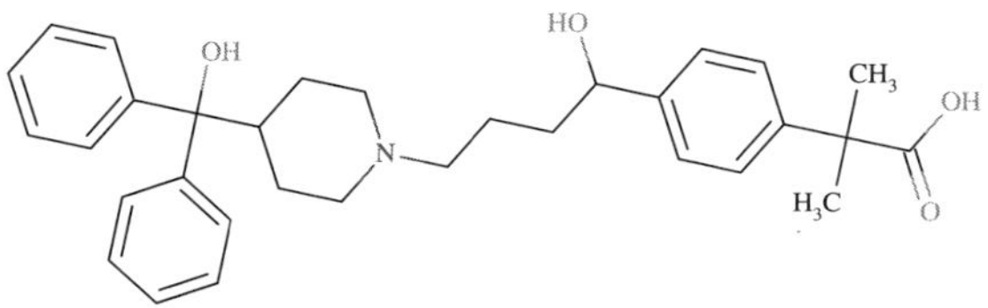	6.4	1.96	Zwitterion
Glycocholate	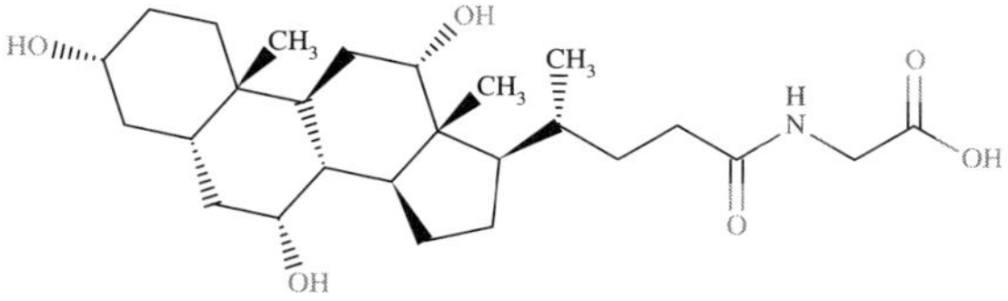			Anion
Imatinib	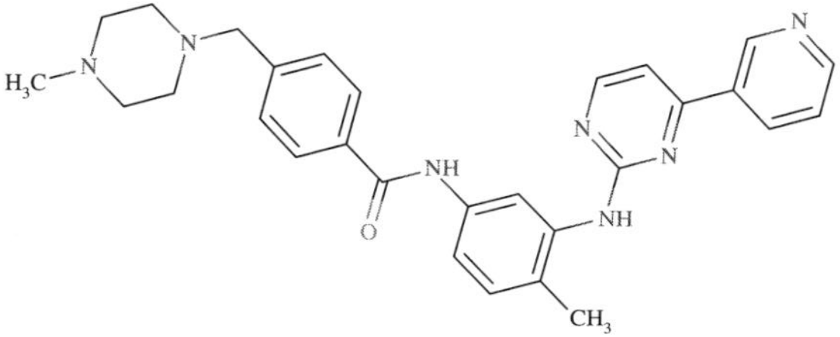			Cation

(*Continued*)

TABLE 7.1 ***(Continued)***

Compound	Structure	K_m (μM)	cLog*P*	Classification
Levofloxacin	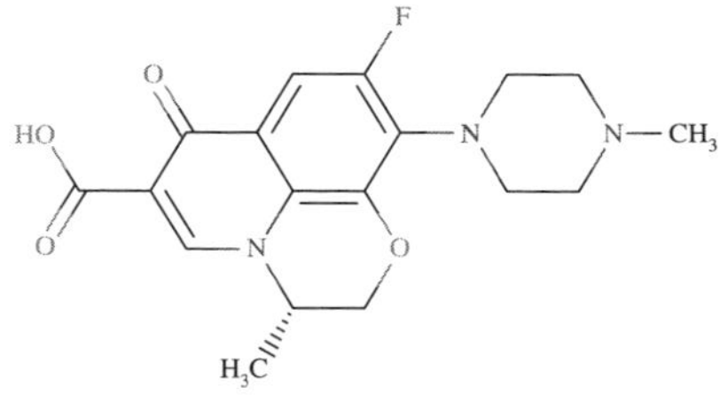	136	–0.51	Zwitterion
Methotrexate	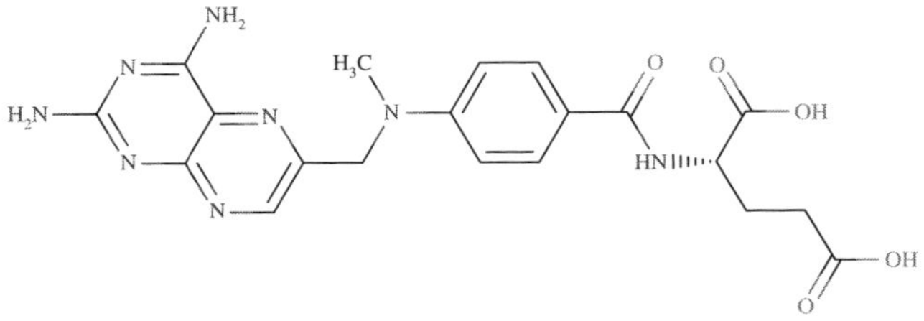	457	–0.53	Multi-anion
N-methylquinine	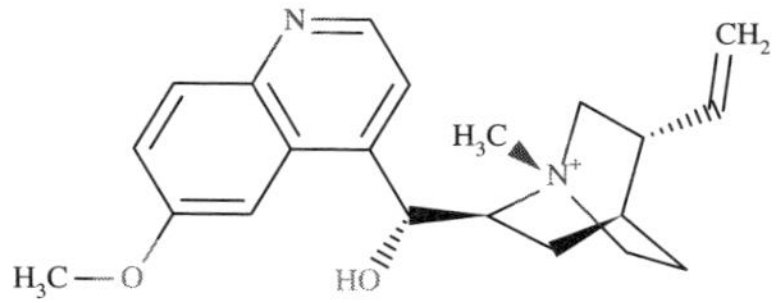	5.1	0.21	Cation
Ouabain	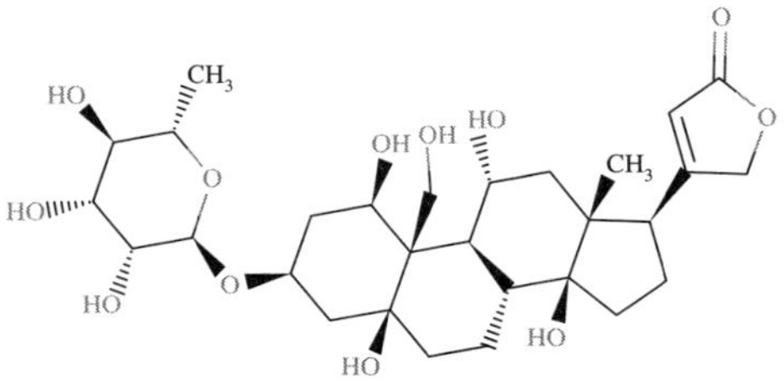	5500	–1.66	Anion

Pitavastatin	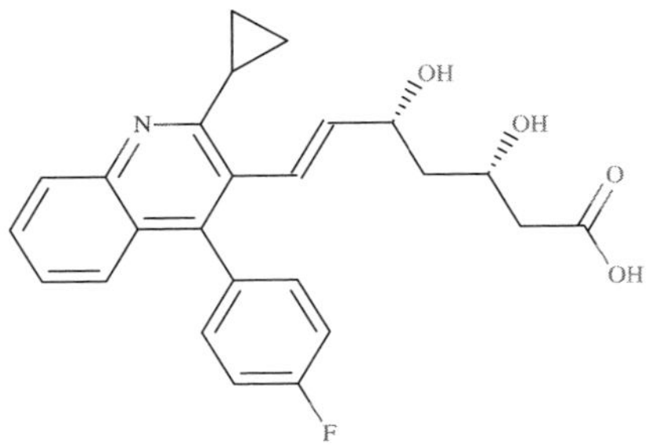	3.4	3.59	Anion
Pravastatin	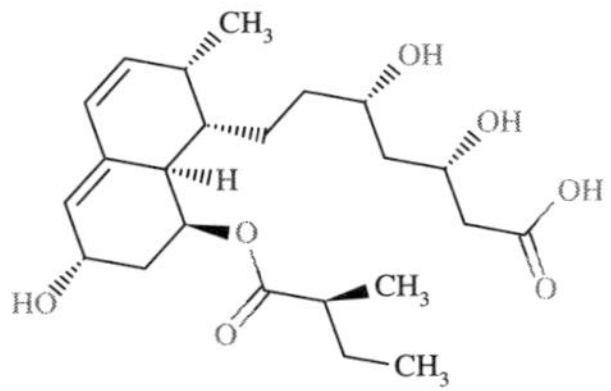			Anion
Prostaglandin E2	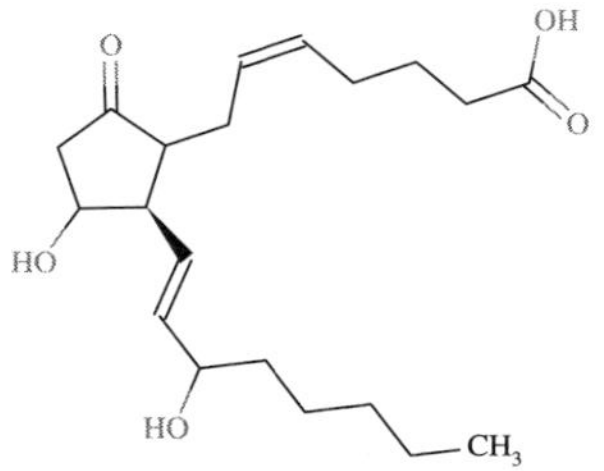			Anion
Rocuronium	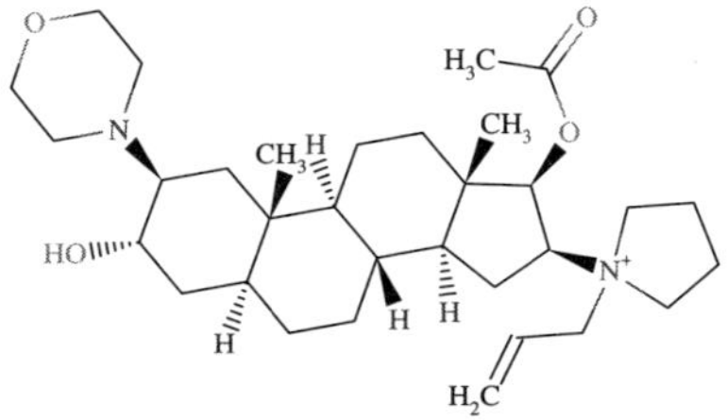			Multi-cation

(*Continued*)

TABLE 7.1 (*Continued*)

Compound	Structure	K_m (μM)	cLog*P*	Classification
Rosuvastatin	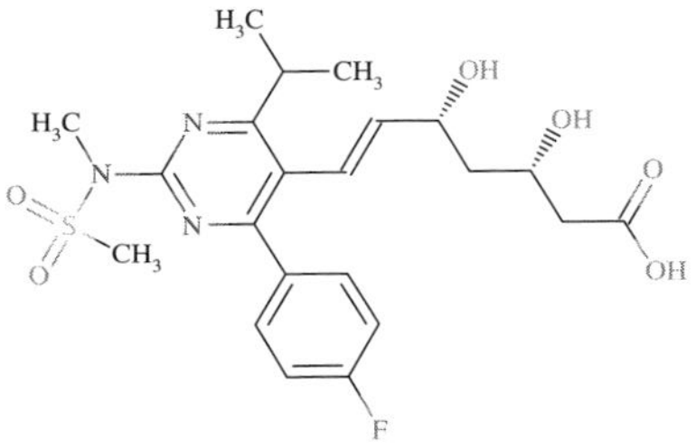	2.6	1.9	Anion
Saquinavir	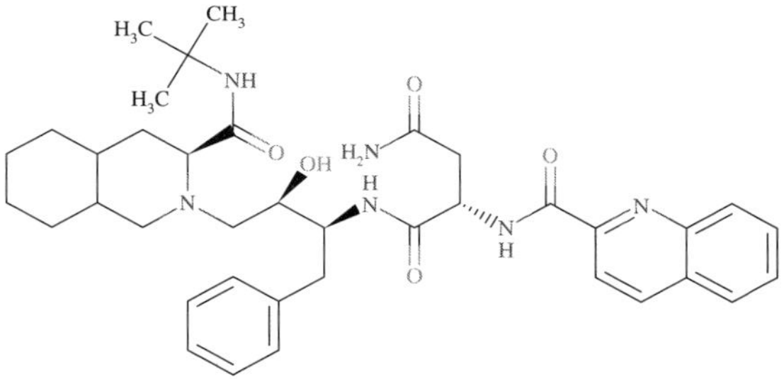	36.4	4.73	Cation
Taurocholate (TCA)	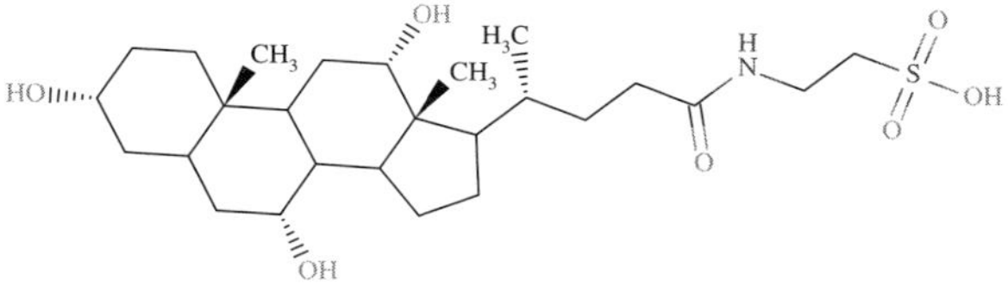	60	–0.01	Anion
Taurourso-deoxycholate (TUDCA)	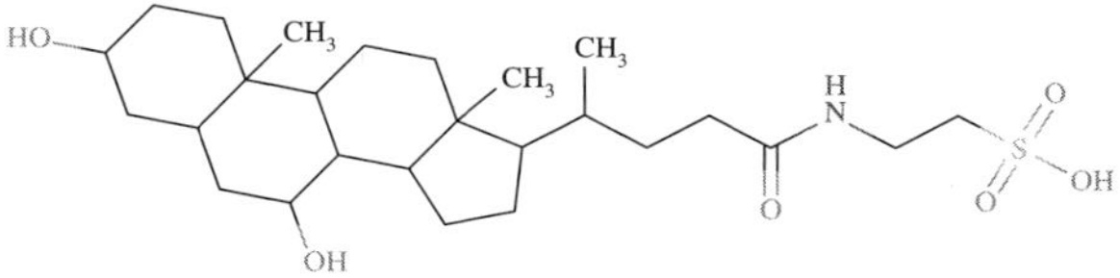	19	2.08	Anion

brain. Cheng et al. [15] have transduced BacMam2-OATP1A2 into Madin–Darby canine kidney cells (MDCKII), which are a commonly used *in vitro* model to study drug transport across the BBB [18], and demonstrated that OATP1A2 was specifically expressed on the apical membrane of the MDCKII monolayer and was capable of facilitating transport of triptans across the MDCKII monolayer from the apical to the basolateral side.

SAR Analysis of Triptan Analogs at OATP1A2

The SAR was studied with triptan structural analogs at OATP1A2 [15]. One common structural moiety observed from triptan marketed drugs is a basic amine atom in the R2 substituents (Table 7.2). To understand the significance of the amine atom in OATP1A2-mediated uptake, we intentionally included two compounds (**17** and **18**) with no amine in R2 while maintaining the same R1 group as that in zolmitriptan. Zolmitriptan had an OATP1A2-mediated uptake rate of 42.2 pmol/min/mg, whereas compounds **17** and **18** showed uptake rates at 0.00 and 3.16 pmol/min/mg protein, respectively. The basic amine atom is protonated at physiological pH. These results indicate that the positively charged basic amine is essential for efficient OATP1A2-mediated uptake. Further comparison of uptake rates for compounds with primary (**5**), secondary (**6**), or tertiary amines (e.g., sumatriptan) suggests that uptake rate is in the order of tertiary > secondary > primary. Visual analysis of R1 substituents shows that electron-rich sulfonamide and amide functional groups exist in all compounds with high OATP1A2-mediated uptake rates. These results reveal that all efficient OATP1A2 substrates bear both an electron-rich functional group and a positively charged basic amine, characterized as an amphipathic pharmacophore. Among the compounds with R1s comprising a sulfonamide group, only two compounds, **12** and **13**, showed an uptake rate higher than 100 pmol/min/mg protein. Both **12** and **13** possess a phenyl ring in addition to the sulfonamide in their R1s. In contrast, other sulfonamide-containing compounds with lower uptake rates, such as **1**, **3**, **5**, **6**, **9**, **10**, and **11**, have only small alkyl groups. The structure–uptake rate relationship indicates that an aromatic ring system in R1 can enhance uptake rate. An amphipathic pharmacophore does not guarantee that a compound will be a good OATP1A2 substrate. A large variation in uptake rate was observed from compound **5**'s 1.14 to compound **13**'s 202 pmol/min/mg protein. To reveal major factors influencing OATP1A2-mediated uptake rate, a training set (compounds **1–16**) was composed for a quantitative SAR study, such that all the compounds contain either *N*-methylmethanesulfonamide in R1 or dimethyl(ethyl)amine in R2. Among over 200 two-dimensional molecular descriptors, OATP1A2-mediated uptake rate forms a linear regression with van der Waals volume (vdw_vol). OATP1A2-mediated uptake rate = 0.94 × vdw_vol − 351 ($N = 16$, $r^{2a} = 0.72$, $s^{b} = 32.4$, $F^{c} = 35.2$, $q^{2d} = 0.61$, $S_{press}{}^{e} = 37.9$). Here, ^{a}is the conventional value of correlation coefficient, ^{b}is standard deviation, ^{c}is F statistic value from an F-test, ^{d}is leave-one-out cross-validated correlation coefficient, and ^{e}is standard deviation for the sum of square predicted errors. The leave-one-out cross-validated correlation coefficient equal to 0.61 reflects good correlation. The model suggests that molecules with larger size (up to 548 Å^{3}) give

TABLE 7.2 QSAR Analysis of Triptans and Its Analogues, OATP1A2 Substrates[a]

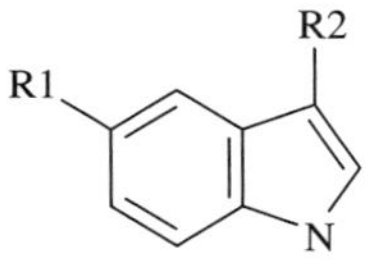

Compound[b]	R1	R2	OATP1A2-mediated uptake rate (data in parentheses were uptake rate from nontransduced HEK293 cells) (pmol/min/mg protein)[c]	vdw_vol (Å^3); MW (Dalton)	cLog$D_{7.4}$[d]
1 (Almotriptan)	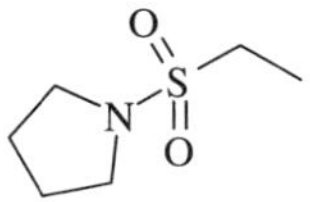	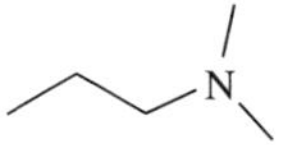	30.0 ± 2.6*** (5.75 ± 0.61)		
2 (Rizatriptan)	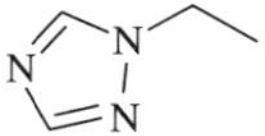	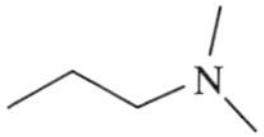	19.6 ± 0.6*** (2.87 ± 0.19)	389; 269	–0.8
3 (Sumatriptan)	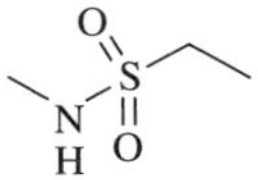	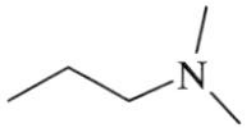	11.7 ± 0.8*** (1.27 ± 0.22)	398; 295	–1.4
4 (Zolmitriptan)	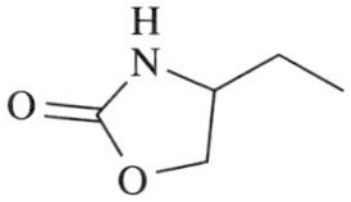	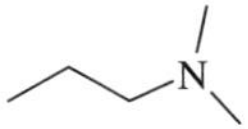	42.1 ± 4.4*** (1.16 ± 0.34)	406; 287	–1.5
5	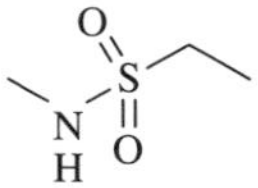		1.14 ± 0.39** (0.402 ± 0.089)	345; 267	–3.1
6	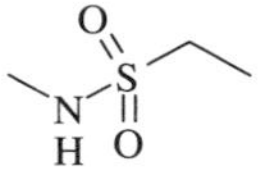	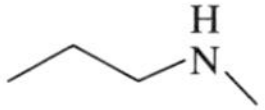	6.80 ± 1.33*** (0.311 ± 0.085)	371; 281	–3.0
7	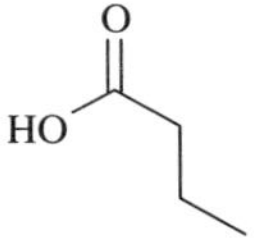	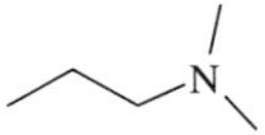	14.1 ± 1.6*** (0.675 ± 0.03)	372; 260	–1.2

8	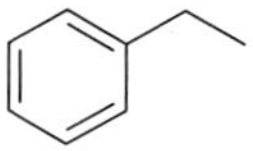	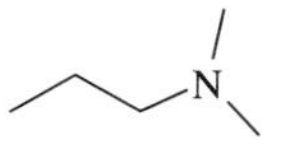	21.0 ± 14.0* (179 ± 13)	433; 278	2.3
9	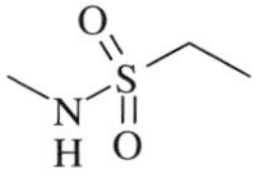	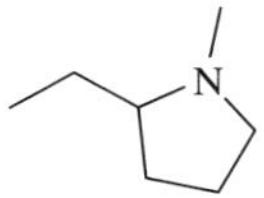	26.6 ± 1.2*** (0.923 ± 0.096)	432; 321	–2.3
10	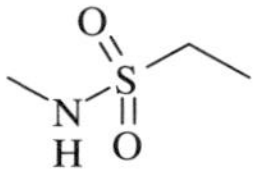	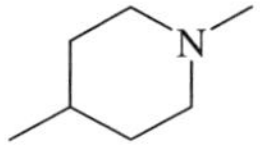	33.8 ± 2.5*** (0.459 ± 0.133)	432; 321	–2.1
11	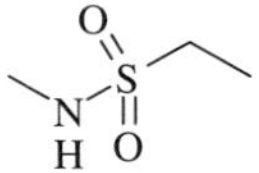	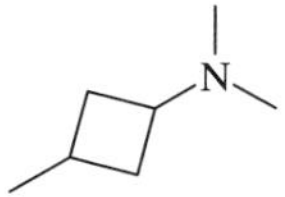	43.4 ± 6.7*** (1.71 ± 0.18)	432; 321	–2.4
12	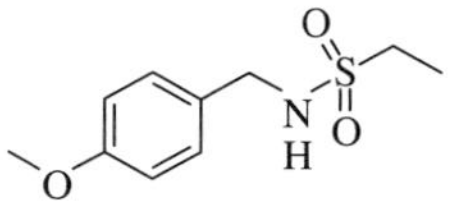	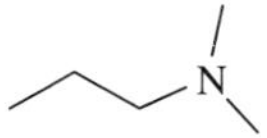	174 ± 13*** (33.9 ± 4.2)	548; 401	–1.0
13	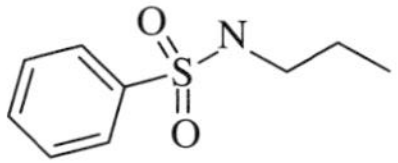	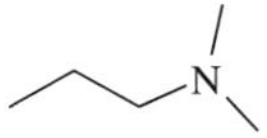	202 ± 22*** (41.8 ± 6.7)	512; 371	1.1
14[e]	—	—	72.7 ± 14.8*** (4.50 ± 0.49)	481; 349	–0.9
15[e]	—	—	82.5 ± 12.7*** (1.01 ± 0.21)	430; 301	–2.1
16[e]	—	—	90.9 ± 3.5*** (2.51 ± 0.57)	432; 287	0.3

(*Continued*)

TABLE 7.2 (*Continued*)

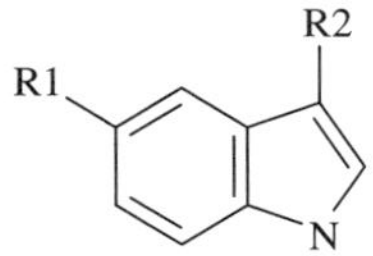

Compound[b]	R1	R2	OATP1A2-mediated uptake rate (data in parentheses were uptake rate from nontransduced HEK293 cells) (pmol/min/mg protein)[c]	vdw_vol (Å^3); MW (Dalton)	cLog$D_{7.4}$[d]
17	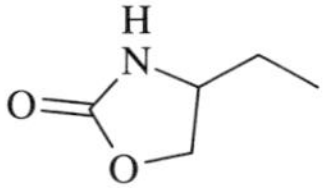	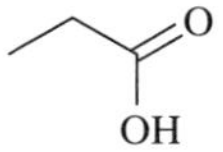	0.00 (0.587 ± 0.296)	340; 274	–3.1
18	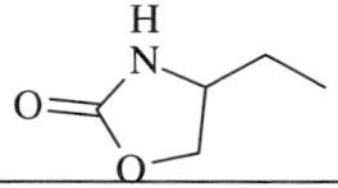	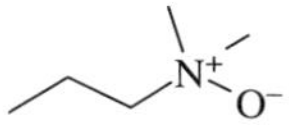	3.16 ± 0.05*** (0.148 ± 0.007)	410; 303	–2.4

[a]Reproduced from Ref. [15]. © 2012 Informa. With kind permission of Informa.
[b]Compounds **1–16** for quantitative SAR and compounds **17** and **18** for qualitative SAR only.
[c]Uptake experiment was conducted with 0.3 μM of drugs for 2 min in triplicate. OATP1A2-mediated uptake rate was obtained by subtracting the uptake rate of nontransduced HEK293 cells from that of BacMam2-OATP1A2 transduced HEK293 cells. Asterisks represent statistically significant uptake in BacMam2-OATP1A2 transduced HEK293 cells compared with nontransduced ones (*$p < 0.05$, **$p < 0.01$, ***$p < 0.001$).
[d]cLog$D_{7.4}$ were calculated using ACD/Labs software (Version 11.0, Advanced Chemistry Development, Inc., Toronto, Canada).
[e]Intellectual Property Department has decided not to disclose the chemical structures for compounds **14**, **15**, and **16**. It will not compromise overall conclusions made.

higher uptake rates. The positive correlation of vdw_vol and OATP1A2-mediated uptake rate for compounds in the training set was illustrated by Cheng et al. [15]. Since larger molecules tend to be more lipophilic, correlation between hydrophobicity (cLog$D_{7.4}$) and OATP1A2-mediated uptake rate was also examined. The result was that r^2 was only 0.14 in the case of cLog$D_{7.4}$ for the same training set, which implies that lipophilicity is irrelevant. Compound **8**, the most lipophilic among six compounds with vdw_vol of 430–433 Å^3, had the lowest uptake rate.

OATP2B1

OATP2B1 is also known as human OATP-B and its gene (*SLCO2B1*) belongs to the *SLCO* family (previously called SLC 21 family). *SLCO2B1* is located on chromosome 11q13 and the encoded 709 amino acid OATP2B1 glycoprotein has 12 transmembrane domains. OATP2B1 mRNA was detected in liver, placenta, brain, heart, lung, kidney, spleen, testis, ovary, and colon with the highest transcript level in the liver [19, 20]. OATP2B1 protein expression was confirmed at the basolateral membrane of hepatocytes, at the apical membrane of enterocytes, at the luminal side of the BBB, at the endothelial cells of the heart, at the myoepithelium of mammary ducts, in the placenta, in keratinocytes, in platelets, and in skeletal muscle. Compared with OATP1A2, OATP2B1 is widely expressed in the human body and the expression at the BBB is probably not the highest. OATP2B1 demonstrates pH-dependent transport activity [13, 21]. When the assays were performed at pH 7.4, OATP2B1 had rather narrow substrate specificity and only transported BSP, estrone-3-sulfate, and dehydroepiandroserone-3-sulfate (DHEAS). When the transport assay was conducted at pH 5–6, OATP2B1 also transported additional substrates, such as taurocholate, fexofenadine, and the loop diuretic M17055. In contrast, OATP1A2 has broad substrate specificity at relevant pH for the BBB, and all the uptake experiments for OATP1A2 substrates in Table 7.1 and 7.2 were performed at pH 7.4. Given that the luminal side of the BBB is facing blood at pH 7.4, the role of OATP2B1 to transport drugs across the BBB may be limited. It makes sense that OATP2B1 can have a significant role in intestinal absorption of drugs because the GI lumen is a more acidic environment [22].

LAT

System L is a transport system that provides cells with large neutral, branched, or aromatic amino acids. The heterodimeric system L is composed of two subunits: the light chain (L-type amino acid transporters LAT1 or LAT2) and heavy chain 4F2hc. The two chains are covalently linked via a disulfide bond. 4F2hc is necessary for plasma membrane trafficking and LAT1 and LAT2 are believed to determine the substrate specificity of the transport system [23]. In addition to endogenous amino acids, LAT1 and LAT2 have been characterized as transporting a few drugs. The structural constraint is severe and all the substrate drugs must resemble amino acids. Therefore, only less than 10 clinically used drugs are substrates of LAT1 or LAT2, including

L-DOPA, melphalan, baclofen, 3-O-methyldopa, alpha-methyltyrosine, gabapentin, alpha-methyldopa, and thyroid hormones. LAT1 is located in the apical and basolateral side of the BBB and the presence of LAT2 in the BBB is controversial.

The substrates for LAT have been elucidated *in vitro* by using mammalian cells and *Xenopus* oocytes; however, the *in vivo* relevance of LAT in drug disposition, including brain penetration, is obscure. The relative importance of the active transporters in pharmacokinetics is not only dependent upon the interaction with the transporter, but also on the passive diffusion across the cell membranes. High passive permeability of the drug often masks the contribution of the active or facilitated transport. L-DOPA, melphalan, baclofen, and gabapentin are hydrophilic molecules with low passive permeability. Theoretically, these drugs are good candidates to elucidate the relative contribution of LAT. However, given the low affinity of the drugs to LAT, therapeutic drug concentrations in the blood are overwhelmed by significant competition from the endogenous large neutral amino acids [23]. As a result, the contribution of LAT to brain penetration and pharmacokinetics *in vivo* for existing drugs as LAT substrates is unlikely.

ENT1

Nucleoside analogs are used clinically to treat cancer and viral infections. Nucleosides are transported across membranes mainly by two types of transporters: the concentrative Na^+-dependent (CNT, SLC28) and equilibrative Na^+ independent (ENT, SLC29) nucleoside transporters. Immunoblotting of rat brain endothelial cells (RBEC) revealed the presence of rENT1, rENT2, and rCNT2 proteins at BBB [24]. Adenosine is a substrate for rENT1, rENT2, and rCNT2. Measurement of [^{14}C] adenosine uptake into primary RBEC grown as monolayers on permeable plastic supports revealed abluminal transport of adenosine by rCNT2 and rENT2, and luminal transport by rENT1 and rENT2 [24]. Both rENT1 and rENT2 are expressed in rat brain microvascular endothelial cells and the human cells show only hENT1-mediated transport [25]. The role of ENT1for brain distribution has been demonstrated *in vivo*. [^{18}F]FHOMP, a C-6 substituted pyrimidine derivative and ENT1 substrate, was developed as a positron emission tomography (PET) tracer. ENT1 knockout mice exhibited significantly lower brain penetration of ^{18}F-3′-deoxy-3′-fluorothymidine (^{18}F-FLT), an ENT1 substrate [26]. Potent ENT1 inhibition by NBMPR-P caused a 40% reduction in brain uptake of [^{18}F] FHOMP in nude mice [27]. These data suggest that ENT1 can be important for CNS drug discovery when nucleoside analogs are the starting point. Existing nucleoside mimetic drugs like cytarabine, gemcitabine, 5-fluoro-5′-deoxyuridine, zidovudine, didanosine, and zalcitabine all have rather minor structural modifications from the nucleoside counterpart, which include (i) using a sugar OH with different chirality; (ii) replacing a sugar OH with other simple fragments; (iii) modifying a nucleobase or sugar moiety; and (iv) eliminating one or two sugar OH groups. Incorporating large structural change like that of NBMPR may increase the chance of delivering a nucleoside transport inhibitor instead of substrates.

MCT1

The monocarboxylate transporter (MCT, SLC16) family is composed of 14 members, of which MCTs 1–4 have been shown to catalyze the proton-linked transport of monocarboxylates, such as L-lactate, pyruvate, and the ketone bodies across the plasma membrane [28]. Besides endogenous compounds, MCTs have been shown to transport a number of therapeutic drugs. Statins are extensively used as cholesterol-lowering agents. Fluvastatin, atorvastatin, lovastatin acid, simvastatin acid, and cerivastatin have moncarboxylate structures within the compounds and have been shown to be MCT4 substrates [28, 29]. Valproic acid is an anticonvulsant and mood-stabilizing drug, whose uptake has been reported to be through MCT4 [28, 30]. Expression of MCT1–4 has been demonstrated in a number of normal tissues and tumor cells and three MCT isoforms that have been reported to be present in the brain. It is generally thought that MCT1 is present in the endothelial cells, MCT2 in neurons, and MCT4 in astrocytes [28]. As a result, MCT1 is perhaps the only candidate among MCTs for drugs transport across the BBB. However, as of today, MCT1 has been only shown to be an important determinant in the uptake of gamma-hydroxybutyrate (GHB) [31, 32]. GHB is an endogenous compound and also a drug of abuse due to its sedative/hypnotic and euphoric effects. The primary role of MCT1 *in vivo*, such as at the BBB, is to facilitate transport of endogenous compounds.

DISCUSSION AND FUTURE PERSPECTIVE

Compounds with high passive permeability that are not efflux transporter substrates and have preferred physicochemical properties of CNS drugs can cross the BBB. In reality, compounds with preferred physicochemical properties may not have potency at the therapeutic target and/or selectivity from off-target effects. The role of uptake transporters in the liver, GI tract, and kidney has been well characterized; however, their role in drug molecules crossing the BBB is less understood. We have reviewed known uptake transporters for drug transport, such as OATP1A2, OATP2B1, LAT, ENT, and MCT. Despite the measured capability of transporting drugs with amino acid structures *in vitro*, LAT is likely saturated by endogenous amino acids in blood and is not available for drug transport *in vivo*. Evidence of drug transport by MCT across the BBB is very limited. hENT1 is likely to mediate transport of some antiviral and anticancer drugs across the BBB; however, potential application of hENT1 for brain penetration is restricted to nucleoside analogs. Transcripts of OATP1A2 and OATP2B1 have been confirmed in the human BBB. Expression of these proteins at the BBB is controversial. Expression has been demonstrated by immunoflurorescence [11], but their concentrations are below the limit of quantification by liquid chromatography/mass spectrometry [33]. OATP2B1 has a broad substrate spectrum at acidic pH; however, it can only transport very limited substrates at pH 7.4, which makes it a less attractive target for CNS drug transport. In addition, OATP2B1 is widely expressed in human tissues and its abundance in the BBB is lower than in many other tissues.

OATP1A2 appears to be a promising uptake transporter to enhance BBB uptake of CNS-active molecules. It is located at the luminal side of the BBB and transcript expression is highest at the BBB compared to other tissues. OATP1A2 can transport a broad spectrum of substrates at pH 7.4, including triptans. Sumatriptan, a relatively hydrophilic triptan (cLog$D_{7.4}$ = −1.40), has very low passive permeability. In a PET study of six migraine patients, it was shown that 6 mg subcutaneous sumatriptan normalizes the migraine attack–related increase in brain serotonin synthesis [34], thus demonstrating that sumatriptan can exert an effect on the brain in migraineurs during an attack. Tfelt-Hansen [35] indicates that sumatriptan can cross the BBB in humans, based on the CNS adverse events of sumatriptan observed in migraine patients and normal volunteers. However, they are not direct evidence of BBB penetration. A human PET study should be conducted with a ^{11}C-labelled triptan to investigate the uptake function of OATP1A2 for triptan brain penetration.

Application of OATP1A2 to CNS drug discovery suffers from a number of gaps. Uchida et al. [33] reported that protein expression of OATP1A2 at the BBB is below the limit of quantitation (LOQ) at 0.695 fmol/µg protein, which is high compared with many other transporters in the same list. A quantitative assay with higher sensitivity is needed to confirm the expression of OATP1A2 at the human BBB. Further exploration of the *in vivo* function of OATP1A2 in CNS penetration is also limited by lack of orthologous rodent protein with similar function and localization. Mouse and rat Oatp1a4, the closest human orthologs, only share between 67% and 73% amino acid sequences with human OATP1A2 [12]. The rodent Oatp1a4 is expressed on both the blood and brain sides of the BBB [9, 36]. Kusuhara and Sugiyama [37] have shown that Oatp1a4 involves both uptake and efflux functions at the rodent BBB; thus the role of Oatp1a4 for net BBB penetration of its substrates can be minimal in rodents. Genetic polymorphism is a useful tool to investigate a protein's function *in vivo*. Franke et al. [12] have summarized the *SLCO1A2* genetic polymorphism and related functional changes by using an *in vitro* system. The impact of *SLCO1A2* genetic polymorphism was recently revealed in a clinical study. Imatinib clearance in chronic myeloid leukemia patients was influenced by the *SLCO1A2*-1105G>A/-1032G>A genotype (p = 0.075) and the *SLCO1A2*-361GG genotype (p = 0.005) [38]. However, the role of the *SLCO1A2* polymorphism for CNS penetration is still unclear. Another proposed experiment is to evaluate OATP1A2 substrates in transgenic mice with human OATP1A2 knock-in and rodent Oatp1a4 knockout at the BBB. Once constructed, the transgenic mouse model may shed light on OATP1A2 function *in vivo*.

In summary, we have reviewed luminal and abluminal expressions of OATP1A2, OATP2B1, ENT1, MCT1, and LAT1 at human BBB and SAR for these uptake transporters. Utilization of BBB uptake transporters have a potential to facilitate uptake of CNS drugs across the BBB. At present, there is insufficient information to utilize any of these uptake transporters to achieve this mission. OATP1A2 appears to be a promising BBB uptake transporter worth further investigation to characterize its function *in vivo*.

REFERENCES

[1] Pardridge, W.M. (2005). The blood-brain barrier: bottleneck in brain drug development. *NeuroRx*, 2, 3–14.

[2] Giacomini, K.M., Huang, S.M., Tweedie, D.J., Benet, L.Z., Brouwer, K.L., Chu, X., Dahlin, A., Evers, R., Fischer, V., Hillgren, K.M., Hoffmaster, K.A., Ishikawa, T., Keppler, D., Kim, R.B., Lee, C.A., Niemi, M., Polli, J.W., Sugiyama, Y., Swaan, P.W., Ware, J.A., Wright, S.H., Yee, S.W., Zamek-Gliszczynski, M.J., Zhang, L., International Transporter Consortium. (2010). Membrane transporters in drug development. *Nature Reviews Drug Discovery*, 9, 215–236.

[3] Enokizono, J., Kusuhara, H., Ose, A., Schinkel, A.H., Sugiyama, Y. (2008). Quantitative investigation of the role of breast cancer resistance protein (Bcrp/Abcg2) in limiting brain and testis penetration of xenobiotic compounds. *Drug Metabolism and Disposition*, 36, 995–1002.

[4] Zhou, L., Schmidt, K., Nelson, F.R., Zelesky, V., Troutman, M.D., Feng, B. (2009). The effect of breast cancer resistance protein and P-glycoprotein on the brain penetration of flavopiridol, imatinib mesylate (Gleevec), prazosin, and 2-methoxy-3-(4-(2-(5-methyl-2-phenyloxazol-4-yl) ethoxy) phenyl) propanoic acid (PF-407288) in mice. *Drug Metabolism and Disposition*, 37, 946–955.

[5] Di, L., Rong, H., Feng, B. (2013). Demystifying brain penetration in central nervous system drug discovery. *Journal of Medicinal Chemistry*, 56, 2–12.

[6] Hitchcock, S.A., Pennington, L.D. (2006). Structure—brain exposure relationships. *Journal of Medicinal Chemistry*, 49, 7559–7583.

[7] Kalvass, J.C., Polli, J.W., Bourdet, D.L., Feng, B., Huang, S-M., Liu, X., Smith, Q.R., Zhang, L.K., Zamek-Gliszczynski, M.J. (2013). Why clinical modulation of efflux transport at the human blood-brain barrier is unlikely: the ITC evidence-based position. *Clinical Pharmacology and Therapeutics*, 94, 80–94.

[8] Kullak-Ublick, G.A., Hagenbuch, B., Stieger, B., Schteingart, C.D., Hofmann, A.F., Wolkoff, A.W., Meier, P.J. (1995). Molecular and functional characterization of an organic anion transporting polypeptide cloned from human liver. *Gastroenterology*, 109, 1274–1282.

[9] Gao, B., Stieger, B., Noé, B., Fritschy, J.M., Meier, P.J. (1999). Localization of the organic anion transporting polypeptide 2 (Oatp2) in capillary endothelium and choroid plexus epithelium of rat brain. *Journal of Histochemistry and Cytochemistry*, 47, 1255–1264.

[10] Lee, W., Glaeser, H., Smith, L.H., Roberts, R.L., Moeckel, G.W., Gervasini, G., Leake, B.F., Kim, R.B. (2005). Polymorphisms in human organic anion-transporting polypeptide 1A2 (OATP1A2): implications for altered drug disposition and central nervous system drug entry. *Journal of Biological Chemistry*, 280, 9610–9617.

[11] Bronger, H., König, J., Kopplow, K., Steiner, H.H., Ahmadi, R., Herold-Mende, C., Keppler, D., Nies, A.T. (2005). ABCC drug efflux pumps and organic anion uptake transporters in human gliomas and the blood-tumor barrier. *Cancer Research*, 65, 11419–11428.

[12] Franke, R.M., Scherkenbach, L.A., Sparreboom, A. (2009). Pharmacogenetics of the organic anion transporting polypeptide 1A2. *Pharmacogenomics*, 10, 339–344.

[13] Hagenbuch, B., Gui, C. (2008). Xenobiotic transporters of the human organic anion transporting polypeptides (OATP) family. *Xenobiotica*, 38, 778–801.

[14] Roth, M., Obaidat, A., Hagenbuch, B. (2012). OATPs, OATs and OCTs: the organic anion and cation transporters of the SLCO and SLC22A gene superfamilies. *British Journal of Pharmacology*, 165, 1260–1287.

[15] Cheng, Z., Liu, H., Yu, N., Wang, F., An, G., Xu, Y., Liu, Q., Guan, C., Aryton, A. (2012). Hydrophilic anti-migraine triptans are substrates for OATP1A2, a transporter expressed at human blood-brain barrier. *Xenobiotica*, 42, 880–890.

[16] Evans, D.C., O'Connor, D., Lake, B.G., Evers, R., Allen, C., Hargreaves, R. (2003). Eletriptan metabolism by human hepatic CYP450 enzymes and transport by human P-glycoprotein. *Drug Metabolism and Disposition*, 31, 861–869.

[17] Moriyama, Y., Hiasa, M., Matsumoto, T., Omote, H. (2008). Multidrug and toxic compound extrusion (MATE)-type proteins as anchor transporters for the excretion of metabolic waste products and xenobiotics. *Xenobiotica*, 38, 1107–1118.

[18] Wang, Q., Rager, J.D., Weinstein, K., Kardos, P.S., Dobson, G.L., Li, J., Hidalgo, I.J. (2005). Evaluation of the MDR-MDCK cell line as a permeability screen for the blood-brain barrier. *International Journal of Pharmaceutics*, 288, 349–359.

[19] Obaidat, A., Roth, M., Hagenbuch, B. (2012). The expression and function of organic anion transporting polypeptides in normal tissues and in cancer. *Annual Review of Pharmacology and Toxicology*, 52, 135–151.

[20] Hagenbuch, B., Stieger, B. (2013). The SLCO (former SLC21) superfamily of transporters. *Molecular Aspects of Medicine*, 34, 396–412.

[21] Varma, M.V., Rotter, C.J., Chupka, J., Whalen, K.M., Duignan, D.B., Feng, B., Litchfield, J., Goosen, T.C., El-Kattan, A.F. (2011). pH-sensitive interaction of HMG-CoA reductase inhibitors (statins) with organic anion transporting polypeptide 2B1. *Molecular Pharmaceutics*, 8, 1303–1313.

[22] Tamai, I. (2012). Oral drug delivery utilizing intestinal OATP transporters. *Advanced Drug Delivery Reviews*, 64, 508–514.

[23] del Amo, E.M., Urtti, A., Yliperttula, M. (2008). Pharmacokinetic role of L-type amino acid transporters LAT1 and LAT2. *European Journal of Pharmaceutical Sciences*, 35, 161–174.

[24] Redzic, Z.B., Biringer, J., Barnes, K., Baldwin, S.A., Al-Sarraf, H., Nicola, P.A., Young, J.D., Cass, C.E., Barrand, M.A., Hladky, S.B. (2005). Polarized distribution of nucleoside transporters in rat brain endothelial and choroid plexus epithelial cells. *Journal of Neurochemistry*, 94, 1420–1426.

[25] Parkinson, F.E., Damaraju, V.L., Graham K., Yao S.Y.M., Baldwin, S.A., Cass, C.E., Young, J.D. (2011). Molecular biology of nucleoside transporters and their distributions and functions in the brain. *Current Topics in Medicinal Chemistry*, 11, 948–972.

[26] Paproski, R.J., Wuest, M., Jans, H-S., Graham, K., Gati, W.P., McQuarrie, S., McEwan, A., Mercer, J., Young, J.D., Cass, C.E. (2010). Biodistribution and uptake of 3′-deoxy-3′-fluorothymidine in ENT1-knockout mice and in an ENT1-knockdown tumor model. *Journal of Nuclear Medicine*, 51, 1447–1455.

[27] Müller, U., Ross, T.L., Ranadheera, C., Slavik, R., Müller, A., Born, M., Trauffer, E., Sephton, S.M., Scapozza, L., Krämer, S.D., Ametamey, S.M. (2013). Synthesis and preclinical evaluation of a new C-6 alkylated pyrimidine derivative as a PET imaging agent for HSV1-tk gene expression. *American Journal of Nuclear Medicine and Molecular Imaging*, 3, 71–84.

[28] Meredith, D., Christian, H.C. (2008). The SLC16 monocaboxylate transporter family. *Xenobiotica*, 38, 1072–1106.

[29] Kobayashi, M., Otsuka, Y., Itagaki, S., Hirano, T., Iseki, K. (2006). Inhibitory effects of stains on human monocarboxylate transporter 4. *International Journal of Pharmaceutics*, 317, 19–25.

[30] Nagasawa, K., Nakai, K., Sumitani, Y., Moriya, Y., Muraki, Y., Takara, K., Ohnishi, N., Yokoyama, T., Fujimoto, S. (2002). Monocarboxylate transporter mediates uptake of lovastatin acid in rat cultures mesangial cells. *Journal of Pharmaceutical Sciences*, 91, 2605–2613.

[31] Roiko, S.A., Felmlee, M.A., Morris, M.E. (2012). Brain uptake of the drug of abuse γ-hydroxybutyric acid in rats. *Drug Metabolism and Disposition*, 40, 212–218.

[32] Bhattacharya, I., Boje, K.M. (2004). GHB (gamma-hydroxybutyrate) carrier-mediated transport across the blood-brain barrier. *Journal of Pharmacology and Experimental Therapeutics*, 311, 92–98.

[33] Uchida, Y., Ohtsuki, S., Katsukara, Y., Ikeda, C., Suzuku, T., Kamiie, J., Terasaki, T. (2011). Quantitative targeted absolute proteomics of human blood-brain barrier transporters and receptors. *Journal of Neurochemistry*, 117, 333–345.

[34] Sakai, Y., Dobson, C., Diksic, M., Aubé, M., Hamel, E. (2008). Sumatriptan normalizes the migraine attack-related increase in brain serotonin synthesis. *Neurology*, 70, 431–439.

[35] Tfelt-Hansen, P.C. (2010). Does sumatriptan cross the blood-brain barrier in animals and man? *The Journal of Headache and Pain*, 11, 5–12.

[36] Ose, A., Kusuhara, H., Endo, C., Tohyama, K., Miyajima, M., Kitamura, S., Sugiyama, Y. (2010). Functional characterization of mouse organic anion transporting peptide 1a4 in the uptake and efflux of drugs across the blood-brain barrier. *Drug Metabolism and Disposition*, 38, 168–176.

[37] Kusuhara, H., Sugiyama, Y. (2005). Active efflux across the blood-brain barrier: role of the solute carrier family. *NeuroRx*, 2, 73–85.

[38] Yamakawa, Y., Hamada, A., Shuto, T., Yuki, K., Uchida, T., Kai, H., Kawaguchi, T., Saito, H. (2011). Pharmacokinetic impact of SLCO1A2 polymorphisms on imatinib disposition in patients with chronic myeloid leukemia. *Clinical Pharmacology and Therapeutics*, 90, 157–163.

8

TRANSPORT OF PROTEIN AND ANTIBODY THERAPEUTICS ACROSS THE BLOOD–BRAIN BARRIER

William M. Pardridge
ArmaGen Technologies, Inc., Calabasas, CA, USA

BIOTECHNOLOGY AND THE BRAIN: HISTORICAL PERSPECTIVE

The biotechnology industry was founded in 1976, and in these nearly 40 years, there is still not a single recombinant protein that is FDA-approved for the treatment of the central nervous system (CNS), wherein drug action in brain requires transport across the blood–brain barrier (BBB). The absence of biologic pharmaceuticals for the brain and spinal cord is not for the lack of trying. As illustrated in Figure 8.1, there have been numerous costly phase III clinical trials of recombinant proteins and monoclonal antibodies (MAb) for CNS diseases, such as amyotrophic lateral sclerosis (ALS), stroke, Parkinson's disease (PD), and Alzheimer's disease (AD). There are several causes for drug failures in clinical trials, but in the case of biologics and the brain, all the drug development efforts shown in Figure 8.1 have a common trait—that in no case was any biologic drug reengineered to cross the (BBB):

- Neurotrophins, such as brain-derived neurotrophic factor (BDNF) or ciliary neurotrophic factor (CNTF), were developed in the 1990s for the treatment of ALS, a chronic neurodegenerative disease, and the phase III trials failed to show benefit [1, 2]. The drugs were administered by subcutaneous (SQ) injection, similar to the administration of insulin for diabetes mellitus. Both BDNF and

Blood–Brain Barrier in Drug Discovery: Optimizing Brain Exposure of CNS Drugs and Minimizing Brain Side Effects for Peripheral Drugs, First Edition. Edited by Li Di and Edward H. Kerns.

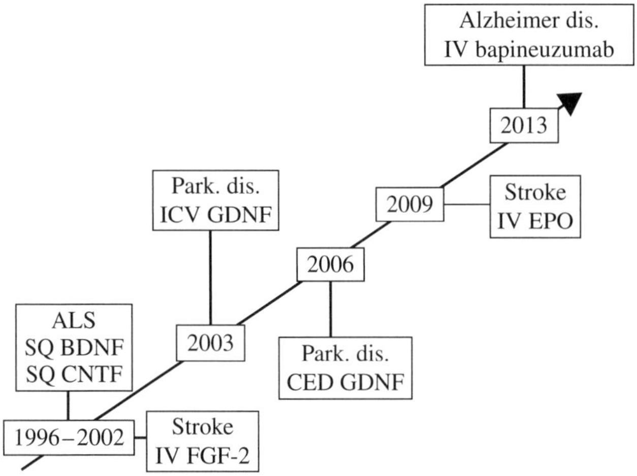

FIGURE 8.1 History of past clinical trials of recombinant proteins or monoclonal antibody drugs for the treatment of CNS disease. ALS, amotrophic lateral sclerosis; BDNF, brain-derived neurotrophic factor; CED, convection-enhanced diffusion; CNTF, ciliary neurotrophic factor; EPO, erythropoietin; FGF, fibroblast growth factor; GDNF, glial-derived neurotrophic factor; ICV, intracerebroventricular; IV, intravenous; SQ, subcutaneous. For primary references of each trial, see Refs. [1, 2, 5, 9, 11, 13, 22].

CNTF are large-molecule drugs that do not cross the BBB [3], and the tight junctions that form the BBB in the spinal cord are intact in ALS [4].

- Neurotrophins are neuroprotective in acute neural disease, such as stoke. However, the intravenous (IV) administration of a potent neurotrophin, fibroblast growth factor (FGF)-2, in acute ischemic stroke in humans was not effective [5]. FGF2 does not cross the BBB in the absence of BBB disruption [6]. Although the BBB becomes disrupted 12–24 h after a stroke in humans [7], the BBB is intact in the early hours after stroke when neuroprotection is still possible [8].
- Glial-derived neurotrophic factor (GDNF) is a potential neuroprotective agent for the nigra-striatal tract of the brain that degenerates in PD. To bypass the BBB, GDNF was administered to PD patients by intracerebroventricular (ICV) injection, but no clinical benefit could be demonstrated [9]. Preclinical work showed that the ICV injection of a neurotrophin results only in drug distribution to the ipsilateral surface of the brain, without significant diffusion from the cerebrospinal fluid (CSF) compartment to the deeper parenchyma of the brain [10].
- So as to bypass the BBB, and to eliminate the need for drug diffusion into the brain, GDNF was administered to PD patients by convection-enhanced diffusion (CED) [11]. In this approach, the GDNF is formulated in a pump that is implanted in the abdomen, and bilateral catheters connect the pump to the striatum of the brain. The clinical trial failed, and parallel studies in primates showed that there is a logarithmic decrease in brain GDNF content with each millimeter removed

from the catheter tip [12]. This finding suggests that the primary process for drug distribution in brain via CED is diffusion, not convection.

- Erythropoietin (EPO) was administered to acute stroke patients by IV injection within 6 h of the stroke, but the clinical trial failed [13]. It was believed that EPO crosses the BBB, because EPO distributes into CSF following IV injection. However, all proteins in plasma distribute into CSF inversely related to molecular weight [14], and drug entry into CSF is not a measure of BBB transport [15]. In fact, EPO does not cross the BBB [16]. The failure of the EPO stroke trial is consistent with (i) the lack of EPO transport across an intact BBB [16], and (ii) the intactness of the BBB in the early hours after stroke when neuroprotection is still possible [8].
- The dementia of AD is correlated with the deposition of amyloid plaque in the brain [17], and the AD plaques are formed from the 40–43 amino acid Abeta amyloid peptide [18, 19]. Anti-amyloid antibodies (AAA), such as bapineuzumab, are potent plaque disaggregation agents. However, the AAAs are large-molecule drugs that do not cross the BBB [20]. The amyloid plaques reside behind the BBB, and the BBB is intact in AD [21]. Therefore, the AAA in blood cannot access the plaque in brain unless there is disruption of the BBB. Chronic treatment of AD patients with an AAA, such as bapineuzumab, failed in a phase III trial [22].

The CNS drug development projects listed in Figure 8.1 all proceeded without any parallel effort in the development of a BBB drug delivery technology. An alternative approach to CNS drug development of biologics is to first produce a functional BBB drug delivery platform, and then reengineer the biologic drug to enable BBB transfer. This is possible with the use of BBB molecular Trojan horses (MTHs) that target endogenous receptor-mediated transport systems within the BBB.

BLOOD–BRAIN BARRIER MOLECULAR TROJAN HORSE TECHNOLOGY

Receptor-Mediated Transport

The BBB expresses multiple carrier-mediated transporters (CMTs), which enable BBB transfer of small-molecule nutrients and vitamins [15]. One such CMT system is the large neutral amino acid transporter type 1, LAT1, which is selectively expressed in the body at the BBB [23]. L-DOPA is an effective treatment of PD because this water-soluble neutral amino acid crosses the BBB via transport on LAT1 [24]. Similarly, the BBB expresses certain receptor-mediated transport (RMT) systems for circulating peptides, such as insulin [25, 26] or transferrin (Tf) [27, 28]. The findings that the BBB expressed endogenous peptide receptors, and that some of these receptors are transport systems, led to the hypothesis that biologics could be made BBB-transportable by linking the protein drug to a ligand that normally crossed the BBB via these endogenous RMT systems [29]. The ligand, such as insulin or Tf, would act as an MTH to ferry the attached biologic drug across the BBB. Insulin would not be an optimal MTH because the insulin domain of the MTH–drug conjugate would cause hypoglycemia. Tf would not be an optimal MTH because the exogenous Tf could not compete with the endogenous Tf in plasma that saturates 99% of the BBB Tf receptor (TfR) binding sites. Alternative MTHs are peptidomimetic MAbs that bind exofacial epitopes on the BBB insulin or TfRs, followed by receptor-mediated transport of the MAb across the BBB [30–33]. An MAb against the TfR is designated TfRMAb, and an MAb against the human insulin receptor (HIR) is designated HIRMAb.

Cell Biology of Receptor-Mediated Transport at the Blood–Brain Barrier

The cell biology of BBB RMT via the TfR has been investigated with light and electron microscopy, emulsion autoradiography, and confocal microscopy (Fig. 8.2). These studies show that (i) the BBB TfR mediates the transcytosis, not the endocytosis, of either Tf or a TfRMAb and that (ii) the TfR is expressed on both luminal and abluminal membranes of the brain capillary endothelial cell. In one study, a conjugate of the OX26 TfRMAb and 5 nm gold (Au) was infused in the carotid artery of rats, followed by saline clearance of the brain microvasculature, and perfusion fixation with glutaraldehyde [34]. The distribution of the TfRMAb at the light microscopic level was examined with silver staining, and the results are shown in Figure 8.2a and 8.2b. The TfRMAb–Au conjugate is shown sequestered within the capillary endothelium, and no TfRMAb–Au is visible in the postvascular compartment of the brain. This observation has been erroneously interpreted as evidence for the BBB TfR mediating only endocytosis, not the transcytosis, of the TfRMAb. However, the contrast between the abundance of the TfRMAb in the intra-endothelial volume versus the postvascular volume is due to dilution effects caused by vast differences in the size of these two volumes in the brain. The volume of the intra-endothelial compartment in the brain is about 1 μl/g brain, but the extravascular volume in the brain is 700 μl/g [15]. Therefore, the TfRMAb undergoes 3 log orders of dilution in transit from the intra-endothelial compartment to the postvascular compartment, and this 700-fold dilution prevents detection of antibody in the parenchyma of brain. Conversely, parenchymal antibody can be detected by the more sensitive emulsion autoradiography, as shown in Figure 8.2 f. [^{125}I]-rat holoTf was infused in the internal carotid artery of the rat for 5 min followed by saline clearance of the brain vasculature [28]. The brain was then removed, frozen, and cryostat sections were applied to emulsion, followed by exposure for 3–4 months. The emulsion autoradiography shows that the Tf rapidly distributes to the entire postvascular volume in the brain (Fig. 8.2f). Similar emulsion autoradiography results have been obtained for insulin in the rabbit brain [26], as well as for HIRMAb fusion proteins in the rhesus monkey brain [35]. The transcytosis of the TfRMAb through the brain capillary endothelium is shown with electron microscopy [34]. The TfRMAb is shown bound to the luminal membrane of the endothelium (Fig. 8.2c). Within the brain endothelial cell, the TfRMAb is observed packaged in 100 nm endosomes, which traverse the intra-endothelial compartment (Fig. 8.2d), followed by exocytosis into the brain interstitial

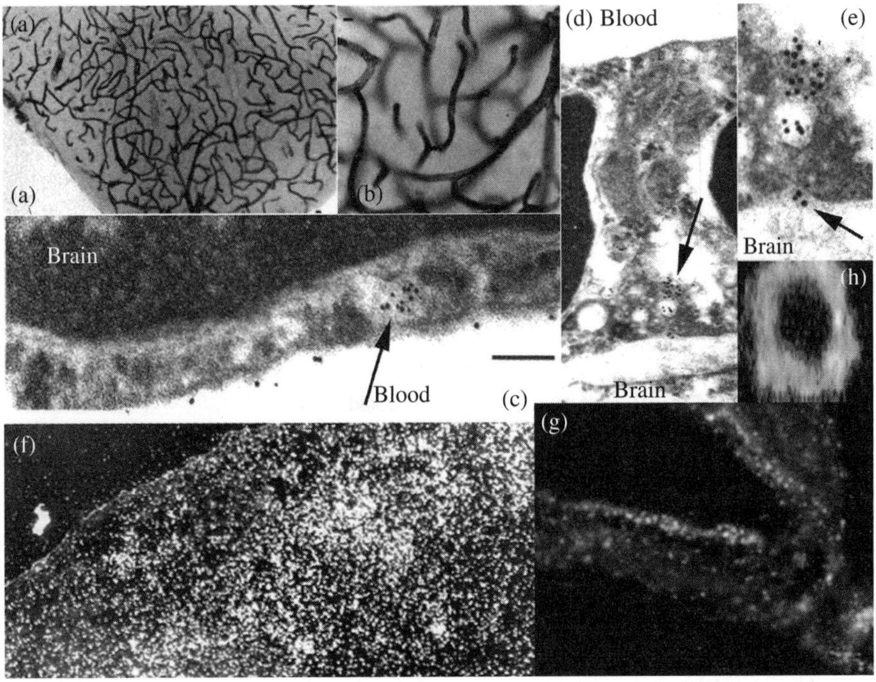

FIGURE 8.2 (a and b) Light microscopic silver staining of rat brain following a 10 min internal carotid artery infusion of a conjugate of 5 nm gold and the OX26 MAb against the rat TfR [34]. Prior to perfusion fixation of rat brain with 2% glutaraldehyde, the brain vasculature was cleared with a saline infusion. (c, d, and e) Electron microscopic examination of rat brain shown in (a) and (b); the OX26 MAb–gold conjugate is observed in the intra-endothelial compartments in 100 nm endosomes (arrows in c and d). The conjugate is observed exocytosed into the brain interstitium in (e) (arrow) [34]. (f) Darkfield light microscopy of rat brain following a 5 min internal carotid artery infusion of [^{125}I]-rat holo transferrin [28]. Prior to removal of the brain for freezing and frozen sectioning, the brain vasculature was cleared with a saline infusion. The autoradiography shows rapid distribution of the holo-transferrin into the entire postvascular compartment of the brain. (g and h) Confocal microscopy of freshly isolated rat brain capillaries showing labeling of the BBB TfR with the OX26 MAb against the rat TfR. In (g), the binding of the OX26 MAb was detected with a fluorescein conjugated secondary antibody; the TfRMAb is observed binding to the TfR on the abluminal membrane of the capillaries, as well as sequestered within intra-endothelial endosomes. In (g), the binding of the TfRMAb to the TfR on both the luminal and abluminal membranes of the brain capillary is demonstrated. Reprinted from Bickel et al. [34]. © 1994, SAGE Publications.

fluid (Fig. 8.2e). The expression of the TfR on both the luminal and the abluminal membranes of the brain capillary endothelium is demonstrated by confocal microscopy of unfixed isolated rat brain capillaries [36], as shown in Figure 8.2 g and h. The sequestration of the TfRMAb within intra-endothelial endosomes is also shown by confocal microscopy (Fig. 8.2 g). The abluminal TfR mediates the reverse transcytosis of Tf from the brain back to blood, which was experimentally demonstrated with the Brain Efflux Index method [37]. Following injection of either apo-Tf or holo-Tf into rat brain, the protein effluxes back to blood with a $T_{1/2}$ of 49±4 min and 170±15 min, respectively [37]. Conversely, 70 kDa dextran, which has no affinity for any BBB receptor, exits the brain with a $T_{1/2}$ of 17 h [37]. The rapid exodus of apo-Tf from the brain back to blood is mediated by reverse transcytosis across the BBB, owing to expression of the TfR on the abluminal membrane of the BBB.

BBB Receptor-Specific Monoclonal Antibodies

MAbs that target the endogenous insulin or TfRs on the BBB are species-specific. A mouse MAb against the rat TfR is transported across the BBB in the rat [30], but not the mouse [32]. A rat MAb against the mouse TfR is transported across the BBB in the mouse [32]. A mouse MAb against the HIR is transported across the BBB in an Old World primate, such as the Rhesus monkey [31], but does not recognize the insulin receptor in the mouse [33]. The genes encoding the variable region of the heavy chain (VH) and the variable region of the light chain (VL) of the murine MAb against the HIR were cloned and fused to genes encoding the constant region of human IgG1 heavy chain and human kappa light chain, respectively, to produce chimeric and humanized forms of the antibody, designated HIRMAb [38]. The genes encoding the VH and the VL of the rat MAb against the mouse TfR were cloned and fused to genes encoding the constant region of mouse IgG1 heavy chain and mouse kappa light chain, respectively, to produce a chimeric MAb against the mouse TfR, designated cTfRMAb [39]. The availability of the genes encoding the heavy chain (HC) and the light chain (LC) of HIRMAb and the cTfRMAb allowed for the genetic engineering of fusion genes encoding fusion proteins of these antibodies for preclinical testing in primates and mice, respectively.

Brain Uptake of IgG Fusion Proteins in the Rhesus Monkey and Mouse

The brain uptake of the HIRMAb-derived fusion proteins in the Rhesus monkey, or the brain uptake of cTfRMAb-derived fusion proteins in the mouse, is high and comparable to the brain uptake of lipid-soluble small molecules. The brain uptake of HIRMAb fusion proteins is 2–3% of injected dose (ID)/brain in the Rhesus monkey [40, 41], and this level of brain uptake is comparable to that of the lipid-soluble small molecule, fallypride [42]. The brain uptake of cTfRMAb fusion proteins is 2–4% ID/g in the mouse [43–45], and this level of brain uptake is comparable to the brain uptake of diazepam after IV injection, which is 2% ID/g in the mouse [46].

REENGINEERING RECOMBINANT PROTEIN THERAPEUTICS AS IGG FUSION PROTEINS

Recombinant protein therapeutics, such as lysosomal enzyme, neurotrophins, or decoy receptors, are large-molecule drugs that cannot enter CNS drug development, because these drugs do not cross the BBB. The BBB problem is best illustrated in the

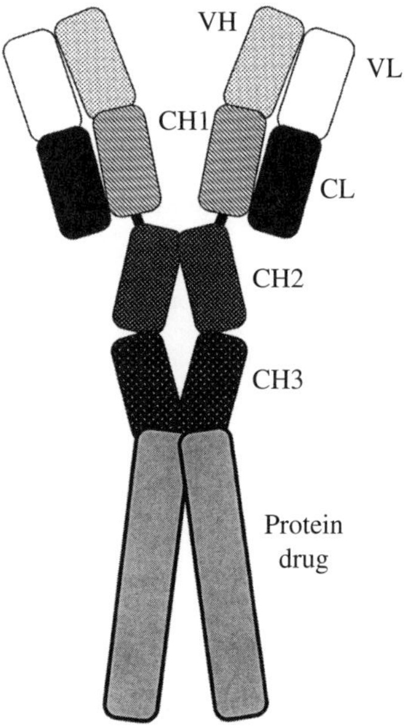

FIGURE 8.3 Structure of IgG fusion protein is shown, and is formed by fusion of a protein drug to the carboxyl terminus of each heavy chain of an MAb against an endogenous BBB receptor. The heavy chain is comprised of the following domains: variable region (VH), CH1, hinge, CH2, CH3, and protein drug. The light chain is comprised of the following domains: variable region (VL) and constant region (CL). The protein drug alone is not transportable across the BBB, and may be a therapeutic enzyme, neurotrophin, or decoy receptor.

case of the biologic tumor necrosis factor alpha (TNFα) inhibitors (TNFI), such as infliximab, adalimumab, or etanercept. The annual market for these three protein drugs exceeds $20 billion, but none of the biologic TNFIs is used to treat CNS disorders, because these drugs do not cross the BBB. Different classes of biologic therapeutics have been reengineered as IgG fusion proteins with the HIRMAb, and the general structure of the fusion protein is shown in Figure 8.3. The protein therapeutic is fused to the carboxyl terminus of the heavy chain of the HIRMAb. This approach has two advantages: (i) the amino terminus of the HIRMAb chains remain free to bind to the target insulin receptor on the BBB and (ii) the protein therapeutic is placed in a dimeric configuration, which replicates the native configuration of many of the protein drugs. The different classes of protein therapeutics that have been reengineered as HIRMAb fusion proteins are summarized in Table 8.1. With one exception, all of the HIRMAb fusion proteins listed in Table 8.1 retain the bifunctionality of the IgG fusion protein: (i) the fusion protein binds the HIR with the same high affinity (KD < 1 nM) as the original HIRMAb, and (ii) the fused protein therapeutic retains biologic activity comparable to the original protein drug prior to fusion. The exception to this rule was observed in the case of the lysosomal enzyme, β-glucuronidase (GUSB). Fusion of GUSB to the carboxyl terminus of the HC of the

TABLE 8.1 HIRMAb fusion proteins engineered for targeted delivery across the human BBB

Category	Protein therapeutic	References
Neurotrophin	Brain-derived neutrophic factor (BDNF)	[48]
	Glial-derived neurotrophic factor (GDNF)	[49]
	Erythropoietin (EPO)	[41]
Enzyme	α-L-iduronidase (IDUA)	[50]
	Iduronate-2-sulfatase (IDS)	[51]
	β-glucuronidase (GUSB)	[47]
	Arylsulfatase A (ASA)	[35]
	Paraoxonase (PON)-1	[52]
Decoy receptor	Tumor necrosis factor receptor (TNFR)	[40]
Monoclonal antibody	Anti-amyloid antibody (AAA)	[20]
Other	Avidin	[53]

HIRMAb caused a >95% reduction in enzyme activity [47]. In contrast, GUSB enzyme activity was retained following fusion of the enzyme to the amino terminus of the HC of the HIRMAb; however, in this case, the affinity of the HIRMAb domain of the fusion protein for the HIR was >95% decreased [47].

IgG FUSION PROTEINS FOR TARGETED BBB DELIVERY

Brain Delivery in the Rhesus Monkey

Lysosomal enzymes, such as iduronidase (IDUA), iduronate 2-sulfatase (IDS), or arylsulfatase A (ASA), have been reengineered as HIRMAb fusion proteins (Table 8.1). There are over 50 lysosomal enzyme storage disorders and about 75% affect the CNS. Enzyme replacement therapy (ERT) with the recombinant enzyme does not treat the brain, because the enzyme does not cross the BBB. The enzyme can be reengineered as an IgG–enzyme fusion protein with the HIRMAb for BBB drug delivery in humans. The effect of this reengineering on the brain uptake of the enzyme is illustrated in the case of IDS, which is the enzyme that is mutated in mucopolysaccharidosis (MPS) Type II, also called Hunters Syndrome. The IDS alone and the HIRMAb–IDS fusion protein were separately labeled with the [^{125}I]-Bolton–Hunter reagent, and injected IV into separate rhesus monkeys [54]. The brain was removed 2h after IV injection and coronal sections were prepared and placed on an X-ray film, which was exposed for 7 days. The brain scans for the monkey injected with the HIRMAb–IDS fusion protein is shown in Figure 8.4a, and the brain scan for the monkey injected with IDS alone is shown in Figure 8.4b. The brain scan of the fusion protein is comparable to a 2-deoxyglucose positron emission tomography (PET) scan, whereas the brain scan of the IDS alone represents background radioactivity. The actual brain uptake of the HIRMAb–IDS fusion protein in the monkey is

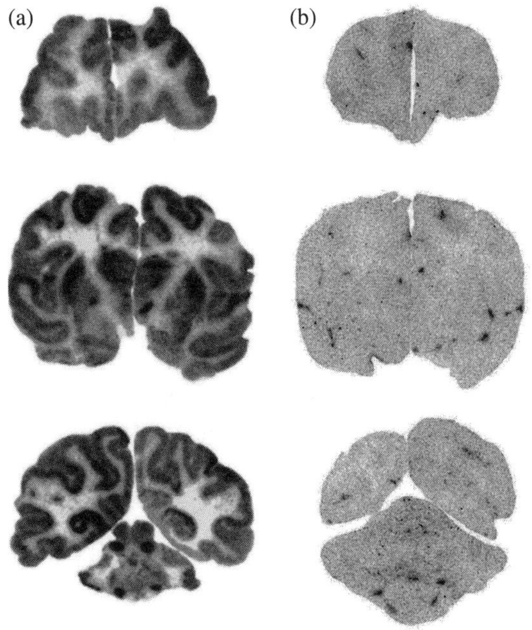

FIGURE 8.4 Film autoradiogram of 20 μm sections of rhesus monkey brain removed 120 min after IV injection of the HIRMAb–IDS fusion protein (a) or IDS alone (b). The forebrain section is on the top, the midbrain section is in the middle, and the hindbrain section with cerebellum is at the bottom. The proteins were separately radiolabeled with the [^{125}I]-Bolton–Hunter reagent. Reprinted with permission from Ref. [54]. © 2013, American Chemical Society.

1.04 ± 0.07% ID/brain, whereas the brain uptake of the IDS is 0.030 ± 0.004% ID/brain. This level of brain uptake of the IDS is effectively zero, as the brain volume of distribution (VD) of the IDS, 9 ± 1 μl/g, is equal to the brain plasma volume [54]. That is, whatever IDS is found in the brain is sequestered within the plasma compartment and has not traversed the BBB. The brain scans in Figure 8.4 illustrate how a recombinant protein, IDS, which does not cross the BBB, can be reengineered as an IgG–IDS fusion protein that can penetrate the BBB.

PHARMACOLOGIC EFFECTS IN MOUSE MODELS OF NEURAL DISEASE

Hurler Mouse Model

The lysosomal enzyme mutated in MPS Type I, Hurler's syndrome, is IDUA. To enable drug testing in the Hurler mouse, a fusion protein of IDUA and the mouse-active cTfRMAb was engineered, and designated the cTfRMAb–IDUA fusion protein [55]. The cTfRMAb–IDUA fusion protein bound the mouse TfR with high affinity, KD = 0.67 ± 0.17 nM, and retained IDUA enzyme activity comparable to recombinant IDUA [55]. Treatment of mouse cells with the cTfRMAb–IDUA fusion protein in culture caused a large increase in intracellular IDUA enzyme activity, which decayed with a $T_{1/2}$ of 2.8 days [55]. Treatment of Hurler mice with the cTfRMAb–IDUA fusion protein increased the organ IDUA enzyme activity to therapeutic levels. Hurler mice were treated with 1 mg/kg of the cTfRMAb–IDUA fusion protein by IV twice-weekly for 8 weeks. Treatment reduced glycosoamino-glycans (GAGs) in peripheral organs, and reduced lysosomal inclusion bodies in the brain of the Hurler mice by 73% [55]. The 6-month-old mice treated in this study were old for a Hurler mouse [55], which means the accumulation of lysosomal inclusion bodies in the brain is reversible, with adequate treatment, even in an older mouse.

Mouse Model of Experimental Parkinson's Disease

PD is a neurodegenerative disease characterized by the dual effects of neural loss and chronic neuroinflammation. Therefore, treatment of PD might be directed at both neuroprotection, with BBB-penetrating neurotrophins, and inflammation blockers, with BBB-penetrating cytokine decoy receptors. A model neurotrophin for PD is glial-derived neurotrophic factor (GDNF), and a model decoy receptor is the extracellular domain (ECD) of the type II TNFα receptor (TNFR). An Fc fusion protein of the TNFR-II ECD is etanercept. Neither GDNF nor etanercept can be developed as a treatment for neurodegenerative diseases, such as PD, because neither GDNF [56] nor etanercept [40] crosses the BBB. BBB-penetrating forms of GDNF and the TNFR ECD were engineered in the form of HIRMAb–GDNF [49] and HIRMAb–TNFR fusion proteins [40]. However, these fusion proteins could not be tested in mouse models of PD, because the HIRMAb does not recognize the mouse insulin receptor [33]. Therefore, a mouse-specific fusion protein of the cTfRMAb and GDNF [43] and a mouse-specific fusion protein of the cTfRMAb and the type II TNFR ECD [57] were genetically engineered. Treatment of a mouse model of PD with the cTfRMAb–GDNF fusion protein caused a 272% increase in striatal tyrosine hydroxylase (TH) enzyme activity, which was correlated with improvement in three models of neuro-behavior [58]. In a separate study, treatment of mice with experimental PD with the cTfRMAb–TNFR fusion protein caused a 130% increase in striatal TH enzyme activity, which was also correlated with improvement in three models of neuro-behavior [59]. Conversely, treatment of the PD mice with etanercept had no therapeutic effect in PD [59], because etanercept does not cross the BBB [40].

Mouse Model of Experimental Stroke

Neurons do not die immediately following an acute ischemic stroke, but take up to 5 h to die following the acute event. During this 5 h window, the neural apoptotic death cycle can be halted by neuroprotective agents, such as neurotrophins, or inflammation blockers, such as the biologic TNFIs. However, the BBB is intact in acute stroke during the period when neuroprotection is still possible [7, 8]. Therefore, neuroprotective neurotrophins, such as GDNF, or biologic TNFIs, such as the TNFR, must be reengineered for BBB penetration. A delayed single IV administration of the cTfRMAb–TNFR fusion protein causes a 45% decrease in hemispheric stroke volume, which was correlated with an improvement in neural deficit, in a reversible middle cerebral artery occlusion (MCAO) stroke model in mice [60]. A delayed single IV injection of the cTfRMAb–GDNF fusion protein caused a 30% decrease in cortical stroke volume in the reversible MCAO model in mice, and the reduction in stroke volume was increased to 69% by the combined IV treatment with both the cTfRMAb–GDNF and cTfRMAb–TNFR fusion proteins [61]. Conversely, the IV administration of etanercept or GDNF alone had no therapeutic effect in acute stroke [60, 61]. The lack of therapeutic effects of the GDNF or etanercept alone is expected, because (i) etanercept [40] and GDNF [56] alone do not cross the BBB and (ii) the BBB is intact in the early hours after stroke when neuroprotection is still possible [7, 8].

REENGINEERING THERAPEUTIC ANTIBODIES AS BBB-PENETRATING BISPECIFIC ANTIBODIES

When the therapeutic protein is an MAb and the BBB MTH is an MAb, then the problem is the engineering of a bispecific antibody (BSA). There are multiple approaches to the genetic engineering of a BSA [62]. The goal is the retention of high affinity binding of both MAb domains of the BSA, that is, retention of high affinity binding for both the BBB receptor/transporter and high affinity binding for the target antigen in the brain. This is possible by fusion of a single chain Fv (ScFv) antibody to the carboxyl terminus of the heavy chain of a second antibody, which results in the engineering of a tetravalent BSA, as shown in Figure 8.5.

REENGINEERING AN ANTI-AMYLOID ANTIBODY AS A BBB-PENETRATING BISPECIFIC ANTIBODY

The model therapeutic MAb is an AAA directed against the amino terminus of the Abeta amyloid peptide of AD [20]. The AAA was reengineered as an ScFv antibody, and the ScFv was fused to the carboxyl terminus of each HC of the HIRMAb, for drug development in humans [20], or to the carboxyl terminus of each HC of the mouse-active cTfRMAb, for drug testing in AD mouse models [44]. The HIRMAb fusion protein is designated the HIRMAb–ScFv fusion protein [20], and the cTfRMAb fusion protein is designated the cTfRMAb–ScFv fusion protein [44]. The HIRMAb–ScFv fusion protein could not be tested in a mouse model of AD, because the HIRMAb domain does not recognize the murine insulin receptor [33]. Therefore, AD transgenic mice were chronically treated with the mouse-active cTfRMAb–ScFv fusion protein [63]. The conventional approach to treatment of AD mouse models with an AAA is called passive immune therapy. The AAA is

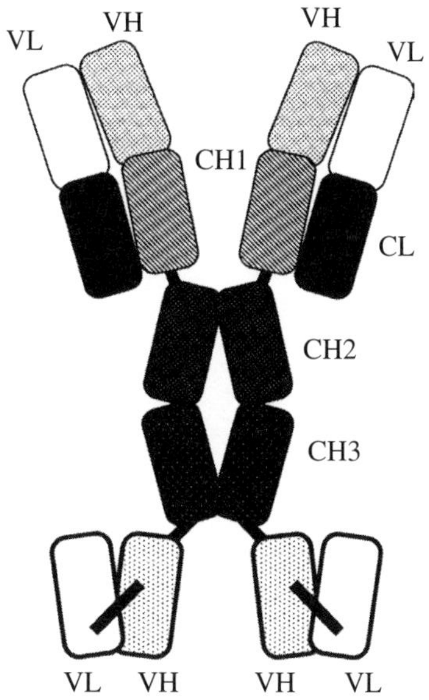

FIGURE 8.5 Structure of a tetravalent bispecific antibody (BSA) formed by fusion of a single chain Fv (ScFv) antibody to the carboxyl terminus of each heavy chain of a second MAb. One antibody domain of the BSA targets an endogenous receptor on the BBB to mediate delivery to the brain, and another antibody domain of the BSA targets an antigen in the brain behind the BBB to mediate the therapeutic effect of the BSA. The heavy chain is comprised of the following domains: variable region (VH), CH1, hinge, CH2, CH3, VH of therapeutic antibody, linker, and VL of therapeutic antibody. The light chain is comprised of the following domains: variable region (VL) and constant region (CL).

administered chronically to AD mice by systemic injection, such as IV or intraperitoneal (IP) administration.The conventional AAAs developed for passive immune therapy of AD have no receptor specificity, do not cross the BBB, and have plasma $T_{1/2}$s in humans of 2–3 weeks [63]. Consequently, the conventional AAA is administered monthly to AD patients (Table 8.2). Owing to the long plasma $T_{1/2}$, the plasma concentration of Abeta increases manyfold with the administration of conventional AAAs [64]. Such elevations of plasma Abeta are associated with BBB disruption and cerebral micro-hemorrhage [65, 66], and a side effect of passive immune therapy of AD with conventional AAAs is vasogenic brain edema [67]. Conventional AAAs do not cross the nondisrupted BBB [20], which is problematic for the treatment of AD, since the amyloid plaque resides in the brain behind the BBB. A pivotal phase III clinical trial of the bapineuzumab AAA in AD failed [22]. This clinical trial failure suggests that alternative strategies should be developed for the delivery of an AAA across the BBB. The goal is the production of an AAA that penetrates the BBB in the absence of BBB disruption or cerebral micro-hemorrhage. This is possible following the reengineering of the AAA as a BBB-penetrating BSA (Fig. 8.5).

TABLE 8.2 Comparison of passive immune therapy and receptor-mediated immune therapy of alzheimer's disease[a]

Parameter	Passive immune therapy	Receptor-mediated immune therapy
Administration	Intravenous, monthly	Subcutaneous, daily
Antibody plasma $T_{1/2}$	2–3 weeks	2–3 h
Brain plaque disaggregation	Yes	Yes
Elevation of plasma Aβ	Yes	No
Cerebral micro-hemorrhage	Yes	No
Penetration of the BBB in absence of BBB disruption	No	Yes

[a] From Ref. 63.

PHARMACOLOGIC EFFECTS IN A MOUSE MODEL OF ALZHEIMER'S DISEASE

The cTfRMAb–ScFv fusion protein was engineered for testing in mouse models of AD [44]. The cTfRMAb–ScFv fusion protein has specificity for the TfR, and is rapidly cleared from plasma in the mouse with a plasma $T_{1/2}$ of 2–3 h [44]. Owing to the pharmacokinetic profile of the cTfRMAb fusion proteins in the mouse, chronic treatment should be administered on a daily basis to insure sustained delivery of the fusion protein to the brain. Fusion proteins have acceptable bioavailability in the mouse following SQ administration [68]. Therefore, aged (12 months of age) double transgenic AD mice were treated daily with SQ injections of 5 mg/kg of the cTfRMAb–ScFv fusion protein for 12 weeks [63]. At the end of the treatment, brain amyloid plaques were quantified with confocal microscopy using both thioflavin-S staining and immunostaining with the 6E10 antibody against Abeta amyloid fibrils. Fusion protein treatment caused a 57 and 61% reduction in amyloid plaque in the cortex and hippocampus, respectively. No increase in plasma immunoreactive Abeta amyloid peptide, and no cerebral micro-hemorrhage, was observed [63]. Chronic daily SQ treatment of the mice with the fusion protein caused no immune reactions and only a low titer antidrug antibody response [63]. This study shows that reengineering AAAs for receptor-mediated BBB transport allows for reduction in brain amyloid plaque without cerebral micro-hemorrhage following daily SQ treatment for 12 weeks.

Passive immune therapy with a conventional AAA is compared to receptor-mediated immune therapy with the cTfRMAb–ScFv fusion protein in Table 8.2. Both conventional AAAs and receptor-mediated BSAs, such as the cTfRMAb–ScFv fusion protein, cause a reduction in brain amyloid plaque [63]. However, chronic treatment with the receptor-mediated BSA causes no elevation in plasma Abeta peptide, and no cerebral micro-hemorrhage (Table 8.2). Therefore, receptor-mediated BSAs for AD may have more favorable therapeutic indices than conventional AAA therapeutics.

SAFETY PHARMACOLOGY OF MOLECULAR TROJAN HORSE FUSION PROTEINS IN MICE AND PRIMATES

Safety Pharmacology of BBB-Penetrating IgG Fusion Proteins in Mice

The safety of MAbs against the TfR was recently questioned following the observation of acute clinical findings in mice administered a single IV injection of an unusual form of TfRMAb [69]. The TfRMAb was comprised of a variable region that had been subjected to site-directed mutagenesis to lower the affinity of the antibody to the mouse TfR, and a constant region derived from human IgG of unknown isotype [70]. A single IV treatment with this antibody, designated anti-TfR^D, resulted in hemoglobinuria and a depressed reticulocyte count in mice [69]. These findings are not corroborated by any mouse models of chronic treatment with cTfRMAb-derived fusion proteins, including chronic treatment of a mouse PD model with the cTfRMAb–GDNF fusion protein [58], chronic treatment of a mouse PD model with the cTfRMAb–TNFR fusion protein [59], or chronic treatment of a mouse MPSI model with the cTfRMAb–IDUA fusion protein [55]. In a model of experimental AD, the mice were treated daily with the cTfRMAb–ScFv fusion protein for 12 consecutive weeks with no evidence of toxicity [63]. In a formal toxicity evaluation, mice were treated with twice-weekly IV injections of the cTfRMAb–GDNF fusion protein for 12 consecutive weeks, and tissue histology was examined in parallel with a clinical chemistry study of 23 different blood tests, including serum iron and total iron-binding capacity [71]. No evidence of toxicity was observed in any parameter [71], validating the safety of chronic treatment of mice with cTfRMAb fusion proteins.

Safety Pharmacology of BBB-Penetrating IgG Fusion Proteins in Primates

The safety of a single, or multiple, IV injections of the HIRMAb–GDNF fusion protein in Rhesus monkeys was evaluated in a Good Laboratory Practice (GLP) investigation of cardiac function, pulmonary function, and behavior, and no findings of toxicity were observed [72]. The HIRMAb–IDUA fusion protein was administered by weekly IV infusion to Rhesus monkeys for 26 consecutive weeks at doses of 0, 3, 9, and 30 mg/kg [73]. The only safety issue observed was hypoglycemia following rapid IV infusion of the HIRMAb fusion protein in saline at the high dose of 30 mg/kg, which is more than 10 times higher than any projected therapeutic dose. Nevertheless, the hypoglycemia at 30 mg/kg was eliminated by the inclusion of dextrose in the infusion vehicle [74]. An IV glucose tolerance test performed at the end of 26 weeks of treatment showed no change in glycemic control in the primates at any dose of the HIRMAb–IDUA fusion protein [74]. BBB-penetrating IgG fusion proteins targeting either the TfR or the insulin receptor have now been administered chronically to over 100 mice and 100 monkeys, respectively, with no indication of safety issues.

ANTIDRUG ANTIBODY RESPONSE TO IgG FUSION PROTEINS

Antidrug Antibody in Mice

The antidrug antibody (ADA) response has been measured in mice treated chronically with the cTfRMAb–GDNF fusion protein [71], the cTfRMAb–EPO fusion protein [75], the cTfRMAb–IDUA fusion protein [55], or the cTfRMAb–TNFR fusion protein [59]. The immune titer (OD/μl) was low, about 1.0, in all studies. The ADA was primarily formed against the variable region of the cTfRMAb domain of the fusion protein. The rate of plasma clearance and the rate of brain uptake of the cTfRMAb–GDNF fusion protein was measured at the end of 12 weeks of twice-weekly chronic treatment, and no changes were observed in either plasma clearance or brain uptake of the fusion protein [71]. Therefore, the low-titer ADAs formed in mice are non-neutralizing and do not interfere with fusion protein binding to the TfR.

Antidrug Antibody in Monkeys

The ADA titer was measured in monkeys treated weekly with IV infusions of the HIRMAb–IDUA fusion protein for 26 weeks [73]. The ADA was directed against both the variable region and the IDUA domain of the HIRMAb–IDUA fusion protein. The ADA titer (OD/μl) in monkeys, about 10, was 10-fold higher than the ADA titer observed in mice [71]. The exaggerated immune response to human proteins in monkeys is known, and is not indicative of potential immune responses in humans [76–78]. The plasma clearance of the HIRMAb–IDUA fusion protein and the plasma IDUA enzyme activity were measured in the monkeys at the end of 26 weeks of chronic weekly IV infusions. The pharmacokinetic parameters of plasma clearance of the fusion protein and plasma IDUA enzyme activity were identical at the start and at the end of the 26 weeks of treatment [73]. Therefore, the ADA formed against the HIRMAb–IDUA fusion protein had no effect on fusion protein binding to the insulin receptor or on IDUA enzyme activity of the fusion protein.

SUMMARY

The biotechnology industry has been attempting to develop recombinant proteins and monoclonal antibodies for neural disorders for over 20 years (Fig. 8.1) but, to date, there is no biopharmaceutical that is FDA-approved for a CNS disease. This failure rate is actually expected, given the following considerations:

- The products of biotechnology are large-molecule drugs that do not cross the BBB.
- The invasive drug delivery strategies, such as intrathecal (IT) drug delivery [9] or convection-enhanced diffusion (CED) [11], are not effective systems for bypassing the BBB; the IT or CED routes do not effectively deliver drug to the parenchyma of brain, as demonstrated in preclinical investigations [10, 12].
- The CNS drug development pathway proceeded in the absence of any parallel effort in BBB drug delivery technology.

BBB drug delivery technology arises from an understanding of the cell biology of BBB transport processes. Targeting endogenous receptor-mediated transport systems within the BBB enables the development of platform technologies, such as BBB MTHs. The engineering of multifunctional IgG fusion proteins, such as those depicted in Figures 8.3 and 8.5, enables the reengineering of protein or antibody therapeutics for the brain as BBB-penetrating neuropharmaceuticals.

REFERENCES

[1] The BDNF Study Group (1999) A controlled trial of recombinant methionyl human BDNF in ALS: (Phase III), *Neurology 52*, 1427–1433.

[2] Miller, R. G., Petajan, J. H., Bryan, W. W., Armon, C., Barohn, R. J., Goodpasture, J. C., Hoagland, R. J., Parry, G. J., Ross, M. A., Stromatt, S. C. (1996) A double-blind placebo-controlled clinical trial of subcutaneous recombinant human ciliary neurotrophic factor (rHCNTF) in amyotrophic lateral sclerosis. ALS CNTF Treatment Study Group, *Neurology 46*, 1244–1249.

[3] Poduslo, J. F. and Curran, G. L. (1996) Permeability at the blood-brain and blood-nerve barriers of the neurotrophic factors: NGF, CNTF, NT-3, BDNF, *Brain research. Molecular brain research 36*, 280–286.

[4] Garbuzova-Davis, S., Haller, E., Saporta, S., Kolomey, I., Nicosia, S. V., and Sanberg, P. R. (2007) Ultrastructure of blood-brain barrier and blood-spinal cord barrier in SOD1 mice modeling ALS, *Brain research 1157*, 126–137.

[5] Bogousslavsky, J., Victor, S. J., Salinas, E. O., Pallay, A., Donnan, G. A., Fieschi, C., Kaste, M., Orgogozo, J. M., Chamorro, A., Desmet, A., and European-Australian Fiblast in Acute Stroke Group. (2002) Fiblast (trafermin) in acute stroke: results of the European-Australian phase II/III safety and efficacy trial, *Cerebrovascular diseases 14*, 239–251.

[6] Fisher, M., Meadows, M. E., Do, T., Weise, J., Trubetskoy, V., Charette, M., and Finklestein, S. P. (1995) Delayed treatment with intravenous basic fibroblast growth factor reduces infarct size following permanent focal cerebral ischemia in rats, *Journal of cerebral blood flow and metabolism 15*, 953–959.

[7] Latour, L. L., Kang, D. W., Ezzeddine, M. A., Chalela, J. A., and Warach, S. (2004) Early blood-brain barrier disruption in human focal brain ischemia, *Annals of neurology 56*, 468–477.

[8] Belayev, L., Busto, R., Zhao, W., and Ginsberg, M. D. (1996) Quantitative evaluation of blood-brain barrier permeability following middle cerebral artery occlusion in rats, *Brain research 739*, 88–96.

[9] Nutt, J. G., Burchiel, K. J., Comella, C. L., Jankovic, J., Lang, A. E., Laws, E. R., Jr., Lozano, A. M., Penn, R. D., Simpson, R. K., Jr., Stacy, M., Wooten, G. F. (2003) Randomized, double-blind trial of glial cell line-derived neurotrophic factor (GDNF) in PD, *Neurology 60*, 69–73.

[10] Yan, Q., Matheson, C., Lopez, O. T., and Miller, J. A. (1994) The biological responses of axotomized adult motoneurons to brain-derived neurotrophic factor, *The journal of neuroscience 14*, 5281–5291.

[11] Lang, A. E., Gill, S., Patel, N. K., Lozano, A., Nutt, J. G., Penn, R., Brooks, D. J., Hotton, G., Moro, E., Heywood, P., Brodsky, M. A., Burchiel, K., Kelly, P., Dalvi, A., Scott, B., Stacy, M., Turner, D., Wooten, V. G., Elias, W. J., Laws, E. R., Dhawan, V., Stoessl, A. J., Matcham, J., Coffey, R. J., and Traub, M. (2006) Randomized controlled trial of intraputamenal glial cell line-derived neurotrophic factor infusion in Parkinson disease, *Annals of neurology 59*, 459–466.

[12] Salvatore, M. F., Ai, Y., Fischer, B., Zhang, A. M., Grondin, R. C., Zhang, Z., Gerhardt, G. A., and Gash, D. M. (2006) Point source concentration of GDNF may explain failure of phase II clinical trial, *Experimental neurology 202*, 497–505.

[13] Ehrenreich, H., Weissenborn, K., Prange, H., Schneider, D., Weimar, C., Wartenberg, K., Schellinger, P. D., Bohn, M., Becker, H., Wegrzyn, M., Jahnig, P., Herrmann, M., Knauth, M., Bahr, M., Heide, W., Wagner, A., Schwab, S., Reichmann, H., Schwendemann, G., Dengler, R., Kastrup, A., Bartels, C., and EPO Stroke Trial Group (2009) Recombinant human erythropoietin in the treatment of acute ischemic stroke, *Stroke 40*, e647–e656.

[14] Reiber, H. (2003) Proteins in cerebrospinal fluid and blood: barriers, CSF flow rate and source-related dynamics, *Restorative neurology and neuroscience 21*, 79–96.

[15] Pardridge, W. M. (2012) Drug transport across the blood-brain barrier, *Journal of cerebral blood flow and metabolism 32*, 1959–1972.

[16] Banks, W. A., Jumbe, N. L., Farrell, C. L., Niehoff, M. L., and Heatherington, A. C. (2004) Passage of erythropoietic agents across the blood-brain barrier: a comparison of human and murine erythropoietin and the analog darbepoetin alfa, *European journal of pharmacology 505*, 93–101.

[17] Naslund, J., Haroutunian, V., Mohs, R., Davis, K. L., Davies, P., Greengard, P., and Buxbaum, J. D. (2000) Correlation between elevated levels of amyloid beta-peptide in the brain and cognitive decline, *JAMA 283*, 1571–1577.

[18] Glenner, G. G. and Wong, C. W. (1984) Alzheimer's disease: initial report of the purification and characterization of a novel cerebrovascular amyloid protein, *Biochemical and biophysical research communications 120*, 885–890.

[19] Masters, C. L., Multhaup, G., Simms, G., Pottgiesser, J., Martins, R. N., and Beyreuther, K. (1985) Neuronal origin of a cerebral amyloid: neurofibrillary tangles of Alzheimer's disease contain the same protein as the amyloid of plaque cores and blood vessels, *The EMBO journal 4*, 2757–2763.

[20] Boado, R. J., Zhang, Y., Zhang, Y., Xia, C. F., and Pardridge, W. M. (2007) Fusion antibody for Alzheimer's disease with bidirectional transport across the blood-brain barrier and abeta fibril disaggregation, *Bioconjugate chemistry 18*, 447–455.

[21] Schlageter, N. L., Carson, R. E., and Rapoport, S. I. (1987) Examination of blood-brain barrier permeability in dementia of the Alzheimer type with [68Ga]EDTA and positron emission tomography, *Journal of cerebral blood flow and metabolism 7*, 1–8.

[22] Vellas, B., Carrillo, M. C., Sampaio, C., Brashear, H. R., Siemers, E., Hampel, H., Schneider, L. S., Weiner, M., Doody, R., Khachaturian, Z., Cedarbaum, J., Grundman, M., Broich, K., Giacobini, E., Dubois, B., Sperling, R., Wilcock, G. K., Fox, N., Scheltens, P., Touchon, J., Hendrix, S., Andrieu, S., Aisen, P., and EU/US/CTAD Task Force Members. (2013) Designing drug trials for Alzheimer's disease: what we have learned from the release of the phase III antibody trials: a report from the EU/US/CTAD Task Force, *Alzheimer's & dementia 9*, 438–444.

[23] Boado, R. J., Li, J. Y., Nagaya, M., Zhang, C., and Pardridge, W. M. (1999) Selective expression of the large neutral amino acid transporter at the blood-brain barrier, *Proceedings of the National Academy of Sciences U.S.A. 96*, 12079–12084.

[24] Kageyama, T., Nakamura, M., Matsuo, A., Yamasaki, Y., Takakura, Y., Hashida, M., Kanai, Y., Naito, M., Tsuruo, T., Minato, N., and Shimohama, S. (2000) The 4F2hc/LAT1 complex transports L-DOPA across the blood-brain barrier, *Brain research 879*, 115–121.

[25] Pardridge, W. M., Eisenberg, J., and Yang, J. (1985) Human blood-brain barrier insulin receptor, *Journal of neurochemistry 44*, 1771–1778.

[26] Duffy, K. R. and Pardridge, W. M. (1987) Blood-brain barrier transcytosis of insulin in developing rabbits, *Brain research 420*, 32–38.

[27] Pardridge, W. M., Eisenberg, J., and Yang, J. (1987) Human blood-brain barrier transferrin receptor, *Metabolism 36*, 892–895.

[28] Skarlatos, S., Yoshikawa, T., and Pardridge, W. M. (1995) Transport of [125I]transferrin through the rat blood-brain barrier, *Brain research 683*, 164–171.

[29] Pardridge, W. M. (1986) Receptor-mediated peptide transport through the blood-brain barrier, *Endocrine reviews 7*, 314–330.

[30] Pardridge, W. M., Buciak, J. L., and Friden, P. M. (1991) Selective transport of an anti-transferrin receptor antibody through the blood-brain barrier in vivo, *The journal of pharmacology and experimental therapeutics 259*, 66–70.

[31] Pardridge, W. M., Kang, Y. S., Buciak, J. L., and Yang, J. (1995) Human insulin receptor monoclonal antibody undergoes high affinity binding to human brain capillaries in vitro and rapid transcytosis through the blood-brain barrier in vivo in the primate, *Pharmaceutical research 12*, 807–816.

[32] Lee, H. J., Engelhardt, B., Lesley, J., Bickel, U., and Pardridge, W. M. (2000) Targeting rat anti-mouse transferrin receptor monoclonal antibodies through blood-brain barrier in mouse, *The journal of pharmacology and experimental therapeutics 292*, 1048–1052.

[33] Zhou, Q. H., Boado, R. J., and Pardridge, W. M. (2012) Selective plasma pharmacokinetics and brain uptake in the mouse of enzyme fusion proteins derived from species-specific receptor-targeted antibodies, *Journal of drug targeting 20*, 715–719.

[34] Bickel, U., Kang, Y. S., Yoshikawa, T., and Pardridge, W. M. (1994) In vivo demonstration of subcellular localization of anti-transferrin receptor monoclonal antibody-colloidal gold conjugate in brain capillary endothelium, *The journal of histochemistry and cytochemistry 42*, 1493–1497.

[35] Boado, R. J., Lu, J. Z., Hui, E. K., Sumbria, R. K., and Pardridge, W. M. (2013) Pharmacokinetics and brain uptake in the rhesus monkey of a fusion protein of arylsulfatase A and a monoclonal antibody against the human insulin receptor, *Biotechnology and bioengineering 110*, 1456–1465.

[36] Huwyler, J. and Pardridge, W. M. (1998) Examination of blood-brain barrier transferrin receptor by confocal fluorescent microscopy of unfixed isolated rat brain capillaries, *Journal of neurochemistry 70*, 883–886.

[37] Zhang, Y. and Pardridge, W. M. (2001) Rapid transferrin efflux from brain to blood across the blood-brain barrier, *Journal of neurochemistry 76*, 1597–1600.

[38] Boado, R. J., Zhang, Y., Zhang, Y., and Pardridge, W. M. (2007) Humanization of anti-human insulin receptor antibody for drug targeting across the human blood-brain barrier, *Biotechnology and bioengineering 96*, 381–391.

[39] Boado, R. J., Zhang, Y., Wang, Y., and Pardridge, W. M. (2009) Engineering and expression of a chimeric transferrin receptor monoclonal antibody for blood-brain barrier delivery in the mouse, *Biotechnology and bioengineering 102*, 1251–1258.

[40] Boado, R. J., Hui, E. K., Lu, J. Z., Zhou, Q. H., and Pardridge, W. M. (2010) Selective targeting of a TNFR decoy receptor pharmaceutical to the primate brain as a receptor-specific IgG fusion protein, *Journal of biotechnology 146*, 84–91.

[41] Boado, R. J., Hui, E. K., Lu, J. Z., and Pardridge, W. M. (2010) Drug targeting of erythropoietin across the primate blood-brain barrier with an IgG molecular Trojan horse, *The journal of pharmacology and experimental therapeutics 333*, 961–969.

[42] Christian, B. T., Vandehey, N. T., Fox, A. S., Murali, D., Oakes, T. R., Converse, A. K., Nickles, R. J., Shelton, S. E., Davidson, R. J., and Kalin, N. H. (2009) The distribution of D2/D3 receptor binding in the adolescent rhesus monkey using small animal PET imaging, *NeuroImage 44*, 1334–1344.

[43] Zhou, Q. H., Boado, R. J., Lu, J. Z., Hui, E. K., and Pardridge, W. M. (2010) Monoclonal antibody-glial-derived neurotrophic factor fusion protein penetrates the blood-brain barrier in the mouse, *Drug metabolism and disposition 38*, 566–572.

[44] Boado, R. J., Zhou, Q. H., Lu, J. Z., Hui, E. K., and Pardridge, W. M. (2010) Pharmacokinetics and brain uptake of a genetically engineered bifunctional fusion antibody targeting the mouse transferrin receptor, *Molecular pharmaceutics 7*, 237–244.

[45] Zhou, Q. H., Boado, R. J., Lu, J. Z., Hui, E. K., and Pardridge, W. M. (2010) Re-engineering erythropoietin as an IgG fusion protein that penetrates the blood-brain barrier in the mouse, *Molecular pharmaceutics 7*, 2148–2155.

[46] Greenblatt, D. J. and Sethy, V. H. (1990) Benzodiazepine concentrations in brain directly reflect receptor occupancy: studies of diazepam, lorazepam, and oxazepam, *Psychopharmacology 102*, 373–378.

[47] Boado, R. J. and Pardridge, W. M. (2010) Genetic engineering of IgG-glucuronidase fusion proteins, *Journal of drug targeting 18*, 205–211.

[48] Boado, R. J., Zhang, Y., Zhang, Y., and Pardridge, W. M. (2007) Genetic engineering, expression, and activity of a fusion protein of a human neurotrophin and a molecular Trojan horse for delivery across the human blood-brain barrier, *Biotechnology and bioengineering 97*, 1376–1386.

[49] Boado, R. J., Zhang, Y., Zhang, Y., Wang, Y., and Pardridge, W. M. (2008) GDNF fusion protein for targeted-drug delivery across the human blood-brain barrier, *Biotechnology and bioengineering 100*, 387–396.

[50] Boado, R. J., Zhang, Y., Zhang, Y., Xia, C. F., Wang, Y., and Pardridge, W. M. (2008) Genetic engineering of a lysosomal enzyme fusion protein for targeted delivery across the human blood-brain barrier, *Biotechnology and bioengineering 99*, 475–484.

[51] Lu, J. Z., Boado, R. J., Hui, E. K., Zhou, Q. H., and Pardridge, W. M. (2011) Expression in CHO cells and pharmacokinetics and brain uptake in the Rhesus monkey of an IgG-iduronate-2-sulfatase fusion protein, *Biotechnology and bioengineering 108*, 1954–1964.

[52] Boado, R. J., Hui, E. K., Lu, J. Z., and Pardridge, W. M. (2011) CHO cell expression, long-term stability, and primate pharmacokinetics and brain uptake of an IgG-paroxonase-1 fusion protein, *Biotechnology and bioengineering 108*, 186–196.

[53] Boado, R. J., Zhang, Y., Zhang, Y., Xia, C. F., Wang, Y., and Pardridge, W. M. (2008) Genetic engineering, expression, and activity of a chimeric monoclonal antibody-avidin

fusion protein for receptor-mediated delivery of biotinylated drugs in humans, *Bioconjugate chemistry 19*, 731–739.

[54] Boado, R. J., Hui, E. K.-W., Lu, J. Z., Sumbria, R. K., and Pardridge, W. M. (2013) Blood-brain barrier molecular Trojan horse enables brain imaging of radioiodinated recombinant protein in the Rhesus monkey. *Bioconjugate chemistry 24*, 1741–1749.

[55] Boado, R. J., Hui, E. K., Lu, J. Z., Zhou, Q. H., and Pardridge, W. M. (2011) Reversal of lysosomal storage in brain of adult MPS-I mice with intravenous Trojan horse-iduronidase fusion protein, *Molecular pharmaceutics 8*, 1342–1350.

[56] Boado, R. J. and Pardridge, W. M. (2009) Comparison of blood-brain barrier transport of glial-derived neurotrophic factor (GDNF) and an IgG-GDNF fusion protein in the rhesus monkey, *Drug metabolism and disposition 37*, 2299–2304.

[57] Zhou, Q. H., Boado, R. J., Hui, E. K., Lu, J. Z., and Pardridge, W. M. (2011) Brain-penetrating tumor necrosis factor decoy receptor in the mouse, *Drug metabolism and disposition 39*, 71–76.

[58] Fu, A., Zhou, Q. H., Hui, E. K., Lu, J. Z., Boado, R. J., and Pardridge, W. M. (2010) Intravenous treatment of experimental Parkinson's disease in the mouse with an IgG-GDNF fusion protein that penetrates the blood-brain barrier, *Brain research 1352*, 208–213.

[59] Zhou, Q. H., Sumbria, R., Hui, E. K., Lu, J. Z., Boado, R. J., and Pardridge, W. M. (2011) Neuroprotection with a brain-penetrating biologic tumor necrosis factor inhibitor, *The journal of pharmacology and experimental therapeutics 339*, 618–623.

[60] Sumbria, R. K., Boado, R. J., and Pardridge, W. M. (2012) Brain protection from stroke with intravenous TNF-alpha decoy receptor-Trojan horse fusion protein, *Journal of cerebral blood flow and metabolism 32*, 1933–1938.

[61] Sumbria, R. K., Boado, R. J., and Pardridge, W. M. (2013) Combination stroke therapy in the mouse with blood-brain barrier penetrating IgG-GDNF and IgG-TNF decoy receptor fusion proteins, *Brain research 1507*, 91–96.

[62] Marvin, J. S. and Zhu, Z. (2005) Recombinant approaches to IgG-like bispecific antibodies, *Acta pharmacologica Sinica 26*, 649–658.

[63] Sumbria, R. K., Hui, E. K., Lu, J. Z., Boado, R. J., and Pardridge, W. M. (2013) Disaggregation of amyloid plaque in brain of Alzheimer's disease transgenic mice with daily subcutaneous administration of a tetravalent bispecific antibody that targets the transferrin receptor and the Abeta amyloid peptide, *Molecular pharmaceutics, 10*, 3507–3513.

[64] Wilcock, D. M., Alamed, J., Gottschall, P. E., Grimm, J., Rosenthal, A., Pons, J., Ronan, V., Symmonds, K., Gordon, M. N., and Morgan, D. (2006) Deglycosylated anti-amyloid-beta antibodies eliminate cognitive deficits and reduce parenchymal amyloid with minimal vascular consequences in aged amyloid precursor protein transgenic mice, *The journal of neuroscience 26*, 5340–5346.

[65] Jancso, G., Domoki, F., Santha, P., Varga, J., Fischer, J., Orosz, K., Penke, B., Becskei, A., Dux, M., and Toth, L. (1998) Beta-amyloid (1–42) peptide impairs blood-brain barrier function after intracarotid infusion in rats, *Neuroscience letters 253*, 139–141.

[66] Su, G. C., Arendash, G. W., Kalaria, R. N., Bjugstad, K. B., and Mullan, M. (1999) Intravascular infusions of soluble beta-amyloid compromise the blood-brain barrier, activate CNS glial cells and induce peripheral hemorrhage, *Brain research 818*, 105–117.

[67] Sperling, R., Salloway, S., Brooks, D. J., Tampieri, D., Barakos, J., Fox, N. C., Raskind, M., Sabbagh, M., Honig, L. S., Porsteinsson, A. P., Lieberburg, I., Arrighi, H. M., Morris, K. A., Lu, Y., Liu, E., Gregg, K. M., Brashear, H. R., Kinney, G. G., Black, R., and Grundman, M. (2012) Amyloid-related imaging abnormalities in patients with Alzheimer's disease treated with bapineuzumab: a retrospective analysis, *Lancet neurology 11*, 241–249.

[68] Sumbria, R. K., Zhou, Q. H., Hui, E. K., Lu, J. Z., Boado, R. J., and Pardridge, W. M. (2013) Pharmacokinetics and brain uptake of an IgG-TNF decoy receptor fusion protein following intravenous, intraperitoneal, and subcutaneous administration in mice, *Molecular pharmaceutics 10*, 1425–1431.

[69] Couch, J. A., Yu, Y. J., Zhang, Y., Tarrant, J. M., Fuji, R. N., Meilandt, W. J., Solanoy, H., Tong, R. K., Hoyte, K., Luk, W., Lu, Y., Gadkar, K., Prabhu, S., Ordonia, B. A., Nguyen, Q., Lin, Y., Lin, Z., Balazs, M., Scearce-Levie, K., Ernst, J. A., Dennis, M. S., and Watts, R. J. (2013) Addressing safety liabilities of TfR bispecific antibodies that cross the blood-brain barrier, *Science translational medicine 5*(183), 183ra57, 1–12.

[70] Yu, Y. J., Zhang, Y., Kenrick, M., Hoyte, K., Luk, W., Lu, Y., Atwal, J., Elliott, J. M., Prabhu, S., Watts, R. J., and Dennis, M. S. (2011) Boosting brain uptake of a therapeutic antibody by reducing its affinity for a transcytosis target, *Science translational medicine 3*, 84ra44.

[71] Zhou, Q. H., Boado, R. J., Hui, E. K., Lu, J. Z., and Pardridge, W. M. (2011) Chronic dosing of mice with a transferrin receptor monoclonal antibody-glial-derived neurotrophic factor fusion protein, *Drug metabolism and disposition 39*, 1149–1154.

[72] Pardridge, W. M. and Boado, R. J. (2009) Pharmacokinetics and safety in rhesus monkeys of a monoclonal antibody-GDNF fusion protein for targeted blood-brain barrier delivery, *Pharmaceutical research 26*, 2227–2236.

[73] Boado, R. J., Hui, E. K., Lu, J. Z., and Pardridge, W. M. (2013) IgG-enzyme fusion protein: pharmacokinetics and anti-drug antibody response in rhesus monkeys, *Bioconjugate chemistry 24*, 97–104.

[74] Boado, R. J., Hui, E. K., Lu, J. Z., and Pardridge, W. M. (2012) Glycemic control and chronic dosing of rhesus monkeys with a fusion protein of iduronidase and a monoclonal antibody against the human insulin receptor, *Drug metabolism and disposition 40*, 2021–2025.

[75] Zhou, Q. H., Hui, E. K., Lu, J. Z., Boado, R. J., and Pardridge, W. M. (2011) Brain penetrating IgG-erythropoietin fusion protein is neuroprotective following intravenous treatment in Parkinson's disease in the mouse, *Brain research 1382*, 315–320.

[76] Ponce, R., Abad, L., Amaravadi, L., Gelzleichter, T., Gore, E., Green, J., Gupta, S., Herzyk, D., Hurst, C., Ivens, I. A., Kawabata, T., Maier, C., Mounho, B., Rup, B., Shankar, G., Smith, H., Thomas, P., and Wierda, D. (2009) Immunogenicity of biologically-derived therapeutics: assessment and interpretation of nonclinical safety studies, *Regulatory toxicology and pharmacology 54*, 164–182.

[77] Baldrick, P. (2011) Safety evaluation of biological drugs: what are toxicology studies in primates telling us? *Regulatory toxicology and pharmacology 59*, 227–236.

[78] Chapman, K. L., Andrews, L., Bajramovic, J. J., Baldrick, P., Black, L. E., Bowman, C. J., Buckley, L. A., Coney, L. A., Couch, J., Maggie Dempster, A., de Haan, L., Jones, K., Pullen, N., de Boer, A. S., Sims, J., and Ian Ragan, C. (2012) The design of chronic toxicology studies of monoclonal antibodies: implications for the reduction in use of non-human primates, *Regulatory toxicology and pharmacology 62*, 347–354.

PART 3

PREDICTING AND MEASURING BRAIN EXPOSURE OF DRUGS

9

IN SILICO TOOLS FOR PREDICTING BRAIN EXPOSURE OF DRUGS

HONGMING CHEN,[1] SUSANNE WINIWARTER,[2] AND OLA ENGKVIST[1]

[1] *Computational Chemistry, Chemistry Innovation Center, AstraZeneca R&D, Mölndal, Sweden*

[2] *Computational ADME/Safety, Drug Safety and Metabolism, AstraZeneca R&D, Mölndal, Sweden*

INTRODUCTION

The blood–brain barrier (BBB) constitutes the primary system to protect the brain from exposure to potentially hazardous xenobiotics. The most important physical structure of the BBB is the brain capillary endothelium, a very tight membrane that hinders paracellular permeation. Transcellular permeation is restricted by the high levels of efflux transporters present in the endothelial cells, for example, P-glycoprotein (P-gp) and multidrug resistance protein (MRP) transporters. This defense mechanism can be utilized by designing peripherally acting drugs with low risk of central side effects. However, for drugs targeting proteins in the central nervous system (CNS), brain exposure may be the biggest hurdle to overcome in the drug discovery process [1]. To be able to understand the likelihood of brain exposure early on in drug discovery, *in silico* tools predicting BBB permeability/brain exposure of drugs are of great interest.

Structure–brain exposure relationships have for many years mainly been derived from rodent total brain-to-plasma concentration ratio values, $K_{p,brain}$ [2] (often expressed in its logarithmic form, log $K_{p,brain}$ or logBB). An alternative measure, also sometimes used to build *in silico* models, is logPS [3], the logarithm of the *in vivo* BBB

Blood–Brain Barrier in Drug Discovery: Optimizing Brain Exposure of CNS Drugs and Minimizing Brain Side Effects for Peripheral Drugs, First Edition. Edited by Li Di and Edward H. Kerns.

permeability surface area product (PS). Recently, it was suggested that the steady-state unbound brain-to-plasma ratio $K_{\mathrm{p,uu,brain}}$ is the most relevant measure to estimate drug exposure in the brain, since the key driving force for drug efficacy in CNS is the free drug concentration in the brain [4–6]. Although publicly available $K_{\mathrm{p,uu,brain}}$ data are rare, recently, some *in silico* $K_{\mathrm{p,uu,brain}}$ models have been published [7, 8].

In silico models based on the parameters mentioned so far can be both quantitative, that is, give a predicted numerical value, or qualitative models, which predict if the compound in question is likely to enter the brain or not. In this chapter we critically review recent *in silico* BBB penetration models available in the literature, including considerations on the experimental data used. Important molecular physicochemical properties which influence the brain exposure are highlighted and potential future directions for developing improved *in silico* BBB penetration prediction tools are discussed.

MEASUREMENTS QUANTIFYING BRAIN EXPOSURE OF DRUGS

LogBB (log $K_{\mathrm{p,brain}}$)

The ratio of a drug's concentration in the brain versus its concentration in the blood plasma at steady state, determined in an *in vivo* animal experiment, has been the most widely used parameter for *in silico* prediction of brain exposure:

$$\mathrm{logBB} = \log \frac{C_{\mathrm{brain}}}{C_{\mathrm{p}}} \quad (9.1)$$

Here, C_{brain} is the total compound concentration in the brain and C_{p} is the total compound concentration in the plasma. LogBB data from rodents are readily available in the literature [9–11] and data from different sources have been compiled and used widely for model building [12–14]. However, it has been argued that logBB, which is based on total concentrations, may be misleading [5, 6, 15], since only the free drug is available for transport across BBB and for binding to target proteins in the brain [15–18].

LogPS

The BBB PS is another measure for brain exposure [19]. PS is determined in *in situ* brain perfusion experiments: the drug's uptake into the brain is measured in an anesthetized laboratory animal over a short time period (tens of seconds to minutes) and PS calculated using Equation 9.2 [20]:

$$\mathrm{PS} = Q_{\mathrm{br}} \times \ln\left(1 - \frac{K_{\mathrm{in}}}{Q_{\mathrm{br}}}\right) \quad (9.2)$$

Q_{br} is the cerebral perfusion fluid flow rate during the experiment and K_{in} denotes the uptake clearance. PS can be regarded as an estimate of the net influx clearance rate of drug into the brain. While this indicates that the free concentration in the brain can thereby be estimated [19], it needs to be considered that the drug concentration is equally influenced by the BBB efflux clearance [16]. Furthermore, PS is a measure of penetration rate and therefore not necessarily correlated with the extent of penetration.

$K_{\mathrm{p,uu,brain}}$

The steady-state unbound brain-to-plasma concentration ratio $K_{\mathrm{p,uu,brain}}$ is defined by the following equation [7]:

$$K_{\mathrm{p,uu,brain}} = \frac{C_{\mathrm{u,brainISF}}}{C_{\mathrm{u,p}}} \tag{9.3}$$

$C_{\mathrm{u,brainISF}}$ is the concentration of unbound compound in the brain interstitial fluid and $C_{\mathrm{u,p}}$ is the concentration of unbound compound in the plasma. $C_{\mathrm{u,brainISF}}$ can be directly measured through microdialysis in the brain [21, 22]. However, this method is experimentally challenging, for example, with technical difficulties when measuring lipophilic drugs, and is therefore only of limited usability in drug discovery projects. Recent developments and validation of methods to measure unbound brain concentrations have contributed to the acceptance of $K_{\mathrm{p,uu,brain}}$ as an important parameter in drug discovery [4, 23–25]. Fridén et al. [7] showed how to assess $K_{\mathrm{p,uu,brain}}$ by combining the total brain-to-plasma ratio $K_{\mathrm{p,brain}}$ determined *in vivo* with estimates of $V_{\mathrm{u,brain}}$ and $f_{\mathrm{u,p}}$ determined *in vitro* in brain slices [26] and by equilibrium dialysis [27], respectively:

$$K_{\mathrm{p,uu,brain}} = \frac{K_{\mathrm{p,brain}}}{\mathrm{V}_{\mathrm{u,brain}} \times f_{\mathrm{u,p}}} \tag{9.4}$$

Here, $V_{\mathrm{u,brain}}$ represents the unbound volume of distribution in the brain and $f_{\mathrm{u,p}}$ is the unbound fraction of drug in plasma. Thus, $K_{\mathrm{p,uu,brain}}$ is experimentally determined from three different measurements: the total brain-to-plasma concentration ratio obtained from an *in vivo* animal experiment and combined *in vitro* determinations of plasma protein binding and binding to brain tissue (Eq. 9.4). Mechanistically, $K_{\mathrm{p,uu,brain}}$ is determined by the relative efficiency of BBB influx and efflux and is independent from plasma protein binding in blood or binding to the tissue components of the brain.

MODELING STRATEGIES FOR BUILDING BBB MODELS

One of the key elements for building *in silico* models is to select relevant molecular descriptors. The descriptor sets employed in BBB modeling evolved over time. In early studies, only a few simple, interpretable descriptors such as log*P* (lipophilicity), polar surface area (PSA), or hydrogen-bonding descriptors were used [9, 12, 28, 29]. Later, when modeling methods to cope with many descriptors were available, descriptor sets could comprise hundreds of descriptors: for example, 2D structure–based Dragon descriptors [30, 31] or 3D structure–based Volsurf [32] descriptors were used for modeling logBB. The rationale for using a large descriptor set was the hope to thereby include specific descriptors that could capture additional subtle structure requirement for BBB penetration alongside the simple physicochemical parameters. Other more complex descriptors were considered as well: for example, solvation free energies based on different solvation models and the compounds' 3D conformations were calculated and utilized [10, 33] as descriptors for building *in silico* BBB models. However, these descriptors require significantly longer computation time compared to 2D descriptors.

Once a relevant descriptor set is chosen, various mathematical or statistical methods can be used to relate the descriptor values to the experimental end point data. In the early logBB modeling studies multiple linear regression (MLR) analysis was used to build models. The biggest advantage of MLR is that the established relationship is totally transparent, that is, the influence of each descriptor is quantified by its coefficient in the MLR equation. However, MLR can only handle a limited number of descriptors. For a larger descriptor set other methods are required. Partial least square [34] (PLS) analysis is a powerful technique to project the descriptors to a few latent variables (also called principal components). Another way to handle a large number of descriptors is to apply a variable selection method during the modeling process. The combination of a genetic algorithm [35], as a global optimization scheme for variable selection, with MLR analysis was reported to build optimal logBB models from a large descriptor set [36, 37]. A drawback for MLR and PLS, both linear methods, is that they cannot handle nonlinear relationships. Recently, computationally more advanced machine-learning algorithms such as support vector machine [38] (SVM), random forest [39] (RF), and neural networks (NN) [40] have been used to build BBB models. These nonlinear methods generally have higher accuracy than linear modeling methods, but with the cost of sacrificing model interpretability due to their nontransparent nature. However, ways to elucidate the relationship between descriptors and the experimental end point in nonlinear models have been suggested [41, 42]. In this review the focus is on global brain exposure models with the goal to cover the whole chemical space relevant for drug discovery. For a drug discovery project in the lead optimization phase it may be more appropriate to create a local model for the chemical series of interest only.

IN SILICO MODELS FOR PREDICTING LOGBB

Given the great importance of studying a compound's brain penetration in drug discovery, prediction of BBB penetration from computed or easily measured experimental parameters has been of interest for a long time. So far most of the reported QSAR models to predict brain exposure are based on logBB data, and representative work is summarized in Table 9.1. The pioneering study by Young et al. [43] showed a good correlation between logBB and $\Delta \text{Log}P$ (the difference between the partition coefficient ($\log P$) in octanol/water versus in cyclohexane/water, thereby describing hydrogen-bonding capacity) for 20 antihistamine compounds. Later, Van de Waterbeemd [28] and Calder [44] investigated the "Young data set" and tried to

TABLE 9.1 Summary of recently published quantitative logBB models

Model descriptors	Modeling method	Number of compounds	R2	sd[a]	References
ΔLogP	MLR	20	0.69	0.44	Young et al. [43]
PSA and molecular volume	MLR	20	0.70	0.45	Van de Waterbeemd and Kansy [28]
Solute descriptors	MLR	57	0.91	0.20	Abraham et al. [12]
Free energy of solvation	MLR	55	0.67	0.41	Lombardo et al. [10]
Molsurf descriptors	PLS	56	0.83	0.31	Norinder et al. [46]
PSA and logP	MLR	55	0.79	0.35	Clark [29]
Free energy of solvation	MLR	55	0.72	0.37	Keseru [33]
Solute descriptors	MLR	148	0.75	0.34	Platts et al. [13]
High charged PSA, SlogP, MW360	MLR	78 (training set)	0.77	0.364	Hou and Xu [47]
		14 (test set 1)	0.88	0.26[b]	
		23 (test set 2)	0.61	0.48[b]	
2D molecular descriptors	Consensus prediction of kNN[c] and SVM models	144 (training set)	0.91	0.21[d]	Zhang et al. [31]
		14 (test set)	0.8	0.29[d]	
2D molecular descriptors	Genetic algorithm–based MLR	193 (training set)	0.74 (0.72[f])	NA[e]	Fan et al. [36]
				NA[e]	
		147 (test set)	0.65		
LogP, ion fraction, plasma protein binding	MLR	329 (training set)	0.52	0.38[b]	Lanevskij et al. [14]
		141 (test set)	0.54	0.39[b]	
2D molecular descriptors	Beam search–based MLR	362	0.59	NA[f]	Muehlbacher et al. [48]

[a]Standard deviation.
[b]Root mean squared error value.
[c]k-nearest neighbors algorithm.
[d]Mean absolute error.
[e]Not available.
[f]Leave-one-out cross validation q^2.

correlate logBB with less experimentally demanding parameters, such as PSA and molecular volume. Abraham et al. [12] extended the "Young data set" with 35 additional compounds to form the so-called Abraham data set, which was subsequently used by various research groups to build BBB models. These early efforts have been extensively reviewed previously [2, 45].

During recent years the size of the publicly available logBB data set has increased gradually as more diverse compounds have been included. So far the biggest public logBB data set comprising around 470 compounds was compiled by Lanevskij et al. [14]. The molecular descriptors utilized have been very diverse: 2D physicochemical descriptors [29, 49–53] representing information about the molecular size, shape, lipophilicity, etc., as well as descriptors derived from 3D molecular structure [10, 33, 46]. For the early logBB models, usually a small number of descriptors were used to build the model and simple MLR statistics were the main modeling strategy utilized. In recent logBB modeling efforts, often large numbers of descriptors and more complex algorithms, to deal with these many variables, are used for building models [31, 54, 48]. The models using nonlinear algorithms [31, 55] have, in general, higher accuracy than the linear models [37, 56, 48], but lack some interpretability.

Overall there is broad agreement [45] on the importance of specific molecular properties and descriptors which have been found in numerous investigations to influence logBB.

- *Hydrogen bonding*: Polarity or hydrogen-bonding capacity related descriptors are those most frequently reported in logBB models, for example, PSA [28, 44], number of hydrogen bond donor/acceptor [29], hydrogen acidity/basicity [13], or the number of oxygen and nitrogen atoms in a molecule [2, 29]. In most cases, such descriptors are negatively correlated to BBB penetration, that is, highly polar compounds or those with strong hydrogen-bonding capacity tend to have low BBB penetration. PSA is one of the most commonly used descriptors for representing the hydrogen-bonding capability. It can be calculated from either 2D or 3D molecular structures [56]. Both definitions were successfully used in logBB models [44, 51, 57].
- *Lipophilicity*: Many MLR models [9, 44, 49, 50, 58] indicate that lipophilicity has a big influence on logBB and, generally, lipophilicity correlates positively with logBB. Both experimental ($\log P_{oct}$, the octanol–water partition coefficient) and calculated lipophilicity, such as Clog*P* [59], ACDlog*P* [60], and Volsurf [61] hydrophobic descriptors, have been used for modeling logBB.
- *Molecular size*: The influence of molecular size on BBB penetration is not entirely clear. Several reports [44, 49, 50] state that molecular volume and molecular weight are negatively related to logBB. This finding is in line with the general understanding of passive diffusion, that is, according to the Stokes–Einstein relation, the diffusion coefficient of a spherical particle is inversely related to its radius. However, there are also investigations indicating that size could be enhancing logBB [13, 62].
- *Ionization states*: It has been observed that basic compounds tend to have higher logBB, while acidic compounds have lower logBB values [8, 63]. Among the known CNS drugs, quite a few are basic amines. Acidic compounds, on the other hand, are known to bind preferably to albumin in plasma, thus limiting the free drug concentration available for brain penetration and thereby contributing to a lower logBB value.

Other descriptors of interest have been studied as well. For example, Lombardo [10] and Keseru [33] both investigated the relation of the solvation free energy to logBB, Gerebtzoff and Seelig [64] suggested cross-sectional area (CSA) as a determinant for logBB, whereas Iyer et al. [37] employed flexibility in a logBB model and found that it is negatively correlated with logBB value. Lanevskij et al. [14] stated that logBB is a direct function of the unbound fraction in the brain and the unbound fraction in plasma, when only passive transport is considered. Their approach was to calculate the unbound fraction of a drug in the brain from log*P* and the unionized fraction at pH 7.4 (as obtained from the compound's p*K*a) and combine them with the free fraction in plasma, which could be either experimentally or computationally determined. This is an interesting approach for predicting logBB. However, considering only passive transportation is somewhat problematic, since most compounds are likely to be dependent on active processes during BBB penetration.

IN SILICO MODELS FOR PREDICTING LOGPS

The concept of PS was originally put forward by Renkin in 1959 [65] as an alternative measure of permeability into tissue. Due to the "free drug hypothesis" the validity of logBB, which is based on total concentrations, was questioned [15]. Pardrige suggested that logPS, describing the uptake clearance into the brain, would be a better measure to estimate the free drug concentration in the brain [19]. However, this parameter does not consider the brain efflux clearance and therefore cannot by itself describe the free drug concentration in the brain. Moreover, PS is a parameter measuring the rate of BBB permeation, not the extent. Gratton et al. [3] proposed several logPS models on a set of 18 compounds using Abraham solvation descriptors [12] and $\log P_{oct}$, respectively:

$$\begin{aligned} &\log \text{PS} = -1.21 + 0.77R_2 - 1.87\pi_2^{\text{H}} - 2.8\sum\beta_2^{\text{H}} + 3.31V_x \\ &N = 18,\ r = 0.976,\ \text{sd} = 0.481,\ F = 65 \end{aligned} \tag{9.5}$$

R_2 is the excess molar refraction, π_2^{H} the dipolarity/polarizability, β_2^{H} the hydrogen bond basicity, and *Vx* the characteristic McGowan volume.

$$\begin{aligned} &\text{logPS} = -2.28 + 0.69\ \log P_{\text{oct}} + 0.69 \log \\ &N = 18,\ r = 0.882,\ \text{sd} = 0.94,\ F = 5 \end{aligned} \tag{9.6}$$

Abraham [66] applied the same solvation descriptors on a data set of 30 compounds, an extension of the Gratton data set, and obtained a model with $R^2 = 0.87$. Around the same time, Liu et al. [67] developed predictive logPS models based on the topological polar surface area (TPSA), Log*D*, and Abraham solvation descriptors. Compared to public logBB data sets, the PS data sets are rather small; thus, the application domain of these logPS models is likely very limited.

BBB+/BBB– (CNS+/CNS–) CLASSIFICATION MODELS

Table 9.2 lists a representative set of classification models based either on *in vivo* brain exposure data or on drug pharmacological indication information. In the latter case drugs acting in the CNS have been used as the active set and drugs acting on periphery targets comprised the inactive set. This is a rather straightforward way to define the active set, since only a compound that enters the brain in a sufficient amount can act centrally. However, the inactive set is less well defined, since drugs with peripheral mode of action may still be able to penetrate the BBB. Ghose et al. [68] addressed this potential problem by ensuring that drugs with CNS-related side effects were not included in the inactive set. Ajay et al. [69] reported the use of CNS+/CNS– classification models based on CNS activity knowledge taken from the *Comprehensive Medicinal Chemistry* (CMC) and the MACCS-II Drug Data Report (MDDR) databases to design compound libraries with high probability of CNS penetration.

Again, various machine-learning techniques were employed for building BBB classification models. Algorithms based on decision trees [74] like recursive-partitioning [75] are particularly interesting due to their capability of building interpretable models that can guide a drug discovery project to improve CNS penetration. For example, Ghose et al. [68] used recursive partitioning to find which parameters increase the likelihood of higher CNS penetration. Martins et al. [73] used both SVM and RF methods and also considered the general probability of a small compound entering the brain via Bayesian logic. Their best-fitted model gave an accuracy of 95% for an independent test set. This model was made available as a free web tool (http://b3pp.lasige.di.fc.ul.pt). Other studies suggested simple rules, similar to Lipinski's "rule of 5" to estimate the likelihood of intestinal absorption [76], to provide guidance on designing CNS-permeable compounds [2, 77, 78]. One such rule, proposed by Norinder and Haeberlein [2], stated that if the sum of the number of nitrogen and oxygen molecules in a compound is less than 5, there is a good chance that the compound can enter the brain. In general, hydrogen-bonding descriptors like PSA or the number of nitrogen/oxygen molecules are to be kept low, whereas lipophilicity enhances a compound's ability to reach the CNS.

PREDICTION OF UNBOUND DRUG EXPOSURE IN THE BRAIN, $K_{P,UU,BRAIN}$

$K_{p,uu,brain}$ has been suggested to be the most relevant parameter to describe drug exposure in the brain [4–7]. To the best of our knowledge, there are only two *in silico* $K_{p,uu,brain}$ modeling studies published so far [7, 8], although several $K_{p,uu,brain}$ data sets are available [17, 25, 79–81].

The first study presenting a computational model based on free brain/plasma ratio was published in 2009 by Fridén et al. [7] The study shows both the comparison of different experimental methods to obtain unbound brain/plasma concentration ratios, species comparison between the animal studies and human cerebrospinal fluid (CSF) data, and a PLS model using simple structural descriptors. The study was based on

TABLE 9.2 Examples of recently published CNS+/CNS– classification models

Descriptor type	Method	Number of compounds (training set/test set)	Experimental data type	Accuracy (test set)	References
1D&2D descriptor	NN	15,000/275	Drugs	92% actives 71% inactives	Ajay et al. [69]
VolSurf descriptor	PLS	110/120	logBB	90%	Crivori et al. [32]
Ghose–Crippen descriptor	NN	7,810/868	Drugs	83% actives 77% inactives	Engkvist et al. [70]
Topological descriptors	NN	108/78	logBB	73%	Guerra et al. [71]
MolconnZ, MOE, Dragon descriptors	kNN[a],SVM consensus model	144/366	logBB	82.5	Zhang et al. [31]
Shape signatures	SVM	351/NA	logBB	83%	Kortagere et al. [54]
Property autocorrelation descriptors	KohNN[b]	1,593/396	Drugs	95%	Wang et al. [72]
Physicochemical descriptors	RP[c]	943/80	Drugs	79% actives 80% inactives	Ghose et al. [68]
Daylight fingerprints, atomic and ring multiplicities, simple molecular parameters, e-Dragon	SVM, RP	1,576/394	logBB	95%	Martins et al. [73]

[a]k-nearest neighbors algorithm.
[b]Kohonen's self-organizing neural network.
[c]Recursive partitioning.

43 compounds, most of which were selected from a review on drugs from five therapeutic areas with known human CSF data [82]. The aim of the selection was to cover drug chemical space as much as possible, but the bioanalytical properties of the selected compounds were considered as well.

$K_{p,uu,brain}$ was determined from the brain/plasma concentration ratio ($K_{p,brain}$) corrected by plasma protein binding ($f_{u,p}$) and unbound brain volume of distribution ($V_{u,brain}$) as shown in Equation 9.4. Sixteen standard 2D molecular descriptors, including PSA, MW, Clog*P*, and hydrogen-bonding descriptors, were calculated for all 43 compounds. Log$K_{p,uu,brain}$ and log $K_{p,brain}$ (=logBB) were used as modeling end points. It was noted that $K_{p,brain}$ spanned a wider range of values, ~0.002–20, than $K_{p,uu,brain}$ with values ranging from 0.006 to 2. The models obtained for log $K_{p,brain}$ were similar to other models for logBB in the literature: lipophilicity was a main factor for $K_{p,brain}$, but hydrogen-bonding descriptors were also of importance. It was seen that basic drugs have higher $K_{p,brain}$ values than acidic drugs. The best model used hydrogen bond acceptor count (HBA), log*D* (ACDlog$D_{7.4}$), acid, and base as descriptors and was rationalized as follows: the HBA provides information on the BBB permeation properties, lipophilicity and basicity give information on binding to phospholipids in tissue (high binding in brain tissue), and acidity provides information on extensive albumin binding in plasma. For log $K_{p,uu,brain}$, on the other hand, lipophilicity was not important. Only hydrogen-bonding descriptors, especially HBA, and size seemed to determine the property. However, $K_{p,uu,brain}$ is more difficult to predict as can be seen from its lower prediction accuracy compared to the $K_{p,brain}$ model (Q^2=0.45 for the $K_{p,uu,brain}$ model versus 0.69 for the $K_{p,brain}$ model). One reason for this difficulty could be that $K_{p,uu,brain}$ is influenced by various transporter interactions rather than by non-specific binding to proteins and lipids. However, it was possible to develop a model with medium accuracy. The simple model based on HBA alone (Q^2=0.43, Fig. 9.1)

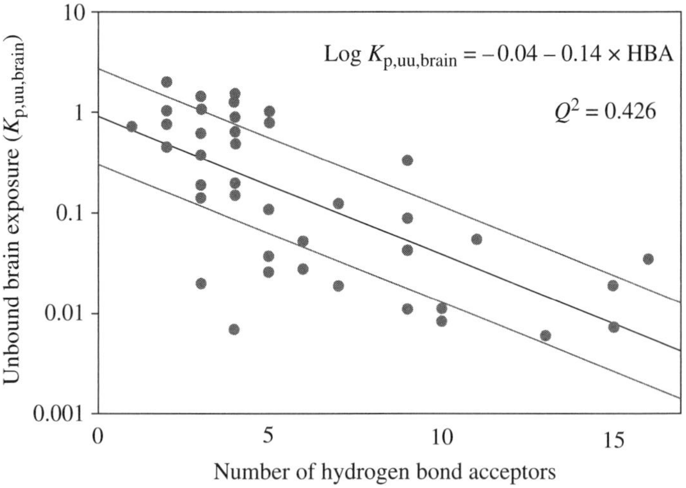

FIGURE 9.1 Correlation between experimental $K_{p,uu,brain}$ and number of hydrogen bond acceptors. Reprinted with permission from Ref. [7]. © 2009, American Chemical Society.

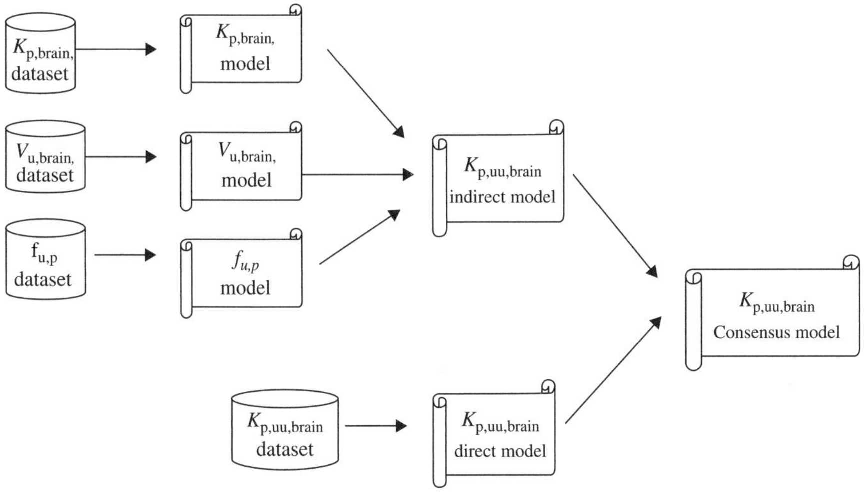

FIGURE 9.2 The workflow applied for the model building. Reprinted with permission from Ref. [8]. © 2011, Elsevier.

gave a predicting power almost as high as the one using all 16 descriptors. The HBA model is useful since it can be easily interpreted: in order to achieve a twofold increase in $K_{p,uu,brain}$ two hydrogen bond acceptors need to be removed from a given molecule.

The $K_{p,uu,brain}$ models were tested on about 140 external compounds with literature values of $K_{p,uu,brain}$ in either rat or mouse. The resulting root mean squared error (RMSE) indicated a likely fourfold error, which was only slightly higher than the RMSE for the training set. It was confirmed that compounds with less than 2 HBAs were likely to have good brain accessibility with a $K_{p,uu,brain}$ value above 0.1, whereas compounds with more than 10 HBAs have a high chance of keeping outside the brain ($K_{p,uu,brain} < 0.1$).

Recently, Chen et al. [8] expanded Fridén's $K_{p,uu,brain}$ data set to 246 compounds by adding proprietary compounds measured with the same experimental protocol as described by Fridén et al. [7]. Besides the $K_{p,uu,brain}$ data set, separate $K_{p,brain,}$ $V_{u,brain}$, and $f_{u,p}$ data sets were also compiled in the study. The modeling workflow is shown in Figure 9.2. Two types of $K_{p,uu,brain}$ models were built: direct models based on the available $K_{p,uu,brain}$ data and indirect models obtained by combining individual models of $K_{p,brain,}$ $V_{u,brain}$, and $f_{u,p}$ data according to Equation 9.4 (Fig. 9.2). The rationale for developing indirect models was to ensure that all relevant experimental data were utilized, since more data points were available for each of $K_{p,brain,}$ $V_{u,brain}$, and $f_{u,p}$ than for $K_{p,uu,brain}$. Thus, the indirect models were built on larger training sets. Consensus models were then built by averaging the predictions for two or more individual $K_{p,uu,brain}$ models. The descriptor set is comprised of 196 2D/3D molecular descriptors, and various modeling methods including RF, SVM, and PLS were employed to build the models. It was found that a consensus model based on three individual nonlinear submodels gave the best prediction on 73 test set compounds ($R^2=0.58$).

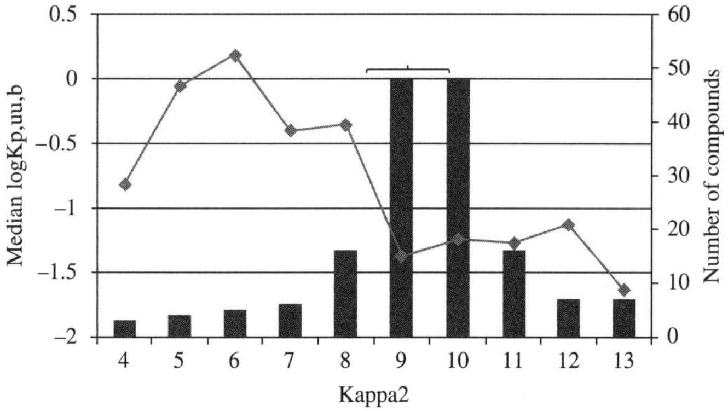

FIGURE 9.3 The relationship between the median log $K_{p,uu,brain}$ and Kappa2. The nonparametric Wilcoxon test shows significance at the $p < 0.05$ level as indicated in the graph. Reprinted with permission from Ref. [8]. © 2011, Elsevier.

Analysis of the descriptor importance showed that several hydrogen bond–related descriptors have significant impact on $K_{p,uu,brain}$. However, the single most important descriptor is Kappa2, which describes molecular shape in terms of its linearity. Highly branched compounds with Kappa2 value larger than 8 have significantly lower $K_{p,uu,brain}$ than more linear compounds (Kappa2 value below 8, see Fig. 9.3). Another interesting finding in that study was that while lipophilicity was positively correlated with $K_{p,brain}$, it did not increase $K_{p,uu,brain}$. A plausible explanation is that the effect of higher passive transport due to increased lipophilicity is offset by the greater efflux caused by the interaction between the lipophilic drug and P-gp or other efflux transporters.

FUTURE DEVELOPMENT AND CHALLENGES

One of the challenges for *in silico* modeling of brain exposure is the choice of the modeling end point. Currently there are quite a few parameters reported in the literature as characterizing brain penetration, such as logBB, $K_{p,uu,brain}$, logPS, and CNS+/CNS−. All these parameters reflect different aspects of brain penetration, logBB still being the most widely used measure. However, debates in recent years [5, 6, 15] have repeatedly highlighted the drawbacks of this parameter. There is an urgent need for the modeling community to make a paradigm shift from logBB to the more relevant $K_{p,uu,brain}$.

Currently, the best reported $K_{p,uu,brain}$ model has only a moderate predictivity (cross-validated $Q^2 = 0.6$ for the training set, $R^2 = 0.58$ for the test set) [8]. Additional $K_{p,uu,brain}$ data for diverse chemical structures are needed to improve the model. However, one needs to keep in mind that the two main mechanisms which determine the level of drug exposure in the brain (in relation to the exposure in blood) are passive diffusion and active influx/efflux due to various transporters. Consequently, any variability in

$K_{p,uu,brain}$ between compounds is not just caused by overall molecular properties (determining passive diffusion), but also by compound-specific, molecular interactions with one or several drug transporters at the BBB, for example, P-gp. Recently, a study [83] on the relationship between efflux ratio measured in the Caco-2 cell line and $K_{p,uu,brain}$ has shown that increasing efflux transportation will decrease $K_{p,uu,brain}$. Therefore, a possible approach to improve predictions on brain exposure can be modeling specific BBB drug transporters to identify structural patterns or pharmacophores for transporter substrates. Advancements in modeling drug transporter interactions have been reviewed elsewhere [84–86]. Species and tissue differences in transporter expression, activity, and substrate specificity will have to be taken into account. Integration of transporter substrate prediction models into a $K_{p,uu,brain}$ model may also help to elucidate if animal model results for a certain compound series can be translated to the human situation with confidence, depending on which transporter is likely to be important.

Combining several QSAR models in a consensus model has been proven to be highly beneficial [8, 31]. Developing consensus models from models built on different descriptor sets and algorithms can be another way to further improve the accuracy of $K_{p,uu,brain}$ models. The use of the bioassay ontology (BAO) [87] to classify the experimental data may also help to collect more relevant data for improving computational models. Recently, efforts to annotate transporter data in such a way have been started [88]. We also think that making models publicly available should be encouraged. Such models can be easily used for benchmarking purposes and thereby help to develop more accurate new models. Additionally such models would be extremely beneficial in areas with low resources, like neglected diseases.

CONCLUSIONS

Despite many years of effort, CNS drug development remains one of the biggest challenges to the pharmaceutical industry, showing the efficient protection mechanism of the BBB. Over recent years, huge amounts of work have been put into generating predictive models for brain penetration. For most of these models, logBB, assessed in *in vivo* animal experiments, was used as a measure of brain exposure. However, there is clear experimental evidence that the total concentration ratio, the determinant for logBB, has little value for understanding drug availability in the brain. A better way to identify drugs that actually reach high enough free concentration in the CNS is to look for compounds known to be active at a CNS target as utilized in CNS+/CNS– classification models. However, compounds without CNS indication may still be able to enter the CNS, resulting in inherent uncertainty in such models. Given the acceptance of the "free drug hypothesis," the steady-state unbound concentration ratio $K_{p,uu,brain}$ appears to be the pharmacologically most meaningful measure for brain exposure. This parameter enables the estimation of the free brain concentration based on the unbound plasma concentration. $K_{p,uu,brain}$ models with modest prediction accuracy have been built recently and these models show that hydrogen-bonding capacity plays an important role for $K_{p,uu,brain}$, while lipophilicity has no correlation with $K_{p,uu,brain}$. This is an interesting difference from the traditional logBB models, which often show a high influence of lipophilicity and may have caused drugs to become more lipophilic than actually needed. To improve the prediction power of $K_{p,uu,brain}$ models, we suggest that the amount of $K_{p,uu,brain}$ data available for modeling needs to be increased and that transporter information should be considered in the models. Robust BBB penetration models can be used for prioritizing virtual sets of compounds for synthesis or testing and for guiding compound design, especially when a large improvement in BBB permeability is desired.

REFERENCES

[1] Reichel, A. The role of blood-brain barrier studies in the pharmaceutical industry. *Curr. Drug Metab.* **2006**, *7*, 183–203.

[2] Norinder, U.; Haeberlein, M. Computational approaches to the prediction of the blood-brain distribution. *Adv. Drug Deliv. Rev.* **2002**, *54*, 291–313.

[3] Gratton, J. A.; Abraham, M. H.; Bradbury, M. W.; Chadha, H. S. Molecular factors influencing drug transfer across the blood-brain barrier. *J. Pharm. Pharmacol.* **1997**, *49*, 1211–6.

[4] Fridén, M.; Gupta, A.; Antonsson, M.; Bredberg, U.; Hammarlund-Udenaes, M. In vitro methods for estimating unbound drug concentrations in the brain interstitial and intracellular fluids. *Drug Metab. Dispos.* **2007**, *35*, 1711–9.

[5] Hammarlund-Udenaes, M. Active-site concentrations of chemicals – are they a better predictor of effect than plasma/organ/tissue concentrations? *Basic Clin. Pharmacol. Toxicol.* **2010**, *106*, 215–20.

[6] Di, L.; Rong, H.; Feng, B. Demystifying brain penetration in central nervous system drug discovery. Miniperspective. *J. Med. Chem.* **2013**, *56*, 2–12.

[7] Fridén, M.; Winiwarter, S.; Jerndal, G.; Bengtsson, O.; Wan, H.; Bredberg, U.; Hammarlund-Udenaes, M.; Antonsson, M. Structure–brain exposure relationships in rat and human using a novel data set of unbound drug concentrations in brain interstitial and cerebrospinal fluids. *J. Med. Chem.* **2009**, *52*, 6233–43.

[8] Chen, H.; Winiwarter, S.; Fridén, M.; Antonsson, M.; Engkvist, O. In silico prediction of unbound brain-to-plasma concentration ratio using machine learning algorithms. *J. Mol. Graph. Model.* **2011**, *29*, 985–95.

[9] Young, R. C.; Mitchell, R. C.; Brown, T. H.; Ganellin, C. R.; Griffiths, R.; Jones, M.; Rana, K. K.; Saunders, D.; Smith, I. R.; Sore, N. E. Development of a new physicochemical model for brain penetration and its application to the design of centrally acting H2 receptor histamine antagonists. *J. Med. Chem.* **1988**, *31*, 656–71.

[10] Lombardo, F.; Blake, J. F.; Curatolo, W. J. Computation of brain-blood partitioning of organic solutes via free energy calculations. *J. Med. Chem.* **1996**, *39*, 4750–5.

[11] Bergström, C. A. S.; Charman, S. A.; Nicolazzo, J. A. Computational prediction of CNS drug exposure based on a novel in vivo dataset. *Pharm. Res.* **2012**, *29*, 3131–42.

[12] Abraham, M. H.; Chadha, H. S.; Mitchell, R. C. Hydrogen bonding. 33. Factors that influence the distribution of solutes between blood and brain. *J. Pharm. Sci.* **1994**, *83*, 1257–68.

[13] Platts, J. A.; Abraham, M. H.; Zhao, Y. H.; Hersey, A.; Ijaz, L.; Butina, D. Correlation and prediction of a large blood-brain distribution data set—an LFER study. *Eur. J. Med. Chem.* **2001**, *36*, 719–30.

[14] Lanevskij, K.; Dapkunas, J.; Juska, L.; Japertas, P.; Didziapetris, R. QSAR analysis of blood-brain distribution: the influence of plasma and brain tissue binding. *J. Pharm. Sci.* **2011**, *100*, 2147–60.

[15] Martin, I. Prediction of blood-brain barrier penetration: are we missing the point? *Drug Discov. Today* **2004**, *9*, 161–2.

[16] Hammarlund-Udenaes, M.; Fridén, M.; Syvänen, S.; Gupta, A. On the rate and extent of drug delivery to the brain. *Pharm. Res.* **2008**, *25*, 1737–50.

[17] Kalvass, J. C.; Maurer, T. S.; Pollack, G. M. Use of plasma and brain unbound fractions to assess the extent of brain distribution of 34 drugs: comparison of unbound concentration ratios to in vivo p-glycoprotein efflux ratios. *Drug Metab. Dispos.* **2007**, *35*, 660–6.

[18] Liu, X.; Chen, C. Strategies to optimize brain penetration in drug discovery. *Curr. Opin. Drug Discov. Devel.* **2005**, *8*, 505–12.

[19] Pardridge, W. M. Log(BB), PS products and in silico models of drug brain penetration. *Drug Discov. Today* **2004**, *9*, 392–3.

[20] Reichel, A. Addressing central nervous system (CNS) penetration in drug discovery: basics and implications of the evolving new concept. *Chem. Biodiv.* **2009**, *6*, 2030–49.

[21] De Lange, E. C.; Danhof, M.; De Boer, A. G.; Breimer, D. D. Critical factors of intracerebral microdialysis as a technique to determine the pharmacokinetics of drugs in rat brain. *Brain Res.* **1994**, *666*, 1–8.

[22] Lindén, K.; Ståhle, L.; Ljungdahl-Ståhle, E.; Borg, N. Effect of probenecid and quinidine on the transport of alovudine (3′-fluorothymidine) to the rat brain studied by microdialysis. *Pharmacol. Toxicol.* **2003**, *93*, 226–32.

[23] Mano, Y.; Higuchi, S.; Kamimura, H. Investigation of the high partition of YM992, a novel antidepressant, in rat brain – in vitro and in vivo evidence for the high binding in brain and the high permeability at the BBB. *Biopharm. Drug Dispos.* **2002**, *23*, 351–60.

[24] Kalvass, J. C.; Maurer, T. S. Influence of nonspecific brain and plasma binding on CNS exposure: implications for rational drug discovery. *Biopharm. Drug Dispos.* **2002**, *23*, 327–38.

[25] Becker, S.; Liu, X. Evaluation of the utility of brain slice methods to study brain penetration. *Drug Metab. Dispos.* **2006**, *34*, 855–61.

[26] Fridén, M.; Ducrozet, F.; Middleton, B.; Antonsson, M.; Bredberg, U.; Hammarlund-Udenaes, M. Development of a high-throughput brain slice method for studying drug distribution in the central nervous system. *Drug Metab. Dispos.* **2009**, *37*, 1226–33.

[27] Wan, H.; Bergström, F. High throughput screening of drug protein binding in drug discovery. *J. Liq. Chromatogr. Related Technol.* **2007**, *30*, 681–700.

[28] Van de Waterbeemd, H.; Kansy, M. Hydrogen-bonding capacity and brain penetration. *CHIMIA Int. J. Chem.* **1992**, *46*, 299–303.

[29] Clark, D. E. Rapid calculation of polar molecular surface area and its application to the prediction of transport phenomena. 2. Prediction of blood-brain barrier penetration. *J. Pharm. Sci.* **1999**, *88*, 815–21.

[30] Todeschini, R.; Lasagni, M.; Marengo, E. New molecular descriptors for 2D and 3D structures. Theory. *J. Chemom.* **1994**, *8*, 263–272.

[31] Zhang, L.; Zhu, H.; Oprea, T. I.; Golbraikh, A.; Tropsha, A. QSAR modeling of the blood-brain barrier permeability for diverse organic compounds. *Pharm. Res.* **2008**, *25*, 1902–14.

[32] Crivori, P.; Cruciani, G.; Carrupt, P.-A.; Testa, B. Predicting blood–brain barrier permeation from three-dimensional molecular structure. *J. Med. Chem.* **2000**, *43*, 2204–2216.

[33] Keseru, G. M.; Molnar, L. High-throughput prediction of blood-brain partitioning: a thermodynamic approach. *J. Chem. Inf. Model.* **2001**, *41*, 120–128.

[34] Wold, S.; Sjöström, M.; Eriksson, L. PLS-regression: a basic tool of chemometrics. *Chemom. Intell. Lab. Syst.* **2001**, *58*, 109–130.

[35] Taherdangkoo, M.; Paziresh, M.; Yazdi, M.; Bagheri, M. H. An efficient algorithm for function optimization: modified stem cells algorithm. *Cent. Eur. J. Eng.* **2012**, *3*, 36–50.

[36] Fan, Y.; Unwalla, R.; Denny, R.a; Di, L.; Kerns, E. H.; Diller, D. J.; Humblet, C. Insights for predicting blood-brain barrier penetration of CNS targeted molecules using QSPR approaches. *J. Chem. Inf. Model.* **2010**, *50*, 1123–33.

[37] Iyer, M.; Mishru, R.; Han, Y.; Hopfinger, A. J. Predicting blood-brain barrier partitioning of organic molecules using membrane-interaction QSAR analysis. *Pharm. Res.* **2002**, *19*, 1611–21.

[38] Meyer, D.; Leisch, F.; Hornik, K. The support vector machine under test. *Neurocomputing* **2003**, *55*, 169–186.

[39] Breiman, L. Random forests. *Machine Learning* **2001**, *45*, 5–32.

[40] Hinton, G. E.; Osindero, S.; Teh, Y.-W. A fast learning algorithm for deep belief nets. *Neural Comput.* **2006**, *18*, 1527–54.

[41] Carlsson, L.; Helgee, E. A.; Boyer, S. Interpretation of nonlinear QSAR models applied to Ames mutagenicity data. *J. Chem. Inf. Model.* **2009**, *49*, 2551–8.

[42] Chen, H.; Carlsson, L.; Eriksson, M.; Varkonyi, P.; Norinder, U.; Nilsson, I. Beyond the scope of Free-Wilson analysis: building interpretable QSAR models with machine learning algorithms. *J. Chem. Inf. Model.* **2013**, *53*, 1324–36.

[43] Young, R. C.; Mitchell, R. C.; Brown, T. H.; Ganellin, C. R.; Griffiths, R.; Jones, M.; Rana, K. K.; Saunders, D.; Smith, I. R.; Sore, N. E. Development of a new physicochemical model for brain penetration and its application to the design of centrally acting H2 receptor histamine antagonists. *J. Med. Chem.* **1988**, *31*, 656–71.

[44] Calder, J. A.; Ganellin, C. R. Predicting the brain-penetrating capability of histaminergic compounds. *Drug Des. Discov.* **1994**, *11*, 259–68.

[45] Clark, D. E. In silico prediction of blood-brain barrier permeation. *Drug Discov. Today* **2003**, *8*, 927–33.

[46] Norinder, U.; Sjöberg, P.; Osterberg, T. Theoretical calculation and prediction of brain-blood partitioning of organic solutes using MolSurf parametrization and PLS statistics. *J. Pharm. Sci.* **1998**, *87*, 952–9.

[47] Hou, T. J.; Xu, X. J. ADME evaluation in drug discovery. 3. Modeling blood-brain barrier partitioning using simple molecular descriptors. *J. Chem. Inf. Comput. Sci.* **2003**, *43*, 2137–52.

[48] Muehlbacher, M.; Spitzer, G. M.; Liedl, K. R.; Kornhuber, J. Qualitative prediction of blood-brain barrier permeability on a large and refined dataset. *J. Comput.-Aided Mol. Des.* **2011**, *25*, 1095–106.

[49] Kaliszan, R. Brain/blood distribution described by a combination of partition coefficient and molecular mass. *Int. J. Pharm.* **1996**, *145*, 9–16.

[50] Salminen, T.; Pulli, A.; Taskinen, J. Relationship between immobilised artificial membrane chromatographic retention and the brain penetration of structurally diverse drugs. *J. Pharm. Biomed. Anal.* **1997**, *15*, 469–77.

[51] Kelder, J.; Grootenhuis, P. D.; Bayada, D. M.; Delbressine, L. P.; Ploemen, J. P. Polar molecular surface as a dominating determinant for oral absorption and brain penetration of drugs. *Pharm. Res.* **1999**, *16*, 1514–9.

[52] Feher, M.; Sourial, E.; Schmidt, J. M. A simple model for the prediction of blood-brain partitioning. *Int. J. Pharm.* **2000**, *201*, 239–47.

[53] Osterberg, T.; Norinder, U. Prediction of polar surface area and drug transport processes using simple parameters and PLS statistics. *J. Chem. Inf. Comput. Sci. 40*, 1408–11.

[54] Narayanan, R.; Gunturi, S. B. In silico ADME modelling: prediction models for blood-brain barrier permeation using a systematic variable selection method. *Bioorg. Med. Chem.* **2005**, *13*, 3017–28.

[55] Kortagere, S.; Chekmarev, D.; Welsh, W. J.; Ekins, S. New predictive models for blood-brain barrier permeability of drug-like molecules. *Pharm. Res.* **2008**, *25*, 1836–45.

[56] Winiwarter, S.; Ridderström, M.; Ungell, A. L.; Andersson, T. B.; Zamora, I. Use of molecular descriptors for ADME predictions. In *Comprehensive Medicinal Chemistry II, Vol. 5 ADME-Tox Approaches*; Testa, B.; Van de Waterbeemd, H., Eds.; Elsevier, Amsterdam, **2007**; pp. 531–554.

[57] Ertl, P.; Rohde, B.; Selzer, P. Fast calculation of molecular polar surface area as a sum of fragment-based contributions and its application to the prediction of drug transport properties. *J. Med. Chem.* **2000**, *43*, 3714–7.

[58] Goodwin, J. T.; Clark, D. E. In silico predictions of blood-brain barrier penetration: considerations to "keep in mind". *J. Pharmacol. Exp. Ther.* **2005**, *315*, 477–83.

[59] ClogP. BioByte Corporation. Claremont, CA.

[60] ACDLogP. Advanced Chemistry Development, Inc., Toronto.

[61] Cruciani, G.; Pastor, M.; Guba, W. VolSurf: a new tool for the pharmacokinetic optimization of lead compounds. *Eur. J. Pharm. Sci.* **2000**, *11* Suppl 2, S29–39.

[62] Kaznessis, Y. N.; Snow, M. E.; Blankley, C. J. Prediction of blood-brain partitioning using Monte Carlo simulations of molecules in water. *J. Comput.-Aided Mol. Des.* **2001**, *15*, 697–708.

[63] Lobell, M.; Molnár, L.; Keserü, G. M. Recent advances in the prediction of blood-brain partitioning from molecular structure. *J. Pharm. Sci.* **2003**, *92*, 360–70.

[64] Gerebtzoff, G.; Seelig, A. In silico prediction of blood-brain barrier permeation using the calculated molecular cross-sectional area as main parameter. *J. Chem. Inf. Model.* **2006**, *46*, 2638–50.

[65] Renkin, E. M. Transport of potassium-42 from blood to tissue in isolated mammalian skeletal muscles. *Am. J. Physiol.* **1959**, *197*, 1205–10.

[66] Abraham, M. H. The factors that influence permeation across the blood-brain barrier. *Eur. J. Med. Chem.* **2004**, *39*, 235–40.

[67] Liu, X.; Tu, M.; Kelly, R. S.; Chen, C.; Smith, B. J. Development of a computational approach to predict blood-brain barrier permeability. *Drug Metab. Dispos.* **2004**, *32*, 132–9.

[68] Ghose, A. K.; Herbertz, T.; Hudkins, R. L.; Dorsey, B. D.; Mallamo, J. P. Knowledge-based, central nervous system (CNS) lead selection and lead optimization for CNS drug discovery. *ACS Chem. Neurosci.* **2012**, *3*, 50–68.

[69] Ajay; Bemis, G. W.; Murcko, M. A. Designing libraries with CNS activity. *J. Med. Chem.* **1999**, *42*, 4942–4951.

[70] Engkvist, O.; Wrede, P.; Rester, U. Prediction of CNS activity of compound libraries using substructure analysis. *J. Chem. Inf. Comput. Sci.* **2003**, *43*, 155–60.

[71] Guerra, A.; Páez, J. A.; Campillo, N. E. Artificial neural networks in ADMET modeling: prediction of blood-brain barrier permeation. *QSAR Comb. Sci.* **2008**, *27*, 586–594.

[72] Wang, Z.; Yan, A.; Yuan, Q. Classification of blood-brain barrier permeation by Kohonen's Self-Organizing Neural Network (KohNN) and Support Vector Machine (SVM). *QSAR Comb. Sci.* **2009**, *28*, 989–994.

[73] Martins, I. F.; Teixeira, A. L.; Pinheiro, L.; Falcao, A. O. A Bayesian approach to in silico blood-brain barrier penetration modeling. *J. Chem. Inf. Model.* **2012**, *52*, 1686–97.

[74] Yuan, Y.; Shaw, M. J. Induction of fuzzy decision trees. *Fuzzy Sets and Systems* **1995**, *69*, 125–139.

[75] Cook, E. F.; Goldman, L. Empiric comparison of multivariate analytic techniques: advantages and disadvantages of recursive partitioning analysis. *J. Chronic Dis.* **1984**, *37*, 721–31.

[76] Lipinski, C. A.; Lombardo, F.; Dominy, B. W.; Feeney, P. J. Experimental and computational approaches to estimate solubility and permeability in drug discovery and development settings. *Adv. Drug Deliv. Rev.* **2001**, *46*, 3–26.

[77] Van de Waterbeemd, H.; Camenisch, G.; Folkers, G.; Chretien, J. R.; Raevsky, O. A. Estimation of blood-brain barrier crossing of drugs using molecular size and shape, and H-bonding descriptors. *J. Drug Target.* **1998**, *6*, 151–65.

[78] Wager, T. T.; Hou, X.; Verhoest, P. R.; Villalobos, A. Moving beyond rules: the development of a central nervous system multiparameter optimization (CNS MPO) approach to enable alignment of druglike properties. *ACS Chem. Neurosci.* **2010**, *1*, 435–49.

[79] Summerfield, S. G.; Read, K.; Begley, D. J.; Obradovic, T.; Hidalgo, I. J.; Coggon, S.; Lewis, A. V; Porter, R. A.; Jeffrey, P. Central nervous system drug disposition: the relationship between in situ brain permeability and brain free fraction. *J. Pharmacol. Exp. Ther.* **2007**, *322*, 205–13.

[80] Maurer, T. S.; Debartolo, D. B.; Tess, D. A.; Scott, D. O. Relationship between exposure and nonspecific binding of thirty-three central nervous system drugs in mice. *Drug Metab. Dispos.* **2005**, *33*, 175–81.

[81] Doran, A.; Obach, R. S.; Smith, B. J.; Hosea, N. A.; Becker, S.; Callegari, E.; Chen, C.; Chen, X.; Choo, E.; Cianfrogna, J.; Cox, L. M.; Gibbs, J. P.; Gibbs, M. A.; Hatch, H.; Hop, C. E. C. A.; Kasman, I. N.; Laperle, J.; Liu, J.; Liu, X.; Logman, M.; Maclin, D.; Nedza, F. M.; Nelson, F.; Olson, E.; Rahematpura, S.; Raunig, D.; Rogers, S.; Schmidt, K.; Spracklin, D. K.; Szewc, M.; Troutman, M.; Tseng, E.; Tu, M.; Van Deusen, J. W.; Venkatakrishnan, K.; Walens, G.; Wang, E. Q.; Wong, D.; Yasgar, A. S.; Zhang, C. The impact of P-glycoprotein on the disposition of drugs targeted for indications of the central nervous system: evaluation using the MDR1A/1B knockout mouse model. *Drug Metab. Dispos.* **2005**, *33*, 165–74.

[82] Shen, D. D.; Artru, A. A.; Adkison, K. K. Principles and applicability of CSF sampling for the assessment of CNS drug delivery and pharmacodynamics. *Adv. Drug Deliver. Rev.* **2004**, *56*, 1825–57.

[83] Plowright, A. T.; Nilsson, K.; Antonsson, M.; Amin, K.; Broddefalk, J.; Jensen, J.; Lehmann, A.; Jin, S.; St-Onge, S.; Tomaszewski, M. J.; Tremblay, M.; Walpole, C.; Wei, Z.; Yang, H.; Ulander, J. Discovery of agonists of cannabinoid receptor 1 with restricted central nervous system penetration aimed for treatment of gastroesophageal reflux disease. *J. Med. Chem.* **2013**, *56*, 220–40.

[84] Winiwarter, S.; Hilgendorf, C. Modeling of drug-transporter interactions using structural information. *Curr. Opin. Drug Discov. Devel.* **2008**, *11*, 95–103.

[85] Ekins, S.; Ecker, G. F.; Chiba, P.; Swaan, P. W. Future directions for drug transporter modelling. *Xenobiotica 37*, 1152–70.

[86] Demel, M. A.; Krämer, O.; Ettmayer, P.; Haaksma, E. E. J.; Ecker, G. F. Predicting ligand interactions with ABC transporters in ADME. *Chem. Biodiv.* **2009**, *6*, 1960–9.

[87] Visser, U.; Abeyruwan, S.; Vempati, U.; Smith, R.P.; Lemmon, V.; Schürer, S.C. BioAssay Ontology (BAO): a semantic description of bioassays and high-throughput screening results. *BMC Bioinf.* **2011**, *12*, 257.

[88] Zdrazil, B.; Pinto, M.; Vasanthanathan, P.; Williams, A. J.; Balderud, L. Z.; Engkvist, O.; Chichester, C.; Hersey, A.; Overington, J. P.; Ecker, G. F. Annotating human P-glycoprotein bioassay data. *Mol. Inf.* **2012**, *31*, 599–609.

10

IN VITRO ASSAYS FOR ASSESSING BBB PERMEABILITY: ARTIFICIAL MEMBRANE AND CELL CULTURE MODELS

ALEX AVDEEF,[1] MÁRIA A. DELI,[2] AND WINFRIED NEUHAUS[3]

[1] *in-ADME Research, New York, USA*
[2] *Institute of Biophysics, Biological Research Centre, Hungarian Academy of Sciences, Szeged, Hungary*
[3] *Department of Pharmaceutical Chemistry, University of Vienna, Vienna, Austria*

INTRODUCTION

This chapter aims at giving an overview about current and future *in vitro* methods to assess blood–brain barrier (BBB) permeability, focusing on artificial membrane and cell culture models. The transport of molecules across the BBB can be mediated or regulated by different mechanisms such as passive diffusion, carrier-mediated uptake, active influx or efflux transport proteins, adsorption, or receptor-mediated transcytosis. These mechanisms are described in previous chapters of this book in detail for small molecules (e.g., drugs) as well as for proteins (e.g., antibodies) (see Chapters 5–8). The BBB and its functionality are highly regulated by the BBB's microenvironment in health as well as in disease. It is known that astrocytes, pericytes, and neurons as well as the shear stress caused by blood flow influence BBB functionality significantly (see Chapters 2 and 4). The challenge of *in vitro* BBB model development is to reproduce the *in vivo* properties, considering these complex

Blood–Brain Barrier in Drug Discovery: Optimizing Brain Exposure of CNS Drugs and Minimizing Brain Side Effects for Peripheral Drugs, First Edition. Edited by Li Di and Edward H. Kerns.

interactions relevant for BBB functionality. In drug discovery there is a huge need for reliable high (at least medium)-throughput methods to screen for BBB permeability, on the one hand, to determine if hit and potential lead substances acting in the periphery do not cross the BBB to avoid unwanted adverse central nervous system (CNS) side effects and, on the other hand, to bring drugs across the BBB to their targeted sites, such as in the treatment of CNS-related diseases (e.g., epilepsy, Alzheimer's disease, stroke, brain tumor, multiple sclerosis, lysosomal storage diseases, and so on—see Chapter 4). In this context, state-of-the-art, but also potential and possible artificial membrane and cell culture models to meet the requirements of drug discovery issues will be discussed in this chapter. Artificial membranes could be established and applied in different test systems (immobilized artificial membrane–high-performance liquid chromatography (IAM-HPLC), bilayer lipid membrane, parallel artificial membrane permeability assay (PAMPA)). Theoretical backgrounds, model types, advantages and disadvantages, application examples, and novel developments, with a special focus on PAMPA for BBB permeability screening, are introduced. Cell culture models provide the possibility to include the investigation of active (energy-dependent) transport processes and molecular mechanisms. An overview of currently used BBB cell culture models in different setups is given (static Transwell, dynamic flow–based hollow-fiber and microfluidic models), their advantages and disadvantages, as well as technical aspects, are discussed. Several used cell sources (primary cells versus immortalized cell lines) for BBB modeling are compared, emphasizing distinct differences between brain-derived models and surrogate models based on cell lines, such as Caco-2, MDCK, or ECV304. Depending on the aims and purposes of BBB cell culture studies, more or less complex models could be applied. In this context, coculture and triple-culture models, based on brain endothelial cells combined with astrocytes and/or pericytes, are described in more detail. The relevance of these models to come closer to properties of the *in vivo* BBB, such as increased paracellular tightness and transport processes, is highlighted. Last, but not least, the fact of species differences has to be considered when choosing a BBB cell culture model.

ARTIFICIAL MEMBRANE MODELS

The sheer size of the surface of the BBB ensures that passive lipoidal transcellular diffusion (or paracellular aqueous diffusion of small hydrophilic molecules, such as atenolol, or in special cases, inulin) represents a possible route for drug permeation across the BBB [1]. As the lead chemical series is assembled for a CNS project, physicochemical property measurements soon follow, for example, pK_a, oil–water partition coefficients (log $PC_{o/w}$), IAM-HPLC retention, PAMPA, plasma unbound fraction ($f_{u,pl}$), and brain homogenate unbound fraction ($f_{u,br}$). BBB penetration models [2, 3] may be constructed from these properties, sometimes in combination with in silico descriptors ("in combo" approach). These models usually cannot predict carrier-mediated transport, and are, thus, mainly confined to passive-diffusion processes (both lipoidal and aqueous) (Table 10.1).

TABLE 10.1 Comparison of artificial membrane and cell culture models

Model type	Advantages	Disadvantages
PAMPA	High throughput	No active transport processes
	Cheaper than cell culture models	
Transwell model	Medium throughput	No flow
	Light microscopically accessible	
	Active transport processes	
	Cheaper than flow-based models	
Flow-based hollow-fiber model	Varied flow rates	No online light microscopically accessibility
	Increased life span	Small throughput
	Active transport processes	Silicon tubings—drug adsorption
	More *in vivo* like properties (increased tightness, changed enzyme, and transporter expression)	Needs gas sterilization (e.g., ethylene dioxide, formaldehyde)
	Online impedance measurement	Technically sophisticated setup
		Expensive
		Pore sizes too small for cell migration studies (leukocytes, bacteria, tumor cells)
Microfluidic model	Same advantages as flow-based hollow-fiber models (increased life span, more *in vivo* like properties (increased tightness, changed enzyme, and transporter expression))	Low analyte volume
		Highly sophisticated
		Expensive
	Light microscopically accessible	
	Less cell and medium consumption	Hardly any two-chamber systems for transport studies developed
	Online impedance measurement	

"Efflux-Minimized" *In Vivo–In Vitro* Correlations (IVIVC) for Passive-Diffusion Processes

To assess the effectiveness of the noncellular passive BBB permeability models described in the following sections, one needs to first identify a reference *in vivo* permeability system that will serve as a validation reference. The selection of such an *in vivo* basis is not a trivial task and can be contentious, due to the possible impact of carrier-mediated processes, which noncellular models cannot address directly. It is thus very important to identify a special subset of *in vivo* measurements which are predominantly passive in character. This section briefly defines such a "pruned" *in vivo* set, before the noncelluar permeability models are described in detail.

The *in vivo* benchmark against which the PAMPA and similar noncellular *in vitro* models have been most often compared is the *in situ* rodent brain perfusion permeability [4–7]. From a survey of the published literature, 602 PS values (permeability-surface area product (PS)) were gathered [8, pp. 635–657], some based on *in vivo* intravenous injection (i.v.), a few from bolus carotid artery injection brain uptake index (BUI), but most from *in situ* brain perfusion methods, from rats, mice, guinea pigs, rabbits, dogs, and cats. Only a portion of the PS values were used in training the PAMPA models. Only rat and mouse data (92% of the collected values) were used. Since plasma protein binding can lower values of PS (in comparison to protein-free perfusion), i.v. data were not used for lipophilic compounds to train the PAMPA models. Compounds that had reported saturable transport were also excluded. Since PAMPA is a passive lipoidal diffusion prediction model, PS values were selected from studies which used some sort of carrier-mediated transport inhibition (e.g., GF120918, PSC833, cyclosporin A, self-inhibition at high concentrations, mdr1a(−/−)/mrp1(−/−)/brcp-knockout (KO) mouse model). Simple amino acids and dipeptides were excluded from the training set, except for those with reported nonsaturable K_d constants from Michaelis–Menten analysis. Out of the starting set of 602 PS values, a total of 197 values were selected as the "efflux-minimized" training set [9]. An additional 85 values were designated as the "external" test set. The latter group was selected to include substrates of carrier-mediated processes, based on the following criteria. The PS values were transformed into charge-corrected intrinsic BBB permeability, $P_0^{in\ situ}$, corresponding to the permeability of the uncharged form of the drug, as described later. In studies where both KO/efflux-inhibited and wild-type (WT)/uninhibited rodent measurements were reported, the KO/efflux-inhibited values were added to the training set ($n=197$), but the corresponding WT/uninhibited paired values were added to the external test set ($n=85$), unless the WT/uninhibited values were either within a factor of 3 of the KO/inhibited or were very high ($P_0^{in\ situ}>0.01$ cm/s), in which case both values were used in training. The external set was not viewed as a rigorous model validation set, but was rather used to test whether actively transported molecules could be recognized by their deviations from the predicted passive values (negative/positive deviations indicating efflux/uptake transport processes, respectively).

In Silico Add-on to *In Vitro* Models: Abraham Solvation Descriptors

Intuitively understandable in silico descriptors, such as those described by Abraham [10], have been used to improve the prediction capability of the artificial membrane models ("in combo" approach, section "In-Combo Double-Sink PAMPA-BBB

Model for Passive BBB Permeability"). The Abraham linear free energy relationship (LFER) applied to BBB intrinsic permeability, $P_0^{in\ situ}$, is defined as

$$\log P_0^{\text{in-situ}} = c_0 + c_1 A + c_2 B + c_3 S + c_4 E + c_5 V \tag{10.1}$$

where c_0, …, c_5 are the multiple linear regression (MLR) coefficients and the five solvation descriptors are as follows:

- *A*—sum of solute H-bond acidity (overall H-bond donor strength).
- *B*—sum of solute H-bond basicity (overall H-bond acceptor strength).
- *S*—solute polarity/polarizability due to solute–solvent interactions between bond dipoles and induced dipoles.
- *E* (dm^3/mol/10)—solute excess molar refractivity, which models dispersion force interaction arising from pi- and *n* electrons of the solute.
- *V* (dm^3/mol/100)—McGowan characteristic molar volume of the solute. The volume term relates to the energy difference in creating a "cavity" in water compared to one in the membrane phase.

Abraham's descriptors are generally calculated from 2D structures (Percepta program from Advanced Chemistry Development, Inc., Toronto, Canada; www.ACDLabs.com).

PAMPA Models

Kansy et al. [11], who invented PAMPA, published a widely circulated study of the permeation of drugs across filters coated with a 10% wt/vol egg lecithin dodecane solution. The investigators were able to relate their measured fluxes to human absorption values with a hyperbolic curve, much like that indicated by cell-based (Caco-2, MDCK, etc.) permeability screening models. The outliers in their assays were drugs known to be actively transported. Since PAMPA is based on passive diffusion and is devoid of active transport systems and metabolizing enzymes, the assay would not be expected to model actively transported molecules.

In the PAMPA assay, a "sandwich" is formed from a 96 well microtitre plate and a 96 well microfilter plate, such that each combined cylinder well is divided into two compartments: donor at the bottom and acceptor at the top, separated by a 125 μm thick microfilter disc (70% porosity PVDF), usually coated with a 1–20% wt/vol lecithin dissolved in dodecane.

Since 1998, many variants of the PAMPA model have been tried, as summarized by Kansy et al. [12] and elaborated in substantial detail by Avdeef [8, Chapter 7]. In the Kansy review, the Di et al. [13] PAMPA-BBB model, based on 2% wt/vol porcine brain lipid extract dissolved in n-dodecane, was compared to the other PAMPA models. More recently, Tsinman et al. [9] developed the Double-Sink™ version of the PAMPA-BBB model, based on 10% wt/vol porcine brain lipid extract dissolved in a viscous alkane, coupled to the use of a pH-gradient universal buffer solution which contained a critically selected surfactant in the receiver wells. Both the receiver and the donor wells were stirred using rotating magnetic discs. In this chapter, all measured PAMPA-BBB values are from the Tsinman et al. [9] model, which may be called Double-Sink PAMPA-BBB, to distinguish it from the original PAMPA-BBB model of Di et al.

PAMPA Models of the BBB

Mensch et al. [14] tested four PAMPA models for predicting the brain/plasma ratio, K_p, which is often defined as $AUC_{TOT,brain}/AUC_{TOT,plasma}$, where AUC is the area under the total drug concentration–time curve in the brain or plasma over a dosing interval. (It should be pointed out that although K_p continues to be widely used in drug discovery as an index of brain penetration, its use in isolation is potentially misleading, since it is generally accepted that it is the unbound drug that exerts the physiological effect.) The CNS+/− discrimination was confirmed with the Di et al. [13] model. The ability to predict K_p was comparable with the porcine brain lipid extract and the much simpler dioleoylphosphatidylcholine-based PAMPA models ($r^2=0.63$ and 0.73, respectively).

Avdeef and Tsinman [15] compared three PAMPA models commonly in use (HDM, DOPC, DS). Figure 10.1 compares Abraham MLR coefficients describing four PAMPA models (HDM, DOPC, DS, and BBB), two partition coefficient models (octanol and 1,9-decadiene), BLM, and rodent *in situ* brain perfusion. The acronyms stand for HDM = hexadecane (intestinal model), DOPC = dioleoylphosphatidylcholine (intestinal model), DS = Double-Sink (intestinal model), BBB = blood–brain barrier model based on 10% wt/vol porcine brain lipid extract (also uses Double-Sink approach), and BLM = black lipid membrane. As can be seen, the Double-Sink PAMPA-BBB model (gray bars) most closely resembles the characteristics of the rodent permeability data (black bars). The Abraham regression coefficients in Figure 10.1 for the Double-Sink PAMPA-BBB model were determined using the 197-molecule training set mentioned earlier.

Dagenais et al. [16] developed a BBB permeability model using the in combo PAMPA-DS intestinal model (in combo refers to the *combination* of measured PAMPA values and one or two Abraham descriptors). Thirty-eight *in situ* PS measurements performed on 19 compounds in WT and P-glycoprotein (P-gp)-deficient (mdr1a(−/−)) CF-1 mice (so-called knockout model) were reported. One aim of the study was to quantify the influence of P-gp on BBB permeability by comparing the mouse genotypes. The second aim was to use permeability values from the P-gp-deficient data to train the in combo PAMPA model. The aim at that time was to develop a model for early screening for passive BBB permeability, which could improve prediction of CNS exposure for test compounds, as part of a broader model incorporating $PS_{passive}$ along with the unbound drug fractions in the brain ($f_{u,br}$), as reviewed by Avdeef (8, pp. 604–607). Although the fit of the *in vivo* data was good using the in combo PAMPA-DS procedure, the correlation between $P_0^{in\ situ}$ and the directly measured log $P_0^{PAMPA\text{-}DS}$ data was not better than that of the octanol–water model, as shown in Figure 10.2a. This prompted the search for a better PAMPA model, described later in the work of Tsinman et al. [9].

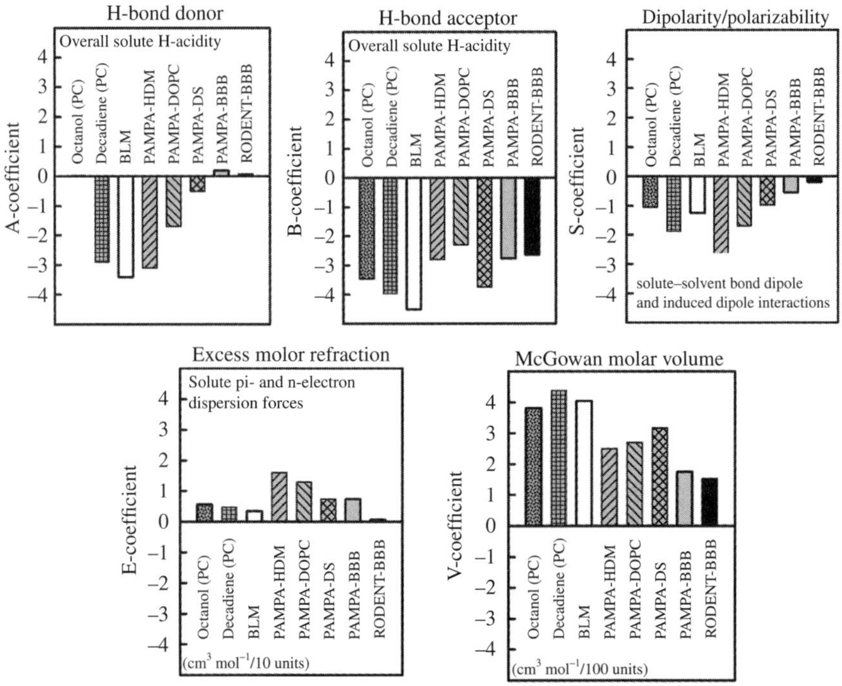

FIGURE 10.1 Graphical depiction of the five Abraham multiple linear regression coefficients for several noncellular permeability models (octanol– and decadiene–water partition coefficients, bilayer lipid membranes (BLM), and four PAMPA models), along with the rodent *in situ* brain prefusion permeability (black bars on the right). The PAMPA-BBB model closely matches the *in vivo* model for most descriptors.

PAMPA-BBB The first purpose-designed PAMPA model for the BBB was described by Di et al. [13], called PAMPA-BBB. It was based on 2% wt/vol porcine brain lipid extract dissolved in dodecane, a novel departure from the use of lecithin. It was demonstrated that drug molecules can be binned into CNS+/− activity classes. A direct comparison of the original PAMPA-BBB model to *in situ* rat brain perfusion permeability coefficients reported by Summerfield et al. [17] suggested appreciable chemical selectivity in the original PAMPA-BBB model [18].

In Combo Double-Sink PAMPA-BBB Model for Passive BBB Permeability In general, PAMPA models have substantially improved since their introduction in 1998, and have become considerably more complex, which in turn has made them more predictive of *in vivo* permeability. The measurements described in this chapter are based on the Double-Sink PAMPA-BBB [9], which uses a 10% wt/vol porcine brain lipid extract model, combined with sink-forming buffers of the Double-Sink approach. For the purpose of keeping clear the distinctions in the PAMPA-BBB variants, the prefix "Double-Sink" is added. When Double-Sink PAMPA-BBB is used in combination with Abraham in silico descriptors, a further prefix is added: in

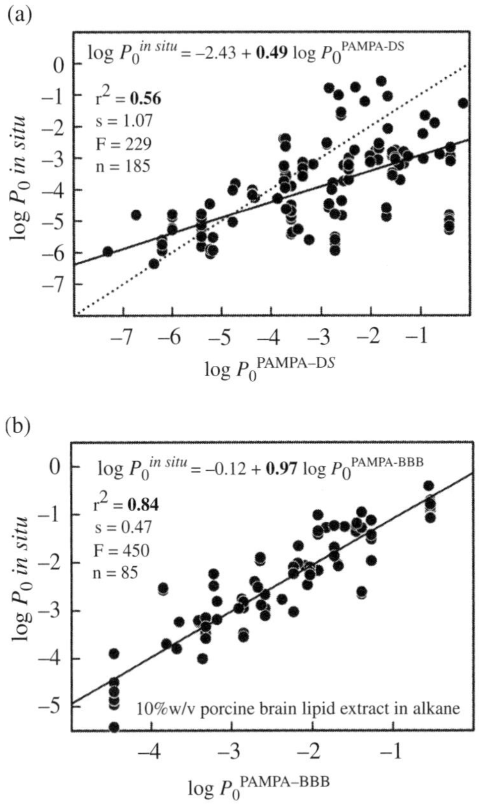

FIGURE 10.2 Correlation between *in situ* brain perfusion intrinsic permeability, log $P_0^{in\ situ}$, and two PAMPA models: (a) Double-Sink intestinal model, log $P_0^{PAMPA\text{-}DS}$, and (b) PAMPA-BBB (10% wt/vol porcine brain lipid extract, PBLE, in alkane), log $P_0^{PAMPA\text{-}BBB}$ for positively charged (weak base) drugs. (Reprinted from Tsinman O, Tsinman K, Sun N, Avdeef A. Physicochemical selectivity of the BBB microenvironment governing passive diffusion—matching with a porcine brain lipid extract artificial membrane permeability model. *Pharm Res.* 2011;28:337–363. Copyright © 2011 Springer. With kind permission of Springer Science+Business Media.)

combo Double-Sink PAMPA-BBB. The prefix "in combo" (originally coined by Han van de Waterbeemd) has been applied in a number of published physical property prediction studies, and describes the "combination" of a *measured* quantity (e.g., PAMPA) and a *calculated* descriptor (e.g., Abraham H-bond acidity or basicity).

An effort was undertaken by Tsinman et al. [9] to develop an improved PAMPA lipid formulation. They described a new formulation based on 10% wt/vol porcine brain lipid extract, using a fivefold higher lipid concentration in a more viscous alkane solvent than dodecane and with thinner membranes, compared to that used by Di et al. [13, 18]. A new sink-forming surfactant was used in the receiver pH 7.4 buffer.

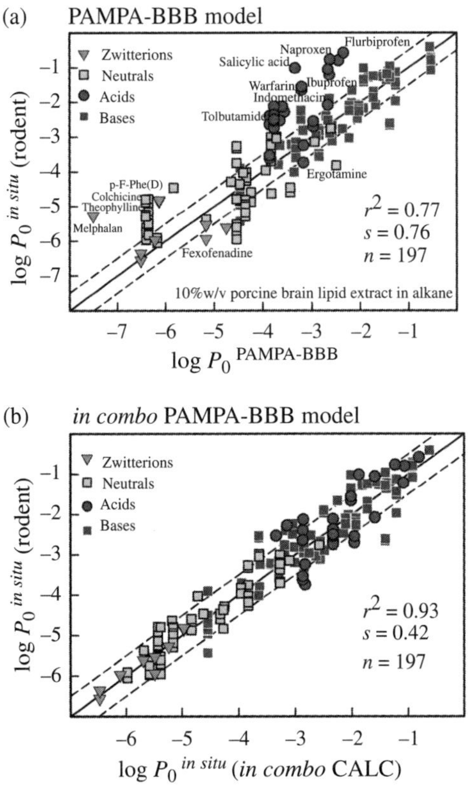

FIGURE 10.3 (a) *In vitro–in vivo* correlation between *in situ* rodent brain perfusion intrinsic permeability (based on the training set of 197 "efflux-minimized" *in situ* brain perfusion) and PAMPA-BBB (10% wt/vol PBLE in alkane) intrinsic permeability. (b) The in combo PAMPA-BBB of the same set of compounds. (Reprinted from Tsinman O, Tsinman K, Sun N, Avdeef A. Physicochemical selectivity of the BBB microenvironment governing passive diffusion—matching with a porcine brain lipid extract artificial membrane permeability model. *Pharm Res*. 2011;28:337–363. Copyright © 2011 Springer. With kind permission of Springer Science + Business Media.)

The 10% porcine brain lipid extract Double-Sink PAMPA-BBB intrinsic permeability values for 108 compounds were correlated to those of 197 "efflux-minimized" published *in situ* rodent brain perfusion PS measurements. Tsinman et al. [9] were able to demonstrate a remarkably high match between the physicochemical selectivity of the new PAMPA-BBB and the *in situ* data, with slope = 0.97 for a series of weak bases thought to permeate passively, as shown in Figure 10.2b. The nature of this physicochemical selectivity was also characterized in terms of the Abraham linear free energy solvation descriptors.

The acids, neutral, and zwitterionic compounds did not fit as well as the bases, as is evident in Figure 10.3a ($r^2 = 0.77$, $s = 0.76$, $n = 197$). To improve the correlation of the acids, neutral, and zwitterionic molecules, the in combo approach [8, pp. 659–663], augmenting the PAMPA-BBB values with one or two Abraham descriptors, resulted in excellent IVIVC as shown in Figure 10.3b ($r^2=0.93$, $s=0.42$, $n=197$) as discussed later.

For the Abraham analysis, PS values were corrected for ionization (using known pK_a values), assuming that *in vivo* permeability was largely free of carrier-mediated effects, to produce intrinsic BBB permeability, $P_0^{in\ situ}$, values. This was done because the Abraham solute descriptors have been developed for uncharged species in the LFER approach. Although it may seem unnecessary, given that the environment of the BBB is very close to pH 7.4, the transformation is primarily a computational strategy to adapt the LFER descriptors to charged drugs [9, 16].

In addition to the 5-parameter in silico LFER models (gray bars for PAMPA-BBB and black bars for rodent *in situ* brain perfusion in Fig. 10.1), it was explored how well PAMPA-BBB measurements, augmented with one (or two) of Abraham's molecular solvation descriptors, can predict passive intrinsic permeability values of the *in situ* data. The combination of measured PAMPA-BBB and a calculated LFER descriptor defines the in combo method:

$$\log P_0^{in\,situ}(\text{in combo}) = c_0 + c_1 \log P_0^{\text{PAMPA-BBB}} + c_2 H \qquad (10.2)$$

where $c_0, \ldots, c_2$ are MLR coefficients and H is a function of Abraham H-bond descriptors (A, $A+B$, or $A-B$). Fewer MLR coefficients are necessary in Equation 10.2, compared to Equation 10.1, because the PAMPA-BBB P_0 already encodes for some of the properties of the microenvironment of the *in vivo* boundary domain (especially well for bases).

In Figure 10.3a, the best-fit slope (selectivity coefficient) = 0.87, already suggesting a highly predictive model. But when the measurements were scrutinized by four charge classes, a more complicated view emerged. The bases (positively charged), indicated by filled squares in Figure 10.3a, were associated with slope = 0.97 ± 0.05 ($r^2=0.84$), which was a near perfect match in selectivity. The acids (negatively charged), indicated by filled circles, showed slope = 1.51 ± 0.32 ($r^2=0.59$). The neutral compounds, represented by unfilled squares, showed slope = 0.56 ± 0.07 ($r^2=0.47$), a selectivity comparable to that of log $PC_{oct/w}$. The zwitterions (triangles) indicated slope ~0 ($r^2 \sim 0$), that is, no dependence on PAMPA permeability. Evidently, the BBB microenvironment affecting passive permeability is not well matched by the neutral and ampholyte drugs. For zwitterions, there was no correlation. The charge-class in combo PAMPA-BBB that will be described later improved the overall fit, and may have shed some light on the presence of carrier-mediated transport undetected in the selection of the training-set *in situ* brain perfusion values.

The 197 "efflux-minimized" *in situ* brain perfusion permeability values (intrinsic form), $P_0^{in\ situ}$, were divided into four predominant-charge groups (at pH 7.4). For each group, the linear regression equation best predicting passive $P_0^{in\ situ}$ was

Positive charge (bases):

$$\begin{aligned} &\text{Log}\, P_0^{in\,situ} = -0.01 + 0.94 \log P_0^{\text{PAMPA-BBB}} - 0.64A \\ &r^2 = 0.86,\ s = 0.46,\ F = 253,\ n = 85 \end{aligned} \qquad (10.3\text{a})$$

Negative charge (acids):

$$\text{Log}\, P_0^{\,in\,situ} = 2.54 + 1.11 \log P_0^{\,\text{PAMPA-BBB}} - 0.65(A+B)$$
$$r^2 = 0.61,\ s = 0.56,\ F = 20,\ n = 28 \tag{10.3b}$$

Uncharged:

$$\text{Log}\, P_0^{\,in\,situ} = -0.40 + 0.63 \log P_0^{\,\text{PAMPA-BBB}} - 0.44(A+B)$$
$$r^2 = 0.88,\ s = 0.33,\ F = 255,\ n = 76 \tag{10.3c}$$

Zwitterionic ± charge:

$$\text{Log}\, P_0^{\,in\,situ} = -4.81 + 0.73(A-B)$$
$$r^2 = 0.86,\ s = 0.22,\ F = 38,\ n = 8 \tag{10.3d}$$

The zwitterions appear to be in a class by themselves, not dependent on lipophilicity or PAMPA. The *in situ* permeability of the zwitterions is strongly correlated to the difference between the H-bond acidity and basicity $(A-B)$ of the solute ("difference" H-bond effect), with PAMPA-BBB playing no role.

The four charge-class analyses were combined into a single equation, using orthonormal indicator indices, I_-, I_+, I_0, and $I_\pm$, each of which has unit value as negatively charged acids, positively charged bases, neutrals, and zwitterions, respectively, and zero otherwise:

$$\begin{aligned}\text{Log}\, P_0^{\,in\,situ} &= \{c_0 + c_1 \log P_0^{\,\text{PAMPA-BBB}} + c_2 A\} I_+ \\ &+ \{c_3 + c_4 \log P_0^{\,\text{PAMPA-BBB}} + c_5(A+B)\} I_- \\ &+ \{c_6 + c_7 \log P_0^{\,\text{PAMPA-BBB}} + c_8(A+B)\} I_0 + \{c_9 + c_{10}(A-B)\} I_\pm\end{aligned} \tag{10.4}$$

The MLR analysis for the combined training set yielded $r^2 = 0.93$, $s = 0.42$, $F = 1454$, $n = 197$. Figure 10.3b shows the IVIVC plot, based on the given combined equation.

Possible Implications of Carrier-Mediated Transport

For new chemical entities (NCEs) with unknown mechanism of transport, having a reliable prediction of passive BBB permeability could serve to indicate the presence of carrier-mediated processes.

Figure 10.4 shows the relationship between the in combo model predictions (calculated using *p*CEL-X v4.0, *in-ADME* Research) and observed BBB permeability values for 85 "external" *in situ* set of measurements not used in the training of the model. Many of the compounds in the external set comparison are known to be substrates for efflux transporters (e.g., quinidine, paclitaxel, fexofenadine, DPDPE), especially the molecules which lie significantly below the identity line in Figure 10.4. The PAMPA-BBB model could suggest that molecules substantially outside of the

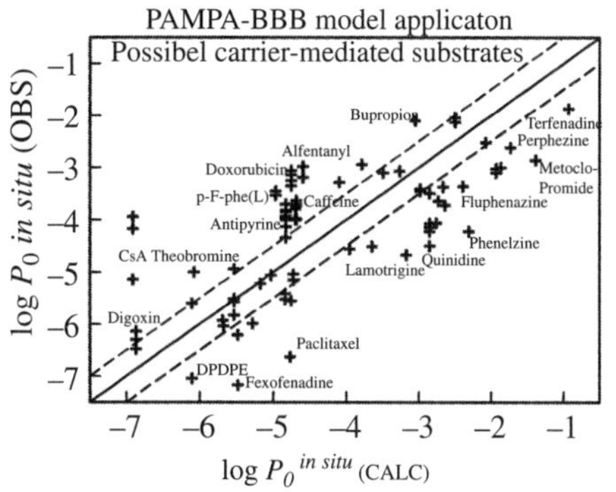

FIGURE 10.4 The 85 measured *in situ* "external" set values of compounds which could potentially be actively transported, compared to those calculated from the in combo PAMPA-BBB model. Values threefold below the identity line (marked off by the dashed line) could be indicative of efflux processes. Values threefold above the identity line could be indicative of carrier-mediated uptake processes. (Reprinted from Tsinman O, Tsinman K, Sun N, Avdeef A. Physicochemical selectivity of the BBB microenvironment governing passive diffusion—matching with a porcine brain lipid extract artificial membrane permeability model. *Pharm Res*. 2011;28:337–363. Copyright © 2011 Springer. With kind permission of Springer Science + Business Media.)

threefold window (dashed lines on both sides of the identity line in Fig. 10.4) might be affected by a carrier-mediated process.

Figure 10.5 shows a diagram of acids arranged vertically according to the $(A+B)$ descriptor sum (left vertical scale). Equation 10.3b was assumed to consist of a passive permeability contribution, 1.11 log $P_0^{\text{PAMPA-BBB}}$, plus an "amplification" term, $2.54 - 0.65\ (A+B)$. The latter term is the calculated "in combo" contribution from the IVIVC analysis of acids. (Note that the superscript PAMPA-BBB refers to the Double-Sink PAMPA-BBB model of Tsinman et al.) The scale on the right indicates values of $10^{+2.54-0.65(A+B)}$. The "amplification" term could indicate a contribution from a carrier-mediated transport mechanism. At the top of Figure 10.5 are valproic acid, ibuprofen, salicylic acid, and flurbiprofen, with amplification factors near 70. That is, if there were a carrier-mediated effect, it would explain why PAMPA-BBB values for these acids are 70 times too low. H-bonding $(A+B)$ sum would mimick the effect. This might be a clue to the nature of the putative carrier-mediated mechanism.

The human whole-brain relative expression of the dozen most prevalent transporters [19] (UCSF-FDA TransPortal, http://bts.ucsf.edu/fdatransportal) follow the rank order: OATP1A2 > MRP5 > PEPT2 > BCRP > MRP4 > OATP2B1 > P-gp > MRP 1 > OCT3 > OCTN1 > OCTN2 > OAT1. On the luminal surface of the BBB, six are efflux transporters (BCRP, P-gp, MRP1,2,4,5) and two are uptake/bidirectional transporters (OATP1A2, OATP2B1). The well-documented substrates of OATP1A2 and OATP2B1 include several statins, fexofenadine, levofloxacin, methotrexate, and saquinavir. The uptake substrates are transported through a positively charged pore by a rocker-switch mechanism, exhibiting the pharmacophore model of two H-bond

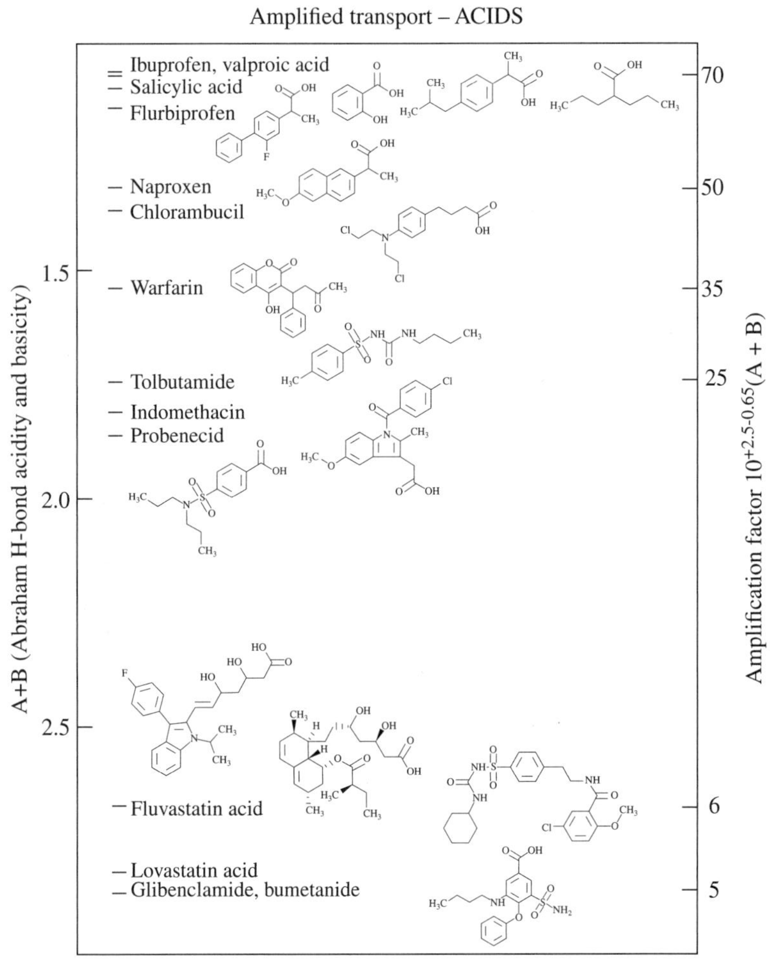

FIGURE 10.5 The *in situ* intrinsic permeability of acids is higher (by up to a factor of 70 for valproic acid, ibuprofen, salicylic acid, and flurbiprofen) than predicted by the PAMPA-BBB model. The differences can be reconciled by the Abraham H-bond sum-descriptor function, $2.54 - 0.65\,(A+B)$. This might be an indication of a carrier-mediated process.

acceptors, one H-bond donor, and two hydrophobic regions [20]. It is not evident how this can be reconciled with the acids "amplification" term in the PAMPA-BBB analysis.

There are few literature indications of nonsteroidal anti-inflammatory drug (NSAID) uptake transporters at the BBB. Although it is well accepted that NSAIDs are potent inhibitors of renal OAT1, NSAIDs themselves are poorly transported [21]. OAT1 is only slightly expressed inside the brain (but not at the BBB). Westholm et al. [22] reported that Oatp1c1 and Oatp1a4 are expressed at the rat BBB. Some NSAIDs were found to be inhibitors of thyroxine uptake by Oatp1c1 (meclofenamic acid > diclofenac > fenamic acid), but indomethacin, ibuprofen, flurbiprofen, and acetylsalicylic acid had no effect. It was suggested that Oatp1c1 could be a transporter of some NSAIDs, but this remains to be further investigated.

CELL CULTURE MODELS

The isolation of brain microvessels, pioneered by Ferenc Joó and his coworkers in 1973, was seminal for the development of BBB culture models [23]. Five years later they observed that endothelial cells grow out of brain capillaries in culture conditions [24]. This early observation was soon followed by a plethora of different culture-based BBB models to study the physiology, pharmacology, and pathology of the BBB [25]. The present review focuses mainly on cell culture–based BBB models developed and used for permeability studies which can be relevant for drug discovery, except human endothelial cell–based *in vitro* BBB models discussed in Chapter 11 of this book [26].

Cell Sources

Primary Cultures of Brain Endothelial Cells versus Cell Lines The first BBB models were monocultures of primary brain microvascular endothelial cells in the early 1980s. These cultures allowed the development of *in vitro* permeability assays by the help of inserts with porous membranes (Table 10.1), but several technical and methodological and other problems arose.

Purity of Cultures One of them was the purity of the cultures, the contamination of endothelial cultures by other cell types, especially brain pericytes, fibroblasts, smooth muscle, or leptomeningeal cells [27, 28], which resulted in loss of monolayer tightness. Methods to solve this problem included complement-mediated cytolysis of Thy1.1 expressing contaminating cells, use of 2–3-week-old rats for brain microvessel endothelial cell isolation, and plasma-derived serum containing low level of platelet-derived growth factor instead of classical fetal bovine sera, all favoring the growth of endothelial cells over pericytes [28]. Treatment of cultures by puromycin, an antibiotic drug and P-gp ligand, to obtain higher endothelial purity, was a big advancement in the field [29, 30]. Brain endothelial cells expressing high amount of P-gp survive puromycin treatment, while the drug selectively kills contaminating cells when applied in the first few days of culture [29, 30]. This selection may also favor capillary endothelial cells expressing the highest level of efflux pumps in the brain vascular tree versus endothelial cells from larger microvessels, thus resulting in tighter monolayers and better BBB models [31]. Another factor contributing to the tightness of intercellular junctions in primary cerebral endothelial cultures is the age-related developmental stage of cerebral capillaries. A recent paper demonstrated that primary cultures from 2-week-old rats mimic newborn BBB unlike cultures from 8-week-old rats showing more differentiated BBB phenotype [32]. This finding is supported by data on new improved BBB in vitro models of porcine and rat origin which use both the puromycin purification method and brain microvessels from adult animals [33–36]. The tightness of these models measured by electric resistance may reach and exceed 1000 $\Omega \cdot cm^2$ and, therefore, is close to the *in vivo* values [37].

Tightness of Junctions The interendothelial junctions of brain capillary endothelial cells form the tightest paracellular pathway in the vascular tree and are tighter than most of the epithelial barriers except urothelium [38, 39]. The tightness of the junctions can be tested morphologically by electron microscopy and functionally by electrical resistance measurements. The morphology of cerebral endothelial tight junctions is seen as specialized contact zones in transmission electron micrographs and as a mesh of interlocking junctional strands made of fine particles in freeze fracture microscopy [39]. The complexity of strands, looking like pearl necklaces, and the degree of association of the particles with the inner (P-face) lipid leaflet of the membrane correlate directly with the observed transepithelial or endothelial resistance [39]. The anatomical structure of primary rat brain endothelial cells in triple coculture including tight junctions is strikingly similar to brain capillaries *in situ* by transmission electron microscopy [40, 41]. By freeze fracture electron microscopy the P-face association of junctional particles of brain endothelial cells in culture models examined so far did not reach that of the brain capillaries [39]. However, culture BBB models with high electric resistance close to *in vivo* values listed in Table 10.2 [33–36, 55, 56, 65] were not examined by this technique. This is an area where more studies are needed to better characterize and validate the tightest BBB models based on cultured cells.

Transporter Functionality The *in situ* BBB expresses a large number of transport systems, including active efflux transporters, solute carriers, and receptor-mediated transport routes. These are indispensable for regulating the homeostasis of the nervous system by protecting it from toxic substances and providing nutrients and hormones (for review see Abbott et al. [1]; also, cf. section "PAMPA Models of the BBB"). Among the transporters, the functionality of efflux pumps are the best studied, and efflux activity by fluorescent or drug substrates and inhibitors were proven on most models. P-gp (ABCB1), the first described and best characterized ABC transporter at the BBB, is present in virtually all primary culture models described so far (for rat models, see, e.g., Perrière et al. [30], Nakagawa et al. [79], Abbott et al. [33], Watson et al. [36]). ABCG2 (BCRP) was only identified at the BBB in 2002 [97]. Its importance is highlighted by recent quantitative proteomic studies indicating that ABCG2 is more abundant than ABCB1 at the human BBB [98]. Its presence has been confirmed in several primary culture BBB models [30, 34, 97, 99]. It is difficult to measure the functionality of individual efflux transporters on BBB culture models, because these transporters have a very broad and overlapping substrate and inhibitor profile and form a robust cooperative efflux system at the BBB *in vivo* [100, 101]. In two comparative permeability studies, primary culture–based BBB models could distinguish between drugs transported by passive and efflux mechanisms [41, 46].

TABLE 10.2 Permeability data obtained on selected *in vitro* BBB models based on primary brain endothelial cells, brain endothelial cell lines, or surrogate cells

In vitro models	References	TEER ($\Omega \cdot cm^2$)	P_e or P_{app} (10^{-6} cm/s)	Correlation
		Primary brain endothelial cells		
BOVINE				
Cloned BBECs + rat ACs	[42]	661 ± 48	Sucrose $P_e \approx 5.4$	$R^2 \approx 0.86^b$ ($n = 12$)
	[43]	NI	Sucrose $P_e \approx 10.5$	$R^2 \approx 0.77^b$ ($n = 10$)
	[44]	NI	Sucrose $P_e \approx 4.2$	$R^2 \approx 0.92^b$ ($n = 9$)
	[45]	NI	Sucrose $P_e \approx 8.3$	$R^2 \approx 0.90^b$ ($n = 13$)
	[46]	NI	Sucrose $P_e \approx 4.0$	$R^2 \approx 0.43$ ($n = 22$)
	[47]	NI	Sucrose $P_e \approx 5.8$	$R^2 \approx 0.81^b$ ($n = 10$)
BBECs	[48]	ND	Sucrose $P_e \approx 85$	$R^2 \approx 0.72^b$ ($n = 13$)
BBECs + ACM + cAMP↑	[49]	625 ± 82	ND	ND
BBECs	[50]	≈150[a]	NI	$R^2 \approx 0.96$ ($n = 7$)
BBECs + rat ACs/ACM	[51]	352–857	SF $P_e \approx 6.0$	ND
BBECs + rat C6 glioma cells	[52]	≈2100	ND	ND
BBECs	[53]	NI	Sucrose $P_{app} \approx 33$	$R^2 \approx 0.38^b$ ($n = 9$)
BBECs + serum-free + rat C6 glioma cells + HC	[54]	689 ± 53	Sucrose $P_{app} \approx 5$	ND
BBECs + rat ACs + DXM + cAMP↑ + buffer	[55]	1638 ± 256	Mannitol $P_e \approx 0.48$	ND
	[56]	1014 ± 70	Mannitol $P_e \approx 0.88$	ND
MURINE				
MBECs + rat C6 glioma cells	[57]	307 ± 15	SF $P_e \approx 30$	ND
MBECs + mouse ACs	[58]	778 ± 15	Sucrose $P_e \approx 4.5$	ND[c]
	[59]	NI	LY $P_e \approx 5.7$	ND
MBECs + HC + serum-free	[60]	≈200	ND	ND
MBECs + rat Acs	[61]	≈200	SF $P_e \approx 3.5$	$R^2 \approx 0.96^b$ ($n = 7$)
MBECs	[62]	NI	$Dextran_{3kDa}$ e $P_e \approx 2.2$	ND
MBECs + Puro (±cAMP↑)	[63]	NI	Albumin $P_e \approx 7.4$	ND
MBECs + Puro + mouse ACs + 20% FBS	[64]	≈200	SF $P_e \approx 1.25$	ND

(*Continued*)

TABLE 10.2 (*Continued*)

In vitro models	References	TEER (Ω·cm²)	P_e or P_{app} (10^{-6} cm/s)	Correlation
PORCINE				
PBECs + serum-free + HC	[65, 66]	700 ± 100	Sucrose $P_e \approx 1$	ND
	[67]	>600	Sucrose $P_e \approx 0.34$	ND
PBECs + serum-free + rat C6 glioma cells + HC	[68]	≈800	Sucrose $P_e \approx 3.3$	ND
	[69]	834 ± 136	Sucrose $P_{app} \approx 1.6$	ND
	[54]	893 ± 89	Sucrose $P_{app} \approx 2$	ND
PBECs + porcine ACs	[70]	139 ± 16	ND	ND
PBECs + rat ACs	[71]	333–550	Sucrose $P_e \approx 92$	$R^2 \approx 0.89$ ($n = 12$)
PBECs + rat ACs (contact)	[72]	1112 ± 43	Sucrose $P_e \approx 0.19$	ND
PBECs + CTX-TNA2 rat AC Line + HC + cAMP↑	[73]	1700–2100	LY $P_e \approx 2.92$	ND
PBECs + Puro + BPDS + HC + cAMP↑ + rat ACs	[34, 35]	779 ± 19	Sucrose $P_{app} \approx 5.7$	ND
RAT				
RBECs + rat ACs + cAMP↑	[74]	300–500	SF $P_e \approx 4.2$	ND
RBECs + rat ACs + ECGF + PDS	[75]	438 ± 75	N.I.	ND
RBECs + Puro + HC + rat ACs	[29]	≈500	SF $P_e \approx 0.75$	ND
	[30]	270 ± 119	Sucrose $P_e \approx 1.43$	$R^2 \approx 0.94^b$ ($n = 19$)
RBECs + Puro + rat ACs	[76]	220–300	Sucrose $P_e \approx 1.86$	ND
RBECs + Puro + rat NPC	[77]	70–120	SF $P_e \approx 5.5$	ND
RBECs + Puro + rat ACs + PCs + HC + cAMP↑	[78, 79]	354 ± 15	SF $P_e \approx 3.9$	$R^2 \approx 0.89$ ($n = 19$)
	[41]	548 ± 125	SF $P_e \approx 2.72$	$R^2 \approx 0.80$ ($n = 10$)
RBECs + Puro + rat ACs + HC + cAMP↑	[33]	>600	ND	ND
RBECs + Puro + rat ACs + EBM-2/EGM-2	[36]	529 ± 14	LY $P_e \approx 2.9$	ND

	Brain endothelial cell lines			
BOVINE				
t-BBEC-117 + rat ACM	[80]	NI	L-Glucose $P_e \approx 1.2$	ND
MURINE				
b.End3 + rat C6 glioma cells + cAMP↑	[68]	≈130	Sucrose $P_e \approx 16$	ND
b.End3 + rat C6 glioma cells	[69]	≈66	Sucrose $P_{app} \approx 19.4$	ND
b.End3 + rat C6 glioma ACM	[81]	≈121	Sucrose $P_{app} \approx 9$	ND
b.End5	[62]	NI	Dextran$_{3\ kDa}$ e $P_e \approx 41$	ND
cEnd + 2% serum + HC	[82–84]	≈800	ND	ND
MCE4	[46]	40–50	Inulin $P_{app} \approx 22.5$	NI ($n = 10$)
TM-BBB1-5	[85]	105–118	Mannitol $P_e \approx 139$	ND
PORCINE				
PBMEC/C1-2 + rat C6 glioma ACM + cAMP↑	[86]	≈300	Sucrose $P_e \approx 5.6$	ND
PBMEC/C1-2 + fibronectin + rat C6 glioma ACM	[87]	101 ± 4	Diazepam $P_e \approx 31$	ND
	[88]	66 ± 2	Dextran$_{APTS}$ $P_e \approx 12.1$	ND
RAT				
GPNT + Puro	[89]	≈66	Sucrose $P_e \approx 123$	ND
RBE4 + rat ACs + 2% FBS	[90]	N.I.	Sucrose $P_e \approx 11$	ND
SV-ARBEC + immortalized rat Acs	[46]	50–70	Sucrose $P_e \approx 16$	NI ($n = 22$)
TR-BBB13	[91, 92]	99–109	NI	NI ($n = 10$)
	Surrogate cells			
HUMAN CACO-2 ADENOCARCINOMA CELL LINE				
Caco-2	[67]	>600	Sucrose $P_e \approx 0.46$	ND
Caco-2	[45]	NI	Sucrose $P_{app} \approx 1.71$	$R^2 \approx 0.46^b$ ($n = 10$)
Caco-2	[46]	600–1000	Sucrose $P_{app} \approx 1.4$	$R^2 \approx 0.34$ ($n = 22$)
Caco-2	[69]	512 ± 155	Sucrose $P_{app} \approx 1.6$	ND
Caco-2	[53]	NI	Sucrose $P_{app} \approx 2.4$	$R^2 \approx 0.37^b$ ($n = 9$)
Caco-2	[41]	1024 ± 184	SF $P_e \approx 1.34$	$R^2 \approx 0.61$ ($n = 10$)
VB-Caco-2	[41]	2012 ± 347	SF $P_e \approx 0.47$	$R^2 \approx 0.72$ ($n = 10$)

(*Continued*)

TABLE 10.2 (*Continued*)

In vitro models	References	TEER (Ω·cm²)	P_e or P_{app} (10^{-6} cm/s)	Correlation
HUMAN ECV 304 CELL LINE[d]				
ECV304 + rat C6 glioma cells + 10% FBS	[93]	221 ± 13	ND	ND
	[94]	120–175	Sucrose $P_{app} \approx 0.13$	ND
ECV304 + rat C6 glioma cells	[46]	≈100	Sucrose $P_{app} \approx 8.1$	NI ($n = 5$)
ECV304 + rat C6 glioma ACM	[88]	150 ± 32	Dextran$_{APTS}$ $P_e \approx 1.98$	ND
ECV304 + rat C6 glioma ACM + cAMP↑	[95]	≈110	Sucrose $P_e \approx 11.3$	ND
MADIN-DARBY CANINE KIDNEY CELL LINES				
MDCK$_{ATCC}$	[46]	200–400	Sucrose $P_{app} \approx 8.1$	$R^2 \approx 0.40$ ($n = 22$)
MDCK$_{wt}$	[46]	130–150	Sucrose $P_{app} \approx 8.1$	$R^2 \approx 0.46$ ($n = 22$)
MDCK	[41]	<100	SF $P_e \approx 0.65$	$R^2 \approx 0.68$ ($n = 10$)
MDCK-MDR1	[46]	120–140	Sucrose $P_{app} \approx 8.1$	$R^2 \approx 0.37$ ($n = 22$)
MDCK-MDR1	[41]	<100	SF $P_e \approx 0.59$	$R^2 \approx 0.78$ ($n = 10$)
MDCK$_{type\ II}$-MDR1	[53]	NI	Sucrose $P_{app} \approx 1.5$	$R^2 \approx 0.36^a$ ($n = 9$)

Molecular weights of paracellular permeability tracers used: albumin: 67,000 Da; dextran$_{APTS}$: 5,779 Da; dextran$_{3k}$: 3,000 Da; diazepam: 290 Da; L-glucose: 180 Da; inulin: 5,000 Da; Lucifer yellow: 444 Da; mannitol: 182 Da; sodium fluorescein: 376 Da; sucrose: 342 Da.

ACs, astrocytes; ACM, astrocyte-conditioned medium; APTS, 8-aminopyrene-1,3,6-trisulfonate; BBECs, bovine brain endothelial cells; BECs, brain endothelial cells; BPDS, bovine plasma–derived serum; cAMP↑, drug combination elevating intracellular adenosine 3′,5′-cyclic monophosphate level; DXM, dexamethasone; EBM-2/EGM-2, EBM-2 microvascular endothelial cell media with the EGM-2 BulletKit SingleQuots without the human recombinant VEGF supplement (from Lonza); ECGF, endothelial cell growth factor; FBS, fetal bovine serum; HC, hydrocortisone; LY, Lucifer yellow; MBECs, murine brain endothelial cells; ND, not determined; NI, not indicated; NPCs, cortical embryonic neural progenitor cells; P_{app}, apparent permeability coefficient; P_e, endothelial permeability coefficient; PBECs, porcine brain endothelial cells; PCs, pericytes; PDS, plasma-derived serum; Puro, puromycin treatment for selection of brain endothelial cells; RBECs, rat brain endothelial cells; SF, sodium fluorescein; TES, *N*-tris(hydroxymethyl)methyl-2-aminoethanesulfonic acid; VB, vinblastine.

[a] Calculated TEER on the basis of surface area and Ω·cm² value published in the paper.

[b] R^2 value for correlation between *in vitro* and *in vivo* permeability data was calculated on the basis of R value published in the paper.

[c] Although no correlation with *in vivo* data was published, $R^2 \approx 0.99$ ($n = 7$) correlation was found when murine *in vitro* results were compared to permeability data obtained on *in vitro* reconstituted model formed by cloned bovine BECs and rat ACs from Dehouck & Cecchelli labs.

[d] Human ECV304 cell line was originally considered as a spontaneously transformed line derived from human umbilical vein endothelial cells (HUVEC) culture; however, genetic analysis by Brown et al. [96] revealed an identical DNA fingerprint pattern of ECV304 and human bladder–derived epithelial cell line T24/83.

A large number of SLC transporters were identified by genomic or proteomic tools at the BBB (for review see Redzic [102]). While the presence of glucose and amino acid transporters have been confirmed in primary cultures of brain endothelial cells [60, 68, 78], there are few BBB culture models which have been characterized for SLC transporter functionality. The kinetics of glucose and L-alanine uptake were described in primary porcine brain endothelial cells indicating functional GLUT-1 and LAT-1 carriers [69]. Active glucose uptake was also described in primary rat brain endothelial cells and was positively modulated by *n*-3 long-chain polyunsaturated fatty acids [103]. Involvement of OCTN2 and amino acid/carnitine transporter *B*(0, +) system in the transport of carnitine and polarized localization of transporter *B*(0, +) was identified in a well-characterized *in vitro* bovine BBB model [104, 105]. There might be several reasons explaining the scarcity of functional studies on SLC transporters on culture BBB models. Influx transporters, like glucose transporter 1, are more sensitive to downregulation by culture conditions than efflux pumps [106]. Garberg et al. [46] suggested that the *in vitro* BBB models tested in their work were not tight enough to measure small-molecule ligands of glucose and amino acid SLC transporters in permeability assay settings. It remains to be seen whether the newly described and paracellularly tighter BBB models will be better applicable in uptake and especially in permeability assays for influx transport studies.

Species Differences Culture morphology does not reveal differences between primary endothelial cells isolated from bovine, human, mouse, porcine, or rat brain microvessels. The same is true for the paracellular tightness of these culture models (Table 10.2). Previously bovine and porcine models were described as the most robust, with the highest transendothelial resistance [44, 65, 107], but recent improvements in culture techniques and conditions have resulted in models with tight barrier function from mouse [58, 59] and rat primary cells [33, 36, 41] (Table 10.2). In a comparative study, both bovine and porcine brain endothelial cell–based models displayed discriminative barrier functionality, but higher resistance and lower permeabilty was measured on the porcine model in identical culture conditions in the same laboratory [54]. An interesting observation on species difference is related to the subcultivation of primary endothelial cells. Bovine brain capillary endothelial cells can be easily cloned and passaged [44], while rodent brain endothelial cells cannot be subcultivated, they not only dedifferentiate, but stop growing.

Recent genomic, proteomic, and functional studies indicate that there are differences between the human and other mammalian BBB, especially at the level of transporters [108] (for more detailed discussion on species differences in BBB functions see Chapter 4). As an example, quantitative proteomic studies indicate that ABCG2 is more abundant than ABCB1 in the human BBB in contrast to mouse BBB [98], which may have relevance for *in vitro* models used for drug permeability testing.

Regulation of the Paracellular Tightness of In Vitro BBB Models Culture conditions, including the composition of culture media, are key factors influencing barrier tightness of culture models. The use of serum in primary brain endothelial cultures is controversial: in a porcine model, serum-free conditions improved barrier tightness [65], but in most primary models, culture media are complemented with animal sera [40]. Sera may contain different levels of lysophosphatidic acid inducing barrier opening [109], but also growth factors like FGF2, which not only induce division of brain endothelial cells, but also barrier tightening (for review see Deli et al. [25]). It should also be noted that in endothelial cells serum deprivation is a strong inducer of oxidative stress and apoptosis [110] and, except one observation, [60] no rodent *in vitro* BBB model is described using serum-free culture conditions. In addition to sera and FGF2, the most effective factors to date to induce tighter barriers are hydrocortisone [111] or other glucocorticoids, cyclic AMP elevation [49, 112], and lipid supplements, which are usually used in combination (for review see Refs. 25 and 113).

Immortalized Brain Endothelial Cell Lines Because the preparation of primary cultures is expensive, time-consuming, and requires technical expertise, immortalized brain endothelial cell lines were established from different species starting from the 1990s. These cell lines are mainly of mouse, rat, porcine, and human origin, and proved to be useful to study several aspects of the physiology or pathology of the BBB including permeability and transport functions [25, 107, 114]. The number of BBB models based on brain endothelial cell lines, which are characterized in detail and used for drug permeability studies, is very low [40]. There are no bovine cell lines with known data on resistance or paracellular marker permeability. From the mouse cells lines MBEC4, b.End3, and b.End5 were tested and compared to the permeability of primary mouse brain endothelial monolayers without coculture [115]. Similar permeability values for fluorescein and expression level of claudin-5 were observed in b.End3 cell line as in primary mouse cells. The limitation of the study is that no resistance data were provided. The comparison of cell lines and primary cells also depends on the complexity of the primary cell–based model. When the permeability of the rat triple coculture model [79] was compared to MBEC4 and SV-ARBEC cells [46], a significantly better *in vitro–in vivo* correlation was found than for the cell lines (Fig. 10.6).

The BBB model from porcine cell line PBMEC/C1-2 was characterized in detail [88, 116]. It was successfully used to assess the functionality of P-gp [117] and for transport ranking of antihistaminic drugs [118] and NSAIDs [119].

While brain endothelial cell lines are valuable research tools their major limitation is the insufficient tightness of the paracellular pathway for pertinent drug penetrability screening [25, 120], as shown in the higher permeability coefficients listed in Table 10.2. The exact reason why immortalization leads to loss of some of the brain endothelial properties [121], especially the tight paracellular barrier [25], is not clear. It is known that angiogenesis is accompanied by weak intercellular junctions, and tight barrier is formed only in nondividing quiescent brain endothelial cells. Tight junction linker proteins, such as ZO-2, serve as nuclear signal factors in angiogenic conditions and are located in the junctional area in confluent, contact-inhibited cultures [122]. If immortalization leads to an "angiogenic" phenotype in brain endothelial cells with continuous cell division and loss of contact inhibition, no tight barrier formation can be expected even if culture conditions are optimized.

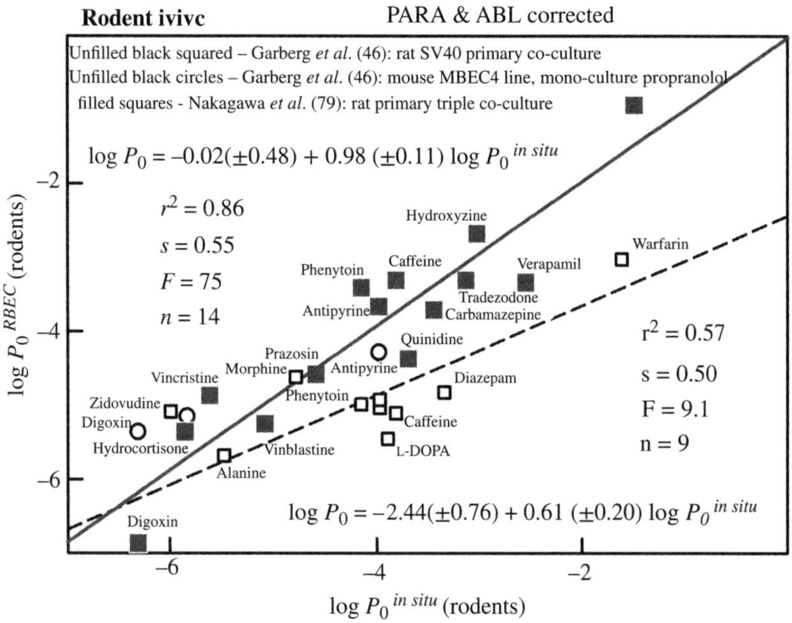

FIGURE 10.6 *In vivo–in vitro* correlation of the apparent permeabilities for drugs: comparison of three rodent BBB culture models: the primary cell–based triple coculture model [79], the mouse MBEC4 cell line, and the SV-ARBEC model, a coculture of SV40 immortalized brain endothelial cells and astrocytes [46].

Brain Endothelial Cells versus Surrogate Models Originally, epithelial cell lines were used as test systems for gastrointestinal absorption by pharmaceutical companies. Because the tight paracellular barrier and the efflux pumps are important characteristics of all biological barriers, including the BBB, these epithelial cell lines possessing both characteristics were also tested as surrogate BBB models [123]. The most popular epithelial surrogate models are the human Caco-2 and the dog MDCK cell lines.

The Caco-2 cell line originates from a human colon adenocarcinoma, and is used primarily as a screen for intestinal absorption. Caco-2 cells give a good correlation for passive diffusion compounds, but the predictive value of this model is less for efflux pump substrates [46]. To overcome the big heterogeneity in cellular morphology and to increase P-glycoprotein expression, Caco-2 cells were selected by vinblastine (VB-Caco-2) [41]. Because cellular morphology is more uniform and P-gp expression is higher, VB-Caco-2 cells can be better used for testing new chemical entities [41].

The dog kidney epithelial MDCK cells are used in primary research on kidney transporters and tight intercellular junctions. They are also applied by pharmaceutical companies as a surrogate BBB model, especially the MDCK-MDR1 cell line, which is a subclone stably expressing the gene of human P-gp [123]. While MDCK cell layers have tight paracellular barrier for small hydrophilic compounds, their resistance can be very low [41, 46]. This is a good example, that low transepithelial electrical resistance (TEER) values do not indicate open paracellular pathways or leaky monolayers if junctional tightness is high by both morphological and functional methods. The high ionic permeability of the MDCK barrier may be related to the presence of ion channels in the plasma membrane and the expression of pore-forming claudin subtypes [41].

Another human cell line used as a surrogate BBB *in vitro* model is ECV304. This cell line was originally introduced by Takahashi et al. [124] as a spontaneously transformed endothelial cell line. Several endothelial markers, such as von Willebrand factor (vWF), the secretion of PAI-1 and endothelin, vitronectin CD51 receptor, ICAM-1, thrombomodulin, VCAM-1, and Ac-LDL uptake were found in ECV304 cells, but PECAM-1 was not proved and the expression of some epithelial markers (desmoglein, cytokeratin) was shown [125]. Comparison of ECV304 cells to urinary bladder carcinoma T24 cells revealed distinct differences between these two cell lines. However, later genetic fingerprint analysis by several groups and cell banks showed identical fingerprints of ECV304 and T24 and, consequently, raised doubt about the endothelial origin of ECV304 cells [96, 126]. Furthermore, they were not able to reproduce the proof of presence of endothelial markers, especially vWF. It was assumed that a cross contamination of original HUVECs had occurred with T24 cells and, consequently, it was recommended not to apply ECV304 cells for endothelial studies and to use T24(ECV304) as the proper term. On the contrary, another study showed that these two cell lines differ by all cytogenetic characteristics [127]. The facts about the usability as BBB model and the origin of ECV304 cells are still controversial, which is probably a result of heterogeneities among cell populations of ECV304 cells. In this regard, Kiessling et al. [128] stated that it was theoretically possible that ECV304 represented embryonal mesothelial cells in transition toward an endothelial phenotype. Considering this, it was proposed that the phenotype is very dependent on the clone and applied culture conditions used and that the properties of ECV304 cells have to be characterized for each single cultivation protocol and subsequent applications. With regard to BBB properties and drug transport studies, several groups were able to confirm the responsiveness of ECV304 cells toward glioma cell line C6, resulting in increased tightness and induced transporter activity (Glut1, P-gp, amino acid transporters) [88, 95]. In comparison to other immortalized BBB cell lines, such as PBMEC/C1-2, RBE4, b.End3, b.End5, RBEC, and EaHy929, ECV304 forms tighter cell layers, which provides a broader measurement window between paracellular and transcellular markers to distinguish between drug permeabilities. ECV304 *in vitro* models were used to investigate and rank the permeability of, for example, sucrose, inulin, amino acids (alanine, leucine), vincristine, benzodiazepines, NSAIDs, or glycine antagonists, which were correlated to *in vivo* logBB data or physicochemical parameters such as log P values [46, 129]. Although BBB properties are inducible in ECV304 cells by astrocytes, which is a major feature of brain endothelial cells, it is not recommended to use them for studies focusing on endothelial properties or signaling pathways, due to distinct doubts about their endothelial origin. Nevertheless, they could be used as a barrier model for investigating drug transport, since they could form significantly tighter barriers than most of the available BBB cell lines. There are very few comparative studies on brain endothelial cell–based and surrogate BBB models. In the study of Lundquist et al. [45], the estimation of BBB penetration was weaker in Caco-2 cells than in the bovine BBB model. Garberg et al. [46] compared the largest number of BBB and epithelial surrogate models so far and the best predictions for passive and efflux compounds were obtained on the bovine BBB, MDCK-MDR1, and Caco-2 models. When compared to bovine brain endothelial or Caco-2 models, MDCK cell monolayers gave a weaker correlation for passively penetrating drugs. Not surprisingly, the MDCK-MDR1 cell line overexpressing P-gp was the best to screen for efflux pump ligands. Although the ECV304 model was included in this study, only 5 compounds were tested from the set of 22, and no conclusion could be drawn on the performance of this model. As compared to the passive and efflux compounds, no model gave a good correlation with in vivo data when all the 22 compounds tested were taken into account, including lipophilic molecules of flow-limited penetration to brain and substrates of nutrient Slc carriers, highlighting the limitations of the tested culture BBB models. (Drugs can only be absorbed into the BBB as fast as the perfusion fluid flow brings them to the site of absorption. If the true permeability is faster than the flow rate, the measurement is said to be "flow-limited.")

In a recent study, Hellinger et al. [41] compared a primary cell-based triple rat coculture model [78, 79] and epithelial cell–based Caco-2, VB-Caco-2, MDCK, and MDCK-MDR1 surrogate BBB models for both cell morphology and functionality. Although it is disregarded in most comparative studies or reviews on BBB models, the cytoarchitecture of epithelial and endothelial models differ strikingly [41]. In brain endothelial cells, anatomically separate adherens junctions, desmosomes and interdigitations, characteristic for epithelium, are missing. In contrast to epithelial cells the luminal surface of brain endothelial cells is smooth and there are no microvilli [40, 41]. Endothelial cells of brain capillaries are very thin both *in vivo* and *in vitro*; the cell thickness of Caco-2 and MDCK cells are 10–20-fold higher. These differences affect the penetration of drugs: the permeability coefficients for passive compounds were 1.3–1.5-fold higher in Caco-2 cells than in the BBB model, which may be related to the higher absorption surface of epithelial cells due to microvilli. All three models showed restrictive paracellular pathway and selective passive permeability. Using a panel of 10 compounds, a good correlation with *in vivo* permeability data was found on the coculture BBB, VB-Caco-2, and MDCK-MDR1 models, with the rat coculture BBB model showing the highest R^2 value [41]. With regard to the P-gp-mediated efflux functionality of the models, the VB-Caco-2 and MDCK-MDR1 were more sensitive than the rat BBB model and identified a significantly higher number of efflux substrate drugs.

Surrogate BBB models may provide a simple and inexpensive alternative for the screening of drug candidates for passive permeability or efflux by transporters. However, the different cytoarchitecture and genetically programmed differences, such as the expression of specific sets of tight junction proteins and BBB-specific transporters, call for the careful use of surrogate models and require alternative approaches for BBB-specific transporter liabilities, or the use of endothelial BBB models.

The Role of the Microenvironment

The other major problem for brain endothelial cells grown in monocultures was the dedifferentiation observed, especially in long-term cultures or following subcultivations. The BBB phenotype in brain microvascular endothelial cells is organ-specific and it results from a cross-talk between endothelial cells and the surrounding cells of the neurovascular unit. When this inductive influence of the CNS microenvironment is lost in culture, the BBB properties of endothelial cells are also lost. Since cerebral microvascular endothelial cells share a common basal membrane with the surrounding pericytes and glial end-feet, the influence of these three elements are key in a culture BBB model.

Glial Cells Astrocytes were the first to be recognized to induce BBB properties in cultured brain endothelial cells in the 1980s. Since then many studies have followed and all have confirmed that glial cells tighten the barrier, as well as induce the expression and functionality of enzymes and transporters in cerebral endothelial cells (for reviews see Refs. 1, 130–132). The mechanism of glial induction is still not known, despite long-time research efforts. Secreted growth factors and small signaling molecules are possible candidates mediating the effect of astrocytes, because not only coculture but even astrocyte-conditioned culture media are effective (Table 10.2). Type I astrocytes and mixed glial cultures are equally efficient for the induction of BBB properties in coculture models (Table 10.2), but some studies indicate that the contact model, where glial cells are on the opposite side of the membrane where endothelial cells are cultured, is the more effective [25, 72, 78]. There seem to be no species differences in glial induction: while some BBB models, especially rat and mouse, are syngeneic, rat glial cells are also used for coculture of bovine or porcine endothelial cells [44, 70]. Glial cell lines can replace primary astrocytes in coculture BBB models; the most widely applied cell line is the rat C6 glioma cells [25] (Table 10.2). However, the inductive capacity of C6 glioma cell line is less effective than that of primary astrocytes, due to the production of the permeability increasing vascular endothelial growth factor (VEGF) [132]. Most models used as permeability screens for drugs are coculture systems of brain endothelial cells or cell lines with either primary astrocytes or with glial cell lines (Table 10.2). Coculture BBB models, based on primary endothelial and glial cells, gave better *in vivo–in vitro* correlations in comparative studies than cell line–based models (Fig. 10.6) [41, 46]. A new study indicates that a well-characterized *in vitro* coculture BBB model, based on bovine brain capillary endothelial and rat glial cells [44], can be used to calculate free brain and free plasma drug concentration ratios (Cu,br/Cu,pl), suggesting that a single *in vitro* model of the BBB can identify compounds with a desirable *in vivo* response in the CNS [133]. From the same laboratory, a bovine brain endothelial cell–based robust, higher-throughput model exhibiting BBB features in a ready-to-use, frozen format was also developed for in-house use by pharmaceutical firms and biotech companies during early-stage drug discovery [134].

Pericytes Pericytes are important elements of the blood vessel wall. They not only stabilize capillary structure and regulate blood flow, but also contribute to low blood vessel permeability [135]. Higher pericytic coverage correlates with the tightness of blood vessels: the ratio of pericytes to endothelial cells is the highest in retina and brain capillaries, which possess the tightest intercellular junctions [136]. It has been recently revealed that pericytes of brain capillaries induce BBB-specific gene expression patterns in endothelial cells and polarization of astrocyte end-feet *in vivo* [137]. They also play a critical role in the regulation of the BBB during development [138]. Despite the early recognition of the importance of pericytes in capillary physiology, there was a 20-year gap between the discovery of the inductive properties of glial cells and pericytes. The effect of brain microvascular pericytes on the tightness of cerebral endothelial monolayers has been demonstrated in a coculture setting using primary rat brain endothelial cells [139]. This observation was confirmed on bovine [140] and immortalized mouse [141] brain endothelial cells. Nakagawa et al. [78, 79] strengthened these observations and developed a new rat syngeneic model where the induction of the BBB properties in endothelial cells was achieved by coculture with both brain pericytes and astrocytes. Since then, several double or triple coculture BBB models have been established with brain pericytes for primary research purposes to study cell–cell interaction or angiogenesis [142, 143], but so far only one model was characterized in detail for drug testing purposes [41, 79].

Neurons It is well accepted that the neural microenvironment plays a key role in inducing BBB function in capillary endothelial cells, but it was difficult to prove the inductive influence of the neuronal environment or the contribution of neurons *in vivo* [144]. The first *in vitro* observation was made on cloned porcine brain endothelial cells, in which incubation with plasma membrane fractions from embryonic brain cells and cortical neurons caused a significant increase in gamma-glutamyl transpeptidase activity [145]. In a coculture system of mouse brain slices and primary rat brain endothelial cells, a tight barrier developed, which restricted the permeability of dopamine and glutamate, but not of L-DOPA across the endothelial monolayers [146]. Immortalized rat neuronal B14 cells could be differentiated into a serotonergic phenotype when added to the abluminal glial surface in a primary rat brain endothelial–glial dynamic coculture model and this neuronal cell line induced the expression of a functional serotonin transporter on brain endothelial cells [147]. Coculture of cortical neurons with RBE4 rat brain endothelial cell line resulted in tighter barrier, junctional localization of occludin [148], and carrier-mediated transport of L-DOPA across brain endothelial monolayers [149]. Cortical embryonic neural progenitor cells (NPCs), which have self-renewal capacity and ability to differentiate into a mixed culture of both neurons and astrocytes, were also tested *in vitro*. Coculture with rat NPCs or treatment with NPC-conditioned medium could induce tighter barrier and the expression of P-gp in primary rat brain microvascular endothelial cells [77, 150]. In a follow-up study both rat and human NPC-derived mixed cultures of astrocytes and neurons (ratio 3:1) could upregulate BBB-related genes and barrier function in rat brain endothelial cells [151], indicating that NPCs can be an alternative to primary neural cells for use in BBB co-culture models. However, no endothelial–neuron coculture BBB model has been tested in a comparative drug penetration study yet.

Basal Membrane The basal membrane of brain capillaries is shared between endothelial cells, pericytes, and astrocytes, which all contribute to its composition. In addition to the influence of the perivascular cells discussed earlier, the barrier function of endothelial cells is also determined by the surrounding basal lamina (for review see Wolburg et al. [39]). The components of the extracellular matrix around brain capillaries include collagen type IV, fibronectin, laminin, and the heparan sulfate proteoglycan, agrin. The importance of extracellular matrix in the culture of brain endothelial cells and in the local control of tight junction formation has long been known. Rat tail collagen gel is an excellent support of monolayer formation in bovine and rat brain endothelial cells [27, 42, 152]. In a comparative study, collagen type IV, fibronectin, and laminin, either alone or in combination, elevated the TEER of low-resistance porcine brain endothelial monolayers [153]. To coat the culture surfaces, either rat tail collagen or a mix of collagen type IV and fibronectin is used for most BBB models [78, 119]. A biological matrix secreted by corneal endothelial cells also improved the growth of cultured brain microvascular endothelial cells [154] and could support the development of tight monolayers of rat brain endothelial cells, but was not used in models for drug testing.

Shear Stress It is known that shear stress is an important factor for BBB functionality. Application of shear stress onto brain endothelial cells resulted in differentiation toward the *in vivo* phenotype indicated by increased expression of tight junction, transport and enzyme proteins, reduced proliferation, changed energy metabolism, increased life span, and changed protein distribution toward the cytoskeleton [155–157] (Table 10.1). The striking role of shear stress is not considered in most of the used BBB *in vitro* models, because probably the model setup and handling are too sophisticated in relation to the expected improvement of obtainable data. Another argument for not implementing flow into BBB *in vitro* models was that blood flow is very low in capillaries and consequently the influence of shear stress is negligible [158]. This is not correct. According to Equation 10.5 [159], flow is indeed directly proportional to the produced shear stress, but the inner radius of capillaries cubed is indirectly proportional:

$$SS = 4\mu Q/\pi r^3 \tag{10.5}$$

SS is the shear stress (units), μ (mPa·s)is the viscosity, Q (ml·s^{-1}) is the flow rate, and r (cm) is the inner radius.

In fact, based on novel shear stress calculations in eye conjunctiva of humans [160], Cucullo et al. [161] proposed that formerly assumed shear stress in brain capillaries of 5 dyne/cm^2 should be corrected to values over 20 dyne/cm^2 in capillaries with 8 μm diameter. The impact of shear stress had been investigated in several different model types, such as conical cylinders, dynamic hollow-fiber flow models from different companies (Cellmax, Fiber-Cell, Flocel) or, more recently, microfluidic devices [162–164]. Although it is well accepted that shear stress is an important inducer of BBB *in vivo* characteristics, the molecular basis for signal transduction is still mainly unresolved. Some publications indicated that cells alter their morphology and elongate by adaption of their cytoskeleton just following the physical force. The rearrangement of the cytoskeleton is linked with Ca^{2+} signaling, which is perhaps then responsible for further assimilations. In other studies, a protein complex was identified, containing molecules known to be of distinct importance for BBB functionality, such as PECAM-1, VE-cadherin, and VEGFR2, which should transform the mechanical stress into a biochemical signal in the endothelial cell. Nonetheless, there have been numerous studies (e.g., flow cessation–reperfusion experiments) underlying the importance of flow-mediated shear stress for BBB functionality [165–168]. How the influence of flow is integrated in current BBB cell culture models is discussed in the section "Different Cell Culture Model Types and Technical Aspects" in more detail.

Physiological Models (Co/Triple-Culture) To mimic the complexity of the neurovascular unit *in vivo*, endothelial–glial coculture BBB models also became more complex. In the first attempts, BBB *in vitro* models were completed with neurons. Using primary neurons or neuronal cell line, both dynamic [147] and static [169], rat triple *in vitro* BBB models were described with improved barrier properties. A rat BBB culture model from NPC-derived mixed cultures of astrocytes and neurons was also introduced [151]. The same laboratory also later described a human pluripotent stem cell–based triple coculture model [170]. These models have a good potential for application in drug testing due to their scalability, but further characterization and more data are needed to prove their reproducibility and applicability by comparative tests.

The aforementioned models, however, miss a crucial brain perivascular cell, pericytes. The triple coculture BBB model developed by Nakagawa et al. [78, 79] consists of primary rat brain capillary endothelial cells grown in the presence of both pericytes and astrocytes mimicking the *in vivo* anatomical position of these cells (Fig. 10.7). Endothelial cells and pericytes share the membrane of the insert, and the astrocytes are cultured in the well holding the culture insert. The cross talk between the three cell types induces barrier properties (Table 10.2) and functional expression of influx and efflux transporters typical to the BBB, including glucose transporter-1, P-gp, and multidrug resistance protein-1 [41, 78, 79, 171]. This rodent triple coculture model gives good *in vitro–in vivo* correlations (Fig. 10.6) [41, 79]. As a primary research tool, the triple rat BBB model was applied in a study on iron transport proteins at the BBB [172]. As a screening system for the BBB transport of drugs and drug candidates, a ready-to-use frozen triple coculture BBB kit (PharmaCo-Cell Co. Ltd.; Nagasaki, Japan) was also used to measure the BBB permeability of alkaloids [173, 174] and nanoparticles [175, 176] *in vitro*. Nakagawa et al. [177] also developed a monkey triple BBB model consisting of primary brain microvascular endothelial cells from *Macaca irus* cocultured with rat pericytes and astrocytes showing barrier properties (TEER $\approx 350\ \Omega\cdot cm^2$, SF $P_{app} \approx 3 \times 10^{-6}$ cm/s) suitable for *in vitro* permeability assays.

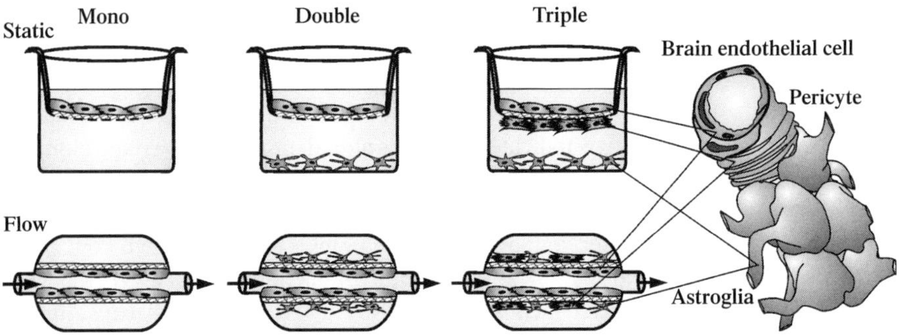

FIGURE 10.7 Schematic picture of *in vitro* BBB models using mono, double, and triple cell cultures under static and flow-based conditions.

TABLE 10.3 ***In vitro*** **models for testing BBB penetrability of compounds**

Cells forming the barrier	Advantages	Disadvantages
Primary brain endothelial cells	Retained morphological, functional, and metabolic BBB characteristics • High TEER values • Low P_e values • Functionally active influx and efflux transporters	Expensive Continuous need for animal tissue Time-consuming Low throughput Sophisticated models need expertise
Brain endothelial cell lines	Partially kept BBB characteristics Feasible for metabolic and receptor studies Restricted suitability for transporter studies	Immortalization: dedifferentiation and partial loss of BBB properties Permeability studies are not feasible in case of high paracellular leak
Surrogate cell types	Tight barrier properties suitable for passive trans- or paracellular assays High throughput	Morphological, functional, and metabolic characteristics are not BBB-like

BBB, blood–brain barrier; P_e, endothelial permeability coefficient; TEER, transendothelial electrical resistance.

A further development of BBB coculture models is expected with even more complexity by the addition of both pericytes and neuronal cells to the endothelial–glial cell duo. The advantage of these complex models is that they mimic the *in vivo* anatomical structure of brain capillaries much better than previous settings (Table 10.3). Their disadvantage is also related to this complexity: the establishment of triple culture models as a new method is technically demanding; however, the ready-to-use kits can solve this problem. Nevertheless, these complex *in vitro* culture models are optimal for testing selected compounds and not as high-throughput assays.

Disease Models In recent years, the importance of astrocytes or pericytes in disease *in vitro* models of the BBB became evident. For example, it was shown that astrocytes contribute significantly to BBB damage in *in vitro* stroke (oxygen/glucose deprivaton (OGD)–treated cells) or HIV models [178, 179]. Integration of astrocytes derived from ApoE4-KO mice in a BBB model, consisting of brain endothelial cells and pericytes, elucidated the crucial role of ApoE4 for BBB functionality [180]. Brain pericytes from stress-susceptible pigs are not able to maintain BBB integrity in comparison with pericytes from wild-type animals [181]. Pericyte damage and pericyte–endothelial cell cross talk–related BBB disruption were demonstrated in various *in vitro* disease models, such as HIV-1 [179, 182], Japanese encephalitis virus [183], and *Escherichia coli* infection [184]. Experiments on the syngeneic triple rat coculture model revealed BBB changes in models of neurological diseases, such as Alzheimer's disease [171] or stroke [185] and BBB protection by edaravone [186] or cilostazol [187]. These studies emphasized the pivotal role of the microenvironment, especially for disease models of the BBB.

Different Cell Culture Model Types and Technical Aspects

The static Transwell™ model is the most widely used system to evaluate drug permeability across cell layers. Although the application principle of this system is quite simple, several technical aspects should be considered to obtain reliable permeability data. First of all, the choice of the plastic membrane is a crucial one. The combination of plastic surface properties and membrane porosity (depending on pore size and number of pores) influences cell growth and differentiation, in particular the tightening process. Recently, it was shown that polyethylene terephthalate (PET) membranes with pores of 0.4 µm size produced the tightest layers formed by mouse brain endothelial cells in comparison to other materials, such as polycarbonate, or to larger available pore sizes, such as 1, 3, or 8 µm. After data evaluation, there was a correlation, especially between porosity and tightness, favoring Transwell™ inserts provided from one company with the lowest porosity. However, if porosity is too low (<1%), the unstirred water layer inside the pores becomes limiting in the transport of lipophilic drugs [8]. Furthermore, using membranes with pores larger than 1 µm holds the danger that cells grow through the membrane to the lower, basolateral side of the membrane and build up an additional cell layer there [188]. With regard to drug transport studies, the membranal design also plays a pivotal role. It is assumed that the larger the pores are, the faster the drug molecules permeate. This can be an important time factor for conducting permeability experiments. On the one hand, a cell layer with high paracellular tightness is wanted, while, on the other hand, a faster experiment is preferred. For this reason, several research groups decided to use membranes with 1 µm pores for permeability studies [88, 116, 118, 119, 189]. Another aspect is certainly the interplay between the physicochemical properties of the membranes and the compounds to be tested. Highly lipophilic compounds could be adsorbed on the plastic or the membrane support, which precludes meaningful, reproducible permeability studies [44]. In one case of derivatives, of pentamidine, PET membrane supports had to be exchanged with membranes made of polytetrafluorethylene (PTFE, teflon) to reduce drug adsorption to a justifiable amount [190]. This change of material led to the need for an additional validation of cell attachment, growth, and tightening processes of the *in vitro* model. Another possibility would be to add bovine serum albumin (BSA) to the basolateral chamber to minimize adsorption to the plastic and cell membranes [191, 192]. As a helpful side effect, added BSA may also help to maintain sink conditions, especially during transport studies with lipophilic, poorly soluble compounds [193]. Consequently, it is recommended to start with permeability tests across the cell-free, but coated, membranes before starting with cell tests, to estimate approximate sampling times the behavior of the compound in the permeability solutions and the recovery rate of the applied substance [44]. In this context, a novel strategy, which is pursued by the Swiss Centre for Electronics and Microtechnology (CSEM), is to minimize the influence of supporting membranes by the usage of porous supports based on ultrathin silicone nitride (Si3N4). These membranes are thinner than commercially available membrane supports (500 nm versus 10 μm) and exhibit an increased percentage of pore area related to the total surface area (15% for 1 μm pores versus 1.3% for 1 μm pores of BD inserts). These properties reduce the interactions of compounds with the membrane support significantly, which should be especially valuable for transport studies of bigger molecules, such as proteins or nanoparticles [194]. Other factors to be considered are the kind of coating of the membranes and the well size. Membranes for BBB models are mostly coated either with collagen I, collagen IV, fibronectin, or layered mixtures of them. The specific coatings should be included in cell-free studies. They also influence cell attachment, differentiation, and tightening processes [195]. The choice of the well format determines the volume of growth medium which is accessible for the cells. On average, around 1.2 ml (0.3 ml apical, 0.9 ml basolateral), 2 ml (0.5 apical, 1.5 basolateral) or 5 ml (2 ml apical, 3 ml basolateral) are used to obtain the same fluid level for the apical and basolateral compartment and prevent hydrostatic pressure within the model. Relating these volumes to the provided surface growth areas of an average of about 0.3 cm^2 (24-well format), 1 cm^2 (12-well format), and 4.2 cm^2 (6-well format) indicates that cells grown in larger well formats have much less growth medium per square centimeter at their disposal than in smaller formats, which may also influence cellular growth and differentiation or increase the frequency of necessary growth medium exchanges. In addition, the chosen well format also influences the magnitude of the TEER, because resistance is inversely proportional to the cross-sectional area of an electrical conductor, which is, in this case, the surface of the membrane. To compensate for the different size of the membranes, therefore, resistance values are given as $\Omega \cdot cm^2$. Certainly, the choice of the well format is also important for subsequent compound analysis and the calculation of permeability coefficients, because the surface area is used to normalize the permeability values. As the well format increases, the volume increases for the permeated test substance that can be collected and, in the case of small absolute amounts, also affects the degree of concentration before analysis. Interestingly, comparison of different culture conditions (monoculture, contact coculture with astrocytes, and noncontact coculture with astrocytes) revealed that the highest tightness in the noncontact coculture setup of brain endothelial cells was with astrocytes on 0.4 μm pores [188]. However, it has to be emphasized that in this study no triple cultures with additional pericytes were tested.

For permeability assessment, the temperature is a very important factor. Following Fick's law of diffusion, the diffusion coefficient in fluids is directly dependent on the temperature. Consequently, this should also be considered for conducting permeability experiments, which implies appropriate prewarming of the plastic ware, the test solutions, and the cells before starting the experiment. In addition, electrical resistance of cell layers increases with decreasing temperature, which has to be considered in the experimental settings for TEER measurements. Conducting experiments at 37 °C and using, for example, chopstick electrodes outside the incubator can require a temperature equilibration phase of a minimum of 30 min (depending on the ambient room temperature) before a temperature-dependent shift in TEER values can be excluded [196]. This could be omitted by measuring at 37 °C, for example, directly in the incubator with an online device (cellZscope, nanoanalytics), or recording the temperature during the measurement and correlating the data to a temperature/electrical resistance calibration curve, which has to be validated for each model system. In another context, the assessment of temperature dependence could be a very helpful tool to elucidate whether active transport processes are involved in the permeability of a specific drug, because active processes (ABC transporters, endocytosis) need energy, which is reduced at 4–8 °C. Comparing permeability experiments at lower temperatures to those at 37 °C could give clear hints about the involvement of active transport processes without the need for specific transport inhibitors. In any case, decreased passive diffusion due to decreased temperature has to be considered in these calculations and, therefore, appropriate cell-free experiments are needed.

The design of the transport study could also significantly influence the measured permeability data. The simplest way to conduct a transport test is to apply the substance in the apical chamber and measure compound concentration in the basolateral chamber after a certain time. Another possibility is to take samples of a certain volume at specific time points to obtain more pharmacokinetic relevant data. In this case, it is important to refill the removed volume with the same amount of prewarmed transport medium to avoid formation of a hydrostatic pressure gradient as an additional driving force ("sip and replace" method). However, with this setup, saturation will occur with time and, consequently, the linearity of permeation will be restricted to a specific time window. Finally, drug-filled inserts could be transferred into new, prewarmed, and prefilled wells at given time intervals. With this technique, also called the "break-sandwich" method, the linearity of drug permeation will be prolonged over the total experimental duration, because sink conditions are maintained [43, 44, 191, 193]. Moreover, this method enables maintenance of sink conditions during the experiment, depending on insert transferring time intervals (sink condition is when lower than 10% permeated substance is found in the basolateral compartment). Depending on the experimental design chosen, appropriate calculation models have to be selected for calculation of permeability coefficients. To assess polarized transport and involvement of an active transport system, comparison of the transport rates in the direction from apical to basolateral to the direction from basolateral to apical

$((P_{a/b})/(P_{b/a}))$ could be used, where a ratio of 1 indicates no preferred net flux in one direction (although it could also be a dynamic equilibration of influx and efflux). However, care must be given to ensure that neither direction of transport is limited by the resistance of the unstirred water layer. Another method is to start with the same drug concentration in the basolateral and apical chambers and monitor the time-dependent establishment of equilibration. Differences in the achieved equilibration concentrations in the apical and basolateral chambers would describe a polarized transport [197]. Certainly, addition of specific transporter inhibitors help to clarify and identify involved mechanisms in detail. Group transport studies, also called cocktail studies, assessing drug permeability of several drugs at the same time have several advantages, such as increased throughput and usage of the same cell layers, which increases experimental reproducibility. Cocktail studies are especially conducted in case of *in vitro* BBB models with modest paracellular tightness, and paracellular and transcellular markers are included to define the measurement window for obtainable permeability coefficients [87, 118, 119, 198–202]. Permeability coefficients are then ranked after normalization to one of the markers. The risk of cocktail studies includes interactions between molecules in the mixture and drug–drug interaction at the level of BBB transporters, not only influencing the transport of active agents, but resulting in BBB breakdown [203]. Consequently, the influence of applied drug cocktails on barrier functionality has to be monitored during transport studies and adverse effects excluded.

As mentioned earlier, hydrostatic pressure, as an additional driving force for drug permeability, has to be excluded by using apical and basolateral volumes ending up with the same fluid levels. A further important parameter is the transport medium in which the experiment is conducted. To simplify the subsequent drug analysis, several groups have used colorless serum-free salt/hepes-buffered solutions, which were supplemented with glucose, because no subsequent protein precipitation step is needed and samples can be directly measured by, for example, HPLC or microplate readers. It has to be kept in mind that the change from growth medium to transport buffer certainly also affects some properties of the cellular models. Of special note, withdrawal of astrocyte-derived factors can lead to significant changes of transport rates within a few hours [119]. Furthermore, due to different ion composition of the solutions, it cannot be excluded that the basis for the TEER measurements (flux of ions) has also been changed significantly. An equilibration phase of about 30 min after changing from growth medium to transport buffer is recommended before starting the actual transport study. Another aspect, especially when correlating *in vitro* to *in vivo* data, is the occurrence of an increased unstirred water layer (UWL) on cell layers *in vitro*. This additional aqueous barrier for lipophilic drugs can lead to significantly lower permeability coefficients in comparison to the *in vivo* situation. To overcome this typical *in vitro* phenomenon, the transport buffer could be stirred during the transport study (e.g., 25–150 rpm) [191, 193]. Rotating discs, developed initially for optimizing PAMPA experiments, could also be applied for monolayer cultures [204]. Another possibility is to modify and further develop the *in vitro* models. For example, the side-by-side diffusion chamber utilizes gas flow to move the growth medium and, thus, minimize the UWL [205]; however, this kind of *in vitro* model has been infrequently applied for BBB *in vitro* studies.

A later generation of BBB *in vitro* models, aiming at simulating the *in vivo* situation more closely, included shear stress by pumping growth medium in a closed circuit. In addition to beneficial effects on BBB *in vitro* properties, due to applied shear stress (see section "The Role of the Microenvironment"), the UWL is reduced in these models at the same time. A flow scheme of this type of model is shown in Figure 10.7. In short, cells are cultivated in plastic capillaries with inner diameters from 300 to 800 µm (wall thickness about 150 µm). The membranes of the capillaries are semipermeable, with pore sizes of around 0.2–0.5 µm. The material is polypropylene, polyethylene, or polysulfone and the number of capillaries per cartridge can vary from 3 to over 300. The capillary ends are sealed together in resins and mounted in polycarbonate cartridges. The lumen of the capillaries represent the blood compartment, which is connected by silicone tubings to a medium reservoir bottle, whereas the extracapillary space (ECS) within the cartridges forms the CNS compartment. Each compartment is accessible by two three-way valves (pre- and postcapillary at the lumen) and the medium is moved by a peristaltic pump in a pulsatile manner. The silicone tubing is also responsible for a proper gas exchange; therefore a sufficient surface area is essential. Although this model type has been used for BBB modeling since the mid 1990s and the effects of applied shear stress are undoubtedly important [156, 157, 161, 163, 166, 206–209], several technical aspects, with regard to drug permeability studies, have still to be considered. The experimental design for the transport study can follow the type of injection or infusion. Permeability studies using one single bolus (injection type) have to be carried out in a short time, since the bolus of the drug will be pumped quite fast through the capillary lumen and samples have to be taken within 15–60 min from the ECS as well as from the capillary lumen. For the infusion transport study type, the experimental duration can be prolonged to several hours until equilibration between luminal and ECS drug concentrations is reached. In this case, the lipophilicity of the silicone tubing can be a critical factor since adsorption loss of lipophilic drugs, such as benzodiazepines, can be around 80% of the initially administered drug amount [129]. Therefore, alternative low adsorbing plastic material for the tubing, which contains less softeners, was recommended, instead of silicone tubing, since repeated addition of the drugs did not yield a saturation of the adsorption process. Sterile filters at the reservoir bottle can be installed to ensure sufficient gas exchange if the new material is not as gas-permeable as silicone. Using either the injection or infusion transport study type, in both cases the assessment of blank values without cells is recommended. These data can give a valuable overview about the behavior of the investigated drug within the model and data for the blank experiment should also be included in the calculation of the resulting permeability coefficients. The mathematical equations used for the calculation of permeability coefficients also differ, depending on the experimental design (injection versus infusion type [157]). Furthermore, permeability of smaller paracellular markers such as mannitol, sucrose, fluorescein, carboxyfluorescein, or Lucifer yellow, or higher–molecular weight markers, such as fluorescent-labeled inulin, dextrans, or albumin, should be assessed, to avoid solely relying on TEER data [201, 202]. Although already integrated TEER modules are available for flow-based hollow-fiber models, it has to be kept in mind that TEER measurement principles are different in these models compared to chopstick electrodes for Transwell models and that TEER values are still related to the inner growth surface area, which can range from 13.5 to 432 cm^2 (Flocel, 19 capillaries per cartridge; Cellmax, Cat. No. 420–007, over 300 capillaries per cartridge). Consequently, a TEER value of 800 $\Omega \cdot cm^2$ in a hollow-fiber model with a total surface area of 13.5 cm^2 was probably obtained from a resistance value (at a distinct frequency) of about 60 Ω. A major disadvantage of these models is that cell attachment and growth can only be monitored indirectly by control of metabolism (glucose consumption), permeability studies, or TEER measurements, since the capillaries are not transparent and, thus, not accessible for light microscopic control. Hence, in order to ensure continuous confluent cell layers in every single capillary, it was recommended not only to seed the number of cells needed to cover the total surface area once, but, sometimes, also to seed the same amount a second time to plug possible leaks within the cell layers. This is because the brain endothelial cells change their metabolism and reduce their proliferation under flow conditions [156]. Confirmation of confluency of cell layers is feasible, for example, by scanning electron microscopy, only after the experiment [157]. Together with the high costs of cells and growth medium, these disadvantages led to the development of novel miniaturized, microfluidic devices of the BBB. By means of these models the beneficial effect of shear stress on BBB properties have been confirmed [164, 210–213]. However, no commercially available model, which is also useful for drug permeability testing, has been developed yet. Although some commercial sources tried to integrate an impedance device, and a few sources provide a two-chamber system for transport studies, the major problem of the small analyte volume in the microliter range remains unresolved. Consequently, the next generation of models have to implement a flow system and an impedance device. They also have to solve the problem of accessibility for online light microscopy, consist of a two-chamber system for transport studies, and provide enough analyte volume for proper subsequent drug analysis. Other *in vitro* models, such as isolated brain capillaries or brain endothelial cells cultured with astrocytes or pericytes in 3D gels (e.g., made of collagen), are still not suitable for drug transport studies, since, in the case of 3D gels, a clear two-chamber system is missing and, in the case of isolated capillaries, the luminal volumes are too small for drug transport studies. Thus, they do not fulfill the demands of drug discovery and development.

CONCLUSION AND OUTLOOK

For the reliable prediction of drug transport through the BBB in preclinical drug screening, a multilevel strategy is suggested. The place of the PAMPA and cell-based BBB models in these serial and parallel processes is between the in silico models and *in vivo* studies.

PAMPA-BBB has shown to be a practical, low-cost, and fast quantitative method, which could be used for early passive BBB permeability screening and for assisting medicinal chemists with structure modification to improve the BBB permeability of test compounds in the CNS drug discovery process. The PAMPA-BBB model, based on porcine brain extract (10% wt/vol PBLE in alkane), can precisely mimic the physicochemical lipoidal microenvironment of the BBB-governing passive permeability of positively charged (basic) drugs, using the rodent *in situ* brain perfusion technique as a benchmark. For negatively charged (acids) and uncharged molecules, the PAMPA-BBB model correlations are weak, but can be substantially improved by the in combo procedure. The zwitterion permeability is best predicted by hydrogen-bonding descriptors and appears not to relate to lipophilicity.

In addition to cell-free PAMPA *in vitro* models, culture models with higher complexity, including cytoarchitecture and transport pathways relevant for drug transport, are needed. Evaluation of the P-gp efflux ratio of drug candidates can be tested on cell line–based BBB and surrogate BBB models, like MDCK-MDR1 and vinblastin-treated Caco-2 cells. Selected drug candidates can be further tested for permeability on *in vitro* reconstituted coculture BBB models, which better mimic the structural and functional organization of the BBB. These double or triple coculture-based BBB models can be considered as added value and medium–low-throughput assays which can also be useful in screening CNS toxicity of non-CNS drugs. While PAMPA models can be part of large-scale screening in early-phase drug discovery, the more complex cell-based methods can be regarded and used as complementary assays.

Further progress is needed in several areas of BBB modeling to reach the goal of better and quicker prediction of BBB penetration. Since species differences exist in the transport proteins of the BBB, human models and human data are very important for further development. A simple, cost-effective, validated human BBB cell culture model is not yet available for drug screening, but the use of human pluripotent stem cells may provide a breakthrough in this field. We can predict that, in the long term, the cell culture–based models described here will be replaced by human models for drug testing, but rodent models will continue to be valuable tools in primary BBB research.

There are two major directions for the development of *in vitro* BBB models. Basic research in the expanding areas of BBB transporters, metabolism, and pathology will help to establish culture models more relevant to clinical situations which are crucial in predicting drug transport to the CNS in diseases. At the same time, development of human BBB culture models suitable for high-throughput screening for CNS drug development is expected. During the development of future models it will be important to find the right balance between the scalability, reproducibility, and simplicity such assays need and the complexity and relevance of the models that are important for correct predictions.

REFERENCES

[1] Abbott NJ, Patabendige AA, Dolman DE, Yusof SR, Begley DJ. Structure and function of the blood-brain barrier. *Neurobiol Discov*. 2010;37:13–25.

[2] Hammarlund-Udenaes M, Fridén M, Syvänen S, Gupta A. On the rate and extent of drug delivery to the brain. *Pharm Res*. 2008;25:1737–1750.

[3] Reichel A. Addressing central nervous system (CNS) penetration in drug discovery: basics and implications of the evolving new concept. *Chem Biodivers*. 2009;6:2030–2049.

[4] Takasato Y, Rapoport SI, Smith QR. An in situ brain perfusion technique to study cerebrovascular transport in the rat. *Am J Physiol*. 1984;247: H484–H493.

[5] Smith QR. A review of blood-brain barrier transport techniques. *Methods Mol Med*. 2003;89:193–208.

[6] Zhao R, Kalvass JC, Pollack GM. Assessment of blood-brain barrier permeability using the in situ mouse brain perfusion technique. *Pharm Res*. 2009;26:1657–1664.

[7] Avdeef A, Sun N. A new in situ brain perfusion flow correction method for lipophilic drugs based on the pH-dependent Crone-Renkin equation. *Pharm Res*. 2011;28:517–530.

[8] Avdeef A. *Absorption and Drug Development*. 2nd ed. Hoboken, NJ: Wiley-Interscience; 2012.

[9] Tsinman O, Tsinman K, Sun N, Avdeef A. Physicochemical selectivity of the BBB microenvironment governing passive diffusion—matching with a porcine brain lipid extract artificial membrane permeability model. *Pharm Res*. 2011;28:337–363.

[10] Abraham MH. The factors that influence permeation across the blood-brain barrier. *Eur J Med Chem*. 2004;39:235–240.

[11] Kansy M, Senner F, Gubernator K. Physicochemical high throughput screening: parallel artificial membrane permeability assay in the description of passive absorption processes. *J Med Chem*. 1998;41:1007–1010.

[12] Kansy M, Avdeef A, Fischer H. Advances in screening for membrane permeability: high-resolution PAMPA for medicinal chemists. *Drug Discov Today Technol*. 2005;1: 349–355.

[13] Di L, Kerns EH, Fan K, McConnell OJ, Carter GT. High throughput artificial membrane permeability assay for blood-brain barrier. *Eur J Med Chem*. 2003;38:223–232.

[14] Mensch J, Melis A, Mackie C, Verreck G, Brewster ME. Evaluation of various PAMPA models to identify the most discriminating method for the prediction of BBB permeability. *Eur J Pharm Sci*. 2010;74:495–502.

[15] Avdeef A, Tsinman O. PAMPA—a drug absorption in vitro model. 13. Chemical selectivity due to membrane hydrogen bonding: in combo comparisons of HDM-, DOPC-, and DS-PAMPA. *Eur J Pharm Sci*. 2006;28:43–50.

[16] Dagenais C, Avdeef A, Tsinman O, Dudley A, Beliveau R. P-Glycoprotein deficient mouse in situ blood-brain barrier permeability and its prediction using an in combo PAMPA model. *Eur J Pharm Sci*. 2009;38:121–137.

[17] Summerfield SG, Read K, Begley DJ, Obradovic T, Hidalgo IJ, Coggon S, Lewis AV, Porter RA, Jeffrey P. Central nervous system drug disposition: the relationship between in situ brain permeability and brain free fraction. *J Pharmacol Exp Ther*. 2007;322: 205–213.

[18] Di L, Kerns EH, Bezar IF, Petusky SL, Huang Y. Comparison of blood-brain barrier permeability assays: in situ brain perfusion, MDR1-MDCKII and PAMPA-BBB. *J Pharm Sci*. 2009;98:1980–1991.

[19] Giacomini KM, Huang S-M, Tweedie DJ, Benet LZ, Brouwer KL, Chu X, Dahlin A, Evers R, Fischer V, Hillgren KM, Hoffmaster KA, Ishikawa T, Keppler D, Kim RB, Lee CA, Niemi M, Polli JW, Sugiyama Y, Swaan PW, Ware JA, Wright SH, Yee SW, Zamek-Gliszczynski MJ, Zhang L; International Transporter Consortium. Membrane transporters in drug development. *Nat Rev Drug Discov*. 2010;9:215–236.

[20] Roth M, Obaidat A, Hagenbuch B. OATPs, OATs and OCTs: the organic anion and cation transporters of the SLCO and SLC22A gene superfamilies. *Br J Pharmacol.* 2012;165: 1260–1287.

[21] Mulato AS, Ho ES, Cihlar T. Nonsteroidal anti-inflammatory drugs efficiently reduce the transport and cytotoxicity of adefovir mediated by the human renal organic anion transporter 1. *J Pharmacol Exp Ther.* 2000;295:10–15.

[22] Westholm DE, Stenehjem DD, Rumbley JN, Drewes LR, Anderson GW. Competitive inhibition of Oatp1c1-mediated transport by the feamate class of nonsteroidal anti-inflammatory drugs. *Endrocrinology.* 2009;150:1025–1032.

[23] Joó F, Karnushina I. A procedure for the isolation of capillaries from rat brain. *Cytobios.* 1973;8(29):41–48.

[24] Panula P, Joó F, Rechardt L. Evidence for the presence of viable endothelial cells in cultures derived from dissociated rat brain. *Experientia.* 1978;34(1):95–97.

[25] Deli MA, Ábrahám CS, Kataoka Y, Niwa M. Permeability studies on in vitro blood-brain barrier models: physiology, pathology, and pharmacology. *Cell Mol Neurobiol.* 2005;25:59–127.

[26] Wilson HK, Shusta EV. Human-based in vitro endothelial cell models. In: Di L, Kerns EH, editors. *Blood-Brain Barrier in Drug Discovery: Optimizing Brain Exposure of CNS Drugs and Minimizing Brain Side Effects.* Hoboken, NJ: Wiley-Interscience; 2014: Chapter 11.

[27] Abbott NJ, Hughes CCW, Revest PA, Greenwood J. Development and characterisation of a rat brain capillary endothelial culture: towards an in vitro blood-brain barrier. *J Cell Sci.* 1992;103(Pt 1):23–37.

[28] Szabó CA, Deli MA, Ngo TK, Joó F. Production of pure primary rat cerebral endothelial cell culture: a comparison of different methods. *Neurobiology (Bp).* 1997;5(1):1–16.

[29] Perrière N, Demeuse PH, Garcia E, Regina A, Debray M, Andreux JP, Couvreur P, Scherrmann JM, Temsamani J, Couraud PO, Deli MA, Roux F. Puromycin-based purification of rat brain capillary endothelial cell cultures. Effect on the expression of blood-brain barrier-specific properties. *J Neurochem.* 2005;93(2):279–289.

[30] Perrière N, Yousif S, Cazaubon S, Chaverot N, Bourasset F, Cisternino S, Decleves X, Hori S, Terasaki T, Deli M, Scherrmann JM, Temsamani J, Roux F, Couraud PO. A functional in vitro model of rat blood-brain barrier for molecular analysis of efflux transporters. *Brain Res.* 2007;1150:1–13.

[31] Ge S, Song L, Pachter JS. Where is the blood-brain barrier … really? *J Neurosci Res.* 2005;79(4):421–427.

[32] Takata F, Dohgu S, Yamauchi A, Matsumoto J, Machida T, Fujishita K, Shibata K, Shinozaki Y, Sato K, Kataoka Y, Koizumi S. *In vitro* blood-brain barrier models using brain capillary endothelial cells isolated from neonatal and adult rats retain age-related barrier properties. *PLoS One.* 2013;8(1):e55166.

[33] Abbott NJ, Dolman DEM, Drndarski S, Fredriksson SM. An improved in vitro blood-brain barrier model: rat brain endothelial cells co-cultured with astrocytes. In: Milner R, editor. *Astrocytes—Methods and Protocols, Methods in Molecular Biology,* Volume 814, New York: Humana Press; 2012:415–430.

[34] Patabendige A, Skinner RA, Abbott JA. Establishment of a simplified *in vitro* porcine blood-brain barrier model with high transendothelial electrical resistance. *Brain Res.* 2013;1521:1–15.

[35] Patabendige A, Skinner RA, Morgan L, Abbott JA. A detailed method for preparation of a functional and flexible blood-brain barrier model using porcine brain endothelial cells. *Brain Res*. 2013;1521:16–30.

[36] Watson PM, Paterson JC, Thom G, Ginman U, Lundquist S, Webster CI. Modelling the endothelial blood-CNS barriers: a method for the production of robust in vitro models of the rat blood-brain barrier and blood-spinal cord barrier. *BMC Neurosci*. 2013;14:59.

[37] Butt AM, Jones HC, Abbott NJ. Electrical resistance across the blood-brain barrier in anaesthetized rats: a developmental study. *J Physiol*. 1990;429:47–62.

[38] Deli MA. Potential use of tight junction modulators to reversibly open membranous barriers and improve drug delivery. *Biochim Biophys Acta*. 2009;1788(4):892–910.

[39] Wolburg H, Noell S, Mack A, Wolburg-Buchholz K, Fallier-Becker P. Brain endothelial cells and the glio-vascular complex. *Cell Tissue Res*. 2009;335(1):75–96.

[40] Veszelka S, Kittel Á, Deli MA. Tools of modelling blood-brain barrier penetrability. In: Tihanyi K, Vastag M, editors. Solubility, Delivery, and ADME Problems of Drugs and Drug Candidates. Washington, DC: Bentham Science; 2011:166–188.

[41] Hellinger É, Veszelka S, Tóth A, Walter F, Kittel Á, Bakk ML, Tihanyi K, Háda V, Nakagawa S, Thuy DH, Niwa M, Deli MA, Vastag M. Comparison of brain capillary endothelial cell based and epithelial cell based (MDCK-MDR1, Caco-2, and VB-Caco-2) surrogate blood-brain barrier penetration models. *Eur J Pharm Biopharm*. 2012;82(2): 340–351.

[42] Dehouck M-P, Méresse S, Delorme P, Fruchart JC, Cecchelli R. An easier, reproducible, and mass-production method to study the blood-brain barrier in vitro. *J Neurochem*. 1990;54(5):1798–1801.

[43] Dehouck MP, Jolliet-Riant P, Brée F, Fruchart JC, Cecchelli R, Tillement JP. Drug transfer across the blood-brain barrier: correlation between *in vitro* and *in vivo* models. *J Neurochem*. 1992;58:1790–1797.

[44] Cecchelli R, Dehouck B, Descamps L, Fenart L, Buée-Scherrer VV, Duhem C, Lundquist S, Rentfel M, Torpier G, Dehouck M-P. In vitro model for evaluating drug transport across the blood-brain barrier. *Adv Drug Deliv Rev*. 1999;36(2–3):165–178.

[45] Lundquist S, Renftel M, Brillault J, Fénart L, Cecchelli R, Dehouck M-P. Prediction of drug transport through the blood-brain barrier in vivo: a comparison between two in vitro cell models. *Pharm Res*. 2002;19(7):976–981.

[46] Garberg P, Ball M, Borg N, Cecchelli R, Fenart L, Hurst RD, Lindmark T, Mabondzo A, Nilsson JE, Raub TJ, Stanimirovic D, Terasaki T, Oberg JO, Osterberg T. In vitro models for the blood-brain barrier. *Toxicol In Vitro*. 2005;19(3):299–334.

[47] Culot M, Lundquist S, Vanuxeem D, Nion S, Landry C, Delplace Y, Dehouck MP, Berezowski V, Fenart L, Cecchelli R. An in vitro blood-brain barrier model for high throughput (HTS) toxicological screening. *Toxicol In Vitro*. 2008;22(3):799–811.

[48] Pardridge WM, Triguero D, Yang J, Cancilla PA. Comparison of *in vitro* and *in vivo* models of drug transcytosis through the blood-brain barrier. *J Pharmacol Exp Ther*. 1990;253(4):884–891.

[49] Rubin LL, Hall DE, Porter S, Barbu K, Cannon C, Horner HC, Janatpour M, Liaw CW, Manning K, Morales J, Tanner LI, Tomaselli KJ, Bard F. A cell culture model of the blood-brain barrier. *J Cell Biol*. 1991;115(6):1725–1735.

[50] Pirro JP, Di Rocco RJ, Narra RK, Nunn AD. Relationship between *in vitro* transendothelial permeability and *in vivo* single-pass brain extraction. *J Nucl Med*. 1994;35:1514–1519.

[51] Gaillard PJ, Voorwinden LH, Nielsen JL, Ivanov A, Atsumi R, Engman H, Ringbom C, de Boer AG. Establishment and functional characterization of an in vitro model of the blood–brain barrier, comprising a coculture of brain capillary endothelial cells and astrocytes. *Eur J Pharm Sci*. 2001;12(3):215–222.

[52] Zenker D, Begley D, Bratzke H, Rübsamen-Waigmann H, von Briesen H. Human blood-derived macrophages enhance barrier function of cultured primary bovine and human brain capillary endothelial cells. *J Physiol*. 2003;551(Pt 3):1023–1032.

[53] Hakkarainen JJ, Jalkanen AJ, Kääriäinen TM, Keski-Rahkonen P, Venäläinen T, Hokkanen J, Mönkkönen J, Suhonen M, Forsberg MM. Comparison of in vitro cell models in predicting in vivo brain entry of drugs. *Int J Pharm*. 2010;402(1–2):27–36.

[54] Nakhlband A, Omidi Y. Barrier functionality of porcine and bovine brain capillary endothelial cells. *Bioimpacts*. 2011;1(3):153–159.

[55] Helms HC, Waagepetersen HS, Nielsen CU, Brodin B. Paracellular tightness and claudin-5 expression is increased in the BCEC/astrocyte blood-brain barrier model by increasing media buffer capacity during growth. *AAPS J*. 2010;12(4):759–770.

[56] Helms MC, Madelung R, Waagepetersen HS, Nielsen CU, Brodin B. In vitro evidence for the brain glutamate efflux hypothesis: brain endothelial cells cocultured with astrocytes display a polarized brain-to-blood transport of glutamate. *Glia*. 2012;60(6):882–893.

[57] Deli MA, Ábrahám CS, Niwa M, Falus A. N,N-diethyl-2-[4-(phenylmethyl)phenoxy] ethanamine increases the permeability of primary mouse cerebral endothelial cell monolayers. *Inflamm Res*. 2003;52(Suppl 1):S39–S40.

[58] Coisne C, Dehouck L, Faveeuw C, Delplace Y, Miller F, Landry C, Morissette C, Fenart L, Cecchelli R, Tremblay P, Dehouck B. Mouse syngenic in vitro blood-brain barrier model: a new tool to examine inflammatory events in cerebral endothelium. *Lab Invest*. 2005;85(6):734–746.

[59] Kuntz M, Mysiorek C, Pétrault O, Pétrault M, Uzbekov R, Bordet R, Fenart L, Cecchelli R, Bérézowski V. Stroke-induced brain parenchymal injury drives blood-brain barrier early leakage kinetics: a combined in vivo/in vitro study. *J Cereb Blood Flow Metab*. 2014;34(1):95–107.

[60] Weidenfeller C, Schrot S, Zozulya A, Galla HJ. Murine brain capillary endothelial cells exhibit improved barrier properties under the influence of hydrocortisone. *Brain Res*. 2005;1053(1–2):162–174.

[61] Shayan G, Choi YS, Shusta EV, Shuler ML, Lee KH. Murine in vitro model of the blood-brain barrier for evaluating drug transport. *Eur J Pharm Sci*. 2011;42(1–2): 148–155.

[62] Steiner O, Coisne C, Engelhardt B, Lyck R. Comparison of immortalized bEnd5 and primary mouse brain microvascular endothelial cells as in vitro blood-brain barrier models for the study of T cell extravasation. *J Cereb Blood Flow Metab*. 2011;31(1):315–327.

[63] Dohgu S, Ryerse JS, Robinson SM, Banks WA. Human immunodeficiency virus-1 uses the mannose-6-phosphate receptor to cross the blood-brain barrier. *PLoS One*. 2012; 7(6):e39565.

[64] Wuest DM, Lee KH. Optimization of endothelial cell growth in murine *in vitro* blood-brain barrier model. *Biotechnol J*. 2012;7(3):409–417.

[65] Franke H, Galla HJ, Beuckmann CT. An improved low-permeability in vitro-model of the blood-brain barrier: transport studies on retinoids, sucrose, haloperidol, caffeine and mannitol. *Brain Res*. 1999;818(1):65–71.

[66] Franke H, Galla H, Beuckmann CT. Primary cultures of brain microvessel endothelial cells: a valid and flexible model to study drug transport through the blood-brain barrier in vitro. *Brain Res Brain Res Protoc*. 2000;5(3):248–256.

[67] Lohmann C, Hüwel S, Galla HJ. Predicting blood-brain barrier permeability of drugs: evaluation of different in vitro assays. *J Drug Target*. 2002;10(4):263–276.

[68] Omidi Y, Campbell L, Barar J, Connell D, Akhtar S, Gumbleton M. Evaluation of the immortalised mouse brain capillary endothelial cell line, b.End3, as an in vitro blood-brain barrier model for drug uptake and transport studies. *Brain Res*. 2003;990(1–2):95–112.

[69] Smith M, Omidi Y, Gumbleton M. Primary porcine brain microvascular endothelial cells: biochemical and functional characterisation as a model for drug transport and targeting. *J Drug Target*. 2007;15(4):253–268.

[70] Jeliazkova-Mecheva VV, Bobilya DJ. A porcine astrocyte/endothelial cell co-culture model of the blood-brain barrier. *Brain Res Brain Res Protoc*. 2003;12(2):91–98.

[71] Zhang Y, Li CS, Ye Y, Johnson K, Poe J, Johnson S, Bobrowski W, Garrido R, Madhu C. Porcine brain microvessel endothelial cells as an in vitro model to predict in vivo blood-brain barrier permeability. *Drug Metab Dispos*. 2006;34(11):1935–1943.

[72] Cohen-Kashi Malina K, Cooper I, Teichberg VI. Closing the gap between the in-vivo and in-vitro blood-brain barrier tightness. *Brain Res*. 2009;1284:12–21.

[73] Cantrill CA, Skinner RA, Rothwell NJ, Penny JI. An immortalised astrocyte cell line maintains the in vivo phenotype of primary porcine in vitro blood-brain barrier model. *Brain Res*. 2012;1479:17–30.

[74] Kis B, Deli MA, Kobayashi H, Abraham CS, Yanagita T, Kaiya H, Isse T, Nishi R, Gotoh S, Kangawa K, Wada A, Greenwood J, Niwa M, Yamashita H, Ueta Y. Adrenomedullin regulates blood-brain barriers functions in vitro. *Neuroreport*. 2001;12(18):4139–4142.

[75] Demeuse P, Kerkhofs A, Struys-Ponsar C, Knoops B, Remacle C, van den Bosch de Aguilar P. Compartmentalized coculture of rat brain endothelial cells and astrocytes: a syngenic model to study the blood-brain barrier. *J Neurosci Methods*. 2002;121(1): 21–31.

[76] Garcia-Garcia E, Gil S, Andrieux K, Desmaële D, Nicolas V, Taran F, Georgin D, Andreux JP, Roux F, Couvreur P. A relevant in vitro rat model for the evaluation of blood-brain barrier translocation of nanoparticles. *Cell Mol Life Sci*. 2005;62(12):1400–1408.

[77] Weidenfeller C, Svendsen CN, Shusta EV. Differentiating embryonic neural progenitor cells induce blood-brain barrier properties. *J Neurochem*. 2007;101(2):555–565.

[78] Nakagawa S, Deli MA, Nakao S, Honda M, Hayashi K, Nakaoke R, Kataoka Y, Niwa M. Pericytes from brain microvessels strengthen the barrier integrity in primary cultures of rat brain endothelial cells. *Cell Mol Neurobiol*. 2007;27(6):687–694.

[79] Nakagawa S, Deli MA, Kawaguchi H, Shimizudani T, Shimono T, Kittel A, Tanaka K, Niwa M. A new blood-brain barrier model using primary rat brain endothelial cells, pericytes and astrocytes. *Neurochem Int*. 2009;54(3–4):253–263.

[80] Sobue K, Yamamoto N, Yoneda K, Hodgson ME, Yamashiro K, Tsuruoka N, Tsuda T, Katsuya H, Miura Y, Asai K, Kato T. Induction of blood–brain barrier properties in immortalized bovine brain endothelial cells by astrocytic factors. *Neurosci Res*. 1999;35(2): 155–164.

[81] Yang T, Roder KE, Abbruscato TJ. Evaluation of bEnd5 cell line as an in vitro model for the blood-brain barrier under normal and hypoxic/aglycemic conditions. *J Pharm Sci*. 2007;96(12):3196–3213.

[82] Förster C, Silwedel C, Golenhofen N, Burek M, Kietz S, Mankertz J, Drenckhahn D. Occludin as direct target for glucocorticoid-induced improvement of blood-brain barrier properties in a murine in vitro system. *J Physiol*. 2005;565(Pt 2):475–486.

[83] Burek M, Salvador E, Förster CY. Generation of an immortalized murine brain microvascular endothelial cell line as an *in vitro* blood brain barrier model. *J Vis Exp*. 2012;66:e4022.

[84] Salvador E, Neuhaus W, Foerster C. Stretch in brain microvascular endothelial cells (cEND) as an *in vitro* traumatic brain injury model of the blood brain barrier. *J Vis Exp*. 2013;80:e50928.

[85] Hosoya K, Tetsuka K, Nagase K, Tomi M, Saeki S, Ohtsuki S, Terasaki T. Conditionally immortalized brain capillary endothelial cell lines established from a transgenic mouse harboring temperature-sensitive simian virus 40 large T-antigen gene. *AAPS PharmSci*. 2000;2(3):E27, pp. 1–11..

[86] Bauer R, Lauer R, Linz B, Pittner F, Peschek G, Ecker GF, Friedl P, Noe CR. An in vitro model for blood brain barrier permeation. *Sci Pharm*. 2002;70(4):317–324.

[87] Neuhaus W, Bogner E, Wirth M, Trzeciak J, Lachmann B, Gabor F, Noe CR. A novel tool to characterize paracellular transport: the APTS-dextran ladder. *Pharm Res*. 2006;23(7): 1491–1501.

[88] Neuhaus W, Plattner VE, Wirth M, Germann B, Lachmann B, Gabor F, Noe CR. Validation of in vitro cell culture models of the blood-brain barrier: tightness characterization of two promising cell lines. *J Pharm Sci*. 2008;97(12):5158–5175.

[89] Romero IA, Radewicz K, Jubin E, Michel CC, Greenwood J, Couraud PO, Adamson P. Changes in cytoskeletal and tight junctional proteins correlate with decreased permeability induced by dexamethasone in cultured rat brain endothelial cells. *Neurosci Lett*. 2003;344(2):112–116.

[90] Blasig IE, Mertsch K, Haseloff RF. Nitronyl nitroxides, a novel group of protective agents against oxidative stress in endothelial cells forming the blood-brain barrier. *Neuropharmacology*. 2002;43(6):1006–1014.

[91] Hosoya KI, Takashima T, Tetsuka K, Nagura T, Ohtsuki S, Takanaga H, Ueda M, Yanai N, Obinata M, Terasaki T. mRNA expression and transport characterization of conditionally immortalized rat brain capillary endothelial cell lines; a new in vitro BBB model for drug targeting. *J Drug Target*. 2000;8(6):357–370.

[92] Terasaki T, Ohtsuki S, Hori S, Takanaga H, Nakashima E, Hosoya K. New approaches to in vitro models of blood-brain barrier drug transport. *Drug Discov Today*. 2003;8(20): 944–954.

[93] Easton AS, Abbott NJ. Bradykinin increases permeability by calcium and 5-lipoxygenase in ECV304/C6 cell culture model of the blood–brain barrier. *Brain Res*. 2002;953(1–2): 157–169.

[94] Youdim KA, Dobbie MS, Kuhnle G, Proteggente AR, Abbott NJ, Rice-Evans C. Interaction between flavonoids and the blood-brain barrier: in vitro studies. *J Neurochem*. 2003;85(1):180–192.

[95] Barar J, Gumbleton M, Asadi M, Omidi Y. Barrier functionality and transport machineries of human ECV304 cells. *Med Sci Monit*. 2010;16(1):BR52–BR60.

[96] Brown J, Reading SJ, Jones S, Fitchett CJ, Howl J, Martin A, Longland CL, Michelangeli F, Dubrova YE, Brown CA. Critical evaluation of ECV304 as a human endothelial cell model defined by genetic analysis and functional responses: a comparison with the human bladder cancer derived epithelial cell line T24/83. *Lab Invest*. 2000;80(1):37–45.

[97] Eisenblätter T, Galla HJ. A new multidrug resistance protein at the blood-brain barrier. *Biochem Biophys Res Commun*. 2002;293(4):1273–1278.

[98] Uchida Y, Ohtsuki S, Katsukura Y, Ikeda C, Suzuki T, Kamiie J, Terasaki T. Quantitative targeted absolute proteomics of human blood-brain barrier transporters and receptors. *J Neurochem*. 2011;117(2):333–345.

[99] Candela P, Gosselet F, Saint-Pol J, Sevin E, Boucau MC, Boulanger E, Cecchelli R, Fenart L. Apical-to-basolateral transport of amyloid-β peptides through blood-brain barrier cells is mediated by the receptor for advanced glycation end-products and is restricted by P-glycoprotein. *J Alzheimers Dis*. 2010;22(3):849–859.

[100] Agarwal S, Elmquist WF. Insight into the cooperation of P-glycoprotein (ABCB1) and breast cancer resistance protein (ABCG2) at the blood-brain barrier: a case study examining sorafenib efflux clearance. *Mol Pharm*. 2012;9(3):678–684.

[101] Lin F, Marchetti S, Pluim D, Iusuf D, Mazzanti R, Schellens JH, Beijnen JH, van Tellingen O. Abcc4 together with abcb1 and abcg2 form a robust cooperative drug efflux system that restricts the brain entry of camptothecin analogues. *Clin Cancer Res*. 2013;19(8):2084–2095.

[102] Redzic Z. Molecular biology of the blood-brain and the blood-cerebrospinal fluid barriers: similarities and differences. *Fluids Barriers CNS*. 2011;8:3.

[103] Pifferi F, Jouin M, Alessandri JM, Roux F, Perrière N, Langelier B, Lavialle M, Cunnane S, Guesnet P. n-3 long-chain fatty acids and regulation of glucose transport in two models of rat brain endothelial cells. *Neurochem Int*. 2010;56(5):703–710.

[104] Berezowski V, Miecz D, Marszałek M, Bröer A, Bröer S, Cecchelli R, Nałecz KA. Involvement of OCTN2 and $B^{0,+}$ in the transport of carnitine through an in vitro model of the blood-brain barrier. *J Neurochem*. 2004;91(4):860–872.

[105] Czeredys M, Mysiorek C, Kulikova N, Samluk Ł, Berezowski V, Cecchelli R, Nałecz KA. A polarized localization of amino acid/carnitine transporter B(0,+) (ATB(0,+)) in the blood-brain barrier. *Biochem Biophys Res Commun*. 2008;376(2):267–270.

[106] Calabria AR, Shusta EV. A genomic comparison of in vivo and in vitro brain microvascular endothelial cells. *J Cereb Blood Flow Metab*. 2008;28(1):135–148.

[107] Cecchelli R, Berezowski V, Lundquist S, Culot M, Renftel M, Dehouck MP, Fenart L. Modelling of the blood-brain barrier in drug discovery and development. *Nat Rev Drug Discov*. 2007;6(8):650–661.

[108] Warren MS, Zerangue N, Woodford K, Roberts LM, Tate EH, Feng B, Li C, Feuerstein TJ, Gibbs J, Smith B, de Morais SM, Dower WJ, Koller KJ. Comparative gene expression profiles of ABC transporters in brain microvessel endothelial cells and brain in five species including human. *Pharmacol Res*. 2009;59(6):404–413.

[109] Nitz T, Eisenblätter T, Psathaki K, Galla HJ. Serum-derived factors weaken the barrier properties of cultured porcine brain capillary endothelial cells in vitro. *Brain Res*. 2003;981(1–2):30–40.

[110] Russell FD, Hamilton KD. Nutrient deprivation increases vulnerability of endothelial cells to proinflammatory insults. *Free Radic Biol Med*. 2014;67:408–415.

[111] Hoheisel D, Nitz T, Franke H, Wegener J, Hakvoort A, Tilling T, Galla HJ. Hydrocortisone reinforces the blood-brain properties in a serum free cell culture system. *Biochem Biophys Res Commun*. 1998;247(2):312–315.

[112] Deli MA, Dehouck MP, Ábrahám CS, Cecchelli R, Joó F. Penetration of small molecular weight substances through cultured bovine brain capillary endothelial cell monolayers:

the early effects of cyclic adenosine 3′,5′-monophosphate. *Exp Physiol*. 1995;80(4): 675–678.

[113] Tóth A, Veszelka S, Nakagawa S, Niwa M, Deli MA. Patented in vitro blood-brain barrier models in CNS drug discovery. *Recent Pat CNS Drug Discov*. 2011;6(2): 107–118.

[114] Roux F, Couraud PO. Rat brain endothelial cell lines for the study of blood-brain barrier permeability and transport functions. *Cell Mol Neurobiol*. 2005;25(1):41–58.

[115] Watanabe T, Dohgu S, Takata F, Nishioku T, Nakashima A, Futagami K, Yamauchi A, Kataoka Y. Paracellular barrier and tight junction protein expression in the immortalized brain endothelial cell lines bEND.3, bEND.5 and mouse brain endothelial cell 4. *Biol Pharm Bull*. 2013;36(3):492–495.

[116] Lauer R, Bauer R, Linz B, Pittner F, Peschek GA, Ecker G, Friedl P, Noe CR. Development of an in vitro blood-brain barrier model based on immortalized porcine brain microvascular endothelial cells. *Farmaco*. 2004;59(2):133–137.

[117] Neuhaus W, Stessl M, Strizsik E, Bennani-Baiti B, Wirth M, Toegel S, Modha M, Winkler J, Gabor F, Viernstein H, Noe CR. Blood-brain barrier cell line PBMEC/C1-2 possesses functionally active P-glycoprotein. *Neurosci Lett*. 2010;469(2): 224–228.

[118] Neuhaus W, Mandikova J, Pawlowitsch R, Linz B, Bennani-Baiti B, Lauer R, Lachmann B, Noe CR. Blood-brain barrier in vitro models as tools in drug discovery: assessment of the transport ranking of antihistaminic drugs. *Pharmazie*. 2012;67(5):432–439.

[119] Novakova I, Subileau E-A, Toegel S, Gruber D, Lachmann B, Urban E, Chesne C, Noe CR, Neuhaus W. Transport rankings of non-steroidal antiinflammatory drugs across blood-brain barrier in vitro models. *PLoS One*. 2014;9(1):e86806.

[120] Reichel A, Begley DJ, Abbott NJ. An overview of in vitro techniques for blood–brain barrier studies. In: Nag S, editor. *The Blood–Brain Barrier: Biology and Research Protocols. Methods in Molecular Medicine*, Volume 89. Totowa, NJ: Humana Press; 2003:307–324.

[121] Kis B, Szabó CA, Pataricza J, Krizbai IA, Mezei Z, Gecse A, Telegdy G, Papp JG, Deli MA. Vasoactive substances produced by cultured rat brain endothelial cells. *Eur J Pharmacol*. 1999;368(1):35–42.

[122] Gonzalez-Mariscal L, Bautista P, Lechuga S, Quiros M. ZO-2, a tight junction scaffold protein involved in the regulation of cell proliferation and apoptosis. *Ann N Y Acad Sci*. 2012;1257:133–141.

[123] Vastag M, Keserű GM. Current in vitro and in silico models of blood-brain barrier penetration: a practical view. *Curr Opin Drug Discov Devel*. 2009;12(1):115–124.

[124] Takahashi K, Sawasaki Y, Hata J, Mukai K, Goto T. Spontaneous transformation and immortalization of human endothelial cells. *In Vitro Cell Dev Biol*. 1990;26(3 Pt 1): 265–274.

[125] Suda K, Rothen-Rutishauser B, Günthert M, Wunderli-Allenspach H. Phenotypic characterization of human umbilical vein endothelial (ECV304) and urinary carcinoma (T24) cells: endothelial versus epithelial features. *In Vitro Cell Dev Biol Anim*. 2001;37(8): 505–514.

[126] Drexler HG, Quentmeier H, Dirks WG, MacLeod RA. Bladder carcinoma cell line ECV304 is not a model system for endothelial cells. *In Vitro Cell Dev Biol Anim*. 2002;38(4):185–186.

[127] Iartseva NM, Fedortseva RF. Characteristics of the spontaneously transformed human endothelial cell line ECV304. I. Multiple chromosomal rearrangements in endothelial cells ECV304 [in Russian]. *Tsitologiia*. 2008;50(7):568–575.

[128] Kiessling F, Kartenbeck J, Haller C. Cell-cell contacts in the human cell line ECV304 exhibit both endothelial and epithelial characteristics. *Cell Tissue Res*. 1999;297(1): 131–140.

[129] Neuhaus W. Development and validation of in-vitro models of the blood-brain barrier [dissertation]. Vienna, Austria: University of Vienna; 2007. Available from the local university library.

[130] Haseloff RF, Blasig IE, Bauer HC, Bauer H. In search of the astrocytic factor(s) modulating blood-brain barrier functions in brain capillary endothelial cells in vitro. *Cell Mol Neurobiol*. 2005;25(1):25–39.

[131] Abbott NJ, Rönnbäck L, Hansson E. Astrocyte-endothelial interactions at the blood-brain barrier. *Nat Rev Neurosci*. 2006;7(1):41–53.

[132] Boveri M, Berezowski V, Price A, Slupek S, Lenfant AM, Benaud C, Hartung T, Cecchelli R, Prieto P, Dehouck M-P. Induction of blood-brain barrier properties in cultured brain capillary endothelial cells: comparison between primary glial cells and C6 cell line. *Glia*. 2005;51(3):187–198.

[133] Culot M, Fabulas-da Costa A, Sevin E, Szorath E, Martinsson S, Ranftel M, Hongmei Y, Cecchelli R, Lundquist S. A simple model for assessing free brain/free plasma ratios using an *in vitro* model of the blood-brain barrier. *PLoS One*. 2013;8(12):e80634.

[134] Vandenhaute E, Sevin E, Hallier-Vanuxeem D, Dehouck MP, Cecchelli R. Case study: adapting in vitro blood-brain barrier models for use in early-stage drug discovery. *Drug Discov Today*. 2012;17(7–8):285–290.

[135] Shepro D, Morel NM. Pericyte physiology. *FASEB J*. 1993;7(11):1031–1038.

[136] Frank RN, Turczyn TJ, Das A. Pericyte coverage of retinal and cerebral capillaries. *Invest Ophthalmol Vis Sci*. 1990;31(6):999–1007.

[137] Armulik A, Genové G, Mäe M, Nisancioglu MH, Wallgard E, Niaudet C, He L, Norlin J, Lindblom P, Strittmatter K, Johansson BR, Betsholtz C. Pericytes regulate the blood-brain barrier. *Nature*. 2010;468(7323):557–561.

[138] Daneman R, Zhou L, Kebede AA, Barres BA. Pericytes are required for blood-brain barrier integrity during embryogenesis. *Nature*. 2010;468(7323):562–566.

[139] Hayashi K, Nakao S, Nakaoke R, Nakagawa S, Kitagawa N, Niwa M. Effects of hypoxia on endothelial/pericytic co-culture model of the blood-brain barrier. *Regul Pept*. 2004;123(1–3):77–83.

[140] Berezowski V, Landry C, Dehouck M-P, Cecchelli R, Fénart L. Contribution of glial cells and pericytes to the mRNA profiles of P-glycoprotein and multidrug resistance-associated proteins in an in vitro model of the blood-brain barrier. *Brain Res*. 2004; 1018(1):1–9.

[141] Dohgu S, Takata F, Yamauchi A, Nakagawa S, Egawa T, Naito M, Tsuruo T, Sawada Y, Niwa M, Kataoka Y. Brain pericytes contribute to the induction and up-regulation of blood-brain barrier functions through transforming growth factor-beta production. *Brain Res*. 2005;1038(2):208–215.

[142] Vandenhaute E, Dehouck L, Boucau MC, Sevin E, Uzbekov R, Tardivel M, Gosselet F, Fenart L, Cecchelli R, Dehouck MP. Modelling the neurovascular unit and the blood-brain barrier with the unique function of pericytes. *Curr Neurovasc Res*. 2011;8(4):258–269.

[143] Katyshev V, Dore-Duffy P. Pericyte coculture models to study astrocyte, pericyte, and endothelial cell interactions. In: Milner R, editor. *Astrocytes—Methods and Protocols. Methods in Molecular Biology*, Volume 814. New York: Humana Press; 2012: 467–481.

[144] Bauer HC, Bauer H. Neural induction of the blood-brain barrier: still an enigma. *Cell Mol Neurobiol*. 2000;20(1):13–28.

[145] Tontsch U, Bauer HC. Glial cells and neurons induce blood-brain barrier related enzymes in cultured cerebral endothelial cells. *Brain Res*. 1991;539(2):247–253.

[146] Duport S, Robert F, Muller D, Grau G, Parisi L, Stoppini L. An in vitro blood-brain barrier model: cocultures between endothelial cells and organotypic brain slice cultures. *Proc Natl Acad Sci U S A*. 1998;95(4):1840–1845.

[147] Stanness KA, Neumaier JF, Sexton TJ, Grant GA, Emmi A, Maris DO, Janigro D. A new model of the blood-brain barrier: co-culture of neuronal, endothelial and glial cells under dynamic conditions. *Neuroreport*. 1999;10(18):3725–3731.

[148] Savettieri G, Di Liegro I, Catania C, Licata L, Pitarresi GL, D'Agostino S, Schiera G, De Caro V, Giandalia G, Giannola LI, Cestelli A. Neurons and ECM regulate occludin localization in brain endothelial cells. *Neuroreport*. 2000;11(5):1081–1084.

[149] Cestelli A, Catania C, D'Agostino S, Di Liegro I, Licata L, Schiera G, Pitarresi GL, Savettieri G, De Caro V, Giandalia G, Giannola LI. Functional feature of a novel model of blood brain barrier: studies on permeation of test compounds. *J Control Release*. 2001;76(1–2):139–147.

[150] Lim JC, Wolpaw AJ, Caldwell MA, Hladky SB, Barrand MA. Neural precursor cell influences on blood-brain barrier characteristics in rat brain endothelial cells. *Brain Res*. 2007;1159:67–76.

[151] Lippmann ES, Weidenfeller C, Svendsen CN, Shusta EV. Blood-brain barrier modeling with co-cultured neural progenitor cell-derived astrocytes and neurons. *J Neurochem*. 2011;119(3):507–520.

[152] Raub TJ, Kuentzel SL, Sawada GA. Permeability of bovine brain microvessel endothelial cells in vitro: barrier tightening by a factor released from astroglioma cells. *Exp Cell Res*. 1992;199(2):330–340.

[153] Tilling T, Korte D, Hoheisel D, Galla H-J. Basement membrane proteins influence brain capillary endothelial barrier function in vitro. *J Neurochem*. 1998;71(3): 1151–1157.

[154] Dömötör E, Sipos I, Kittel Á, Abbott NJ, Ádám-Vizi V. Improved growth of cultured brain microvascular endothelial cells on glass coated with a biological matrix. *Neurochem Int*. 1998;33(6):473–478.

[155] Chang E, O'Donnell ME, Barakat AI. Shear stress and 17beta-estradiol modulate cerebral microvascular endothelial Na-K-Cl cotransporter and Na/H exchanger protein levels. *Am J Physiol Cell Physiol*. 2008;294(1):C363–C371.

[156] Cucullo L, Hossain M, Puvenna V, Marchi N, Janigro D. The role of shear stress in Blood-Brain Barrier endothelial physiology. *BMC Neurosci*. 2011;12:40.

[157] Neuhaus W, Lauer R, Oelzant S, Fringeli UP, Ecker GF, Noe CR. A novel flow based hollow-fiber blood-brain barrier in vitro model with immortalised cell line PBMEC/C1-2. *J Biotechnol*. 2006;125(1):127–141.

[158] de Boer AG, Gaillard PJ, Breimer DD. The transference of results between blood-brain barrier cell culture systems. *Eur J Pharm Sci*. 1999;8(1):1–4.

[159] Redmond EM, Cahill PA, Sitzmann JV. Perfused transcapillary smooth muscle and endothelial cell co-culture—a novel in vitro model. *In vitro Cell Dev Biol Anim.* 1995;31(8):601–609.

[160] Koutsiaris AG, Tachmitzi SV, Batis N, Kotoula G, Karabatsas CH, Tsironi E, Chatzoulis DZ. Volume flow and wall shear stress quantification in the human conjunctival capillaries and post-capillary venules in vivo. *Biorheology.* 2007;44(5–6):375–386.

[161] Cucullo L, Hossain M, Tierney W, Janigro D. A new dynamic in vitro modular capillaries-venules modular system: cerebrovascular physiology in a box. *BMC Neurosci.* 2013;14:18.

[162] Kaiser D, Freyberg MA, Schrimpf G, Friedl P. Apoptosis induced by lack of hemodynamic forces is a general endothelial feature even occurring in immortalized cell lines. *Endothelium.* 1999;6(4):325–334.

[163] Desai SY, Marroni M, Cucullo L, Krizanac-Bengez L, Mayberg MR, Hossain MT, Grant GG, Janigro D. Mechanisms of endothelial survival under shear stress. *Endothelium.* 2002;9(2):89–102.

[164] Griep LM, Wolbers F, de Wagenaar B, ter Braak PM, Weksler BB, Romero IA, Couraud PO, Vermes I, van der Meer AD, van den Berg A. BBB on chip: microfluidic platform to mechanically and biochemically modulate blood-brain barrier function. *Biomed Microdevices.* 2013;15(1):145–150.

[165] Conway DE, Breckenridge MT, Hinde E, Gratton E, Chen CS, Schwartz MA. Fluid shear stress on endothelial cells modulates mechanical tension across VE-cadherin and PECAM-1. *Curr Biol.* 2013;23(11):1024–1030.

[166] Krizanac-Bengez L, Kapural M, Parkinson F, Cucullo L, Hossain M, Mayberg MR, Janigro D. Effects of transient loss of shear stress on blood-brain barrier endothelium: role of nitric oxide and IL-6. *Brain Res.* 2003;977(2):239–246.

[167] Orsenigo F, Giampietro C, Ferrari A, Corada M, Galaup A, Sigismund S, Ristagno G, Maddaluno L, Koh GY, Franco D, Kurtcuoglu V, Poulikakos D, Baluk P, McDonald D, Grazia Lampugnani M, Dejana E. Phosphorylation of VE-cadherin is modulated by haemodynamic forces and contributes to the regulation of vascular permeability in vivo. *Nat Commun.* 2012;3:1208. doi:10.1038/ncomms2199.

[168] Walsh TG, Murphy RP, Fitzpatrick P, Rochfort KD, Guinan AF, Murphy A, Cummins PM. Stabilization of brain microvascular endothelial barrier function by shear stress involves VE-cadherin signaling leading to modulation of pTyr-occludin levels. *J Cell Physiol.* 2011;226(11):3053–3063.

[169] Schiera G, Sala S, Gallo A, Raffa MP, Pitarresi GL, Savettieri G, Di Liegro I. Permeability properties of a three-cell type in vitro model of blood-brain barrier. *J Cell Mol Med.* 2005;9(2):373–379.

[170] Lippmann ES, Azarin SM, Kay JE, Nessler RA, Wilson HK, Al-Ahmad A, Palecek SP, Shusta EV. Derivation of blood-brain barrier endothelial cells from human pluripotent stem cells. *Nat Biotechnol.* 2012;30(8):783–791.

[171] Veszelka S, Tóth AE, Walter FR, Datki Z, Mózes E, Fülöp L, Bozsó Z, Hellinger É, Vastag M, Orsolits B, Környei Z, Penke B, Deli MA. Docosahexaenoic acid reduces amyloid β-induced toxicity in cells of the neurovascular unit. *J Alzheimers Dis.* 2013; 36(3):487–501.

[172] Yang WM, Jung KJ, Lee MO, Lee YS, Lee YH, Nakagawa S, Niwa M, Cho SS, Kim DW. Transient expression of iron transport proteins in the capillary of the developing rat brain. *Cell Mol Neurobiol.* 2011;31(1):93–99.

[173] Imamura S, Tabuchi M, Kushida H, Nishi A, Kanno H, Yamaguchi T, Sekiguchi K, Ikarashi Y, Kase Y. The blood-brain barrier permeability of geissoschizine methyl ether in uncaria hook, a galenical constituent of the traditional Japanese medicine Yokukansan. *Cell Mol Neurobiol.* 2011;31(5):787–793.

[174] Tabuchi M, Imamura S, Kawakami Z, Ikarashi Y, Kase Y. The blood-brain barrier permeability of 18β-glycyrrhetinic acid, a major metabolite of glycyrrhizin in *Glycyrrhiza* root, a constituent of the traditional Japanese medicine *Yokukansan. Cell Mol Neurobiol.* 2012;32(7):1139–1146.

[175] Hanada S, Fujoka K, Inoue Y, Kanaya F, Manome Y, Yamamoto K. Application of *in vitro* BBB model to measure permeability of nanoparticles. *J Phys Conf Ser.* 2013;429(1):012028.

[176] Hanada S, Fujioka K, Inoue Y, Kanaya F, Manome Y, Yamamoto K. Cell-based *in vitro* blood-brain barrier model can rapidly evaluate nanoparticles' brain permeability in association with particle size and surface modification. *Int J Mol Sci.* 2014;15(2): 1812–1825.

[177] Nakagawa S, Deli MA, Thuy DHD, Sagara M, Yamada N, Tanaka K, Niwa M. Development of in vitro monkey BBB model. *J Pharmacol Sci.* 2010;112(Suppl I): 91P.

[178] Mysiorek C, Culot M, Dehouck L, Derudas B, Staels B, Bordet R, Cecchelli R, Fenart L, Berezowski V. Peroxisome-proliferator-activated receptor-alpha activation protects brain capillary endothelial cells from oxygen-glucose deprivation-induced hyperpermeability in the blood-brain barrier. *Curr Neurovasc Res.* 2009;6(3):181–193.

[179] Nakagawa S, Castro V, Toborek M. Infection of human pericytes by HIV-1 disrupts the integrity of the blood-brain barrier. *J Cell Mol Med.* 2012;16(12):2950–2957.

[180] Nishitsuji K, Hosono T, Nakamura T, Bu G, Michikawa M. Apolipoprotein E regulates the integrity of tight junctions in an isoform-dependent manner in an in vitro blood-brain barrier model. *J Biol Chem.* 2011;286(20):17536–17542.

[181] Vandenhaute E, Culot M, Gosselet F, Dehouck L, Godfraind C, Franck M, Plouët J, Cecchelli R, Dehouck M-P, Ruchoux M-M. Brain pericytes from stress-susceptible pigs increase blood-brain barrier permeability in vitro. *Fluids Barriers CNS.* 2012;9:11.

[182] Dohgu S, Banks WA. Brain pericytes increase the lipopolysaccharide-enhanced transcytosis of HIV-1 free virus across the in vitro blood-brain barrier: evidence for cytokine-mediated pericyte-endothelial cell crosstalk. *Fluids Barriers CNS.* 2013; 10(1):23.

[183] Chen C-J, Ou Y-C, Li J-R, Chang C-Y, Pan H-C, Lai C-Y, Liao S-L, Raung S-L, Chang C-J. Infection of pericytes in vitro by Japanese encephalitis virus disrupts the integrity of endothelial barrier. *J Virol.* 2014;88(2):1150–1161.

[184] Salmeri M, Motta C, Anfuso CD, Amodeo A, Scalia M, Toscano MA, Alberghina M, Lupo G. VEGF receptor-1 involvement in pericyte loss induced by *Escherichia coli* in an in vitro model of blood brain barrier. *Cell Microbiol.* 2013;15(8):1367–1384.

[185] Ceruti S, Colombo L, Magni G, Viganò F, Boccazzi M, Deli MA, Sperlágh B, Abbracchio MP, Kittel Á. Oxygen-glucose deprivation increases ectonucleotidase activity in the cells composing the blood-brain barrier. *Neurochem Int.* 2011;59(2):259–271.

[186] Lukic-Panin V, Deguchi K, Yamashita T, Shang J, Zhang X, Tian F, Liu N, Kawai H, Matsuura T, Abe K. Free radical scavenger edaravone administration protects against tissue plasminogen activator induced oxidative stress and blood brain barrier damage. *Curr Neurovasc Res.* 2010;7(4):319–329.

[187] Takeshita T, Nakagawa S, Tatsumi R, So G, Hayashi K, Tanaka K, Deli MA, Nagata I, Niwa M. Cilostazol attenuates ischemia-reperfusion-induced blood-brain barrier dysfunction enhanced by advanced glycation endproducts via transforming growth factor-β1 signaling. *Mol Cell Neurosci*. 2014;60:1–9 in press.

[188] Wuest DM, Wing AM, Lee KH. Membrane configuration optimization for a murine in vitro blood-brain barrier model. *J Neurosci Methods*. 2013;212(2):211–221.

[189] Neuhaus W, Trauner G, Gruber D, Oelzant S, Klepal W, Kopp B, Noe CR. Transport of a GABAA receptor modulator and its derivatives from Valeriana officinalis L. s. l. across an in vitro cell culture model of the blood-brain barrier. *Planta Med*. 2008;74(11): 1338–1344.

[190] Peters C. Untersuchungen von trypanoziden Pentamidin Derivaten mit Hilfe eines in vitro—Zellkulturmodells der Blut—Hirn Schranke [master thesis]. Kiel, Germany: Christian-Albrechts University of Kiel; 2008. Available from the local university library.

[191] Youdim KA, Avdeef A, Abbott NJ. In vitro trans-monolayer permeability calculations: often forgotten assumptions. *Drug Discov Today*. 2003;8(21):997–1003.

[192] Lakeram M, Lockley DJ, Pendlington R, Forbes B. Optimisation of the Caco-2 permeability assay using experimental design methodology. *Pharm Res*. 2008;25(7): 1544–1551.

[193] Buckley ST, Fischer SM, Fricker G, Brandl M. In vitro models to evaluate the permeability of poorly soluble drug entities: challenges and perspectives. *Eur J Pharm Sci*. 2012;45(3):235–250.

[194] Leonard F. Novel cell based in vitro models to study nanoparticle interaction with the inflamed intestinal mucosa [dissertation]. Saarbrücken, Germany: University of Saarbrücken, 2012. Available from the local university library.

[195] Tilling T, Engelbertz C, Decker S, Korte D, Hüwel S, Galla HJ. Expression and adhesive properties of basement membrane proteins in cerebral capillary endothelial cell cultures. *Cell Tissue Res*. 2002;310(1):19–29.

[196] Blume LF, Denker M, Gieseler F, Kunze T. Temperature corrected transepithelial electrical resistance (TEER) measurement to quantify rapid changes in paracellular permeability. *Pharmazie*. 2010;65(1):19–24.

[197] Römermann K, Wanek T, Bankstahl M, Bankstahl JP, Fedrowitz M, Müller M, Löscher W, Kuntner C, Langer O. (R)-[(11)C]verapamil is selectively transported by murine and human P-glycoprotein at the blood-brain barrier, and not by MRP1 and BCRP. *Nucl Med Biol*. 2013;40(7):873–878.

[198] Hakala KS, Laitinen L, Kaukonen AM, Hirvonen J, Kostiainen R, Kotiaho T. Development of LC/MS/MS methods for cocktail dosed Caco-2 samples using atmospheric pressure photoionization and electrosprayionization. *Anal Chem*. 2003;75(21):5969–5977.

[199] Koljonen M, Hakala KS, Ahtola-Sätilä T, Laitinen L, Kostiainen R, Kotiaho T, Kaukonen AM, Hirvonen J. Evaluation of cocktail approach to standardise Caco-2 permeability experiments. *Eur J Pharm Biopharm*. 2006;64(3):379–387.

[200] Smalley J, Kadiyala P, Xin B, Balimane P, Olah T. Development of an on-line extraction turbulent flow chromatography tandem mass spectrometry method for cassette analysis of Caco-2 cell based bi-directional assay samples. *J Chromatogr B Analyt Technol Biomed Life Sci*. 2006;830(2):270–277.

[201] Avdeef A. Leakiness and size exclusion of paracellular channels in cultured epithelial cell monolayers—interlaboratory comparison. *Pharm Res*. 2010;27:480–489.

[202] Avdeef A. How well can *in vitro* barrier microcapillary endothelial cell models predict *in vivo* blood-brain barrier permeability? *Eur J Pharm Sci.* 2011;43:109–124.

[203] Berezowski V, Landry C, Lundquist S, Dehouck L, Cecchelli R, Dehouck MP, Fenart L. Transport screening of drug cocktails through an in vitro blood-brain barrier: is it a good strategy for increasing the throughput of the discovery pipeline? *Pharm Res.* 2004; 21(5):756–760.

[204] Avdeef A, Nielsen PE, Tsinman O. PAMPA—a drug absorption in vitro model 11. Matching the in vivo unstirred water layer thickness by individual-well stirring in microtitre plates. Eur J Pharm Sci. 2004;22(5):365–374.

[205] Hidalgo IJ, Hillgren KM, Grass GM, Borchardt RT. Characterization of the unstirred water layer in Caco-2 cell monolayers using a novel diffusion apparatus. *Pharm Res.* 1991;8(2):222–227.

[206] Cucullo L, McAllister MS, Kight K, Krizanac-Bengez L, Marroni M, Mayberg MR, Stanness KA, Janigro D. A new dynamic in vitro model for the multidimensional study of astrocyte-endothelial cell interactions at the blood-brain barrier. *Brain Res.* 2002; 951(2):243–254.

[207] Colgan OC, Ferguson G, Collins NT, Murphy RP, Meade G, Cahill PA, Cummins PM. Regulation of bovine brain microvascular endothelial tight junction assembly and barrier function by laminar shear stress. *Am J Physiol Heart Circ Physiol.* 2007;292(6): H3190–H3197.

[208] Santaguida S, Janigro D, Hossain M, Oby E, Rapp E, Cucullo L. Side by side comparison between dynamic versus static models of blood-brain barrier in vitro: a permeability study. *Brain Res.* 2006;1109(1):1–13.

[209] Stanness KA, Guatteo E, Janigro D. A dynamic model of the blood-brain barrier "in vitro." *Neurotoxicology.* 1996;17(2):481–496.

[210] Ma SH, Lepak LA, Hussain RJ, Shain W, Shuler ML. An endothelial and astrocyte co-culture model of the blood-brain barrier utilizing an ultra-thin, nanofabricated silicon nitride membrane. *Lab Chip.* 2005;5(1):74–85.

[211] Booth R, Kim H. Characterization of a microfluidic in vitro model of the blood-brain barrier (μBBB). *Lab Chip.* 2012;12(10):1784–1792.

[212] Yeon JH, Na D, Choi K, Ryu SW, Choi C, Park JK. Reliable permeability assay system in a microfluidic device mimicking cerebral vasculatures. *Biomed Microdevices.* 2012; 14(6):1141–1148.

[213] Prabhakarpandian B, Shen MC, Nichols JB, Mills IR, Sidoryk-Wegrzynowicz M, Aschner M, Pant K. SyM-BBB: a microfluidic Blood Brain Barrier model. *Lab Chip.* 2013;13(6):1093–1101.

11

HUMAN-BASED *IN VITRO* BRAIN ENDOTHELIAL CELL MODELS

Hannah K. Wilson and Eric V. Shusta

Department of Chemical and Biological Engineering, University of Wisconsin–Madison, Madison, WI, USA

INTRODUCTION

In vitro models of the blood–brain barrier (BBB) have the potential to make a large impact on the neuroscience field. There is significant interest in understanding the physiology and pathology of the BBB, both to develop new and better therapeutics for cerebral disorders and to deliver these therapeutics more effectively. Such information is often gained through detailed mechanistic studies, including those of BBB development, maintenance, and disease, as well as by high-throughput drug screening. All of these studies can be facilitated through the use of *in vitro* BBB models.

There are many advantages of using *in vitro* BBB models as complements to *in vivo* systems. First, *in vitro* models offer the ability to exert precise control over experimental variables, while *in vivo* experiments are often limited to evaluating basic changes in phenotype. *In vitro* models are thus more amenable to investigations at the cellular and molecular level. Second, they tend to be less expensive than *in vivo* experiments, particularly for high-throughput screens that would otherwise be cost-prohibitive. Third, *in vitro* models allow the study of the human system, as opposed to *in vivo* animal research. While *in vitro* models are inherently simplified compared to the *in vivo* context, they still maintain much of the cellular complexity that is responsible for the BBB phenotype, giving them an advantage over *in silico* models. Taken together, the ability to examine the BBB in isolation gives *in vitro* brain endothelial cell models significant value.

Blood–Brain Barrier in Drug Discovery: Optimizing Brain Exposure of CNS Drugs and Minimizing Brain Side Effects for Peripheral Drugs, First Edition. Edited by Li Di and Edward H. Kerns.

The ideal *in vitro* BBB model would mimic the *in vivo* BBB in terms of its physical, transport, and metabolic barriers. While no *in vitro* model can fully reproduce all of the intricate functions of the BBB, there are several major properties that an ideal BBB model would possess: (i) a physical barrier comprising well-organized tight junction proteins, high transendothelial electrical resistance (TEER), and low permeability to paracellular tracers; (ii) a transport barrier comprising efflux transporters, carrier-mediated transporters, and receptor-mediated transporters expressed at *in vivo* levels; (iii) a metabolic barrier comprising relevant drug-metabolizing enzymes; (iv) the ability to respond appropriately to physiological stimuli such as shear force, inflammatory activation, or disease state; (v) the capacity for facile, reliable, and reproducible implementation; and (vi) relevancy to the human condition. The human BBB exhibits important differences compared with the BBB of other species, including altered expression and function of efflux transporters [1–3], differential response to inflammatory mediators [4], and altered clearance of β-amyloid [5]. Thus, the ideal *in vitro* BBB model would be of human origin.

This chapter describes the current state of the art for *in vitro* modeling of the human BBB and discusses how well these models reflect the idealized attributes listed earlier. A discussion of the many studies that have employed human *in vitro* BBB models follows as a demonstration of their utility.

SOURCES OF HUMAN BRAIN MICROVASCULAR ENDOTHELIAL CELLS (hBMECs)

Human *in vitro* BBB models are composed of human brain microvascular endothelial cells (hBMECs). Three available sources of hBMECs have been described: primary cells, immortalized cell lines, and human pluripotent stem cells (hPSCs). While no single *in vitro* model currently meets all the requirements of an ideal model as described earlier, *in vitro* models do display many key BBB properties. Importantly, depending on the cell source and model implementation, different human BBB models display different subsets of BBB properties. Therefore, care must be taken during human BBB model selection to address the specific experimental needs. To help guide the choice of human BBB model, the current strengths and limitations of each hBMEC source will be discussed in detail and are summarized in Table 11.1.

Primary Cells

Primary hBMECs are generally isolated from resected brain tissue obtained from patients undergoing neurosurgery, but can also be obtained from autopsied brain tissue and fetal brain tissue [14, 15]. Primary hBMECs are generated by outgrowth of human BBB from isolated human tissue, and their principal advantage is that they tend to maintain the closest resemblance to *in vivo* hBMECs in terms of the cellular and structural characteristics. For example, primary hBMECs have been demonstrated to express high levels of glucose transporter Glut-1, efflux transporters BCRP and P-glycoprotein (P-gp), and metabolic enzyme CYP1B1 (Table 11.1). However,

TABLE 11.1 Comparison of hBMEC sources

Model	TEER	Paracellular tracers	Drug-related efflux transporters	Transporters and receptors	Metabolic enzymes	Scalable?
Primary	Max. 339 $\Omega \times cm^2$ monoculture [6]	Sucrose $P_e = 1.02 \times 10^{-3}$ cm min^{-1} [7]	BCRP, P-gp protein most highly expressed, MRP4 expressed at ~10-fold lower level, MRP1 protein not detected [8]	Highest ABC transporter proteins = ABCA2, ABCA8; ABCA3 not expressed. Highest SLC transporter = Glut-1. LAT1, LRP1, insulin receptor, and transferrin receptor expressed [8]	CYP1B1 transcript most highly expressed, CYP2J2, 2U1, 2E1, 1A1 also expressed [9]	No
Immortalized hCMEC/d3 cells	<40 $\Omega \times cm^2$ monoculture [10], 140 $\Omega \times cm^2$ with astrocyte coculture [11]	Sucrose $P_e = 1.65 \times 10^{-3}$ cm min^{-1} [10]	P-gp, BCRP, MRP1, MRP4 proteins highly expressed [8]	Highest ABC transporter proteins = ABCA2, ABCA3; ABCA8 not expressed. Highest SLC transporter = Glut-1. Insulin and transferrin receptors expressed. LAT1 and LRP1 protein not detected [8]	CYP2U1 transcript most highly expressed. CYP2S1 and CYP2R1 transcripts also expressed. Low expression of CYP1B1 [12]	Yes
Stem cells	250 $\Omega \times cm^2$ monoculture, 1450 $\Omega \times cm^2$ with astrocyte coculture [13]	Sucrose $P_e = 3.4 \times 10^{-5}$ cm min^{-1} [13]	P-gp, BCRP, MRP1, MRP4 protein expressed and active [13]	Protein expression of Glut-1. mRNA expression of LAT1, LRP1, STRA6, insulin receptor, and transferrin receptor [13]	No data	Yes

primary hBMECs are well known to de-differentiate and lose their BBB phenotype in tissue culture, and these changes are exacerbated if signals from surrounding cells of the neurovascular unit are absent. Primary cells can only be used for 3–4 passages at most before experiencing significant phenotypic de-differentiation [15].

While primary hBMECs express many key proteins such as transporters and those involved in tight junctions, they lose the ability to form an *in vivo*-like barrier. For instance, TEER of primary hBMECs can be highly variable as a function of the source material, and typically only reaches 30–50 $\Omega \times cm^2$ (reviewed in [16]), compared to *in vivo* values between 1000 and 2000 $\Omega \times cm^2$ or higher [17, 18]. The highest published TEER using a monoculture of primary hBMECs is 339 ± 107 $\Omega \times cm^2$ [6] (Table 11.1), and upon monocyte coculture, TEER could reach 500 $\Omega \times cm^2$ [19].

In addition to limitations in barrier formation, there are also practical concerns when using primary hBMECs, the foremost of which is the paucity of material. Moreover, given that cells are primarily obtained through neurosurgery, there are concerns regarding the health and preservation of the tissue. Even when hBMECs are isolated from a healthy region of the brain (i.e., outside the glioma or epileptic lesion), one cannot necessarily consider it a source of normal BBB. On the other hand, these isolates can also be used advantageously should one wish to study the BBB in a specific pathology, such as glioma [20], epilepsy [21], or cerebral malaria [22], or from a specific patient subset, such as children [23, 24].

Another practical concern is that hBMECs are technically challenging to isolate, given the generally small amounts of tissue and issues with cellular contamination (e.g., pericytes and smooth muscle cells) that can interrupt the hBMEC monolayer. However, hBMECs are also available for purchase from commercial vendors, which circumvents the challenge of tissue procurement and hBMEC isolation, but the cells are still subject to rapid de-differentiation and senescence. Primary hBMECs, therefore, do not meet the idealized requirement of being facile and scalable. Despite these practical challenges, primary hBMECs hold great value in basic and translational research as discussed in the Section "*In Vitro* Model Applications".

Immortalized Cell Lines

Immortalized hBMEC lines have been developed to overcome the practical challenges of working with primary hBMECs. Immortalization is commonly accomplished via transfection with SV40 large T antigen or human telomerase, which allows appropriately genetically modified cells to evade senescence [25, 26]. Immortalized hBMECs are simpler to culture and can be passaged extensively while maintaining a fairly stable BBB phenotype.

The most widely characterized immortalized hBMEC line is the hCMEC/d3 cell line, which has been used in over 100 studies since it was established in 2005 (reviewed in [27]). The hCMEC/d3 cells express key BBB properties, including normal endothelial markers, tight junction proteins, and active efflux transporters. Several studies have addressed the question of how closely hCMEC/d3 cells recapitulate the *in vivo* BBB by using genomics and proteomics approaches on specific sets of BBB markers such as transporters, receptors, and junctional proteins [8, 12, 28].

High expression of Glut-1, P-gp, and BCRP is maintained in hCMEC/d3 cells, although they have low expression of metabolic enzyme CYP1B1 compared to primary hBMECs (Table 11.1). Additional studies have addressed the utility of hCMEC/d3 cells for specific applications including response to cytokines [29], immune cell migration [11], and drug transport studies [30]. In general these studies have highlighted the value of hCMEC/d3 cells, although one study cautioned that they were not suitable to study response to ischemia [31].

While hCMEC/d3 cells are quite representative of the *in vivo* BBB in terms of marker expression, they lack an appreciable barrier phenotype. The TEER of an hCMEC/d3 cell monolayer is approximately 40 $\Omega \times cm^2$, which is significantly lower than necessary to study permeability or drug transport [32] (Table 11.1). It should be noted that introduction of astrocyte coculture and shear forces as discussed later can increase TEER to 1000 $\Omega \times cm^2$ [33]. As expected, given the aberrantly low TEER, the paracellular permeability to hydrophilic tracers is elevated (sucrose $P_e = 1.65 \times 10^{-3}$ cm/min), but is on par with values obtained from primary hBMECs [7, 34] (Table 11.1).

Besides barrier phenotype, a second disadvantage is the nonnegligible effect of the immortalization procedure. The introduction of immortalizing genes can affect a wide variety of cellular functions, possibly altering basic physiological and biological responses [35]. A genomics study comparing hCMEC/d3 cells to primary cells noted differences in genes known to be affected by SV40 integration, as well as genes linked to the immune system and interferon pathway, presumably due to the process of viral transduction [36].

Several other immortalized hBMEC lines have been developed, including BB19 [37], hCMEC-5i [38], TY08 [39], and hBMEC-ciβ [40], but most have only been minimally characterized. Overall, the advantage of immortalized hBMECs lies in their ease of culture, scalability, and reproducibility. The cells express many transporters and receptors that reside at the *in vivo* BBB, although to date, they lack significant barrier properties and, thus, are not ideal for small-molecule drug screening. Optimizing barrier tightness remains a major challenge to further increase the utility of immortalized hBMEC lines.

Human Pluripotent Stem Cells (hPSCs)

hPSCs are defined by their ability to both self-renew and differentiate into any somatic cell type, which makes them an attractive source for constructing an *in vitro* BBB model, provided one can identify the appropriate differentiation conditions. There are two types of hPSCs: (1) embryonic stem cells, which are derived from the inner cell mass of a preimplantation stage blastocyst [41] and (2) induced pluripotent stem cells (iPSCs), which are derived from terminally differentiated cells forced to express key pluripotency transcription factors that revert them to an undifferentiated state [42, 43]. Stem cells have been differentiated into many different cell types, including endothelial cells, but until recently had not been shown to express organ-specific phenotypes or gene signatures [44–46]. Recently, our research group developed a protocol by which hPSCs could be differentiated into hBMEC-like cells via codifferentiation with neural cells [13] (reviewed in [47]). This codifferentiation strategy provides an embryonic-like brain microenvironment that instructs the developing endothelial cells to gain brain-specific attributes, such as elevated glucose transporter (Glut-1) expression, expression and localization of tight junction proteins occludin and claudin-5, and expression of efflux transporters such as P-gp. At the end of differentiation, hBMECs are purified from the surrounding neural cells by selective adhesion to a collagen-IV/fibronectin matrix, where the population is essentially 100% positive for BBB and vascular markers.

One particularly attractive feature of this model is its resultant barrier properties. Of the three sources for human endothelial cell models, hPSC-derived hBMECs are the only model that achieves a TEER that is well matched to drug permeability screens. For instance, a monoculture of hPSC-derived BMECs achieves a baseline TEER of 250 $\Omega \times cm^2$, which can be increased up to 1450 $\Omega \times cm^2$ with the addition of coculture with rat astrocytes. In terms of batch-to-batch reproducibility, hPSC-derived BMECs reached a TEER of 860 ± 260 $\Omega \times cm^2$ across 30 individual differentiation experiments. Accordingly, these cells display relatively low permeability to sucrose ($P_e = 3.4 \times 10^{-5}$ cm/min) compared to hCMEC/d3 ($P_e = 1.65 \times 10^{-3}$ cm/min) or primary hBMEC ($P_e = 1.02 \times 10^{-3}$ cm/min) (Table 11.1). They also display a 40-fold dynamic range of permeability between diazepam (BBB permeable) and sucrose (BBB impermeable), which is greater than the 10-fold dynamic range reported for hCMEC/d3 cells [10]. While encouraging, the initial cohort of test compounds was not exhaustive enough to determine the model's true predictive power. Finally, efflux transporters are also an important part of the transport barrier. At minimum, P-gp, BCRP, MRP1, MRP2, MRP4, and MRP5 are all expressed at the transcript level in the hPSC-derived BMECs, and functional assays using P-gp, BCRP, and pan-MRP inhibitors suggest functional polarized activity (Table 11.1). Overall, hPSC-derived BMECs display a reasonable physical and transport barrier.

Stem cell–derived hBMECs offer several advantages compared to primary hBMECs. Like immortalized cells, stem cells are highly scalable, which makes the issue of cellular yields practically inconsequential. In fact, a single vial of hPSCs is enough to create thousands of filters for drug screening purposes. Second, the newly differentiated nature of stem cell–derived hBMECs helps to avoid the issue of de-differentiation that is encountered when passaging primary hBMECs. However, to date each batch of hPSC-derived hBMECs is derived from a new differentiation of hPSCs, and thus requires the operators to have a working knowledge of stem cell culture. Further definition and standardization of the protocol is currently under way to enable more widespread dissemination of the technique.

Moving forward, there are several unexplored areas where the stem cell–derived BBB model could help make further contributions. First, recent success creating patient-specific pluripotent stem cell lines suggests that skin cells can be biopsied from patients, converted into iPSC lines, and potentially differentiated into hBMECs to conduct central nervous system (CNS) studies *in vitro*. Second, advances in genetic manipulation of hPSCs using zinc finger nucleases [48] and TAL effector nucleases [49] could allow for the exploration of genetic contributions to disease. Finally, stem cell–derived hBMECs offer the capability to examine BBB development. For example, during the hPSC differentiation process, endothelial progenitors lacking tight junctions and BBB transporters are instructed to become barrier-forming hBMECs. Thus, it is believed that the hPSC-derived system will allow for the detailed mechanistic study of BBB specification in complement with *in vivo* transgenic studies. Indeed, there is evidence suggesting that the acquisition of BBB phenotype in the stem cell–derived hBMECs occurs in part through the *in vivo* relevant Wnt pathway [50–52]. These data suggest that the stem cell model mimics certain aspects of *in vivo* development. However, it is unlikely that the model represents *in vivo* development entirely, as it may be missing important factors or may express irrelevant factors that drive the BBB phenotype in ways not observed *in vivo*.

For those researchers weighing whether to employ animal or human *in vitro* BBB models, it should be noted that human and animal models are similarly implemented. Animal models also require a primary or immortalized source of BMECs that have properties that are subject to the same limitations as their human counterparts, as discussed earlier. Moreover, human and animal BBB models require similar resources in terms of culture medium and culture-ware. The most substantial difference would be the less widespread access to primary human tissue, an issue that will likely be mitigated by continued advances in hBMEC immortalization and stem cell sourcing.

MODEL REFINEMENT

The foundation of human *in vitro* BBB models is clearly formed by the choice of hBMECs; however, hBMEC response to the *in vitro* microenvironment is also key. Creating a more *in vivo*-like microenvironment through structural, cellular, and molecular cues can help to upregulate BBB properties. Such interventions include coculturing with astrocytes or other cells of the neurovascular unit, and the addition of shear force. Each strategy can be adapted to any hBMEC source (primary, immortalized, or stem cell–derived) in an effort to bring the cells closer to an *in vivo* phenotype. To demonstrate the impact of microenvironmental manipulation on human BBB models, hBMEC coculture models, flow-based models, and microfabricated systems will be discussed.

Effects of Coculture

The culture of hBMECs on Transwells® is extremely valuable for *in vitro* BBB modeling. In the Transwell setup, hBMECs are seeded onto a porous membrane residing in a cell culture insert, establishing a blood compartment and a brain compartment (Fig. 11.1). Such a two-compartment system possesses significant utility for measuring compound and cell permeability across the BBB as discussed in the Section "*In Vitro* Model Applications". While a monoculture system is the simplest model, it is widely accepted that culturing hBMECs alone will lead to a loss of BBB phenotype [53, 54]. Conveniently, the two-compartment model also makes it possible to establish cocultures of hBMECs with other cells from the neurovascular unit (Fig. 11.1). Therefore, features of the *in vivo* neurovascular cellular microenvironment have been introduced into human hBMEC-based models. Most common is the use of astrocyte coculture, in which astrocytes are seeded either in contact with hBMECs on the opposite side of the

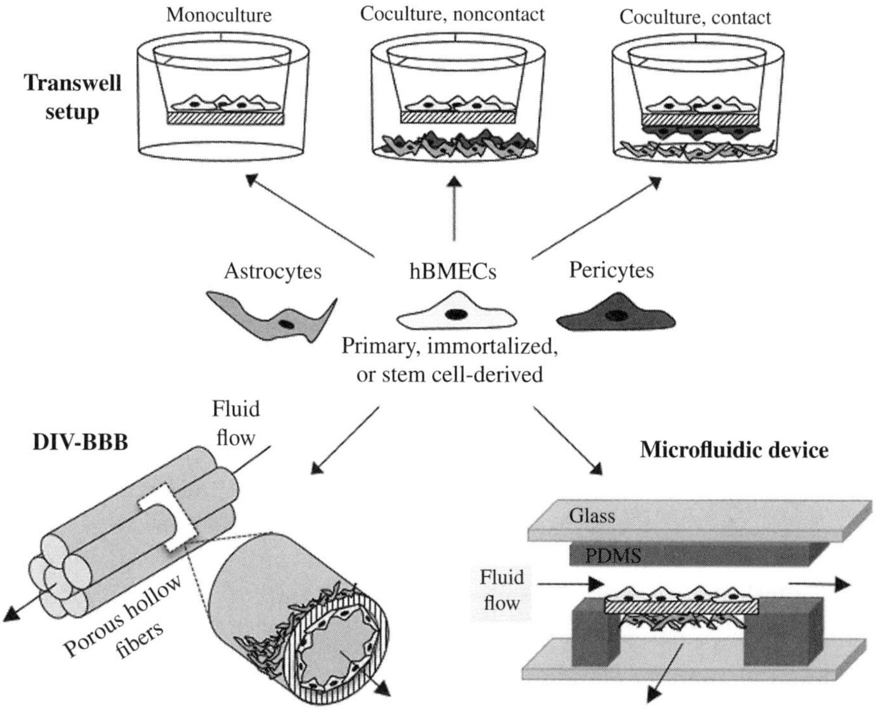

FIGURE 11.1 hBMEC model refinement. hBMECs from any source (primary, immortalized, or stem-cell derived) can be incorporated into Transwell® inserts, dynamic *in vitro* BBB models (DIV-BBB), or microfluidic devices. Each strategy is amenable to coculture with astrocytes and/or pericytes, which enhance BBB properties.

Transwell filter or in a noncontact configuration at the base of the culture well (Fig. 11.1). Astrocytes derived from various species (often rat) have been used in hBMEC cocultures, although the most relevant source for human BBB studies is human astrocytes.

Most notably, astrocyte coculture elevates the TEER of hBMEC monolayers in primary, immortalized, and stem cell–derived hBMECs. The magnitude of TEER increase upon astrocyte coculture was greatest for hPSC-derived hBMECs (from 222 ± 51 to $860 \pm 260\ \Omega \times cm^2$ [13]), whereas hCMEC/d3 cells modestly increased from 30 to 60 $\Omega \times cm^2$ [55], and primary hBMEC from 180 to 230 $\Omega \times cm^2$ [56]. Accordingly, astrocyte coculture also decreases paracellular permeability [56] and increases expression of tight junctions in primary hBMECs [57]. Because it is known that astrocytes release soluble factors, employing astrocyte-conditioned media to culture hBMEC monolayers also boosts TEER, but the effect is diminished compared to astrocyte coculture [57, 58]. Besides their effect on the physical barrier, astrocytes affect the transport barrier as well by upregulating efflux transporter expression of P-gp and MRPs in primary hBMECs [57]. There is also evidence that astrocytes may participate in the regulation of immune surveillance of the CNS. A global gene expression study of primary and hCMEC/d3 cells uncovered a small subset of genes downregulated upon astrocyte coculture which were involved in cell–cell adhesion, cellular extravasation, and cell migration [36]. Because astrocytes have such a profound effect on hBMEC phenotype, much research has focused on elucidating the molecular mechanisms of cross talk between the two cell types (for review, see [59, 60]). Astrocyte-derived factors that affect BBB function *in vitro* include transforming growth factor-β (TGFβ), glial-derived neurotrophic factor (GDNF), basic fibroblast growth factor (bFGF), angiopoeitin-1 (ANG1), and sonic hedgehog (Shh) [61–63], although much of this work was developed using animal models.

Pericytes are also being increasingly used in BBB coculture models, as *in vivo* experiments have indicated an important role for pericytes in regulating barrier function [64]. More often than not, pericytes are cocultured in addition to astrocytes to create a triple coculture (Fig. 11.1), and a partial additive effect of pericytes with astrocytes has been noted in nonhuman BBB models [65]. However, pericyte coculture did not elevate TEER over astrocytes alone in an hCMEC/d3 cell model, although this may be a result of the low baseline TEER in this system [55]. Research on the effects of other physiologically relevant cell types, such as neurons [66–68], neural progenitor cells [66, 69], and oligodendrocytes [70], indicates these cells can also affect BBB phenotype in nonhuman *in vitro* BBB models.

As a final note, Transwells can facilitate but are not absolutely necessary for the study of interactions between hBMECs and other cells of the neurovascular unit. hBMECs can also be plated on a gel matrix along with astrocytes and pericytes in a setup termed a gliovascular complex [71]. Another setup utilizes hanging drop culture to construct a spheroidal model of hBMECs, astrocytes, and pericytes that self-assemble into defined structures recapitulating their *in vivo* morphological arrangement [72].

Effects of Shear Force

Static coculture, as described in the Section "Effects of Coculture", is the most widely used embodiment of *in vitro* BBB models. However, endothelial cells *in vivo* experience shear stress as a result of blood flow. Shear stress is a major pleiotropic modulator, critically influencing endothelial differentiation and tight junction formation (for review, see [35]). So-called dynamic BBB systems incorporate the aspect of shear stress by the introduction of fluid flow using specially constructed culture chambers.

The first human BBB model to incorporate shear stress, termed "dynamic *in vitro* BBB model" (DIV-BBB), involved seeding of hBMECs inside porous hollow fibers to create a capillary-like structure allowing for intraluminal flow and abluminal astrocyte coculture (Fig. 11.1) [73]. Seeding hCMEC/d3 cells in the DIV-BBB model in combination with astrocyte coculture dramatically elevated TEER to 1000 $\Omega \times cm^2$ from just 70 $\Omega \times cm^2$ in static culture [33]. In addition to increased TEER, other important features of DIV-BBB include low permeability to paracellular tracers [33], expression of specialized transporters and efflux proteins [73, 74], and the ability to introduce relevant cell types such as red or white blood cells into the fluid circulation [75].

Several interesting adaptations to the DIV-BBB model have been developed. The addition of microholes in the hollow-fiber supports makes the model suitable for immune cell transmigration studies [76]. DIV-BBB capillary modules can also be serially connected to venule modules, which incorporate coculture of vascular smooth muscle cells and mimic *in vivo* rheological characteristics (i.e., shear stress, transmural pressure, and flow rate) to study distinct segments of the cerebrovasculature [77]. Finally, incorporation of hBMECs isolated from epileptic patients into the DIV-BBB model was used to screen for antiepileptic drugs [73].

As an alternative to the DIV-BBB, microfluidic devices also provide shear stress stimulation to hBMECs through the use of microfabricated channels rather than hollow fibers. In this setup, a microporous membrane is sandwiched between polydimethylsiloxane (PDMS) layers and anchored on a glass substrate, establishing patterns of microchannels and microvalves (Fig. 11.1) [78]. hBMECs are seeded onto the membrane to generate a monolayer of cells across which fluids can flow. Where the microfluidic model has an advantage over the DIV-BBB is the capability for high-throughput permeability screening of drugs, as parallelism is inherent in the microfabricated design and fluid volumes are significantly lower [78, 79]. Microfluidic membranes have a surface area of approximately 0.25 mm^2 for cell growth compared to the intraluminal surface area of 67 cm^2 in the DIV-BBB, and on-chip functional volumes are on the order of microliters rather than milliliters [33, 79].

The hCMEC/d3 cell line has been utilized in the microfluidic device with astrocytes cocultured on the opposing side of the microporous membrane. Astrocyte stimulation increased TEER from 37 to 120 $\Omega \times cm^2$ in hCMEC/d3 cells [79]. As with the DIV-BBB, the dynamic presentation of cells, proteins, and small molecules is more physiologically relevant than the static Transwell setup. Accordingly, microfluidic systems have been used to study adhesion of *Plasmodium falciparum*–infected erythrocytes to hBMECs [22] and the effect of the addition of inflammatory cytokine TNFα [79]. However, due to the technical challenges associated with constructing the models, both the DIV-BBB and microfluidic systems require substantial user investment. Increasing model complexity through the methods mentioned previously increases the utility and fidelity of human *in vitro* BBB models, thus widening the array of potential applications. A summary of those applications is provided in the Section "*In Vitro* Model Applications".

IN VITRO MODEL APPLICATIONS

While hBMEC sourcing and model format are constantly evolving, human *in vitro* BBB models have been widely used for both basic science and translational research. Major areas of research include mechanistic studies of drug action and evaluation of delivery strategies, studies of BBB development and maintenance, and studies of BBB pathology under various disease conditions. In the following sections we will briefly highlight these research areas with more comprehensive human BBB model usage compiled in Tables 11.2, 11.3, 11.4, and 11.5.

Drug Development and Delivery

As discussed previously, a major motivation for the development of human *in vitro* BBB models is therapeutic screening. Because the BBB is located at the interface between the brain and the circulation, there are many ways in which it influences

TABLE 11.2 Use of human *in vitro* BBB models to evaluate therapeutics and delivery strategies

Topic	hBMEC source	References
	Drug delivery	
Nanoparticle delivery	Primary	[80–86]
	Immortalized (hCMEC/d3)	[87–89]
	Immortalized (other)	[90, 91]
Gene delivery	Immortalized (hCMEC/d3)	[92]
	Immortalized (other)	[93]
Therapeutic alone	Primary	[7]
	Immortalized (hCMEC/d3)	[94]
	Mechanistic activity of drugs targeting the BBB to treat disease	
Neuroinflammation	Primary	[95, 96]
	Immortalized (hCMEC/d3)	[97]
	Immortalized (other)	[98, 99]
Ischemic stroke	Primary	[100, 101]
	Immortalized (hCMEC/d3)	[102, 103]
	Immortalized (other)	[104, 105]
Brain hemorrhaging	Primary	[106]
Glioma	Primary	[107]
	Immortalized (other)	[108]
	Mechanistic activity of drugs not specifically targeted to BBB	
HIV-1 infection	Primary	[109]
	Immortalized (hCMEC/d3)	[110, 111]

TABLE 11.3 Human *in vitro* BBB models in development and maintenance

Topic	hBMEC source	References
	Development	
Signaling pathways	Immortalized (hCMEC/d3)	[112–114]
	Primary (fetal)	[24]
Receptor expression	Primary (children aged 4–7)	[23]
Bacterial invasion	Primary (fetal/children aged 4–7)	[115]
	Maintenance	
Transporters, receptors, tight junctions	Primary	[116–118]
	Immortalized (hCMEC/d3)	[119–123]
	Immortalized (other)	[124]
Immune cell interactions	Immortalized (hCMEC/d3)	[125]
	Immortalized (Other)	[126–128]

TABLE 11.4 Human *in vitro* BBB models of disease.

Disease condition	hBMEC source	References
	Neurodegenerative diseases	
Alzheimer's disease (AD)	Primary	[129–132]
	Immortalized (hCMEC/d3)	[5, 133–136]
	Immortalized (other)	[137, 138]
Multiple sclerosis (MS)	Primary	[139, 140]
	Primary (isolated from MS patients)	[141, 142]
	Immortalized (hCMEC/d3)	[141–143]
	Immortalized (other)	[144]
Epilepsy	Primary (normal) and primary (isolated from epileptic brain)	[21, 73, 145, 146]
	Hypoxia and glucose deprivation	
Hypoxia alone	Primary	[147]
	Immortalized (hCMEC/d3)	[31, 148]
	Immortalized (other)	[149, 150]
Hypoxia and glucose deprivation	Primary	[151, 152]
	Immortalized (hCMEC/d3)	[103, 153, 154]
	Inflammation	
TNF-α only	Primary	[23, 34, 155]
	Immortalized (hCMEC/d3)	[156]
TNF-α, IFN-γ, and/or interleukins	Primary	[157–159]
	Immortalized (hCMEC/d3)	[29, 159–161]
	Immortalized (other)	[157, 160]
TNF family member TWEAK	Primary and immortalized (hCMEC/d3)	[162]
	Cancer	
Glioma/glioblastoma	Primary	[86, 163–166]
	Primary (isolated from glioma)	[20]
	Immortalized (hCMEC/d3)	[167]
Breast cancer	Primary	[168–170]
	Stem cell-derived	[171]
Leukemia	Primary	[172]
	Environmental poisons	
Methylmercury	Primary	[173]
Diesel exhaust particles	Immortalized (hCMEC/d3)	[174]
Perfluorooctane sulfonate (PFOS)	Immortalized (other)	[175]

drug interactions in the CNS. Here we will discuss how human *in vitro* BBB models are being utilized in drug discovery and delivery (summarized in Table 11.2).

Since the BBB often prevents small- and large-molecule therapeutics from entering the brain, human *in vitro* BBB models have been used to help predict whether

TABLE 11.5 Human *in vitro* BBB models of bacterial and viral pathogenesis

Infectious agent	hBMEC source	References
	Bacteria/bacterial toxins	
Escherichia coli	Primary	[115, 176–184]
	Primary (children aged 4–7)	[185–187]
	Immortalized (other)	[188–190]
Cronobacter sakazakii	Immortalized (other)	[191–193]
Group B *Streptococcus*	Immortalized (other)	[194–196]
Borrelia burgdorferi	Primary	[197, 198]
Borrelia turicatae	Immortalized (other)	[199]
Listeria monocytogenes	Immortalized (other)	[200]
Neisseria meningitides	Immortalized (other)	[201]
Salmonella enteric	Immortalized (other)	[202]
LPS (*S. enterica, Pseudomonas aeruginosa)*	Immortalized (hCMEC/d3)	[4]
AB5 toxin, B subunit (*S. typhi*)	Immortalized (other)	[203]
Pertussis toxin (*B. pertussis*)	Immortalized (other)	[188]
Shiga toxins (*E. coli)*	Primary	[204]
	Yeast	
Cryptococcus neoformans	Primary	[205–211]
	Immortalized (hCMEC/d3)	[212–214]
	Immortalized (other)	[205]
Candida albicans	Primary	[215, 216]
	Primary (children aged 4–7)	[216]
	Viruses/viral proteins	
HIV-1	Primary	[155, 217–223]
	Immortalized (hCMEC/d3)	[224]
HIV-1 Tat protein	Primary	[226, 226]
HIV-1 glycoprotein gp120	Primary	[227]
Hepatitis C	Immortalized (hCMEC/d3 and other)	[228]
Sindbis virus	Immortalized (other)	[229]
Dengue virus	Immortalized (other)	[230]
Enteroviruses coxsackievirus B and poliovirus	Primary	[231]
	Parasites	
Plasmodium falciparum	Primary	[118, 232]
	Immortalized (hBEC-5i)	[232–235]
	Immortalized (hCMEC/d3)	[233, 236]

a drug crosses the BBB and to help design drugs and drug carriers that may pass the BBB more effectively. One drug delivery method that has been widely studied using human *in vitro* BBB models is the use of nanoparticles to potentially deliver drug payloads across the BBB [80, 87], and gene delivery strategies using viral and polymer carrier systems have also been investigated [92, 93] (Table 11.2). Screening for drug penetration across the human BBB has been suboptimal to date because of the poor physical barrier properties of primary and immortalized cells as discussed earlier. That being said, there have been several studies investigating the use of hBMECs to evaluate drug efflux activity and to look at small-molecule delivery [7, 110] (Table 11.2). The development of stem cell–derived hBMECs possessing tight barrier properties may better accommodate these research directions in the future.

Second, as many cerebral pathologies are associated with BBB opening or dysregulation of the BBB phenotype, the BBB itself can serve as a therapeutic target. Potential therapeutics that have been evaluated on human *in vitro* BBB models include growth factors [104] and free radical scavengers [100, 105] to treat symptoms of stroke, and compounds such as flavonoids [98] and phosphodiesterases [106] to treat neuroinflammation and brain hemorrhaging, respectively. Thus, hBMEC-derived *in vitro* BBB models can be used to evaluate therapeutics that could potentially tighten the BBB under normal or pathological conditions and investigate the mechanism by which tightening occurs.

Finally, drugs reaching the BBB through the circulatory system can have off-target effects and modulate hBMEC properties. For example, HIV antiretrovirals can induce efflux transporter expression and affect drug uptake *in vitro* [110, 111]. Therefore, human *in vitro* BBB models can also be used to evaluate whether a drug that is not specifically targeted to hBMECs can modulate its properties.

BBB Development and Maintenance

The understanding of BBB development is rapidly evolving. Several recent *in vivo* studies have implicated involvement of several signaling pathways, including Wnt [50–52], retinoic acid [112], Hedgehog [63], and the orphan G-protein-coupled receptor GPR124 [237]. The study of development presents several challenges for *in vitro* models, because it is a dynamic process involving immature hBMECs, yet most of the *in vitro* development work has employed primary hBMECs from mature, adult brain tissue. Alternatively, primary hBMECs can be obtained from fetal tissue [24, 115], although yield, developmental timing, and availability remain as issues. In addition, once grown in culture, fetal hBMECs would undergo the same de-differentiation process as adult hBMECs. Thus, primary hBMECs have had a limited use in developmental studies.

As an alternative to primary hBMEC study of human BBB development, immortalized hBMECs have been employed. First, hCMEC/d3 cells have been used to demonstrate relevancy of signaling pathways such as retinoic acid signaling to the human system [112]. They have also been used to further probe pathway involvement such as Wnt/β-catenin signaling in hBMEC phenotype [113, 114]. However, the latter situation is nonideal from a developmental standpoint because immortalized hBMECs sourced from adult brain tissue may not respond to signaling components in the same manner as immature hBMECs undergoing BBB specification.

Stem cell–derived hBMECs are poised to make an impact in the field of BBB development. They are the only *in vitro* model that allows the study of developmental progression from an immature endothelial population to an hBMEC population having significant barrier properties and BBB phenotype. As mentioned earlier, the differentiation process occurred through the physiologically relevant Wnt signaling pathway [13]. Therefore, it is possible that this model could be further examined to identify additional signaling cascades involved in the conversion of nascent endothelial cells to hBMECs. Stem cell–derived hBMECs could also be used for detailed evaluation of pathways identified using animal models.

The *in vitro* study of hBMECs as mature cells has been more thoroughly explored. *In vitro* models are often used to determine whether a particular protein or pathway is expressed at the human BBB and/or how it is regulated. Studies of this kind include investigation of BBB solute and ion transporters [119, 120], tight junctions [121], efflux transporters [113, 122, 123, 238], chemokine receptors [116, 117], receptor tyrosine kinases [124], and growth factors [239]. *In vitro* human BBB models have also been used to study processes such as iron transport and accumulation [240, 241], the microstructure of BBB-resident P-gp [242], and the identification of proteins released in microsomes [243].

Another important hBMEC function is interaction with immune cells. It was once believed that the BBB was an immunoprivileged site and that immune cell interaction was mainly a symptom of disease. While immune cells certainly contribute to the progression of many diseases, recent studies have suggested that even during healthy BBB function, hBMECs interact with monocytes, T-cells, and B-cells [126–128], and can even support T-cell function and proliferation [125].

BBB Physiology in Disease

BBB dysfunction has been implicated in a variety of diseases, such as stroke, neurodegenerative diseases, and neurological infections. In order to develop appropriate therapies, it is necessary to understand the role of the BBB in each disease. To this end, human *in vitro* BBB models have been widely used to study diseases having BBB involvement.

Neurodegenerative Diseases, Hypoxia/Ischemia, Inflammation Disease modeling has been extensively adapted to human *in vitro* BBB studies. As stated earlier, a key motivation for using *in vitro* models is the ability to control the microenvironment. Disease modeling involves manipulating the microenvironment by adding a toxic or inflammatory component (i.e., β-amyloid fibrils and inflammatory cytokines) or removing a crucial component (i.e., oxygen and glucose) to recapitulate a certain facet of disease. The effects of the disease insult can then be studied in a controlled manner. The ways in which human *in vitro* BBB models are currently being used to study disease are compiled in Table 11.4.

Examples where components can be added to *in vitro* hBMECs to recapitulate disease symptoms include Alzheimer's disease (AD) and inflammation. The primary component of AD is the presence of amyloid-β (A-β) fibrils that accumulate in the brain and have neurotoxic effects. A number of studies have described adding various isoforms of A-β to hBMECs to determine mechanisms of A-β transport [129, 133] or the resultant changes in hBMEC phenotype, such as increased permeability and tight junction relocalization [130], efflux transporter expression [134], and apoptosis [135]. Inflammation is caused by a variety of sources, including infection, injury, and autoimmune disease, and inflammatory effects can often be reproduced *in vitro* by the addition of soluble cytokines in the culture media. Stimulation of *in vitro* hBMECs with cytokines, such as TNF-α and IFN-γ, induces a variety of phenotypic changes, including increased permeability and altered expression of cell adhesion molecules, tight junctions, and efflux transporters [23, 156, 157].

Aspects of stroke can also be relatively straightforward to model *in vitro*. Stroke is principally modeled using hypoxia and glucose deprivation, which can be achieved *in vitro* by manipulating culture conditions. Glucose deprivation and/or hypoxic conditioning of hCMEC/d3 cells increases their metallomatrix protease (MMP) expression and can increase immune cell transmigration [153, 154], although it was not found to regulate ABC transporter expression [148].

Multiple sclerosis (MS) and epilepsy are more difficult to model *in vitro*. MS is a complex autoimmune disease characterized by loss of oligodendrocytes, followed by T-cell migration into the brain. To date, human *in vitro* BBB models have been used to examine T-cell adhesion and migration through hBMECs, and to assess the potential role of deregulated targets identified through analysis of MS patients, such as miRNAs, glucocorticoid ANXA-1, and somatostatin [141–143]. *In vitro* studies of epilepsy have employed primary hBMECs derived from epileptic patients to model the disease. Primary epileptic hBMECs have been used to screen for therapeutic penetration or differences in drug-metabolizing enzymes compared to "normal" BBB (in this case, usually from stroke or trauma victims) [21, 73, 145].

Microbial and Viral Pathogenesis The pathogenesis of bacterial and viral meningitis, encephalitis, and HIV-1 infection is, in part, mediated by the interaction of infectious agents with brain endothelial cells. In some cases, pathogens interact directly with hBMECs, crossing the BBB and infecting cells within the brain (i.e., meningitis). In other cases, pathogens infect peripheral cells (i.e., malaria or HIV-1 infection), and the resultant interaction between infected cells and/or activated immune cells with hBMECs contributes to the pathogenesis. Studies that employed human *in vitro* BBB models to study microbial and viral pathogenesis are listed in Table 11.5.

In the case of bacterial or fungal meningitis, which is common among neonates and immunocompromised individuals, the pathogen must adhere, invade, and transmigrate across hBMECs in order to infect the meninges [115]. Thus many studies are focused on the molecular mechanisms at each of these crucial steps (binding [194, 197], invasion [176, 177, 191], and transmigration [188, 205]). In addition, hBMECs dynamically respond to the presence of microbes, including cytoskeletal rearrangements, production of cytokines, increased immune cell transmigration, and apoptosis [200, 203, 212].

HIV-1 and cerebral malaria are two diseases where infection of peripheral cells can lead to neurological disorders. HIV-1 encephalitis is characterized by brain infiltration of virus-infected monocytes and macrophages [217]. Several studies utilizing human *in vitro* BBB models show that hBMECs do not become infected with HIV-1 virus [218], but HIV-1 virus and HIV-1-infected monocytes can cross hBMEC monolayers [155, 219]. The mechanism by which transmigration occurs is an active area of research. Cerebral malaria is caused by the parasitic infection of red blood cells by *P. falciparum* which bind and become internalized by hBMECs [233]. In response to parasitized red blood cells, hBMECs exhibit increased permeability, increased cytokine production, and become apoptotic, a response that is mediated by platelets [234, 236].

Ultimately, the role of the BBB in health and disease is complex. Achieving a comprehensive understanding will likely require integrating knowledge from all of the facets described here, that is, mechanisms of BBB development, maintenance, and disease pathology, and investigation of effective therapeutic delivery strategies. Each of these efforts can benefit from the use of human *in vitro* BBB models.

CONCLUSION

The quality of human *in vitro* BBB models has progressed substantially since the first hBMECs were isolated. Alternative sources of hBMECs, including immortalized and stem cell–derived hBMECs, have overcome certain practical challenges associated with primary culture and opened the door for a wide range of studies on the human BBB, including drug screening and developmental studies. Refinement of hBMEC models, using either astrocyte and pericyte coculture or addition of shear force, has dramatically improved hBMEC phenotype and functionality. Moreover, the pool of *in vitro* models continues to grow as new adaptations are added. However, it is the authors' opinion that no "gold standard" hBMEC model currently exists. Rather, model suitability for a particular investigation depends on the properties required, and the cohort of human *in vitro* BBB models affords researchers a significant breadth of choices. Since no model faithfully recapitulates every aspect of the *in vivo* BBB, *in vitro* models should be used as a complement to *in vivo* studies.

ACKNOWLEDGMENTS

This work was funded in part by the US National Science Foundation (NSF) Graduate Research Fellowship Program (H.K.W.) and the US National Institutes of Health (NIH) grant NS083688 (E.V.S.).

REFERENCES

[1] Warren, M. S.; Zerangue, N.; Woodford, K.; Roberts, L. M.; Tate, E. H.; Feng, B.; Li, C.; Feuerstein, T. J.; Gibbs, J.; Smith, B.; de Morais, S. M.; Dower, W. J.; Koller, K. J. Comparative gene expression profiles of ABC transporters in brain microvessel endothelial cells and brain in five species including human. *Pharmacological Research: The Official Journal of the Italian Pharmacological Society* **2009**, *59* (6), 404–413.

[2] Syvanen, S.; Lindhe, O.; Palner, M.; Kornum, B. R.; Rahman, O.; Langstrom, B.; Knudsen, G. M.; Hammarlund-Udenaes, M. Species differences in blood-brain barrier transport of three positron emission tomography radioligands with emphasis on P-glycoprotein transport. *Drug Metabolism and Disposition: The Biological Fate of Chemicals* **2009**, *37* (3), 635–643.

[3] Hoshi, Y.; Uchida, Y.; Tachikawa, M.; Inoue, T.; Ohtsuki, S.; Terasaki, T. Quantitative atlas of blood-brain barrier transporters, receptors, and tight junction proteins in rats and common marmoset. *Journal of Pharmaceutical Sciences* **2013**, *102* (9), 3343–3355.

[4] Jin, L.; Nation, R. L.; Li, J.; Nicolazzo, J. A. Species-dependent blood-brain barrier disruption of lipopolysaccharide: Amelioration by colistin *in vitro* and *in vivo*. *Antimicrobial Agents and Chemotherapy* **2013**, *57* (9), 4336–4342.

[5] Qosa, H.; Abuasal, B. S.; Romero, I. A.; Weksler, B.; Couraud, P. O.; Keller, J. N.; Kaddoumi, A. Differences in amyloid-β clearance across mouse and human blood-brain barrier models: Kinetic analysis and mechanistic modeling. *Neuropharmacology* **2014**, *79*, 668–678.

[6] Rubin, L. L. The blood-brain barrier in and out of cell culture. *Current Opinion in Neurobiology* **1991**, *1* (3), 360–363.

[7] Megard, I.; Garrigues, A.; Orlowski, S.; Jorajuria, S.; Clayette, P.; Ezan, E.; Mabondzo, A. A co-culture-based model of human blood-brain barrier: Application to active transport of indinavir and *in vivo–in vitro* correlation. *Brain Research* **2002**, *927* (2), 153–167.

[8] Ohtsuki, S.; Ikeda, C.; Uchida, Y.; Sakamoto, Y.; Miller, F.; Glacial, F.; Decleves, X.; Scherrmann, J.-M.; Couraud, P.-O.; Kubo, Y.; Tachikawa, M.; Terasaki, T. Quantitative targeted absolute proteomic analysis of transporters, receptors and junction proteins for validation of human cerebral microvascular endothelial cell line hCMEC/D3 as a human blood-brain barrier model. *Molecular Pharmaceutics* **2013**, *10* (1), 289–296.

[9] Dauchy, S.; Dutheil, F.; Weaver, R. J.; Chassoux, F.; Daumas-Duport, C.; Couraud, P. O.; Scherrmann, J. M.; De Waziers, I.; Decleves, X. ABC transporters, cytochromes P450 and their main transcription factors: Expression at the human blood-brain barrier. *Journal of Neurochemistry* **2008**, *107* (6), 1518–1528.

[10] Weksler, B. B.; Subileau, E. A.; Perriere, N.; Charneau, P.; Holloway, K.; Leveque, M.; Tricoire-Leignel, H.; Nicotra, A.; Bourdoulous, S.; Turowski, P.; Male, D. K.; Roux, F.; Greenwood, J.; Romero, I. A.; Couraud, P. O. Blood-brain barrier-specific properties of a human adult brain endothelial cell line. *Federation of American Societies for Experimental Biology* **2005**, *19* (11), 1872–1874.

[11] Daniels, B. P.; Cruz-Orengo, L.; Pasieka, T. J.; Couraud, P.-O.; Romero, I. A.; Weksler, B.; Cooper, J. A.; Doering, T. L.; Klein, R. S. Immortalized human cerebral microvascular endothelial cells maintain the properties of primary cells in an *in vitro* model of immune migration across the blood brain barrier. *Journal of Neuroscience Methods* **2013**, *212* (1), 173–179.

[12] Dauchy, S.; Miller, F.; Couraud, P. O.; Weaver, R. J.; Weksler, B.; Romero, I. A.; Scherrmann, J. M.; De Waziers, I.; Decleves, X. Expression and transcriptional regulation of ABC transporters and cytochromes P450 in hCMEC/D3 human cerebral microvascular endothelial cells. *Biochemical Pharmacology* **2009**, *77* (5), 897–909.

[13] Lippmann, E. S.; Azarin, S. M.; Kay, J. E.; Nessler, R. A.; Wilson, H. K.; Al-Ahmad, A.; Palecek, S. P.; Shusta, E. V. Derivation of blood-brain barrier endothelial cells from human pluripotent stem cells. *Nature Biotechnology* **2012**, *30* (8), 783–791.

[14] Navone, S. E.; Marfia, G.; Invernici, G.; Cristini, S.; Nava, S.; Balbi, S.; Sangiorgi, S.; Ciusani, E.; Bosutti, A.; Alessandri, G.; Slevin, M.; Parati, E. A. Isolation and expansion of human and mouse brain microvascular endothelial cells. *Nature Protocols* **2013**, *8* (9), 1680–1693.

[15] Bernas, M. J.; Cardoso, F. L.; Daley, S. K.; Weinand, M. E.; Campos, A. R.; Goncalves Ferreira, A. J.; Hoying, J. B.; Witte, M. H.; Brites, D.; Persidsky, Y.; Ramirez, S. H.;

Brito, M. A. Establishment of primary cultures of human brain microvascular endothelial cells to provide an *in vitro* cellular model of the blood-brain barrier. *Nature Protocols* **2010**, *5* (7), 1265–1272.

[16] Deli, M. A.; Abraham, C. S.; Kataoka, Y.; Niwa, M. Permeability studies on *in vitro* blood-brain barrier models: Physiology, pathology, and pharmacology. *Cellular and Molecular Neurobiology* **2005**, *25* (1), 59–127.

[17] Butt, A. M.; Jones, H. C.; Abbott, N. J. Electrical resistance across the blood-brain barrier in anaesthetized rats: A developmental study. *The Journal of Physiology* **1990**, *429*, 47–62.

[18] Crone, C.; Olesen, S. P. Electrical resistance of brain microvascular endothelium. *Brain Research* **1982**, *241* (1), 49–55.

[19] Zenker, D.; Begley, D.; Bratzke, H.; Rubsarrien-Waigmann, H.; von Briesen, H. Human blood-derived macrophages enhance barrier function of cultured primary bovine and human brain capillary endothelial cells. *Journal of Physiology–London* **2003**, *551* (3), 1023–1032.

[20] Dong, B.; Mu, L.; Qin, X.; Qiao, W.; Liu, X.; Yang, L.; Xue, L.; Rainov, N. G.; Liu, X. Stathmin expression in glioma-derived microvascular endothelial cells: A novel therapeutic target. *Oncology Reports* **2012**, *27* (3), 714–718.

[21] Ghosh, C.; Hossain, M.; Puvenna, V.; Martinez-Gonzalez, J.; Alexopolous, A.; Janigro, D.; Marchi, N. Expression and functional relevance of UGT1A4 in a cohort of human drug-resistant epileptic brains. *Epilepsia* **2013**, *54* (9), 1562–1570.

[22] Herricks, T.; Seydel, K. B.; Turner, G.; Molyneux, M.; Heyderman, R.; Taylor, T.; Rathod, P. K. A microfluidic system to study cytoadhesion of *Plasmodium falciparum* infected erythrocytes to primary brain microvascularendothelial cells. *Lab on a Chip* **2011**, *11* (17), 2994–3000.

[23] Stins, M. F.; Gilles, F.; Kim, K. S. Selective expression of adhesion molecules on human brain microvascular endothelial cells. *Journal of Neuroimmunology* **1997**, *76* (1–2), 81–90.

[24] Kasa, P.; Pakaski, M.; Joo, F.; Lajtha, A. Endothelial-cells from human fetal brain microvessels may be cholinoceptive, but do not synthesize acetylcholine. *Journal of Neurochemistry* **1991**, *56* (6), 2143–2146.

[25] Yang, J. W.; Chang, E.; Cherry, A. M.; Bangs, C. D.; Oei, Y.; Bodnar, A.; Bronstein, A.; Chiu, C. P.; Herron, G. S. Human endothelial cell life extension by telomerase expression. *Journal of Biological Chemistry* **1999**, *274* (37), 26141–26148.

[26] O'Hare, M. J.; Bond, J.; Clarke, C.; Takeuchi, Y.; Atherton, A. J.; Berry, C.; Moody, J.; Silver, A. R. J.; Davies, D. C.; Alsop, A. E.; Neville, A. M.; Jat, P. S. Conditional immortalization of freshly isolated human mammary fibroblasts and endothelial cells. *Proceedings of the National Academy of Sciences of the United States of America* **2001**, *98* (2), 646–651.

[27] Weksler, B.; Romero, I. A.; Couraud, P.-O. The hCMEC/D3 cell line as a model of the human blood brain barrier. *Fluids and Barriers of the CNS* **2013**, *10* (1), 16.

[28] Carl, S. M.; Lindley, D. J.; Couraud, P. O.; Weksler, B. B.; Romero, I.; Mowery, S. A.; Knipp, G. T. ABC and SLC transporter expression and pot substrate characterization across the human CMEC/D3 blood-brain barrier cell line. *Molecular Pharmaceutics* **2010**, *7* (4), 1057–1068.

[29] Fasler-Kan, E.; Suenderhauf, C.; Barteneva, N.; Poller, B.; Gygax, D.; Huwyler, J. Cytokine signaling in the human brain capillary endothelial cell line hCMEC/D3. *Brain Research* **2010**, *1354*, 15–22.

[30] Poller, B.; Gutmann, H.; Krahenbuhl, S.; Weksler, B.; Romero, I.; Couraud, P. O.; Tuffin, G.; Drewe, J.; Huwyler, J. The human brain endothelial cell line hCMEC/D3 as a human blood-brain barrier model for drug transport studies. *Journal of Neurochemistry* **2008**, *107* (5), 1358–1368.

[31] Lindner, C.; Sigruner, A.; Walther, F.; Bogdahn, U.; Couraud, P. O.; Schmitz, G.; Schlachetzki, F. ATP-binding cassette transporters in immortalised human brain microvascular endothelial cells in normal and hypoxic conditions. *Experimental & Translational Stroke Medicine* **2012**, *4* (1), 9.

[32] Toth, A.; Veszelka, S.; Nakagawa, S.; Niwa, M.; Deli, M. A. Patented *in vitro* blood-brain barrier models in CNS drug discovery. *Recent Patents on CNS Drug Discovery* **2011**, *6* (2), 107–118.

[33] Cucullo, L.; Couraud, P.-O.; Weksler, B.; Romero, I.-A.; Hossain, M.; Rapp, E.; Janigro, D. Immortalized human brain endothelial cells and flow-based vascular modeling: A marriage of convenience for rational neurovascular studies. *Journal of Cerebral Blood Flow and Metabolism* **2008**, *28* (2), 312–328.

[34] Didier, N.; Romero, I. A.; Creminon, C.; Wijkhuisen, A.; Grassi, J.; Mabondzo, A. Secretion of interleukin-1β by astrocytes mediates endothelin-1 and tumour necrosis factor-α effects on human brain microvascular endothelial cell permeability. *Journal of Neurochemistry* **2003**, *86* (1), 246–254.

[35] Naik, P.; Cucullo, L. *In vitro* blood-brain barrier models: Current and perspective technologies. *Journal of Pharmaceutical Sciences* **2012**, *101* (4), 1337–1354.

[36] Urich, E.; Lazic, S. E.; Molnos, J.; Wells, I.; Freskgard, P.-O. Transcriptional profiling of human brain endothelial cells reveals key properties crucial for predictive *in vitro* blood-brain barrier models. *Public Library of Science* **2012**, *7* (5), e38149.

[37] Prudhomme, J. G.; Sherman, I. W.; Land, K. M.; Moses, A. V.; Stenglein, S.; Nelson, J. A. Studies of *Plasmodium falciparum* cytoadherence using immortalized human brain capillary endothelial cells. *International Journal for Parasitology* **1996**, *26* (6), 647–655.

[38] Wassmer, S. C.; Combes, V.; Candal, F. J.; Juhan-Vague, L.; Grau, G. E. Platelets potentiate brain endothelial alterations induced by *Plasmodium falciparum*. *Infection and Immunity* **2006**, *74* (1), 645–653.

[39] Sano, Y.; Shimizu, F.; Abe, M.; Maeda, T.; Kashiwamura, Y.; Ohtsuki, S.; Terasaki, T.; Obinata, M.; Kajiwara, K.; Fujii, M.; Suzuki, M.; Kanda, T. Establishment of a new conditionally immortalized human brain microvascular endothelial cell line retaining an *in vivo* blood-brain barrier function. *Journal of Cellular Physiology* **2010**, *225* (2), 519–528.

[40] Kamiichi, A.; Furihata, T.; Kishida, S.; Ohta, Y.; Saito, K.; Kawamatsu, S.; Chiba, K. Establishment of a new conditionally immortalized cell line from human brain microvascular endothelial cells: A promising tool for human blood-brain barrier studies. *Brain Research* **2012**, *1488*, 113–122.

[41] Thomson, J. A.; Itskovitz-Eldor, J.; Shapiro, S. S.; Waknitz, M. A.; Swiergiel, J. J.; Marshall, V. S.; Jones, J. M. Embryonic stem cell lines derived from human blastocysts. *Science* **1998**, *282* (5391), 1145–1147.

[42] Takahashi, K.; Tanabe, K.; Ohnuki, M.; Narita, M.; Ichisaka, T.; Tomoda, K.; Yamanaka, S. Induction of pluripotent stem cells from adult human fibroblasts by defined factors. *Cell* **2007**, *131* (5), 861–872.

[43] Takahashi, K.; Yamanaka, S. Induction of pluripotent stem cells from mouse embryonic and adult fibroblast cultures by defined factors. *Cell* **2006**, *126* (4), 663–676.

[44] Levenberg, S.; Golub, J. S.; Amit, M.; Itskovitz-Eldor, J.; Langer, R. Endothelial cells derived from human embryonic stem cells. *Proceedings of the National Academy of Sciences of the United States of America* **2002**, *99* (7), 4391–4396.

[45] James, D.; Nam, H.-S.; Seandel, M.; Nolan, D.; Janovitz, T.; Tomishima, M.; Studer, L.; Lee, G.; Lyden, D.; Benezra, R.; Zaninovic, N.; Rosenwaks, Z.; Rabbany, S. Y.; Rafii, S. Expansion and maintenance of human embryonic stem cell-derived endothelial cells by TGF-β inhibition is Id1 dependent. *Nature Biotechnology* **2010**, *28* (2), 161–166U15.

[46] Choi, K.-D.; Yu, J.; Smuga-Otto, K.; Salvagiotto, G.; Rehrauer, W.; Vodyanik, M.; Thomson, J.; Slukvin, I. Hematopoietic and endothelial differentiation of human induced pluripotent stem cells. *Stem Cells* **2009**, *27* (3), 559–567.

[47] Lippmann, E. S.; Al-Ahmad, A.; Palecek, S. P.; Shusta, E. V. Modeling the blood-brain barrier using stem cell sources. *Fluids and Barriers of the CNS* **2013**, *10* (1), 2.

[48] Hockemeyer, D.; Soldner, F.; Beard, C.; Gao, Q.; Mitalipova, M.; DeKelver, R. C.; Katibah, G. E.; Amora, R.; Boydston, E. A.; Zeitler, B.; Meng, X.; Miller, J. C.; Zhang, L.; Rebar, E. J.; Gregory, P. D.; Urnov, F. D.; Jaenisch, R. Efficient targeting of expressed and silent genes in human ESCs and iPSCs using zinc-finger nucleases. *Nature Biotechnology* **2009**, *27* (9), 851–857.

[49] Hockemeyer, D.; Wang, H.; Kiani, S.; Lai, C. S.; Gao, Q.; Cassady, J. P.; Cost, G. J.; Zhang, L.; Santiago, Y.; Miller, J. C.; Zeitler, B.; Cherone, J. M.; Meng, X.; Hinkley, S. J.; Rebar, E. J.; Gregory, P. D.; Urnov, F. D.; Jaenisch, R. Genetic engineering of human pluripotent cells using TALE nucleases. *Nature Biotechnology* **2011**, *29* (8), 731–734.

[50] Daneman, R.; Agalliu, D.; Zhou, L.; Kuhnert, F.; Kuo, C. J.; Barres, B. A. Wnt/β-catenin signaling is required for CNS, but not non-CNS, angiogenesis. *Proceedings of the National Academy of Sciences of the United States of America* **2009**, *106* (2), 641–646.

[51] Stenman, J. M.; Rajagopal, J.; Carroll, T. J.; Ishibashi, M.; McMahon, J.; McMahon, A. P. Canonical Wnt signaling regulates organ-specific assembly and differentiation of CNS vasculature. *Science* **2008**, *322* (5905), 1247–1250.

[52] Liebner, S.; Corada, M.; Bangsow, T.; Babbage, J.; Taddei, A.; Czupalla, C. J.; Reis, M.; Felici, A.; Wolburg, H.; Fruttiger, M.; Taketo, M. M.; von Melchner, H.; Plate, K. H.; Gerhardt, H.; Dejana, E. Wnt/β-catenin signaling controls development of the blood-brain barrier. *Journal of Cell Biology* **2008**, *183* (3), 409–417.

[53] Debault, L. E.; Cancilla, P. A. γ-glutamyl-transferase transpeptidase in isolated brain endothelial-cells—induction by glial-cells *in vitro*. *Science* **1980**, *207* (4431), 653–655.

[54] Dehouck, B.; Dehouck, M. P.; Fruchart, J. C.; Cecchelli, R. Upregulation of the low density lipoprotein receptor at the blood-brain barrier: Intercommunications between brain capillary endothelial cells and astrocytes. *The Journal of Cell Biology* **1994**, *126* (2), 465–473.

[55] Hatherell, K.; Couraud, P.-O.; Romero, I. A.; Weksler, B.; Pilkington, G. J. Development of a three-dimensional, all-human *in vitro* model of the blood-brain barrier using mono-, co-, and tri-cultivation transwell models. *Journal of Neuroscience Methods* **2011**, *199* (2), 223–229.

[56] Kuo, Y.-C.; Lu, C.-H. Effect of human astrocytes on the characteristics of human brain-microvascular endothelial cells in the blood-brain barrier. *Colloids and Surfaces B-Biointerfaces* **2011**, *86* (1), 225–231.

[57] Kuo, Y.-C.; Lu, C.-H. Regulation of endocytosis into human brain-microvascular endothelial cells by inhibition of efflux proteins. *Colloids and Surfaces B-Biointerfaces* **2011**, *87* (1), 139–145.

[58] Siddharthan, V.; Kim, Y. V.; Liu, S.; Kim, K. S. Human astrocytes/astrocyte-conditioned medium and shear stress enhance the barrier properties of human brain microvascular endothelial cells. *Brain Research* **2007**, *1147*, 39–50.

[59] Haseloff, R. F.; Blasig, I. E.; Bauer, H. C.; Bauer, H. In search of the astrocytic factor(s) modulating blood-brain barrier functions in brain capillary endothelial cells *in vitro*. *Cellular and Molecular Neurobiology* **2005**, *25* (1), 25–39.

[60] Abbott, N. J.; Ronnback, L.; Hansson, E. Astrocyte-endothelial interactions at the blood-brain barrier. *Nature Reviews Neuroscience* **2006**, *7* (1), 41–53.

[61] Lee, S. W.; Kim, W. J.; Choi, Y. K.; Song, H. S.; Son, M. J.; Gelman, I. H.; Kim, Y. J.; Kim, K. W. SSeCKS regulates angiogenesis and tight junction formation in blood-brain barrier. *Nature Medicine* **2003**, *9* (7), 900–906.

[62] Igarashi, Y.; Utsumi, H.; Chiba, H.; Yamada-Sasamori, Y.; Tobioka, H.; Kamimura, Y.; Furuuchi, K.; Kokai, Y.; Nakagawa, T.; Mori, M.; Sawada, N. Glial cell line-derived neurotrophic factor induces barrier function of endothelial cells forming the blood-brain barrier. *Biochemical Biophysical Research Communication* **1999**, *261* (1), 108–112.

[63] Alvarez, J. I.; Dodelet-Devillers, A.; Kebir, H.; Ifergan, I.; Fabre, P. J.; Terouz, S.; Sabbagh, M.; Wosik, K.; Bourbonniere, L.; Bernard, M.; van Horssen, J.; de Vries, H. E.; Charron, F.; Prat, A. The hedgehog pathway promotes blood-brain barrier integrity and CNS immune quiescence. *Science* **2011**, *334* (6063), 1727–1731.

[64] Daneman, R.; Zhou, L.; Kebede, A. A.; Barres, B. A. Pericytes are required for blood-brain barrier integrity during embryogenesis. *Nature* **2010**, *468* (7323), 562–566.

[65] Nakagawa, S.; Deli, M. A.; Kawaguchi, H.; Shimizudani, T.; Shimono, T.; Kittel, A.; Tanaka, K.; Niwa, M. A new blood-brain barrier model using primary rat brain endothelial cells, pericytes and astrocytes. *Neurochemistry International* **2009**, *54* (3–4), 253–263.

[66] Lippmann, E. S.; Weidenfeller, C.; Svendsen, C. N.; Shusta, E. V. Blood-brain barrier modeling with co-cultured neural progenitor cell-derived astrocytes and neurons. *Journal of Neurochemistry* **2011**, *119* (3), 507–520.

[67] Savettieri, G.; Di Liegro, I.; Catania, C.; Licata, L.; Pitarresi, G. L.; D'Agostino, S.; Schiera, G.; De Caro, V.; Giandalia, G.; Giannola, L. I.; Cestelli, A. Neurons and ECM regulate occludin localization in brain endothelial cells. *Neuroreport* **2000**, *11* (5), 1081–1084.

[68] Schiera, G.; Bono, E.; Raffa, M. P.; Gallo, A.; Pitarresi, G. L.; Di Liegro, I.; Savettieri, G. Synergistic effects of neurons and astrocytes on the differentiation of brain capillary endothelial cells in culture. *Journal of Cellular and Molecular Medicine* **2003**, *7* (2), 165–170.

[69] Weidenfeller, C.; Svendsen, C. N.; Shusta, E. V. Differentiating embryonic neural progenitor cells induce blood-brain barrier properties. *Journal of Neurochemistry* **2007**, *101* (2), 555–565.

[70] Miyamoto, N.; Pham, L. D.; Seo, J. H.; Kim, K. W.; Lo, E. H.; Arai, K. Crosstalk between cerebral endothelium and oligodendrocyte. *Cellular and Molecular Life Sciences* **2013**, *71* (6), 1055–1066.

[71] Itoh, Y.; Toriumi, H.; Yamada, S.; Hoshino, H.; Suzuki, N. Astrocytes and pericytes cooperatively maintain a capillary-like structure composed of endothelial cells on gel matrix. *Brain Research* **2011**, *1406*, 74–83.

[72] Urich, E.; Patsch, C.; Aigner, S.; Graf, M.; Iacone, R.; Freskgard, P.-O. Multicellular self-assembled spheroidal model of the blood brain barrier. *Scientific Reports* **2013**, *3*, 1–8.

[73] Cucullo, L.; Hossain, M.; Rapp, E.; Manders, T.; Marchi, N.; Janigro, D. Development of a humanized *in vitro* blood-brain barrier model to screen for brain penetration of antiepileptic drugs. *Epilepsia* **2007**, *48* (3), 505–516.

[74] McAllister, M. S.; Krizanac-Bengez, L.; Macchia, F.; Naftalin, R. J.; Pedley, K. C.; Mayberg, M. R.; Marroni, M.; Leaman, S.; Stanness, K. A.; Janigro, D. Mechanisms of glucose transport at the blood-brain barrier: An *in vitro* study. *Brain Research* **2001**, *904* (1), 20–30.

[75] Krizanac-Bengez, L.; Mayberg, M. R.; Cunningham, E.; Hossain, M.; Ponnampalam, S.; Parkinson, F. E.; Janigro, D. Loss of shear stress induces leukocyte-mediated cytokine release and blood-brain barrier failure in dynamic *in vitro* blood-brain barrier model. *Journal of Cellular Physiology* **2006**, *206* (1), 68–77.

[76] Cucullo, L.; Marchi, N.; Hossain, M.; Janigro, D. A dynamic *in vitro* BBB model for the study of immune cell trafficking into the central nervous system. *Journal of Cerebral Blood Flow and Metabolism* **2011**, *31* (2), 767–777.

[77] Cucullo, L.; Hossain, M.; Tierney, W.; Janigro, D. A new dynamic *in vitro* modular capillaries-venules modular system: Cerebrovascular physiology in a box. *BMC Neuroscience* **2013**, *14* (18), 1–12.

[78] Booth, R.; Kim, H. Characterization of a microfluidic *in vitro* model of the blood-brain barrier (mu BBB). *Lab on a Chip* **2012**, *12* (10), 1784–1792.

[79] Griep, L. M.; Wolbers, F.; de Wagenaar, B.; ter Braak, P. M.; Weksler, B. B.; Romero, I. A.; Couraud, P. O.; Vermes, I.; van der Meer, A. D.; van den Berg, A. BBB on chip: Microfluidic platform to mechanically and biochemically modulate blood-brain barrier function. *Biomedical Microdevices* **2013,** *15* (1), 145–150.

[80] Kuo, Y.-C.; Chen, I. C. Evaluation of surface charge density and surface potential by electrophoretic mobility for solid lipid nanoparticles and human brain-microvascular endothelial cells. *Journal of Physical Chemistry B* **2007**, *111* (38), 11228–11236.

[81] Kuo, Y.-C.; Yu, H.-W. Transport of saquinavir across human brain-microvascular endothelial cells by poly(lactide-co-glycolide) nanoparticles with surface poly-(gamma-glutamic acid). *International Journal of Pharmaceutics* **2011,** *416* (1), 365–375.

[82] Gil, E. S.; Wu, L.; Xu, L.; Lowe, T. L. β-Cyclodextrin-poly(β-amino ester) nanoparticles for sustained drug delivery across the blood-brain barrier. *Biomacromolecules* **2012,** *13* (11), 3533–3541.

[83] Kuo, Y.-C.; Lee, C.-L. Methylmethacrylate-sulfopropylmethacrylate nanoparticles with surface RMP-7 for targeting delivery of antiretroviral drugs across the blood-brain barrier. *Colloids and Surfaces B-Biointerfaces* **2012,** *90*, 75–82.

[84] Kuo, Y.-C.; Chung, C.-Y. Transcytosis of CRM197-grafted polybutylcyanoacrylate nanoparticles for delivering zidovudine across human brain-microvascular endothelial cells. *Colloids and Surfaces B: Biointerfaces* **2012,** *91*, 242–249.

[85] Kuo, Y.-C.; Shih-Huang, C.-Y. Solid lipid nanoparticles carrying chemotherapeutic drug across the blood-brain barrier through insulin receptor-mediated pathway. *Journal of Drug Targeting* **2013,** *21* (8), 730–738.

[86] Kenzaoui, B. H.; Angeoni, S.; Overstolz, T.; Niedermann, P.; Bernasconi, C. C.; Liley, M.; Juillerat-Jeanneret, L. Transfer of ultrasmall iron oxide nanoparticles from human brain-derived endothelial cells to human glioblastoma cells. *ACS Applied Materials & Interfaces* **2013,** *5* (9), 3581–3586.

[87] Chattopadhyay, N.; Zastre, J.; Wong, H.-L.; Wu, X. Y.; Bendayan, R. Solid lipid nanoparticles enhance the delivery of the HIV protease inhibitor, atazanavir, by a human brain endothelial cell line. *Pharmaceutical Research* **2008,** *25* (10), 2262–2271.

[88] Freese, C.; Uboldi, C.; Gibson, M. I.; Unger, R. E.; Weksler, B. B.; Romero, I. A.; Couraud, P.-O.; Kirkpatrick, C. J. Uptake and cytotoxicity of citrate-coated gold nanospheres: Comparative studies on human endothelial and epithelial cells. *Particle and Fibre Toxicology* **2012,** *9* (23), 1–11.

[89] Ye, D.; Raghnaill, M. N.; Bramini, M.; Mahon, E.; Aberg, C.; Salvati, A.; Dawson, K. A. Nanoparticle accumulation and transcytosis in brain endothelial cell layers. *Nanoscale* **2013,** *5* (22), 11153–11165.

[90] Hemmelmann, M.; Metz, V. V.; Koynov, K.; Blank, K.; Postina, R.; Zentel, R. Amphiphilic HPMA-LMA copolymers increase the transport of rhodamine 123 across a BBB model without harming its barrier integrity. *Journal of Controlled Release* **2012,** *163* (2), 170–177.

[91] Thomsen, L. B.; Linemann, T.; Pondman, K. M.; Lichota, J.; Kim, K. S.; Pieters, R. J.; Visser, G. M.; Moos, T. Uptake and transport of superparamagnetic iron oxide nanoparticles through human brain capillary endothelial cells. *ACS Chemical Neuroscience* **2013,** *4* (10), 1352–1360.

[92] Laakkonen, J. P.; Engler, T.; Romero, I. A.; Weksler, B.; Couraud, P.-O.; Kreppel, F.; Kochanek, S. Transcellular targeting of fiber- and hexon-modified adenovirus vectors across the brain microvascular endothelial cells *in vitro*. *Public Library of Science* **2012,** *7* (9), e45977.

[93] Thomsen, L. B.; Lichota, J.; Kim, K. S.; Moos, T. Gene delivery by pullulan derivatives in brain capillary endothelial cells for protein secretion. *Journal of Controlled Release* **2011,** *151* (1), 45–50.

[94] Li, T.; Bourgeois, J.-P.; Celli, S.; Glacial, F.; Le Sourd, A.-M.; Mecheri, S.; Weksler, B.; Romero, I.; Couraud, P.-O.; Rougeon, F.; Lafaye, P. Cell-penetrating anti-GFAP VHH and corresponding fluorescent fusion protein VHH-GFP spontaneously cross the blood-brain barrier and specifically recognize astrocytes: Application to brain imaging. *Federation of American Societies for Experimental Biology* **2012,** *26* (10), 3969–3979.

[95] Mahajan, S. D.; Aalinkeel, R.; Reynolds, J. L.; Nair, B.; Sykes, D. E.; Bonoiu, A.; Roy, I.; Yong, K.-T.; Law, W.-C.; Bergey, E. J.; Prasad, P. N.; Schwartz, S. A. Suppression of MMP-9 expression in brain microvascular endothelial cells (BMVEC) using a gold nanorod (GNR)-siRNA nanoplex. *Immunological Investigations* **2012,** *41* (4), 337–355.

[96] Ramirez, S. H.; Hasko, J.; Skuba, A.; Fan, S.; Dykstra, H.; McCormick, R.; Reichenbach, N.; Krizbai, I.; Mahadevan, A.; Zhang, M.; Tuma, R.; Son, Y.-J.; Persidsky, Y. Activation of cannabinoid receptor 2 attenuates leukocyte-endothelial cell interactions and blood-brain barrier dysfunction under inflammatory conditions. *Journal of Neuroscience* **2012,** *32* (12), 4004–4016.

[97] Li, J.; Ye, L.; Wang, X.; Liu, J.; Wang, Y.; Zhou, Y.; Ho, W. (–)-Epigallocatechin gallate inhibits endotoxin-induced expression of inflammatory cytokines in human cerebral microvascular endothelial cells. *Journal of Neuroinflammation* **2012,** *9* (161), 1–13.

[98] Tahanian, E.; Sanchez, L. A.; Shiao, T. C.; Roy, R.; Annabi, B. Flavonoids targeting of I kappa B phosphorylation abrogates carcinogen-induced MMP-9 and COX-2 expression in human brain endothelial cells. *Drug Design Development and Therapy* **2011,** *5*, 299–309.

[99] Annabi, B.; Lord-Dufour, S.; Vezina, A.; Beliveau, R. Resveratrol targeting of carcinogen-induced brain endothelial cell inflammation biomarkers MMP-9 and COX-2 is Sirt1-independent. *Drug Target Insights* **2012,** *6*, 1–11.

[100] Onodera, H.; Arito, M.; Sato, T.; Ito, H.; Hashimoto, T.; Tanaka, Y.; Kurokawa, M. S.; Okamoto, K.; Suematsu, N.; Kato, T. Novel effects of edaravone on human brain microvascular endothelial cells revealed by a proteomic approach. *Brain Research* **2013,** *1534*, 87–94.

[101] Cavdar, Z.; Egrilmez, M. Y.; Altun, Z. S.; Arslan, N.; Yener, N.; Sayin, O.; Genc, S.; Genc, K.; Islekel, H.; Oktay, G.; Akdogan, G. G. Resveratrol reduces matrix metalloproteinase-2 activity induced by oxygen-glucose deprivation and reoxygenation in human cerebral microvascular endothelial cells. *International Journal for Vitamin and Nutrition Research* **2012,** *82* (4), 267–274.

[102] Horai, S.; Nakagawa, S.; Tanaka, K.; Morofuji, Y.; Couraud, P.-O.; Deli, M. A.; Ozawa, M.; Niwa, M. Cilostazol strengthens barrier integrity in brain endothelial cells. *Cellular and Molecular Neurobiology* **2013,** *33* (2), 291–307.

[103] Lapergue, B.; Bao Quoc, D.; Desilles, J.-P.; Ortiz-Munoz, G.; Delbosc, S.; Loyau, S.; Louedec, L.; Couraud, P.-O.; Mazighi, M.; Michel, J.-B.; Meilhac, O.; Amarenco, P. High-density lipoprotein-based therapy reduces the hemorrhagic complications associated with tissue plasminogen activator treatment in experimental stroke. *Stroke* **2013,** *44* (3), 699–707.

[104] Shimizu, F.; Sano, Y.; Saito, K.; Abe, M.-a.; Maeda, T.; Haruki, H.; Kanda, T. Pericyte-derived glial cell line-derived neurotrophic factor increase the expression of claudin-5 in the blood-brain barrier and the blood-nerve barrier. *Neurochemical Research* **2012,** *37* (2), 401–409.

[105] Li, W.; Xu, H.; Hu, Y.; He, P.; Ni, Z.; Xu, H.; Zhang, Z.; Dai, H. Edaravone protected human brain microvascular endothelial cells from methylglyoxal-induced injury by inhibiting ages/RAGE/oxidative stress. *Public Library of Science* **2013,** *8* (9), e76025–e76025.

[106] Liu, S.; Yu, C.; Yang, F.; Paganini-Hill, A.; Fisher, M. J. Phosphodiesterase inhibitor modulation of brain microvascular endothelial cell barrier properties. *Journal of the Neurological Sciences* **2012,** *320* (1–2), 45–51.

[107] Kavitha, C. V.; Agarwal, C.; Agarwal, R.; Deep, G. Asiatic acid inhibits pro-angiogenic effects of VEGF and human gliomas in endothelial cell culture models. *Public Library of Science* **2011,** *6* (8), e22745.

[108] Tahanian, E.; Lord-Dufour, S.; Das, A.; Khosla, C.; Roy, R.; Annabi, B. Inhibition of tubulogenesis and of carcinogen-mediated signaling in brain endothelial cells highlight the antiangiogenic properties of a mumbaistatin analog. *Chemical Biology & Drug Design* **2010,** *75* (5), 481–488.

[109] Roy, U.; Bulot, C.; Honer Zu Bentrup, K.; Mondal, D. Specific increase in MDR1 mediated drug-efflux in human brain endothelial cells following co-exposure to HIV-1 and saquinavir. *Public Library of Science* **2013,** *8* (10), e75374.

[110] Zastre, J. A.; Chan, G. N. Y.; Ronaldson, P. T.; Ramaswamy, M.; Couraud, P. O.; Romero, I. A.; Weksler, B.; Bendayan, M.; Bendayan, R. Up-regulation of P-glycoprotein by HIV protease inhibitors in a human brain microvessel endothelial cell line. *Journal of Neuroscience Research* **2009,** *87* (4), 1023–1036.

[111] Chan, G. N. Y.; Patel, R.; Cummins, C. L.; Bendayan, R. Induction of P-glycoprotein by antiretroviral drugs in human brain microvessel endothelial cells. *Antimicrobial Agents and Chemotherapy* **2013,** *57* (9), 4481–4488.

[112] Mizee, M. R.; Wooldrik, D.; Lakeman, K. A. M.; van het Hof, B.; Drexhage, J. A. R.; Geerts, D.; Bugiani, M.; Aronica, E.; Mebius, R. E.; Prat, A.; de Vries, H. E.; Reijerkerk,

A. Retinoic acid induces blood-brain barrier development. *Journal of Neuroscience* **2013,** *33* (4), 1660–1671.

[113] Lim, J. C.; Kania, K. D.; Wijesuriya, H.; Chawla, S.; Sethi, J. K.; Pulaski, L.; Romero, I. A.; Couraud, P. O.; Weksler, B. B.; Hladky, S. B.; Barrand, M. A. Activation of β-catenin signalling by GSK-3 inhibition increases p-glycoprotein expression in brain endothelial cells. *Journal of Neurochemistry* **2008,** *106* (4), 1855–1865.

[114] Paolinelli, R.; Corada, M.; Ferrarini, L.; Devraj, K.; Artus, C.; Czupalla, C. J.; Rudini, N.; Maddaluno, L.; Papa, E.; Engelhardt, B.; Couraud, P. O.; Liebner, S.; Dejana, E. Wnt activation of immortalized brain endothelial cells as a tool for generating a standardized model of the blood brain barrier *in vitro*. *Public Library of Science* **2013,** *8* (8), e70233.

[115] Stins, M. F.; Nemani, P. V.; Wass, C.; Kim, K. S. Escherichia coli binding to and invasion of brain microvascular endothelial cells derived from humans and rats of different ages. *Infection and Immunity* **1999,** *67* (10), 5522–5525.

[116] Andjelkovic, A. V.; Pachter, J. S. Characterization of binding sites for chemokines MCP-1 and MIP-1α on human brain microvessels. *Journal of Neurochemistry* **2000,** *75* (5), 1898–1906.

[117] Berger, O.; Gan, X. H.; Gujuluva, C.; Burns, A. R.; Sulur, G.; Stins, M.; Way, D.; Witte, M.; Weinand, M.; Said, J.; Kim, K. S.; Taub, D.; Graves, M. C.; Fiala, M. CXC and CC chemokine receptors on coronary and brain endothelia. *Molecular Medicine* **1999,** *5* (12), 795–805.

[118] Avril, M.; Tripathi, A. K.; Brazier, A. J.; Andisi, C.; Janes, J. H.; Soma, V. L.; Sullivan, D. J. Jr.; Bull, P. C.; Stins, M. F.; Smith, J. D. A restricted subset of var genes mediates adherence of *Plasmodium falciparum*-infected erythrocytes to brain endothelial cells. *Proceedings of the National Academy of Sciences of the United States of America* **2012,** *109* (26), E1782–E1790.

[119] Shimomura, K.; Okura, T.; Kato, S.; Couraud, P.-O.; Schermann, J.-M.; Terasaki, T.; Deguchi, Y. Functional expression of a proton-coupled organic cation (H+/OC) antiporter in human brain capillary endothelial cell line hCMEC/D3, a human blood-brain barrier model. *Fluids and Barriers of the CNS* **2013,** *10* (8), 1–10.

[120] Okura, T.; Kato, S.; Deguchi, Y. Functional expression of organic cation/carnitine transporter 2 (OCTN2/SLC22A5) in human brain capillary endothelial cell line hCMEC/D3, a human blood-brain barrier model. *Drug Metabolism and Pharmacokinetics* **2013,** *29* (1), 69–74.

[121] Luissint, A.-C.; Federici, C.; Guillonneau, F.; Chretien, F.; Camoin, L.; Glacial, F.; Ganeshamoorthy, K.; Couraud, P.-O. Guanine nucleotide-binding protein G-α i2: A new partner of claudin-5 that regulates tight junction integrity in human brain endothelial cells. *Journal of Cerebral Blood Flow and Metabolism* **2012,** *32* (5), 860–873.

[122] Durk, M. R.; Chan, G. N. Y.; Campos, C. R.; Peart, J. C.; Chow, E. C. Y.; Lee, E.; Cannon, R. E.; Bendayan, R.; Miller, D. S.; Pang, K. S. 1a,25-dihydroxyvitamin D3-liganded vitamin D receptor increases expression and transport activity of P-glycoprotein in isolated rat brain capillaries and human and rat brain microvessel endothelial cells. *Journal of Neurochemistry* **2012,** *123* (6), 944–953.

[123] Hoque, M. T.; Robillard, K. R.; Bendayan, R. Regulation of breast cancer resistant protein by peroxisome proliferator-activated receptor-α in human brain microvessel endothelial cells. *Molecular Pharmacology* **2012,** *81* (4), 598–609.

[124] Zhou, N.; Zhao, W.-D.; Liu, D.-X.; Liang, Y.; Fang, W.-G.; Li, B.; Chen, Y.-H. Inactivation of EphA2 promotes tight junction formation and impairs angiogenesis in brain endothelial cells. *Microvascular Research* **2011,** *82* (2), 113–121.

[125] Wheway, J.; Obeid, S.; Couraud, P.-O.; Combes, V.; Grau, G. E. R. The brain microvascular endothelium supports T cell proliferation and has potential for alloantigen presentation. *Public Library of Science* **2013,** *8* (1), e52586.

[126] Man, S.; Ubogu, E. E.; Williams, K. A.; Tucky, B.; Callahan, M. K.; Ransohoff, R. M. Human brain microvascular endothelial cells and umbilical vein endothelial cells differentially facilitate leukocyte recruitment and utilize chemokines for T cell migration. *Clinical & Developmental Immunology* **2008,** *2008*, 384982–384982.

[127] Man, S.; Tucky, B.; Cotleur, A.; Drazba, J.; Takeshita, Y.; Ransohoff, R. M. CXCL12-Induced monocyte-endothelial interactions promote lymphocyte transmigration across an *in vitro* blood-brain barrier. *Science Translational Medicine* **2012,** *4* (119).

[128] Callahan, M. K.; Williams, K. A.; Kivisak, P.; Pearce, D.; Stins, M. F.; Ransohoff, R. M. CXCR3 marks CD4+ memory T lymphocytes that are competent to migrate across a human brain microvascular endothelial cell layer. *Journal of Neuroimmunology* **2004,** *153* (1–2), 150–157.

[129] Mackic, J. B.; Stins, M.; McComb, J. G.; Calero, M.; Ghiso, J.; Kim, K. S.; Yan, S. D.; Stern, D.; Schmidt, A. M.; Frangione, B.; Zlokovic, B. V. Human blood-brain barrier receptors for Alzheimer's amyloid-β 1-40—Asymmetrical binding, endocytosis, and transcytosis at the apical side of brain microvascular endothelial cell monolayer. *Journal of Clinical Investigation* **1998,** *102* (4), 734–743.

[130] Gonzalez-Velasquez, F. J.; Kotarek, J. A.; Moss, M. A. Soluble aggregates of the amyloid-β protein selectively stimulate permeability in human brain microvascular endothelial monolayers. *Journal of Neurochemistry* **2008,** *107* (2), 466–477.

[131] Bachmeier, C.; Mullan, M.; Paris, D. Characterization and use of human brain microvascular endothelial cells to examine β-amyloid exchange in the blood-brain barrier. *Cytotechnology* **2010,** *62* (6), 519–529.

[132] Singh, I.; Sagare, A. P.; Coma, M.; Perlmutter, D.; Gelein, R.; Bell, R. D.; Deane, R. J.; Zhong, E.; Parisi, M.; Ciszewski, J.; Kasper, R. T.; Deane, R. Low levels of copper disrupt brain amyloid-β homeostasis by altering its production and clearance. *Proceedings of the National Academy of Sciences of the United States of America* **2013,** *110* (36), 14771–14776.

[133] Tai, L. M.; Loughlin, A. J.; Male, D. K.; Romero, I. A. P-glycoprotein and breast cancer resistance protein restrict apical-to-basolateral permeability of human brain endothelium to amyloid-β. *Journal of Cerebral Blood Flow and Metabolism* **2009,** *29* (6), 1079–1083.

[134] Kania, K. D.; Wijesuriya, H. C.; Hladky, S. B.; Barrand, M. A. β-amyloid effects on expression of multidrug efflux transporters in brain endothelial cells. *Brain Research* **2011,** *1418*, 1–11.

[135] Fossati, S.; Ghiso, J.; Rostagno, A. Insights into Caspase-Mediated Apoptotic Pathways Induced by Amyloid-β in Cerebral Microvascular Endothelial Cells. *Neurodegenerative Diseases* **2012,** *10* (1–4), 324–328.

[136] Hernandez-Guillamon, M.; Mawhirt, S.; Fossati, S.; Blais, S.; Pares, M.; Penalba, A.; Boada, M.; Couraud, P.-O.; Neubert, T. A.; Montaner, J.; Ghiso, J.; Rostagno, A. Matrix metalloproteinase 2 (MMP-2) degrades soluble vasculotropic amyloid-β E22Q and L34V mutants, delaying their toxicity for human brain microvascular endothelial cells. *Journal of Biological Chemistry* **2010,** *285* (35), 27144–27158.

[137] Man, S.-M.; Ma, Y.-R.; Shang, D.-S.; Zhao, W.-D.; Li, B.; Guo, D.-W.; Fang, W.-G.; Zhu, L.; Chen, Y.-H. Peripheral T cells overexpress MIP-1α to enhance its transendothelial migration in Alzheimer's disease. *Neurobiology of Aging* **2007,** *28* (4), 485–496.

[138] Zhang, K.; Tian, L.; Liu, L.; Feng, Y.; Dong, Y.-B.; Li, B.; Shang, D.-S.; Fang, W.-G.; Cao, Y.-P.; Chen, Y.-H. CXCL1 contributes to β-amyloid-induced transendothelial migration of monocytes in alzheimer's disease. *Public Library of Science* **2013,** *8* (8), e72744.

[139] Prat, A.; Biernacki, K.; Lavoie, J. F.; Poirier, J.; Duquette, P.; Antel, J. P. Migration of multiple sclerosis lymphocytes through brain endothelium. *Archives of Neurology* **2002,** *59* (3), 391–397.

[140] Liu, K. K. Y.; Dorovini-Zis, K. Differential regulation of CD4+ T cell adhesion to cerebral microvascular endothelium by the β-chemokines CCL2 and CCL3. *International Journal of Molecular Sciences* **2012,** *13* (12), 16119–16140.

[141] Cristante, E.; McArthur, S.; Mauro, C.; Maggioli, E.; Romero, I. A.; Wylezinska-Arridge, M.; Couraud, P. O.; Lopez-Tremoleda, J.; Christian, H. C.; Weksler, B. B.; Malaspina, A.; Solito, E. Identification of an essential endogenous regulator of blood-brain barrier integrity, and its pathological and therapeutic implications. *Proceedings of the National Academy of Sciences of the United States of America* **2013,** *110* (3), 832–841.

[142] Reijerkerk, A.; Lopez-Ramirez, M. A.; van het Hof, B.; Drexhage, J. A. R.; Kamphuis, W. W.; Kooij, G.; Vos, J. B.; Kraan, T. C. T. M. v. d. P.; van Zonneveld, A. J.; Horrevoets, A. J.; Prat, A.; Romero, I. A.; de Vries, H. E. MicroRNAs Regulate human brain endothelial cell-barrier function in inflammation: Implications for multiple sclerosis. *Journal of Neuroscience* **2013,** *33* (16), 6857–6863.

[143] Basivireddy, J.; Somvanshi, R. K.; Romero, I. A.; Weksler, B. B.; Couraud, P.-O.; Oger, J.; Kumar, U. Somatostatin preserved blood brain barrier against cytokine induced alterations: Possible role in multiple sclerosis. *Biochemical Pharmacology* **2013,** *86* (4), 497–507.

[144] Mueller, M.; Frese, A.; Nassenstein, I.; Hoppen, M.; Marziniak, M.; Ringelstein, E. B.; Kim, K. S.; Schaebitz, W.-R.; Kraus, J. Serum from interferon-β-1b-treated patients with early multiple sclerosis stabilizes the blood-brain barrier *in vitro*. *Multiple Sclerosis Journal* **2012,** *18* (2), 236–239.

[145] Ghosh, C.; Gonzalez-Martinez, J.; Hossain, M.; Cucullo, L.; Fazio, V.; Janigro, D.; Marchi, N. Pattern of P450 expression at the human blood-brain barrier: Roles of epileptic condition and laminar flow. *Epilepsia* **2010,** *51* (8), 1408–1417.

[146] Kubota, H.; Ishihara, H.; Langmann, T.; Schmitz, G.; Stieger, B.; Wieser, H. G.; Yonekawa, Y.; Frei, K. Distribution and functional activity of P-glycoprotein and multidrug resistance-associated proteins in human brain microvascular endothelial cells in hippocampal sclerosis. *Epilepsy Research* **2006,** *68* (3), 213–228.

[147] Zhang, W.-j.; Feng, J.; Zhou, R.; Ye, L.-y.; Liu, H.-l.; Peng, L.; Lou, J.-n.; Li, C.-h. Tanshinone IIA protects the human blood-brain barrier model from leukocyte-associated hypoxia-reoxygenation injury. *European Journal of Pharmacology* **2010,** *648* (1–3), 146–152.

[148] Patak, P.; Jin, F.; Schafer, S. T.; Metzen, E.; Hermann, D. M. The ATP-binding cassette transporters ABCB1 and ABCC1 are not regulated by hypoxia in immortalised human brain microvascular endothelial cells. *Experimental & Translational Stroke Medicine* **2011,** *3*, 12–12.

[149] Zhang, Y. H.; Zhang, X. C.; Park, T. S.; Gidday, J. M. Cerebral endothelial cell apoptosis after ischemia-reperfusion: Role of PARP activation and AIF translocation. *Journal of Cerebral Blood Flow and Metabolism* **2005,** *25* (7), 868–877.

[150] Busu, C.; Li, W.; Caldito, G.; Aw, T. Y. Inhibition of glutathione synthesis in brain endothelial cells lengthens S-phase transit time in the cell cycle: Implications for proliferation in recovery from oxidative stress and endothelial cell damage. *Redox Biology* **2013,** *1* (1), 131–139.

[151] Zhang, Y.; Park, T. S.; Gidday, J. N. Hypoxic preconditioning protects human brain endothelium from ischemic apoptosis by Akt-dependent survivin activation. *American Journal of Physiology–Heart and Circulatory Physiology* **2007,** *292* (6), H2573–H2581.

[152] Allen, C.; Srivastava, K.; Bayraktutan, U. Small GTPase rhoA and its effector rho kinase mediate oxygen glucose deprivation evoked *in vitro* cerebral barrier dysfunction. *Stroke* **2010,** *41* (9), 2056–2063.

[153] Reuter, B.; Rodemer, C.; Grudzenski, S.; Couraud, P.-O.; Weksler, B.; Romero, I. A.; Meairs, S.; Bugert, P.; Hennerici, M. G.; Fatar, M. Temporal profile of matrix metalloproteinases and their inhibitors in a human endothelial cell culture model of cerebral ischemia. *Cerebrovascular diseases (Basel, Switzerland)* **2013,** *35* (6), 514–520.

[154] Quoc Bao, D.; Lapergue, B.; Alexy, T.-D.; Diallo, D.; Moreno, J.-A.; Mazighi, M.; Romero, I. A.; Weksler, B.; Michel, J.-B.; Amarenco, P.; Meilhac, O. High-density lipoproteins limit neutrophil-induced damage to the blood-brain barrier *in vitro*. *Journal of Cerebral Blood Flow and Metabolism* **2013,** *33* (4), 575–582.

[155] Fiala, M.; Looney, D. J.; Stins, M.; Way, D. D.; Zhang, L.; Gan, X. H.; Chiappelli, F.; Schweitzer, E. S.; Shapshak, P.; Weinand, M.; Graves, M. C.; Witte, M.; Kim, K. S. TNF-α opens a paracellular route for HIV-1 invasion across the blood-brain barrier. *Molecular Medicine* **1997,** *3* (8), 553–564.

[156] Foerster, C.; Burek, M.; Romero, I. A.; Weksler, B.; Couraud, P.-O.; Drenckbahni, D. Differential effects of hydrocortisone and TNF-α on tight junction proteins in an *in vitro* model of the human blood-brain barrier. *Journal of Physiology-London* **2008,** *586* (7), 1937–1949.

[157] Lee, N.-Y.; Rieckmann, P.; Kang, Y.-S. The changes of P-glycoprotein activity by interferon-gamma and tumor necrosis factor-α in primary and immortalized human brain microvascular endothelial cells. *Biomolecules & Therapeutics* **2012,** *20* (3), 293–298.

[158] Wong, D.; Dorovini-Zis, K.; Vincent, S. R. Cytokines, nitric oxide, and cGMP modulate the permeability of an *in vitro* model of the human blood-brain barrier. *Experimental Neurology* **2004,** *190* (2), 446–455.

[159] Haarmann, A.; Deiss, A.; Prochaska, J.; Foerch, C.; Weksler, B.; Romero, I.; Couraud, P.-O.; Stoll, G.; Rieckmann, P.; Buttmann, M. Evaluation of soluble junctional adhesion molecule-A as a biomarker of human brain endothelial barrier breakdown. *Public Library of Science* **2010,** *5* (10), e13568.

[160] Chaitanya, G. V.; Cromer, W. E.; Wells, S. R.; Jennings, M. H.; Couraud, P. O.; Romero, I. A.; Weksler, B.; Erdreich-Epstein, A.; Mathis, J. M.; Minagar, A.; Alexander, J. S. Gliovascular and cytokine interactions modulate brain endothelial barrier *in vitro*. *J Neuroinflammation* **2011,** *8*, 162.

[161] Lopez-Ramirez, M. A.; Male, D. K.; Wang, C.; Sharrack, B.; Wu, D.; Romero, I. A. Cytokine-induced changes in the gene expression profile of a human cerebral microvascular endothelial cell-line, hCMEC/D3. *Fluids Barriers CNS* **2013,** *10* (1), 27.

[162] Stephan, D.; Sbai, O.; Wen, J.; Couraud, P.-O.; Putterman, C.; Khrestchatisky, M.; Desplat-Jego, S. TWEAK/Fn14 pathway modulates properties of a human microvascular endothelial cell model of blood brain barrier. *Journal of Neuroinflammation* **2013,** *10* (9), 1–14.

[163] Sun, L. X.; Hui, A. M.; Su, Q.; Vortmeyer, A.; Kotliarov, Y.; Pastorino, S.; Passaniti, A.; Menon, J.; Walling, J.; Bailey, R.; Rosenblum, M.; Mikkelsen, T.; Fine, H. A. Neuronal and glioma-derived stem cell factor induces angiogenesis within the brain. *Cancer Cell* **2006,** *9* (4), 287–300.

[164] Zagzag, D.; Lukyanov, Y.; Lan, L.; Ali, M. A.; Esencay, M.; Mendez, O.; Yee, H.; Voura, E. B.; Newcomb, E. W. Hypoxia-inducible factor 1 and VEGF upregulate CXCR4 in glioblastoma: Implications for angiogenesis and glioma cell invasion. *Laboratory Investigation* **2006,** *86* (12), 1221–1232.

[165] Zhu, T. S.; Costello, M. A.; Talsma, C. E.; Flack, C. G.; Crowley, J. G.; Hamm, L. L.; He, X.; Hervey-Jumper, S. L.; Heth, J. A.; Muraszko, K. M.; DiMeco, F.; Vescovi, A. L.; Fan, X. Endothelial cells create a stem cell niche in glioblastoma by providing NOTCH ligands that nurture self-renewal of cancer stem-like cells. *Cancer Research* **2011,** *71* (18), 6061–6072.

[166] Wang, D.; Olman, M. A.; Stewart, J. Jr.; Tipps, R.; Huang, P.; Sanders, P. W.; Toline, E.; Prayson, R. A.; Lee, J.; Weil, R. J.; Palmer, C. A.; Gillespie, G. Y.; Liu, W. M.; Pieper, R. O.; Guan, J.-L.; Gladson, C. L. Downregulation of FIP200 induces apoptosis of glioblastoma cells and microvascular endothelial cells by enhancing pyk2 activity. *Public Library of Science* **2011,** *6* (5), e19629.

[167] Dwyer, J.; Hebda, J. K.; Le Guelte, A.; Galan-Moya, E.-M.; Smith, S. S.; Azzi, S.; Bidere, N.; Gavard, J. Glioblastoma cell-secreted interleukin-8 induces brain endothelial cell permeability via CXCR2. *Public Library of Science* **2012,** *7* (9), e45562.

[168] Lee, T. H.; Avraham, H. K.; Jiang, S. X.; Avraham, S. Vascular endothelial growth factor modulates the transendothelial migration of MDA-MB-231 breast cancer cells through regulation of brain microvascular endothelial cell permeability. *Journal of Biological Chemistry* **2003,** *278* (7), 5277–5284.

[169] Lee, B. C.; Lee, T. H.; Avraham, S.; Avraham, H. K. Involvement of the chemokine receptor CXCR4 and its ligand stromal cell-derived factor 1α in breast cancer cell migration through human brain microvascular endothelial cells. *Molecular Cancer Research* **2004,** *2* (6), 327–338.

[170] Rodriguez, P. L.; Jiang, S.; Fu, Y.; Avraham, S.; Avraham, H. K. The proinflammatory peptide substance P promotes blood-brain barrier breaching by breast cancer cells through changes in microvascular endothelial cell tight junctions. *International Journal of Cancer (Journal International du Cancer)* **2013,** *134* (5), 1034–1044.

[171] Malin, D.; Strekalova, E.; Petrovic, V.; Deal, A. M.; Al Ahmad, A.; Adamo, B.; Miller, C. R.; Ugolkov, A.; Livasy, C.; Fritchie, K.; Hamilton, E. P.; Blackwell, K.; Geradts, J.; Ewend, M.; Carey, L. A.; Shusta, E. V.; Anders, C. K.; Cryns, V. L. αB-crystallin: A novel regulator of breast cancer metastasis to the brain. *Clinical Cancer Research: An Official Journal of the American Association for Cancer Research* **2013,** *20* (1), 56–67.

[172] Akers, S. M.; O'Leary, H. A.; Minnear, F. L.; Craig, M. D.; Vos, J. A.; Coad, J. E.; Gibson, L. F. VE-cadherin and PECAM-1. enhance ALL migration across brain microvascular endothelial cell monolayers. *Experimental Hematology* **2010,** *38* (9), 733–743.

[173] Hirooka, T.; Fujiwara, Y.; Shinkai, Y.; Yamamoto, C.; Yasutake, A.; Satoh, M.; Eto, K.; Kaji, T. Resistance of human brain microvascular endothelial cells in culture to methylmercury: Cell-density-dependent defense mechanisms. *Journal of Toxicological Sciences* **2010,** *35* (3), 287–294.

[174] Tobwala, S.; Zhang, X.; Zheng, Y.; Wang, H.-J.; Banks, W. A.; Ercal, N. Disruption of the integrity and function of brain microvascular endothelial cells in culture by exposure to diesel engine exhaust particles. *Toxicology Letters* **2013,** *220* (1), 1–7.

[175] Wang, X.; Li, B.; Zhao, W.-D.; Liu, Y.-J.; Shang, D.-S.; Fang, W.-G.; Chen, Y.-H. Perfluorooctane sulfonate triggers tight junction "opening" in brain endothelial cells via phosphatidylinositol 3-kinase. *Biochemical and Biophysical Research Communications* **2011,** *410* (2), 258–263.

[176] Huang, S. H.; Wass, C.; Fu, Q.; Prasadarao, N. V.; Stins, M.; Kim, K. S. *Escherichia coli* invasion of brain microvascular endothelial-cells *in vitro* and *in vivo*—Molecular-cloning and characterization of invasion gene IBE10. *Infection and Immunity* **1995,** *63* (11), 4470–4475.

[177] Huang, S. H.; Chen, Y. H.; Fu, Q.; Stins, M.; Wang, Y.; Wass, C.; Kim, K. S. Identification and characterization of an Escherichia coli invasion gene locus, ibeB, required for penetration of brain microvascular endothelial cells. *Infection and Immunity* **1999,** *67* (5), 2103–2109.

[178] Chi, F.; Wang, L.; Zheng, X.; Jong, A.; Huang, S.-H. Recruitment of α 7 nicotinic acetylcholine receptor to caveolin-1-enriched lipid rafts is required for nicotine-enhanced *Escherichia coli* K1 entry into brain endothelial cells. *Future Microbiology* **2011,** *6* (8), 953–966.

[179] Kim, K. J.; Chung, J. W.; Kim, K. S. 67 kDa laminin receptor promotes internalization of cytotoxic necrotizing factor 1-expressing *Escherichia coli* K1 into human brain microvascular endothelial cells. *Journal of Biological Chemistry* **2005,** *280* (2), 1360–1368.

[180] Zhao, W.-D.; Liu, W.; Fang, W.-G.; Kim, K. S.; Chen, Y.-H. Vascular endothelial growth factor receptor 1 contributes to *Escherichia coli* K1 invasion of human brain microvascular endothelial cells through the phosphatidylinositol 3-kinase/Akt signaling pathway. *Infection and Immunity* **2010,** *78* (11), 4809–4816.

[181] Zhu, L.; Maruvada, R.; Sapirstein, A.; Malik, K. U.; Peters-Golden, M.; Kim, K. S. Arachidonic acid metabolism regulates *Escherichia coli* penetration of the blood-brain barrier. *Infection and Immunity* **2010,** *78* (10), 4302–4310.

[182] Che, X. J.; Chi, F.; Wang, L.; Jong, T. D.; Wu, C. H.; Wang, X. N.; Huang, S. H. Involvement of IbeA in meningitic *Escherichia coli* K1-induced polymorphonuclear leukocyte transmigration across brain endothelial cells. *Brain Pathology* **2011,** *21* (4), 389–404.

[183] Maruvada, R.; Kim, K. S. IbeA and OmpA of *Escherichia coli* K1 exploit Rac1 activation for invasion of human brain microvascular endothelial cells. *Infection and Immunity* **2012,** *80* (6), 2035–2041.

[184] Zhou, Y.; Tao, J.; Yu, H.; Ni, J.; Zeng, L.; Teng, Q.; Kim, K. S.; Zhao, G.-P.; Guo, X.; Yao, Y. Hcp family proteins secreted via the type VI secretion system coordinately regulate *Escherichia coli* K1 interaction with human brain microvascular endothelial cells. *Infection and Immunity* **2012,** *80* (3), 1243–1251.

[185] Pascal, T. A.; Abrol, R.; Mittal, R.; Wang, Y.; Prasadarao, N. V.; Goddard, W. A. III, Experimental validation of the predicted binding site of *Escherichia coli* K1 outer membrane protein A to human brain microvascular endothelial cells identification of critical mutations that prevent *E. coli* meningitis. *Journal of Biological Chemistry* **2010,** *285* (48), 37753–37761.

[186] Krishnan, S.; Chen, S.; Turcatel, G.; Arditi, M.; Prasadarao, N. V. Regulation of toll-like receptor 2 interaction with Ecgp96 controls *Escherichia coli* K1 invasion of brain endothelial cells. *Cellular Microbiology* **2013,** *15* (1), 63–81.

[187] Krishnan, S.; Fernandez, G. E.; Sacks, D. B.; Prasadarao, N. V. IQGAP1 mediates the disruption of adherens junctions to promote *Escherichia coli* K1 invasion of brain endothelial cells. *Cellular Microbiology* **2012,** *14* (9), 1415–1433.

[188] Seidel, G.; Boecker, K.; Schulte, J.; Wewer, C.; Greune, L.; Humberg, V.; Schmidt, M. A. Pertussis toxin permeabilization enhances the traversal of *Escherichia coli* K1, macrophages, and monocytes in a cerebral endothelial barrier model *in vitro*. *International Journal of Medical Microbiology* **2011,** *301* (3), 204–212.

[189] Stins, M. F.; Badger, J.; Kim, K. S. Bacterial invasion and transcytosis in transfected human brain microvascular endothelial cells. *Microbial Pathogenesis* **2001,** *30* (1), 19–28.

[190] Khan, N. A.; Iqbal, J.; Siddiqui, R. Escherichia coli K1-induced cytopathogenicity of human brain microvascular endothelial cells. *Microbial Pathogenesis* **2012,** *53* (5–6), 269–275.

[191] Li, Q.; Zhao, W.-D.; Zhang, K.; Fang, W.-G.; Hu, Y.; Wu, S.-H.; Chen, Y.-H. PI3K-dependent host cell actin rearrangements are required for *Cronobacter sakazakii* invasion of human brain microvascular endothelial cells. *Medical Microbiology and Immunology* **2010,** *199* (4), 333–340.

[192] Liu, D.-X.; Zhao, W.-D.; Fang, W.-G.; Chen, Y.-H. cPLA(2) α-mediated actin rearrangements downstream of the Akt signaling is required for *Cronobacter sakazakii* invasion into brain endothelial cells. *Biochemical and Biophysical Research Communications* **2012,** *417* (3), 925–930.

[193] Giri, C. P.; Shima, K.; Tall, B. D.; Curtis, S.; Sathyamoorthy, V.; Hanisch, B.; Kim, K. S.; Kopecko, D. J. *Cronobacter* spp. (previously *Enterobacter sakazakii*) invade and translocate across both cultured human intestinal epithelial cells and human brain microvascular endothelial cells. *Microbial Pathogenesis* **2012,** *52* (2), 140–147.

[194] Seo, H. S.; Mu, R.; Kim, B. J.; Doran, K. S.; Sullam, P. M. Binding of glycoprotein Srr1 of *Streptococcus agalactiae* to fibrinogen promotes attachment to brain endothelium and the development of meningitis. *PLoS Pathogens* **2012,** *8* (10), e1002947.

[195] Nizet, V.; Kim, K. S.; Stins, M.; Jonas, M.; Chi, E. Y.; Nguyen, D.; Rubens, C. E. Invasion of brain microvascular endothelial cells by group B streptococci. *Infection and Immunity* **1997,** *65* (12), 5074–5081.

[196] Magalhaes, V.; Andrade, E. B.; Alves, J.; Ribeiro, A.; Kim, K. S.; Lima, M.; Trieu-Cuot, P.; Ferreira, P. Group B streptococcus hijacks the host plasminogen system to promote brain endothelial cell invasion. *Public Library of Science* **2013,** *8* (5), e63244.

[197] Gandhi, G.; Londono, D.; Whetstine, C. R.; Sethi, N.; Kim, K. S.; Zueckert, W. R.; Cadavid, D. Interaction of variable bacterial outer membrane lipoproteins with brain endothelium. *Public Library of Science* **2010,** *5* (10), e13257.

[198] Brissette, C. A.; Kees, E. D.; Burke, M. M.; Gaultney, R. A.; Floden, A. M.; Watt, J. A. The multifaceted responses of primary human astrocytes and brain microvascular endothelial cells to the lyme disease spirochete, *Borrelia burgdorferi. American Society for Neurochemistry* **2013,** *5* (3), 221–229.

[199] Londono, D.; Carvajal, J.; Strle, K.; Kim, K. S.; Cadavid, D. IL-10 prevents apoptosis of brain endothelium during bacteremia. *Journal of Immunology* **2011,** *186* (12), 7176–7186.

[200] Wang, C.; Chou, C.-H.; Tseng, C.; Ge, X.; Pinchuk, L. M. Early gene response of human brain microvascular endothelial cells to *Listeria monocytogenes* infection. *Canadian Journal of Microbiology* **2011,** *57* (5), 441–446.

[201] Sokolova, O.; Heppel, N.; Jagerhuber, R.; Kim, K. S.; Frosch, M.; Eigenthaler, M.; Schubert-Unkmeir, A. Interaction of *Neisseria meningitidis* with human brain microvascular endothelial cells: Role of MAP- and tyrosine kinases in invasion and inflammatory cytokine release. *Cellular Microbiology* **2004,** *6* (12), 1153–1166.

[202] van Sorge, N. M.; Zialcita, P. A.; Browne, S. H.; Quach, D.; Guiney, D. G.; Doran, K. S. Penetration and activation of brain endothelium by *Salmonella enterica Serovar Typhimurium*. *Journal of Infectious Diseases* **2011,** *203* (3), 401–405.

[203] Wang, H.; Paton, J. C.; Herdman, B. P.; Rogers, T. J.; Beddoe, T.; Paton, A. W. The B subunit of an AB5 toxin produced by *Salmonella enterica Serovar typhi* up-regulates chemokines, cytokines, and adhesion molecules in human macrophage, colonic epithelial, and brain microvascular endothelial cell lines. *Infection and Immunity* **2013,** *81* (3), 673–683.

[204] Ramegowda, B.; Samuel, J. E.; Tesh, V. L. Interaction of Shiga toxins with human brain microvascular endothelial cells: Cytokines as sensitizing agents. *Journal of Infectious Diseases* **1999,** *180* (4), 1205–1213.

[205] Kim, J.-C.; Crary, B.; Chang, Y. C.; Kwon-Chung, K. J.; Kim, K. J. Cryptococcus neoformans activates RhoGTPase proteins followed by protein kinase C, focal adhesion kinase, and ezrin to promote traversal across the blood-brain barrier. *Journal of Biological Chemistry* **2012,** *287* (43), 36147–36157.

[206] Chen, S. H. M.; Stins, M. F.; Huang, S. H.; Chen, Y. H.; Kwon-Chung, K. J.; Chang, Y.; Kim, K. S.; Suzuki, K.; Jong, A. Y. *Cryptococcus neoformans* induces alterations in the cytoskeleton of human brain microvascular endothelial cells. *Journal of Medical Microbiology* **2003,** *52* (11), 961–970.

[207] Huang, S.-H.; Long, M.; Wu, C.-H.; Kwon-Chung, K. J.; Chang, Y. C.; Chi, F.; Lee, S.; Jong, A. Invasion of *Cryptococcus neoformans* into human brain microvascular endothelial cells is mediated through the lipid rafts-endocytic pathway via the dual specificity tyrosine phosphorylation-regulated kinase 3 (DYRK3). *Journal of Biological Chemistry* **2011,** *286* (40), 34761–34769.

[208] Huang, S.-H.; Wu, C.-H.; Jiang, S.; Bahner, I.; Lossinsky, A. S.; Jong, A. Y. HIV-1 gp41 ectodomain enhances *Cryptococcus neoformans* binding to human brain microvascular endothelial cells via gp41 core-induced membrane activities. *Biochemical Journal* **2011,** *438*, 457–466.

[209] Jong, A.; Wu, C.-H.; Prasadarao, N. V.; Kwon-Chung, K. J.; Chang, Y. C.; Ouyang, Y.; Shackleford, G. M.; Huang, S.-H. Invasion of *Cryptococcus neoformans* into human brain microvascular endothelial cells requires protein kinase C-α activation. *Cellular Microbiology* **2008,** *10* (9), 1854–1865.

[210] Liu, T.-B.; Kim, J.-C.; Wang, Y.; Toffaletti, D. L.; Eugenin, E.; Perfect, J. R.; Kim, K. J.; Xue, C. Brain inositol is a novel stimulator for promoting cryptococcus penetration of the blood-brain barrier. *PLoS Pathogens* **2013,** *9* (4), e1003247.

[211] Long, M.; Huang, S.-H.; Wu, C.-H.; Shackleford, G. M.; Jong, A. Lipid raft/caveolae signaling is required for *Cryptococcus neoformans* invasion into human brain microvascular endothelial cells. *Journal of Biomedical Science* **2012,** *19* (19), 1–14.

[212] Vu, K.; Eigenheer, R. A.; Phinney, B. S.; Gelli, A. *Cryptococcus neoformans* promotes its transmigration into the central nervous system by inducing molecular and cellular changes in brain endothelial cells. *Infection and Immunity* **2013,** *81* (9), 3139–3147.

[213] Vu, K.; Weksler, B.; Romero, I.; Couraud, P.-O.; Gelli, A. Immortalized human brain endothelial cell line HCMEC/D3 as a model of the blood-brain barrier facilitates *in*

vitro studies of central nervous system infection by *Cryptococcus neoformans*. *Eukaryotic Cell* **2009,** *8* (11), 1803–1807.

[214] Sabiiti, W.; May, R. C. Capsule independent uptake of the fungal pathogen *Cryptococcus neoformans* into brain microvascular endothelial cells. *Public Library of Science* **2012,** *7* (4), e35455.

[215] Jong, A. Y.; Stins, M. F.; Huang, S. H.; Chen, S. H. M.; Kim, K. S. Traversal of *Candida albicans* across human blood-brain barrier *in vitro*. *Infection and Immunity* **2001,** *69* (7), 4536–4544.

[216] Liu, Y.; Mittal, R.; Solis, N. V.; Prasadarao, N. V.; Filler, S. G. Mechanisms of *Candida albicans* trafficking to the brain. *PLoS Pathogens* **2011,** *7* (10), e1002305.

[217] Persidsky, Y.; Heilman, D.; Haorah, J.; Zelivyanskaya, M.; Persidsky, R.; Weber, G. A.; Shimokawa, H.; Kaibuchi, K.; Ikezu, T. Rho-mediated regulation of tight junctions during monocyte migration across the blood-brain barrier in HIV-1 encephalitis (HIVE). *Blood* **2006,** *107* (12), 4770–4780.

[218] Liu, N. Q.; Lossinsky, A. S.; Popik, W.; Li, X.; Gujuluva, C.; Kriederman, B.; Roberts, J.; Pushkarsky, T.; Bukrinsky, M.; Witte, M.; Weinand, M.; Fiala, M. Human immunodeficiency virus type 1 enters brain microvascular endothelia by macropinocytosis dependent on lipid rafts and the mitogen-activated protein kinase signaling pathway. *Journal of Virology* **2002,** *76* (13), 6689–6700.

[219] Persidsky, Y.; Stins, M.; Way, D.; Witte, M. H.; Weinand, M.; Kim, K. S.; Bock, P.; Gendelman, H. E.; Fiala, M. A model for monocyte migration through the blood-brain barrier during HIV-1 encephalitis. *Journal of immunology (Baltimore, Md.: 1950)* **1997,** *158* (7), 3499–3510.

[220] Dhawan, S.; Weeks, B. S.; Soderland, C.; Schnaper, H. W.; Toro, L. A.; Asthana, S. P.; Hewlett, I. K.; Stetlerstevenson, W. G.; Yamada, S. S.; Yamada, K. M.; Meltzer, M. S. HIV-1 infection alters monocyte interactions with human microvascular endothelial-cells. *Journal of Immunology* **1995,** *154* (1), 422–432.

[221] Nottet, H.; Persidsky, Y.; Sasseville, V. G.; Nukuna, A. N.; Bock, P.; Zhai, Q. H.; Sharer, L. R.; McComb, R. D.; Swindells, S.; Soderland, C.; Gendelman, H. E. Mechanisms for the transendothelial migration of HIV-1-infected monocytes into brain. *Journal of Immunology* **1996,** *156* (3), 1284–1295.

[222] Ramirez, S. H.; Fan, S.; Dykstra, H.; Reichenbach, N.; Del Valle, L.; Potula, R.; Phipps, R. P.; Maggirwar, S. B.; Persidsky, Y. Dyad of CD40/CD40 ligand fosters neuroinflammation at the blood-brain barrier and is regulated via jnk signaling: Implications for HIV-1 encephalitis. *Journal of Neuroscience* **2010,** *30* (28), 9454–9464.

[223] Yang, B.; Singh, S.; Bressani, R.; Kanmogne, G. D. Cross-talk between STAT1 and PI3K/AKT signaling in HIV-1-induced blood-brain barrier dysfunction: Role of CCR5 and implications for viral neuropathogenesis. *Journal of Neuroscience Research* **2010,** *88* (14), 3090–3101.

[224] Li, J.; Wang, Y.; Wang, X.; Ye, L.; Zhou, Y.; Persidsky, Y.; Ho, W. Immune activation of human brain microvascular endothelial cells inhibits HIV replication in macrophages. *Blood* **2013,** *121* (15), 2934–2942.

[225] Xu, R.; Feng, X.; Xie, X.; Zhang, J.; Wu, D.; Xu, L. HIV-1 Tat protein increases the permeability of brain endothelial cells by both inhibiting occludin expression and cleaving occludin via matrix metalloproteinase-9. *Brain Research* **2012,** *1436*, 13–19.

[226] Mishra, R.; Singh, S. K. HIV-1 Tat C modulates expression of miRNA-101 to suppress VE-cadherin in human brain microvascular endothelial cells. *Journal of Neuroscience* **2013,** *33* (14), 5992–6000.

[227] Kanmogne, G. D.; Schall, K.; Leibhart, J.; Knipe, B.; Gendelman, H. E.; Persidsky, Y. HIV-1 gp120 compromises blood-brain barrier integrity and enhances monocyte migration across blood-brain barrier: Implication for viral neuropathogenesis. *Journal of Cerebral Blood Flow and Metabolism: Official Journal of the International Society of Cerebral Blood Flow and Metabolism* **2007,** *27* (1), 123–134.

[228] Fletcher, N. F.; Wilson, G. K.; Murray, J.; Hu, K.; Lewis, A.; Reynolds, G. M.; Stamataki, Z.; Meredith, L. W.; Rowe, I. A.; Luo, G. X.; Lopez-Ramirez, M. A.; Baumert, T. F.; Weksler, B.; Couraud, P. O.; Kim, K. S.; Romero, I. A.; Jopling, C.; Morgello, S.; Balfe, P.; McKeating, J. A. Hepatitis C virus infects the endothelial cells of the blood-brain barrier. *Gastroenterology* **2012,** *142* (3), 634–U326643.

[229] Rust, N. M.; Papa, M. P.; Scovino, A. M.; Carneiro da Silva, M. M.; Calzavara-Silva, C. E.; de Azevedo Marques, E. T.Jr.; Torres Pecanha, L. M.; Scharfstein, J.; Arruda,L. B. Bradykinin enhances Sindbis virus infection in human brain microvascular endothelial cells. *Virology* **2012,** *422* (1), 81–91.

[230] da Conceicao, T. M.; Rust, N. M.; Egypto Rosa Berbel, A. C.; Martins, N. B.; do Nascimento Santos, C. A.; Da Poian, A. T.; de Arruda, L. B. Essential role of RIG-I in the activation of endothelial cells by dengue virus. *Virology* **2013,** *435* (2), 281–292.

[231] Coyne, C. B.; Bozym, R.; Morosky, S. A.; Hanna, S. L.; Mukherjee, A.; Tudor, M.; Kim, K. S.; Cherry, S. Comparative RNAi screening reveals host factors involved in enterovirus infection of polarized endothelial monolayers. *Cell Host & Microbe* **2011,** *9* (1), 70–82.

[232] Claessens, A.; Adams, Y.; Ghumra, A.; Lindergard, G.; Buchan, C. C.; Andisi, C.; Bull, P. C.; Mok, S.; Gupta, A. P.; Wang, C. W.; Turner, L.; Arman, M.; Raza, A.; Bozdech, Z.; Rowe, J. A. A subset of group A-like var genes encodes the malaria parasite ligands for binding to human brain endothelial cells. *Proceedings of the National Academy of Sciences of the United States of America* **2012,** *109* (26), E1772–E1781.

[233] Jambou, R.; Combes, V.; Jambou, M.-J.; Weksler, B. B.; Couraud, P.-O.; Grau, G. E. *Plasmodium falciparum* adhesion on human brain microvascular endothelial cells involves transmigration-like cup formation and induces pening of intercellular junctions. *PLoS Pathogens* **2010,** *6* (7), e1001021.

[234] Barbier, M.; Faille, D.; Loriod, B.; Textoris, J.; Camus, C.; Puthier, D.; Flori, L.; Wassmer, S. C.; Victorero, G.; Alessi, M.-C.; Fusai, T.; Nguyen, C.; Grau, G. E.; Rihet, P. Platelets alter gene expression profile in human brain endothelial cells in an *in vitro* model of cerebral malaria. *Public Library of Science* **2011,** *6* (5), e19651.

[235] Khaw, L. T.; Ball, H. J.; Golenser, J.; Combes, V.; Grau, G. E.; Wheway, J.; Mitchell, A. J.; Hunt, N. H. Endothelial cells potentiate interferon-gamma production in a novel tripartite culture model of human cerebral malaria. *Public Library of Science* **2013,** *8* (7), e69521.

[236] Zougbede, S.; Miller, F.; Ravassard, P.; Rebollo, A.; Ciceron, L.; Couraud, P.-O.; Mazier, D.; Moreno, A. Metabolic acidosis induced by *Plasmodium falciparum* intraerythrocytic stages alters blood-brain barrier integrity. *Journal of Cerebral Blood Flow and Metabolism* **2011,** *31* (2), 514–526.

[237] Cullen, M.; Elzarrad, M. K.; Seaman, S.; Zudaire, E.; Stevens, J.; Yang, M. Y.; Li, X.; Chaudhary, A.; Xu, L.; Hilton, M. B.; Logsdon, D.; Hsiao, E.; Stein, E. V.; Cuttitta, F.; Haines, D. C.; Nagashima, K.; Tessarollo, L.; St Croix, B. GPR124, an orphan G protein-coupled receptor, is required for CNS-specific vascularization and establishment of the blood-brain barrier. *Proceedings of the National Academy of Sciences of the United States of America* **2011,** *108* (14), 5759–5764.

[238] Avemary, J.; Salvamoser, J. D.; Peraud, A.; Remi, J.; Noachtar, S.; Fricker, G.; Potschka, H. Dynamic regulation of P-glycoprotein in human brain capillaries. *Molecular Pharmacology* **2013,** *10* (9), 3333–3341.

[239] Lok, J.; Sardi, S. P.; Guo, S.; Besancon, E.; Ha, D. M.; Rosell, A.; Kim, W. J.; Corfas, G.; Lo, E. H. Neuregulin-1 signaling in brain endothelial cells. *Journal of Cerebral Blood Flow and Metabolism* **2009,** *29* (1), 39–43.

[240] McCarthy, R. C.; Kosman, D. J. Mechanistic analysis of iron accumulation by endothelial cells of the BBB. *Biometals* **2012,** *25* (4), 665–675.

[241] McCarthy, R. C.; Kosman, D. J. Ferroportin and exocytoplasmic ferroxidase activity are required for brain microvascular endothelial cell iron efflux. *Journal of Biological Chemistry* **2013,** *288* (24), 17932–17940.

[242] Huber, O.; Brunner, A.; Maier, P.; Kaufmann, R.; Couraud, P.-O.; Cremer, C.; Fricker, G. Localization microscopy (SPDM) reveals clustered formations of P-glycoprotein in a human blood-brain barrier model. *Public Library of Science* **2012,** *7* (9), e44776.

[243] Haqqani, A. S.; Delaney, C. E.; Tremblay, T.-L.; Sodja, C.; Sandhu, J. K.; Stanimirovic, D. B. Method for isolation and molecular characterization of extracellular microvesicles released from brain endothelial cells. *Fluids and Barriers of the CNS* **2013,** *10* (1), 4.

12

METHODS FOR ASSESSING BRAIN BINDING

Li Di and Cheng Chang

Pharmacokinetics, Dynamics and Metabolism, Pfizer Inc., Groton, CT, USA

INTRODUCTION

Neuroscience is one of the largest therapeutic areas in drug discovery research and development. Central nervous system (CNS) diseases, such as stroke, Parkinson's, Alzheimer's, multiple sclerosis, and schizophrenia, affect millions of people and many of these devastating diseases still do not have adequate treatment. One of the major challenges in CNS drug discovery is the blood–brain barrier (BBB). BBB is a highly protective barrier to the CNS due to its tight cellular junctions, limited endocytosis, and high P-glycoprotein (P-gp) efflux activity. It prevents toxic or foreign substances and certain drug molecules from entering into the brain. Many potent drug candidates fail to demonstrate clinical efficacy because they do not possess the physicochemical properties necessary to enter the brain.

The ability to cross the BBB is a prerequisite for CNS drugs. CNS pharmacology is related to the free drug concentration at the site of action in the brain (C_{bu}). Unbound brain drug concentration divided by unbound plasma drug concentration (K_{puu}) determines whether distribution equilibrium between plasma and brain has been achieved. $K_{puu} > 1$ suggests influx into the brain by active uptake processes; $K_{puu} = 1$ indicates distribution equilibrium; and $K_{puu} < 1$ reveals efflux mechanisms by efflux transporters. C_{bu} and K_{puu} are some of the most important parameters to be optimized for CNS drug candidates or minimized for drugs with peripheral targets. Typically, C_{bu} and K_{puu} are difficult to measure directly *in vivo* due to technical challenges. Instead, total brain and

Blood–Brain Barrier in Drug Discovery: Optimizing Brain Exposure of CNS Drugs and Minimizing Brain Side Effects for Peripheral Drugs, First Edition. Edited by Li Di and Edward H. Kerns.

plasma drug concentrations are measured in neuro pharmacokinetics (neuroPK) studies and free drug concentrations are calculated using the following equations: $C_{b,u} = f_{u,b} \times C_b$; $C_{p,u} = f_{u,p} \times C_p$, where $C_{b,u}$ and $C_{p,u}$ are unbound brain and plasma concentrations, C_b and C_p are total brain and plasma concentrations, $f_{u,b}$ and $f_{u,p}$ are fraction unbound of brain tissue and plasma. For large animals and humans, brain samples are usually not available. Other approaches can be used to estimate $C_{b,u}$ and K_{puu} [1, 2]. Fraction unbound values ($f_{u,b}$ or $f_{u,p}$) are important parameters to measure, so that free drug concentrations can be derived from total concentration. The approach has been validated against *in vivo* brain microdialysis and has been demonstrated as a viable strategy [3, 4].

Although fraction unbound of brain and plasma is essential for converting total drug concentration to free drug concentration, the parameters ($f_{u,b}$ and $f_{u,p}$) themselves have no impact on *in vivo* efficacy [5]. There are no "good" or "bad" fraction unbound values and increasing fraction unbound is not going to increase free drug concentration in the brain. Hence, structure modifications should not be made to optimize fraction unbound. One will only need to measure fraction unbound when *in vivo* total drug concentration is available to be converted to free concentration [1]. Screening or rank-ordering of fraction unbound is of limited value as it has no relevance for *in vivo* efficacy, namely, it does not affect unbound drug concentration in the brain or K_{puu} at steady state. Free fraction and free concentration are two very different concepts and much confusion comes from using the two terms interchangeably. For detailed discussions on free drug hypothesis, free fraction, and misconceptions on plasma protein binding, readers can refer to Chapter 3 and Refs. 1, 5, and 6.

SPECIES AND REGIONAL INDEPENDENCE OF BRAIN TISSUE BINDING

Plasma protein binding (PPB), though not very commonly, can be species-dependent due to binding to different proteins in plasma or differences in binding sites of a protein among different species. PPB can also change with disease state, due to changes in the concentrations of certain plasma proteins. Brain tissue binding, however, has been shown to be independent of species and strain [7–10]. A study with about 50 structurally diverse compounds and 7 different species/strains (Sprague–Dawley and Wistar Han rat, CD-1 mouse, Hartley guinea pig, Beagle dog, Cynomolgus monkey, and humans) showed that brain tissue binding is independent of species and strain [7]. Brain tissue binding is mainly governed by nonspecific binding to brain lipids rather than specific binding to brain proteins, due to the higher lipid content in brain compared to plasma [11] and consistency of brain lipid composition across species [12]. Binding to brain tissue is mostly driven by lipophilicity [8]. Because of the lack of difference in brain tissue binding for different species/strains, it is recommended to measure brain fraction unbound in a single species (e.g., Wistar Han rat) and use it to predict brain tissue binding of any preclinical species and strains, as well as humans. This can greatly reduce the cost and resources needed for $f_{u,b}$ determination. It has also been shown that brain tissue binding lacks regional differences [13] and, therefore, brain tissue binding studies with different regions of the brain are unnecessary.

IMPACT OF CONTAMINATED BLOOD ON BRAIN TISSUE BINDING

Brain tissues are typically contaminated with a small amount of blood if they are not perfused with buffer. The brain plasma volume varies significantly depending on the euthanasia methods and the tracers used to determine the brain vascular spaces [14]. Typically, the plasma fraction (D) in brain tissue of ~3% is used to evaluate its effect. The simulation of the ratio of fraction unbound between brain tissue containing 3% plasma ($f_{u,final}$) and $f_{u,b}$ as a function of $f_{u,b}/f_{u,p}$ is shown in Table 12.1 and Figure 12.1 using Equation 12.1. Theoretically, when $f_{u,p}$ is similar or greater than $f_{u,b}$, contaminated blood in brain tissue has no or minimal impact on $f_{u,final}$. Only when $f_{u,p}$ is much smaller (>10-fold) than $f_{u,b}$, $f_{u,final}$ starts to deviate from $f_{u,b}$. For example, when $f_{u,p}$ is 10 times smaller than $f_{u,b}$, about 20% lower $f_{u,final}$ can be observed compared to $f_{u,b}$ without blood contamination, which is still within the variability of the assay. Since $f_{u,p}$ and $f_{u,b}$ track pretty well for most compounds [15], the impact of contaminated blood on $f_{u,b}$ measurement is expected to be minimal. There is also little concern of overestimation of $C_{b,u}$, since the maximum overestimation is ~3%.

$$\frac{f_{u,final}}{f_{u,b}} = \frac{1}{\left[(1-D)+D\times\left(f_{u,b}/f_{u,p}\right)\right]} \tag{12.1}$$

IN VITRO BRAIN BINDING METHODS

A number of brain binding assays have been developed and some of them are widely applied in CNS research. A few common brain binding methods are highlighted in this chapter. Many of the methods are similar to PPB methods with slight modifications. Since brain tissue binding is unlikely to be saturated, a single concentration (e.g., 2 μM) is tested for $f_{u,b}$ determination. The exception is the brain slice uptake method, where ≤0.1 μM test compound concentration is recommended to avoid nonlinearity in binding (Section "Brain Slice Uptake Method") [16].

Equilibrium Dialysis with Brain Homogenates

Equilibrium dialysis with brain homogenate to measure brain tissue binding is the gold standard method and is widely applied in drug discovery research and development [7, 17]. In order to generate consistent data, a large batch of brain homogenate is typically prepared or purchased to minimize batch-to-batch variability. Homogenization to fine suspension is necessary to generate reliable and reproducible data. Usually a two-stage homogenization is applied to produce brain homogenate [7]. Dilution of brain tissues is required to produce brain homogenate for equilibrium dialysis. A fivefold dilution is used regularly for brain binding studies. When a dilution factor is too high, it can be problematic for weakly bound compounds (high $f_{u,b}$) leading to high uncertainty in binding values. When a dilution factor is too low, brain homogenate is too viscous to generate reproducible data.

TABLE 12.1 Theoretical values on the effect of contaminated plasma in brain tissue on brain fraction unbound

	$f_{u,p}$								
$f_{u,b}$	0.0001	0.0005	0.001	0.005	0.01	0.05	0.1	0.5	1.0
0.0001	0.0001	0.00010	0.00010	0.00010	0.00010	0.00010	0.00010	0.00010	0.00010
0.0005	0.00045	0.0005	0.00051	0.00051	0.00052	0.00052	0.00052	0.00052	0.00052
0.001	0.00079	0.00097	0.001	0.0010	0.0010	0.0010	0.0010	0.0010	0.0010
0.005	0.0020	0.0039	0.0045	0.005	0.0051	0.0051	0.0051	0.0052	0.0052
0.01	0.0025	0.0064	0.0079	0.0097	0.01	0.010	0.010	0.010	0.010
0.05	0.0031	0.013	0.020	0.039	0.045	0.05	0.051	0.051	0.051
0.1	0.0032	0.014	0.025	0.064	0.079	0.097	0.1	0.10	0.10
0.5	0.0033	0.016	0.031	0.13	0.20	0.39	0.45	0.5	0.51
1.0	0.0033	0.016	0.032	0.14	0.25	0.64	0.79	0.97	1.0

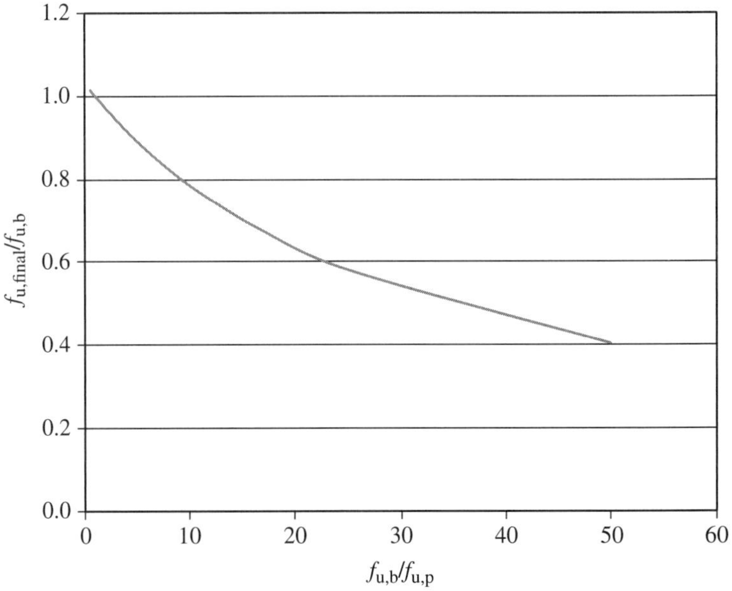

FIGURE 12.1 Impact of contaminated plasma in brain tissue on brain fraction unbound.

Therefore, selection of an appropriate dilution factor for homogenate preparation is important to generate reliable data.

For highly bound compounds, the time required for the experiment to reach equilibrium may be longer than that in the standard protocol. The time to reach equilibrium will need to be determined for these types of compounds to ensure equilibrium has been fully achieved for the experiment. This is done by determining $f_{u,b}$ at different time points (e.g., 4, 6, 8, 18, and 24 h). $f_{u,b}$ values will increase with time and gradually reach equilibrium. If volume shift is significant due to the difference in osmotic pressure between donor and receiver and/or Donnan ion effect, it will need to be corrected for $f_{u,b}$ calculation [18, 19].

Both the rapid equilibrium dialysis (RED) [20] device and the 96-well high-throughput dialysis (HTD) [21] device are widely used in the pharmaceutical industry for measuring binding. RED is about 20 times more expensive than HTD, but it is more amendable to automation. Leakage can sometimes happen for HTD as a result of wear and tear of the device. The two devices, in general, give comparable results for fraction unbound determination [22, 23]. Cassette assay can be applied to increase assay throughput and has been shown to have a similar coefficient of variation to single compound assay [8].

Recovery of the assay is an indicator of potential issues of the assay, such as high nonspecific binding to the dialysis device, instability, or precipitation. However, recovery has been shown to have no impact on fraction unbound determination as long as the system has reached equilibrium [22]. Scientific judgment has to be made to fail or pass a run when recovery is low. Other factors (e.g., stability or solubility issues) need to be considered as well.

Reiterated Stepwise Equilibrium Dialysis

Reiterated stepwise equilibrium dialysis has recently been developed to measure PPB of compounds with high nonspecific binding, such as peptides [24]. The method can also be applied to brain tissue binding of sticky compounds. It works by presaturating the dialysis device with the test compound before the test compound is added in both donors and receivers at different ratios of concentrations based on the anticipated f_u value. The final result with the smallest differences between the starting donor/receiver ratio to the final ratio suggests equilibrium has been achieved. This method has been successfully applied to liraglutide [24], a peptide with high nonspecific binding, and it is expected to work for brain binding measurement of very sticky or highly bound compounds.

Brain Slice Uptake Method

Drug distribution in brain tissues has various mechanisms, including nonspecific binding to brain lipids and proteins, specific binding to pharmacologic targets, pH partitioning into subcellular compartments (e.g., cytosol and lysosomes), carrier-mediated uptake, and membrane potential effect for charged species [25]. Binding measurements using brain homogenates only determine nonspecific binding to brain tissue, while binding with brain slice captures multiple mechanisms of intrabrain distribution for drug molecules. The major differences of binding between using brain homogenate and brain slice are seen in basic compounds, due to pH-partitioning and transporter uptake of substrates [25].

In the brain slice uptake method, freshly prepared brain slices are incubated with test compounds at low concentration (≤1 μM to avoid saturation of brain slice and nonlinearity) for 5 h [3, 16, 26–28]. At the end of incubation, the concentrations of test compounds in the medium and the brain slice are determined and $V_{u,b}$ (unbound volume of distribution in brain) is calculated: $V_{u,b} = A_{brain}/C_{buffer}$, where A_{brain} is the amount of test compound in the brain slice (nanomoles/g of brain slice), C_{buffer} is the final test compound concentration in the buffer (μM), and $V_{u,b}$ is calculated in ml/g of brain. $V_{u,b}$ and $f_{u,b}$ are inversely related ($V_{u,b} \times f_{u,b} = K_{puu,cell}$) and, for compounds where tissue binding is the only distribution mechanism, $V_{u,b} = 1/f_{u,b}$. However, for compounds that partition into subcellular compartments (e.g., lysosomal trapping (pH 5.0) [29] for basic compounds) due to pH gradient or active influx into brain cells by uptake transporters (e.g., gabapentin transported by LAT1 transporter) [25], $V_{u,b}$ correlates to $1/f_{u,b}$ through K_{puu} ($V_{u,b} = K_{puu,cell} \times 1/f_{u,b}$).

The brain slice uptake method is more applicable for basic compounds and transporter substrates, where intracellular free drug concentration is different than free medium concentration. Even though the high-throughput brain slice uptake method has been developed and cassette assaying has been validated, the method requires freshly isolated brain slice, which is more demanding than using brain homogenate. When $V_{u,b}$ is not available from a brain slice uptake study, it is recommended to use default K_{puu} values (~3 for amines, ~0.6 for carboxylic acids, and 1 for neutrals or zwitterions) to convert $f_{u,b}$ to $V_{u,b}$.

TRANSIL Brain Absorption Method

The TRANSIL method is based on membrane affinity of drugs to porcine brain lipid bilayer that is noncovalently absorbed onto surface-modified silica beads [30]. The method was found to give good correlation to $f_{u,b}$ determined by equilibrium dialysis [12]. It is a high-throughput method in 96-well format and can also be used in cassette assaying to further increase throughput. The method has 2 min incubation time and is not based on biomatrices. It is a useful approach to determine brain tissue binding of unstable compounds and compounds that have high nonspecific binding. Since the system only measures binding to phospholipids, for compounds that are highly bound to albumin, the method could potentially overestimate $f_{u,b}$ [12].

Ultracentrifugation with Brain Homogenates

Ultracentrifugation is another method that has been used to measure $f_{u,b}$ [31]. However, the method is not commonly applied in the pharmaceutical industry due to high nonspecific binding to the filter membrane and the device, leading to artificially low $f_{u,b}$. Strategies trying to reduce nonspecific binding did not significantly improve the results. The method is not recommended for compounds with high nonspecific binding.

Microemulsion Retention Factor

The retention factor generated from microemulsion-based capillary electrophoresis (MEEKC k') with mass spectrometry detection has been shown to have high correlation with $f_{u,b}$ [32]. The method performed better than using calculated log P or LC retention-based lipophilicity. Lauric acid–based micellar electrokinetic chromatography was used for the application. The method can be applied to diverse chemical structures, is amendable for automation, has short runtime, and consumes a very small amount of material. However, the error in $f_{u,b}$ estimation appears to be too large to provide true quantitative $f_{u,b}$ prediction necessary to support *in vitro–in vivo* extrapolation (IVIVE) or pharmacokinetic–pharmacodynamics (PK–PD) understanding [15].

IN SILICO MODELS FOR BRAIN FRACTION UNBOUND

A quantitative structure–activity relationship (QSAR) model was built based on 70 compounds with mouse $f_{u,b}$ data [8]. The analysis suggested that the mechanisms regulating $f_{u,b}$ were mainly nonspecific. $c\log P$ was the most important parameter in the model. The number of aromatic atoms were found to have negative influence on $f_{u,b}$, while the solvent accessible polar surface area was positively contributing to $f_{u,b}$. The model was expected to be continuously updated with newly available data to ensure better predictive power and adequate coverage of chemical space.

Another QSAR model was developed for $f_{u,b}$ using descriptors of log P and pK_a as a nonlinear ionization-specific model based on ~500 compounds [33]. The results of internal and external validation demonstrated good predictive power of the model and the statistical parameters were similar among training and validation sets, indicating that the model is not likely to be overfitted.

A simple physical model has been developed to predict $V_{u,b}$ from chemical structure [34]. The model is based on predicted lipid binding and pH partitioning in interstitial fluid, intracellular, and lysosomal compartments. No fitted parameters were included and the model is not dependent on the dataset used. The model confirms that lipid binding and pH partioning are important aspects of brain distribution.

CONCLUSIONS

Brain binding ($f_{u,b}$) is an important parameter to estimate C_{bu} and K_{puu}, which could subsequently help establish PK–PD relationships for brain targets or build physiologically based pharmacokinetic (PBPK) models for the brain compartment. Many methodologies are available to measure $f_{u,b}$, including equilibrium dialysis with brain homogenate, iterated stepwise equilibrium dialysis, uptake with brain slice, TRANSIL brain absorption, ultracentrifugation with brain homogenates, microemulsion retention factor (MEEKC k'), and in silico approaches. Equilibrium dialysis with brain homogenate is the most common method applied in the pharmaceutical industry for $f_{u,b}$ determination. Brian tissue binding is mostly driven by nonspecific binding to brain lipids and is highly correlated to lipophilicity. Consequently, brain fraction unbound is independent of species, strain, and brain region. Single species brain binding (e.g., Wistar Han rat) can be used to predict brain binding of all preclinical species/strains and humans. The primary utility of $f_{u,b}$ is to convert total drug concentration in brain to unbound concentration. There is no mechanistic link between $f_{u,b}$ and brain penetration and $f_{u,b}$ is not an indicator of brain penetrability. Although $f_{u,b}$ is critical for PBPK modeling and for developing PK–PD relationships, structure modification to improve $f_{u,b}$ is not recommended. There are no "good" or "bad" $f_{u,b}$ values and $f_{u,b}$-based screening or rank-ordering could be misleading in drug discovery. $F_{u,b}$ will only need to be determined when *in vivo* total brain concentration is available to be converted to unbound concentration.

REFERENCES

[1] Di, L.; Rong, H.; Feng, B. Demystifying brain penetration in central nervous system drug discovery. *J. Med. Chem.* **2013,** 56, 2–12.

[2] Rong, H.; Feng, B.; Di, L. Integrated approaches to blood-brain barrier. *Encycl. Drug Metab. Interact.* **2012,** 3, 563–590.

[3] Friden, M.; Gupta, A.; Antonsson, M.; Bredberg, U.; Hammarlund-Udenaes, M. *In vitro* methods for estimating unbound drug concentrations in the brain interstitial and intracellular fluids. *Drug Metab. Dispos.* **2007,** 35, 1711–1719.

[4] Liu, X.; Van Natta, K.; Yeo, H.; Vilenski, O.; Weller, P. E.; Worboys, P. D.; Monshouwer, M. Unbound drug concentration in brain homogenate and cerebral spinal fluid at steady state as a surrogate for unbound concentration in brain interstitial fluid. *Drug Metab. Dispos.* **2009,** 37, 787–793.

[5] Smith, D. A.; Di, L.; Kerns, E. H. The effect of plasma protein binding on *in vivo* efficacy: Misconceptions in drug discovery. *Nat. Rev. Drug Discov.* **2010,** 9, 929–939.

[6] Liu, X.; Chen, C.; Hop, C. E. C. A. Do we need to optimize plasma protein and tissue binding in drug discovery? *Curr. Top. Med. Chem. (Sharjah, United Arab Emirates)* **2011,** 11, 450–466.

[7] Di, L.; Umland, J. P.; Chang, G.; Huang, Y.; Lin, Z.; Scott, D. O.; Troutman, M. D.; Liston, T. E. Species independence in brain tissue binding using brain homogenates. *Drug Metab. Dispos.* **2011,** 39, 1270–1277.

[8] Wan, H.; Rehngren, M.; Giordanetto, F.; Bergstroem, F.; Tunek, A. High-throughput screening of drug-brain tissue binding and *in silico* prediction for assessment of central nervous system drug delivery. *J.Med. Chem.* **2007,** 50, 4606–4615.

[9] Read, K. D.; Braggio, S. Assessing brain free fraction in early drug discovery. *Expert Opin. Drug Metab. Toxicol.* **2010,** 6, 337–344.

[10] Summerfield, S. G.; Lucas, A. J.; Porter, R. A.; Jeffrey, P.; Gunn, R. N.; Read, K. R.; Stevens, A. J.; Metcalf, A. C.; Osuna, M. C.; Kilford, P. J.; Passchier, J.; Ruffo, A. D. Toward an improved prediction of human *in vivo* brain penetration. *Xenobiotica* **2008,** 38, 1518–1535.

[11] Di, L.; Kerns, E. H.; Carter, G. T. Strategies to assess blood-brain barrier penetration. *Expert Opin. Drug Discov.* **2008,** 3, 677–687.

[12] Longhi, R.; Corbioli, S.; Fontana, S.; Vinco, F.; Braggio, S.; Helmdach, L.; Schiller, J.; Boriss, H. Brain tissue binding of drugs: evaluation and validation of solid supported porcine brain membrane vesicles (TRANSIL) as a novel high-throughput method. *Drug Metab. Dispos.* **2011,** 39, 312–321.

[13] Hong, L.; Jiang, W.; Pan, H.; Jiang, Y.; Zeng, S.; Zheng, W. Brain regional pharmacokinetics of p-aminosalicylic acid and Its N-acetylated metabolite: Effectiveness in chelating brain manganese. *Drug Metab. Dispos.* **2011,** 39, 1904–1909.

[14] Friden, M.; Ljungqvist, H.; Middleton, B.; Bredberg, U.; Hammarlund-Udenaes, M. Improved measurement of drug exposure in the brain using drug-specific correction for residual blood. *J. Cereb. Blood Flow Metab.* **2010,** 30, 150–161.

[15] Zamek-Gliszczynski, M. J.; Sprague, K. E.; Espada, A.; Raub, T. J.; Morton, S. M.; Manro, J. R.; Molina-Martin, M. How well do lipophilicity parameters, MEEKC microemulsion capacity factor, and plasma protein binding predict CNS tissue binding? *J. Pharm. Sci.* **2012,** 101, 1932–1940.

[16] Friden, M.; Ducrozet, F.; Middleton, B.; Antonsson, M.; Bredberg, U.; Hammarlund-Udenaes, M. Development of a high-throughput brain slice method for studying drug distribution in the central nervous system. *Drug Metab. Dispos.* **2009,** 37, 1226–1233.

[17] Kalvass, J. C.; Maurer, T. S. Influence of nonspecific brain and plasma binding on CNS exposure: implications for rational drug discovery. *Biopharm. Drug Dispos.* **2002,** 23, 327–338.

[18] Tozer, T. N.; Gambertoglio, J. G.; Furst, D. E.; Avery, D. S.; Holford, N. H. G. Volume shifts and protein binding estimates using equilibrium dialysis: Application to prednisolone binding in humans. *J. Pharm. Sci.* **1983,** 72, 1442–1446.

[19] Giacomini, K. M.; Wong, F. M.; Tozer, T. N. Correction for volume shift during equilibrium dialysis by measurement of protein concentration. *Pharm. Res.* **1984**, 179–181.

[20] http://www.piercenet.com/product/rapid-equilibrium-dialysis-red.

[21] http://htdialysis.com/.

[22] Di, L.; Umland, J. P.; Trapa, P. E.; Maurer, T. S. Impact of recovery on fraction unbound using equilibrium dialysis. *J. Pharm. Sci.* **2012,** 101, 1327–1335.

[23] Waters, N. J.; Jones, R.; Williams, G.; Sohal, B. Validation of a rapid equilibrium dialysis approach for the measurement of plasma protein binding. *J. Pharm. Sci.* **2008,** 97, 4586–4595.

[24] Plum, A.; Jensen, L. B.; Kristensen, J. B. *In vitro* protein binding of liraglutide in human plasma determined by reiterated stepwise equilibrium dialysis. *J. Pharm. Sci.* **2013,** 102, 2882–2888.

[25] Friden, M.; Bergstroem, F.; Wan, H.; Rehngren, M.; Ahlin, G.; Hammarlund-Udenaes, M.; Bredberg, U. Measurement of unbound drug exposure in brain: Modeling of pH partitioning explains diverging results between the brain slice and brain homogenate methods. *Drug Metab. Dispos.* **2011,** 39, 353–362.

[26] Kakee, A.; Terasaki, T.; Sugiyama, Y. Brain efflux index as a novel method of analyzing efflux transport at the blood-brain barrier. *J. Pharmacol. Exp. Ther.* **1996**, 277(3), 1550–1559.

[27] Becker, S.; Liu, X. Evaluation of the utility of brain slice methods to study brain penetration. *Drug Metab. Dispos.***2006,** 34, 855–861.

[28] Loryan, I.; Friden, M.; Hammarlund-Udenaes, M. The brain slice method for studying drug distribution in the CNS. *Fluids Barriers CNS* **2013**, 10, 6.

[29] Ohkuma, S.; Poole, B. Fluorescence probe measurement of the intralysosomal pH in living cells and the perturbation of pH by various agents. *Proc. Natl. Acad. Sci. U.S.A.* **1978**, 75, 3327–3331.

[30] http://www.sovicell.com/scitech-faq-transil.asp.

[31] Mano, Y.; Higuchi, S.; Kamimura, H. Investigation of the high partition of YM992, a novel antidepressant, in rat brain—*in vitro* and *in vivo* evidence for the high binding in brain and the high permeability at the BBB. *Biopharm. Drug Dispos.* **2002,** 23, 351–360.

[32] Wan, H.; Aahman, M.; Holmen, A. G. Relationship between brain tissue partitioning and microemulsion retention factors of CNS drugs. *J. Med. Chem.* **2009,** 52, 1693–1700.

[33] Lanevskij, K.; Dapkunas, J.; Juska, L.; Japertas, P.; Didziapetris, R. QSAR analysis of blood-brain distribution: The influence of plasma and brain tissue binding. *J. Pharm. Sci.* **2011,** 100, 2147–2160.

[34] Spreafico, M.; Jacobson, M. P. *In silico* prediction of brain exposure: Drug free fraction, unbound brain to plasma concentration ratio and equilibrium half-life. *Curr. Top. Med. Chem. (Sharjah, United Arab Emirates)* **2013**, 13, 813–820.

13

IN VIVO STUDIES OF BRAIN EXPOSURE IN DRUG DISCOVERY

Edward H. Kerns
Laytonsville, MD, USA

INTRODUCTION

Brain exposure research has resulted in various methodologies: in silico predictions and models, *in vitro* measurements, and *in vivo* investigations. In the past, the methodologies were typically aligned sequentially (i.e., in silico first followed by *in vitro* then *in vivo*), but are now integrated across the discovery/development continuum in new ways that increase value. This integration enhances understanding of the disease and target, drug design, and safety and predicts and determines human pharmacokinetics/pharmacodynamics (PKPD) for the development of superior clinical drug candidates. The high value of brain exposure characterization is clear when we consider one of three "pillar" research questions to be answered in drug research and development: *What is the "free drug exposure at the target site of action*" [1]?

In vivo studies of brain exposure are discussed in this chapter. First, the core methods used for these studies are summarized. Next, examples of the research questions for which *in vivo* brain studies provide data, and how these data are integrated with other data, are highlighted. The main product of *in vivo* brain studies is the total brain concentrations of a drug candidate at time points, which, in combination with *in vitro* and *in silico* methods, provides unbound drug concentrations ($C_{u,b}$), brain PK, and a PK/PD model.

Blood–Brain Barrier in Drug Discovery: Optimizing Brain Exposure of CNS Drugs and Minimizing Brain Side Effects for Peripheral Drugs, First Edition. Edited by Li Di and Edward H. Kerns.

CORE METHODS

Most *in vivo* brain studies include core methods [2–5]. These are incorporated or modified in the experimental protocol according to the research question under investigation.

NeuroPK

In vivo brain exposure studies (neuroPK) involve dosing and sacrificing a separate animal for every time point. For example, a study would involve 20 animals for 10 time points in duplicate. This is unlike *in vivo* plasma studies, for which blood samples can be drawn at multiple time points from one animal. Thus, more resources must be anticipated for an *in vivo* brain exposure study (e.g., number of animals, lab space, animal associate time). For studies involving efflux transporters, knockout animals are available from vendors [6]. Blood samples are also saved from the sacrificed animals to assess the extent ($K_{p,uu,brain}$) of partitioning into the brain from the blood. Various routes of administration can be used for neuroPK studies, including intravenous (IV), subcutaneous, intramuscular, oral (PO), or intraperatoneal (IP). Dosing solutions have also been administered using continuous subcutaneous osmotic minipump for steady-state infusion [6].

Prior to sampling, the animal is anesthetized. Blood is sampled by cardiocentesis (cardiac puncture) or from the abdominal aorta with anticoagulant (e.g., heparin, ethylenediaminetetraacetic (EDTA)), centrifuged to produce plasma and stored frozen until analysis. Cerebrospinal fluid (CSF) is sampled by cisterna magna puncture in small animals (from the lumbar CSF in humans) and frozen. (Lumbar CSF may not have rapid equilibrium with CSF in the brain.) Some investigators perfuse the brain of the anesthetized animal with buffer while the heart is beating to remove residual blood that contains the test compound and minimize blood contamination of brain samples, especially if the brain concentration is expected to be relatively low compared to the blood concentration, which is relatively high. Alternately, the contribution of residual blood in the brain capillaries can be corrected for [7]. The brain is carefully removed from the animal's skull immediately after euthanasia. The brain is cleaned of external blood by washing with buffer, quickly frozen, placed in an individual pre-tared storage vial, weighed, and stored frozen (e.g., $\leq -80\,°C$) until sample preparation.

In Situ Brain Perfusion

To determine the rate of brain penetration, the *in situ* brain perfusion method has been used [8, 9]. This experimental format can also be used to affect the *in vivo* conditions on the luminal side of the blood–brain barrier (BBB) by substituting a surrogate fluid matrix containing selected additives (e.g., transporter inhibitors) for the circulating blood. *In situ* perfusion experiments are resource-consuming and are usually only performed for a few compounds or to answer a specific question.

In this experiment, the rat is anesthetized, the carotid artery is severed and cannulated, and artery branches that do not perfuse the brain are quickly tied off. The cannula is immediately attached to an infusion pump that supplies buffer (e.g., Krebs Ringer Bicarbonate, pH 7.4, saturated with 95% oxygen/5% carbon dioxide) fluid at the physiological flow rate (e.g., 20 ml/min in rat) and pressure (e.g., 80–120 mm Hg) opposite to normal blood flow. The perfusion fluid contains the test compound (e.g., 10 μM), internal control compounds, and any additives. The internal control compounds have well characterized BBB permeability values (e.g., atenolol is a negative to low BBB permeability control, antipyrine is a positive BBB passive permeability control, loperamide is a P-glycoprotein (P-gp) efflux control). Infusion continues for a short time period (e.g., 30 s), after which the fluid is changed to blank buffer to wash the test compound from the brain capillaries for 30 s. The brain is then immediately removed. One brain hemisphere, on the side of the animal where the carotid artery was perfused, is separated and analyzed using liquid chromatography–mass spectrometry (LC/MS). This separation is because each brain hemisphere is perfused by only the carotid artery on that side of the body. Because of the short experimental time, *in vivo* BBB permeability is unaffected by nonspecific binding of brain tissue or brain metabolism.

Brain Uptake Assay

In the brain uptake assay [10] mice are injected via the tail vein, and at 5 or 60 m post injection the mice were sacrificed. The blood and a 200–400 mg portion of brain, from the same lobe for all animals, are removed for assay. The 5 m time point provides an assessment of BBB permeability. The 60 m time point provides brain PK insights, such as the test compound loss from the brain relative to loss from the plasma. Consistent brain and plasma loss indicates rapid equilibration and BBB permeability and faster loss from plasma indicates slow equilibration or brain tissue accumulation potential. The assay enables assessment of factors, such as initial brain uptake rate, active uptake and efflux, and serum binding, which limit brain exposure of drug discovery candidates, thus supporting hit identification and lead optimization. In a similar manner [11], the test compound is injected into rats via the femoral vein.

IV Infusion

For the purpose of determining $K_{p,uu}$, one lab administers the test compound IV to rats for 4 h [2]. Measurement of the brain and plasma concentrations provides the K_p. Using *in vitro* methods, unbound fraction in plasma (f_u) and unbound brain volume ($V_{u,b}$) in brain slices is measured. This enables the following calculation: $K_{p,uu} = K_p/(f_u \times V_{u,b})$.

Microdialysis

Microdialysis uses a dialysis probe to directly sample and measure unbound test compound interstitial fluid (ISF) concentration *in vivo*. This is detailed in another chapter of this book.

Brain Tissue and Plasma Sample Preparation

At the time of brain sample preparation, the samples are thawed and then processed rapidly to avoid degradation. All or part of each time point tissue sample is transferred to a vial and weighed. Plasma samples are vortexed and then a consistent volume is transferred for all samples.

The brain tissue is disrupted to obtain complete extraction of the test compound, metabolites, and biomarkers of interest. Disruption has been performed using enzymatic tissue digestion [12] or various homogenization techniques [13–15] and cryo-pulverization [16] (CryoPrep Pulverizer). Homogenization disruption involves adding a measured volume of buffer (e.g., potassium phosphate at pH 7.4) or water, at a multiple (e.g., 3- to 10-fold) of the weight of the tissue (e.g., 300 μl of buffer with 100 mg of brain tissue). Homogenization has been performed with an individual ultrasonic or rotor stator homogenizer probe. Apparatus with multiple homogenizer probes, for parallel processing of samples, have also been used (e.g., Tomtec Autogizer). A more recent disruption procedure uses a bead homogenizer (e.g., Precellys®24, SPEX® SamplePrep® Geno/Grinder), which pulverizes the tissue very rapidly into a viscous suspension. A stainless-steel bead is placed in each vial in a rack or well in a plate with the buffer and tissue, wells are sealed and then the entire assembly is vigorously agitated. This allows rapid parallel processing of multiple tissue samples.

The test compound or biomarker is typically extracted from the tissue homogenate/suspension or from plasma samples using multiple volumes (e.g., threefold the sample volume) of organic solvent containing an internal standard that either precipitates the protein material (e.g., acetonitrile) or extracts the analytes into an immiscible solvent (e.g., ethyl acetate) for liquid/liquid extraction. The solutions are agitated or vortexed for a period (e.g., Glas-Col Pulsing Vortex Mixer) for extraction. The solutions are then centrifuged to precipitate the suspended denatured protein or separate the layers, and then the supernatant is transferred to a separate microtiter plate for analysis. The organic solvent is sometimes evaporated and the extract is resuspended in the high-performance liquid chromatography (HPLC) mobile phase.

A standard curve is extracted in parallel with the experiment's samples. This is prepared by adding known quantities of analyte to known amounts of blank brain tissue (for tissue analysis) or plasma (for plasma analysis) obtained from undosed naive animals of a comparable strain in-house or from a vendor (e.g., Bioreclamation). Various standard curve wells/vials are prepared for various concentrations to prepare a standard curve over a wide concentration range that brackets the sample concentrations.

LC/MS Analysis

A small volume of each sample extract is injected into an LC/MS for quantitation of the test compound concentration in the brain sample. Different MS techniques have been used. A single-stage MS might be used with selected ion monitoring of the pseudomolecular ion of the analyte; however, this may not have sufficient selectivity to separate the analyte signal from the background signal, especially at low analyte concentrations. The most commonly used MS technique is tandem MS, in which the pseudomolecular ion of the analyte ("precursor" or "parent" ion) is selectively passed by the first MS stage (e.g., quadrupole analyzer operated at unit mass resolution) and other ions are rejected, followed by collisional activation by an inert gas in the second MS stage to form fragment ions specific to the analyte ("product" or "daughter" ions) that are then selectively passed by the third MS stage operated at unit mass resolution. Instrument manufacturers have different names for this tandem MS technique (e.g., multiple reaction monitoring (MRM), selected reaction monitoring (SRM)). In recent years, increasing numbers of investigators have been using high-resolution MS techniques, in which the pseudomolecular ion of the analyte is selectively passed by a high-resolution MS stage (e.g., time of flight (TOF), orbitrap) with a narrow mass window and other ions are rejected, even those of the same nominal unit mass, but with a different accurate mass owing to a different molecular formula. For high-resolution MS quantitation, collisional activation and third MS stage separation are not required. The analyte concentration in the experimental sample (either unit or high-resolution technique) is determined by comparison of the relative response (analyte response/internal standard response) in the sample extract versus the relative response from the standard curve. The usual bioanalytical validation and quality control statistical values for plasma analysis should be achieved.

The measured values are the total brain concentrations (C_b) corresponding to the respective animal sacrifice times. C_b can be corrected for the contribution of residual blood in the brain capillaries [7]. Correction is particularly important when $K_{p,uu,brain}$ is low (i.e., $C_b < C_p$), because the residual blood will significantly increase $K_{p,uu,brain}$. The unbound brain concentration ($C_{u,b}$) is calculated by multiplying C_b by the fraction unbound in brain ($f_{u,b}$), which was determined using an *in vitro* brain tissue binding experiment [17], for example, involving equilibrium dialysis. The $C_{u,b}$ and unbound concentration in plasma ($C_{u,p}$) values are loaded into software (e.g., Phoenix® WinNonlin®) to fit the data to a PK noncompartmental model and calculate the PK parameters. The unbound brain/unbound plasma ratio ($K_{p,uu,brain}$) is also calculated by dividing $C_{u,b}$ (usually at C_{max} or C_{ss}) by $C_{u,p}$ or by dividing the unbound exposure in brain ($AUC_{u,b}$) by unbound exposure in plasma ($AUC_{u,p}$).

Brain Exposure Localization

For some studies there is a need to know the exposure of a test compound in a specific brain substructure. This is a more resource-consuming study, but can provide more definitive results. Several approaches have been used: (i) a radioactive drug can be administered and then detected by autoradiography to determine the location and concentration in a brain slice, (ii) microdialysis sampling of live animals performed in a specific brain region or CSF, (iii) fluorescence imaging of a fluorescent test compound, and (iv) mass spectrometry imaging.

IN VIVO BRAIN EXPOSURE RESEARCH QUESTIONS AND APPROACHES FOR THEIR STUDY

To address various drug discovery project research questions, the core methods described earlier are modified as appropriate to understand the factors that affect brain exposure. Examples of these brain exposure questions and the information they provide are discussed later. A consistent topic of discussion is the amount of resources needed for such studies versus the useful data and valuable research answers they provide. In the past, drug discovery research was driven by the pillars of target engagement and biology, but the important impact of the third pillar, *in vivo* target delivery/exposure, has increased the reliance on information from *in vivo* brain exposure studies.

What Was the Brain Exposure During an Early Proof-of-Concept Efficacy Study?

Following early *in vitro* biology studies with receptor- or cell-based assays, drug discovery teams may obtain an initial assessment of the capability of an early hit or lead compound to produce efficacious results (e.g., minimum efficacious dose (MED)) *in vivo*. After dosing in the efficacy model, biological or biomarker end points are measured. During this study, a small number of animals might be sacrificed at one or more time points to measure the brain concentrations. In this way, the project team can obtain an initial idea of the brain concentration at the time that the efficacious response was observed to develop the PK/PD relationship.

Do Some Chemical Series Examples Have Higher Brain Exposure?

In a fast-moving drug discovery program, many analogs of a chemical series may be synthesized within a short time period. Different analogs may have different outstanding characteristics, such as target binding, bioavailability, or selectivity. In addition, different analogs may have been synthesized to increase BBB permeation by enhancing passive BBB permeability or reducing efflux. In these cases, the question may be: What are the comparable brain exposures of the analogs? To address this question, the identical protocol for dosing vehicle, level, and route are repeated for the same animal species and strain and the comparable analog brain PK values are determined.

In order to speed up and conserve resources for this type of study [18], the use of a cassette dosing approach, in which multiple analogs are combined into one vehicle prior to dosing, was studied. Eleven model compounds were compared by discrete (i.e., single) compound dosing versus cassette subcutaneous dosing in wild-type, Mdr1a/1b(–/–), Bcrp1(–/–), and Mdr1a/1b(–/–)/Bcrp1(–/–) mice at 1 and 3 mg/kg. The results were that the K_p values from discrete and cassette dosing were within twofold for the nine compounds with high enough concentrations to measure. The ratios of K_p values for wild-type versus three efflux transporter–deficient mice strains were also consistent with the ones from the literature.

Other investigators have previously expressed concern about the possibility of drug–drug interactions of the multiple analogs in cassette dosing experimental protocols. To address this concern, the authors [18] purposefully selected study compounds that are known P-gp and Bcrp inhibitors and substrates; however, the cassette dosing data were nevertheless comparable to the discrete dosing for all the compounds. In addition, they purposefully selected a low dose level for analogs in the cassette to make drug–drug interactions less likely. This study indicates that a cassette protocol is effective at 1 and 3 mg/kg with subcutaneous dosing.

Cassette protocols reduce the number of sacrificed animals for humane treatment. The number of expensive transgenic animals can also be reduced. Furthermore, with the same resource budget, more information can be provided to the discovery program for better decision making. For example, instead of discrete studies of four analogs in wild-type mice, cassette studies of four analogs in wild-type and efflux transporter–deficient mice can be performed with the same resources. These protocols also reduce the supplies, instrumentation, and human resources consumed and increase the speed at which such data can be made available to the program investigators. Cassette protocols also facilitate analog side-by-side comparisons for decision making. The analogs are studied in the same animals, on the same day, and with the same investigator.

Do the Chemical Series Structural Modifications Yield Different Brain Exposures?

In some cases, a therapeutic target may be found in both the brain and the peripheral tissues. It may be advantageous to modify a lead compound to stay out of the brain and reduce CNS side effects. In some cases, the therapeutic target is expressed in both the brain and peripheral tissues.

As described in other chapters of this book, a reduction in BBB permeation can be produced by the following methods:

a) Increasing efflux transport (e.g., increase hydrogen-bonding acceptors, increase molecular weight) [19]
b) Adding one or more substructures that reduce BBB passive diffusion (e.g., increase hydrogen bonding and TPSA, increase molecular weight, increase acidity, reduce lipophilicity)

One group [20] studied rimonabant, which has desirable antiobesity activity as a pheripheral CB1 GPCR receptor antagonist, but also has undesirable CNS side effects through CB1 GPCR receptors in the brain. Rimonabant was structurally modified by increasing the molecular weight and hydrogen bond receptors to produce AM6545 for the purpose of limited BBB penetrance. *In vivo* studies showed that rimonabant reached similar plasma and brain concentrations at 1 h after a single PO or IP dose and after 28 days of IP dosing at 10 mg/kg. In contrast, AM6545 reached a plasma concentration similar to rimonabant in plasma, but approximately one-tenth of its concentration in brain after single or 28-day dosing. Further testing in

Mdr1a/b$^{-/-}$ and Mdr1a/b$^{-/-}$Bcrp$^{-/-}$ mice and wild-type controls demonstrated a dramatic increase in brain levels of AM6545 in the knockout strains, supporting the conclusion that the brain concentration reduction was owing to efflux transport at the BBB. The involvement of efflux transport was supported by *in vitro* studies in Caco-2 which showed an efflux ratio of 1.1 and good apical to basolateral and basolateral to apical permeability. Furthermore, AM6545 demonstrated desirable peripheral CB1 effects without the undesirable CNS behavioral effects.

In another example, it was demonstrated [21] that levocetirizine H1 receptor occupancy in (peripheral) ileum is correlated to plasma concentration, whereas H1 receptor occupancy in the brain is poorly correlated to plasma concentration, but is instead correlated to brain concentration. This is in agreement with the observed human clinical peripheral H1 efficacy without having CNS side effects at low (10 mg) dose. Brain H1 receptor occupancy is lower and increases slowly with time compared to ileum occupancy. At a dose of 0.1 mg/kg, levocetirizine occupied less than 20% of central H1 receptors, but at the same time occupied 80% of peripheral receptors. These results are consistent with the much lower brain concentration versus plasma ($K_p = 0.06$). Thus, compounds that are limited in BBB permeation might be clinically dosed at a lower level at which peripheral receptors are sufficiently occupied for efficacy, without occupying enough CNS receptors to produce side effects. These results are in agreement with clinical human efficacy.

Another example suggests the possibility of structural modification to increase BBB permeability by conjugation to a substrate for an uptake transporter. In one study [22] L-tyrosine was conjugated to ketoprofen and enhanced brain uptake of the prodrug via the LAT1 (large neutral amino acid) uptake transporter was observed. The authors used *in situ* perfusion and a coadministered LAT1 inhibitor to confirm the uptake. The same group also conjugated glucose to ketoprofen and indomethacin and observed enhanced brain uptake of the prodrug via the GluT1 (glucose) uptake transporter [23].

Is Brain Exposure Limited or Enhanced by a Transporter?

In vivo experiments can indicate the predominant BBB permeation mechanisms for a candidate compound. *In vivo* dosing studies allow measurement of plasma and brain concentrations, which, in combination with *in vitro* $f_{u,p}$ and $f_{u,b}$, allow calculation of $K_{p,uu,brain}$ [7]. This value indicates whether the compound is affected by BBB efflux transport ($K_{p,uu,brain} < 1$), uptake transport ($K_{p,uu,brain} > 1$), or primarily passive diffusion permeation ($K_{p,uu,brain} \sim 1$).

Efflux and uptake transport can greatly influence the brain concentration of transporter substrates. The extent of this reduction or enhancement is useful information for the project team. As an example of such *in vivo* studies [21], quinacrine, an antimalarial drug, was investigated for treatment of the brain disease Creutzfeldt–Jakob disease. While quinacrine was active *in vitro*, it was ineffective when administered *in vivo*. Therefore, authors investigated whether BBB permeability was limiting brain exposure. Quinacrine was administered orally at 40 mg/kg/day to mice for 29 days using wild-type, *Mdr1*$^{0/0}$-deficient (*mdr1a* and *mdr1b* genes ablated) mice. The resulting steady-state concentrations were 1.6 μM for wild-type mice and 84 μM for Mdr1-deficient mice. In another experiment, the P-gp inhibitor cyclosporine A was administered orally at 100 mg/kg to wild-type mice 1 or 2 h prior to quinacrine administration. Brain concentrations were sixfold higher in the inhibitor-treated mice than in untreated mice. Thus, the two experiments, comparison of wild-type and P-gp-deficient animals and comparison of wild-type and wild-type plus P-gp inhibitor, were comparable in providing confirmation of the mechanism limiting brain exposure of the subject compound. This information can be useful to project teams to understand the mechanism causing discrepancies between *in vitro* activity and *in vivo* brain efficacy, to select candidates, or to guide optimization of the chemical series to increase brain exposure by reducing P-gp efflux.

In a similar manner, brain penetration enhancement of other efflux transporter substrates has been increased by predosing with efflux transporter inhibitors. Paclitaxel and docetaxel brain penetrations were enhanced using valspodar, cyclosporine A, elacridar, and zosuquidar. Imatinib was increased by zosuquidar [24]. Reduced brain accumulation of dasatinib was observed in wild-type mice [25], but higher accumulation in $Abcb1a/1b^{-/-}$ and even higher in $Abcb1a/1b;Abcg2^{-/-}$ mice, implicating P-gp efflux as the primary factor reducing accumulation and Abcg2 as a secondary factor. Elacridar, which inhibits both efflux transporters, had a similar accumulation as $Abcb1a/1b;Abcg2^{-/-}$.

The plasma to brain distribution ($K_{p,uu}$) of seven CNS drugs in both wild-type and Mdr1a knockout mice and rats was studied [6]. Compounds affected by P-gp efflux in mice were also affected in rats. Furthermore, the knockout models were supported as viable alternatives to chemical efflux transporter inhibitor codosing models.

Where Is a Compound Localized in the Brain?

When the question is "what is the test compound exposure in various brain regions?", a specialized MS technique has been used [26]. A rat was dosed with clozapine, and slices of the brain were prepared and placed on a slide. The slices were sprayed with a "matrix" compound. The slide was placed in a device that bombards the sprayed slices with small-diameter pulses of laser light at specific positions. The laser light was absorbed by the matrix compound, which transferred its energy to the tissue and test compound, causing it to become charged and desorb from the slice surface. This is called matrix-assisted laser desorption ionization (MALDI). The test compound molecules were detected by a sensitive quadrupole time-of-flight (QTOF) MS. By rastering the laser pulse across the surface of the slice the clozapine concentrations associated with the brain tissue structures were measured. The spatial localization of clozapine in the tissue agreed with radioautographic analysis. The technique is not as sensitive as whole-brain bioanalysis techniques, especially for low–molecular weight test compounds, but, when it works, it can provide a map of the concentrations of the test compound in various brain structures for specialized studies.

What Is the Predicted Free Drug Concentration in Brain in Large Animals?

Later drug discovery and predevelopment studies focus on large animal species, such as dog and nonhuman primate, for studies of neuropharmacology and toxicity. For example, $C_{b,u}$ is related to receptor occupancy. In planning such large animal studies, it would be beneficial to reliably project the large animal unbound brain concentration ($C_{b,u}$) using a smaller species.

It was found [4] that $C_{b,u}/C_{p,u}$ from single dosing in rats is a reliable means of projecting steady-state $C_{b,u}$ in dogs and nonhuman primates for compounds that permeate the BBB primarily via passive or active uptake mechanisms, but not for efflux substrates (for which CSF concentration is better). $C_{b,u}$ was used as a surrogate for brain ISF compound concentration (C_{ISF}), which is most predictive of neuropharmacodynamics. The predicted unbound brain concentration in a large animal species was calculated using the following equation, where $K_{p,uu,rat}$ was calculated from measured $C_{b,u}$ and $C_{p,u}$ in the rat and C_p and $f_{u,p}$ in larger species:

$$C_{b,u,human} \approx K_{p,uu,rat} \times C_{p,u,human}$$

Thus, large animal steady-state $C_{b,u}$ can be predicted from single-dose rat brain and plasma studies in combination with large animal plasma data. For efflux substrates (e.g., resperidone), $C_{b,u}/C_{p,u}$ values were about fourfold higher in dogs and nonhuman primates than in rats, and C_{CSF} is recommended for such compounds. This study also measured rat and large animal $f_{u,b}$ values that were equivalent (≤1.5-fold different), which was consistent with previous findings [17, 27]; this supports the use of one $f_{u,b}$ value across species.

CONCLUSIONS

In vivo brain studies offer important opportunities for drug discovery teams to obtain the information needed for informed decisions in advancing their projects toward a strong clinical candidate. In a higher-throughput cassette format, they can rapidly provide brain exposure $C_{b,u}$ and plasma–brain partitioning $K_{p,uu,brain}$ results for lead selection and optimization. In a more definitive format, they can provide data for discovery-level PK/PD relationship development as the time course of the relationship between $C_{b,u}$ and efficacy values are carefully measured. The design of late discovery or predevelopment studies can be facilitated as $C_{b,u}$ in large animals is predicted. The involvement of transporters can be better understood as the discovery team seeks to assess the extent of the effects of efflux, to reduce it by compound design, or explore enhancement of brain exposure by increasing BBB uptake transporter affinity. All of these research questions are addressed by tailoring *in vivo* brain study experiments to the research question. While brain exposure was formerly the concern of neuroscientists, the viability of peripheral drugs is increasingly affected by their purposeful design to reduce CNS side effects by reducing CNS exposure.

REFERENCES

[1] Morgan, P.; Van Der Graaf, P. H.; Arrowsmith, J.; Feltner, D. E.; Drummond, K. S.; Wegner, C. D.; Street, S. D. A. Can the Flow of Medicines be Improved? Fundamental Pharmacokinetic and Pharmacological Principles toward Improving Phase II Survival. *Drug Discov. Today* **2012**, *17*, 419–424.

[2] Hammarlund-Udenaes, M.; Bredberg, U.; Friden, M. Methodologies to Assess Brain Drug Delivery in Lead Optimization. *Curr. Topics Med. Chem.* **2009**, *9*, 148–162.

[3] Ahn, M.; Ghaemmaghami, S.; Huang, Y.; Phuan P. W.; May, B. C. H.; Giles, K.; DeArmond, S. J.; Prusiner, S. B.; Pharmacokinetics of Quinacrine Efflux from Mouse Brain via the P-glycoprotein Efflux Transporter. *PLoS One* **2012**, *7*, e3912–e39112.

[4] Doran, A. C.; Osgood, S. M.; Mancuso, J. Y., Shaffer, C. L. An Evaluation of Using Rat-Derived Single-Dose Neuropharmacokinetic Parameters to Project Accurately Large Animal Unbound Brain Drug Concentrations. *Drug Metab. Dispos.* **2012**, *40*, 2162–2173.

[5] Tamvakopoulos, C. S.; Colwell, L. F.; Karakat, K.; Fenyk-Melody, J.; Griffin, P. R.; Nargund, R.; Palucki, B.; Sebhat, I.; Shen, X.; Stearns, R. A. Determination of Brain and Plasma Drug Concentrations by Liquid Chromatography/Tandem Mass Spectrometry. *Rapid Commun. Mass Spectrom.* **2000**, *14*, 1729–1735.

[6] Bundgaard, C.; Jensen, C. J. N.; Garmer, M. Species Comparison of In Vivo P-Glycoprotein-Mediated Brain Efflux Using mdr1a-Deficient Rats and Mice. *Drug Metab. Dispos.* **2012**, *40*, 461–466.

[7] Friden, M.; Ljundqvist, H.; Middleton, B.; Bredberg, U.; Hammarlund-Udenaes, M. Improved Measurement of Drug Exposure in the Brain Using Drug-Specific Correction for Residual Blood. *J. Cereb. Blood Flow Metab.* **2010**, *30*, 150–161.

[8] Smith, Q. R. Brain Perfusion Systems for Studies of Drug Uptake and Metabolism in the Central Nervous System, In *Models for Assessing Drug Absorption and Metabolism;* Borchardt, R. T., Smith, P. L., Wilson, G., Eds.; Plenum Press: New York, 1996, pp. 285–307.

[9] Obradovic, T.; Dobson, G. G.; Shingaki, T.; Kungu, T.; Hidalgo, I. J. Assessment of the First and Second Generation Antihistamines Brain Penetration and Role of P-Glycoprotein. *Pharm. Res.* **2007**, *24*, 318–327.

[10] Raub, T. J.; Sutzke, B. S.; Andrus, P. K.; Sawada, G. A.; Staton, B. A. Early Preclinical Evaluation of Brain Exposure in Support of Hit Identification and Lead Optimization, In *Optimizing the "Drug-Like" Properties of Leads in Drug Discovery;* Borchardt, R. T.; Kerns, E. H.; Hageman, M. J.; Thakker, D. R.; Stevens, J. L., Eds.; Springer: New York, 2006, pp. 355–410.

[11] Ohno, K.; Pettigrew, K. D.; Rapoport, S. I. Lower Limites of Cerebrovascular Permeability to Nonelectrolytes in the Conscious Rat. *Am. J. Physiol.* **1978**, *172*, 354–359.

[12] Yu, C.; Penn, L. D.; Hollembaek, J.; Li, W.; Cohen, L. H. Enzymatic Tissue Digestion as an Alternative Sample Preparation Approach for Quantitative Analysis Using Liquid Chromatography – Tandem Mass Spectrometry. *Anal. Chem.* **2004**, *76*, 1761–1767.

[13] Liang, X.; Ubhayakar, S.; Liederer, B. M.; Dean, B.; Qin, A. R. R.; Shahidi-Latham, S.; Deng, Y. Evaluation of Homogenization Techniques for the Preparation of Mouse Tissue Samples to Support Drug Discovery. *Bioanalysis* **2011**, *3*, 1923–1933.

[14] Want, E. J.; Masson, P.; Michopoulos, F.; Wilson, I. D.; Theodoridis, G.; Plumb, R. S.; Shockcor, J.; Loftus, N.; Holmes, E; Nicholson, J. K. Global Metabolic Profiling of Animal and Human Tissues via UPLC-MS. *Nat. Protoc.* **2013**, *8*, 17–32.

[15] Watson, J.; Wright, S.; Lucas, A.; Clarke, K. L.; Viggers, J.; Cheetham, S.; Jeffrey, P.; Porter, R.; Read, K. D. Receptor Occupancy and Brain Free Fraction. *Drug Metab. Dispos.* **2009**, *37*, 753–760.

[16] Xiao, G.; Black, C.; Hetu, G.; Sands, E.; Wang, J.; Caputo, R.; Rohde, E.; Gan, L. S. L. Cerebrospinal Fluid Can be Used as a Surrogate to Assess Brain Exposures of Breast Cancer Resistance Protein and P-Glycoprotein Substances. *Drug Metab. Dispos.* **2012**, *40*, 779–787.

[17] Di, L.; Rong, H.; Feng, B. Demystifying Brain Penetration in Central Nervous System Drug Discovery. Miniperspective. *J. Med. Chem.* **2013**, *56*, 2–12.

[18] Liu, X.; Ding, X.; Deshmukh, L.; Bianca, M.; Hop, C. E. C. A. Use of Cassette-Dosing Approach to Assess Brain Penetration in Drug Discovery. *Drug Metab. Dispos.* **2012**, *40*, 963–969.

[19] Cole, S.; Bagal, S.; El-Kattan, A.; Fenner, A.; Hay, T.; Kempshall, S.; Lunn, G.; Varma, M.; Supple, P.; Speed, W. Full Efficacy With No CNS Side-Effects: Unachievable Panacea or Reality? DMPK Considerations in Design of Drugs With Limited Brain Penetration. *Xenobiotica* **2012**, *42*, 11–27.

[20] Tam, J.; Vemuri, V. K.; Liu, J.; Bátkai, S.; Mukhopadhyay, B.; Godlewski, G.; Osei-Hyiaman, D.; Ohnuma, S.; Ambudkar, S. V.; Pickel, J.; Makriyannis, A.; Kunos, G. Peripheral CB1 Cannabinoid Receptor Blockade Improves Cardiometabolic Risk in Mouse Models of Obesity. *J. Clin. Investig.* **2012**, *120*, 2953–2966.

[21] Gupta, A.; Gillard, M.; Christophe, B.; Chatelain, P.; Massingham, R.; Hammarlund-Udenaes, M. Peripheral and Central H1 Histamine Receptor Occupancy by Levocetirizine, A Non-Sedating Antihistamine; A Time Course Study in the Guinea Pig. *Br. J. Pharmacol.* **2007**, *151*, 1129–1136.

[22] Gynther, M.; Laine, K.; Ropponen, J.; Leppänen, J.; Mannila, A.; Nevalainen, T.; Savolainen, J.; Järvinen, T.; Rautio, J. Large Neutral Amino Acid Transporter Enables Brain Drug Delivery via Prodrugs. *J. Med. Chem.* **2008**, *51*, 932–993.

[23] Gynther, M.; Ropponen, J.; Laine, K.; Leppänen, J.; Haapakoski, P.; Peura, L.; Järvinen, T.; Rautio, J. Glucose Promoiety Enables Glucose Transporter Mediated Brain Uptake of Ketoprofen and Indomethacin Prodrugs in Rats. *J. Med. Chem.* **2009**, *52*, 3348–3353.

[24] Breedveld, P.; Beijnen, J. H.; Schellens, J. H. M. Use of P-glycoprotein and BCRP Inhibitors to Improve Oral Bioavailability and CNS Penetration of Anticancer Drugs. *Trends Pharmacol. Sci.* **2006**, *27*, 17–24.

[25] Lagas, J. S.; van Waterschoot, R. A. B.; van Tilburg, V. A. C. J.; Hillebrand, M. J.; Lankheet, N.; Rosing, H.; Beijnen, J. H.; Schinkel, A. H. Brain Accumulation of Dasatinib Is Restrictedby P-Glycoprotein (ABCB1) and Breast Cancer Resistance Protein (ABCG2) and Can Be Enhanced by Elacridar Treatment. *Clin. Cancer Res.* **2009**, *15*, 2344–2351.

[26] Hsieh, Y.; Casale, R.; Fukuda, E.; Chen, J.; Knemeyer, I.; Wingate, J.; Morrison, R.; Korfmacher, W. Matrix-Assisted Laser Desorption/Ionization Imaging Mass Spectrometry for Direct Measurement of Clozapine in Rat Brain Tissue. *Rapid Commun. Mass Spectrom.* **2006**, *20*, 965–972.

[27] Summerfield, S. G.; Lucas, A. J.; Porter, R. A.; Jeffrey, P.; Gunn, R. N.; Read, K. R.; Stevens, A. J.; Metcalf, A. C.; Osuna, M. C.; Kilford, P. J.; Passchier, J.; Ruffo, A. D. Toward an Improved Prediction of Human *In Vivo* Brain Penetration. *Xenobiotica* **2008**, *38*, 1518–1535.

14

PBPK MODELING APPROACH FOR PREDICTIONS OF HUMAN CNS DRUG BRAIN DISTRIBUTION

ELIZABETH C.M. DE LANGE

Division of Pharmacology, Leiden Academic Center of Drug Research, Gorlaeus Laboratories, Leiden University, Leiden, The Netherlands

PROBLEMS IN CNS DRUG DEVELOPMENT

Central nervous system (CNS) drug development is facing a very high attrition rate and a number of pharmaceutical companies have decided to reduce or even close their CNS Research & Development sites. On the other hand, we face a huge need for CNS therapies and the question is how to find a way to improve prediction of adequate CNS drug effects in human, too many gaps in our current understanding exist for reasonable prediction of CNS drug effects in humans. Among these gaps is the lack of adequate knowledge of CNS target site drug distribution.

As information on CNS target site drug distribution cannot (readily) be obtained directly from human brain, indirect approaches should be used, including *in vitro* and *in vivo* preclinical studies. With CNS drug discovery and development aiming at rapid evaluation of the attribute of a compound, the following parameters have been typically measured for CNS drug candidates:

1. The "BBB permeability" (the rate of passing the blood–brain barrier (BBB), mostly obtained in an *in vitro* system), as a measure of brain penetration, with high values considered to be good.

Blood–Brain Barrier in Drug Discovery: Optimizing Brain Exposure of CNS Drugs and Minimizing Brain Side Effects for Peripheral Drugs, First Edition. Edited by Li Di and Edward H. Kerns.

2. The “Kp” value (the ratio of total brain concentrations divided by total plasma concentrations, mostly obtained in vivo at assumed steady-state conditions), as another measure of brain penetration, where a larger value is considered to indicate better penetration.
3. The “P-glycoprotein (P-gp) efflux ratio” (the extent of P-gp mediated polarized transport), typically obtained in vitro in a monolayer of cells that express the P-gp where a value larger than 1.5–2 indicates the compound to be a P-gp substrate, with higher values being considered as worse.

However, it has become clear that the aforementioned parameters, judged upon more or less in isolation, have not provided insight into the understanding of the impact of these individual parameters on CNS target site drug concentrations as interpretation of the meaning of the parameters has not been unambiguously:

1. “BBB permeability” is only an indication for the rate of entrance into the brain, and does not inform on, but has often been mixed up with the extent of brain distribution.
2. The “Kp value” does not inform on the equilibration between plasma and brain as it is based on total concentrations, while to that end we would need the unbound concentrations at either side of the BBB. So, the Kp_{uu} value should be used. In addition, for brain effects, the unbound concentrations is what should be focused on [1].
3. The “P-gp efflux ratio” is not the only factor to be taken into account in active drug transport between blood and brain.

It should be realized that many factors are involved in CNS target site drug distribution. There is the presence of the BBBs, which include the BBB but also the blood–cerebrospinal fluid barrier (BCSFB). Both the BBB and BCSFB, made up of endothelial and epithelial cells respectively, have tight junctions that restrict the passive exchange of hydrophilic drug molecules. Furthermore, these barriers possess multiple active efflux and influx transport systems that recognize many drugs, while, also at the level of the brain parenchymal cells, active transport processes may be working. Moreover, the bulk flow of brain extracellular fluid ($brain_{ECF}$) and turnover of CSF might further contribute to differences between the pharmacokinetics in plasma, and at different sites within the brain.

In vitro cellular assays based on real BBB cellular anatomy and physiology provide information on transporters, metabolic enzymes, and how to modulate these [2]. In situ methods like brain homogenate and brain slice methods have allowed insight into the relationship between the physicochemical properties of drugs and their main tissue subtype distribution characteristics [3, 4]. However, information on regional differences in drug concentrations, brain ECF bulk flow, and CSF flow, and, therewith, on-target site concentrations cannot be assessed from *in vitro or in situ* studies. This indicates the need for *in vivo* studies in preclinical species.

It has to be realized that it is the combination of drug properties and biological system characteristics that determine the target site pharmacokinetics and resulting pharmacodynamics of a drug. Drug properties will remain unchanged in different systems, but, for example, clear differences exist between the rat and human biological systems. Thus, research approaches that can distinguish between drug- and systems-specific properties are of special interest, as, in principle, physiological parameters from the rat can be replaced by those of humans (Table 14.1) to predict the target site pharmacokinetics and resulting pharmacodynamics in humans. However, not only can the biological system differ between species. Gender, genetic background, age, diet, disease conditions, drug treatment, etc. also contribute to differences between one and another biological system [34–44]. So, the rate and extent of processes in the body are context-dependent and contribute to variability in CNS target site pharmacokinetics and pharmacodynamics. This indicates the need for cross-compared designed experimental approaches to provide information on contributions of the (main) individual processes, in terms of rate and extent, as well as their interplay [45].

This chapter will address gross anatomy and physiology of the brain, and how physiological processes play a role in CNS drug target site distribution. This is followed by physiologically based pharmacokinetic model (PBPK) characteristics and how PBPK models will have better translational predictive power for CNS target site distribution in humans, as it explicitly distinguishes between drug and system properties. Knowledge of human brain target site concentrations will then be useful for further development of PBPK-PD models in healthy and disease conditions [43, 46], to further pave the way to predict the right drug at the *right location, right time, and right concentrations* [45].

TABLE 14.1 Human and rat approximate values for brain physiological parameters

Parameter	Rat value	Human value
Blood volume	20 ml	5000 ml
Plasma volume	10.6 ml [5]	2900 ml [6]
Brain weight	1.8 g (own observations)	1400 g [7]
Cerebral blood flow	1.1 ml/min [8, 9]	40 ml/min/g brain [10], 700 ml/min [11]
$Brain_{ECF}$ volume	290 µl [12, 13]	240–280 ml [14, 15]
$Brain_{ECF}$ bulk flow	0.2–0.5 µl/min [12, 16]	0.15–0.20 ml/min [17, 18]
CSF production	2.2 µl/min [12, 13]	0.35–0.4 ml/min [19, 20]
CSF turnover	11 times/day [20]	4 times/day [20]
CSF volume	250–300 µl [21]	140–150 ml [18, 20, 22]
CSF volume lateral ventricle	50 µl [23, 24]	22.5 ml [23–25]
CSF volume cisterna magna	17 µl [26, 27]	7.5 ml [26, 27]
CSF volume third and fourth ventricle	50 µl [28]	22.5 ml [23, 24]
CSF volume subarachnoidal space (SAS)	180 µl [21, 28]	90 ml [29]
BBB surface area	155 cm^2 [30, 31]	10–20 m^2 [32]
Choroid plexus surface area	75 cm^2 [31]	0.021 m^2 [33]

PHYSIOLOGY OF THE BRAIN

Physiological Compartments

The brain consists of different physiological compartments. These include the $brain_{ECF}$, the brain parenchyma cells, and the different spaces of CSF being the lateral ventricles, third ventricle, fourth ventricle, cisterna magna, and the subarachnoid spaces [47]. The brain is separated from direct contact with blood by the presence of barriers. The BBB is situated between the blood and the $brain_{ECF}$ and is made up from endothelial cells of brain capillaries joined by tight junctions. The BCSFB is mainly situated at the epithelium of the choroid plexuses [48–50]. The CSF is separated from the brain parenchyma cells by an ependymal layer without barrier function [51].

Barrier Functions

With respect to the vasculature of the brain, there are more than 100 billion capillaries in the human brain comprising a total length of approximately 400 miles, indicating that the microvasculature in the human brain is dense. With tight junctions present between the adjacent cerebral endothelial cells, the BBB forms a continuous cerebral blood vessel wall. As a result, the BBB has a high endothelial electrical resistance, in the range of 1500–2000 $\Omega \times cm^2$ (pial vessels), as compared to 3–33 $\Omega \times cm^2$ in other tissues. The net result of this elevated resistance is low paracellular permeability. The resistance may even increase to approximately 8000 $\Omega \times cm^2$ in nonpial capillaries [32, 52, 53]. The surface area of the brain endothelial cells that make up the BBB is approximately 100 cm^2/g tissue, with the capillary volume and endothelial cell volume constituting approximately 1 and 0.1% of the tissue volume, respectively [54]. The intercapillary distance in the brain is about 40 μm, which provides space for two neurons [55]. The choroid plexus and arachnoid epithelial cells provide the BCSFB and also have restricted and highly controlled exchange for compounds, which is comparable but not quantitatively and qualitatively equal to the BBB [56–59].

Cerebrospinal Fluid

The principal sources of the CSF are the choroid plexus epithelia of the lateral, third, and fourth ventricles [30]. The volume of CSF in humans is 140–150 ml, with only about 30–40 ml actually in the ventricular system. The production rate is approximately 21 ml/h. The turnover time of the total CSF is approximately 5 h for humans. This rate is species-dependent and is approximately 1 h for rat [20, 48, 60]. The majority of the CSF is in the subarachnoid space. CSF moves within the ventricles and subarachnoid spaces under the influence of hydrostatic pressure generated by its production. In addition, there are indications that drainage of the brain ECF contributes to CSF formation [16]. The CSF flows from the lateral ventricles via the third ventricle to the fourth ventricle to the cranial and spinal subarachnoid spaces. Finally, the CSF is absorbed into the peripheral bloodstream across the arachnoid villi. This production, continuous flow, and elimination of CSF serve as a washout system, especially affecting hydrophilic and large molecules [60, 61].

Circumventricular Organs

Moreover, the circumventricular organs (CVOs) that border the third and fourth ventricles, have permeable fenestrated capillaries. CVOs include the pineal gland, median eminence, neurohypophysis, subfornical organ, area postrema, subcommissural organ, organum vasculosum of the lamina terminalis, and the intermediate and neural lobes of the pituitary. These BBB-deficient areas are recognized as important sites for "sensing" blood components that inform the brain on the status of the body [60]. The choroid plexus is also devoid of a BBB, but as already discussed, the epithelium provides the barrier function in the form of the BCSFB.

Barrier Surfaces

The surface area of the human BBB is between 10 and 20 m^2, and is approximately 5000-fold greater than that of the CVOs [32]. The estimated surface area of the human choroid plexus is approximately 0.021 m^2 [33], and for rats it is approximately 75 cm^2 [62, 63]. This rat choroid plexus surface at the apical side takes into account the apical microvilli. It is much greater than older estimates and is similar to the surface area of the cerebral capillaries in rat (~155 cm^2), and suggests that the choroid plexuses may play a more important role in the regulation of the brain microenvironment than previously thought [62–64]. Table 14.1 summarizes the reported physiological values for rats and humans.

Metabolic Enzymes

A number of metabolic enzymes have been identified in the BBB and BCSFB [65–70]. These enzymes include cytochrome P-450–dependent monooxygenases, epoxide hydrolases, and several conjugating enzymes. Enzymes such glutathione S-transferase, alkaline phosphatase, and aromatic acid decarboxylase are in elevated concentration in cerebral capillaries (i.e., BBB), yet often in low concentration or absent in non-neuronal capillaries. The activity of several drug-metabolizing enzymes is especially high in the choroid plexus (i.e., BCSFB). Evidence that these enzyme activities influence the brain concentrations of drugs has been shown only for some enzymes in the choroid plexus. In the choroid plexus, glucuronic acid or glutathione conjugation occurs, which is coupled to basolateral efflux of the formed conjugates, likely mediated by multidrug resistance–related proteins (MRPs), and, thereby, makes up another barrier function that limits distribution into the CSF.

Disease Conditions

Several disease-related processes result in enhanced BBB permeability to fluid and/or solutes. These include hypertension, radioactive exposure, edema, inflammation, ischemia, and reperfusion (reoxygenation) [71, 72]. Changes in cerebral blood flow, BBB functionality, BCSFB functionality, plasma protein binding, brain tissue binding, CSF flow, and enzyme functionality may all have their effects on drug brain distribution and elimination.

PHYSIOLOGICAL PROCESSES INVOLVED IN CNS DRUG DISTRIBUTION

A number of factors play a role in the relationship between CNS drug dose and resulting CNS effects, and it is important to realize that these processes occur in parallel and do influence each other [45, 73]. The main processes that determine CNS target site distribution include plasma pharmacokinetics and plasma protein binding [74], cerebral blood flow, transport across the BBB, transport across the BCSFB, $brain_{ECF}$ bulk flow, CSF turnover, brain tissue binding, brain metabolism, and brain degradation [75, 76].

Plasma Pharmacokinetics

The pharmacokinetics of the (unbound) drug in plasma and the unbound concentration on the other side of the BBB (and BCSFB) are the driving forces for drug distribution between plasma and CNS [77] (Fig. 14.1).

Cerebral Blood Flow

For drugs that have no problem in crossing membranes, blood flow rate will determine the rate of membrane crossing. If a membrane has a barrier function for a particular drug, then the barrier crossing rate will be determined by membrane permeability of the drug [75, 77] (Fig. 14.2).

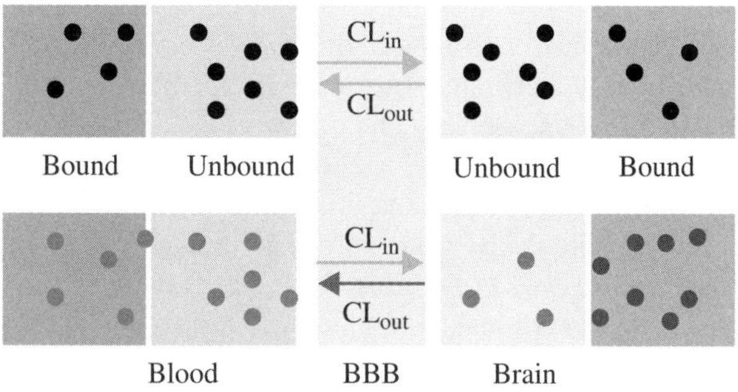

FIGURE 14.1 The unbound concentrations at either side of a membrane strive toward concentration equilibrium and the concentration difference is the driving force for transport. In the upper panel, the transport is based only on passive transport. At equilibrium the unbound concentrations at the blood and brain side are equal and net transport is zero. In the lower panel, at equilibrium, the unbound concentrations are different. This is due to active efflux from the brain (largest arrow) that maintains a lower unbound concentration at the brain side. The bound concentrations in blood and brain will not influence membrane transport of the unbound drug.

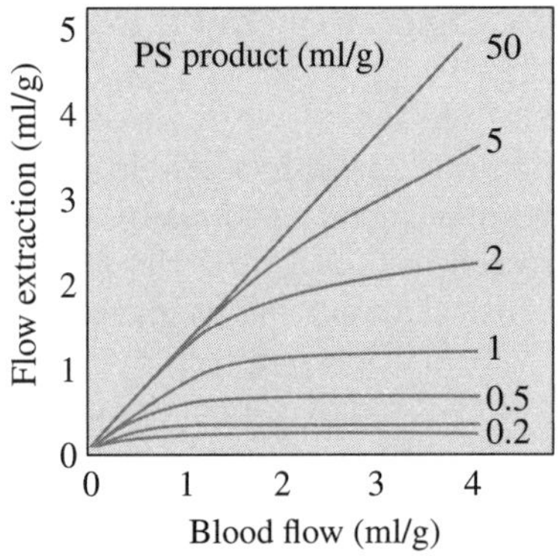

FIGURE 14.2 Transport of low permeability surface area (PS) product value drugs are membrane permeability–limited. Drugs with higher PS products are more sensitive to cerebral blood flow. Transport of high PS product value drugs are cerebral blood flow–limited (Redrawn from Ref. 77).

Drug Transport across the Blood–Brain Barriers

A number of main transport modes across the BBBs can be distinguished [60, 70, 75].

Passive simple diffusion is driven by the concentration gradient, from high to low concentrations. The cerebral capillary endothelial cells (BBB) and choroid plexus epithelial cells (BCSFB) contain tight junctions, which seal cell-to-cell contacts. This limits the diffusion of hydrophilic molecules. For transport across the lipid bilayer of the cell membranes the lipophilicity of the drug is an important factor [78, 79]. Another passive transport mode is the *facilitated diffusion* where drug transport also occurs from high to low concentrations, but is speeded up by a helper molecule (transport proteins imbedded within the cellular membrane, such as glucose transport proteins). *Fluid phase transport* occurs by vesicles, formed out of the membrane enters the cell and fuses with the membrane at the other side of the cell. This transport requires energy and occurs in the direction of the brain. It is nonselective for pinocytosis, little more specific for adsorption-mediated transcytosis (which is based on formation of the vesicle following interaction of a positively charge macromolecule with the BBB membrane at particular sites), and specific for receptor-mediated transcytosis (where binding of a ligand to plasma membrane-spanning receptors leads to the vesicle formation). These routes are useful for larger molecules.

Active influx and efflux transport occurs through transporter proteins for which the molecule should have a relatively high and specific binding site. It requires energy and is able to occur against the concentration gradient. In the last decade it has become clear that drug distribution between blood and tissues is significantly influenced for many drugs [80–87]. Starting with insights on the impact of the P-gp efflux transporter being present at the brain endothelial cells that make up the BBB [88], many studies followed in which information was provided on the expression or functionality of other efflux, as well as influx transporters, including MRPs [89], breast cancer resistance protein (BCRP), and organic anion (transporting) polypeptides (OA(T)Ps) [83]. This has made clear that exchange of drugs between blood and brain is more often complex than simple. With time, the potential contribution of the BCSFB in drug transport into and out of the brain has become clear. The BCSFB is based in the epithelial cells of the choroid plexus. In these cells many transporters are expressed [56, 90, 91]. An overview of the transporters identified at the level of the BBB and the BCSFB has been nicely visualized by Othsuki and Terasaki [83].

Intrabrain Distribution

The $\text{brain}_{\text{ECF}}$ bulk flow and the CSF production and elimination contribute to elimination clearance of drugs from the brain, especially for the drugs with low permeability, and should explicitly be distinguished from actual BBB transport. Intra-extracellular exchange of drugs should also be considered. This may include both passive and active transport. As to our current knowledge, passive membrane transport only occurs for unionized molecules, and the pH gradient from plasma (pH=7.4), the extracellular space (pH=7.3), and the intracellular space (pH=7.0) is of importance for weak bases and acids. In addition, transport from $\text{brain}_{\text{ECF}}$ and brain cells may be governed by active transport processes [92, 93].

Brain Metabolism and Degradation

A final process to be considered is actual intrabrain metabolism or degradation. A number of metabolic enzymes have been identified in the BBB and BCSFB (as described earlier). Furthermore, drugs may end up in lysosomes and degradation of the drug may result.

Integration

An important feature is that the BBB is under continuous physiologic control by astrocytes, pericytes, neurons, and plasma components. All together, these factors determine the delicate homeostasis of the brain environment. This dynamic regulation of the BBB indicates that different situations may result in different BBB functionalities and changes in pathological conditions [37, 38, 94–102]. BBB functionality changes may influence drug transport across the BBB and, therefore, they may have important implications for the target site kinetics. Thus, not only BBB functionality, but also many other brain processes are subjected to changes. All these mentioned processes occur in parallel and are interconnected. Thus, it can be seen that CNS target site drug delivery includes a complex combination of processes. As indicated earlier, oversimplification of these processes has significantly contributed to the very high attrition rate in the development of CNS drugs. Thus, we need to put additional effort into performing the type of investigations that will provide data from which we can learn to have the right CNS drug "*at the right place, at the right time, and at the right concentration*" [45].

PHYSIOLOGICALLY BASED PHARMACOKINETIC MODEL CHARACTERISTICS

Basic Considerations

The PBPK modeling approach does not only use concentration–time profiles of drugs described for virtual compartments that are just based on rate of equilibrium with the plasma compartment [103, 104]. In conjunction with physicochemical properties, these physiological and biochemical parameter values determine the pharmacokinetics of the drug. To that end, the PBPK modeling mathematically describes mass transport of the drug between true body (physiological) compartments, using quantitative parameter values of physiological volumes of tissues, tissue components, tissue blood flow, interstitial bulk flow, as well as expression of transporters [105], expression of enzymes, pH values, etc. All together this results in values for tissue permeability of a drug (PS value, which is an expression for rate of transport) and tissue distribution (Kp value, expressed as a ratio of total drug concentrations in tissue divided by total drug concentrations in blood or plasma at equilibrium, which is an expression for extent) [106]. In principle, this is the strongest approach to derive and to predict the impact of a change in a physiological value on the pharmacokinetics of a drug, but a lot of data and, therewith, time is needed for the development of a PBPK model. Besides, there is room for improvement.

Classical PBPK models have been typically based on total drug concentration measurements in body compartments, drug elimination processes, and passive diffusion between blood and tissues, including the pH partition theory, to derive fully mass-balanced equations. To that end, many data were included on total concentrations measured in multiple parts of the (animal) body [8].

Because the unbound drug concentrations drive membrane transport and interaction with targets and enzymes [1, 107, 108], PBPK models should include unbound drug concentrations. PBPK models of drug distribution into the brain today typically include the unbound fraction, or, better, the unbound drug concentrations in the brain.

In processes, there are two main aspects to consider. One is the rate at which a process occurs (time to steady-state) and the other is the extent to which concentration equilibration occurs (ratio of concentrations) (Fig. 14.3). These parameters have been made explicit by [1], also clarifying the multiple parameters in the BBB transport community that had been used in a confusing manner, enabling us to work now with clear definitions.

The aspect of time to equilibrium between plasma and brain concentrations (rate) has been specifically addressed by Liu et al. [109]. These authors investigated the combination of different values for BBB permeability (permeability–surface area product (PS), determined by *in situ* brain perfusion), plasma protein binding (unbound fraction in plasma), and brain tissue binding (unbound fraction in brain tissue), using equilibrium dialysis of brain homogenates), on the time to reach equilibrium between brain and plasma for seven model compounds with distinctively different physicochemical properties. They proposed the intrinsic brain equilibrium half-life as a parameter, to be equal to $V_b \times \ln 2/(PS \cdot f_{u,brain})$ as a parameter,

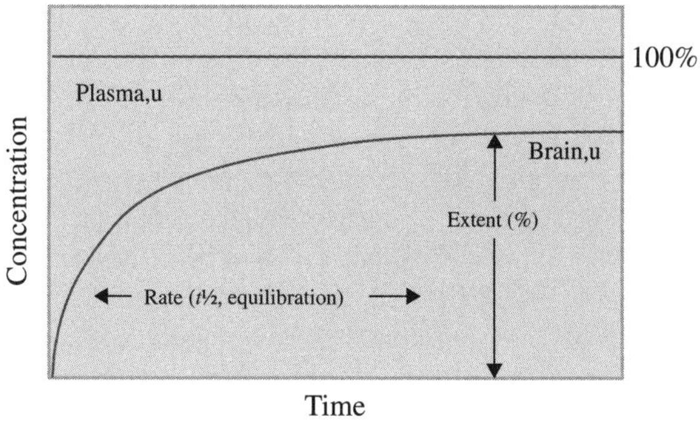

FIGURE 14.3 Relationship between unbound plasma and unbound brain concentrations as a function of time, for the theoretical case that plasma unbound concentration is instantaneously at steady state. Brain unbound concentration will rise until a steady-state condition is reached. The time to steady state is governed by the rate of transport across the BBB (equilibration half-life, *t*½ equilibration), while the extent of brain equilibration is governed by the ratio of AUCbrain,u/AUCplasma,u, or, under steady-state conditions, the ratio of the values of CLbrain,u-plasma,u and CLplasma,u-brain,u [1].

where V_b is the physiological volume of brain, and $f_{u,brain}$ is the unbound fraction in brain. It was demonstrated that a high BBB permeability alone does not necessarily result in a rapid brain equilibration, but actually requires a combination of high BBB permeability and low brain tissue binding. So, if looking for a drug with rapid brain equilibration, drug discovery should look for compounds with high BBB permeability and low nonspecific binding in brain tissue.

The extent of concentration equilibration is often viewed between plasma and brain. Moreover, here it is the unbound concentration that should be taken as the basic input (Figs. 14.1 and 14.3). The extent of concentration equilibration between two compartments is expressed as the ratio of the $AUC_{0-\infty}$ values of both compartments, or, if at steady state, the ratio of the clearance from the one to the other compartment divided by the clearance from the other to the one compartment (CL_{1-2}/CL_{2-1}) can be used. Differences in these clearance values may be due to active transport, metabolism, or other elimination routes, such as $brain_{ECF}$ bulk flow, CSF turnover, or degradation in lysosomes.

Type of Data Needed

Unbound and Total Drug Concentrations The transport of drugs between compartments is governed by the concentration gradient of the unbound drug across the membrane that separates the compartments. This indicates that unbound concentrations should explicitly be taken into account. Total concentrations remain important though, as they indicate how much has been transported from one compartment to the other.

Time Resolution Data Information on time dependency is crucial [1, 37]. It is often thought that only steady-state concentrations are of interest, as chronic dosing should result in steady-state concentrations. Drug discovery and development studies have, therefore, focused on measuring drug concentrations in brain and plasma under (assumed) steady-state conditions. However, for drugs with a desired rapid onset of action, the rate of target site distribution is also relevant.

Physiological Data The brain consists of brain parenchymal cells, surrounded by the $brain_{ECF}$, and CSF spaces. These compartments have their physiological volumes. Exchange of drugs may occur by diffusion, as well as by active transport processes for crossing cell membranes. BrainECF bulk flow and CSF production and elimination are an additional mode of drug transport. In PBPK models, of course, physiological data are key. The combination of the drug properties and physiological processes govern the CNS target site pharmacokinetics. For interspecies extrapolation, for example, rat to human, the physiological values of rats should be replaced by those of humans. For the purpose of PBPK models, values that have been reported on a number of brain physiological compartments in rats and in humans are presented in Table 14.1.

Distinction between Drug- and System-Specific Characteristics Drug properties include lipophilicity (hydrophobicity, hydrogen-bonding potential), size, charge, conformation, polar surface area, and rotatable bonds. Drug properties remain drug properties, but system conditions may be different. Clearly system-specific properties are different between rats and humans, but they are even different between rats with the same gender and genetic background, but living in different situations, for example, on different food. Therefore, all characteristics should be considered. As the rate and extent of individual processes may differ between different conditions, and may be interdependent, it is of importance to investigate condition dependency of these parameters. This can be done by cross-compared designed experiments, measuring multiple parameters (as many as possible) in each setting. A systematic variation of extent and rate of specific factors on the causal chain between dose and drug effect may then be extremely useful to investigate the impact it has on brain distribution in a time-dependent manner.

Integration—Mastermind Approach The wealth of data that results from cross-compared multilevel studies should be subjected to advanced mathematical modeling to derive a useful set of parameters that provide insight into the interrelationship of the contributing factors between dose and CNS target site pharmacokinetics and maybe even (biomarkers) of the effects. This is called the Mastermind Approach [45], which will be useful in explicit distinction of the role of drug-specific properties and the characteristics of the system and "stored" in physiologically based PK(PD) models.

TOWARDS DEVELOPMENT OF PREDICTIVE PBPK BRAIN DISTRIBUTION MODELS

With the use of PBPK modeling approaches, the aim is to develop models with higher predictive power of CNS target site distribution in humans. Information on species- and/or condition-dependent differences in abundance levels and activities of the different active transport proteins and drug-metabolizing enzymes at the BBB and BCSFB, as well as at the liver and kidney, under healthy or diseased conditions, is essential for extrapolation purposes. With the use of advanced PBPK modeling, the contributions of individual mechanisms in animals can be revealed, to serve as links to the human situation. The physiological values of the brain volumes, surfaces, and flows have been summarized in Table 14.1. Thus, PBPK models integrate drug-specific and system-specific physiological parameters that vary between species, subjects, or within subjects with different age and/or disease states [110–113]. There are a number of studies on brain distribution that do not really include physiological parameters but, nevertheless, provide considerable insight into multiple brain compartmental distribution processes [114–124]. There are only a few studies that have truly used the physiological modeling approach. These will be discussed later.

Atomoxetine and Duloxetine

Kielbasa and Stratford [125] explored the potential utility of PBPK modeling using rat brain microdialysis data to predict human brain$_{ECF}$ PK of atomoxetine and duloxetine. Using an intravenous maintenance dose of 1.25 mg/kg/h for atomoxetine [126] and 4.2 mg/kg/h for duloxetine [125], plasma and brain microdialysates were obtained from the rat as a function of time, and were used in combination with end-of-experiment concentrations in total brain and CSF. These data were included in a model based on four brain compartments (plasma (not PBPK), brain$_{ECF}$, brain cells (BC), and CSF) and clearances between these compartments (Fig. 14.4).

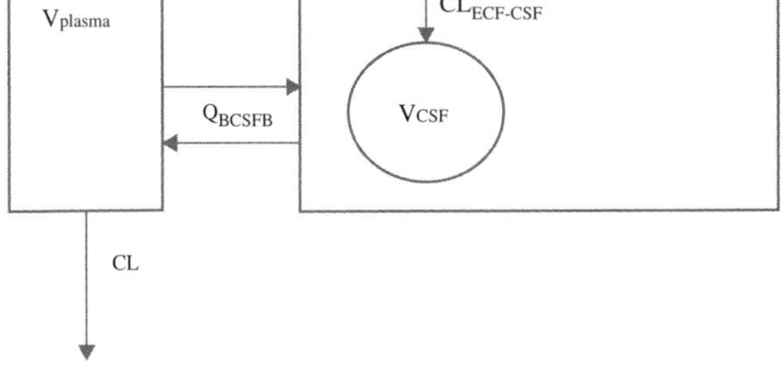

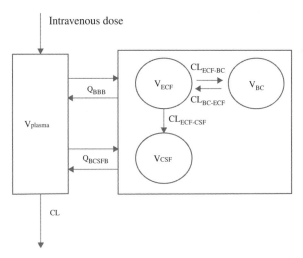

FIGURE 14.4 PBPK model for brain distribution by Kielbasa and Stratford [125], in which physiological volumes of the different brain compartments were used. Exchange of drug between the different compartments was expressed as bidirectional clearances between plasma and brain$_{ECF}$ (Q_{BBB}), and plasma and CSF (Q_{BCSFB}), clearances between brain$_{ECF}$ and brain cells (CL_{ECF-BC}) and vice versa (CL_{BC-ECF}), and a unidirectional clearance for flow-mediated transport from brain$_{ECF}$ to CSF ($CL_{ECF-CSF}$).

TABLE 14.2 Pharmacokinetic parameters of atomoxetine and duloxetine used in translational models to predict human brain$_{ECF}$ pharmacokinetics

	Atomoxetine		Duloxetine	
Parameter	Clinical	Scaled from rat	Clinical	Scaled from rat
K (h^{-1})	3.1		0.168	
CL/F (l/h)	20.6		45.1	
V/F (l)	121		814	
V_{CSF} (l)	0.16		0.16	
V_{ECF} (l)	0.31		0.31	
BBC (l)	1.04		1.04	
$C_{PL\text{-}CSF}$ (l/h)	0.00825		NE	
$CL_{CSF\text{-}PL}$ (l/h)	0.0205		NE	
Q_{BCSFB} (l/h)	NE	0.015		0.009
Q_{BBB} (l/h)		0.181		0.026
$CL_{ECF\text{-}CSF}$ (l/h)		0.021		0.004
$CL_{ECF\text{-}BC}$ (l/h)		0.355		0.279
$CL_{BC\text{-}ECF}$ (l/h)		0.153		0.412
ω-CL/F (%)	90.8		58.9	
ω-V/F (%)	65.6		96.6	

Data from Kielbasa and Stratford [125].
ω, intersubject variability; NE, not estimated.

Nonlinear mixed-effects modeling was performed using NONMEM. The rat model was converted into a human model by using human physiological values for the brain compartments and rat PK parameters were scaled to human values by allometric principles according to the equation $P_h = P_r \times (Wt_h/Wt_r)^{0.75}$, in which P_h is the scaled human parameter, P_r is the model-predicted parameter in the rat, Wt_h is the average human brain weight (1.35 kg), and Wt_r is the average rat brain weight (0.0015 kg). On the basis of an adult brain volume of 1.35 l, the estimated human VECF was 0.31 l and VBC was 10.4 l. Human V_{CSF} was fixed at 0.16 l.

Predictions of human brain$_{ECF}$, brain cells, and CSF PK were made and the authors concluded that these results may support the clinical development of CNS-mediated drug candidates by enhancing the ability to predict pharmacologically relevant doses in humans (Table 14.2).

Acetaminophen, Quinidine, and Methotrexate—Toward a Generic Brain Distribution Model

Westerhout et al. [44, 127, 128] worked on PBPK brain models where the data were produced in-house on (unbound) concentrations in plasma, brain$_{ECF}$, brain cells, CSF_{LV} (CSF in lateral ventricle), and CSF_{CM} (CSF in cisterna magna) obtained as far as possible in parallel from single animals (Fig. 14.5) following short infusion of model drugs with distinct physicochemical properties (Table 14.3).

In the different studies, one of the model drugs, acetaminophen, quinidine, or methotrexate, was administered intravenously to the rat as a short infusion, with or without concomitant administration of blockers of active transport at the BBBs.

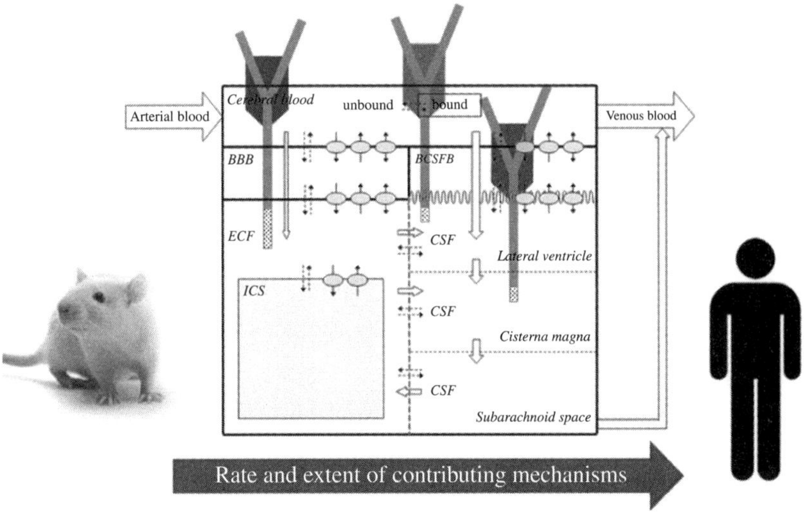

FIGURE 14.5 Locations of microdialysis measurements of unbound concentrations in the rat brain. In figure on left side = brain$_{ECF}$, in the middle = CSF in lateral ventricle, and on the right side = CSF in cisterna magna. Serial samples were also obtained from blood, from which unbound plasma concentrations were derived. The resulting data on unbound concentrations in plasma, brain$_{ECF}$, CSF$_{LV}$, and CSF$_{CM}$, in combination with end-of-experiment total brain concentrations were used in brain distribution model development.

The data were used to define the time-dependent parameters for passive and active exchange between plasma, brain$_{ECF}$, and CSF concentration. Physiological parameters for volumes, surfaces, fluid flows, and active transport processes were obtained from the literature. This was all performed using nonlinear mixed-effects modeling using the NONMEM software package. With the use of the same structural model for all three paradigm compounds (Fig. 14.6), it is important to note that this will allow investigation, in a mechanistic manner, of the impact of drug characteristics on brain kinetics in the different physiological compartments. The three compounds are only the start of this approach; the brain distribution of many more drugs with different physicochemical properties should be measured to develop a generic brain distribution model that only needs the input from *in vitro* assessed properties of a (new) drug to predict its brain distribution kinetics. For the drugs used so far (acetaminophen, quinidine, and methotrexate) the brain compartment that can be entered after passing the BBB and the brain$_{ECF}$ ("deep" brain compartment) remained difficult to characterize due to the limited number of data points (1 per animal; end-of-experiment).

Acetaminophen—Validation by Human Data

Given that CSF concentrations are considered to be the best available surrogate for brain$_{ECF}$ concentrations in humans [73, 129–131], we focused on predicting human brain$_{ECF}$ concentrations. Acetaminophen is only subjected to passive transport

TABLE 14.3 Physicochemical properties of three different compounds

Compound	MW	PSA	log P	log D (7.4)	pK_a1 (acid)	pK_a2 (acid)	pK_a1 (base)	pK_a1 (base)	Ionized at physiological pH (%)	Substrate for
Acetaminophen	151	49.3	0.25	0.23				10.2	0 (neutral)	—
Quinidine	324	45.6	2.29	1.4	4.2		8		99.8 (positive)	P-gp
Methotrexate	454	210	−1.55	−5.49	3.4	5.4			99.9 (negative)	MRP's

log P, measure of lipophilicity determined as log of partition of unionized compound in octanol versus water; log D (7.4), measure of lipophilicity at physiological pH, determined as log of distribution of the compound in octanol versus buffer pH = 7.4; MW, molecular weight; PSA, polar surface area.

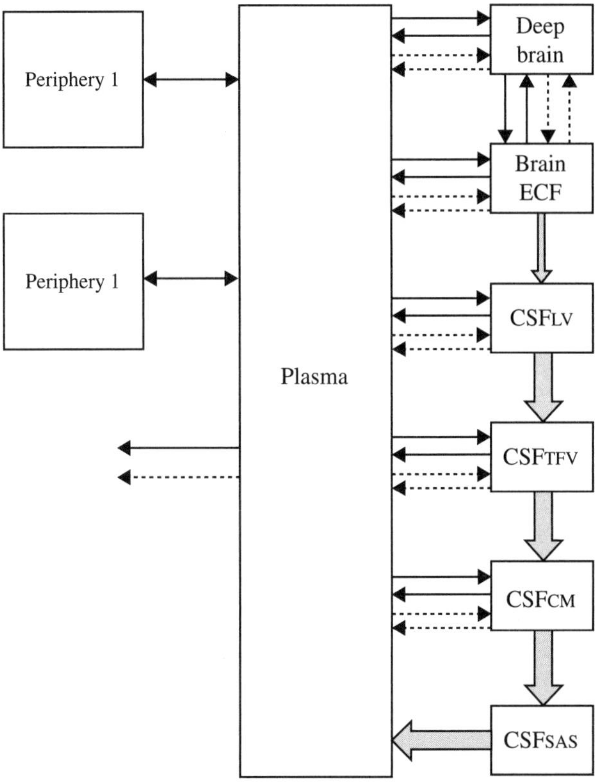

FIGURE 14.6 Structural model that is used for drugs with different physicochemical properties. The different brain compartments have physiological volumes and flows that have been reported in the literature (Table 14.1). The lines represent passive transport, and the dashed lines represent the active transport processes. The big arrows indicate fluid flow. CM, cisterna magna; LV, lateral ventricle; SAS, subarachnoidal space; TFV, third and fourth ventricle.

(diffusion and flow). Rat data and physiological parameters were used for the preclinical PBPK model [44]. By changing the different values of the physiological parameters of the rat to their corresponding human values, and by fitting the human plasma data to our model while extrapolating the plasma–brain exchange in a systems-based manner, we were able to adequately predict human lumbar CSF concentrations as observed by Bannwarth et al. [132] (Fig. 14.7). For acetaminophen in humans, it was predicted that $brain_{ECF}$ concentrations are on average approximately twofold higher than unbound plasma concentrations, whereas the $brain_{ECF}$-to-CSF (from the subarachnoid space) concentration relationship is highly dependent on the time after dose. The fact that the data, as predicted for human CSF lumbar concentrations, are in line with observed lumbar concentrations by Bannwarth et al. [132] provides confidence in the usefulness of the model to predict human $brain_{ECF}$ concentrations as a function of time.

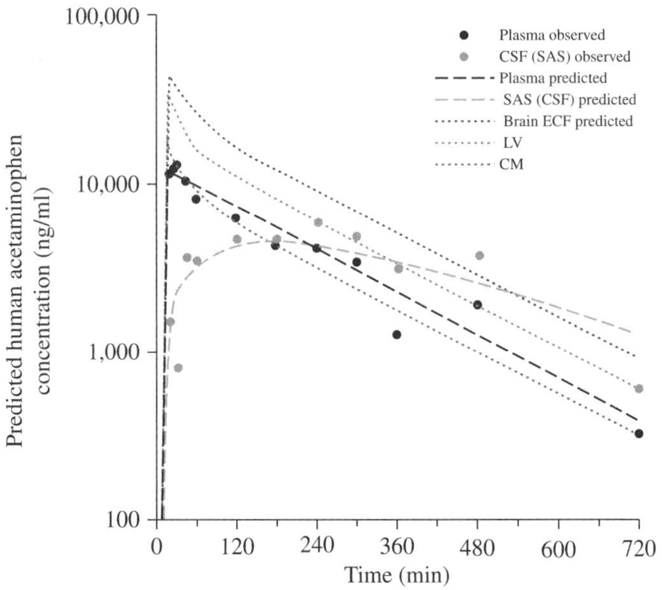

FIGURE 14.7 Prediction of human concentration–time profiles in plasma (unbound) and different brain compartments on the basis of the preclinically derived translation brain distribution model for acetaminophen [44]. The predicted profiles in plasma (black bold dashed line) and CSF SAS (light grey bold dashed line) nicely correspond to the observed concentrations, as measured by Bannwart et al. [132] (Republished with permission of Springer from Ref. 44, with permission conveyed through © Clearance Center, Inc.).

Quinidine—The Impact of P-gp Functionality

For quinidine, the inclusion of the influence of P-gp-mediated transport at the blood–brain barriers was taken into account. Data were obtained for quinidine doses of 10 and 20 mg/kg, without or with coadministration of the P-gp blocker tariquidar (Fig. 14.8). It was clear that P-gp functionality is an important factor in the relationship between CSF and $\text{brain}_{\text{ECF}}$ exposure, given the fact that the relative order of distribution of quinidine over the brain compartments changes with blocking P-gp-mediated transport by coadministration of tariquidar [127]. Substantial changes in P-gp functionality might, therefore, underlie mispredictions of human brainECF concentrations on the basis of CSF measurements.

Methotrexate—Influence of Disease

Methotrexate data were obtained in rats at doses of 40 and 80 mg/kg, without or with coadministration of probenecid as a blocker of MRPs and OATPs as methotrexate is a substrate for these transporters [128]. For methotrexate, the relative order of the concentrations in the different brain compartments upon probenecid cotreatment did not change; all concentrations only increased, in part, also, because of larger exposure by increased plasma concentrations (Fig. 14.9).

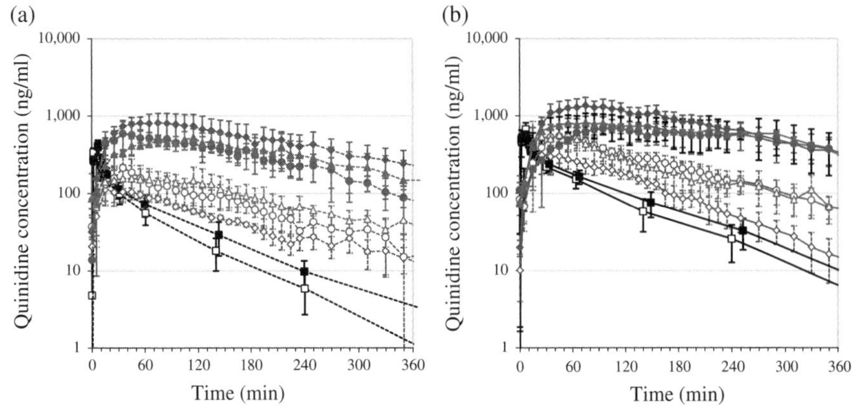

FIGURE 14.8 Average (geometric mean ± S.E.M.) unbound quinidine concentration–time profiles in plasma (squares), $\text{brain}_{\text{ECF}}$ (diamonds), CSFLV (circles), and CSFCM (triangles) following (a) 10 mg/kg quinidine with vehicle (open symbols) or 10 mg/kg quinidine with coadministration of 15 mg/kg tariquidar (closed symbols) and (b) 20 mg/kg quinidine with vehicle (open symbols) or 20 mg/kg quinidine with coadministration of 15 mg/kg tariquidar (closed symbols).

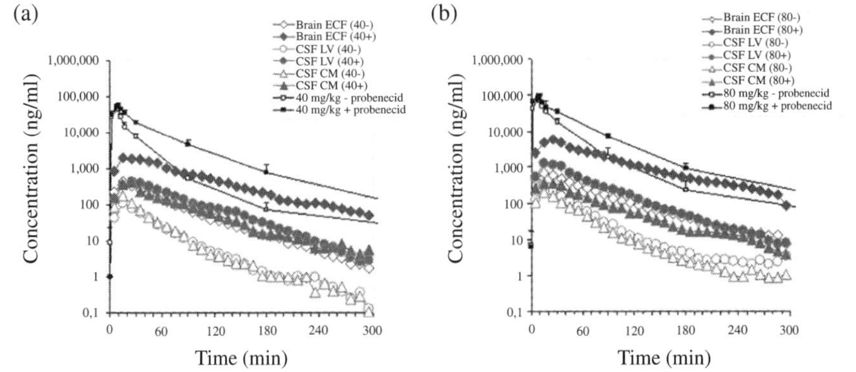

FIGURE 14.9 Average (geometric mean ± S.E.M.) unbound methotrexate concentration–time profiles in plasma (squares), $\text{brain}_{\text{ECF}}$ (diamonds), CSFLV (circles), and CSFCM (triangles) following (a) 40 mg/kg methotrexate with vehicle (open symbols) or 40 mg/kg methotrexate with coadministration of probenecid (closed symbols) and (b) 80 mg/kg methotrexate with vehicle (open symbols) or 80 mg/kg methotrexate with coadministration of probenecid (closed symbols).

For methotrexate, there is quite some clinical data available, including $\text{brain}_{\text{ECF}}$ concentrations in humans [133]. However, all published human data (children and adults) have been obtained in different disease states. It is, therefore, not logical to expect proper prediction of diseased human concentrations in different brain compartments on the basis of a preclinical model developed on data obtained in healthy rats. Westerhout et al. [128] applied the PBPK model on literature data for methotrexate brain distribution, first to predict data obtained in other healthy rats (plasma and brain$_{ECF}$ data), and then to investigate the impact of disease status on the PK of methotrexate. By using the same PK parameter values that were estimated based on our data, we were able to predict the methotrexate plasma and brain$_{ECF}$ concentrations in other healthy rats reasonably well. For earlier reported brain$_{ECF}$ concentrations of methotrexate in brain tumor–bearing rats [134] the predictions by the preclinical brain distribution model were found to be significantly lower, indicating increased distribution of methotrexate at the brain tumor site. The next step was to use the PBPK model to predict plasma and CSF concentrations in healthy dogs. When taking into account that the hepatic elimination of methotrexate in dogs is only a fraction of the renal clearance [135], whereas in rats the hepatic elimination of methotrexate is estimated to be over fivefold higher than the renal clearance, the predictions of plasma and CSF concentrations were reasonable.

In humans, methotrexate undergoes extensive enterohepatic circulation, effectively reducing the hepatic elimination rate to the same level as the renal elimination rate [136]. With this information incorporated into the model, the prediction of human unbound methotrexate plasma concentrations was reasonable. In the human disease conditions, the brain$_{ECF}$ and CSF concentrations were significantly higher than predicted for healthy human conditions. Comparison of simulations of the model with the actual measured data in the patients indicates a possible decreased active efflux from the brain$_{ECF}$ as well as a lower CSF flow to cause these higher brain$_{ECF}$ and CSF concentrations in the patients. Actually, a reduced CSF flow as "suspect" contributor to the higher methotrexate brain PK measured in the patients might indeed be a real possibility as it is in line with the reported observation that several adult patients had an obstruction of normal CSF flow [137]. So, interestingly, apart from blood–brain exchange, the CSF flow seems to play an important role in the brain$_{ECF}$–CSF relationship [138].

CONCLUSIONS

To be able to predict CNS drug effects in humans on the basis of preclinical data, it is essential to (i) study the underlying processes and mechanisms that govern the ultimate concentration–effect relationship as a function of time, and (ii) include information on unbound drug concentrations. The value of intracerebral microdialysis in this prediction is clearly exemplified here.

FUTURE PERSPECTIVES

For the ultimate development of a generic brain distribution model, in which prediction of brain distribution in humans can be made on the basis of physicochemical properties of drugs measured *in vitro*, further detailed time-course data of a number of other drugs will be needed with other distinct combinations of physicochemical properties.

Further development of the preclinical CNS PBPK to a CNS PBPKPD model lies in the improvement of the quality of the CNS effect data. Often, the focus has been on a single biomarker to reflect the CNS drug effect. However, given the complexity of brain diseases, it can be seen that the search for a single biomarker to explain the disease relative to the healthy condition and/or changes in the disease condition by (drug) treatment will never lead to success. Actually, we do not deal with "the" effect, but a composite of effects. The search should therefore be for "fingerprints" of multiple biomarkers, in a time-dependent manner, for investigations on the "effect spectrum." With metabolomics as an emerging scientific tool, many more compounds in brain fluids and in plasma can be measured in parallel, in a quantitative and time-dependent manner. Furthermore, the emphasis should lie on measures that can be obtained both preclinically and clinically to enhance translational insights and, therewith, predictive power of preclinically obtained information [139].

Knowledge of human brain target site concentrations will then be useful for further development of PBPKPD models in health and disease conditions [43, 46], to further pave the way to predict the right drug at the *right location, right time, and right concentrations* [45].

REFERENCES

[1] Hammarlund-Udenaes M, Fridén M, Syvänen S, Gupta A. On the rate and extent of drug delivery to the brain. Pharm Res 2008;25:1737–1750.

[2] Abbott NJ, Dolman DE, Patabendige AK. Assays to predict drug permeation across the blood-brain barrier, and distribution to brain. Curr Drug Metab 2008 Nov;9(9):901–910.

[3] Fridén M, Gupta A, Antonsson M, Bredberg U, Hammarlund-Udenaes M. In vitro methods for estimating unbound drug concentrations in the brain interstitial and intracellular fluids. Drug Metab Dispos 2007;35:1711–1719.

[4] Loryan I, Fridén M, Hammarlund-Udenaes M. The brain slice method for studying drug distribution in the CNS. Fluids Barriers CNS 2013 Jan 21;10(1):6.

[5] Lee HB, Blaufox MD. Blood volume in the rat. J Nucl Med 1985;26:72–76.

[6] Frank H, Gray SJ. The determination of plasma volume in man with radioactive chromic chloride. J Clin Invest 1953;32:991–999.

[7] Dobbing J, Sands J. Quantitative growth and development of human brain. Arch Dis Childhood 1973;48:757.

[8] Harashima H, Sawada Y, Sugiyama Y, Iga T, Hanano M. Analysis of nonlinear tissue distribution of quinidine in rats by physiologically based pharmacokinetics. J Pharmacokin Biopharm 1985;13:425–550.

[9] Sandor P, Cox-van Put J, de Jong W, de Wied D. Continuous measurement of cerebral blood volume in rats with the photoelectric technique: effect of morphine and naloxone. Life Sci 1986 Nov 3;39(18):1657–1665.

[10] Forster A, Juge O, Morel D. Effects of midazolam on cerebral blood flow in human volunteers. Anesthesiology 1982 Jun;56(6):453–455.

[11] Ito H, Inoue K, Goto R, Kinomura S, Taki Y, Okada K, Sato K, Sato T, Kanno I, Fukuda H. Database of normal human cerebral blood flow measured by SPECT: I. Comparison

between I-123-IMP, Tc-99m-HMPAO, and Tc-99m-ECD as referred with O-15 labeled water PET and voxel-based morphometry. Ann Nucl Med 2006;20:131–138.

[12] Cserr HF. Potassium exchange between cerebrospinal fluid, plasma, and brain. Am J Physiol 1965;209:1219–1226.

[13] Cserr HF, Cooper DN, Suri PK, Patlak CS. Efflux of radiolabeled polyethylene glycols and albumin from rat brain. Am J Physiol 1981;240:F319–F328.

[14] Dekaban AS, Sadowsky D. Changes in brain weights during the span of human life: relation of brain weights to body heights and body weights. Ann Neurol 1978;4:345–356.

[15] Thorne RG, Hrabětová S, Nicholson C. Diffusion of epidermal growth factor in rat brain extracellular space measured by integrative optical imaging. J Neurophysiol 2004;92:3471–3481.

[16] Abbott NJ. Evidence of bulk flow of brain interstitial fluid: significance for physiology and pathology. Neurochem Int 2004;45:545–552.

[17] Abbott NJ. Dynamics of CNS barriers: evolution, differentiation, and modulation. Cell Mol Neurobiol 2005 Feb;25(1):5–23.

[18] Kimelberg HK. Water homeostasis in the brain: basic concepts. Neuroscience 2004;129: 851–760.

[19] Nilsson C, Stahlberg F, Thomsen C, Henriksen O, Herning M, Owman C. Circadian variation in human cerebrospinal fluid production measured by magnetic resonance imaging. Am J Physiol 1992;262:R20–R24.

[20] Johanson CE, Duncan JA 3rd, Klinge PM, Brinker T, Stopa EG, Silverberg GD. Multiplicity of cerebrospinal fluid functions: new challenges in health and disease. Cerebrospinal Fluid Res 2008;14:5–10.

[21] Bass NH, Lundborg P. Postnatal development of bulk flow in the cerebrospinal fluid system of the albino rat: clearance of carboxyl-(14C)inulin after intrathecal infusion. Brain Res 1973;30:52:323–332.

[22] Oldendorf WH. Cerebrospinal fluid formation and circulation. Prog Nucl Med 1972;1: 336–358.

[23] Condon P, Patterson J, Wyper D, Hadley D, Grant R, Teasdale G. Use of magnetic resonance imaging to measure intracranial cerebrospinal fluid volume. Lancet 1986;327: 1355–1357.

[24] Kohn MI, Tanna NK, Herman GT, Resnick SM, Mozley PD, Gur RE, Alavi A, Zimmerman RA, Gur RC. Analysis of brain and cerebrospinal fluid volumes with MR imaging. Part I. Methods, reliability, and validation. Radiology 1991;178:115–122.

[25] Dickey CC, Shenton ME, Hirayasu Y, Fischer I, Voglmaier MM, Niznikiewicz MA, Seidman LJ, Fraone S, McCarley RW. Large CSF volume not attributable to ventricular volume in schizotypical personality disorder. Am J Psychiatry 2000;157:48–54.

[26] Adam R, Greenberg JO. The mega cisterna magna. J Neurosurg 1978;48:190–192.

[27] Robertson EG. Developmental defects of the cisterna magna and dura mater. J Neurol Neurosurg Psychiatry 1949;12:39–51.

[28] Levinger IM. The cerebral ventricles of the rat. J Anat 1971;108:447–451.

[29] Pardridge WM. Drug transport in brain via the cerebrospinal fluid. Fluids Barriers CNS 2011;8:7.

[30] Bradbury MW. The Concept of a Blood–Brain Barrier. Chichester: Wiley; 1979.

[31] Keep RF, Jones HC. A morphometric study on the development of the lateral ventricle choroid plexus, choroid plexus capillaries and ventricular ependyma in the rat. Brain Res Dev Brain Res 1990 Oct 1;56(1):47–53.

[32] Crone C, Christensen O. Electrical resistance of a capillary endothelium. J Gen Physiol 1981;77:349–371.

[33] Dohrmann GJ. The choroid plexus: a historical review. Brain Res 1970;18:197–218.

[34] Letrent SP, Pollack GM, Brouwer KR, Brouwer KL. Effects of a potent and specific P-glycoprotein inhibitor on the blood-brain barrier distribution and antinociceptive effect of morphine in the rat. Drug Metab Dispos 1999;27:827–834.

[35] Karssen AM, Meijer OC, van der Sandt ICJ, Lucassen P J, de Lange ECM, de Boer AG, de Kloet ER. Multidrug resistance P-glycoprotein hampers the access of cortisol but not of corticosterone to mouse and human brain. Endocrinology 2001;142:2686–2694.

[36] Kooij G, van Horssen J, de Lange EC, Reijerkerk A, van der Pol SM, van Het Hof B, Drexhage J, Vennegoor A, Killestein J, Scheffer G, Oerlemans R, Scheper R, van der Valk P, Dijkstra CD, de Vries HE. T lymphocytes impair P-glycoprotein function during neuroinflammation. J Autoimmun.2010; 34(4):416–425.

[37] De Lange ECM, Ravenstijn PGM, Groenendaal D, van Steeg TS. Toward the prediction of CNS drug effect profiles in physiological and pathological conditions using microdialysis and mechanism-based pharmacokinetic-pharmacodynamic modeling. AAPS J 2005;7:article 54.

[38] Mulder M, Blokland A, van den Berg DJ, Schulten H, Bakker AH, Terwel D, Honig W, de Kloet ER, Havekes LM, Steinbusch HW, de Lange EC. Apolipoprotein E protects against neuropathology induced by a high-fat diet and maintains the integrity of the blood-brain barrier during aging. Lab Invest 2001;81(7):953–960.

[39] Danhof M, de Jongh J, de Lange ECM, Della Pasqua OE, Ploeger BA, Voskuyl RA. Mechanism-based pharmacokinetic-pharmacodynamic modeling: biophase distribution, receptor theory, and dynamical systems analysis. Annu Rev Pharmacol Toxicol 2007;47: 357–400.

[40] Ravenstijn PG, Merlini M, Hameetman M, Murray TK, Ward MA, Lewis H, Ball G, Mottart C, de Ville de Goyet C, Lemarchand T, van Belle K, O'Neill MJ, Danhof M, De Lange EC. The exploration of rotenone as a toxin for inducing Parkinson's disease in rats, for application in BBB transport and PKPD experiments. J Pharmacol Toxicol Methods 2007;57(2):114–130.

[41] Ravenstijn PGM, Drenth H, Baatje MS, O'Neill MJ, Danhof M, de Lange ECM. Evaluation of BBB transport and CNS drug metabolism in diseased and control brain after intravenous L-DOPA in a unilateral rat model of Parkinson's disease. Fluids Barriers CNS 2012;9:4.

[42] Syvänen S, Lindhe Ö, Palner M, Kornum BR, Rahman O, Långström B, Knudsen GM, Hammarlund-Udenaes M. Species differences in blood-brain barrier transport of three positron emission tomography radioligands with emphasis on P-glycoprotein transport. Drug Metab Dispos 2009;37:635–643.

[43] Westerhout J, Danhof M, de Lange EC. Preclinical prediction of human brain target site concentrations: Considerations in extrapolating to the clinical setting. J Pharm Sci 2011; 100(9):3577–3593.

[44] Westerhout J, Ploeger B, Smeets J, Danhof M, de Lange ECM. Physiologically based pharmacokinetic modeling to investigate regional brain distribution kinetics in rats. AAPS J 2012;14(3):543–553.

[45] De Lange ECM. The mastermind approach to CNS drug therapy: translational prediction of human brain distribution, target site kinetics, and therapeutic effects. Fluids Barriers CNS 2013b;10:12.

[46] Stevens J, Ploeger B, Hammarlund-Udenaes M, Osswald G, vd Graaf PH, Danhof M, de Lange ECM. Mechanism-based PK–PD model for the prolactin biological system response following an acute dopamine inhibition challenge: quantitative extrapolation to humans. J Pharmacokin Pharmacodyn 2012;39(5):463–477.

[47] Segal MB. Extracellular and cerebrospinal fluids. J Inher Metab Dispos 1993;16:617–638.

[48] Cserr HF. Convection of brain interstitial fluid. In: Shapiro K, Marmarou A, Portnoy H, eds. Hydrocephalus. New York: Raven Press; 1984:59–68.

[49] Abbott NJ, Rönnbäck L, Hansson E. Astrocyte-endothelial interactions at the blood-brain barrier. Nat Rev Neurosci 2006;7:41–53.

[50] Bernacki J, Dobrowolska A, Nierwińska K, Małecki, A. Physiology and pharmacological role of the blood-brain barrier. Pharmacol Rep 2008;60:600–622.

[51] Del Bigio MR. The ependyma: a protective barrier between brain and cerebrospinal fluid. *Glia* 1995;14:1–13.

[52] Crone C, Olesen SP. Electrical resistance of a brain microvascular endothelium. Brain Res 1982;241:49–55.

[53] Smith QR, Rapoport SI. Cerebrovascular permeability coefficients to sodium, potassium, and chloride. J Neurochem 1986;46(6):1732–1742.

[54] Pardridge WM. Peptide Drug Delivery to the Brain. New York: Raven Press; 1991.

[55] Duvernoy H, Delon S, Vannson JL. The vascularization of the human cerebellar cortex. Brain Res Bull 1983;11:419–480.

[56] De Lange EC. Potential role of ABC transporters as a detoxification system at the blood-cerebrospinal fluid-barrier. Adv Drug Del Rev 2004;56(12);1793–1809.

[57] Ghersi-Egea JF, Strazielle N, Murat A, Jouvet A, Buénerd A, Belin MF. Brain protection at the blood-cerebrospinal fluid interface involves a glutathione-dependent metabolic barrier mechanism. J Cereb Blood Flow Metab 2006;26(9):1165–1175.

[58] Ghersi-Egea JF, Mönkkönen KS, Schmitt C, Honnorat J, Fèvre-Montange M, Strazielle N. Blood-brain interfaces and cerebral drug bioavailability. Rev Neurol (Paris) 2009; 165(12):1029–1038.

[59] Redzic Z. Molecular biology of the blood-brain and the blood-cerebrospinal fluid barriers: similarities and differences. Fluids Barriers CNS 2011 Jan 18;8(1):3.

[60] Davson H, Segal MB. Physiology of the CSF and Blood-Brain Barriers. Boca Raton: CRC Press; 1996.

[61] Davson H. The cerebrospinal fluid. Handbook Neurochem 1969;2:23–48.

[62] Keep RF, Jones HC. Cortical microvessels during brain development: a morphometric study in the rat. Microvasc Res 1990a;40:412–426.

[63] Keep RF, Jones HC. A morphometric study on the development of the lateral ventricle choroid plexus, choroid plexus capillaries and ventricular ependymal in the rat. Dev Brain Res 1990b;56:47–53.

[64] Johanson CE, Stopa EG, McMillan PN. The blood-cerebrospinal fluid barrier: structure and functional significance. Methods Mol Biol 2011;686:101–131.

[65] Betz AL, Firtha JA, Goldstein GW. Polarity of the blood-brain barrier: distribution of enzymes between the luminal and antiluminal membranes of brain capillary endothelial cells. Brain Res 1980;192(1):17–28.

[66] Tayarani I, Cloez I, Clément M, Bourre JM. Antioxidant enzymes and related trace elements in aging brain capillaries and choroid plexus. J Neurochem 1989;53:817–824.

[67] Fukushima H, Fujimoto M, Ide M. Quantitative detection of blood-brain barrier-associated enzymes in cultured endothelial cells of procine brain microvessels. J In Vitro Cell Dev Biol 1990;26(6):612–620.

[68] Volk B, Hettmansperger U, Papp TH, Amelizad Z, Oesch F, Knoth R. Mapping of phenytoin-inducible cytochrome P450 immunoreactivity in the mouse central nervous system. Neuroscience 1991;42:215–235.

[69] Minn A, Ghersi-Egea JF, Perrin R, Leininger B, Siest G. Drug metabolizing enzymes in the brain and cerebral microvessels. Brain Res 1991;16:65–82.

[70] Strazielle N, Ghersi-Egea JF. Physiology of blood-brain interfaces in relation to brain disposition of small compounds and macromolecules. Mol Pharm 2013 May 6;10(5): 1473–1491.

[71] Banks WA, Kastin AJ. Passage of peptides across the blood-brain barrier: pathophysiological perspectives. Life Sci 1996;59(23):1923–1943.

[72] Abbott NJ, Patabendige AA, Dolman DE, Yusof SR, Begley DJ. Structure and function of the blood-brain barrier. Neurobiol Dis 2010 Jan;37(1):13–25.

[73] De Lange ECM. Utility of CSF in translational neuroscience. J Pharmacokinet Pharmacodyn 2013a;40(3):315–326.

[74] Schmidt S, Gonzalez D, Derendorf H. Significance of protein binding in pharmacokinetics and pharmacodynamics. J Pharm Sci 2010 Mar;99(3):1107–1122.

[75] Fenstermacher JD, Patlak CS, Blasberg RG. Transport of material between brain extracellular fluid, brain cells and blood. Fed Proc 1974;33:2070–2074.

[76] Spector R, Spector AZ, Snodgrass SR. Model for transport in the central nervous system. Am J Physiol 1977;1:R73–R79.

[77] Fenstermacher JD, Wei L, Acuff V, Lin SZ, Chen JL, Bereczki D, Otsuka T, Nakata H, Tajima A, Hans FJ, Ghersi-Egea JF, Finnegan W, Richardson G, Haspel H, Patlak C. The dependency of influx across the blood-brain barrier on blood flow and the apparent flow-independence of glucose influx during stress. In: Greenwood, Begley, Segal, eds. New Concepts of a Blood-Brain Barrier; Plenum, NY, 1995:89–101.

[78] Oldendorf WH. Lipid solubility and drug penetration of the blood-brain barrier. Proc Exp Biol Med 1974;14:813–816.

[79] Levin VA. Relationship of octanol/water partition coefficient and molecular weight to rat brain capillary permeability. J Med Chem 1980;23:682–684.

[80] De Boer AG, van der Sandt I, Gaillard PJ. The role of drug transporters at the blood-brain barrier. Annu Rev Pharmacol Toxicol 2003;43:629–656.

[81] Kusuhara H, Sugiyama Y. Efflux transport systems for organic anions and cations at the blood-CSF barrier. Adv Drug Del Rev 2004;56:1741–1763.

[82] Kusuhara H, Sugiyama Y. Active efflux across the blood-brain barrier: role of the solute carrier family. NeuroRx 2005;2:73–85.

[83] Ohtsuki S, Terasaki T. Contribution of carrier-mediated transport systems to the blood–brain barrier as a supporting and protecting interface for the brain; Importance for CNS drug discovery and development. Pharm Res 2007;24(9):1745–1758.

[84] Uchida Y, Ohtsuki S, Katsukura Y, Ikeda C, Suzuki T, Kamiie J, Terasaki T. Quantitative targeted absolute proteomics of human blood-brain barrier transporters and receptors. J Neurochem 2011;117(2):333–345.

[85] Uchida Y, Ohtsuki S, Kamiie J, Terasaki T. Blood-brain barrier (BBB) pharmacoproteomics (PPx): reconstruction of in vivo brain distribution of 11 P-glycoprotein substrates

based on the BBB transporter protein concentration, in vitro intrinsic transport activity, and unbound fraction in plasma and brain in mice. J Pharmacol Exp Ther 2012;339(2): 579–588.

[86] Hillgren KM, Keppler D, Zur AA, Giacomini KM, Stieger B, Cass CE, Zhang L. Emerging transporters of clinical importance: an update from the international transporter consortium. Clin Pharmacol Ther 2013;94:52–63.

[87] Zamek-Gliszczynski MJ, Lee CA, Poirier A, Bentz J, Chu X, Ellens H, Ishikawa T, Jamei M, Kalvass JC, Nagar S, Pang KS, Korzekwa K, Swaan PW, Taub ME, Zhao P, Galetin A. International transporter consortium. ITC recommendations for transporter kinetic parameter estimation and translational modeling of transport-mediated PK and DDIs in humans. Clin Pharmacol Ther 2013 Jul;94(1):64–79.

[88] Schinkel AH, Smit JJM, van Tellingen O, Beijnen JH, Wagenaar E, van Deemter L, Mol CAAM, van der Valk MA, Robanus-Maandag EC, te Riele HPJ, Berns AJM, Borst P. Disruption of the Mouse mdr1a P-glycoprotein gene leads to a deficiency in the blood-brain barrier and to increased sensitivity to drugs. Cell 1994;77:491–502.

[89] Borst P, Evers R, Kool M, Wijnholds. A family of drug transporters: the multidrug resistance-associated proteins. J Natl Cancer Inst. 2000 Aug 16;92(16):1295–302.

[90] Nishino J, Suzuki H, Sugiyama D, Kitazawa T, Ito K, Hanano M, Sugiyama Y. Transepithelial transport of organic anions across the choroid plexus: possible involvement of organic anion transporter and multidrug resistance-associated protein. J Pharmacol Exp Ther 1999;290(1):289–294.

[91] Wijnholds J, de Lange ECM, Scheffer GL, van den Berg D-J, Mol CAAM, van der Valk M, Schinkel AH, Scheper RJ, Breimer DD, Borst P. Multidrug resistance protein 1 protects the choroid plexus epithelium and contributes to the blood-cerebrospinal fluid barrier. J Clin Invest 2000;105:279–285.

[92] Lee G, Dallas S, Hong M, Bendayan R. Drug transporters in the central nervous system: brain barriers and brain parenchyma considerations. Pharmacol Rev 2001;53:569–596.

[93] Scism JL, Powers KM, Artru AA, Lewis L, Shen DD. Probenecid-inhibitable efflux transport of valproic acid in the brain parenchymal cells of rabbits: a microdialysis study. Brain Res 2000 Nov 24;884(1–2):77–86.

[94] Zlokovic BV, Skundric DS, Segal MB, Colover J, Jankov RM, Pejnovic N, Lackovic V, Mackic J, Lipovac MN, Davson H. Blood-brain barrier permeability changes during acute allergic encephalomyelitis induced in the guinea pig. Metab Brain Dispos 1989; 4(1):33–40.

[95] Oztaş B, Küçük M. Influence of acute arterial hypertension on blood-brain barrier permeability in streptozocin-induced diabetic rats. Neurosci Lett 1995;188(1):53–56.

[96] Oztaş B, Akgül S, Arslan FB. Influence of surgical pain stress on the blood-brain barrier permeability in rats. Life Sci 2004;74(16):1973–1979.

[97] Oztas B, Akgul S, Seker FB. Gender difference in the influence of antioxidants on the blood-brain barrier permeability during pentylenetetrazol-induced seizures in hyperthermic rat pups. Biol Trace Elem Res 2007;118(1):77–83.

[98] Ederoth P, Tunblad K, Bouw R, Lundberg CJ, Ungerstedt U, Nordström CH, Hammarlund-Udenaes M. Blood-brain barrier transport of morphine in patients with severe brain trauma. Br J Clin Pharmacol 2004;57(4):427–435.

[99] Langford D, Grigorian A, Hurford R, Adame A, Ellis RJ, Hansen L, Masliah E. Altered P-gp expression in AIDS patients with HIV encephalitis. J Neuropathol Exp Neurol 2004;63:1038–1047.

[100] Bell RD, Zlokovic BV. Neurovascular mechanisms and blood-brain barrier disorder in Alzheimer's disease. Acta Neuropathol 2009;118(1):103–113.

[101] Bengtsson J, Ederoth P, Ley D, Hansson S, Amer-Wåhlin I, Hellström-Westas L, Marsál K, Nordström CH, Hammarlund-Udenaes M. The influence of age on the distribution of morphine and morphine-3-glucuronide across the blood-brain barrier in sheep. Br J Pharmacol 2009;157(6):1085–1096.

[102] Zlokovic BV. Neurodegeneration and the neurovascular unit. Nat Med 2010;16(12): 1370–1371.

[103] Dedrick RL, Bisschoff KB. Species similarities in pharmacokinetics. Fed Proc 1980; 39:54–59.

[104] Davies B, Morris T. Physiological parameters in laboratory animals and humans. Pharm Res 1993;10:1093–1095.

[105] Shawahna R, Uchida Y, Declèves X, Ohtsuki S, Yousif S, Dauchy S, Jacob A, Chassoux F, Daumas-Duport C, Couraud PO, Terasaki T, Scherrmann JM. Transcriptomic and quantitative proteomic analysis of transporters and drug metabolizing enzymes in freshly isolated human brain microvessels. Mol Pharm 2011 Aug 1;8(4):1332–1341.

[106] Rowland M, Peck C, Tucker G. Physiologically-based pharmacokinetics in drug development and regulatory science. Annu Rev Pharmacol Toxicol 2011;51:45–73.

[107] Watson J, Wright S, Lucas A, Clarke KL, Viggers J, Cheetham S, Jeffrey P, Porter R, Read KD. Receptor occupancy and brain free fraction. Drug Metab Dispos 2009;37: 753–760.

[108] Hammarlund-Udenaes M. Active-site concentrations of chemicals—are they a better predictor of effect than plasma/organ/tissue concentrations? Basic Clin Pharmacol Toxicol 2009;106:215–220.

[109] Liu X, Smith BJ, Chen C, Callegari E, Becker SL, Chen X, Cianfrogna J, Doran AC, Doran SD, Gibbs JPN, Hosea J, Liu, Nelson FR, Szewc MA, Van Deusen J. Use of a physiologically based pharmacokinetic model to study the time to reach brain equilibrium: an experimental analysis of the role of blood–brain barrier permeability, plasma protein binding, and brain tissue binding. J Pharmacol Exp Ther 2005;313:1254–1262.

[110] Collins JM, Dedrick LD. Distributed model for drug delivery to CSF and brain tissue. J Am Physiol 1983;14:R303–R310.

[111] Colburn WA. Physiologic pharmacokinetic modeling. J Clin Pharmacol 1988;28: 673–677.

[112] Espié P, Tytgat D, Sargentini-Maier M-L, Poggesi I, Watelet J-P. Physiologically based pharmacokinetics (PBPK). Drug Metab Rev 2009;41:391–407.

[113] Ings RMJ. Interspecies scaling and comparisons in drug development and toxicogenetics. Xenobiotica 1990;20:1201–1231.

[114] Mayer S, Maickel RP, Brodie BB. Kinetics of penetration of drugs and other foreign compounds into cerebrospinal fluid and brain. J Pharmacol Exp Ther 1959;127: 205–211.

[115] Ooie T, Suzuki H, Terasaki T and Sugiyama Y. Kinetic evidence for active efflux transport across the blood-brain barrier of quinolone antibiotics. J Pharmacol Exp Ther 1997;283:293–304.

[116] Wang YF, Welty DF. The simultaneous estimation of the influx and efflux blood-brain barrier permeabilities of gabapentin using a microdialysis-pharmacokinetic approach. Pharm Res 1996;13:398–403.

[117] Proescholdt MG, Hutto B, Brady LS, Herkenham M. Studies of cerebrospinal fluid flow and penetration into brain following lateral ventricle and cisterna magna injections of the tracer [^{14}C]inulin in rat. Neurosci 2000;95:577–592.

[118] Lindberger M, Tomson T, Wallstedt L, Stahle L. Distribution of valproate to subdural cerebrospinal fluid, subcutaneous extracellular fluid, and plasma in humans" a microdialysis study. Epilepsia 2001;42:256–261.

[119] Kalvass JC, Maurer TS. Influence of nonspecific brain and plasma binding of CNS exposure: implications for rational drug discovery. Biopharm Drug Dispos 2002;23:327–338.

[120] Kamiie J, Ohtsuki S, Iwase R, Ohmine K, Katsukura Y, Yanai K, Sekine Y, Uchida Y, Ito S, Terasaki T. Quantitative atlas of membrane transporter proteins: development and application of a highly sensitive simultaneous LC/MS/MS method combined with novel in-silico peptide selection criteria. Pharm Res 2008;25:1469–1483.

[121] Maurer TS, DeBartolo DB, Tess DA, Scott D. Relationship between exposure and nonspecific binding of thirty-three central nervous system drugs in mice. Drug Metab Dispos 2005;33:175–181.

[122] Vladić A, Klarica M, Bulat M. Dynamics of distribution of 3H-inulin between the cerebrospinal fluid compartments. Brain Res 2009;1248:127–135.

[123] Liu X, Van Natta K, Yeo H, Vilenski O, Weller PE, Worboys PD, Monshouwer M. Unbound drug concentration in brain homogenate and cerebral spinal fluid at steady state as a surrogate for unbound concentration in brain interstitial fluid. Drug Metab Dispos 2009;37:787–793.

[124] Ball K, Bouzom F, Scherrmann JM, Walther B, Declèves X. Physiologically based pharmacokinetic modelling of drug penetration across the blood-brain barrier-towards a mechanistic IVIVE-based approach. AAPS J 2013;15(4):913–932.

[125] Kielbasa W, Stratford RE Jr. Exploratory translational modeling approach in drug development to predict human brain pharmacokinetics and pharmacologically relevant clinical doses. Drug Metab Dispos 2012 May;40(5):877–883.

[126] Kielbasa W, Kalvass JC, Stratford R. Microdialysis evaluation of atomoxetine brain penetration and central nervous system pharmacokinetics in rats. Drug Metab Dispos 2009;37(1):137–142.

[127] Westerhout J, Smeets J, Danhof M, de Lange ECM. The impact of P-gp functionality on non-steady state relationships between CSF and brain extracellular fluid. J Pharmacokinet Pharmacodyn 2013;40:327–342.

[128] Westerhout J, van den Berg DJ, Hartman R, Danhof M, de Lange ECM. Prediction of methotrexate CNS distribution in different species and the influence of disease conditions. Eur J Pharmacol Sci 2014;57:11–24.

[129] De Lange EC, Danhof M. Considerations in the use of cerebrospinal fluid pharmacokinetics to predict brain target concentrations in the clinical setting: implications of the barriers between blood and brain. Clin Pharmacokinet 2002;41:691–703.

[130] Lin JH. CSF as a surrogate for assessing CNS exposure: an industrial perspective. Curr Drug Metab 2008;9:46–59.

[131] Shen DD, Artru AA, Adkison KK. Principles and applicability of CSF sampling for the assessment of CNS drug delivery and pharmacodynamics. Adv Drug Deliv Rev 2004;56:1825–1857.

[132] Bannwarth B, Netter P, Lapicque F, Gillet P, Péré P, Boccard E, Royer RJ, Gaucher A. Plasma and cerebrospinal fluid concentrations of paracetamol after a single intravenous dose of propacetamol. Br J Clin Pharmacol 1992;34:79–81.

[133] Blakeley JO, Olson J, Grossman SA, He X, Weingart J, Supko JG. Effect of blood brain barrier permeability in recurrent high grade gliomas on the intratumoral pharmacokinetics of methotrexate: a microdialysis study. J Neurooncol 2009;91:51–58.

[134] De Lange ECM, Hesselink MB, Danhof M, De Boer AG, Breimer DD. The use of intracerebral microdialysis to determine changes in blood-brain barrier transport characteristics. Pharm Res 1995;12:129–133.

[135] Henderson ES, Adamson RH, Denham C, Oliverio VT. The metabolic fate of tritiated methotrexate I. absorption, excretion, and distribution in mice, rats, dogs and monkeys. Cancer Res 1965;25:1008–1017.

[136] Hendel J, Brodthagen H. Entero-hepatic cycling of methotrexate estimated by use of the D-isomer as a reference marker. Eur J Clin Pharmacol 1984;26:103–107.

[137] Glantz MJ, Cole BF, Recht L, Akerley W, Milles P, Saris S, Hochberg F, Calabresi P, Egorin MJ. High-dose intravenous methotrexate for patients with nonleukemic leptomeningeal cancer: is intrathecal chemotherapy necessary? J Clin Oncol 1998;16: 1561–1567.

[138] Reiber H. Flow rate of cerebrospinal fluid (CSF)—a concept common to normal blood-CSF barrier function and to dysfunction in neurological diseases. J Neurol Sci 1994 Apr;122(2):189–203.

[139] De Lange ECM. Pharmacometrics in psychiatric diseases. In: Schmidt S, Derendorf H, eds. Applied Pharmacometrics. Springer, New York; 2014.

15

PK/PD MODELING OF CNS DRUG CANDIDATES

JOHAN GABRIELSSON,[1] STEPHAN HJORTH,[2] AND LAMBERTUS A. PELETIER[3]

[1]*Department of Biomedical Sciences and Veterinary Public Health, Division of Pharmacology and Toxicology, Swedish University of Agricultural Sciences, Uppsala, Sweden*
[2]*Department of Bioscience, AstraZeneca R&D, Pepparedsleden, Mölndal, Sweden*
[3]*Mathematical Institute, Leiden University, Leiden, The Netherlands*

INTRODUCTION

The increased emphasis on mechanistic and quantitative biomarker approaches in central nervous system (CNS) pharmacology presents new challenges to the design and analysis of kinetic/dynamic models used in drug discovery. The central objective in pharmacology is to gain an understanding of physiological processes and find drugs to avoid undesirable effects. The first step in achieving this objective is to obtain precise measurements of the processes involved. For some responses this may be done in a direct manner, for example, blood pressure or heart rate. For others, especially CNS-elicited drug effects, it may be necessary to resort to indirect methods, and identify appropriate biomarkers (Fig. 15.1). Having obtained data about the process implicated, the questions that arise are how to interpret alterations in the biomarker responses and how to use these in improving our understanding of the underlying mechanisms. The ultimate goal is to identify the determinants of onset, intensity, and duration of the pharmacological response. We have collected four case studies, each

Blood–Brain Barrier in Drug Discovery: Optimizing Brain Exposure of CNS Drugs and Minimizing Brain Side Effects for Peripheral Drugs, First Edition. Edited by Li Di and Edward H. Kerns.

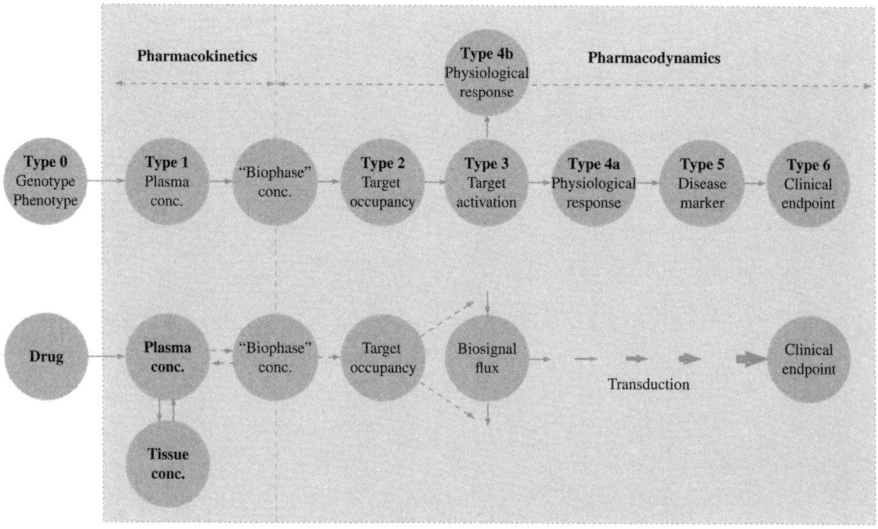

FIGURE 15.1 Schematic diagram of the components of pharmacodynamic models and the comparison with classification of biomarkers. Adapted from Jusko and Ebling [1] and Danhof et al. [2].

of which focuses on a different issue and points to challenges that have not yet been fully explored and would benefit from mathematical analysis.

What are the typical trademarks of a CNS response/biomarker? The CNS is a well-perfused but still protected space where distributional processes such as target-specific partitioning, diffusion, and sometimes active (carrier-mediated, saturable) transport may play a role as determinants for delayed onset of response. Moreover, particulars of the CNS physiological milieu and operational properties (e.g., endogenous ligands and tone, adaptational mechanisms, regional differences and interplay), as well as the potential influence of drug-metabolizing capacity in the brain [3] are factors that need to be considered when dealing with CNS-mediated responses. Several of these complexities will be touched upon in the examples discussed here, and suggestions on how to approach them will be given.

All datasets and figures used in this chapter were originally analyzed by Gabrielsson and Weiner [4], Gabrielsson and Peletier [5], and Gabrielsson and Hjorth [6]. The dataset of Case Study 1 was digitized from Lundström et al. [7].

In this chapter we focus on four challenges:

1. How to proceed when exposure data are in instantaneous equilibrium with the CNS response but the relationship displays multiphasic behavior.
2. How the advent of turnover models in mechanism-based pharmacodynamic modeling resulted in a new approach to study the interaction of drug and target.
3. How to tackle functional tolerance by extended drug exposure.
4. How to deal with functionally adaptational mechanisms offsetting drug response.

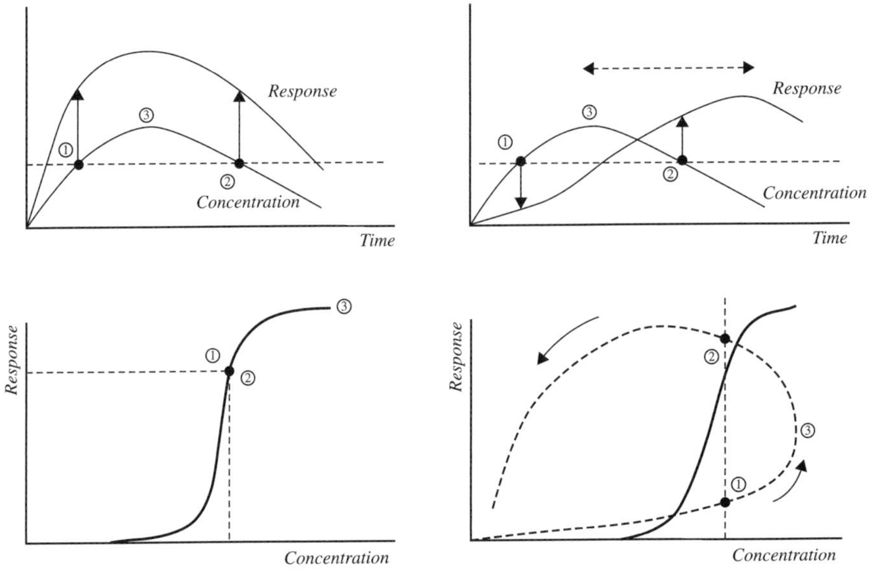

FIGURE 15.2 Schematic illustration of concentration and response versus time for instantaneous equilibrium (upper left panel) and delayed equilibrium (upper right panel). The lower plots illustrate the response versus concentration relationship for the instantaneous (left-hand panel) and delayed (right-hand panel) equilibrium. In the latter case, the same concentration (exposure) results in two different responses. The profile shown in the lower right panel is an example of a hysteresis loop. Adapted from Gabrielsson and Hjorth [6].

We illustrate each of these challenges with case studies. We would also like to refer to the recently published *Themed Issue on Translational Modeling in Neuroscience* (*Journal of Pharmacokinetics and Pharmacodynamics* [8]). This issue contains several timely examples of modeling CNS biomarkers such as "Disease progression and neuroscience," "Modeling and simulation of placebo effect: application to drug development in schizophrenia," "Exposure–response modeling of antidepressant treatments: the confounding role of placebo effect," and "Translational PK-PD modeling in pain" to mention just a few.

When instantaneous equilibrium occurs between the drug plasma concentration and the pharmacological response, we have a situation as shown in Figure 15.2 (left). Thus, there is a direct correspondence between plasma concentration and response amplitude, whether concentrations are ascending or descending. However, if there is a delay between the plasma concentration-time course and the pharmacological effect-time course, a situation arises that is illustrated by the right-hand plots in Figure 15.2. The same plasma concentration occurring at two different time points (rise and fall phase) now coincides with two different response amplitudes. In general, the response is lagging behind the plasma concentration both on the upswing and washout. There are temporal differences between the plasma concentration and response—equilibrium does not occur instantaneously. Exposure/response divergences may lead to the erroneous conclusion that the drug is not efficacious, or that there is no relationship between the concentration of a drug and the response. When it comes to CNS-elicited responses, temporal departures may be introduced by simple distribution delays for a drug to access its action biophase, including, for example, active metabolite formation and carrier-mediated limitations. However, more complex events, linked to drug mechanisms and/or triggering of secondary processes and factors controlling onset and offset of the response, may also be involved. As an example, we may consider gamma secretase inhibitors, intended to lessen Alzheimer plaque formation in the brain via a decrease in the formation of amyloid peptide fragments. Even if the drug inhibition of relevant enzymatic activity is quite rapid, the formation and deposit of plaques occurs over a weeks-to-months time frame, and it is changes in this readout that represent the actual antidisease progression response of interest (further discussed later in this chapter).

PHARMACODYNAMIC RESPONSE MODELS

Instantaneous Equilibrium—*The Model Gallery*

Typical pharmacodynamic models used to capture the equilibrium concentration–response relationships are given by Equations 15.1–15.4. Figure 15.3 shows the pharmacodynamic model gallery used to capture instantaneous or equilibrium concentration–response relationships. The simplest functional form of these models is the linear concentration–response mode (Eq. 15.1):

$$E = E_0 + S \cdot C \tag{15.1}$$

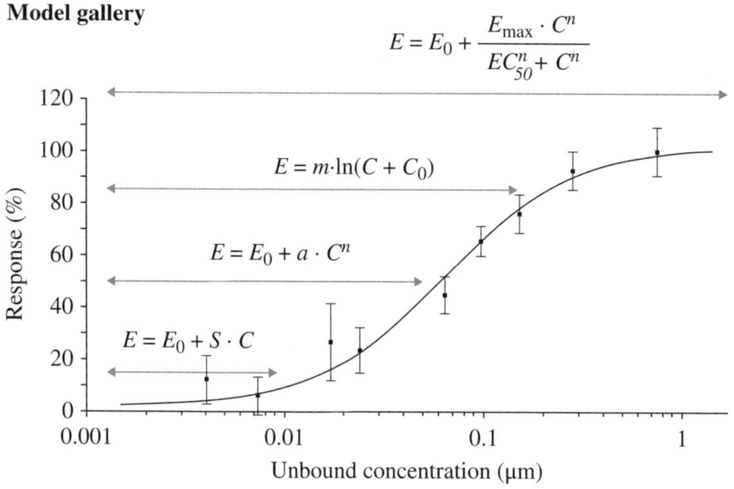

FIGURE 15.3 Pharmacodynamic model gallery of instantaneous or equilibrium concentration–response relationships commonly used [6]. The models from bottom to top represent the linear, exponential, log-linear, and the Hill model, respectively. The horizontal double arrows indicate the relative ranges over which they capture concentration–response data. Note that the linear model is a subset of the exponential model when the exponent n is set to unity.

where E, E_0, S, and C are the response, baseline response, slope parameter, and the plasma concentration, respectively. This model can then be expanded to capture a wider concentration–response range (display more flexibility) by either the exponential model (Eq. 15.2)

$$E = E_0 + a \cdot C^n \tag{15.2}$$

where a and n are the pharmacodynamic regression parameters to be used, or the log-concentration model (Eq. 15.3). The log-linear approach has traditionally been used for exploratory purposes to facilitate graphical representation over a large range of concentrations:

$$E = m \cdot \ln(C + C_0) \tag{15.3}$$

The m parameter is the slope of the log-linear portion of the concentration–response data when C is much greater than C_0, and C_0 is a hypothetical parameter allowing the model to also capture baseline response E_0. Note that C_0 has nothing to do with the plasma concentration at time zero, but is rather added to increase the numerical robustness of the model. The baseline response of the log-linear model is then given by Equation 15.4:

$$E_0 = m \cdot \ln(C_0) \tag{15.4}$$

Equations 15.1–15.3 should be used carefully because they do not display a maximum plateau at higher concentrations. To remedy this, the *Hill model* is more common, particularly since it covers a wider concentration range and inherently mimics the other models during limited concentration intervals as indicated in Figure 15.3. The *Hill model* contains, in addition to the baseline effect E_0, the efficacy parameter E_{max}, the potency EC_{50} and the Hill exponent n:

$$E = E_0 \pm \frac{E_{max} \cdot C^n}{EC_{50}^n + C^n} \tag{15.5}$$

For further information about the behavior of the Hill model see Gabrielsson and Weiner [4]. The thermodynamic measure of the tendency of the ligand and receptor to stay together or to have affinity for one another is also called the potency of the ligand, commonly expressed as the dissociation constant K_d (defined by the ratio of k_{off}/k_{on} rate constants) in receptor-binding studies. With regard to functional responses, it is desirable to correlate the pharmacological response to plasma or tissue measurements and then the potency parameter is denoted EC_{50}. When EC_{50} is measured, it is commonly confounded by target expression (receptor number, transduction efficiency, spare receptors, etc.) and, therefore, often falls to the left of the true affinity (potency) K_d on the concentration axis. In other situations EC_{50} may also shift to the right as a result of endogenous substrate competition. So, the EC_{50} parameter corresponds to the plasma concentration at half-maximal drug-induced response, which may or may not correspond to 50% target occupancy. In fact, due to very efficient receptor stimulus–response coupling (transduction), it is not uncommon among high-efficacy receptor agonists to produce a maximal effect even if only a small fraction of the target receptor is occupied (say, 10–15%). The Hill n-exponent is a positive term, which does not necessarily have a strict biological translation when analyzing functional data.

In Case Study 1 we will utilize an extension of the Hill function to also capture multiple receptor systems by means of a composite E_{max}/I_{max} model (Eq. 15.6).

Composite E_{max} Model—Case Study 1

The situation of combined drug action arises either when two or more active compounds exert an effect on a certain biological system or when a single drug acts simultaneously at two or more receptors. Combination approaches have been introduced or proposed in several treatment frameworks in the hope of enhancing efficacy through offsetting counter-regulatory mechanisms triggered by single drug action. While such approaches may well turn out more effective, as compared to monotherapy from a PD perspective, they also introduce pharmacokinetic challenges and the need to control for disposition and PD effects of more than one drug simultaneously (recent examples include, e.g., fixed-dose combination (FDC) topiramate/phentermine in obesity and various drug combination regimens in major depression; cf. Refs. 9 and 10, respectively).

Case Study 1 on composite E_{max} models involves studies of the racemate of a centrally acting dopamine (DA) ligand, 3-(3-hydroxyphenyl)-*N*-*n*-propylpiperidine HCl (3-PPP), in which both its (+) and (−) enantiomers display significant effects on DA neurotransmission: (+)3-PPP acting as a "classical" pre- and postsynaptic DA agonist and its (−) counterpart acting mainly as an agonist and antagonist at these sites, respectively [11]. The racemate, therefore, in essence, represents a kind of "combination" treatment in a single drug entity. The example is based on data from Lundström et al. [7] and analysis of Gabrielsson and Weiner [4] (Fig. 15.4). After subcutaneous administration of racemic 3-PPP to rats, plasma and brain levels were monitored in relation to the amount of spontaneous locomotor activity—a well-known and often-used rodent behavioral readout of central DA activity. Plotting brain concentrations of (−)3-PPP against locomotor activity resulted in a good fit to a declining two-phase curve, in all probability reflecting preferential actions at pre- and postsynaptic DA receptors, respectively. These two underlying mechanisms could also be identified from the biphasic effects produced by (+)3-PPP on locomotor activity: suppression followed by stimulation. Together, these observations provide further support for the contention that at low doses both enantiomers have sedative actions due to stimulation of inhibitory DA autoreceptors (cf., e.g., Ref. [13]). With increasing doses, however, a postsynaptic DA receptor blockade will predominate for a partial agonist like (−)3-PPP, producing suppression of locomotion, whereas the full agonist (+)3-PPP will produce behavioral activation due to mounting stimulation of postsynaptic DA D_2 receptors [7, 11]. Lundström et al. [7] therefore proposed the biphasic composite E_{max} function (Eq. 15.6):

$$E = E_0 - \frac{I_{max1} \cdot C^{n_1}}{IC_{51}^{n_1} + C^{n_1}} - \frac{I_{max2} \cdot C^{n_2}}{IC_{52}^{n_2} + C^{n_2}} \tag{15.6}$$

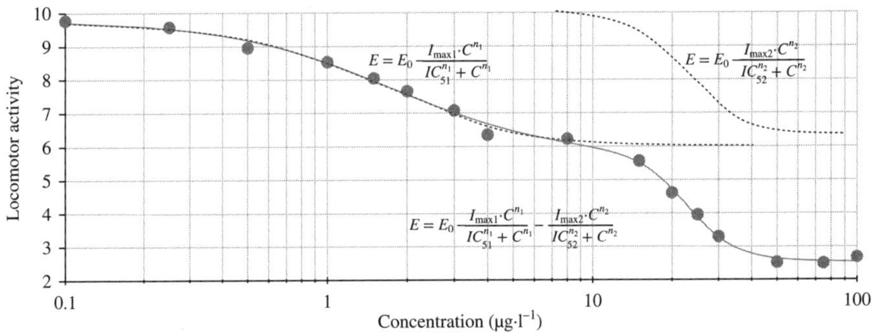

FIGURE 15.4 The relationship between observed (solid circles) and model-predicted (solid line, Equations 15.6 or 15.7) response and the steady-state concentration of an experimental agent, on a Cartesian scale (*Y*-axis) and a logarithmic scale (*X*-axis). Note the three functions of the first and second receptor interaction (upper left and right equations) and the combined one (bottom equation). Data digitized from Lundström et al. [7] and analyzed by Gabrielsson and Weiner [4]. For further information on modeling mixture dynamics see Gabrielsson and Peletier [12].

TABLE 15.1 Final parameter estimates and their CV%

Equation 15.6—additive terms			Equation 15.7—multiplicative terms		
Parameter	Estimate	CV%	*Parameter*	Estimate	CV%
I_{max1}	4.13	9	I_{max1}	0.42	9
IC_{51} (μg l⁻¹)	1.76	16	IC_{51}	1.75	16
n_1	1.37	14	n_1	1.38	14
I_{max2}	3.15	12	I_{max2}	0.55	6
IC_{52} (μg l⁻¹)	22.9	5	IC_{52}	23.1	5
n_2	4.72	21	n_2	4.67	20

E_0 was set to 9.8 in both models.

and showed that it offered a good fit to the data for the parameter values listed in Table 15.1. E_0, I_{max1}, I_{max2}, IC_{51}, IC_{52}, n_1, and n_2 denote the baseline response, maximum inhibitory efficacy of receptor population 1 and 2, their corresponding potency values, and the Hill exponents, respectively.

An alternative model based on a multiplicative combination of inhibitory I_{max} functions (Eq. 15.7) proposed by Gabrielsson and Peletier [12] on *mixture dynamics* can also be shown to capture the concentration–response data in Figure 15.4:

$$E = E_0 \cdot \left(1 - \frac{I_{max1} \cdot C^{n_1}}{IC_{51}^{n_1} + C^{n_1}}\right) \cdot \left(1 - \frac{I_{max2} \cdot C^{n_2}}{IC_{52}^{n_2} + C^{n_2}}\right) \tag{15.7}$$

The corresponding parameter values are also listed in Table 15.1.

Typical examples where composite E_{max}/I_{max} models have been successfully applied are apomorphine, clonidine, haloperidol, sulpiride, and (+/–)3-PPP [4, 7, 12, 14–20]. For review of bell- or U-shaped concentration–response relationships, particularly in the toxicological data context, see Calabrese and Baldwin [21].

This case study offers an example of how to model drug interaction with multiple receptor sites, which appears as a composite concentration–response relationship. Provided the functional potency values are well separated, the model parameters will have low correlation and high precision. The real value of this analysis is, of course, not the modeling by itself, but what the results are intended for, that is, design of new studies, translation across species, etc.

TIME DELAYS BETWEEN PLASMA CONCENTRATION AND CNS RESPONSE

This section deals with temporal differences between concentration-time and response-time data and the concept of *hysteresis* (time delay between plasma concentration and response). The half-life of biological response, which will impact the extent of hysteresis, will be examined from a pharmacodynamic perspective. The pharmacodynamic model gallery relating to time delays is exemplified by three case studies, which include the *receptor binding on/off model* (Case Study 2), the *ordinary turnover model* (Case Study 3), and the *feedback turnover model* (Case Study 4).

So, the question arises: What available modeling approaches do we have at hand to tackle temporal divergences? Figure 15.5 shows the three major classes of models that typically depict time delays between plasma concentration and the pharmacological CNS response. We intentionally skip further elaborating the *effect-compartment model*, also known as the *distributional model* (A), as a first-choice tool of CNS drugs, because in our own experience it confounds the interpretation of model parameters from a mechanistic point of view. A case study is therefore not given for the distributional model in the present review.

Temporal divergences like the aforementioned (Fig. 15.2) are deceptive as they may mislead the investigator to conclude that there is no relationship between the concentration of a drug in blood or plasma and the time course of response. Actually, there is a relationship between plasma exposure and pharmacological response, but it is a long-term one that will be more obvious at equilibrium. However, as will be further discussed, the pharmacological effects of many drugs lag behind the drug concentration in plasma (Fig. 15.2, right).

An example with substantial temporal differences between plasma concentration and response is the impact of cannabinoid receptor type 1 (CB1r) antagonists on body weight and composition in diet-induced obese mice (Fig. 15.6). CB1r antagonists decrease appetite and increase the energy turnover and therefore in turn reduce body weight and fat stores. The apparent steady state is established in plasma after 2–3 days of repeated subcutaneous dosing. However, the time to pharmacodynamic steady state, with regard to body weight loss, is about 3–4 weeks due to the long

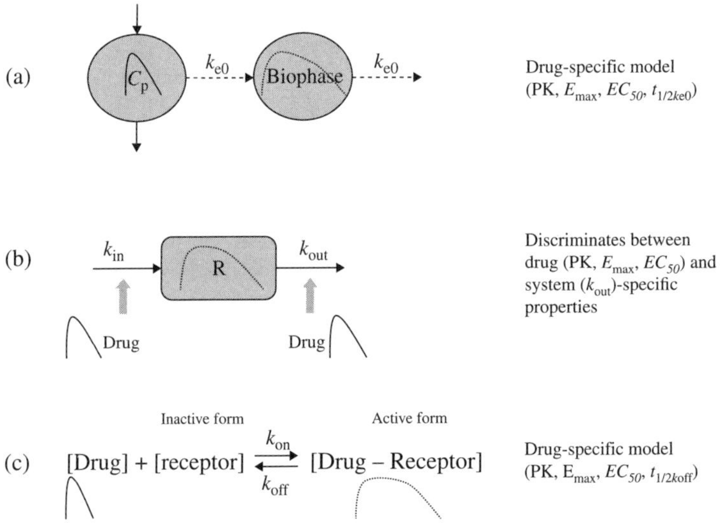

FIGURE 15.5 Schematic illustration of the (a) distributional (effect-compartment or link) model, (b) turnover model, and (c) receptor on/off binding model—models that are commonly used to capture time delays between plasma concentration and the pharmacological response. The shapes of the concentration-time profiles and response-time profiles illustrate the temporal differences schematically. The inactive and active forms in (c) denote the two stages of receptor function. Adapted from Gabrielsson and Hjorth [6]. Feedback as a time delay will be further discussed under Section "Time Delays—Feedback Turnover Models—Case Study 4 on SSRI".

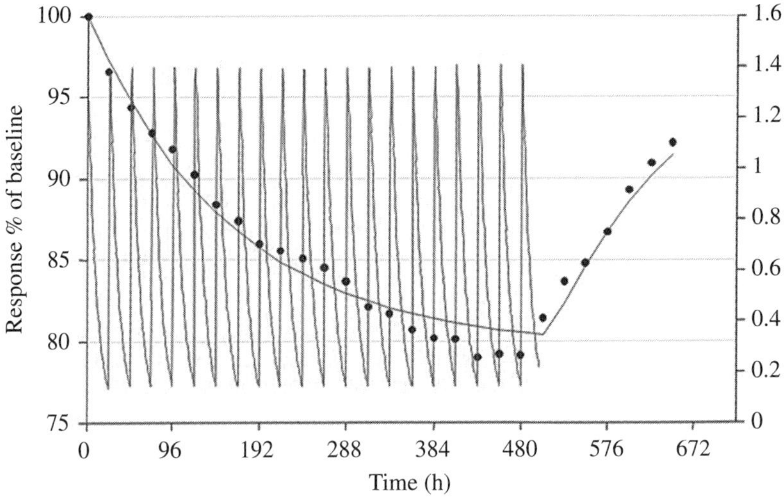

FIGURE 15.6 Plasma concentration-time (CB1r antagonist X, right-hand axis, rapid oscillating curve) and response-time (body weight reduction, left hand axis, solid symbols superimposed on model predictions) Data after repeated daily subcutaneous dosing of a CB1r antagonist during 3 weeks in obese mice.

half-life of adipocyte turnover (body weight loss in obese mice is almost entirely explained by loss of adipose tissue, and various compensatory changes).

The half-life of the adipocytes (fat cells) falls in the range of several weeks in experimental animals and several counterbalancing adjustments to the CB1r antagonist–induced lowered food intake (and hence relative "energy deficit state") are simultaneously triggered. Pharmacodynamic steady state is established after months of dosing when rates of production and removal of adipocytes balance each other. The body weight reduction now fully reflects the increased removal of adipocytes. This apparent discrepancy between plasma concentration- and response-time curves can be rationalized by the use of some of the approaches shown in Figure 15.5. The advent of the turnover and receptor occupancy models has given us tools to better explain and quantitatively tackle temporal differences (Fig. 15.5b and c). Using a model allows a more stringent design, analysis, and interpretation of kinetic dynamic experiments. It also allows predictions to be made and the model is further challenged with new experimental data. The temporal differences between plasma concentration and response (Fig. 15.2, right) may be due to distribution from plasma to biophase, active metabolites, initiation of changes of factors controlling the onset, and offset of response (cf. e.g., the CB1r example), and/or a slow on/offset of receptor occupancy. A further complexity for PK/PD analysis arises with drugs like the CB1r antagonists. This is because their mechanism of action (MoA) with respect to body weight and lipid homeostasis may relate not only to a centrally elicited appetite reduction, but also to an interaction with peripheral CB1r in, for example, the liver (see, e.g., Ref. [22]). When a drug's MoA encompasses CNS as well as peripheral tissue compartments, it is to be expected that different degrees of accessibility and counter-regulatory capacities will be encountered, with implications for how best to capture and describe the relation between exposure and PD response.

One of the primary goals in kinetic/dynamic assessment is therefore to identify the rate-limiting step along the concentration–response axis, and to judge whether this rate-limiting event is due to distributional, turnover, or other (e.g., receptor on/off; cf. further below) processes. Thus, the pharmacokinetic requirements for an optimal dynamic effect may vary substantially depending on the type of target and the position of the target in the biological system eliciting the effect. A generic way of working is, therefore, not possible and an integrated PK/PD strategy has to be adopted using a case-by-case approach.

We will further discuss how to assess *potency*, *efficacy*, and *steepness* and factors determining the time delay, based on a combination of the concentration-time, response-time, and concentration–response relationships in Case Studies 15.2–4.

Of note is also that development of PD tolerance or sensitization may vary substantially despite being triggered by one and the same receptor (sub)type. One example is 5-HT_{1A} receptor agonist–induced PD responses (e.g., Ref. [23]) that may develop marked tachyphylaxis already after a single administration (hypothermia), gradually (behavioral syndrome components) or only after protracted administration regimens, if at all (5-HT synthesis/turnover indices). This again illustrates the prominent adaptive physiological capacity of the CNS, and the importance of choosing and matching PK to a contextually appropriate PD response marker.

Time-Delays—Receptor On/Off Binding Models—Case Study 2

The third class of models that illustrates temporal differences between plasma concentration and the drug-induced response comprises the receptor binding on/off model (Fig. 15.5c). Typical occupancy-time profiles for three different intravenous infusion doses of a compound given to dogs are shown in Figure 15.7. Three observed occupancy-time courses are superimposed on their model-predicted values. Note that with increasing doses there is a shift to the left in the peak level of occupancy because the receptors are occupied more rapidly with increasing doses. This is not captured by the turnover model (usually a peak shift toward the right).

Because the total amount of receptor, free, and bound is conserved ([R] + [RC] = B_{max}), the on/off-binding model can be formulated mathematically as

$$\begin{aligned}\frac{d[RC]}{dt} &= k_{on} \cdot C \cdot R - k_{off} \cdot [RC] \\ &= k_{on} \cdot C \cdot (B_{max} - [RC]) - k_{off} \cdot [RC]\end{aligned} \tag{15.8}$$

where k_{on}, k_{off}, B_{max}, R, [RC], and C are the second-order on-rate binding constant, first-order off-rate binding constant, maximum binding capacity, concentration of free or unbound receptors, concentration of the receptor-ligand binding complex, and the total plasma concentration, respectively. By simultaneously fitting Equation 15.8 to the three occupancy-time courses shown in Figure 15.7, we obtain final parameter estimates of k_{on}, k_{off}, and B_{max} and their corresponding precision (Table 15.2). Ideally

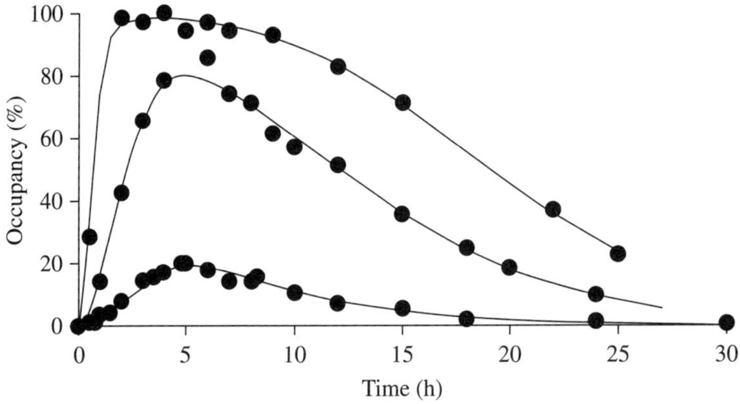

FIGURE 15.7 Observed (filled symbols) and predicted (solid lines, Equation 15.8) response-time data after three 4 h constant rate infusions of test compound X with a total dose of 10, 100, and 800 μmol·kg^{-1}. The bottom, middle, and upper time courses represent the low-, intermediate-, and high-dose groups, respectively. Note the leftward peak shifts in response with increasing doses. Adapted from Gabrielsson and Weiner [4]. The original functional response data were presented on a 0–7 response unit scale, and were then converted to 0–100% in order to mimic partial and full occupancy since there was an instantaneous equilibrium between occupancy and response.

the drug concentration at the receptor target site should have been used for the binding interaction in order to make the model fully mechanistic. Needless to say, *in vivo* examples with known actual biophase concentrations at the target (or even in the regional tissue of interest) in the CNS are rare (particularly in the clinic, if ever encountered). Thus, we deliberately applied the total plasma concentration instead, simply to demonstrate that this concentration could be used to approximate the "driving force" of rise and decline of response.

The concentration–binding relationship at equilibrium $[\mathrm{RC}]_{\mathrm{ss}}$ can then be derived as

$$[\mathrm{RC}]_{\mathrm{ss}} = \frac{k_{\mathrm{on}} \cdot C \cdot B_{\mathrm{max}}}{k_{\mathrm{off}} + k_{\mathrm{on}} \cdot C} = \frac{C \cdot B_{\mathrm{max}}}{(k_{\mathrm{off}} / k_{\mathrm{on}}) + C} = \frac{B_{\mathrm{max}} \cdot C}{K_{\mathrm{d}} + C} \qquad (15.9)$$

where K_{d} is the binding dissociation constant (affinity parameter and ratio of k_{off}-to-k_{on}).

The final parameter estimates (Table 15.2) from regressing Equation 15.8 are plugged into Equation 15.9 to give the equilibrium concentration–occupancy relationship in Figure 15.8.

TABLE 15.2 Final parameter estimates and their CV%

Equation 15.8		
Parameter	Estimate	CV%
k_{on} ($\mathrm{conc}^{-1} \cdot \mathrm{h}^{-1}$)	0.96	3
k_{off} (h^{-1})	0.19	3
B_{max} (conc)	101	1

The k_{on} and k_{off} parameters were estimated with high precision (low relative standard deviations CV%) and displayed very low correlation.

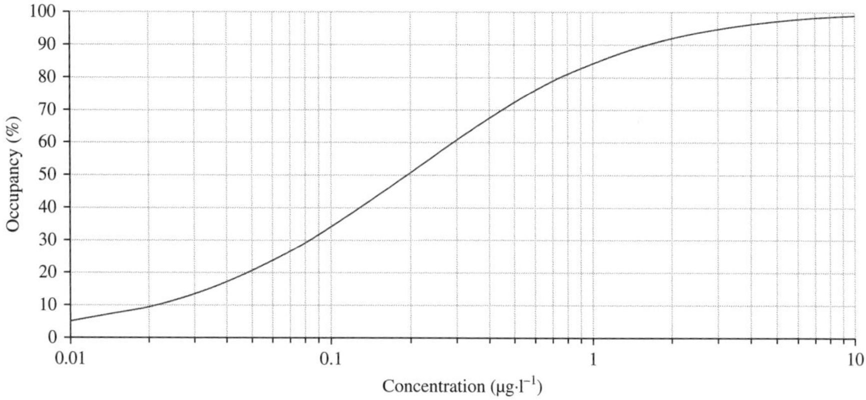

FIGURE 15.8 Simulated concentration–occupancy graph derived from Equation 15.9 and the final parameter estimates in Table 15.2. Adapted from Gabrielsson and Hjorth [6].

We have quantitatively analyzed and summarized the occupancy-time data in Figure 15.7, derived the equilibrium concentration–occupancy relationship and graphically displayed the effective concentration range against target occupancy in Figure 15.8. The latter can then be communicated to and used in the project team and further serve as a quantitative tool of experimental design.

From the PD point of view, the slow off-rate example may have a number of significant ramifications too (cf. Ref. [24]). Thus, a long residence time of the drug at its target pharmacodynamic efficacy can extend quite some time beyond when plasma levels have fallen several orders of magnitude. With regard to selectivity and safety aspects, a slow off-rate property displayed by a drug may be desirable or not, depending on the profile and context. Thus, a long residence time at the primary target relative to other sites would be consistent with an accentuated selectivity and favorable drug safety profile. On the other hand, a slow off-rate from nonprimary drug interaction sites, or when side effects *as well as* desired pharmacodynamic effects are mediated via the primary target site, could be clearly more problematic with regard to desensitization and safety. Needless to say, these aspects also need to be factored into the benefit/risk assessment for a particular target and candidate drug ligand that project teams have to consider.

Time Delays—Turnover Models—Case Study 3 on γ-Secretase (GSECR) Inhibitors

Numerous drugs have been developed using an acute response measurement. However, many are then administered to patients requiring long-term treatment. In some notable cases, evidence has emerged that the downstream therapeutic mechanism of action is probably not the mechanism used to assist selection in the discovery process (although the latter might be responsible for initiating or sustaining the long-term response).

An example of temporal difference between plasma concentration and pharmacological response was found in studies on a compound inhibiting gamma secretase (GSECR) in the brain to treat production of senile plaques in Alzheimer's disease (for refs to this principle, see, e.g., Ref. [25]). The compound decreases the activity of GSECR, reflected in the levels of its product: soluble Aβ, an amyloid precursor protein, APP, fragment (Fig. 15.9; thereby in turn indirectly reducing the formation of amyloid plaques). The peak plasma concentration C_{max} occurred at 0.5 h, whereas the effect of GSECR inhibition on soluble Aβ was seen after a delay of more than 2 h—thus recording a temporal difference of about 2 h. By only sampling the pharmacological response at 0.5 h after different doses of the compound, little information would have been obtained about the onset of action, maximum intensity and duration of response. The C_{max} approach would have resulted in biased estimates of potency of test compound (Gabrielsson and Weiner, 4th ed., 2010).

There is a temporal delay between the maximal plasma concentration of test compound X and the maximum decrease in ensuing response (Fig. 15.10). The peak plasma concentration occurs before the time of maximal intensity of response. The direct action of GSECR inhibitors is on the production of response, Aβ (turnover rate k_{in}).

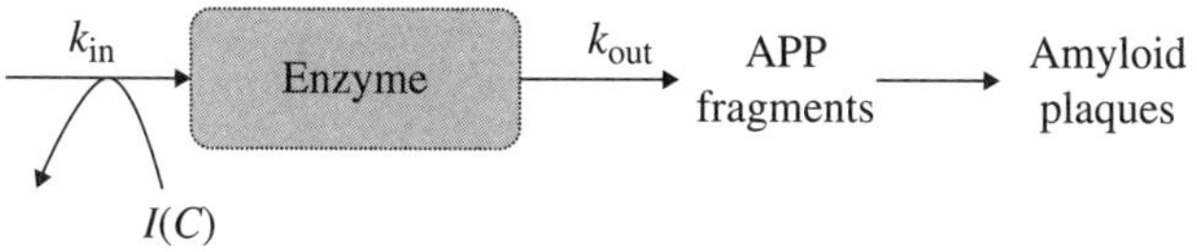

FIGURE 15.9 Basic structure of a turnover model. The test compound inhibits the enzyme activity or the buildup of response. The turnover rate k_{in} and fractional turnover rate k_{out} are respectively the zero-order and first-order rate constants that govern the enzyme activity. The formation of amyloid plaques is a slow process that occurs over weeks or months. The model is the most parsimonious with experimental data although alternative parameterizations may be elaborated provided several doses and a wider time range are available [12].

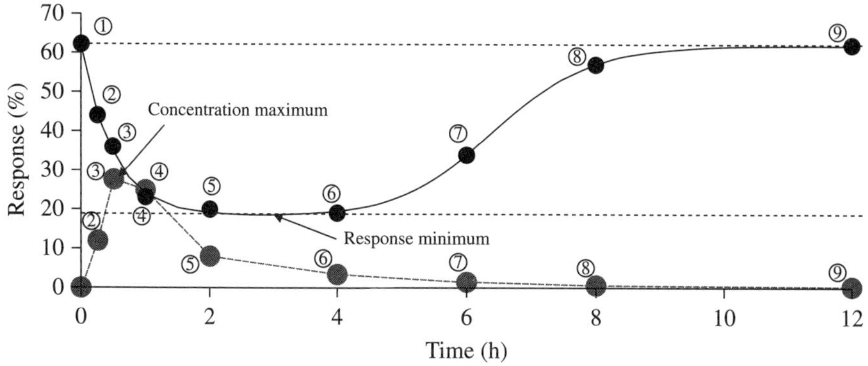

FIGURE 15.10 Response-time course of Aβ production (filled circles in solid upper line curve) and concentration-time (filled circles in lower dashed curve) data of GSECR-inhibiting test compound *X* after an oral dose of 40 μg·kg^{-1} to mice. Note that plasma concentration peaks at 30 min, but the time of maximal effect (trough) occurs at about 2.5 h. Stacking up the response measurements at the same time point as the plasma peak concentration is not optimal. Note that the pre-dose level starts at a fraction (62%) of the normal population. Data not yet published.

$$\frac{dR}{dt} = k_{in} \cdot I(C) - k_{out} \cdot R \tag{15.10}$$

The drug "mechanism function" can be written as an inhibitory I_{max} model.

$$I(C) = 1 - \frac{I_{max} \cdot C^{n_H}}{IC_{50}^{n_H} + C^{n_H}} \tag{15.11}$$

where I_{max}, IC_{50}, and n_H are the maximum drug-induced response, the plasma concentration at half-maximal drug-induced response, and the Hill exponent, respectively. Regressing Equations 15.10 and 15.11 to data in Figure 15.10 generates a good fit and model parameters with high precision (Table 15.3).

The concentration–response relationship shown in Figure 15.11 demonstrates a clear time delay in terms of hysteresis when data are plotted in time order, a flat minimum response region in spite of an eightfold concentration range (3–25 μM),

TABLE 15.3 Final parameter estimates and their CV%

Equations 15.6 and 15.7		
Parameter	Estimate	CV%
k_{out} (h^{-1})	2.0	7
R_0 (%)	62	1
I_{max}	0.71	2
IC_{50} ($\mu g \cdot l^{-1}$)	0.97	7
n_H	2.5	16

Original response-time data were generously shared by AstraZeneca, Wilmington, DE.

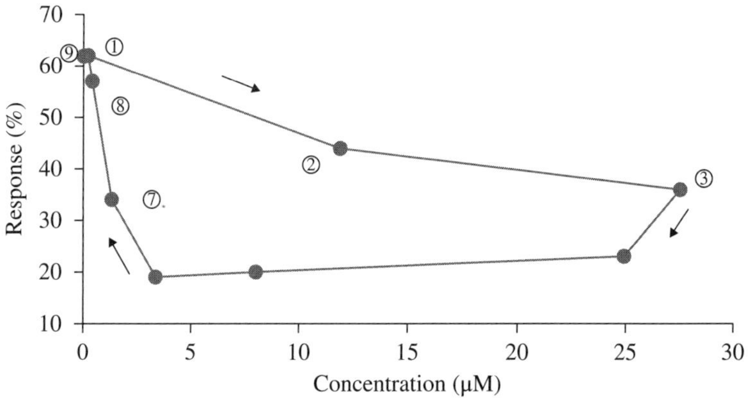

FIGURE 15.11 Concentration–response plot of data from Figure 15.10. Data plotted in time order (clockwise hysteresis) indicated by the circled numbers and arrows. C_{max} occurs at ③(~30 min). The clockwise hysteresis is due to the fact that response decreases from a basal level to lower levels when animals are exposed to a test compound. Normally, a counterclockwise hysteresis is observed for responses increasing above the baseline.

and then a steep return toward the baseline response. When the final parameter estimates in Table 15.3 are inserted into Equation 15.12, one gets the concentration–response relationship shown in Figure 15.12.

$$\left.\begin{aligned} R_{ss} &= R_0 \cdot I(C) = R_0 \cdot \left(1 - \frac{I_{max} \cdot C^{n_H}}{IC_{50}^{n_H} + C^{n_H}}\right) \\ R_0 &= \frac{k_{in}}{k_{out}} \end{aligned}\right\} \tag{15.12}$$

Simulations of the response-time course following a low and a high dose are superimposed over the original data from the analysis in Figure 15.13. Note that the onset of action is not affected by a change in dose as the time of observation was at C_{max}.

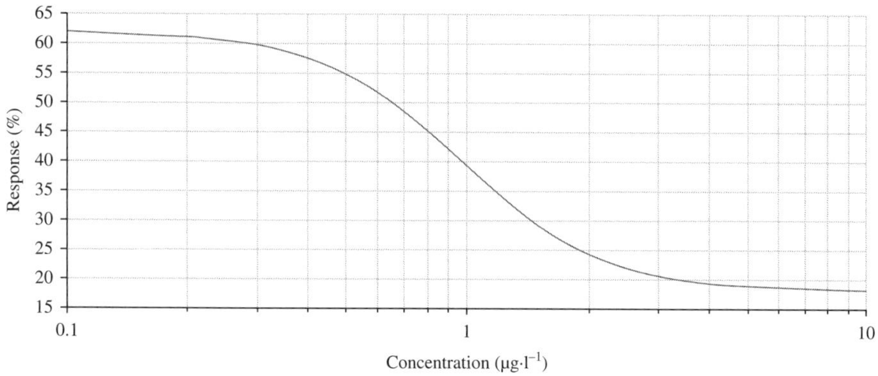

FIGURE 15.12 Semi-logarithmic plot of the concentration–response relationship of compound *X*. The final parameter estimates of the system and drug parameters together with their individual precisions (CV %) are also included. Note that this equilibrium function (Equation 15.12) lacks the hysteresis loop when plotted. That is because Equation 15.12 mimics the steady-state situation where time is no longer an issue.

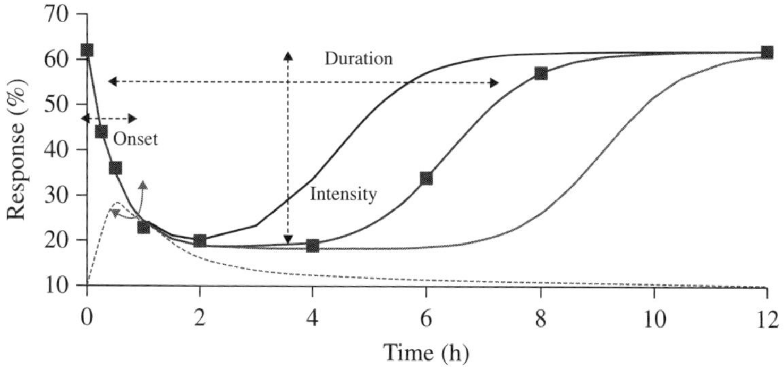

FIGURE 15.13 Simulations of the response-time courses after three different doses of 8, 40, and 200 μg·kg−1 (left to right) With the inhibitory action of the drug on factors controlling the production of response there is very little change of the onset of response with a change in dose. The downswing of the response-time course is governed by the fractional rate constant k_{out}. The duration of response will be affected by a change in dose.

There is very little change in intensity with the other simulated doses. However, the duration of response increases proportionally to the log(Dose) [13].

We have fitted Equations 15.10 and 15.11 to the data in Figure 15.10, derived the equilibrium concentration–response relationship (Fig. 15.12) and finally simulated different experimental designs as response-time courses (Fig. 15.13). The latter approach shows the lack of utility of the single point C_{max} approach.

While the time delay example still refers to the semi-acute drug treatment and PD response situation, obviously the desired therapeutic effect is a reduction of the amyloid plaque buildup in Alzheimer's, which is a process that takes much longer to achieve, may involve rebound, and is more difficult to detect—and thus model—in PD terms (e.g., Ref. [26]).

Time Delays—Feedback Turnover Models—Case Study 4 on SSRI

Separations between plasma concentration curves and pharmacodynamic responses are not uncommon, particularly in contexts where the latter emanate from targets within the BBB. Reasons for such temporal disconnect of course include distribution and accessibility factors (e.g., physicochemical properties of drugs, carriers, etc.), but the basic physiology of the CNS is a major determinant that needs to be taken into account as well. This includes redundancies in important basic processes, recruitment of secondary and cascade mechanisms, autoregulatory responses, tolerance, and sensitisation phenomena. Some examples will be discussed here.

In some cases, the pharmacological response peaks *before* rather than *after* plasma exposure, or disappears much faster than the plasma exposure. This is the case for the subjective "*feel drug*" effect relative to the plasma concentration of *d*-amphetamine (Fig. 15.14, left), which results in a clockwise hysteresis loop (Fig. 15.14, right) [27]. In this example there is a well-defined relationship between plasma concentration and pharmacological response, but it may not be evident from an acute dose. Thus, the mismatch between the subjective PD ("*feel drug*") response to *d*-amphetamine and plasma exposure levels likely involves an extent of functional adaptation that greatly outlasts the stimulatory action of *d*-amphetamine. The actual level of "*feel drug*" is determined not only by initiation but also by counteracting homeostatic processes resulting in the rapid decline of PD effects despite sustained levels of drug. Blunted effects of indirectly acting psychostimulants are often associated with rapid depletion of a releaseable pool of transmitter, down-regulation of receptor responsiveness, and/or other more or less undefined adaptational processes. The exact mechanism underlying the apparent tachyphylaxis of the subjective response to *d*-amphetamine in this example has yet to be fully understood. Nevertheless, the noticeable dissociation of exposure and PD readout time frames serves to illustrate the powerful buffering capacity of brain circuits to drug-induced effects. In turn, this necessitates proper PK/PD modeling approaches to gain insight into the relationship between drug concentrations and target response.

Our fourth case study is another example of functional adaptation, but one where the PD response is delayed relative to plasma exposure. Many tricyclic antidepressants, serotonin selective reuptake inhibitors (SSRIs), and monoamine oxidase inhibitors were developed on the basis of an acute action on monoamine neurotransmitters. However, the time delay in their clinical antidepressant effect (typically 2–3 weeks) and subsequent preclinical biochemical pharmacology studies led to the suggestion that the drugs produced neuroadaptive effects and that it was these longer-term changes that were crucial to their therapeutic effect (cf. e.g., Ref. [28]). The example discussed here stems from a recently developed feedback turnover model for SSRI ([29]; see also Refs. 5 and 6). Adaptational mechanisms are the rule rather than the exception in biological systems, even when drug-induced responses are concerned.

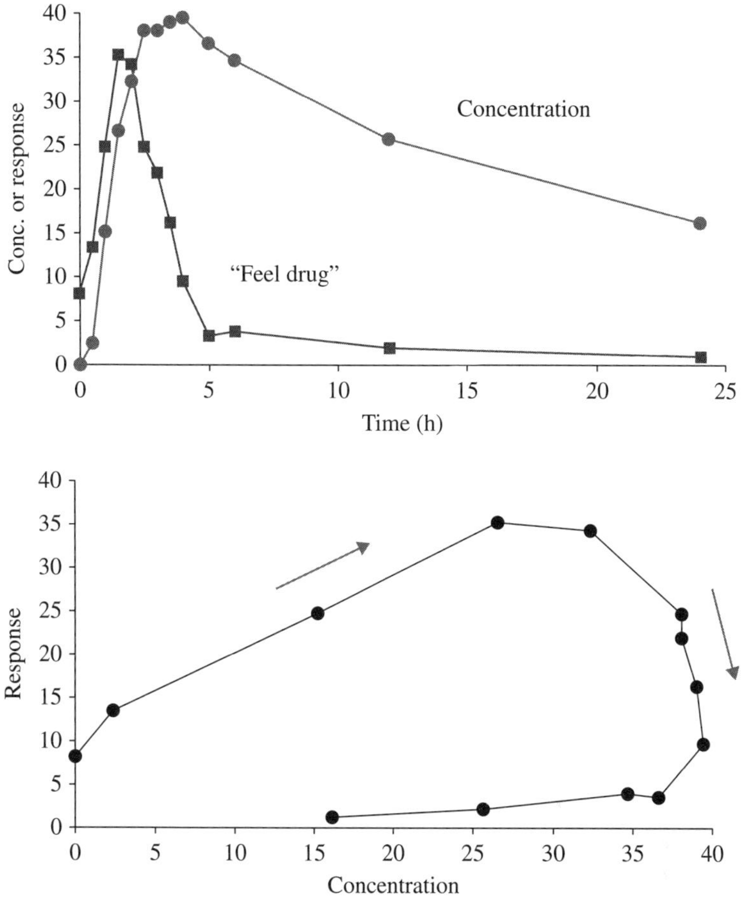

FIGURE 15.14 Left: Concentration (d-amphetamine)- and response (feel drug)-time data of d-amphetamine in healthy male volunteers (data digitized from Brauer et al. [27]). Right: Concentration (d-amphetamine)–response ("feel drug") data plotted in time order. Note the clockwise hysteresis (time order indicated by the arrows) which indicates that response declines over time in spite of stable or even increasing plasma concentrations of d-amphetamine.

The endogenous physiological modulatory mechanisms will therefore have to be considered as well, in the description and understanding of PD properties of drugs and relations to exposure metrics. For example, the acute serotonin-elevating action of antidepressant SSRIs such as citalopram or its S-enantiomer (escitalopram) is offset by 5-HT receptor-mediated feedback inhibition (cf., Refs. [30–35]). This autoregulatory impact thereby limits the SSRI-induced effect on 5-HT neurotransmission, and a modeling approach was therefore introduced to describe the relation between exposure and observed PD response (Fig. 15.15).

The model mimics drug-induced effects on brain extracellular levels of serotonin (5-HT) after constant rate infusions of the SSRI escitalopram in rats (Fig. 15.16).

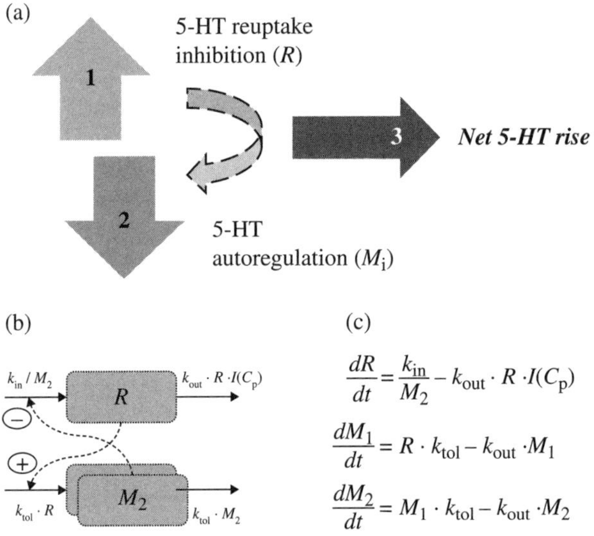

$$\frac{dR}{dt} = \frac{k_{in}}{M_2} - k_{out} \cdot R \cdot I(C_p)$$

$$\frac{dM_1}{dt} = R \cdot k_{tol} - k_{out} \cdot M_1$$

$$\frac{dM_2}{dt} = M_1 \cdot k_{tol} - k_{out} \cdot M_2$$

FIGURE 15.15 (a) Schematic illustration of SSRI effects on serotonergic neurotransmission: reuptake inhibition (R) results in increased synaptic levels of 5-HT (1), in turn this leads to enhanced autoregulatory impact (dashed curve arrow) through somatodendritic, nerve terminal, and postsynaptic 5-HT receptor–mediated countermodulatory processes (M_1 and M_2) aiming to decrease levels of 5-HT (2). The relative balance between these processes determines the net rise in 5-HT levels (3). (b) Conceptual illustration of the impact on 5-HT by means of a feedback buffering model. (c) Mathematical description of the conceptual model in terms of a system of differential equations.

In Figure 15.17 we show the model fits (three lower curves) superimposed on experimental data (observations: symbols). In addition, in Figure 15.17 we show a model simulation (upper curve) for a system lacking autoregulatory feedback modulation ("nonbuffered").

The dynamics of escitalopram-evoked changes of 5-HT response were characterized by a turnover model, which included an inhibitory feedback moderator component:

$$\left.\begin{aligned} \frac{dR}{dt} &= \frac{k_{in}}{M_n^{p}} - k_{out} \cdot R \cdot I(C) \\ \frac{dM_1}{dt} &= k_{tol} \cdot R - k_{tol} \cdot M_1 \\ &\cdots \\ \frac{dM_n}{dt} &= k_{tol} \cdot M_{n-1} - k_{tol} \cdot M_n \end{aligned}\right\} \qquad (15.13)$$

Here k_{in}, k_{out}, k_{tol}, and M denote the zero-order release of serotonin into the synapse, the first-order loss of serotonin (response) from the synapse, the first-order rate constant for buildup, and decay of receptor-mediated feedback (moderator), respectively. The drug mechanism function $I(C)$ is given by

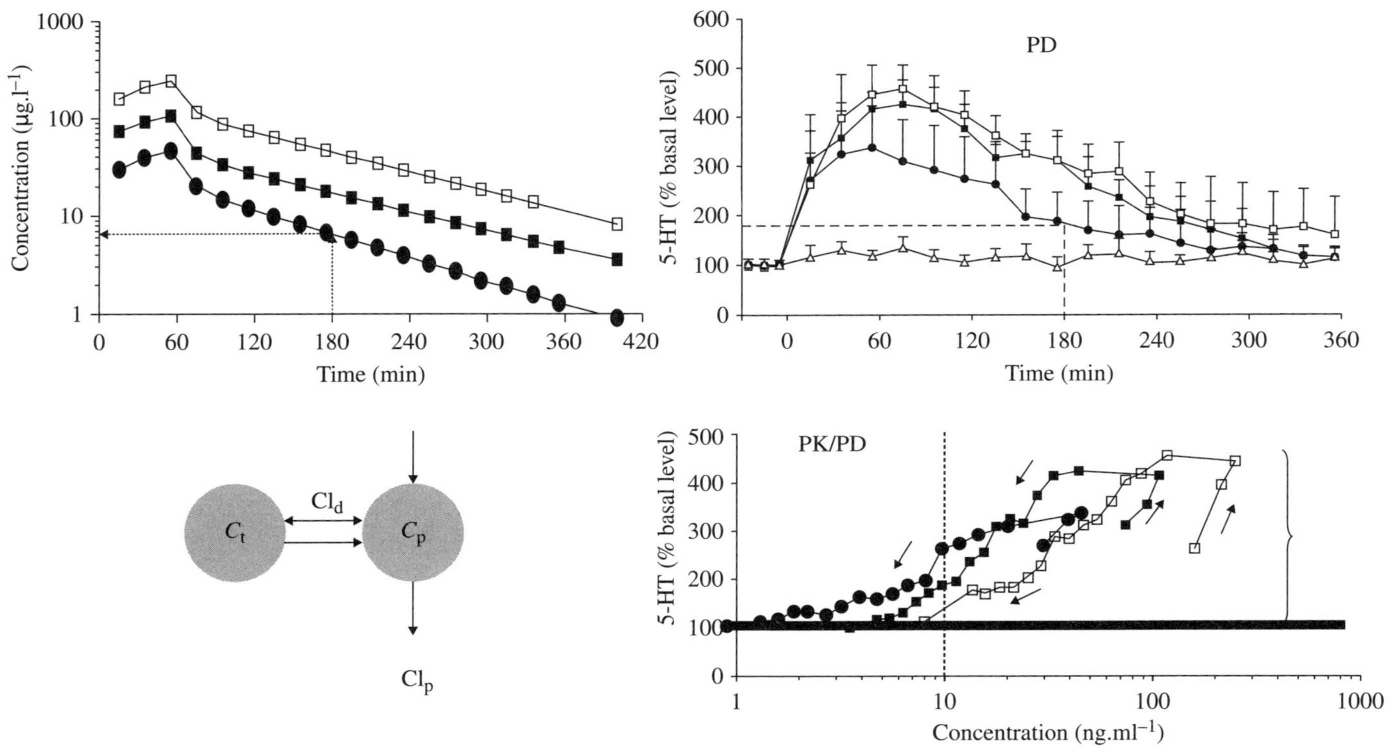

FIGURE 15.16 Upper left: Semi-logarithmic plot of model-predicted plasma concentrations of escitalopram after administration of 2.5 (solid circles), 5 (solid squares), and 10 (open squares) mg·kg^{-1} of escitalopram as an intravenous infusion over 60 min to three groups of rats. Upper right: Observed (filled symbols) response versus time from the three dose groups of rats. Lower left: The two-compartment model fitted to plasma concentration-time data. Lower right: Semi-logarithmic plot–observed concentration versus response loops of the three dose groups. The small arrows indicate the time order. Note that the higher-dose groups display lower responses (see, e.g., vertical dashed line at 10 ng·ml^{-1}), implying stronger impact of autoregulation.

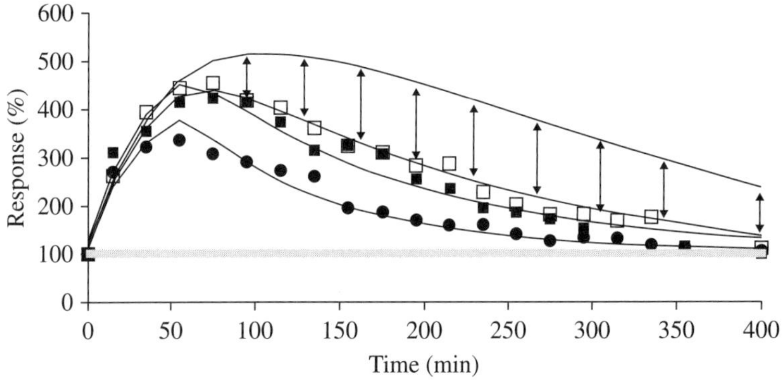

FIGURE 15.17 Plot of observed (filled symbols) and superimposed model response-time data of % 5-HT of the hippocampal area after administration of 2.5 (filled circles), 5 (black squares), and 10 (open squares) mg·kg^{-1} of escitalopram as an intravenous infusion over 60 min to three groups of rats. Note that the intermediate- and high-dose curves almost superimpose due to the offsetting impact of 5-HT autoregulatory buffering.

$$I(C) = 1 - \frac{I_{max} \cdot C^n}{IC_{50}^n + C_n} \tag{15.14}$$

where $0 < I_{max} \leq 1$, and IC_{50} and n are potency and Hill exponent, respectively. Estimates of the model parameters are given in Table 15.4.

Fitting the three response-time courses enables one to mimic the experimental data and to obtain robust and precise parameter estimations. Typical characteristics of the dataset are (i) a rapid upswing without overshoot, (ii) almost superimposable response-time curves for the higher doses due to counter-regulatory impact, and (iii) no rebound during the terminal response-time course due to the extended decline in the exposure to escitalopram. The value of this analysis is then to derive the concentration–response relationship at equilibrium, which is given by Equation 15.15 of a buffered (tolerant) system

$$R_{ss} = \sqrt{\frac{k_{in}}{k_{out}} \cdot \frac{1}{I(C)}} = R_0 \cdot \sqrt{\frac{1}{I(C)}} = R_0 \cdot \sqrt{\frac{1}{1 - \frac{I_{max} \cdot C^{n_H}}{IC_{50}^{n_H} + C^{n_H}}}} \tag{15.15}$$

and Equation 15.16 for a nonbuffered (nontolerant) system

$$R_{ss} = \frac{k_{in}}{k_{out}} \cdot \frac{1}{I(C)} = R_0 \cdot \frac{1}{I(C)} = R_0 \cdot \frac{1}{1 - \frac{I_{max} \cdot C^{n_H}}{IC_{50}^{n_H} + C^{n_H}}} \tag{15.16}$$

and shown graphically in Figure 15.18.

TABLE 15.4 Final parameter estimates and their corresponding precision (CV%) of escitalopram

Parameter	Final estimate	CV%
R_0 (%)	100	6
k_{out} (min^{-1})	0.18	11
I_{max}	0.84	2
IC_{50} ($\mu g \cdot l^{-1}$)	4.1	25
N	0.87	11
k_{tol} (min^{-1})	0.003	18

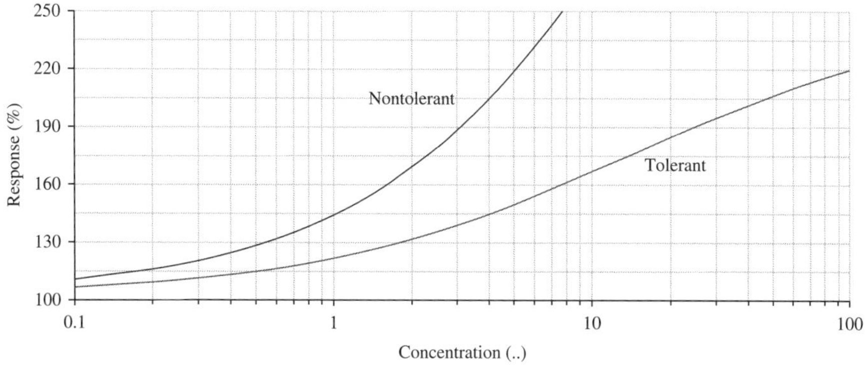

FIGURE 15.18 Semi-logarithmic plot of the concentration–response relationship of compound *X*. The final parameter estimates of the system and drug parameters are included. Note that this equilibrium function (Eqs. 15.15 and 15.16) lacks the hysteresis loop when plotted. That is because Equations 15.15 and 15.16 mimic equilibrium where time is no longer an issue. The buffered and nonbuffered curves are given by Equations 15.15 and 15.16, respectively.

We have quantitatively analyzed and summarized the response-time data in Figure 15.16 displaying functional adaptation, derived the equilibrium concentration–response relationship of both a tolerant and a hypothetical nontolerant system, and graphically displayed the effective concentration range against response in Figure 15.17. The response-time courses can then be projected at different doses by means of model simulations, and further serve as a quantitative tool of experimental design.

Interestingly, *in vivo* data on the effect of citalopram in the presence of 5-HT autoreceptor antagonists demonstrate almost a doubling of the 5-HT levels versus the SSRI given alone [31, 34]. This is in agreement with the modeled prediction of increased 5-HT response size in a nonbuffered system (Figs. 15.17 and 15.18). The observations argue in favor of the proposed modeling approach and emphasize the temporal PD response profile of a CNS-acting agent like this. In turn, it may also help devising novel means of treating conditions associated with altered 5-HT neurotransmission. As evident, integrated PK/PD strategies thus represent a significant advantage with respect to optimization of therapeutic principles toward man. One long-lived and prominent hypothesis for the delayed onset of antidepressant drug action of SSRIs has been the chronic administration-induced desensitization of the acute autoregulatory impact discussed and modeled earlier. This said, it should be noted that enhanced serotonergic neurotransmission resulting from chronic treatment with such agents also leads to other cellular, neurochemical, neuroadaptive, and neuroplastic changes, which may or may not be connected to the clinically antidepressant effects of these drugs (see, e.g., Ref. [36]). This case study clearly illustrates the need to take any biological buffering mechanisms and their relative impact into account when attempting to decipher the PD effect of a given drug versus its observed exposure. Additionally, it emphasizes how modeling approaches may be used in projects to direct and optimize experimental design when alterations in the dynamics of such systems may be expected to occur upon long-term drug treatment of CNS conditions.

CONCLUSIONS—PROPOSALS FOR THE PK/PD MODELER OF CNS SYSTEMS

As discussed in this chapter, examining CNS drug action *in vivo* may introduce several intricacies to be accounted for in order to generate a suitable and useful PK/PD analysis.

We have attempted to illustrate some of these challenges and how they may be approached, by means of the four case studies presented. The real value of a kinetic and dynamic analysis is of course not the modeling by itself, but what the results are intended for, that is, design of new studies, translation across species, etc. Needless to say, ascertaining that data are reliable is also of major importance prior to any data analysis. However, a close integration of insights into the *modus operandi* of relevant CNS processes and circuits for a particular target, paired with mathematical treatment and analysis of *in vivo* PD and PK study data, is required to establish the best foundation for optimal modeling. A particular challenge with regard to CNS drugs is to define biologically relevant PD responses, as truly predictive nonhuman animal models of neuro- and psychopharmacological agents are very scarce and indirect biomarkers may therefore have to be used. An ample compilation of *in vivo* data generated in animal models of drug action and discussions of their extrapolative utility in various psychiatric and neurological disorders was recently published in a themed issue of *British Journal of Pharmacology* [37].

To summarize, the four case studies discussed in this chapter display examples of a wide range of different response-time curves, pointing toward typical CNS phenomena such as multiple receptor sites, saturation, delay, and feedback. They also highlight the difficulties involved in extracting an adequate biophase concentration. A modeler has to elaborate on the pattern of onset, intensity, and duration of PD response. If no prior information is available about the mechanism of action, model building may start with some points to consider addressing baseline behavior, time delays, peak shifts with increasing doses, saturation of response at higher doses, and decline toward baseline (Fig. 15.19). For optimal outcome, however, study data should include monitoring of directly target-related and translatable biomarkers across several species, concomitant with measurements of plasma drug concentrations across time and dose ranges, plasma protein binding of the drug *ex vivo* (or, if not possible,

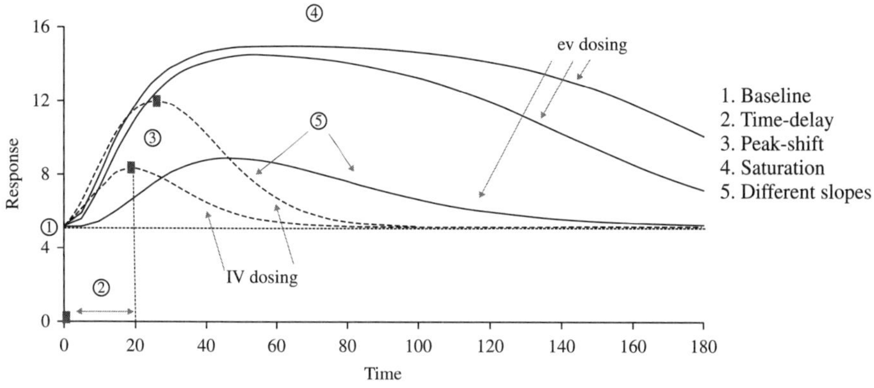

FIGURE 15.19 Schematic presentation of the pivotal characteristics of pharmacodynamic data. Data contain a baseline ①, time delay between peak exposure ② (not measured), and peak of response, peak shift with increasing doses ③, saturation at the highest dose ④, and different slopes of decline post peak depending on route of administration ⑤. Since response-time data were available after both intravenous (dashed lines) and subcutaneous (solid lines) dosing, the biophase availability could be estimated as a model parameter.

in vitro), and knowledge of the presence and activity of potential active metabolites that may influence the results (cf. Ref. [38]). Moreover, despite significant advances in human PK prediction consideration should also be given to ensure that the resulting plasma exposure is relevant to the way the drug will be used in humans, thereby further enhancing the translational value of the modeling.

In conclusion, as demonstrated by the examples in this chapter, a basic understanding of the dynamics of CNS physiology will greatly facilitate the kinetic–dynamic modeling when temporal delays between plasma exposure and a CNS-triggered response of a compound are observed. It is our firm belief that whenever feasible an integrated approach is superior to a reductionist approach. We therefore hope that the examples given will prove useful and accessible not only to the PK/PD modeler community but to support among project leaders, "hard-core" biologists, and chemists alike novel thoughts and ideas on how best to design future studies and drugs. Finally, with regard to kinetic–dynamic interactions in the CNS it should always be kept in mind that "*the brain is no democracy*" (Nobel laureate A. Carlsson, personal communication). Thus, if sometimes perplexing, and even seemingly anarchy-derived, sources of complexity in connecting plasma concentrations to brain drug response parameters may be revealed, identified, and accounted for with the help of high-quality kinetic–dynamic modeling.

REFERENCES

[1] Jusko WJ, and Ebling W. (1995) Components of PKPD models. *J Pharmacokin Pharmacodyn* **23**: 1–2.

[2] Danhof M, Alvan G, Dahl SG, Kuhlmann J, and Paintaud G (2005) Mechanism-based pharmacokinetic-pharmacodynamic modeling—a new classification of biomarkers, *Pharm Res* **22**:1432–1437.

[3] Miksys SL, and Tyndale RF (2002) Drug-metabolizing cytochrome P450s in the brain. *Rev Psychiatr Neurosci* **27**: 406–415.

[4] Gabrielsson J, and Weiner D (1994, 2000, 2010) Pharmacokinetic–Pharmacodynamic Data Analysis: Concepts and Applications (4th edn, 2nd print), Swedish Pharmaceutical Press, Stockholm, ISBN 13 978 91 9765 100 4.

[5] Gabrielsson J, and Peletier LA (2007) A nonlinear feedback model capturing different patterns of tolerance and rebound. *Eur J Pharm Sci* **32**: 85–104.

[6] Gabrielsson J, and Hjorth S (2012) Quantitative Pharmacology: An Introduction to Integrative Pharmacokinetic–Pharmacodynamic Analysis (1st edn), Swedish Pharmaceutical Press, Stockholm, ISBN 9 789197 945233.

[7] Lundström J, Lindgren JE, Ahlenius S, and Hillegaart V (1992) Relationship between brain level of 3-(3-hydroxyphenyl)-N-n-propylpiperidine HCl enantiomers and effects on locomotor activity in rats. *J Pharmacol Exp Ther* **262**: 41–47.

[8] Bonate PL (2013) Themed issue on translational modeling in neuroscience. *J Pharmacokinet Pharmacodyn* **40**: 255–418.

[9] Bays HE, and Gadde KM (2011) Phentermine/topiramate for weight reduction and treatment of adverse metabolic consequences in obesity. *Drugs Today (Barc).* **47**: 903–914.

[10] Rocha FL, Fuzikawa C, Riera R, and Hara C (2012) Combination of antidepressants in the treatment of major depressive disorder: A systematic review and meta-analysis. *J Clin Psychopharmacol* **32**: 278–281.

[11] Hjorth S, Carlsson A, Clark D, Svensson K, Wikström H, Sanchez D, Lindberg P, Hacksell U, Arvidsson LE, Johansson A, and Nilsson JLG (1983) Central dopamine receptor agonist and antagonist actions of the enantiomers of 3-PPP. *Psychopharmacol* **81**: 89–99.

[12] Gabrielsson J, and Peletier LA (2013) Mixture dynamics: Dual action of inhibition and stimulation. *Eur J Pharm Sci* **50**: 215–226.

[13] Carlsson A (1977) Dopaminergic autoreceptors: Background and implications. *Adv Biochem Psychopharmacol* **16**: 439–441.

[14] Paalzow LK, and Edlund PO (1979) Multiple receptor responses: A new concept to describe the relationship between pharmacological effects and pharmacokinetics of a drug: Studies on clonidine in the rat and cat. *J Pharmacokin Biopharm* **7**: 495–510.

[15] Paalzow GH, and Paalzow LK (1983) Opposing effects of Apomorphine on pain in rats. Evaluation of dose-response curve. *Eur J Pharmacol* **88**: 27–35.

[16] Paalzow LK, and Paalzow GH (1986) Concentration-response relations for apomorphine effects on heart rate in conscious rats. *J Pharm Pharmacol* **38**: 28–34.

[17] Bredberg E, and Paalzow LK (1991) Effects of apomorphine on heart rate during simultaneous administration of sulpiride: A challenge of the composed concentration effect model. *J Pharmacol Exp Ther* **258**: 1055–1060.

[18] Cheng VF, and Paalzow LK (1990) A pharmacodynamic model to predict the time-dependent adaptation of dopaminergic activity during constant concentrations of haloperidol. *J Pharm Pharmacol* **422**: 566–571.

[19] Levy G (1993) The case for preclinical pharmacodynamics. *In*: *Integration of Pharmacokinetics, Pharmacodynamics and Toxicokinetics in Rational Drug Development.* Ed. by Yacobi A, Shah VP, Skelly JP, and Benet LZ. pp. 7–13, Plenum Press, New York.

[20] Levy G (1998) Predicting effective drug concentrations for individual patients: Determinants of pharmacodynamic variability. *Clin Pharmacokin* **34**: 323–333.

[21] Calabrese E, and Baldwin LA (2003) Special issue: HORMESIS: Environmental and biomedical perspectives. *Crit Rev Toxicol* **33**: 213–463.

[22] Kirilly E, Gonda X, and Bagdy G (2012) CB1 receptor antagonists: New discoveries leading to new perspectives. *Acta Physiol (Oxf)* **205**: 41–60.

[23] Larsson LG, Rényi L, Ross SB, Svensson B, and Ängeby-Möller K (1990) Different effects on the responses of functional pre- and postsynaptic 5-HT1A receptors by repeated treatment of rats with the 5-HT1A receptor agonist 8-OH-DPAT. *Neuropharmacology* **29**: 85–91.

[24] Tummino PJ, and Copeland RA (2008) Residence time of receptor-ligand complexes and its effect on biological function. *Biochemistry* **47**: 5481–5492.

[25] Panza F, Frisardi V, Imbimbo BP, Capurso C, Logroscino G, Sancarlo D, Seripa D, Vendemiale G, Pilotto A, and Solfrizzi V (2010) REVIEW: γ-Secretase inhibitors for the treatment of Alzheimer's disease: The current state. *CNS Neurosci Ther* **16**: 272–284.

[26] Cook JJ, Wildsmith KR, Gilberto DB, Holahan MA, Kinney GG, Mathers PD, Michener MS, Price EA, Shearman MS, Simon AJ, Wang JX, Wu G, Yarasheski KE, and Bateman RJ (2010) Acute gamma-secretase inhibition of nonhuman primate CNS shifts amyloid precursor protein (APP) metabolism from amyloid-beta production to alternative APP fragments without amyloid-beta rebound. *J Neurosci* **30**: 6743–6750.

[27] Brauer LH, Ambre J, and De Wit H (1996) Acute tolerance to subjective but not cardiovascular effects of d-amphetamine in normal, healthy men. *J Clin Psychopharmacol* **16**: 72–76.

[28] Grahame-Smith DG (1997) 'Keep on taking the tablets': Pharmacological adaptation during long-term drug therapy. *Br J Clin Pharmacol* **44**: 227–238.

[29] Bundgaard C, Larsen F, Jørgensen M, and Gabrielsson J (2006) Mechanistic model of acute autoinhibitory feedback action after administration of SSRIs in rats: application to escitalopram-induced effects on brain serotonin levels. *Eur J Pharm Sci* **29**: 394–404.

[30] Blier P (1991) Terminal serotonin autoreceptor function in the rat hippocampus is not modified by pertussis and cholera toxins. *Naunyn-Schmiedeberg's Arch Pharmacol* **344**: 160–166.

[31] Hjorth S (1993) Serotonin 5 HT1A autoreceptor blockade potentiates the ability of the 5-HT reuptake inhibitor citalopram to increase nerve terminal 5 HT output in vivo: A micro-dialysis study. *J Neurochem* **60**: 776–779.

[32] Hjorth S, and Auerbach SB (1995) 5-HT1A autoreceptors and the mode of action of selective serotonin reuptake inhibitors (SSRI). *Behav Brain Res* **73**: 281–283.

[33] Hjorth S, Bengtsson HJ, and Milano S (1996) Raphe 5-HT1A autoreceptors, but not postsynaptic 5-HT1A receptors or β-adrenoceptors, restrain the citalopram-induced increase in extracellular 5-HT in vivo. *Eur J Pharmacol* **316**: 43–47.

[34] Hjorth S, Bengtsson HJ, Kullberg A, Carlzon D, Peilot H, and Auerbach SB (2000) Serotonin autoreceptor function and antidepressant drug action. *J Psychopharmacol* **14**: 177–185.

[35] Sharp T, Boothman L, Raley J, and Quérée P (2007) Important messages in the 'post': Recent discoveries in 5-HT neurone feedback control. *Trends Pharmacol Sci* **28**: 629–636.

[36] Zhong H, Haddjeri N, and Sánchez C (2012) Escitalopram, an antidepressant with an allosteric effect at the serotonin transporter—a review of current understanding of its mechanism of action. *Psychopharmacology (Berl)* **219**: 1–13.

[37] *British Journal of Pharmacology* (2011) *Special Issue: Themed Issue:* Translational Neuropharmacology—Using Appropriate Animal Models to Guide Clinical Drug Development. Guest Editor: A Richard Green. http://onlinelibrary.wiley.com/doi/10.1111/bph.2011.164.issue-4/issuetoc (Accessed August 28, 2013).

[38] Gabrielsson J, and Green R (2009) Quantitative pharmacology or pharmacokinetic pharmacodynamic integration should be a vital component in integrative pharmacology. *J Pharmacol Exp Ther* **331**:767–774.

16

MICRODIALYSIS TO ASSESS FREE DRUG CONCENTRATION IN BRAIN

William Kielbasa[1] and Robert E. Stratford, Jr.[2]

[1] *Eli Lilly and Company, Indianapolis, IN, USA*
[2] *Xavier University of Louisiana, New Orleans, LA, USA*

INTRODUCTION

Whether in a preclinical or clinical setting, the potential for disconnection between plasma and central nervous system (CNS) exposures is well recognized for its ability to confound CNS drug discovery and development. Microdialysis, herein referred to as quantitative microdialysis (QMD), is the only technique that enables sampling of brain extracellular fluid (bECF) in conscious animals to determine the absolute unbound bECF drug concentration. Determination of unbound drug concentration in bECF provides direct evidence of exposure at extracellular target sites and supports development of physiologically based pharmacokinetic (PBPK) models that can characterize the disposition of drugs. In pharmacodynamic (PD) research, relationships between drug concentrations and pharmacological effects are sought. The unbound drug PK profile is an important determinant of the time course and intensity of target binding and, accordingly, the potential for an ensuing pharmacologic effect. As such, QMD has the potential to fulfill the first two pillars of a recently advocated drug-hunting paradigm [1]. By extension, translation of preclinical PBPK models to humans has been advocated as a tool to support clinical drug development [2–5].

The intent of this chapter is to provide the reader a general overview of the QMD technique and its application in advancing CNS molecules in the context of model-based drug development. Microdialysis can also be used to measure change in

Blood–Brain Barrier in Drug Discovery: Optimizing Brain Exposure of CNS Drugs and Minimizing Brain Side Effects for Peripheral Drugs, First Edition. Edited by Li Di and Edward H. Kerns.

concentration of an endogenous substance, such as a neurotransmitter, in response to sensory stimulation or drug treatment, or of a putative biomarker of a disease process, such as amyloid beta protein [6]. This pharmacologic application of microdialysis, which measures relative concentration change rather than absolute concentrations, will be discussed only in the context of the historical development of the technique. Approaches related to microdialysis that are also capable of sampling bECF but which are tuned more toward pharmacologic applications, such as electrochemical sensors for specific neurotransmitters [7, 8] and open-flow microperfusion for endogenous high–molecular weight substances [9], will not be discussed.

OVERVIEW OF THE BRAIN MICRODIALYSIS TECHNIQUE

Microdialysis involves stereotaxic surgical implantation of a probe that possesses a semipermeable membrane. The principle of solute transport in microdialysis is the same as the diffusional process of solute exchange across a capillary blood vessel. Namely, the semipermeable membrane in a probe is perfused with a fluid that is physiologically compatible with bECF to support diffusion of solutes between this fluid and bECF. Figure 16.1 illustrates the principle of solute exchange across a concentric probe design commonly used in brain microdialysis studies. Because solute diffusion across the dialysis membrane takes time, the concentration of solute "recovered" in the collected dialysate will be lower than in the bECF. How much lower depends on the rate of perfusion of the probe, with solute recovery decreasing as perfusate flow increases. When perfusate flow rate is zero, equilibrium across the dialysis membrane is achieved, and measured concentrations in the sampled dialysate are equivalent to bECF concentrations. In such cases, solute recovery is considered

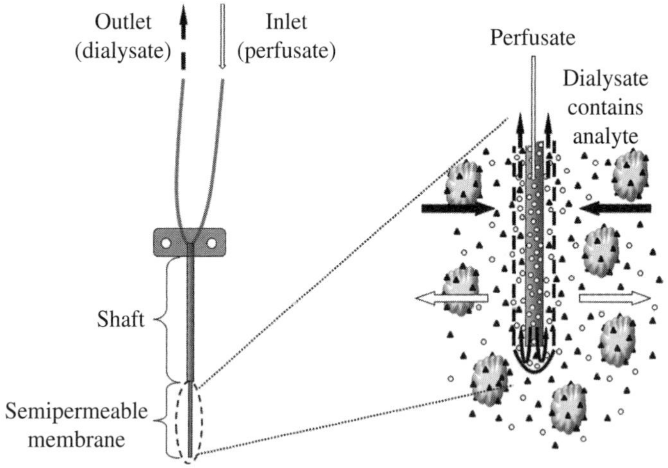

FIGURE 16.1 Diagram of a concentric probe design (Reprinted as free media from http://en.wikipedia.org/wiki/File:Schematic_illustration_of_a_microdialysis_probe.png).

to be 100%. Flow rates of 1 or 2 μl/min are commonly used and represent a balance between the competing factors of recovery and collection time, the latter to generate sufficient volume to support analysis yet achieve reasonable temporal resolution.

Use of a brain atlas and stereotaxic device ensures precise location of probe implantation. This ability to control probe placement allows for discrete brain regions to be sampled. Specific brain regions are implicated in different brain diseases; thus the probe can be placed in the region most relevant to the disease under investigation to measure unbound drug concentration at the intended site of action. Conceivably, lower unbound concentrations could exist at a target site relative to a nontarget site for a drug that possesses high specific binding. Region-specific distribution of the target, a receptor, for example, could conceivably lead to significant contribution of specific receptor-mediated binding in a region of high receptor density relative to nonspecific binding only at nontarget sites, thus resulting in different unbound concentrations. Invoking contribution of specific binding to dopamine D_2 and serotonin $5HT_{2A}$ receptors was recently shown to be necessary to model risperidone and paliperidone brain concentrations [10].

The sampling technique employed in brain microdialysis does not change the net fluid balance of the surrounding ECF matrix, and for that reason samples can be collected continuously for several hours. Implantation of a guide cannula enables probes to be removed and, perhaps several days later, another probe to be reinserted to implement once again sampling from the same animal. Damage to surrounding tissue can occur during probe implantation; however, evidence indicates that neurons remain functional because they retain the ability to respond to systemic and local pharmacologic manipulation [11]. Moreover, the blood–brain barrier (BBB) remains functionally intact after probe implantation because it is responsive to carrier-mediated transport inhibition [12] and passive transport continues to depend on partition coefficient and molecular weight [13]. Following probe insertion, a range of 18–24 h before initiating sample collection is generally regarded as an acceptable recovery time for the tissue. After this period, probes are considered useful for sampling for the next 24 h; beyond that time, decay in probe recovery due to tissue responses to the probe can become problematic [13].

At the heart of sample generation in microdialysis is the probe, which provides the structural interface of the dialysis membrane with the tissue. While there are several types of commercially available probes, the concentric probe design shown in Figure 16.1 has proven most useful for CNS application due to its rigid design, which enables precise implantation that is stable over time. Key variables in probe configuration are probe length, diameter, and dialysis membrane type. Various combinations of these three provide an array of choices to support fit-for-purpose application. For example, in selection of a smaller diameter for use in mice, or a longer probe tip to increase drug recovery to support bioanalysis, or a higher–molecular weight cutoff (MWCO) membrane to detect endogenous peptides or RNA fragments as potential biomarkers.

Detection of neurotransmitters in the brain of living organisms was one of the first successful research applications of microdialysis [14–16]. Demonstration that relative concentrations of endogenous chemicals respond in a consistent manner to pharmacologic input or sensory stimuli has, in a sense, validated the technique as a tool to improve our understanding of brain function. Improvements in probe design and coupled bioanalytical methods have enabled greater sophistication in its use as a pharmacologic tool with high spatial and temporal resolution. Kennedy [17] provides an excellent review with numerous references of the application of microdialysis to support the discovery of peptidic disease-related biomarkers, and its use to study interactions among neurotransmitter systems. Not surprisingly, these improvements in probe design and more selective and sensitive bioanalytical methods also apply to QMD, and have led to its increasingly larger application as a tool to support PK–PD investigations [5]. The capability to collect multiple samples from the same animal combined with more powerful bioanalytical approaches also makes it possible to measure biomarker levels and drug concentrations in the same animal. This approach was used to investigate serotonin response and escitalopram concentrations simultaneously in brain microdialysis samples [18]. Notwithstanding this unique capability of microdialysis, the remainder of this chapter will emphasize the QMD-specific application of the technique.

TECHNICAL IMPLEMENTATION OF QUANTITATIVE MICRODIALYSIS

Evolution of probe design has been continuous over the past 30 years. Early probes relied on cellulose membranes and had limited spatial resolution due to their size. Temporal resolution was also limited due to analytical sensitivity, which necessitated longer collection times. While cellulose-based membranes are still available, other membranes are now also available and provide flexibility in surface properties and MWCO to maximize recovery of a given solute. Commercially available probe diameters range from 0.2 to 0.5 mm, while the length of the dialysis membrane ranges from 1 to 4 mm; thus different combinations of the two provide a range of membrane surface area. As this surface area declines, spatial resolution increases, but the absolute mass of analyte that can be recovered also declines. Achieving the right balance needs to be considered within the context of live phase study design, primarily administration of pharmacologically relevant doses and analytical method capability. A summary of the factors relevant to commercially available probes that influence the dependent variable of solute recovery, and their effect on spatial or temporal resolution, is provided in Table 16.1. A comprehensive review of how these factors and others, such as perfusate composition and temperature, influence recovery is given by de Lange et al. [13].

For PK assessments, the absolute drug concentration needs to be calculated by taking recovery of the microdialysis probe into account. Drug recovery is less than 100% because the flow rate of perfusate through the probe does not allow sufficient time for the solute to equilibrate between the perfusate and bECF across the dialysis membrane. To account for the difference between the measured concentration in the dialysate and the actual bECF concentration, several correction methods have been developed. Examples are no-net-flux [19], dynamic no-net-flux [20], and

TABLE 16.1 Factors that influence recovery in standard commercially available probes and their impact on spatial and temporal resolution

	Membrane structure	
Polymer composition	Cellulose	Polyethers
MWCO (kDa)	30	6–100
	Probe configuration	
Probe diameter range (mm)	0.2–0.5	
Membrane length range (mm)	1–4	
Surface area range (mm^2)	0.2–2	
Solute recovery	Increases as surface area increases	
Spatial resolution	Decreases as surface area increases	
	Volume flow	
Perfusate flow rate range (μL/min)	0.5–2	
Dialysate collection volume range (μL/min)	10–30	
Solute recovery	Increases as flow rate decreases	
Temporal resolution	Decreases as collection time increases	

retrodialysis [13, 21]. With a newer approach—the modified ultraslow microdialysis technique—the dialysate drug concentration measured is used without the need for probe recovery correction [22].

Given the multiplicity of probe-related variables that can influence solute recovery (Table 16.1), it is unreasonable to attempt to optimize these *in vivo*. Thus, as a general strategy, *in vitro* assessments of solute recovery, because of their relatively lower cost and higher throughput, are used first to define variables of probe configuration and perfusate flow rate that optimize recovery. Importantly, these variables are independent of the matrix the probe is sampling, whether buffer for *in vitro* work or actual brain tissue for *in vivo* work; thus, knowledge obtained *in vitro* is directly relevant to the *in vivo* situation.

The objectives for *in vitro* recovery experiments are to achieve steady-state solute recovery quickly (ideally within the first collection interval) and to maximize the extent of this recovery. To support attainment of these objectives, recovery should be measured in both directions. That is, measurement of recovery in the collected dialysate of solute *loss* from perfusate containing a known concentration of solute with the probe placed in a solute-free vessel of the perfusate ("recovery-by-loss"); and measurement of solute *gain* into the collected dialysate following perfusion of solute-free solvent into a probe placed into a vessel that contains a known concentration of solute ("recovery-by-gain"). The equation to calculate "recovery-by-loss" is

$$\%\text{loss} = \left(\frac{C_{\text{perfusate}} - C_{\text{dialysate}}}{C_{\text{perfusate}}} \right) \cdot 100$$

where $C_{\text{perfusate}}$ is the drug concentration in the perfusate and $C_{\text{dialysate}}$ is the drug concentration in the dialysate.

The equation to calculate "recovery-by-gain" is

$$\%\text{gain} = \left(\frac{C_{\text{dialysate}}}{C_{\text{reservoir}}}\right)\cdot 100$$

where $C_{\text{reservior}}$ is the drug concentration in the vessel.

A reasonable starting point for initiating an *in vitro* experiment is to select a membrane type (cellulose versus noncellulose), a moderate surface area (2 mm length × 0.3–0.4 mm diameter), a flow rate of 1 µl/min with 30 min collection intervals over a 2–3 h duration, and a drug concentration range of 10–100 nM. The same probe can be used to evaluate recovery in both directions. To gain confidence in conducting the experiment, use of a drug that does not bind readily to microdialysis tubing, dialysis membranes, and collection vessels is advisable (e.g., antipyrine). Many drug candidates have physicochemical properties conducive to binding to polymeric surfaces (adsorption), and because the bECF concentrations can often be in the single-digit nM range, this low concentration can be problematic for accurate bioanalytical assessments with reasonable variability. In fact, by anecdote, high nonspecific binding to brain tissue, such as the case with duloxetine [3], portends problematic adsorption to the dialysis membrane. Poor *in vitro* recovery (<20%), asymmetric recovery of gain versus loss, and slow progression to the final (steady-state) recovery value are diagnostic of problematic adsorption. In such cases, incorporation of a binding agent in the perfusate, such as 0.2–0.5% bovine serum albumin, can improve the kinetics and extent of recovery to acceptable levels (>20% asymmetric recovery within 30 min). A recent study systematically evaluated various factors for several compounds ranging in $c \log D$, pH 7.4 from 1.5 to 3.5 [23].

Once probe-related variables influencing recovery have been optimized through *in vitro* work, it is still necessary to measure recovery *in vivo* due to tissue-related factors [13, 24, 25]. As mentioned previously, several *in vivo* recovery approaches exist [13, 22]. *In vivo* recovery (R) relates $C_{\text{dialysate}}$ to bECF concentration (C_{ecf}) through the following equation:

$$C_{\text{ecf}} = \frac{C_{\text{dialysate}}}{R}$$

Of the *in vivo* recovery approaches available, retrodialysis with a calibrator is the most efficient because it provides measurement of recovery in the same animal over the identical time course of sample collection following drug administration. A retrodialysis calibrator is a molecule that has similar physicochemical properties to the molecule of interest, and, therefore, similar dialysis membrane diffusion properties so that recovery of both is similar. Apart from the use of radiolabelled drug, which is usually not available in a discovery or early development setting, an excellent candidate for a retrodialysis calibrator is the stable label form of the drug, which is used when quantitation is by high-performance liquid chromatography (HPLC) with mass spectrometric detection. The stable label form is the drug molecule that is synthesized in such a way as to incorporate either deuterium or, preferably, carbon-13, into the structure rather than the corresponding common isotope. If stable label is not available, use of a structurally related molecule from the same SAR library may suffice. The actual drug of interest could also be used as a retrodialysis calibrator. However, the downside of this approach is that the duration of the experiment is extended because it requires two additional phases prior to the PK experiment: (1) measurement of recovery by loss, and (2) a washout period to allow clearance of the drug from the bECF.

Application of the retrodialysis calibrator approach involves perfusion of the calibrator into the probe, to determine its recovery by loss in the collected dialysate, coincident with administration of the drug of interest to achieve systemic and bECF exposure. Thus, probe and drug are contained in the collected dialysate samples. A concern with the use of the retrodialysis with calibrator approach is that the time required to achieve steady-state recovery of the calibrator may be different from that of the solute (drug) of interest. This can be problematic for compounds that equilibrate slowly across the dialysis membrane; accordingly, the desire to reach steady-state recovery quickly reinforces optimization of recovery conditions based on prior *in vitro* work. Another aspect to consider is that it is useful to minimize the retrodialysis calibrator concentration, but yet remain above bioanalytical sensitivity, so any potential competition of calibrator with drug for physiological binding or transport sites is diminished.

Two other approaches to determine *in vivo* recovery are the zero-no-net-flux and dynamic no-net-flux approaches [13]. Unlike the retrodialysis approach, these approaches require separate experiments to determine an average drug recovery across animals, and then apply this average recovery to subsequent experiments. Thus, more animals are needed and there are no means to account for intersubject variability in recovery with these two approaches. A third approach that has been recently developed [22] uses a modified probe design that enables near zero flow of perfusate to approximate 100% drug recovery. While this approach is obviously attractive by removing the need to use a calibrator, one needs to keep in mind the approximate 10-fold dilution of dialysate inherent with this approach.

The practical issues that need consideration to successfully implement QMD creates the potential for researchers to consider QMD as a demanding and intensive technique that is not worth investing in. However, for laboratories that may not have the resources to conduct QMD experiments, several fee-for-service companies are available to support the gamut of activities such as *in vitro* probe recovery, surgical implantation of guide cannula, and *in vivo* sample generation with subsequent bioanalysis and PKPD analysis. In addition to several seminal review articles that provide theoretical background, valuable insights and knowledge on QMD can be gained through microdialysis user groups that can be accessed readily through the Internet. For example, the American Association of Pharmaceutical Sciences (AAPS) has a Microdialysis Focus Group that has members with long-standing experience and is accessible via social media portals.

APPLICATION OF QUANTITATIVE MICRODIALYSIS IN DRUG RESEARCH

A key question in drug discovery, particularly neuroscience research, is whether a drug is at a sufficient concentration to interact with a target to be effective [1]. Besides microdialysis, researchers have used several preclinical and clinical approaches including brain homogenates [26–29], brain slices [30–32], and cerebrospinal fluid (CSF) measurements [33, 34] to inform the unbound drug concentration at the target site. While microdialysis is the only technique that can sample bECF, it is not a high-throughput technique that can be used efficiently to drive a drug discovery program. Alternatively, it is better suited to support candidate drugs selected for clinical investigation or to discriminate between a smaller set of compounds that have been identified as potential clinical candidates. Microdialysis-derived concentrations have served as a standard to lend confidence to other approaches [35] and to highlight when precautions regarding their ability to extrapolate need to be taken [2, 33, 36].

The invasive nature of microdialysis precludes widespread use in clinical trials of CNS candidate drugs, and CSF is considered the best surrogate matrix for bECF [37]. In animals, multiple probes can be implanted to enable simultaneous sampling of bECF and CSF, which allows for a determination of the relationship between the bECF and CSF drug exposure. Knowledge of this relationship in an animal is useful for translating an observed human CSF concentration to a predicted bECF concentration [5]. The use of receptor occupancy through positron emission tomography (PET) or single photon emission computed tomography (SPECT) imaging is a useful technique to infer the time course of drug concentration in brain [38, 39], but the availability of a PET or SPECT tracer to provide such information is a common limiting factor in applying the technique.

By virtue of its capability to measure unbound brain concentrations *in vivo* with high temporal and spatial resolution, QMD provides an excellent preclinical substrate for the development of PBPK models for translational (animal to human) research. An application of QMD in translational research to support dose selection for clinical trials was described by Kielbasa and Stratford [3] using atomoxetine, a norepinephrine transporter (NET) inhibitor, and duloxetine, a NET inhibitor and serotonin transporter (SERT) inhibitor, as probe drugs. The approach aims to predict the human dose–bECF concentration relationship for drugs to identify clinical doses that should engage the brain target. For each drug, a PBPK model was developed based on QMD experiments conducted in rats. The rat PBPK models were subsequently translated to humans using allometric principles and human physiology to predict the human bECF PK profile at clinically efficacious doses for atomoxetine and duloxetine. Target site engagement was assessed by linking the predicted human bECF PK profile to *in vitro* measures of receptor-binding kinetics (K_i, inhibition constant). The results demonstrated that for clinically approved doses of atomoxetine and duloxetine the predicted human bECF PK profiles were suggestive of target inhibition based on the human NET K_i for atomoxetine (Fig. 16.2) and the human NET and SERT K_i for duloxetine (Fig. 16.3). Putting these results in clinical context, according to the atomoxetine product label in the United States, dosing adults, children, and adolescents

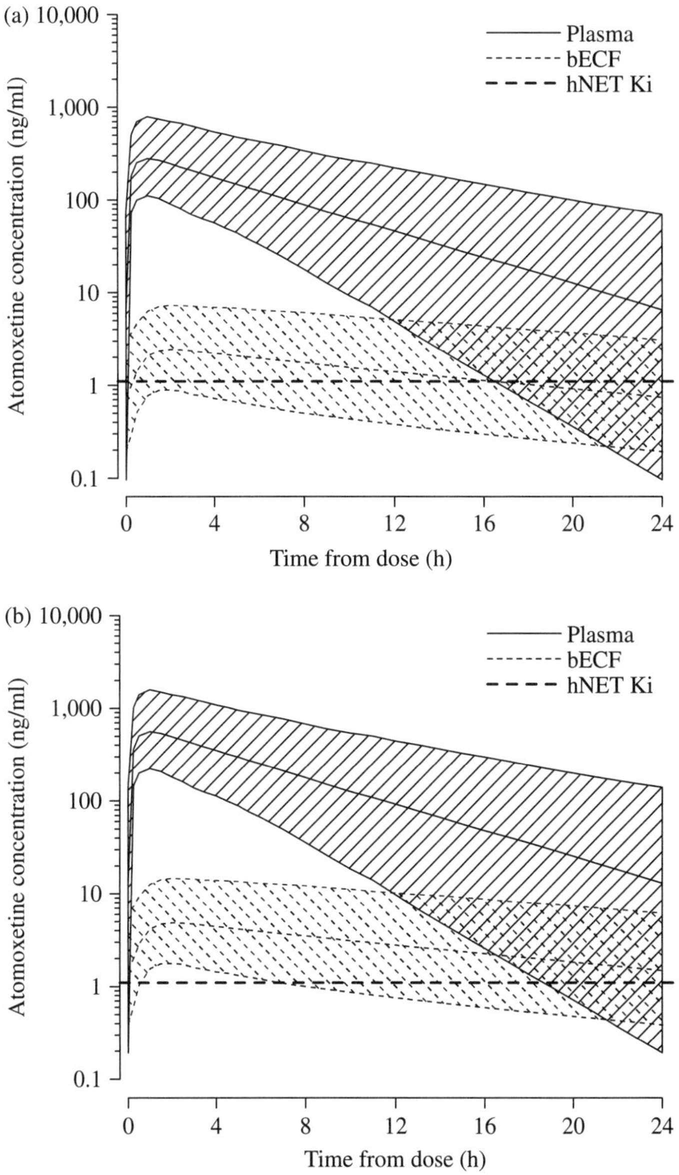

FIGURE 16.2 Model-predicted human plasma and bECF concentrations of atomoxetine after once-daily administration of (a) 40 mg and (b) 80 mg. Shown are the predicted median and 90% confidence interval of plasma and bECF atomoxetine concentrations. The dashed horizontal line represents the atomoxetine inhibition constant determined for the human norepinephrine reuptake transporter (Reprinted from Ref. [3]. With kind permission from ASPET).

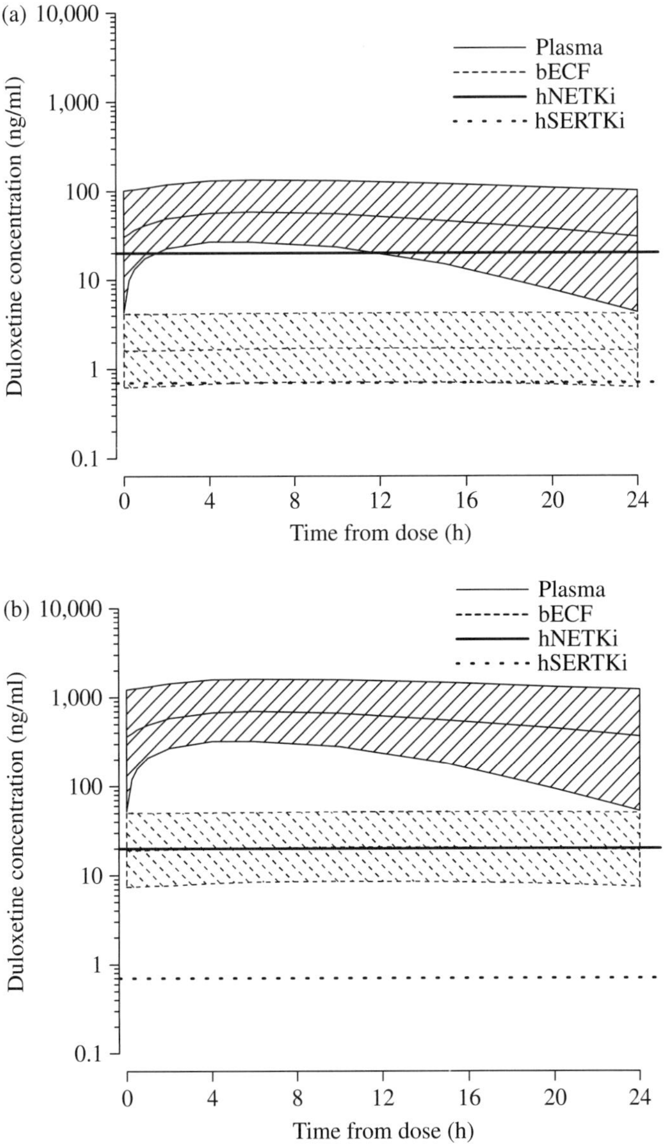

FIGURE 16.3 Model-predicted human plasma and bECF concentrations of duloxetine after once-daily administration of (a) 5 mg and (b) 60 mg. Shown are the predicted median and 90% confidence interval of plasma and bECF duloxetine concentrations. The solid and dashed horizontal lines represent the duloxetine inhibition constant determined for the human norepinephrine and serotonin reuptake transporters, respectively (Reprinted from Ref. [3]. With kind permission from ASPET).

over 70 kg body weight should be initiated at 40 mg QD and increased after a minimum of 3 days to approximately 80 mg QD. After 2–4 additional weeks, the dose may be increased to a maximum of 100 mg QD in patients who have not achieved an optimal response. These predictions were also corroborated with reports that assessed NET target activity in humans via measurement of norepinephrine and 3,4-dihydroxy-phenylglycine in plasma, CSF, and/or urine for duloxetine [40] and atomoxetine [41, 42]. The model prediction for duloxetine was also consistent with the dose-dependent SERT receptor occupancy findings clinically [43], including the ability to discriminate a noneffective 5 mg dose from the clinically effective 60 mg dose. Overall, these results provided confidence that the translational approach can be applied prospectively to identify human efficacious doses of future drug candidates.

CONCLUSIONS

QMD provides important PK–PD data that serve to strengthen predictions of drug candidate performance based on traditional higher-throughput lead optimization pharmacologic and drug disposition assays. Quantitative microdialysis in animals supports translational research and model-based predictions of dose–target site exposure–effect relationships in humans.

REFERENCES

[1] Morgan P, Van der Graaf PH, Arrowsmith J, Feltner DE, Drummond KS, Wegner CD and Street SDA (2012). Can the flow of medicines be improved? Fundamental pharmacokinetic and pharmacologic principles toward improving Phase II survival. Drug Discov Today 17(9/10):419–424.

[2] Doran AC, Osgood SM, Mancuso JY and Shaffer CL (2012). An evaluation of using rat-derived single-dose neuropharmacokinetic parameters to project accurately large animal unbound brain drug concentrations. Drug Metab Dispos 40:2162–2173.

[3] Kielbasa W and Stratford R (2012). Exploratory translational modeling approach in drug development to predict human brain pharmacokinetics and pharmacologically relevant clinical doses. Drug Metab Dispos 40:877–883.

[4] Ball K, Bouzom F, Scherrmann J, Walther B and Declèves X (2013). Physiologically based pharmacokinetic modeling of drug penetration across the blood-brain barrier—towards a mechanistic understanding of IVIVE-based approach. AAPS J 15:913–932.

[5] de Lange ECM (2013). The mastermind approach to CNS drug therapy: translational prediction of human brain distribution, target site kinetics, and therapeutic effects. Fluids Biol Barriers CNS 10:12–28.

[6] Bero AW, Yan P, Roh JH, Cirrito JR, Stewart FR, Raichle ME, Lee JM and Holtzman DM (2011). Neuronal activity regulates the regional vulnerability to amyloid-beta deposition. Nat Neurosci 14:750–756.

[7] Kita JM and Wightman RM (2008). Microelectrodes for studying neurobiology. Curr Opin Chem Biol 12:491–496.

[8] Hashemi P, Walsh PL, Guillot TS, Gras-Najjar J, Takmakov P, Crews FT and Wightman RM (2011). Chronically implanted, nafion-coated Ag/AgCl reference electrodes for neurochemical applications. ACS Chem Neurosci 2:658–666.

[9] Pieber TR, Birngruber T, Bodenlenz M, Höfferer C, Mautner S, Tiffner K and Sinner F (2012). Open flow microperfusion: an alternative method to microdialysis? In *Microdialysis in Drug Development* (pp. 283–302). ed. Müller, M. Springer, New York

[10] Kozielska M, Johnson M, Reddy VP, Vermeulen A, Li C, Grimwood S, de Greef R, Goothuis GMM, Danhof M and Proost JH (2012). Pharmacokinetic-pharmacodynamic modeling of the D_2 and 5-HT_{2A} receptor occupancy of risperidone and paliperidone in rats. Pharm Res 29:1932–1948.

[11] Watson C, Venton B, Kennedy R (2006). In vivo measurements of neurotransmitters by microdialysis sampling. Anal Chem 78(5):1391–1399.

[12] Sun H, Bungay PM and Elmquist WF (2001). Effect of capillary efflux transport inhibition on the determination of probe recovery during in vivo microdialysis in the brain. J Pharmacol Exp Therap 297:991–1000.

[13] de Lange ECM, de Boer AG and Breimer DD (2000). Methodological issues in microdialysis sampling for pharmacokinetic studies. Adv Drug Deliv Rev 45:125–148.

[14] Ungerstedt U and Pycock C (1974). Functional correlates of dopamine neurotransmission. Bull Schweiz Akad Med Wiss 30(1–3):44–55.

[15] Westerink BH, Damsma G, Rollema H, de Vries JB and Horn AS (1987). Scope and limitations of in vivo brain dialysis: a comparison of its application to various neurotransmitter systems. Life Sci 41(15):1763–1776.

[16] Benveniste H and Hüttemeier PC (1990). Microdialysis-theory and application. Prog Neurobiol. 35:195–215.

[17] Kennedy R (2013) Emerging trends in in vivo neurochemical monitoring by microdialysis. Curr Opin Chem Biol 17(5):860–867.

[18] Bundgaard C, Jørgensen M and Mørk A (2007). An integrated microdialysis rat model for multiple pharmacokinetic/pharmacodynamic investigations of serotonergic agents. J Pharmacol Toxicol Methods 55:214–223.

[19] Lönnroth P, Jansson PA and Smith U (1987). A microdialysis method allowing characterization of intercellular water space in humans. Am J Physiol 253(Endocrinol Metab 16): E228–E231.

[20] Olson RJ and Justice JB (1993). Quantitative microdialysis under transient conditions. Anal Chem 65:1017–1022.

[21] Bouw MR and Hammarlund-Udenaes M (1998). Methodological aspects of the use of a calibrator in in vivo microdialysis-further improvement of the retrodialysis method. Pharm Res 15:1673–1679.

[22] Cremers TIFH, de Vries MG, Huinink KD, van Loon JP, v d Hart M, Ebert B, Westerink BHC and de Lange ECF (2009). Quantitative microdialysis using modified ultraslow microdialysis: direct rapid and reliable determination of free brain concentrations with the MetaQuant technique. J Neurosci Methods 178:249–254.

[23] Nirogi R, Kandikere V, Bhyrapuneni G, Benade V, Saralaya R, Irappanavar S, Muddana N and Ajjala DR (2012). Approach to reduce the non-specific binding in microdialysis. J Neurosci Methods 209 (2):379–387.

[24] Bungay PM, Morrison PF and Dedrick RL (1990). Steady state theory for quantitative microdialysis of solutes and water in vivo and in vitro. Life Sci 46:105–119.

[25] Morrison PF, Bungay PM, Hsiao JK, Ball BA, Mefford IN and Dedrick RL (1991). Quantitative microdialysis: analysis of transients and application to pharmacokinetics in brain. J Neurochem 57:103–119.

[26] Kalvass JC and Maurer TS (2002). Influence of nonspecific brain and plasma binding of CNS exposure: implications for rational drug discovery. Biopharm Drug Dispos 23:327–338.

[27] Maurer TS, Debartolo DB, Tess DA and Scott DO (2005). Relationship between exposure and nonspecific binding of thirty-three central nervous system drugs in mice. Drug Metab Dispos 33:175–181.

[28] Summerfield S, Stevens AJ, Cutler L, Del Carmen Osuna M, Hammond B, Tang SP, Hersey A, Spalding DJ and Jeffrey P (2006). Improving the in vitro prediction of in vivo CNS penetration: integrating permeability, Pgp efflux and free fractions in blood and brain. J Pharmacol Exp Ther 316:1282–1290.

[29] Kalvass JC, Maurer TS and Pollack GM (2007). Use of plasma and brain unbound fractions to assess the extent of brain distribution of 34 drugs: comparison of unbound concentration ratios to in vivo P-glycoprotein efflux ratios. Drug Metab Dispos 35:660–666.

[30] Kakee A, Terasaki T and Sugiyama Y (1996). Brain efflux index as a novel method of analyzing efflux transport at the blood brain barrier. J Pharmacol Exp Ther 277:1550–1559.

[31] Becker S and Liu X (2006). Evaluation of the utility of brain slice methods to study brain penetration. Drug Metab Dispos 34:855–861.

[32] Hammarlund-Udenaes M, Bredberg U and Fridén M (2009). Methodologies to assess brain drug delivery in lead optimization. Curr Topics Med Chem 9:148–162.

[33] Shen DD, Artru AA and Adkison KK (2004). Principles and applicability of CSF sampling for the assessment of CNS drug delivery and pharmacodynamics. Adv Drug Deliv Rev 56:1825–1857.

[34] Lin JH (2008). CSF as a surrogate for assessing CNS exposure: an industrial perspective. Curr Drug Metab 9:46–59.

[35] Fridén M, Gupta A, Antonsson M, Bredberg U and Hammarlund-Udenaes M (2007). In vitro methods for estimating unbound drug concentrations in the brain interstitial and intracellular fluids. Drug Metab Dispos 35:1711–1719.

[36] Liu X, Van Natta K, Yeo H, Vilenski O, Weller PE, Worboys PD and Monshouwer M (2009). Unbound drug concentration in brain homogenate and cerebral spinal fluid at steady state as a surrogate for unbound concentration in brain interstitial fluid. Drug Metab Dispos 37:787–793.

[37] Liu X, Smith BJ, Chen C, Callegari E, Becker SL, Chen X, Cianfrogna J, Doran AC, Doran SD, Gibbs JP, Hosea N, Liu J, Nelson FR, Szewc MA and Van Deusen J (2006). Evaluation of cerebrospinal fluid concentration and plasma free concentration as a surrogate measurement for brain free concentration. Drug Metab Dispos 34:1443–1447.

[38] Tauscher J, Jones C, Remington G, Zipursky RB and Kapur S (2002). Significant dissociation of brain and plasma kinetics with antipsychotics. Mol Psychiatry 7:317–321.

[39] Westerhout J, Danhof M and De Lange ECM (2011). Preclinical prediction of human brain target site concentrations: considerations in extrapolating to the clinical setting. J Pharm Sci 100:3577–3593.

[40] Chappell JC, Eisenhofer G, Owens MJ, Haber H, Lachno DR, Dean RA, Knadler MP, Nemeroff CB, Mitchell MI, Detke MJ, Iyengar S, Pangallo B and Lobo ED (2014). Effects of duloxetine on norepinephrine and serotonin transporter activity in healthy subjects. J Clin Psychopharmacol 34(1):9–16.

[41] Kielbasa W, Bingham J and Bieck P (2006). Pharmacokinetic-pharmacodynamic modeling of atomoxetine in adults: characterization of dihydroxyphenylglycol (DHPG), a pharmacological marker of norepinephrine transport (NET) inhibition, in plasma and cerebrospinal fluid. American Association of Pharmaceutical Sciences Annual Meeting; November 1, 2006; San Antonio, TX. Poster Presentation Abstract 002685, American Association of Pharmaceutical Sciences, Arlington, VA.

[42] Montoya A, Escobar R, García-Polavieja MJ, Lachno DR, Alda JÁ, Artigas J, Cardo E, García M, Gastaminza X and Gilaberte I (2011). Changes of urine dihydroxyphenylglycol to norepinephrine ratio in children with attention-deficit hyperactivity disorder (ADHD) treated with atomoxetine. J Child Neurol 26:31–36.

[43] Takano A, Suzuki K, Kosaka J, Ota M, Nozaki S, Ikoma Y, Tanada S and Suhara T (2006). A dose-finding study of duloxetine based on serotonin transporter occupancy. Psychopharmacology 185:395–399.

17

IMAGING TECHNIQUES FOR CENTRAL NERVOUS SYSTEM (CNS) DRUG DISCOVERY

Lei Zhang and Anabella Villalobos

Pfizer Worldwide Research and Development, Neuroscience Medicinal Chemistry, Cambridge, MA, USA

INTRODUCTION

The complexity of the human central nervous system (CNS) presents a daunting challenge to the development of effective treatments for a vast spectrum of neurological disorders that cause a tremendous burden for society and personal caregivers. Despite significant strides made in the past several decades in the understanding of the possible causes of CNS disorders and the availability of therapeutic options, unmet medical needs continue to be high [1]. For example, Alzheimer's disease (AD) is estimated to affect 5.4 million Americans with roughly one in eight people aged 65 and older living with this highly debilitating disease [2]. More alarmingly, without disease-modifying therapy, AD prevalence is predicted to grow rapidly to impact over 106 million people worldwide, reaching a pandemic status in 2050 [3]. While current treatment options for AD patients alleviate symptoms to help daily activity management, they offer no impact on disease onset and progression. Situations like this clearly highlight the urgent need for new and more effective therapeutics for various neurological disorders. It is essential to incorporate new strategies that could accelerate CNS drug discovery and reduce the current high attrition rate of clinical candidates. Noninvasive imaging techniques can help to address some of the key challenges facing the current

Blood–Brain Barrier in Drug Discovery: Optimizing Brain Exposure of CNS Drugs and Minimizing Brain Side Effects for Peripheral Drugs, First Edition. Edited by Li Di and Edward H. Kerns.

CNS drug discovery process, including the inaccessibility of the human brain, uncertainty around target engagement by the drug molecules, and a lack of predictive functional biomarkers for efficacy readout. In this chapter, we will provide an overview of different types of noninvasive imaging techniques that can be integrated into the CNS drug discovery process and their impact from informing basic pharmacology to enabling critical decision-making in clinical evaluations of novel CNS therapeutics. We will primarily focus on positron emission tomography (PET) and single photon emission computed tomography (SPECT) techniques.

APPLICATIONS OF IMAGING TECHNIQUES IN THE DRUG DISCOVERY PROCESS

Strengths and Limitations of Commonly Used Imaging Techniques

Commonly used noninvasive imaging techniques can be broadly divided into two major categories: (1) morphological/anatomical imaging techniques to detect tissue structure changes, such as magnetic resonance imaging (MRI) and X-ray computed tomography (CT); (2) molecular imaging techniques to detect molecular and cellular events utilizing appropriate fluorescence probes (optical imaging) or radiotracers (PET and SPECT) [4]. Each of these imaging techniques has its own strengths and limitations, as summarized in Table 17.1 [5].

MRI provides excellent soft-tissue contrast and high spatial resolution (10–100 μm), which allow sensitive detection of pathological changes in the soft tissues. MRI can also be used to map functional neural activities (fMRI) by detecting changes in blood flow related to energy use by brain cells [6]. Since its introduction in the early 1980s, MRI has evolved into a highly versatile neuroimaging technology extensively used in the clinical setting to provide qualitative diagnostic and quantitative parametric information for CNS disorders such as AD [7], Parkinson's diseases (PD) [8], and stroke [9]. Its preclinical usage in early drug discovery and small animal imaging, however, is somewhat limited partly due to its relatively high cost [10]. CT is a lower-cost anatomic imaging technology, particularly suitable for skeletal structures and lung imaging. In the clinical setting, it is often used in combination with other imaging methods such as PET [11] and SPECT [12] to allow detections of pathological changes with higher anatomic resolution. In terms of limitations, neither MRI nor CT is able to detect diseases prior to tissue structural changes or provide information related to a specific target.

Such limitations can be addressed by using molecular imaging techniques, which employ target-specific probes or radiotracers to detect molecular and cellular events related to a given target. In optical imaging, fluorescence probes are used, which are either conjugated to biocompatible fluorescent dyes [13] or bioluminescent proteins such as green fluorescent protein (GFP) [14] or luciferin/luciferase system [15]. Fluorescence emission triggered by external light source (fluorescence dyes) or generated endogenously (bioluminescence) can be captured by a detection unit, such as a charged-coupled device (CCD) camera, to give information ranging from microscopic live-cell tracking [16] to macroscopic small animal *in vivo* distribution [17].

TABLE 17.1 Overview of commonly used imaging techniques and their strengths and limitations

	Morphological/anatomical imaging		Molecular imaging		
Imaging techniques	MRI	CT	Optical Imaging	PET	SPECT
Strengths	High soft tissue resolution; No depth limit; Clinical usage	High bone/lung resolution; No depth limit; Clinical usage	High sensitivity; Low cost	High sensitivity; No depth limit; Clinical transaction	High sensitivity; No depth limit; Clinical transaction
Limitations	High cost; No target-specific information	No target-specific information	Limited depth penetration; Limited clinical transaction	High cost	Limited spatial resolution

Compared to other imaging methods, optical imaging has distinct advantages in cost and throughput, and has been used extensively in early-stage preclinical studies from high-throughput screening to *in vivo* compound characterization. Newer optical imaging methods, such as fluorescence molecular tomography (FMT), can provide quantification at femtomolar level and submillimeter spatial resolution in tissues of small animals (e.g., rodents). However, the usage of optical imaging techniques in larger animals (e.g., nonhuman primates (NHPs)) and human studies has been hampered due to the limitation in their tissue depth penetration (currently micrometers to single digit centimeters depending on technologies and light sources used). New technological advances in light source and detector system are necessary to enable optical imaging in larger animals and eventually in humans for clinical translation. Complementary to optical imaging, radiotracer–based nuclear imaging techniques, such as PET and SPECT, have no depth limit and offer similar sensitivity; thus they are well suited for brain imaging in preclinical species (from rodents to NHP) and humans. This feature of PET and SPECT offers a unique opportunity for the translation of *in vivo* outcomes from preclinical animals to humans, bridging the preclinical and clinical space. Compared to optical imaging, PET and SPECT incur higher cost. Therefore, they are typically used to assess more advanced leads of near-clinical candidate status.

Integration of Imaging Techniques in the CNS Drug Discovery Process

Based on what they can offer, various imaging techniques can be integrated in the drug discovery process in a complementary rather than competitive manner as illustrated in Scheme 17.1 [18]. Optical imaging methods can play a significant role in the early discovery phases at the preclinical stage, for example, target identification

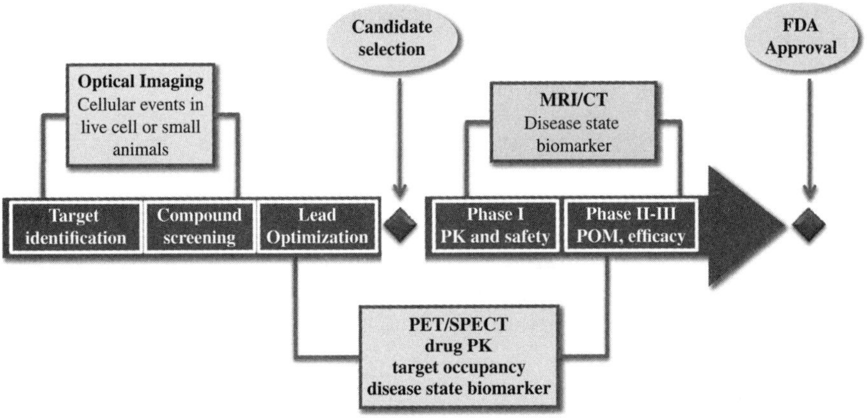

SCHEME 17.1 Integration of various imaging techniques in CNS drug discovery.

and compound screening, to inform *in vitro* pharmacology and target characteristics through live cell imaging and *in vivo* behaviors of early leads in small animals [19]. Conversely, MRI and CT can be used in later clinical stages (Phase I–III) to understand pathological changes for the diseases of interest and serve as diagnostic tools and disease state biomarkers for patient selection and disease progression monitoring [20]. Across preclinical and clinical stages, PET and SPECT can serve as powerful translational tools to bridge the preclinical research to clinical study design and decision-making [21]. PET and SPECT can be used at the lead optimization stage to establish correlation between target occupancy (TO) and efficacy of advanced leads. In conjunction with animal safety studies, PET and SPECT can also inform optimization and differentiation strategies to ensure the best molecule is advanced to costly clinical studies. Furthermore, the preclinically established TO/efficacy correlation can also be used in the clinical setting to achieve proof of mechanism (drug on target), optimize clinical doses (efficacy/safety separation), and enable clinical go/no go decisions. Besides target-specific information, PET and SPECT can also be used to assess human brain exposure and blood–brain barrier (BBB) permeability, which would be otherwise difficult to measure. Finally, if a given neurological target is up-regulated or down-regulated in a given CNS disorder, a target-specific PET or SPECT tracer may be used as a disease-state biomarker to assist in diagnosis, and disease progression monitoring. With new advances in all of these technologies, the impact of various imaging methods could expand beyond what has been described.

Incorporation of these noninvasive imaging techniques into the appropriate stages of the CNS drug discovery process could thus provide valuable decision-making information to key topics: what is the right target (optical imaging), which is the right molecule (optical imaging, PET/SPECT), what is the right patient group (PET/SPECT, MRI, CT), and how to monitor disease progression (PET/SPECT, MRI, CT). Right choices in these key aspects should undoubtedly lead to reduction in the attrition rate and accelerate the introduction of new and improved therapeutics for CNS disorders which still face high unmet medical need.

In this chapter, we dedicate the following sections to an in-depth discussion of PET and SPECT in light of their multifaceted roles in the CNS drug discovery process, covering the following topics: (i) basic principles of PET and SPECT, (ii) discovery process of PET and SPECT radiotracers, and (iii) applications in CNS drug discovery illustrated by specific literature examples.

PET AND SPECT: BASIC PRINCIPLES AND RADIOTRACER DEVELOPMENT

PET and SPECT Comparison

PET and SPECT are noninvasive nuclear imaging techniques that utilize tomographic reconstruction methods to provide three-dimensional (3D) images of radiotracer distribution in the regions of interest [22, 23]. Both technologies have been extensively used in brain imaging to assist decision-making at different stages of the CNS drug discovery process. These technologies, however, differ on the properties of their corresponding radionuclides and the resolution and quality of their images. In terms of radionuclides, PET requires positron-emitting radionuclides while SPECT requires single gamma-emitting radionuclides (Table 17.2) [24]. Routinely used PET radionuclides, such as [^{11}C] and [^{18}F], are isotopes of elements commonly found in CNS drug-like molecules. Incorporation of these radionuclides typically results in little to no impact on physicochemical and pharmacological properties, thus allowing more flexibility in tracer design. SPECT radionuclides, on the other hand, are rarely found in drug-like small molecules and may lead to considerable changes in compound properties. In terms of radiotracer preparation, PET requires rapid synthesis/purification as well as proximity to a cyclotron facility due to the short half-lives of PET radionuclides, [^{11}C] ($T_{1/2}$ = 20 min) and [^{18}F] ($T_{1/2}$ = 110 min). In comparison, SPECT radionuclides have much longer half-lives, [^{123}I] ($T_{1/2}$ = 13 h) and [^{99m}Tc] ($T_{1/2}$ = 6 h), and can be bought commercially or synthesized onsite with low-cost generators, thus offering greater flexibility in tracer synthesis and better access to the technology.

In terms of image quality, PET allows concurrent detection of high-energy collinear gamma photon pairs resulting from the annihilation events of emitted positrons

TABLE 17.2 PET and SPECT radionuclides

PET radionuclides	$T_{1/2}$	SPECT radionuclides	$T_{1/2}$
^{15}O	2 min	^{99m}Tc	6 h
^{13}N	10 min	^{123}I	13 h
^{11}C	20 min	^{111}In	68 h
^{18}F	110 min	^{67}Ga	78 h
^{64}Cu	12.8 h	^{201}TI	73 h
^{68}Ga	68 min	^{133}Xe	127 h
^{82}Rb	1.3 min	^{131}I	192 h

with electrons from the surrounding media, leading to high precision in locating the original emission events in the regions of interest. This translates into images with high sensitivity and spatial resolution (1–3 mm), consequently allowing quantification of tracer concentration in brain regions of interest with a high degree of accuracy. In contrast, the collimators used as the SPECT detector system detect photons within a small angular range, while SPECT radionuclides emit photons in all directions. Thus, only a smaller percentage of emission events are recorded, leading to lower spatial resolution (~10–14 mm) and lower sensitivity (~2–3 magnitudes) than PET. In addition, the spatial resolution and sensitivity of SPECT are also position-dependent, with increased attenuation and scattering of the photons as they travel through dense tissues. Therefore, SPECT in general requires a larger number of attenuation corrections and longer scan times for quantification.

Attributes for Effective PET/SPECT Radiotracers

In order to enable PET and SPECT imaging, the availability of radiotracers with suitable attributes is essential [25]. A demanding list of prerequisite attributes need to be considered when designing or selecting target-specific PET and SPECT tracers. First of all, the tracer candidates should contain structural moieties that allow late-stage incorporation of PET or SPECT radionuclides. In particular for PET, rapid synthesis and purification are essential considering the short half-lives of PET radionuclides. Second, a viable PET/SPECT tracer must be potent and selective toward its intended pharmacological target. The maximum concentration of target binding sites (B_{max}) is a key parameter for consideration, as it defines the level of affinity (K_d) required for a successful radiotracer ($B_{max}/K_d \geq 10$) [26]. In general, radiotracers are more potent than typical drug candidates, often in the subnanomolar (nM) range considering the low expression levels of most brain targets [27]. The requisite off-target selectivity is dependent not only on relative affinities, but also on the brain distribution and expression levels of competing targets. A high level of selectivity (>30- to 100-fold) is recommended, particularly against targets that are highly expressed in the brain or colocated with the target of interest. Third, a suitable radiotracer should be brain-permeable and has low nonspecific binding (NSB) to achieve acceptable signal-to-noise ratio. Finally, a suitable radiotracer should not form brain-permeable radioactive metabolites, which would compromise imaging quantification since PET and SPECT detector systems are unable to distinguish signals from different radioactive chemical entities.

Novel CNS PET/SPECT Tracer Development Process and Recent Advances

In many ways, the development process of novel CNS PET/SPECT tracers bears great similarity to the drug discovery process. As illustrated in Scheme 17.2, the process starts with an understanding of the expression level (B_{max}) and brain biodistribution of the target of interest (stage 1), which would inform subsequent lead selection criteria and imaging study planning. At stage 2, the design, synthesis, and pharmacology/PK profiling of potential tracer leads with the aforementioned attributes are carried out. These first two stages are interconnected. If no previous

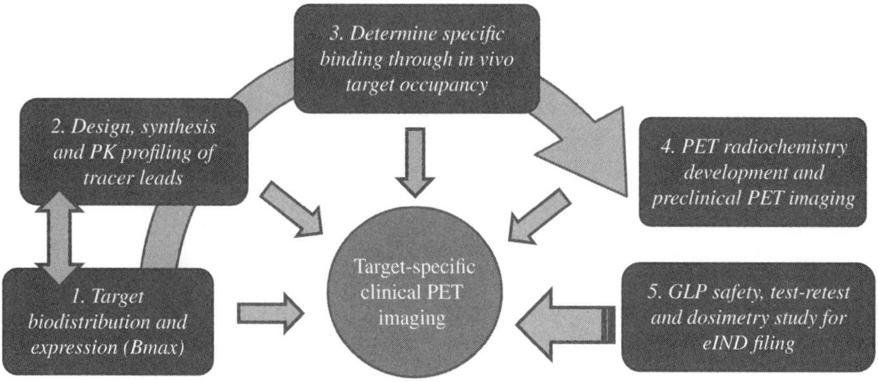

SCHEME 17.2 Novel PET/SPECT tracer development process.

B_{max} and brain biodistribution knowledge exists for a neuro-target, one can start with stage 2 by identifying a lead that might not meet all desirable attributes, but with good potency and selectivity. Such a lead can then be tritiated and used in an autoradiography study to provide the requisite B_{max} and biodistribution information. At stage 3, suitable tracer leads are assessed *in vivo* (typically in rodents) to ensure that specific binding can be achieved, usually in an *in vivo* binding study either with a high dose of blocking compound or in wild-type (WT)/knockout (KO) animals to define the specific binding window. If an acceptable level of specific binding is achieved, the lead will be advanced to stage 4, wherein the tracer lead is labeled with a suitable radionuclide and assessed for viability in PET or SPECT imaging studies in preclinical species. Although stages 3 and 4 share a similar tracer assessment function, stage 3 often serves as a more cost-effective way of gaining an early read on *in vivo* performance of potential tracers, offering the ability to prioritize tracer leads prior to initiating more expensive and labor-intensive PET/SPECT imaging studies.[1] As an added benefit, if *in vivo* specificity is established with a tritiated ligand at stage 3, it can be used to set up *in vivo* TO studies in rodents to facilitate the evaluation of drug candidates. At stage 5, requisite data are collected for exploratory investigational new drug (eIND) application filing, including reproducibility in test/retest studies, organ exposure of radioactivity in dosimetry studies, and compound safety in rat good laboratory practice (GLP) toxicity studies. At the final stage, a tracer will undergo clinical validation in human PET or SPECT imaging studies and, if favorable, will be used to measure target engagement and facilitate the progression of drug candidates.

As described earlier, development of a novel target-specific PET or SPECT tracer can be a fairly lengthy process, due to challenges in meeting a demanding list of requisite tracer attributes and the relatively empirical nature of the tracer design and selection. Recent advances in tracer discovery strategies, particularly in the realm of rational design and effective evaluation, have offered encouraging improvement in the tracer discovery process. Two representative strategies are highlighted herein. The first strategy was disclosed by our group, focusing on predictive tools and *in vitro* assays for rational tracer design and selection [28]. Recognizing that poor brain permeability and high NSB are the most frequent causes for failure in CNS PET tracer development, our group aimed to gain a better understanding of molecular properties that are required for brain permeability and low NSB, and identify predictive methods for these two important *in vivo* parameters. An analysis of a PET tracer database consisting of 62 clinically validated CNS PET tracers and 15 failed tracers as negative control was carried out. Physicochemical and absorption, distribution, metabolism, and excretion (ADME) properties were examined, from which we identified a set of preferred properties to inform the design and selection of novel PET tracers including a new design tool, CNS PET multi-parameter optimization (MPO) (Fig. 17.1a). In addition to the existing criteria around *in vitro* binding potential ($B_{max}/K_d > 10$) and selectivity, we recommend that novel tracers should be designed and selected based on suitable physicochemical properties (CNS PET MPO > 3), permeability (RRCK Papp AB[2] [29] > 5 × 10^{-6} cm/s), and P-glycoprotein (P-gp) efflux (MDR1 BA/AB [30] < 2.5) for good brain permeability, and appropriate fraction unbound in brain ($F_{u\_b} > 0.05$) to minimize NSB. In addition, by targeting CNS PET MPO scores greater than 3, a higher probability of aligning all three key ADME parameters in one molecule can be achieved. Subsequent analysis of a collection of 10 high-performing PET tracers showed greater alignment with the criteria defined earlier, suggesting that such criteria could be used to steer tracer development efforts toward higher performing tracers. The prospective use of this set of PET design and selection parameters was further illustrated by the identification of [^{18}F]PF-05270430, the first highly selective PDE2A PET tracer. Guided by the parameters mentioned, [^{18}F]PF-05270430 was identified within a single design cycle with a total of six analogs, and demonstrated an uptake pattern (high in striatum and low in cerebellum) consistent with the PDE2A enzyme biodistribution in monkey (Fig. 17.1a).

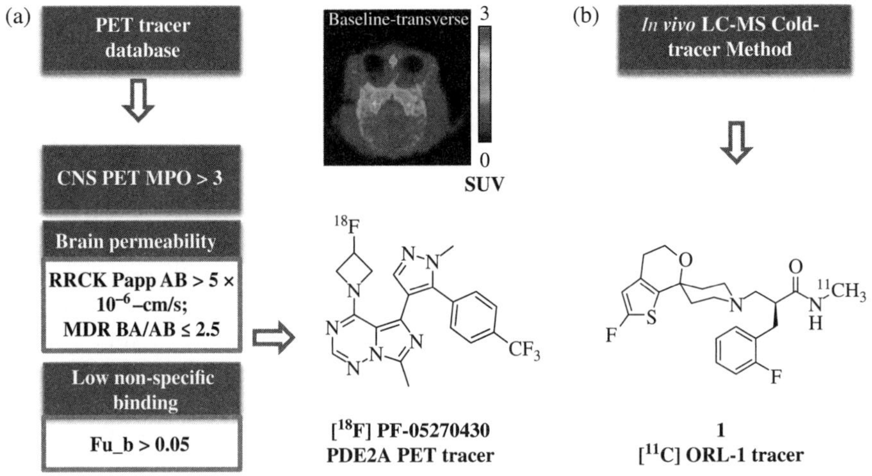

FIGURE 17.1 CNS PET MPO and *in vivo* LC-MS cold tracer method. For color detail, please see color plate section.

In another strategy, a liquid chromatography–mass spectral (LC-MS) cold tracer method was proposed to serve as a simplified and cost-effective technique for *in vivo* tracer evaluation [31]. In this method, a low dose of the tracer (≤5 μg/kg) in a nonradiolabeled "cold" form is injected into a group of rats. After a set time, the animals are sacrificed and brain tissue regions of interest are dissected and analyzed. Instead of scintillation spectroscopy (radio counts), the distribution of the "cold" tracer in various brain regions is quantified by high-sensitivity LC-MS. The specific binding is determined by co-dosing the cold tracer with a high dose of target-selective blocking compound or using KO animals. This method offers many advantages over the traditional "hot" radiotracer method. First, it eliminates the need for tritiation, saving time and cost as well as avoiding environmental concerns. Second, it can offer PK measurements for both tracer and the blocking compound, yielding target engagement together with exposure information in a single experiment. Finally, it allows concurrent use of multiple "cold" tracers in the same animals for pharmacology studies. All these advantages mount to a faster and more cost-effective way to test new radiotracer candidates *in vivo*. The usage of this methodology has been demonstrated recently by the discovery of novel nociceptin/orphanin FQ peptide (NOP) PET tracer [^{11}C]-(*S*)-3-(2′-fluoro-4′,5-dihydrospiro[piperidine-4,7′-thieno-[2,3-*c*] pyran-1-yl)-2(2-fluorobenzyl)-*N*-methylpropanamide (**1**) (Fig. 17.1b), in which three potential tracer candidates were tested in the LC-MS cold tracer methodology, from which compound **1** emerged as the best lead [32]. Compound **1** was subsequently radiolabeled with carbon-11 and demonstrated specific binding in NHP with an estimated specific-to-nonspecific ratio of 1.28, consistent with the outcome from the cold tracer method.

PET AND SPECT IMAGING IN CNS DRUG DISCOVERY

With the rapid expansion of PET and SPECT imaging applications in preclinical research and clinical drug development, effective CNS PET and SPECT tracers have been developed for a wide range of targets including G-protein coupled receptors (GPCRs), ion channels, transporters, enzymes, and amyloid plaques. Their applications in CNS drug discovery can be broadly divided into three categories: (1) biodistribution studies to inform drug brain permeability, (2) target occupancy studies to enable proof of mechanism (POM) and inform clinical decisions, and (3) disease state imaging to monitor disease progression and drug-induced changes. In this section, recent examples are highlighted to illustrate PET and SPECT applications in each of these three areas.

Biodistribution Studies

In CNS drug discovery, one key requirement for drug candidates is that they must penetrate the BBB and reach the target of interest in the brain. The human brain permeability of a drug candidate is typically predicted based on preclinical neuroPK studies [33] and is not measured in the clinic. The noninvasive nature of PET and

SPECT allows direct measurement of human brain permeability. For this type of study, a radiolabeled version of the exact drug molecule is prepared with a radionuclide (e.g., ^{11}C and ^{18}F) that induces no structural change. The biodistribution of the radiotracer in plasma and brain can then be quantified by PET or SPECT imaging to give a direct read on the brain permeability of the drug molecule [34]. In a recent example, a radiolabeled muscarinic acetylcholine receptor 1 (mAchR1) allosteric agonist clinical candidate [^{11}C]GSK1034702 (**2**) was prepared [35] and advanced to an open-label[3] PET study for brain penetration in healthy volunteers (Fig. 17.2).[4] A tracer dose of **2** was administered intravenously and the subjects underwent baseline PET scans. The time activity curves (TACs) of **2** illustrated good brain uptake and heterogeneous distribution with increased uptake in medial temporal lobe (MTL) and thalamus. The peak concentration was achieved in plasma at approximately 3 h post dose and slightly later in brain. The overall equilibrium partition coefficient of the tracer between brain and plasma is 2.63, consistent with preclinical data in baboons. The results confirmed the passive diffusion of the clinical candidate across the human BBB.

The human brain permeability from PET studies can also be used to enhance our understanding of the human BBB [36]. For example, clinically measured human brain permeability data can be used to assess the predictive power of various *in vitro* cell-based BBB models [37]. In a recent publication by Mabondzo et al., 6 PET tracers were evaluated in an *in vitro* coculture-based model of human BBB (cold form) in parallel to PET imaging studies [38]. Within the small sample size, a good correlation ($r^2 = 0.90$) was demonstrated between this coculture-based *in vitro* assay and PET-measured permeability coefficient, suggesting the potential of this assay as a high-throughput model to predict human brain permeability. In another study, PET imaging was used to directly measure Pgp transport activity in BBB across several preclinical

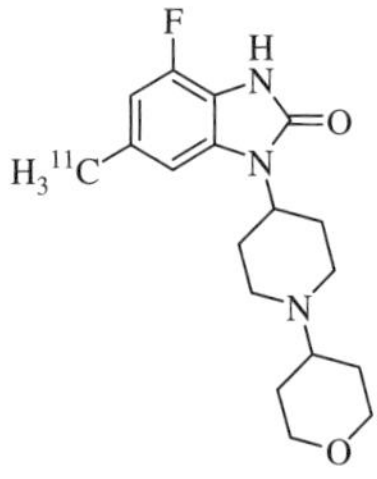

2
[^{11}C] GSK1034702

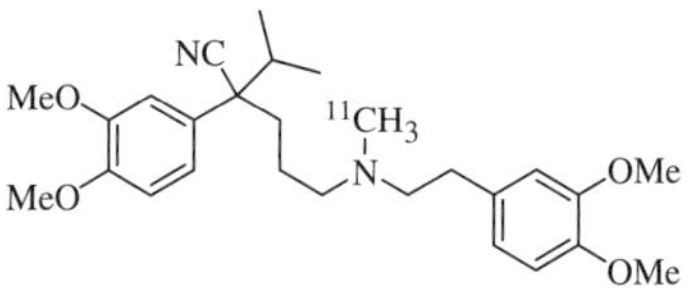

3
[^{11}C] Verapamil

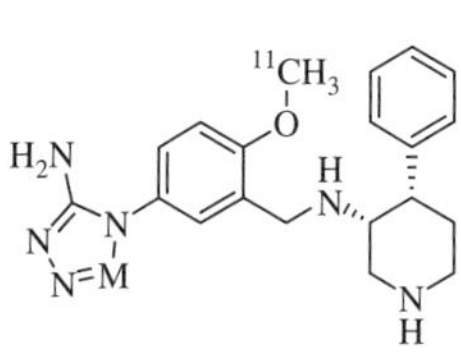

4
[^{11}C] GR205171

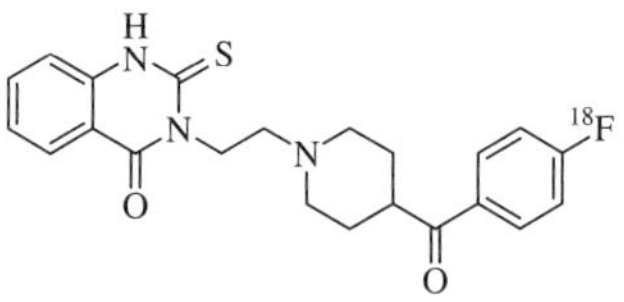

5
[^{18}F] Altanserin

FIGURE 17.2 Examples of PET tracers used in human biodistribution studies.

species and humans [39]. Three radiolabeled P-gp substrates (shown in Fig. 17.2), [^{11}C] verapamil (**3**), [^{11}C] GR205171 (**4**), and [^{18}F] altanserin (**5**), were used to quantify brain-to-plasma ratios in various preclinical species and humans with or without a P-gp inhibitor cyclosporine A (CsA). Pronounced species differences were found in the brain concentrations of these three tracers, with higher brain distribution in humans, monkeys, and mini pigs than in rats and guinea pigs, suggesting species differences in BBB P-gp activity. For example, the brain-to-plasma ratio of [^{11}C] GR205171 in humans was almost ninefold higher than the ratio in rats. Such differences should be considered when extrapolating neuroPK data in preclinical species (particularly rat) to humans to avoid underestimation of human brain permeability for drug molecules that might be P-gp substrates.

Target Occupancy (TO) Measurement

In CNS drug discovery, PET and SPECT are most frequently used to quantify the concentration of drug binding to a specific pharmacological target, confirming target engagement to enable POM [40] and providing TO measurement [41]. There are several ways of using TO effectively to inform critical drug discovery and development decisions. First of all, the TO information can be used to guide clinical dose selection to ensure adequate testing of the therapeutic effect of a drug molecule. This was demonstrated by the clinical evaluation of D2 partial agonist aripiprazole, a highly successful treatment for psychosis. While previously developed antipsychotics based on D2 antagonism suggested that efficacy was typically associated with 60–80% D2 TO [42], a full dose–response PET TO study of aripiprazole using [^{11}C] raclopride (Fig. 17.3a) revealed that this drug required much higher TO (~90–95%) to elicit antipsychotic efficacy [43]. Prior to this knowledge, some of the early clinical studies of aripiprazole in healthy volunteers were run with underestimated therapeutic doses. Without the key D2 TO information, the subsequent proof of concept efficacy studies with aripiprazole could have targeted lower exposures, thus missing the therapeutic efficacy window of this successful drug. Second, the TO information can be used to show efficacy and side effect (therapeutic index, TI) differentiation among multiple clinical leads in similar mechanisms. This is illustrated by recent work published by Atack et al. [44] in which the PET TO at the benzodiazepine binding site for TPA-023, a GABAa receptor α2/α3 subtype-selective agonist, was measured across species using [^{18}F] flumazenil (Fig. 17.3b), showing a good alignment of EC_{50} (50% occupancy) values across rats, baboons, and humans. In humans, a single 2 mg dose of TPA-023 produced 50–60% TO without adverse sedative side effects, a clear improvement over previously evaluated nonselective GABAa agonists, which typically produced the sedative side effects at TO less than 30%. These results demonstrated the potential TI benefit associated with the profile of TPA-023 and thus differentiation from previous clinical entries. Finally, TO can be used to establish well-defined clinical “no go” criteria, if a clinical candidate cannot achieve targeted TO due to PK or safety reasons or cannot demonstrate clinical efficacy even with complete TO. For example, early trials of arepitant, an NK-1 receptor antagonist, showed promise for treating depression.

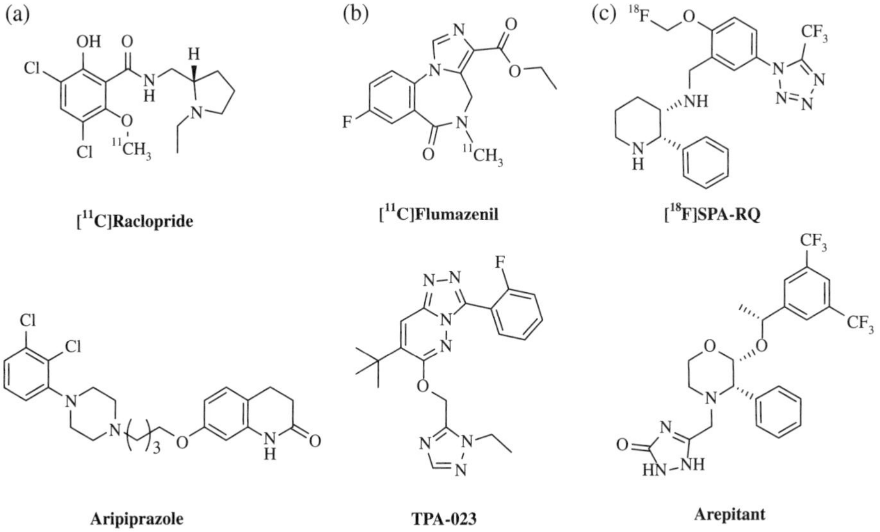

FIGURE 17.3 (a) D2 TO measurement of aripiprazole by [^{11}C]raclopride; (b) GABAa benzodiazepine site TO measurement of TPA-023 by [^{11}C]flumazenil; (c) NK1 TO measurement of arepitant by [^{18}F]SPA-RQ.

However, subsequent trials failed to demonstrate clinical efficacy [45] even with doses (80 and 160 mg) that gave near complete TO as measured with [^{18}F]SPA-RQ [46] (Fig. 17.3c), which ultimately led to the termination of clinical development of arepitant for this indication. A clear "no go" decision of an ineffective therapy or mechanism is important as this will stop unnecessary resource drain and redirect focus to more promising targets and molecules.

In light of the clinical significance of the TO information, incorporation of a PET or SPECT strategy early in the preclinical program to align TO, exposures, and efficacy can often lead to significant acceleration of project progression. This is exemplified by a recent publication in the development of an mGluR5 negative allosteric modulator (NAM) as a treatment for L-DOPA-induced dyskinesia in Parkinson's disease patients (PD-LID) [47]. Evaluation of efficacy in preclinical species required a translatable NHP PD-LID model that was lengthy, expensive, and labor-intensive (6-month data turnaround, >$500K). Therefore routine screening of leads using this assay was impractical and expensive. To avoid a significant bottleneck in compound evaluation and progression, a research strategy driven by TO was pursued. Analysis of the existing clinical data indicated that PD-LID efficacy generally associated with greater than 80% projected mGluR5 TO. Assessment of *in vitro* mGluR5 binding affinity, *in vivo* target occupancy (IVTO) in rats using [^{3}H]MPEPy and PET TO in NHP using [^{18}F]FPEB showed excellent alignment, which allowed facile lead development and compound prioritization driven by *in vitro* binding assay and rat IVTO studies. From this effort, PF470 emerged as a promising lead and was evaluated in an NHP PET TO study, yielding an IC_{50} (50% occupancy) of 0.43 nM in alignment with its rat IVTO IC_{50} (1.04 nM) as well as the *in vitro* binding affinity (0.9 nM).

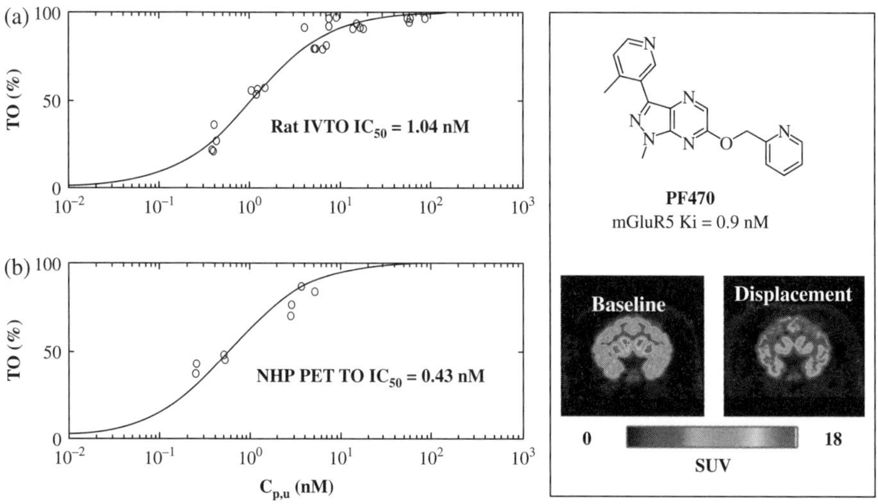

FIGURE 17.4 (a) Rat IVTO and PET TO of **PF470**; (b) the chemical structure of **PF470** and the coronal views of [^{18}F]FPEB-dependent PET images from an NHP-receiving vehicle (baseline) and **PF470** corresponding to 76% TO. For color detail, please see color plate section.

Based on the NHP PET TO data, three doses of PF470 (0.1, 0.32, and 1.0 mg/kg) were selected for the NHP PD-LID model. As expected, robust efficacy was demonstrated with the two top doses which showed TO exceeding 80% (corresponding to 90 and 97% TO, respectively) (Fig. 17.4).

Disease State Biomarkers

One of the fastest-growing fields within neuroimaging is the use of PET and SPECT imaging as disease state biomarkers to enable early disease detection and disease progression. A distinct advantage of PET and SPECT compared to other diagnostic tools (e.g., MRI and CT) is that they may detect neurotarget changes prior to the emergence of clinical symptoms and brain structural changes, thus allowing early diagnosis and therapeutic intervention. For example, until recently the diagnosis of AD was heavily dependent on the presence of symptoms of dementia. However, brain alterations, such as accumulation of Aβ protein aggregates (Aβ plaques), could occur years before the initial manifestation of dementia. Efforts in developing PET or SPECT tracers that specifically bind to Aβ plaques proved to be highly fruitful, yielding multiple diagnostic tools to detect AD pathology in pre-dementia stages [48]. This advance in neuroimaging may be highly relevant to the development of new experimental AD therapies, as they are predicted to have the greatest therapeutic benefit at the early stages of AD. Florbetapir, a fluorine-18-labeled stilbene derivative (Fig. 17.5), became the first FDA-approved AD diagnostic tool for Aβ plaque density estimation in adult patients with cognitive impairment in 2012.[5] This was quickly followed by the approval of two other Aβ plaque tracers, flutemetamol in October

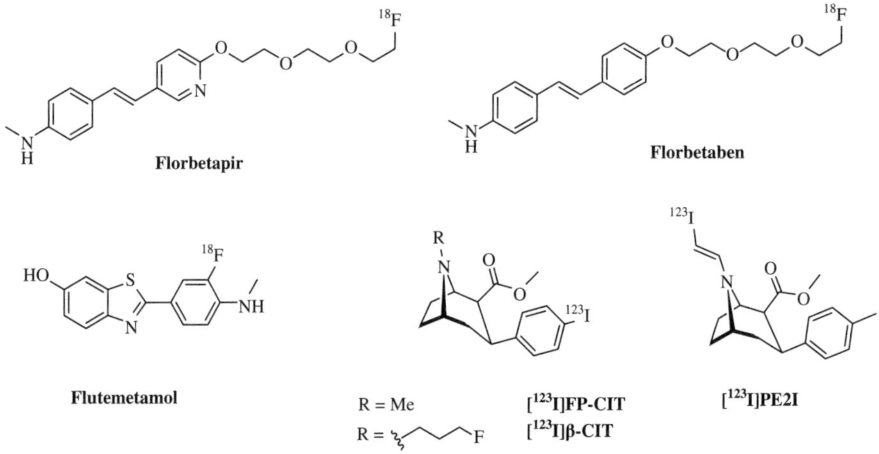

FIGURE 17.5 Examples of PET and SPECT tracers as disease state biomarkers for AD or PD.

2013[6] and florbetaben, a structurally similar stilbene analog of florbetapir, in March 2014 (Fig. 17.5).[7] Extensive use of these Aβ plaque tracers is expected in future clinical studies of experimental AD therapies for selecting the appropriate patient group and monitoring disease progression. Similarly, dopamine transporter SPECT tracers, such as [^{123}I]FP-CIT and [^{123}I]β-CIT (Fig. 17.5), have also been used as sensitive markers of dopamine neurodegeneration in clinical studies to assess PD, based on the correlation between reduction in nigrostriatal binding and disease severity [49]. In a recent report, SPECT imaging using [^{123}I]FP-CIT was used to generate a striatal asymmetry index as a potential predictor of responsiveness to L-DOPA in patients with PD [50]. A structurally similar tracer, [^{123}I]PE2I, has also shown promise as a highly sensitive and specific diagnostic tool to detect striatal neurodegeneration in patients with minor Parkinsonian symptoms [51].

CONCLUSIONS

The significant unmet medical need in CNS disorders calls for new strategies to improve the overall efficiency and success rate in drug discovery. Imaging techniques allow noninvasive detection of pathological, cellular, and molecular events in living subjects, from cells to humans. Each type of imaging technology presents its own unique strength and can be incorporated into various stages of the drug discovery process to provide valuable information for key preclinical and clinical decision-making, which in turn would allow for a more focused drug and efficient discovery effort. Furthermore, recent advances in imaging technology have started to address the specific limitations associated with each of them. For example, newer optical imaging methods, for example, diffuse optical tomography (DOT) and FMT, can image relatively deeper tissues up to 10 cm utilizing near-infrared light (NIR) and higher-density detector systems, which open the possibility of imaging studies in larger animals [52, 53]. Recent advances in PET and SPECT tracer design and evaluation have shown promise in expediting the discovery of novel PET tracers to enable preclinical to clinical translation studies on neuro-targets. New technical advancements in SPECT have led to better resolution/sensitivity [54] and an expanded role of SPECT in small animal preclinical imaging. The combination of imaging techniques such as PET/MRI allows for better localization and quantification of target binding in the brain regions of interest [55]. Such improvements will undoubtedly broaden the use of imaging techniques in the CNS drug discovery process, enable high-quality preclinical and clinical studies, and ultimately result in the identification of novel treatment of neurological disorders of high unmet medical need.

NOTES

1. In certain cases wherein the compound series consistently show rodent and human/NHP potency disconnect, one would need to go directly to PET or SPECT imaging in higher species in order to have an accurate read on the tracer viability.
2. RRCK cells were generated in-house (Pfizer Inc., Groton, CT, USA) as a subclone of Mardin-Darby canine kidney wild-type (MDCK-WT) cells that displayed low expression of endogenous P-gp (~1–2% of MDCK-WT cells, based on mRNA level). For detailed information, see: Ref. [29].
3. An open-label trial is a type of clinical trial that both the researcher and the participant know the treatment that the participant is receiving.
4. GSK clinical study register, study ID 110771. An open label positron emission tomography (PET) study to investigate brain penetration by [^{11}C]GSK1034702 in healthy subjects, 2009. Available from: http://www.gsk-clinicalstudyregister.com/study/110771#rs.
5. Eli Lilly Pharmaceuticals Press Release. FDA approves Amyvid™ (florbetapir F-18 injection) for use in patients being evaluated for Alzheimer's disease and other causes of cognitive decline. Indianapolis: 2012. Available from: https://investor.lilly.com/releasedetail2.cfm?ReleaseID=662647.
6. GE Healthcare Press Release. FDA approval of Viazmyl™ (flutemetamol F-18 injection) for detection of beta amyloid. Bethesda: 2013. Available from: http://www3.gehealthcare.com/en/News_Center/Press_Kits/FDA_Approves_Vizamyl.
7. Piramal Imaging. FDA approves Piramal Imaging's Neuraceq™ (florbetaben F-18 injection) for PET imaging of beta-amyloid neuritic plaques in the brain. Berlin/Boston: 2014. Available from: http://www.piramal.com/imaging/pdf/FDA-Approval-Press-Release.pdf.

REFERENCES

[1] Manji, K. H.; DeSouza, E. B. "CNS drug discovery and development: when will we rescue Tantalus?" CNS drug discovery and development: challenges and opportunities. *Neuropsychopharmacol. Rev.* **2008**, *2*, 1–4.

[2] Alzheimer's Disease facts and figures, Alzheimer's Association, Washington, DC: **2012**. Available from https://www.alz.org/downloads/facts_figures_2012.pdf (accessed on September 12, 2014).

[3] Brookmeyer, R.; Johnson, E.; Ziegler-Graham, K.; Arrighi, H. M. Forecasting the global burden of Alzheimer's disease. *Alzheimers Dement.* **2007**, *3*, 186–191.

[4] Willmann, J. K.; van Bruggen, N.; Dinkelborg, L. M.; Gambhir, S. S. Molecular imaging in drug development. *Nat. Rev.* **2008**, *7*, 591–607.

[5] Rudin, M.; Weissleder, R. Molecular imaging in drug discovery and development. *Nat. Rev. Drug Discov.* **2003**, *2*, 123–131.

[6] Paulus, M. P.; Stein, M. B. Role of functional magnetic resonance imaging in drug discovery. *Neuropsychol. Rev.* **2007**, *17*, 179–188.

[7] Filippi, M.; Agosta, F.; Frisoni, G. B.; De Stefano, N.; Bizzi, A.; Bozzali, M.; Falini, A.; Rocca, M. A.; Sorbi, S.; Caltagirone, C.; Tedeschi, G. Magnetic resonance imaging in Alzheimer's disease: from diagnosis to monitoring treatment effect. *Curr. Alzheimer Res.* **2012**, *10*, 1198–1209.

[8] Yekhlef, F.; Ballan, G.; Macia, F.; Delmer, O.; Sourgen, C.; Tison, F. Routine MRI for the differential diagnosis of Parkinson's disease, MSA, PSP, and CBD. *J. Neural Transm.* **2003**, *110*, 151–169.

[9] Burke, J. F.; Gelb, D. J.; Quint, D. J.; Morgenstern, L. B.; Kerber, K. A. The impact of MRI on stroke management and outcomes: a systematic review. *J. Eval. Clin. Pract.* **2013**, *19*, 987–993.

[10] Beckmann, N.; Mueggler, T.; Allegrini, P. R.; Laurent, D.; Rudin, M. From anatomy to the target: contributions of magnetic resonance imaging to preclinical pharmaceutical research. *Anat. Rec.* **2001**, *265*, 316–333.

[11] Schoder, H.; Erdi, Y. E.; Larson, S. M.; Yeung, H. W. PET/CT: a new imaging technology in nuclear medicine. *Eur. J. Nucl. Med. Mol. Imaging* **2003**, *30*, 1419–1437.

[12] Buck, A. K.; Nekolla, S.; Ziegler, S.; Beer, A.; Krause, B. J.; Hermann, K.; Scheidhauer, K.; Wester, H.-J.; Rummeny, E. J.; Schwaiger, M.; Drzezga, A. SPECT/CT. *J. Nucl. Med.* **2008**, *49*, 1305–1319.

[13] Glepmans, B. N.; Adams, S. R.; Ellisman, M. H.; Tsien, R. Y. Fluorescent toolbox for assessing protein location and function. *Science* **2006**, *312*, 217–224.

[14] Yang, F.; Moss, L. G.; Phillips Jr., G. N. The molecular structure of green fluorescence protein. *Nat. Biotechnol.* **1996**, *14*, 1246–1251.

[15] Greer III, L. F.; Szalay, A. A. Imaging of light emission from the expression of luciferases in living cells and organisms: a review. *Luminescence* **2002**, *17*, 43–74.

[16] Isherwood, B.; Timpson, P.; McGhee, E. J.; Anderson, K.; Canel, M.; Serrels, A.; Brunton, V. G.; Carragher, N. O. Live cell in vitro and in vivo imaging applications: accelerating drug discovery. *Pharmaceutics* **2011**, *3*, 141–170.

[17] Rao, J.; Dragulescu-Andrasi, A.; Yao, H. Fluorescence imaging in vivo: recent advances. *Curr. Opin. Biotechnol.* **2007**, *18*, 17–25.

[18] Wong, D. F.; Tauscher, J.; Gründer, G. The role of imaging in proof of concept for CNS drug discovery and development. *Neuropsychopharmacology* **2009**, *34*, 187–203.

[19] Licha, K.; Olbrich, C. Optical imaging in drug discovery and diagnostic applications. *Adv. Drug. Deliv. Rev.* **2005**, *57*, 1087–1108.

[20] Barentsz, J.; Takahashi, S.; Oyen, W.; Mus, R.; De Mulder, P.; Reznek, R.; Oudkerk, M.; Mali, W. Commonly used imaging techniques for diagnosis and staging. *J. Clin. Oncol.* **2006**, *24*, 3234–3244.

[21] Brooks, D. J. Positron emission tomography and single-photon emission computed tomography in central nervous system drug development. *NeuroRx* **2005**, *2*, 226–236.

[22] Ametamy, S. M.; Honer, M.; Schubiger, P. A. Molecular imaging with PET. *Chem. Rev.* **2008**, *108*, 1501–1516.

[23] Sharma, S.; Ebadi, M. SPECT neuroimaging in translational research of CNS disorders. *Neurochem. Int.* **2008**, *52*, 352–362.

[24] Lecomte, R. Biomedical imaging: SPECT and PET. *AIP Conference Proc.* **2007**, *958*, 115–122.

[25] Zhang, L.; Villalobos, A. Recent advances in the development of PET and SPECT tracers for brain imaging. *Ann. Rep. Med. Chem.* **2012**, *47*, 105–119.

[26] Patel, S.; Gibson, R. In vivo site-directed radiotracers: a mini-review. *Nucl. Med. Biol.* **2008**, *35*, 805–815.

[27] Laruelle, M.; Slifstein, M.; Huang, Y. Y. Positron emission tomography: imaging and quantification of neurotransporter availability. *Methods* **2002**, *27*, 287–299.

[28] Zhang, L.; Villalobos, A.; Beck, E. M.; Bocan, T.; Chappie, T. A.; Chen, L.; Grimwood, S.; Heck, S. D.; Helal, C. J.; Hou, X.; Humphrey, J. M.; Lu, J.; Skaddan, M. B.; McCarthy, T. J.; Verhoest, P. R.; Wager, T. T.; Zasadny, K. Design and selection parameters to accelerate the discovery of novel central nervous system positron emission tomography (PET) ligands and their application in the development of a novel phosphodiesterase 2A PET ligand. *J. Med. Chem.* **2013**, *56*, 4568–4579.

[29] Callegari, E.; Malhotra, B.; Bungay, P. J.; Webster, R.; Fenner, K. S.; Kempshall, S.; LaPerle, J. L.; Michel, M. C.; Kay, G. G. A comprehensive non-clinical evaluation of the CNS penetration potential of antimuscarinic agents for the treatment of overactive bladder. *Br. J. Clin. Pharmacol.* **2011**, *72*, 235–246.

[30] Feng, B.; Mills, J. B.; Davidson, R. E.; Mireles, R. J.; Janiszewski, J. S.; Troutman, M. D.; de Morais, S. M. In vitro P-glycoprotein assays to predict the in vivo interactions of P-glycoprotein with drugs in the central nervous system. *Drug Metab. Dispos.* **2008**, *36*, 268–275.

[31] Chernet, E.; Martin, L. J.; Li, D.; Need, A. B.; Barth, V. N.; Rash, K. S.; Phebus, L. A. Use of LC/MS to assess brain tracer distribution in preclinical in vivo receptor occupancy studies: dopamine D2, serotonin 2A and NK-1 receptors as examples. *Life Sci.* **2005**, *78*, 340–346.

[32] Pike, V. W.; Rash, K. S.; Chen, Z.; Pedregal, C.; Statnick, M. A.; Kimura, Y.; Hong, J.; Zoghbi, S. S.; Fujita, M.; Toledo, M. A.; Diaz, N.; Gackenheimer, S. L.; Tauscher, J. T.; Barth, V. N.; Innis, R. B. Synthesis and evaluation of radioligands for imaging brain nociceptin/orphanin FQ peptide (NOP) receptors with positron emission tomography. *J. Med. Chem.* **2011**, *54*, 2687–2700.

[33] Shaffer, C. L. Defining neuropharmacokinetic parameters in CNS drug discovery to determine cross-species pharmacologic exposure-response relationships. *Ann. Rep. Med. Chem.* **2010**, *45*, 55–70.

[34] Fischman, A. J.; Alpert, N. M.; Rudin, R. H. Pharmacokinetic imaging: a non-invasive method for determining drug distribution and action. *Clin. Phamacokinet.* **2002**, *41*, 581–602.

[35] Huiban, M.; Pampols-Maso, S.; Passchier, J. Fully automated synthesis of the M_1 receptor agonist [^{11}C]GSK1034702 for clinical use on an Eckert & Ziegler Modular Lab system. *App. Rad. Isot.* **2011**, *69*, 1390–1394.

[36] Syvanen, S.; Eriksson, J. Advances in PET imaging of p-glycoprotein function in blood-brain barrier. *ACS Chem. Neurosci.* **2013**, *4*, 225–237.

[37] Josserand, V.; Pelerin, H.; de Bruin, B.; Jego, B.; Kuhnast, B.; Hinnen, F.; Duconge, F.; Boisgard, R.; Beuvon, F.; Chassoux, F.; Daumas-Duport, C.; Ezan, E.; Dolle, F.; Mabonzo, A.; Tavitan, B. Evaluation of drug penetration into the brain: a double study by in vivo imaging with positron emission tomography and using an in vitro model of the human blood-brain barrier. *Pharmacol. Exp. Ther.* **2006**, *316*, 79–86.

[38] Mabondzo, A.; Bottlaender, M.; Guyot, A. C.; Tsaouin, K.; Devere, J. R.; Balimane, P. V. Validation of in vitro cell-based human blood-brain barrier model using clinical positron emission tomography radioligands to predict in vivo human brain penetration. *Mol. Pharm.* **2010**, *7*, 1805–1815.

[39] Syvanen, S.; Lindhe, O.; Palner, M.; Kornum, B. R.; Rahman, O.; Langstrom, B.; Knudsen, G. M.; Hammerand-Udenaes, M. Species differences in blood-brain barrier transport of three positron emission tomography radioligands with emphasis on P-glycoprotein transport. *Drug Metab. Dispo.* **2009**, *37*, 635–643.

[40] Morgan, P.; Van Der Graaf, P. H.; Feltner, D. E.; Drummond, K. S.; Wegner, C. D.; Street, A. D. A. Can the flow of medicines be improved? Fundamental pharmacokinetic and pharmacological principles toward improving phase II survival. *Drug Discov. Today* **2012**, *17*, 419–424.

[41] Grimwood, S.; Hartig, P. R. Target site occupancy: emerging generalization from clinical and preclinical studies. *Pharmacol. Ther.* **2009**, *122*, 281–301.

[42] Kapur, S.; Zipurski, R.; Jones, C.; Remington, G.; Houle, S. Relationship between dopamine D(2) occupancy, clinical response, and side effects: a double-blind PET study of first-episode schizophrenia. *Am. J. Psychiatry* **2000**, *157*, 514–520.

[43] Yokoi, F.; Grunder, G.; Biziere, K.; Stephane, M.; Dogan, A. S.; Dannals, R. F.; Ravert, H.; Suri, A.; Bramer, S.; Wong, D. F. Dopamine D2 and D3 receptor occupancy in normal humans treated with the antipsychotic drug aripiprazole (OPC 14597): a study using positron emission tomography and [^{11}C]raclopride. *Neuropsychopharmacology* **2002**, *27*, 248–259.

[44] Atack, J. R.; Wong, D. F.; Fryer, T. D.; Ryan, C.; Sanabria, S.; Zhou, Y.; Dannals, R. F.; Eng, W.; Gibson, R. E.; Burns, H. D.; Vega, J. M.; Vessy, L.; Scott-Stevens, P.; Beech, J. S.; Baron, J.-C.; Sohal, B.; Schrag, M. L.; Aigbirhio, F. I.; McKernan, R. M.; Hargreaves, R. J. Benzodiazepine binding site occupancy by the novel $GABA_A$ receptor subtype-selective drug 7-(1,1-dimethylethyl)-6-(2-ethyl-2*H*-1,2,4-triazol-3-ylmethoxy)-3-(2-fluorophenyl)-1,2,4-triazolo[4,3-*b*]pyridazine (TPA023) in rats, primates, and humans. *J. Pharmacol. Exp. Ther.* **2010**, *332*, 17–25.

[45] Keller, M.; Montgomery, S.; Ball, W.; Morrison, M.; Snavely, D.; Liu, G.; Hargreaves, R.; Hietala, J.; Lines, C.; Beebe, K.; Reines, S. Lack of efficacy of the substance p (neurokinin1 receptor) antagonist aprepitant in the treatment of major depressive disorder. *Biol. Psychiatry* **2006**, *59*, 216–223.

[46] Solin, O.; Eskila, O.; Hamill, T. G.; Bergman, J.; Lehikoinen, P.; Gronroos, T.; Forsback, S.; Haaparanta, M.; Viljanenen, T.; Ryan, C.; Gibson, R.; Kieczykowski, G.; Hietala, J.; Hargreaves, R.; Burns, H. D. Synthesis and characterization of a potent, selective, radiolabeled substance-P antagonist for NK_1 receptor quantification: ([^{18}F]SPA-RQ). *Mol. Imaging Biol.* **2004**, *6*, 373–384.

[47] Zhang, L.; Balan, G.; Barreiro, G.; Boscoe, B. P.; Chenard, L. K.; Cianfrogna, J.; Claffey, M. M.; Chen, L.; Coffman, K. J.; Drozda, S. E.; Dunetz, J. R.; Fonseca, K. R.; Galatsis, P.; Grimwood, S.; Lazzaro, J. T.; Mancuso, J. Y.; Miller, E. L.; Reese, M. R.; Rogers, B. N.; Sakurada, I.; Skaddan, M.; Smith, D. L.; Stepan, A. F.; Trapa, P.; Tuttle, J. B.; Verhoest, P. R.; Walker, D. P.; Wright, A. S.; Zaleska, M. M.; Zasadny, K.; Shaffer, C. L.

Discovery and preclinical characterization of 1-methyl-3-(4-methylpyridin-3-yl)-6-(pyridin-2-ylmethoxy)-1H-pyrazolo-[3,4-b]pyrazine (PF470): a highly potent, selective, and efficacious metabotropic glutamate receptor 5 (mGluR5) negative allosteric modulator. *J. Med. Chem.* **2014**, *57*, 861–877.

[48] Dubois, B.; Feldman, H. H.; Jacova, C.; Dekosky, S. T.; Barberger-Gateau, P.; Cummings, J.; Delacourte, A.; Galasko, D.; Gauthier, S.; Jicha, G.; Meguro, K.; O'brien, J.; Pasquier, F.; Robert, P.; Rossor, M.; Salloway, S.; Stern, Y.; Visser, P. J.; Scheltens, P. Research criteria for the diagnosis of Alzheimer's disease: revising the NINCDS-ADRDA criteria. *Lancet Neurol.* **2007**, *6*, 734–746.

[49] Wang, L.; Zhang, Q.; Li, H.; Zhang, H. SPECT molecular imaging in Parkinson's disease. *J. Biomed. Biotechnol.* **2012**, *2012*, 1–11.

[50] Contrafatto, D.; Mostile, G.; Nicoletti, A.; Raciti, L.; Luca, A.; Dibilio, V.; Lanzafame, A.; Distefano, A.; Drago, F.; Zappia, M. Single photon emission computed tomography striatal asymmetry index may predict dopaminergic responsiveness in Parkinson's disease. *Clin. Neuropharm.* **2011**, *34*, 71–73.

[51] Ziebell, M.; Andersen, B. B.; Thomsen, G.; Pinborg, L. H.; Karlsborg, M.; Hasselbalch, S. G.; Knudsen, G. M. Predictive value of dopamine transporter SPECT imaging with [^{123}I]PE2I in patients with subtle Parkinsonian symptoms. *Eur. J. Nucl. Mol. Imaging* **2012**, *39*, 242–250.

[52] Durduran, T.; Choe, R.; Baker, W. B.; Yodh, A. G. Diffuse optics for tissue monitoring and tomography. *Rep. Prog. Phys.* **2010**, *73*, 1–43.

[53] Solomon, M.; Nothdruft, R. E.; Akers, W.; Edwards, W. B.; Liang, K.; Xu, B.; Suddlow, G. P.; Deghani, H.; Tai, Y. C.; Eggebrecht, A. T.; Achilefu, S.; Culver, J. P. Multimodal fluorescence-mediated tomography and SPECT/CT for small-animal imaging. *J. Nucl. Med.* **2013**, *54*, 639–646.

[54] DePuey, E. G. Advances in SPECT camera software and hardware: currently available and new on the horizon. *J. Nucl. Cardiol.* **2012**, *19*, 551–581.

[55] Pichler, B. J.; Wehrl, H. F.; Judenhofer, M. S. Latest advances in molecular imaging instrumentation. *J. Nucl. Med.* **2008**, *49*, 5S–25S.

PART 4

MODULATING BRAIN PENETRATION OF LEADS DURING DRUG DISCOVERY

18

DESIGNING CNS DRUGS FOR OPTIMAL BRAIN EXPOSURE

ZORAN RANKOVIC

Eli Lilly and Company, Indianapolis, IN, USA

INTRODUCTION

The human brain relies on a sophisticated protection system to preserve its physiological environment and greatly reduce injury from external insults and toxins. An important component of this system is the blood–brain barrier (BBB) at the interface between the blood capillaries of the brain and brain tissue. It is composed of astrocytes, pericytes, and endothelial cells that line the blood vessels and form the brain capillary endothelium [1]. The main role of the BBB is to enable selective access of required nutrients, ions, hormones, solutes, and proteins to the brain while diminishing exposure to potentially harmful xenobiotics. The BBB is very effective in this role; more than 98% of small-molecule drugs and approximately 100% of large-molecule drugs are excluded, often via selective active transport processes [2]. The blood–cerebrospinal fluid barrier (BCSFB) is a similar barrier, albeit with a several thousand times smaller surface area, between the blood capillaries and the cerebrospinal fluid (CSF), which is a clear liquid that fills the ventricles and canals of the brain and bathes the external surface of the brain to provide buoyancy and mechanical protection inside the skull [3]. In brain tissue, the CSF is exchanged with the brain's interstitial fluid (ISF), which washes metabolic wastes out of the CNS and regulates the chemical environment of the brain.

Active transport and passive diffusion are the two principal mechanisms by which molecules can enter the brain. In contrast to the endothelial cells of capillaries

Blood–Brain Barrier in Drug Discovery: Optimizing Brain Exposure of CNS Drugs and Minimizing Brain Side Effects for Peripheral Drugs, First Edition. Edited by Li Di and Edward H. Kerns.

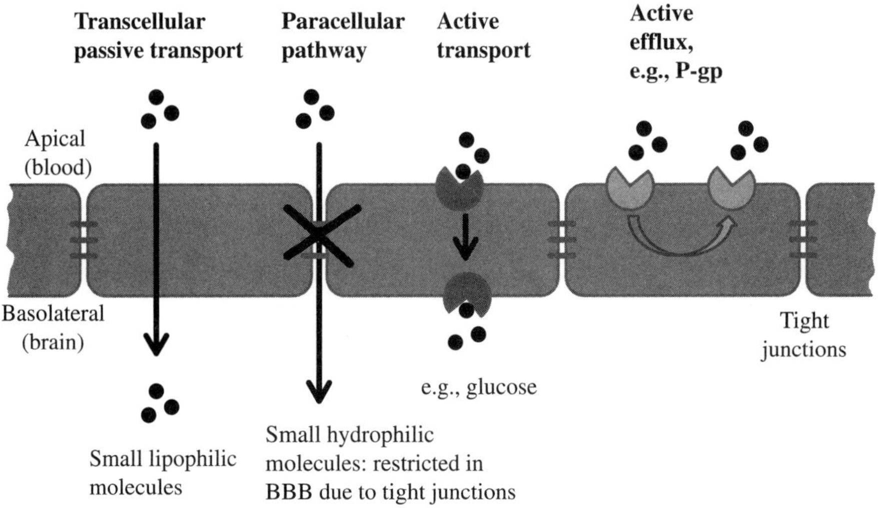

FIGURE 18.1 Principal mechanisms by which small molecules cross the BBB [1].

elsewhere in the body, paracellular diffusion is effectively precluded by the presence of tight junctions that characterize the BBB and BCSFB. For example, brain capillary endothelial cells have far higher transendothelial electrical resistance than do peripheral capillaries (>1000 versus <20 ohm/cm^2) due to the tight junctions that restrict the movement of ions such as Na^+ and Cl^- [4]. This increased resistance is thought to protect the brain from fluctuations in ionic composition that may occur due to, for example, exercise or a meal.

The great majority of CNS drugs are small molecules that are designed to cross the BBB via the transcellular passive diffusion route (Fig. 18.1). Achieving optimal brain exposure is a unique and major challenge for medicinal chemists working on the CNS targets [4]. Design strategies toward such molecules and their physicochemical properties are the main foci of this chapter. Readers are referred to earlier chapters in this book for the state of the art in targeting active transport mechanisms and CNS delivery strategies for large molecules such as proteins and antibodies.

IMPORTANT PARAMETERS AND CONSIDERATIONS IN THE DESIGN OF CNS DRUGS

The existence of the BBB renders classic pharmacokinetic (PK) parameters, such as oral bioavailability and plasma concentration, insufficient for assessing CNS drug candidates due to uncertainty regarding whether such parameters will reflect the time course or exposure levels in the brain. To provide better understandings of the pharmacokinetic to pharmacodynamic (PK/PD) relationships of CNS drug candidates, additional data and more sophisticated PK methods are required [5–10]. The commonly reported measures of brain penetration, along with the most commonly reported acronyms, are listed in Table 18.1.

TABLE 18.1 Pharmacokinetic parameters commonly used in CNS drug discovery [4]

Parameter	Definition	Description	Units
RO	Receptor occupancy	Percentage of available targeted receptors occupied by a drug	%
K_p B/P B:P Br:Bl	$\text{LogBB} = \text{Log}\left(\frac{C_{\text{brain}}}{C_{\text{plasma}}}\right)$	Total brain-to-plasma concentration ratio (includes nonspecific plasma protein and brain tissue bound drug)	No units (ratio)
LogBB	$\text{LogBB} = \text{Log}(K_p)$	Brain-to-plasma distribution coefficient of total drug between brain and plasma on a logarithmic scale	No units (ratio)
LogPS	$\text{Log PS} = \text{Log}(-F \ln(1 - \text{Cl}_{up}/F))$ F = the regional flow rate Cl_{up} = uptake clearance (ml/s/g)	The BBB permeability, expressed as permeability surface area product (PS, quantified as log PS)	No units
$C_{b,u}$ $C_{free,br}$		Concentration of unbound drug in brain	ng/g, nM
$C_{p,u}$ $C_{free,pl}$		Concentration of unbound drug in plasma	ng/g, nM
C_{CSF}		Concentration of drug in cerebrospinal fluid (CSF)	ng/g, nM
$K_{p,uu}$ $K_{p,free}$	$K_{p,uu} = \frac{C_{u,brain}}{C_{u,plasma}}$	Brain-to-plasma concentration ratio of unbound drug (often reported as surrogate measure: CSF to free plasma distribution coefficient; $C_{CSF}/C_{u,plasma}$)	No units (ratio)
$f_{u,b}$ $f_{u,brain,}$	$f_{u,b} = C_{b,u}/C_b$	Fraction unbound or free drug fraction in brain tissue	No units (ratio)
$f_{u,p}$ $f_{u,plasma}$	$f_{u,p} = C_{p,u}/C_p$	Fraction unbound or free drug fraction in plasma	No units (ratio)
P_{app}		Apparent permeability; total permeability through a specific membrane	nm/s, or 10^{-6} cm/s
ER	$\text{ER} = P_{app}$ (B to A)/P_{app} (A to B)	Efflux ratio (A = apical; B = basolateral)	No units (ratio)

Receptor occupancy studies provide the most direct information about exposure at the site of action; however, due to the technical complexities, time, and costs involved, these studies tend to be performed only for the most advanced compounds. Due to the relative ease of measurement, drug total brain concentration (C_b) and the whole brain-to-plasma ratio K_p (C_b/C_p, also referred to as the B/P ratio) have historically been used as the main drivers of decision making in CNS drug discovery.

However, relying solely on these parameters has proven insufficient to enable adequate understanding of the CNS exposures needed to establish robust PK/PD correlations. This inadequacy is primarily because only a fraction of the total drug in the *in vivo* plasma or tissues, such as brain, is unbound to proteins or lipids and free to diffuse across tissues, thus affecting the free drug concentration ($C_{b,u}$) that interacts with the intended therapeutic target. In addition, the unbound fraction is usually different between plasma and tissue, thus affecting the unbound brain-to-plasma ratio ($K_{p,uu}$) [11, 12]. The potential for misleading conclusions derived from the reliance on total brain concentrations is exemplified well by studies of KA-672, **1** (ensaculin), developed for the treatment of cognitive deficits associated with Alzheimer's disease (AD) [12]. In rats at a dose of 1 mg/kg, the compound exhibited a total brain concentration equivalent to its IC_{50} value as determined *in vitro* (0.36 µM). However, doses up to 10 mg/kg produced no pharmacological effect. Analyses of hippocampal microdialysates indicated that free concentration of the compound was below 0.01 µM, which readily explained the lack of activity. This low concentration was probably a consequence of the compound's high unbound drug clearance. Total B/P ratio (K_p) data can be misleading even when used to rank structurally related compounds. For example, the K_p value of 0.22 for S-cetirizine **2** indicates that this compound apparently penetrates the CNS significantly better than its enantiomer R-cetirizine, which has a K_p value of 0.04 [12]. Only when the concentrations of the unbound drugs in the brain ($C_{b,u}$) and plasma ($C_{p,u}$) are used in the calculation does it become apparent that S- and R-cetirizine in fact have very similar if not identical brain penetration properties with $K_{p,uu}$ ($C_{b,u}/C_{p,u}$) of 0.17 and 0.14, respectively. Further investigation revealed that the observed "K_p enantioselectivity" is related to the lower plasma protein binding of S-cetirizine (PPB = 50%) compared to R-cetirizine (PBB = 85%) [12] (Fig. 18.2).

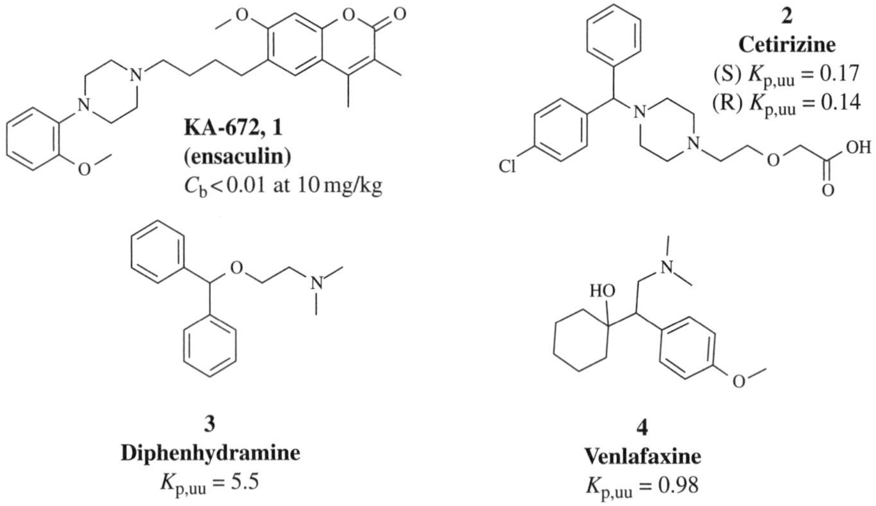

FIGURE 18.2 Examples of drugs with different levels of brain exposure.

The importance of free drug concentrations is highlighted by the free drug hypothesis, which states that the free (unbound) drug concentration (C_u) at the site of action is responsible for the pharmacological activity *in vivo* and that, at steady state and in absence of active transport, the free drug concentration is the same on both sides of any biomembrane (e.g., the BBB) [13]. Consequently, the drug unbound brain concentration ($C_{b,u}$) is often equated to its unbound plasma concentration ($C_{p,u}$) at steady state, which can then be used to calculate the exposure required for *in vivo* efficacy based on the known *in vitro* EC_{50} or K_i. However, it is important to note that this assumption is not valid for compounds with the following properties:

- A low rate of passive permeability across the BBB/BCSFB (i.e., long time period is required to reach equilibrium across membranes)
- Compounds that are actively pumped *out of* the brain by efflux transporters (e.g., P-glycoprotein (P-gp))
- Compounds that are actively transported *into* the brain by uptake transporters

In these cases, there is a disconnect between a compound's free drug concentrations in the plasma and brain; for example, $C_{b,u} \neq C_{p,u}$ ($K_{p,uu} < 1$ or $K_{p,uu} > 1$). $K_{p,uu}$ values below 1, for example, that of cetirizine, indicate that the compound is a substrate for an efflux transporter and/or has low passive permeability across the BBB [14]. This profile is desirable for cetirizine, which is a second generation antihistamine that was intentionally designed to minimize brain exposure to avoid the CNS side effects, such as sedation, that are generally associated with the first-generation antihistamines [15]. Pursuing compounds with low $K_{p,uu}$ values as CNS targets presents significant risks due to the difficulties in estimating human brain unbound drug concentrations and the consequently low confidence in human dose prediction [16]. However, $K_{p,uu}$ values greater than 1 suggest active uptake processes mediated by influx transporters as exemplified by the first-generation antihistamine diphenhydramine, **3**, which has a $K_{p,uu} = 5.5$ (in rats) [17]. Intentionally leveraging active transport processes could be an effective design approach for challenging CNS targets for which the ligands are nonconducive to passive BBB penetration. L-dopa is used to treat Parkinson's disease and is a frequently quoted example of exploiting an active transport mechanism [18]. However, the research surrounding the targeted specific transport proteins or transcytosis mechanisms for CNS drug delivery is still in its infancy (for additional information see dedicated chapters in this book). A successful small-molecule CNS drug candidate typically complies with the free drug hypothesis, and the ratio of the unbound drug in brain to the unbound drug in the plasma (i.e., the $K_{p,uu}$ value) should be close to 1 [9]. With a $K_{p,uu}$ in mouse of 0.98, the antidepressant venlafaxine **4** is such an example [19].

Direct measurements of free drug concentrations in the brain ISF, which bathes the neurons and neuroglia and is also known as the extracellular fluid (ECF), is technically challenging, and the number of datasets that provide such information is small. Compromises are found in the use of a range of surrogate techniques such as CSF sampling. The barrier separating the ISF from the CSF is a single layer of ependyma, which is a cell layer that does not have tight junctions, and microdialysis studies suggest that drug concentrations in the CSF are good approximations of those in the brain ISF [20]. Furthermore, because the CSF contains very low levels of proteins (~0.2 versus 70 mg/ml in the plasma), CSF concentrations (C_{CSF}s) of compounds that are not subject to active efflux can be approximated to their $C_{b,u}$s. Indeed, a number of reports in the literature have described good correlations between CSF concentrations and efficacies in both preclinical and clinical studies [21].

Another parameter that is frequently used in drug discovery is the fraction unbound ratio in the brain or plasma ($f_{u,tissue} = C_{u,tissue}/C_{tissue}$). Notably, however, the free drug hypothesis implies that, irrespective of differential $f_{u,plasma}$ and $f_{u,tissue}$, it is $C_{u,plasma}$ that reflects the unbound drug concentration in the tissue of interest in the steady state; that is, $C_{u,plasma} = C_{u,tissue}$. For example, following either single or multiple doses of fluconazole (an antifungal drug with a rapid rate of membrane permeation), similar free drug concentrations were found in various body fluids, including vaginal secretions, breast milk, saliva, sputum, prostatic and seminal vesicle fluid, CSF, and plasma [22]. Because f_u does not affect the unbound drug concentration at the therapeutic target and, therefore, has no clinical relevance for oral drugs, drug discovery programs should be driven by the free drug concentration rather than the f_u [23].

BBB PERMEABILITY AND P-GLYCOPROTEIN EFFLUX

In both *in vitro* and *in vivo* systems, drug permeability can be considered to be the sum of passive (diffusion-driven) and active (transporter-mediated) processes. The *in vitro* models of BBB permeability, including the parallel artificial membrane permeability assay (PAMPA), Caco2 (heterogeneous human epithelial colorectal adenocarcinoma cells), and MDR1-MDCK (Madin-Darby canine kidney cells expressing P-gp (from MDR-1gene), are identical to those used to estimate gut wall permeability for oral absorption evaluations. However, for CNS activity, the implications of low permeability and active efflux are more profound, because the concentration of drug compound in the brain blood capillaries is likely to be significantly lower than in the gut; thus, efflux processes are unlikely to be saturated. Consequently, although the influences of physicochemical properties exhibit similar trends for oral absorption and CNS permeability, the requirements for BBB permeability are more stringent. For example, in a study of 48 marketed CNS drugs and 45 marketed non-CNS drugs, 96% of the CNS drugs exhibited passive permeabilities with $P_{app} > 150$ nm/s in an MDR1-MDCK assay, and only 76% of the non-CNS drugs met this criterion [24]. Thus, the guideline of $P_{app} > 150$–200 nm/s was suggested for compound with CNS exposure. In contrast, Veber suggested the widely quoted lower cutoff of Caco2 A (apical) to B (basolateral) $P_{app} > 100$ nm/s as a guideline for obtaining good oral bioavailability [25].

Due to the high expression levels of transporters at the BBB, active efflux is a common issue for CNS drug discovery programs. There are a number of transporters expressed in the endothelial cells of the BBB that can affect drug transport across the BBB in both directions, through active efflux or uptake (Table 18.2) [26].

TABLE 18.2 Transporters found at the human BBB [26]

Transport direction	Transporter name	Aliases	Gene
Efflux	P-glycoprotein	P-gp MDR1 ABCB1	*ABCB1*
Efflux	Breast cancer resistance protein	BCRP MXR	*ABCG2*
Efflux	Multidrug resistance protein 4	MRP4 ABCC4	*ABCC4*
Efflux	Multidrug resistance protein 5	MRP5 ABCC5	*ABCC5*
Uptake	Organic anion transporting polypeptide 1A2	OATP1A2 OATP-A	*SLCO1A2*
Uptake	Organic anion transporting polypeptide 2B1	OATP2B1 OATP-B	*SLCO2B1*

However, the P-gp (or MDR1, ABCB1) is by far the most commonly reported transporter involved in drug active efflux [27]. P-gp is a membrane protein of the adenosine triphosphate (ATP)-binding cassette transporter superfamily that is expressed in a variety of human tissues including the intestine and the BBB. However, the effect of P-gp efflux on intestinal absorption is generally much lower compared to the effect it can have on the CNS entry, mainly due to the fact that, in contrast to the intestine, drug plasma concentrations at the BBB rarely reach P-gp saturation levels. Indeed, Polli et al. showed that CNS drugs have a threefold lower incidence of P-gp-mediated efflux than non-CNS drugs, which emphasizes the greater importance of the P-gp efflux for CNS compared to other therapeutic areas [28].

P-gp acts as a gatekeeper for tissue exposure and, as such, has evolved to be highly promiscuous and able to recognize a large variety of molecules. Attempts to predict whether a compound will be a P-gp substrate based on structure have either examined pharmacophore models using known substrates [29] or utilized physicochemical property analyses [30–32]. However, both approaches have flaws. Pharmacophore generation is hampered by the fact that there are several binding domains within the large P-gp-binding cavity, and the physicochemical property approach is equally challenging because it is unlikely to have the subtlety required to account for micro-phenomena such as secretion-associated and Ras-related (SAR) protein. The recent publication of low-resolution crystal structures of P-gp and the related homology models allows for the possibility of ligand-docking approaches, although such approaches have not yet been routinely reported [33]. However, the limited utility of global P-gp models is mitigated by the ready availability of *in vitro* models of P-gp efflux including the MDR1-MDCK and Caco-2 cell lines.

Recently a rule of 4 was formulated to predict P-gp efflux liabilities, which suggests that compounds with molecular weights (MWs) less than 400, total numbers of nitrogen and oxygen atoms less than 4, and basic pK_as less than 8 are unlikely to be

TABLE 18.3 Morphine analogues: an effect of HBD on P-gp efflux [35]

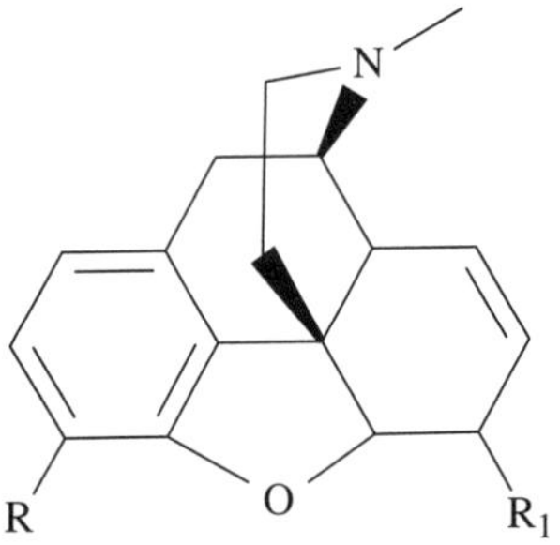

Compound	R	R_1	MOR IC_{50} (nM)	P-gp[a] substrate
5	OH	OH	1.7	Yes
6	MeO	MeO	1910	No
7	OH	MeO	1.1	No
8	OH	H	2.9	No

[a]P-gp Glo ATP assay with 200 μM test compound [35].

P-gp substrates [34]. Although a significant number of CNS-active molecules exceed one of these criteria, this rule of 4 is a useful indicator of strategies that can be applied to reduce efflux by P-gp. Functional groups that specifically favor P-gp recognition are typically those that contain hydrogen-bond donors (HBDs), such as alcohols, phenols, carboxylic acids, primary and secondary amides, sulfonamides, ureas, and N-heterocycles bearing uncapped NH groups. Lowering pK_a and reducing the number of heteroatoms, particularly HBDs, in a molecule have been demonstrated to reduce P-gp activity in different chemical series [27].

Morphine **5** is a P-gp substrate, and Coop and coworkers were interested in structural modifications that could diminish its P-gp efflux while maintaining its μ-opioid receptor (MOR) potency and selectivity [35]. While the capping of the 3-OH group was not tolerated, analogues with modified 6-OH groups retained the MOR potency. For example, a di-methoxy analogue **6** exhibited a significant loss in MOR potency (K_i = 1910 nM), whereas 6-methoxy **7** and deletion analogues **8** both retained their MOR activities. Interestingly, the reduction in total HBD count rendered all three of these compounds nonsubstrates for P-gp (Table 18.3).

In cases in which reducing HBD counts by capping groups such as NH or OH with an alkyl or aryl group is not tolerated, more creative approaches, such as masking the HBD group by intramolecular hydrogen bonding (H-bonding), can be explored (Table 18.4).

For example, derivatives of the MOR antagonist naltrexamine, NAP (**9**), and NAQ (**10**) have similar subnanomolar MOR potencies; however, in a mouse tail immersion test of nociception, NAP is 10-fold less potent (AD_{50} = 4.98 mg/kg) than NAQ (AD50 = 0.46 mg/kg) in antagonizing the activity of morphine [36]. Permeability measurements in Caco-2 cells have revealed both that NAP has low permeability

TABLE 18.4 Masking HBDs by intramolecular H-bonding can reduce P-gp efflux [36]

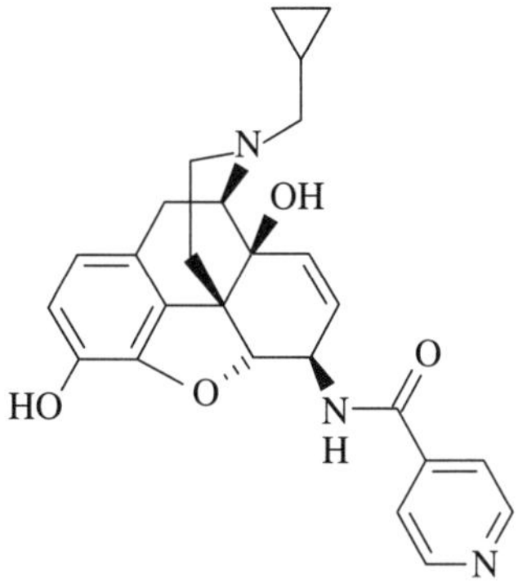

9

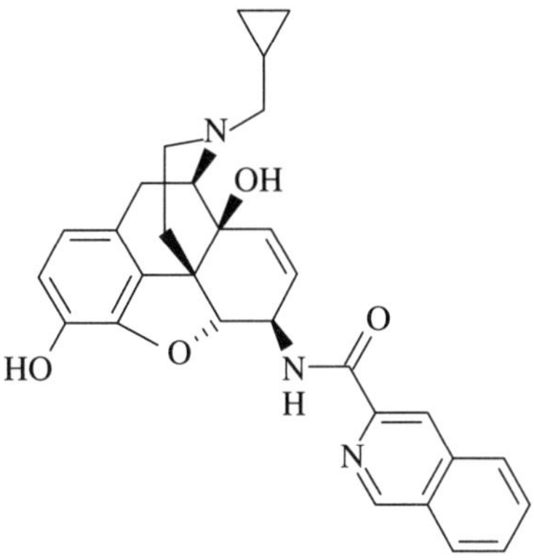

10

Compound	MOR IC50 (nM)	Caco-2 ER	P_{app} (nm/s); ER	cLogP	HBD
9	0.37	>10	<10	1.2	3
10	0.55	1.3	30	2.6	3

(P_{app} (A to B) $<1\times10^{-6}$ cm/s) and that it is an efflux substrate (ER > 10). When NAP was tested in the presence of a P-gp inhibitor, the P_{app} (A to B) increased to approximately 4×10^{-6} cm/s. In contrast, NAQ displayed P_{app} (A to B) of 3×10^{-6} cm/s and did not appear to be an efflux substrate. Despite the facts that NAP and NAQ have identical HBD counts and topological polar surface area (TPSA) values and the MW of NAQ is higher, the latter compound has superior permeability and lower efflux. These differences are likely due to a combination of the more favorable cLogP of 2.6 for NAQ relative to 1.2 for NAP and the intramolecular masking of the amide NH via H-bonding to the isoquinoline nitrogen atom [27, 37].

Also notable is that small structural changes can have significant effects on P-gp recognition, particularly when the molecule is already within the physicochemical space consistent with brain penetration. For example, Whitlock et al. at Pfizer tackled the issue of efflux from the CNS by P-gp during the discovery of the dual serotonin and noradrenaline reuptake inhibitor (SNRI) clinical candidate PF-184,298 **12** [38, 39]. The starting compound **11** exhibited many physicochemical properties consistent with CNS space, including low MW, but exhibited no efficacy due to poor CNS penetration ($C_{CSF}/C_{u,p}$ ~0.1, a surrogate measure of $K_{p,uu}$) and a significant degree of efflux in the MDR1-MDCK cell line *in vivo* model (ER 4.3) (Fig. 18.3). The pK_a of the secondary amine was high (pK_a 9.4), which is consistent with many molecules that target monoamine transporters and is a key part of the pharmacophore. Thus, rather than addressing the global physicochemical properties, the authors elected to rearrange the key pharmacophoric elements in **11** in an attempt to disrupt P-gp recognition while maintaining the desired pharmacology. Indeed, a subtle change in **12**,

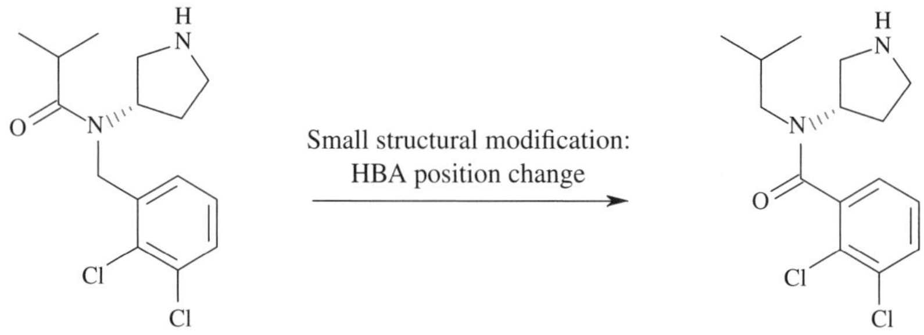

11
SERT K_i 9 nM
NET K_i 52 nM
$C_{CSF}/C_{u,p}$ ~0.1
P_{app} 15×10^{-6} cm/s
ER 4.3

12, PF-184,298
SERT K_i 6 nM
NET K_i 21 nM
$C_{CSF}/C_{u,p}$ ~0.45
P_{app} 16×10^{-6} cm/s
ER 2.7

FIGURE 18.3 SNRIs: small structural changes can affect P-gp [39].

wherein the carbonyl of the amide was moved to the benzylic position, proved sufficient to reduce the P-gp efflux (ER 2.7) and improve brain exposure ($C_{CSF}/C_{u,p}$ ~0.45) and efficacy in the preclinical *in vivo* model.

In contrast to efflux, active influx can have a positive effect on brain exposure, particularly for compounds with poor inherent passive permeability. In these cases, the challenges lie in understanding the transport mechanism and how that mechanism is related to compound structure to inform the optimization process. Even if a project is successful in delivering a clinical candidate, concerns related to cross-species and interpatient variability are likely to be raised. For these reasons, the generally preferred optimization strategy for overcoming the poor permeability of small-molecule drugs is to address inherent passive permeability rather than exploiting active transport.

PHYSICOCHEMICAL PROPERTIES—CNS EXPOSURE RELATIONSHIPS

Understanding how physicochemical properties affect each of the parameters discussed earlier and how they influence each other enables the design of molecules with improved CNS exposure. By analogy to the now widely accepted Lipinski guidelines for the design of successful oral drugs [40], the physicochemical properties for brain penetration have been studied and several groups have defined the characteristics of successful CNS drug candidates using a variety of approaches (Table 18.5).

Inevitably, the choice of the dataset greatly influences the conclusions drawn from these analyses, and this has spawned much debate in the CNS medicinal chemistry community regarding the best methods for evaluating brain penetration. Many

TABLE 18.5 Summary of the literature describing physicochemical properties required for optimal CNS exposure [4]

Author	Year	Dataset	Conclusions/criteria	Ref.
Hitchcock	2012	Literature review	General guidelines for minimizing P-gp recognition include maximizing ligand efficiency, keeping the HBD count < 2 and the TPSA < 90 Å^2 (preferably <70 Å^2). The presence of a basic amino group can further lower the threshold for tPSA tolerance.	[27]
Ghose et al.	2011	317 CNS and 626 non-CNS oral drugs	CNS drug discovery property profile guidelines: TPSA < 76 Å^2 (25–60 Å^2); at least one N atom; <7 (2–4) linear chains outside of rings; <3 (0–1) HBD; 740–970 Å^3 volume; solvent accessible surface of 460–580 Å^2; positive QikProp parameter CNS. One violation may be tolerated.	[41]
Wager	2010	119 marketed CNS drugs and 108 Pfizer CNS clinical candidates	Multiparameter optimization tool based on weighted physicochemical properties including cLogP, cLogD, PSA, MW, HBD, pK_a.	[42]
Wager	2010	119 marketed CNS drugs and 108 Pfizer CNS clinical candidates	Median values for CNS drugs: MW 305.3, cLogP 2.8, cLogD 1.7; MW 305.3; PSA 44.8 Å^2, HBD = 1, pK_a 8.4.	[43]
Waring	2009	9571 AstraZeneca Caco-2 measurements	Guidelines to achieve >50% chance of high permeability for a given MW: MW < 300 AZ LogD > 0.5; MW 300–350; AZ LogD > 1.1; MW 350–400; AZ LogD > 1.7; MW 400–450; AZ LogD > 3.1, MW 450–500; AZ LogD > 3.4; MW > 500; AZ LogD > 4.5.	[44]
Gleeson	2008	3059 rat CNS Log BB datapoints; 1975 P-gp efflux ratio datapoints 50641 P_{app} artificial membrane assay datapoints. 986 brain tissue binding datapoints	↑ MW leads to: ↓P_{app}, ↑P-gp ER and ↓ LogBB. cLogP has a weak nonlinear effect on ER. The optimal cLogP is <3 or >5. Permeability of neutral molecules shows a nonlinear dependence on cLogP. Basic, acid, and zwitterionic molecules show ↑ P_{app} with ↑ cLogP. ↑ MW ↑ nonspecific brain tissue binding. ↑ cLogP ↑ brain tissue binding.	[45]

(*Continued*)

TABLE 18.5 ***(Continued)***

Author	Year	Dataset	Conclusions/criteria	Ref.
Hitchcock and Pennington	2006	Literature review	Suggested physicochemical property ranges for increasing the potential for BBB penetration: PSA <90 Å^2; HBD <3; cLogP 2–5; cLogD (pH 7.4) 2–5; MW < 500.	[46]
Pajouhesh and Lenz	2005	Literature review	Attributes of a successful CNS drug candidate: MW < 450; cLogP < 5; HBD < 3; HBA < 7; Rot. bond < 8; H-bonds < 8; pK_a 7.5–10.5; PSA < 60–70 Å^2	[47]
Leeson and Davis	2004	329 launched oral drugs from the period 1983–2002	CNS drugs showed significantly different (mean/median) MW (310/307); polar properties (O + N (4.32/4); HBA (2.12/2); Rot. bond (4.7/4.5)) relative to other therapeutic area categories.	[48]
Petrauskas et al.	2003	1000 P-gp datapoints	Proposed cutoffs to avoid P-gp efflux liability: MW < 400; N + O < 4; pK_a < 8.	[34]
Norinder and Haeberlein	2002	Literature review	For a good chance of CNS penetration: O + N ≤ 5; cLogP – (O + N) > 0.	[49]
Polli et al.	2002	48 CNS and 45 non-CNS drugs	Physicochemical properties with significant differences between the CNS and non-CNS set. CNS set had ↑ cLogP (CNS mean 3.43); ↑ cLogD (CNS mean 2.08); ↓HBD (CNS mean 0.67); ↓ PSA (CNS mean 40.5 Å^2); and were less flexible.	[28]
Van der Waterbeemd et al.	1998	125 CNS and non-CNS drugs	MW < 450; PSA < 90 Å^2; LogD between 1 and 4; principle axis length/width ratio < 5.	[50]
Gratton et al.	1997	18 chemically diverse compounds with LogPS data	LFER equation relating Log PS to solute excess molar refraction and solute volume. ↑ solute dipolarity/polarizability and hydrogen-bond basicity leads to↓ Log PS.	[51]
Van der Waterbeemd and Kansy	1992	20 compounds	Propose that LogP alkane/water and calculated molar volume are suitable predictors of brain uptake.	[52]

LFER, linear free energy relationship; MW, molecular weight; PSA, polar surface area; HBD, H-bond donors; HBA, H-bond acceptors; Rot. bond., rotatable bonds; ↑ increased; ↓ decreased.

studies have been based on the physicochemical properties of marketed CNS drugs, and consequently their conclusions reflect not only brain penetration but also the overall absorption, distribution, metabolism, and excretion (ADME) and toxicological requirements for a successful CNS drug candidate [50]. Alternatively, researchers have analyzed large datasets of surrogate *in vitro* BBB permeation data (e.g., Caco-2, PAMPA, MDR1-MDCK) or crude *in vivo* measures of brain penetration such as brain/plasma ratios (K_p or Log BB). The former approach provides information on the *rate of permeation* of the BBB, whereas the latter approach provides information on the *extent of penetration*, in this case, the total drug concentration in brain. Used in isolation, neither technique provides insight into the more relevant brain free drug concentration ($C_{u,b}$); hence the associated physicochemical guidelines should be interpreted with care.

Several key physicochemical properties have been identified that influence the rate of brain permeability and the extent of brain penetration including H-bonding potential, MW, lipophilicity, polar surface area (PSA), ionization state, and rotatable bond count. It is often the case that no single physicochemical property can be used to predict a given parameter such as P_{app}; rather, multivariate models that incorporate several descriptors are required. These multivariate models are very useful when applied, often via an online tool, as a filter to prioritize ideas for synthesis. However, these models can be difficult to intuit when designing a molecule "in one's head." In the latter case, it is useful to have an idea of how each of the individual physicochemical properties influences CNS penetration and the rate of permeation.

One of the critical challenges for the medicinal chemist in drug discovery is balancing multiple physicochemical properties with the SAR to address deficits in one parameter without negatively affecting another. For example, increasing lipophilicity can improve BBB permeability but may also negatively impact blood clearance and increase nonspecific binding and, thus, lead to an overall reduction in $C_{u,b}$. To achieve an optimal balance, one should understand how individual physicochemical properties may influence the extent of CNS penetration and the rate of permeation. For the purposes of this review, key physicochemical properties are discussed sequentially with examples from the medicinal chemistry literature [4]. However, it should be noted that it is not possible to alter one parameter in isolation, which may confound interpretation of the controlling factor mitigating CNS exposure. Therefore, the categorization applied here is best regarded a mnemonic rather than a rigorous classification.

Lipophilicity

Lipophilicity is widely considered to be one of the most, if not the most, important physicochemical properties, the control of which is critical for ultimate success in drug discovery and development. This property reflects the critical events of molecular desolvation in the transfer from aqueous phases to cell membranes and to protein-binding sites, which are mostly hydrophobic in nature. Increases in ligand lipophilicity often result in improved *in vitro* potency, which makes increasing lipophilicity a relatively straightforward and tempting medicinal chemistry optimization strategy. However, if lipophilicity is too high, there is not only increased likelihood of poor solubility and high metabolic clearance, but also greater risks of nonspecific polypharmacology and consequent off-target related toxicology [53]. Hence, the surrogate measures cLogP and cLogD are commonly monitored parameters from the earliest stages of the discovery process.

The recognition that lipophilicity is an important determinant in brain penetration predates modern computational approaches to QSAR. Early studies using the Hansch approach, which was developed in the early 1970s, demonstrated that increase in octanol–water partition coefficients (LogP) are associated with increases in LogBB and CNS activity [54]. This observation was subsequently confirmed by studies of marketed CNS and CNS-inactive drugs, which also introduced a more physiologically relevant lipophilicity indicator, the pH-dependent octanol–water distribution coefficient (LogD). Good correlations were reported and values in the range of cLogD 1–4 were proposed for optimal LogBB [50]. However, as discussed earlier in this chapter, the pharmacological relevance of LogBB and its correlation with lipophilicity has been questioned in recent years [5–10, 16]. Gleeson's analysis of LogBB data for over 3000 diverse molecules from GlaxoSmithKline (GSK) showed that cLogP is well correlated with brain tissue binding. This suggests that the higher LogBB values observed for more lipophilic compounds reflect increased nonspecific brain tissue binding, which will not necessarily correlate with the therapeutic effect [45]. Importantly, Gleeson's study uncovered the negative correlation between $f_{u,b}$ and LogP, which supports the lowering of LogP as a strategy to improve brain exposure.

Lipophilicity has been shown to affect not only the *extent* of brain penetration, but also the *rate* of the BBB permeation. Numerous reports in the literature suggest that in addition to lipophilicity, the BBB permeability is also influenced by ionization state, molecular volume, and H-bonding potential as represented by descriptors such as the pK_a, PSA, H-bond donors and acceptors count, and MW [45, 55, 56]. Since some of these descriptors correlate well with each other (e.g., MW, PSA), a number of reported CNS permeability models rely on an indicator of lipophilicity and only one additional descriptor. An analysis of over 50,000 measurements from an artificial membrane (PAMPA) assay from GSK showed that increased LogP is associated with an increased passive permeability of molecules containing ionisable groups such as acidic, basic, or zwitterions; while no such linear correlation was observed for neutral molecules [45]. In line with these findings, Waring's analysis of a large Caco-2 dataset at AstraZeneca indicated that LogD is a better permeability predictor than LogP [44]. He also demonstrated that, in addition to the LogD, MW needs to be considered. As MW increases, LogD also needs to be increased in order to maintain a 50% chance of high permeability, for example, a compound of 300–350 Da requires a cLogD >1.1, but a compound of 450–500 Da requires a cLogD >3.4 [44, 57]. Similar dependencies of the BBB permeability on LogD, MW, as well as on PSA, have been reported by others [57–63].

An example of how the control of lipophilicity can improve brain exposure was reported by Johnson et al. [13]. In the course of a program seeking to identify selective muscarinic M1 agonists for the treatment of the cognitive deficits

TABLE 18.6 M1 agonist: reduce lipophilicity to improve $C_{u,b}$ [13]

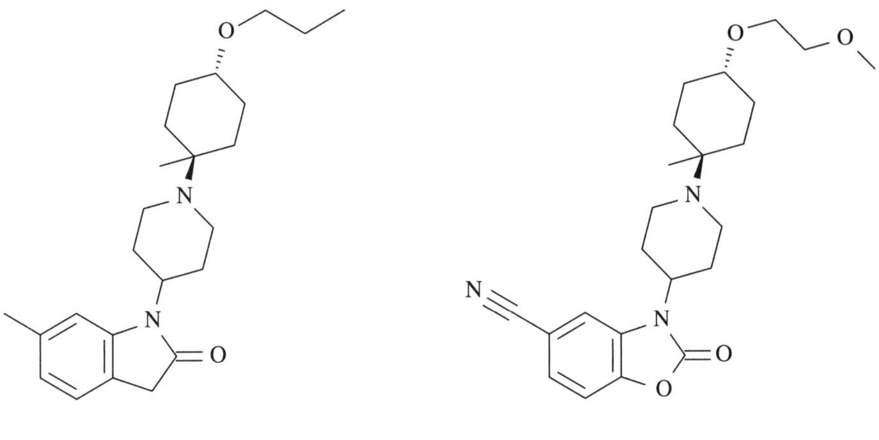

13 **14**

Compound	cLogP	CL[a] ml/min/kg	K_p (B:P)	$f_{u,b}$ (%)	$f_{u,p}$ (%)	$C_{u,b}$[a] (nM)	$C_{u,p}$[a] (nM)
13	3.5	85	5.7	6	20	2.5	2.6
14	1.5	23	1.7	39	38	261	265

[a]Estimated from a 3 mg/kg po. dose.

associated with schizophrenia and AD, compound **13** was identified (Table 18.6). This compound has a moderate cLogP of 3.5 and a whole brain/plasma ratio K_p of 5.7; however, the unbound drug concentrations were low at $C_{u,b}$ 2.5 nM and $C_{u,p}$ 2.6 nM, likely because of high clearance. Encouragingly, the ratio of unbound drug in the brain and plasma $K_{p,uu}$ was approximately 1, which is indicative of a lack of active efflux and contrasts the earlier analogues within the series. To address the metabolic instability, a series of compounds with reduced cLogPs was prepared, and this process resulted in the identification of compound **14**, which exhibits reduced plasma and brain tissue binding ($f_{u,p}$, $f_{u,b}$), lower clearance, and much improved $C_{u,b}$. Notably, the reduced cLogP of **14** (1.53 versus 3.48 for **13**) led to a reduction in K_p; however, due to the increased $f_{u,b}$ and reduced clearance, the overall effect was an increase in the unbound concentration in the brain $C_{u,b}$. This example highlights the fact that strategies that seek to lower clearance from the plasma by reducing cLogP may also increase the $f_{u,b.}$ The parameters thus work synergistically to beneficially affect $C_{u,b}$.

Another example that highlights the importance of looking beyond total brain levels to assess CNS exposure has been reported by Smith et al. at GSK [64]. The authors discussed the optimization of a series of neurokinin-3 (NK_3) receptor antagonists beginning with the lead compound **15** (Table 18.7). Following the first round optimization, the authors selected two antagonists, **16** and **17**, for an *ex vivo* NK_3 receptor occupancy study in gerbils.

TABLE 18.7 NK3 antagonists: total brain C_{max} levels do not necessarily reflect $C_{u,b}$ [64]

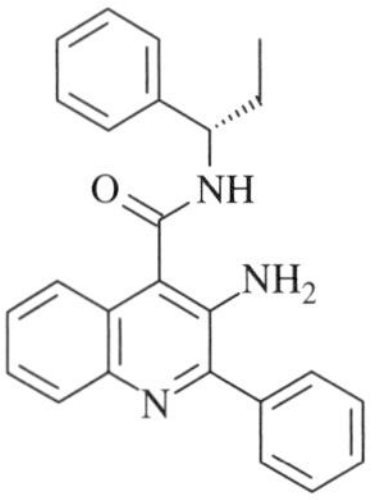

15

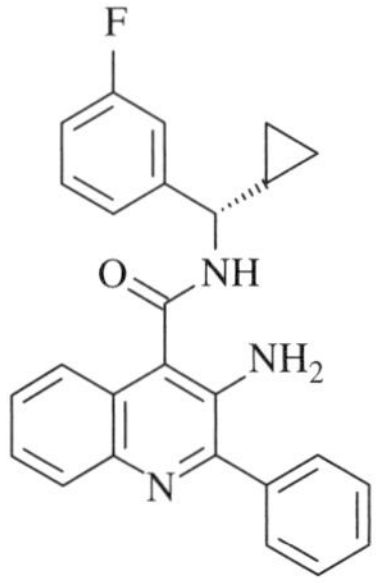

16, GSK172981

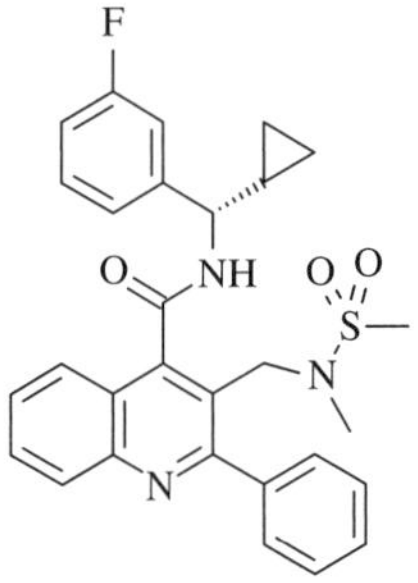

17, GSK256471

Compound	cLogP	hNK$_3$ pK_i	Rat brain C_{max} (ng/g)[a]	Gerbil dose mg/kg (route)	Mean C_b ng/g (nM)	$f_{u,b}$ (%)	Mean RO (%)
15	6.6	8.7	80	—	—	—	—
16	6.8	8.7	464	30 (ip)	2062 (5011)	0.7	60
17	4.5	9.0	43	10 (po)	61 (118)	3.3	61

[a]Sprague–Dawley male rats. Dose: 3 mg/kg po.

As shown in Table 18.7, despite the higher whole brain exposure observed for compound **16**, the receptor occupancy (RO) at 30 mg/kg ip was equivalent to that observed with compound **17** at 10 mg/kg po. A combination of slightly increased potency and reduced nonspecific binding to brain tissue of compound 17 ($f_{u,b}$ 3.3% versus $f_{u,b}$ 0.7% for **16**) compensated for the reduced whole brain level and thus predicted equivalent efficacy *in vivo*. As compound **17** has the lower cLogP (4.5 versus 6.8 for **16**), it also had the lower intrinsic risk with respect to development.

Controlling lipophilicity to achieve optimal brain exposure was also a critical aspect of the medicinal chemistry strategy for optimizing class IIa histone deacetylase (HDAC) inhibitors as a potential therapy for Huntington's disease that was described by Burli et al. [65]. The starting point of the project was cyclopropyl hydroxamic acid **18**, which was derived from molecular modeling and an HDAC4 biostructural–based design. Compound **18** has a moderate potency for class IIa HDAC isoforms; for example, HDAC4 = 340 nM (Table 18.8). Based on the modeling information and with the objective of further improving the activity while achieving balanced physicochemical properties that are suitable for optimal CNS exposure, the authors investigated a range of analogues containing small heteroaromatic "capping" groups. This approach led to the discovery of oxazole **19**, which has high HDAC4 biochemical (25 nM) and cellular (220 nM) potency, and excellent ligand efficiency (LE) and ligand-lipophilicity efficiency (LLE) (0.43 and 5.4,

TABLE 18.8 HDAC4 inhibitors: lower CLogP to improve P_{app}, P-gp, and $f_{u,b}$ [65]

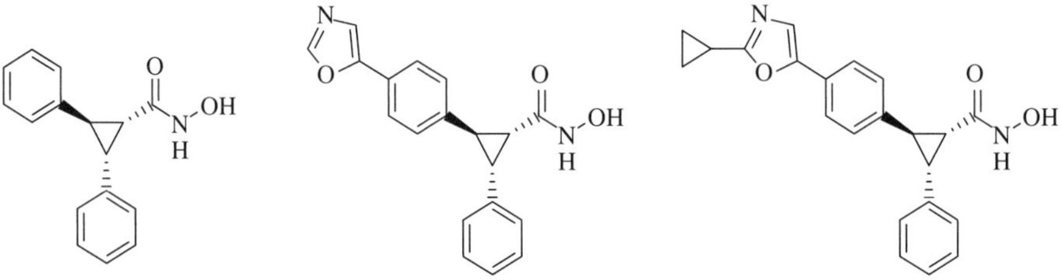

15 **16** **17**

Compound	cLogP	HDAC4 IC_{50} (nM)	MDR1-MDCK P_{app} (nm/s)	ER	K_p[a] (B/P)	$f_{u,b}$[a,b]
15	2.6	340	—	—	—	—
16	2.1	25	442	2.4	0.2–1.0	0.026
17	3.0	40	255	4.0	0.3–0.6	0.0037

[a]C57Bl/6 mice; dose: 10 mg/kg.
[b]Estimation based on *in vitro* equilibrium dialysis data.

respectively). The compound also exhibited high permeability (P_{app} = 442 nm/s) and a low efflux ratio, which suggest that it has no significant interaction with P-gp (2.4).

However, **19** was found to be a potent inhibitor of human hepatic P450 isoforms CYP2C9 (IC_{50} = 0.4 μM), CYP2C19 (IC_{50} = 0.06 μM), and CYP2D6 (IC_{50} = 2.7 μM). To address the P450 inhibition, the authors explored a range of substituted isoxazole analogues of which **20** displayed the best overall properties. While retaining high HDAC4 potency (40 nM), this compound exhibited no significant P450 inhibition (IC_{50} values: CYP2C9 = 23 μM; CYP2C19 > 100 μM; CYP2D6 = 45 μM). However, despite the increased lipophilicity (AlogP = 3.2), this compound showed lower permeability (255 nm/s) and a higher efflux ratio (ER = 4) compared to the parent **19** (AlogP = 2.2; P_{app} = 442 nm/s; ER = 2.4). The MDR1-MDCK data translated well *in vivo* where more lipophilic compound **20** was found to have slightly lower K_p and $f_{u,b}$ (0.3–0.6 and 0.0037) than **19** (0.2–1.0 and 0.026). Consequently, when dosed at 10 mg/kg, the free concentration of compound **19** in the brain is estimated to reach its cell-based IC_{50} value, whereas **20** requires administration at a higher dose to achieve its free brain concentration at or above its IC_{50}.

These examples focus on lowering the lipophilicities of candidate compounds and highlight the fact that a high whole brain to plasma level is not a prerequisite of a successful CNS drug candidate. However, high nonspecific binding to brain tissue has not precluded some compounds from becoming successful drugs. Indeed, consistent with the observation that CNS drugs, on average, have higher LogPs than non-CNS drugs [28], many commercially successful CNS therapeutics have high brain tissue binding, such as the antidepressant sertraline (**21a**) and the antipsychotic chlorpromazine (**21b**) with $f_{u,b}$ values in mouse of 0.00066 and 0.00076, respectively [10].

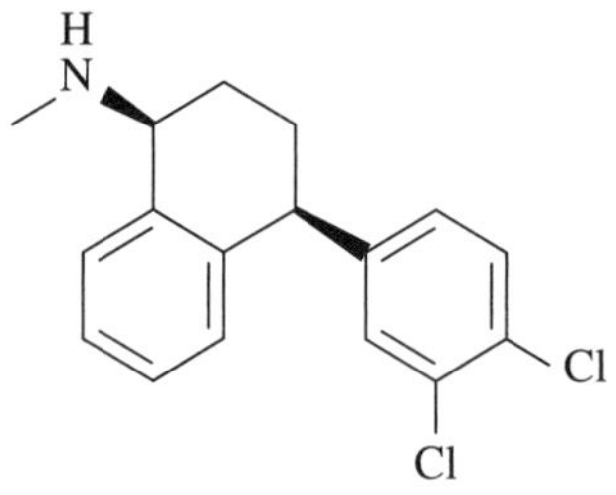

21a, sertraline
clogP = 5.1
MW = 306

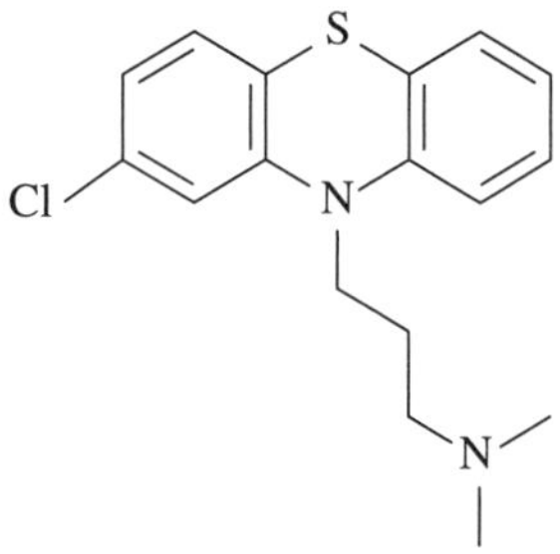

21b, chloropromazine
clogP = 3.9
MW = 284

Hydrogen Bonding

H-bonding capacity is arguably one of the most critical physicochemical properties in the context of the CNS penetration. Increased H-bonding potential is correlated with increased PSA and has been associated with reduced BBB penetration. Leeson and Davis and Polli et al. have shown in comparative analyses of CNS and non-CNS drugs that the CNS agents have significantly reduced numbers of H-bond donors (CNS mean 0.67) [28] and H-bond acceptors (CNS mean 2.12) [48]. Österberg showed that H-bond descriptors combined with cLogP were sufficient to generate a BBB penetration model with good predictivity [66]. This work was subsequently extended to provide two simple rules for predicting LogBB values that are indicative of good brain penetration [49]:

- N + O atoms < 5
- cLogP-(N + O) > 0

H-bonding also has an effect on the rate of permeation [23]. It has been shown that H-bond acidity and basicity both decrease the rate of BBB permeation [67]. Studies using a PAMPA devoid of efflux transporters suggest that increased H-bonding potential can lower passive permeability [23]; however, it is likely that the role of H-bonding is twofold because increased H-bonding potential also increases the risk of efflux via P-gp. Consequently, the reduction of H-bond donors and acceptors is often a successful strategy for the optimization of targeting the CNS. Pajouhesh and Lenz proposed the following frequently quoted guidelines for successful CNS drug candidates: H-bond donors < 3; H-bond acceptors < 7; and total H-bonds < 8 [47].

A group at GSK has described a successful optimization of the poorly brain-penetrant 5-HT_6 antagonist SB-271046 **22** by focusing on the H-bond count [68]. Compound **22** had good oral bioavailability; however, its brain to plasma ratio, K_p, was low, and the compound was shown to be a P-gp substrate (Table 18.9). To improve brain penetration, compounds such as **23** and **24** in which the acidic NH

TABLE 18.9 5-HT6 antagonist: reduce HBD count to improve brain exposure [68]

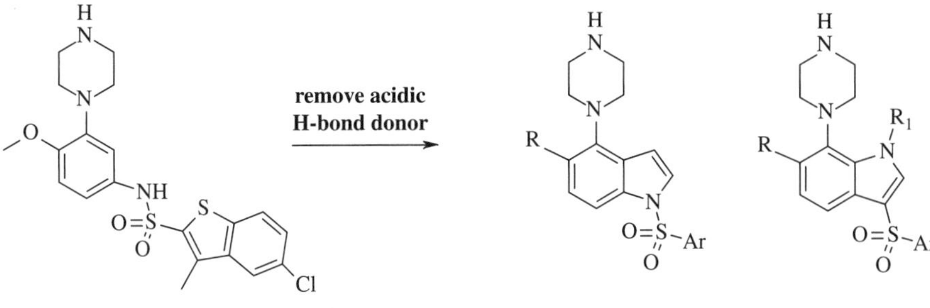

22, SB-271046 **23a-c** **24a-b**

Compound	Ar	R	R¹	5-HT_6 pK_i	Rat CL_b ml/min/kg	K_p
23a	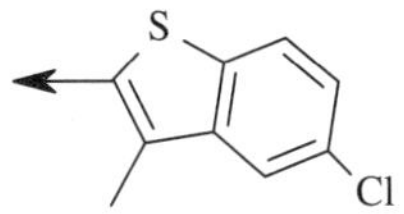	H	—	8.5	—	—
23b	C_6H_4(3-Cl)	H	—	9.6	—	—
23c	C_6H_4(3-Cl)	Cl	—	8.6	44	3
24a	C_6H_4(3-Cl)	H	H	9.5	41	0.7
24b	C_6H_4(3-Cl)	H	Me	8.6	34	2.6

donor group of sulfonamide had been removed by cyclization onto the adjacent phenyl ring were prepared. This process also reduced the overall flexibility of the molecule, which was predicted to further benefit BBB permeability.

This approach resulted in the 5-HT_6 ligands **23a–b**; however, these compounds suffered from high *in vivo* blood clearance. Resolution of this issue led to the discovery of compound **23c**, which was shown to have maintained 5-HT_6 affinity and reduced efflux liability in an MDR1-MDCK assay. Alternative cores, such as compounds **24a–b,** were also investigated, and the removal of the additional H-bond donor from **24a** to afford **24b** led to an increase in the brain/plasma ratio (K_p 2.6 versus 0.7). To confirm that the elevated K_p values were indicative of improved $C_{u,br}$, compounds **23c** and **24b** were selected for evaluation in a rat *ex vivo* binding assay. Compounds **23c** and **24b** had ED_{50} values of 3 and 5 mg/kg po, respectively. In comparison, the initial compound SB-271046 **22** was significantly less potent with an ED_{50} of 11 mg/kg po.

Reduction of the H-bond donor count was also a successful strategy for Hu and colleagues at Amgen who reported on the optimization of PDE10A inhibitors for potential treatment of schizophrenia [69]. To establish the minimum structural requirements for the PDE10A inhibition of HTS hit **25** and to address its high efflux in both human (ER = 24.8) and rat (ER = 13.1) P-gp assays, systematic deletion studies were performed. Interestingly, the truncated analogue **26**, which lacks methyl-amino-pyridine functionality and two fluorine atoms, lost its PDE10A potency by only fivefold, while its efflux ratio significantly improved to 4.7 in the human and 4.9 in the rat assays. Replacing the pyridine with a morpholine moiety in **27** resulted in improved PDE10A potency (IC_{50} = 1.1 nM) and a decreased efflux ratio in human (ER = 11.1) but an increased ratio in rat (ER = 32.8) assays. Because a cocrystal structure of the HTS hit (**25**) in PDE10A suggested that the NH linker was not essential, the authors decided to replace it with a carbonyl group to reduce the H-bond count; this approach was also expected to maintain the planarity of the phenyl and the benzimidazole rings that was observed in the cocrystal structure. Indeed, the keto analogue **28** not only retained low nanomolar potency in the PDE10A assay but also exhibited both low efflux ratios (human ER = 0.9; rat ER = 1.0) and improved *in vivo* rat clearance (Cl = 0.53 l/h/kg) compared to the screening hit (Cl = 3.9 l/h/kg), albeit the oral bioavailability was still low (F = 10%). Compound **28** was advanced into the LC-MS/MS RO assay where it produced 21% RO at 10 mg/kg and 55% RO at 30 mg/kg (po). The modest CNS target engagement event at the higher dose was clearly due to the compound's poor systemic exposure, which was attributed mainly to the oxidative metabolism of the morpholine ring (Table 18.10).

TABLE 18.10 PDE10A inhibitors: reduce HBD capacity to reduced ER [69]

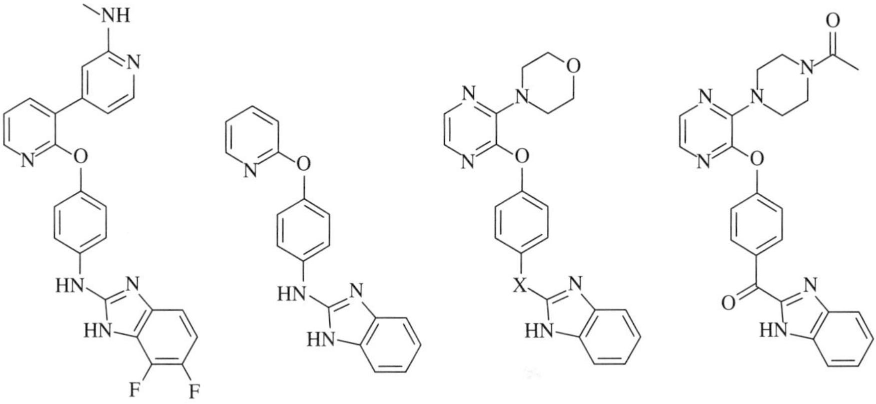

25 **26** **27, X = NH** **28, X = CO** **29**

Compound	PDE10A IC_{50} (nM)	ER human/rat	CL (l/h/kg)	F^a (%)	RO^b (%)
25	9.7	24.8/13.1	3.90	—	—
26	45	4.9/4.7	—	—	—
27	1.1	11.1/32.8	—	—	—
28	4.5	0.9/1.0	0.53	10	21.3
29	5.1	—	0.07	56	57.1

[a]Fed Sprague–Dawley male rats. Dose: 5 mg/kg po.
[b]Dose: 10 mg/kg.

Consequently, in order to reduce the plasma clearance and preserve the favorable CNS properties, a range of structurally diverse but in terms of physicochemical properties similar analogues of **28** with variations around the morpholine region were prepared. This strategy led to the identification of compound **29**, which exhibited superior *in vivo* ADME profile and achieved 57% RO at 10 mg/kg (po) in the LC-MS/MS RO assay.

Achieving optimal physicochemical properties that are conducive to CNS exposure is one of the major challenges in the design of β-Secretase (BACE-1) inhibitors for the treatment of AD. Early drug discovery efforts focused primarily on the transition state isosteres of the peptide substrate, a strategy that proved to be successful in the design of inhibitors for other members of the aspartic protease family, such as HIV and rennin. Indeed, this approach led to highly potent BACE-1 inhibitors, but due to their peptidic nature they were also characterized by high numbers of H-bond donors and acceptors, PSAs, and MWs, and consequent poor CNS exposures [70]. The more recent discoveries of small cyclic amidine-, guanidine-, and sulfamide-containing heterocycles that efficiently interact with the two catalytic aspartates marked a major breakthrough in the development of brain penetrant BACE-1 inhibitors. For example, Brodney et al. [71] described a structure-based drug discovery (SBDD)-driven optimization of a series of spirocyclic sulfamides (represented by compound **30**) in which one of the critical issues was P-gp-driven poor CNS exposure (Table 18.11). Based on the postulation that the high P-gp efflux was due to the presence of the H-bond donor, the optimization

TABLE 18.11 BACE-1 inhibitors: eliminate or "mask" HBD to improve brain penetration [71]

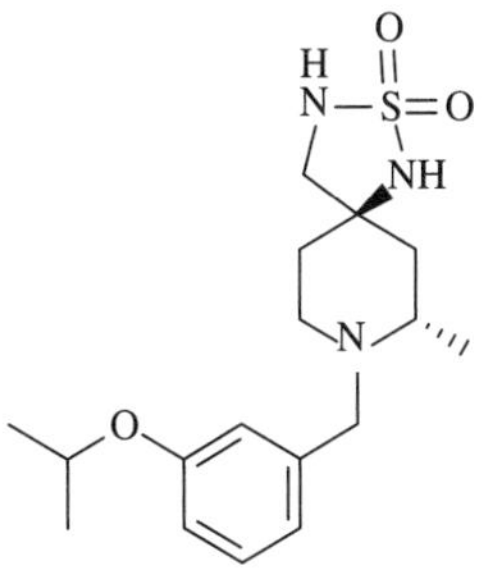

30

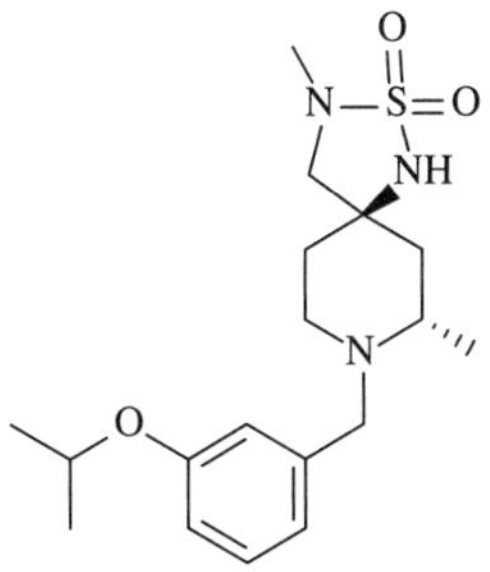

31

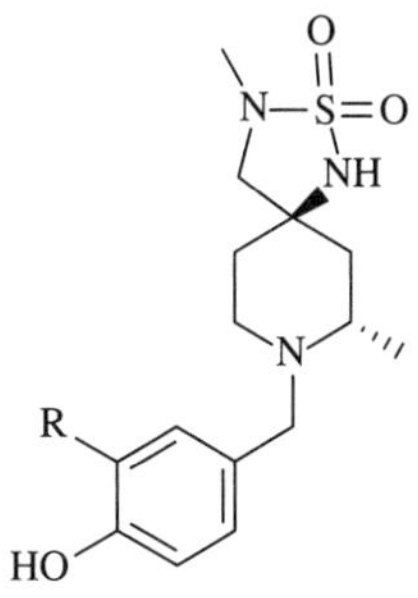

32a R = iPrO-
32b R = H

Compound	BACE-1[a] IC_{50} (μM)	MDCK AB (10^{-6} cm/s)	ER	$f_{u,p}$[b]	$f_{u,b}$[b]	$K_{p,uu}$
30	2.9	10.2	3.8	0.041	0.018	0.22
31	1.8	7.5	1.7	0.036	0.024	0.65
32a	0.1	17.6	3.4	0.17	0.12	0.27

[a]Whole cell assay.
[b]Determined from equilibrium analysis (mice).

strategy was to cap the sulfamide to reduce the HBD count. Indeed, the methyl analogue **31** showed comparable potency and a reduced P-gp-mediated efflux compared to the NH analogue **30** (ER = 1.7 versus 3.8). Compound **31** possesses greatly improved physicochemical properties that include a lack of HBDs, a moderate PSA of 69, and a LogP of 3.9; consequently, this compound displays favorable unbound brain to unbound plasma ratios ($K_{p,uu}$ = 0.65) that suggest that no efflux transporters limit access to the central compartment.

However, further improvements of the potencies of BACE-1 s were needed (**31** IC_{50} = 1.8 μM). Analysis of the X-ray structure of **31** co-crystallized with BACE-1 suggested that an H-bond-donating group attached to the benzyl moiety might interact with one of the backbone carbonyl groups in the enzyme S1 pocket and result in improved potency. Because all other attempts to deliver improvements failed, the authors pursued this strategy despite the concern that incorporating an HBD into the scaffold may reintroduce P-gp issues. However, this approach was found to be successful and resulted in the discovery of isopropoxyphenol **32a**, which not only exhibits improved potency over the parent **31** (BACE-1 IC_{50} = 0.1 and 1.8 μM, respectively) but also exhibits an improved metabolic stability profile. As anticipated, **32a** also proved to be a P-gp substrate with an MDCK ER of 3.4 and consequently displays an asymmetry between the unbound concentrations in the plasma and brain compartments in mice ($K_{p,uu}$ = 0.27). Interestingly, isopropoxyphenol **32a** exhibits a slightly lower P-gp-mediated efflux relative to the phenol analogue **32b** (ER = 4.3). The authors speculated that this difference was a consequence of the HBD being "masked" by an engagement into an intramolecular H-bonding with the iPro group in **32a** [37]. Despite the reduced brain penetration relative to **31**, the higher free fraction, excellent potency, and selectivity of **32a** resulted in its advancement into the PK–PD studies. The compound displayed significant dose-dependent reductions in both the brain and CSF Aβ40 levels at the two highest doses (100 and 300 mg; s.c.). In an agreement with the MDR1-MDCK data, the brain drug exposures of **32a** at the 10–300 mg/kg dose range were approximately 7- to 12-fold higher in the P-gp KO mice compared to the wild types.

Polar Surface Area

PSA is often described as a surrogate measure of H-bonding capacity and polarity. Early computational QSAR studies of LogBB data by van de Waterbeemd and Kansy identified PSA as a key descriptor for determining the extent of BBB penetration [52]. The initial studies were confirmed upon the analyses of larger datasets [50, 51, 69, 72, 73]; however, many of these studies produced QSAR equations that rely on access to special computational software and are not easily interpreted outside of the cyber environment by medicinal chemists. Analyses of non-CNS versus CNS drugs resulted in the groups of both Kelder and van de Waterbeemd translating their results into guidelines for optimal PSA; van de Waterbeemd proposed a PSA cutoff of less than 90 Å^2 [50], and Kelder proposed a more stringent cutoff of less than 60–70 Å^2 [72]. The effect of PSA on the extent of brain penetration could be due to changes in brain tissue binding or differences in permeability through the BBB; in reality, however, PSA probably influences both. In an analysis of permeability LogPS data for 23 compounds, Liu et al. demonstrated that PSA is one of three key descriptors that can be used to generate a linear model of brain permeation [62]. However, PSA alone has been shown to have only modest predictive power for even simpler datasets, such as the Caco-2 permeability dataset [57]. Egan's "confidence ellipse" based on ALogP98 and PSA, which was used at Pharmacopeia to predict Caco-2 permeability, is perhaps one of the simplest-to-apply composite models of PSA that can be used to predict permeability [58].

Pinard at Roche described the discovery of the selective GlyT1 inhibitor RG1678, which has demonstrated efficacy in a Phase II clinical study in schizophrenic patients [74]. Using their proprietary SAR analyzer (ROSARA), this group determined that PSA and cLogP were the sole descriptors required to predict brain penetration within the series. The starting point for their work was the lead compound **34** (Table 18.12), which has low brain penetration ($K_p \sim 0.1$) and

TABLE 18.12 Glyt-1 inhibitors: reduce PSA to improve brain penetration [74]

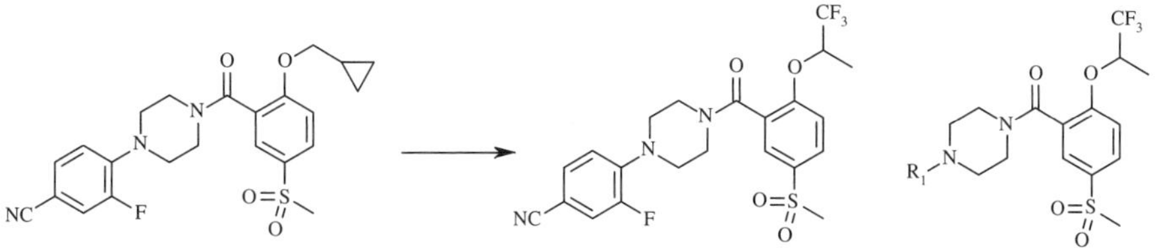

34 **35** **36a-c**

Compound	R^1	GlyT1 EC_{50} μM	hERG IC_{50} μM	cLogP	PSA Å^2	*In vivo* ED_{50} mg/kg[a]	Mouse K_p
34	—	0.016	0.6	2.64	80	3.0	0.10
35	—	0.044	3.0	2.77	80	1.0	0.25
36a	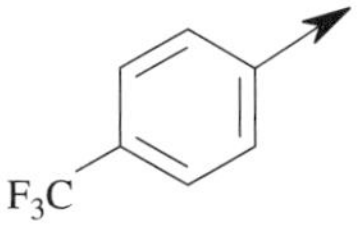 F_3C	0.030	10	3.92	59	2.0	1.10
36b	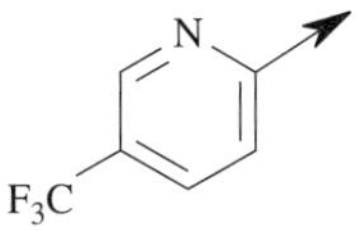 N, F_3C	0.037	20	2.82	69	5.0	0.20
36c	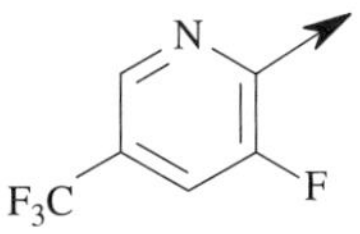 N, F, F_3C	0.040	20	3.00	69	1.0	0.50
(*S*)-**36c** (RG1678)		0.030	17	3.00	69	0.5	0.5

[a]Reversal of L-687,414-induced hyperlocomotion in mouse.

an unacceptable hERG potency (IC_{50} 0.6 μM, in patch clamp assay). The optimization challenge was thus to balance the conflicting requirements of a PSA that was sufficiently low for acceptable brain permeation and a PSA that was sufficiently high to reduce hERG liability. The authors commented that they observed a correlation between the improvements in the efficacy in their *in vivo* model and the improvements in C_b in this series (which implies that $f_{u,b}$ remained relatively constant between the analogues); thus, during the optimization they used this parameter in combination with K_p. The trifluoroisopropyloxy group in derivative **35** proved to be a favorable substitute. Despite inducing minimal changes in PSA and cLogP, this group improved the brain exposure and *in vivo* efficacy as well as the selectivity over hERG. Replacement of the nitrile group with a trifluoromethyl group afforded **36a** (cLogP 3.92, PSA 59 Å^2) with a lower PSA and a higher cLogP compared to the lead **34** (cLogP 2.64, PSA 80 Å^2); this replacement also resulted in an improved K_p ratio. The incorporation of the pyridyl nitrogen in **36b** was successful in further improving the hERG selectivity; however, it also proved detrimental to the brain exposure and *in vivo* efficacy (Table 18.12).

To address this issue, the authors followed the well-documented strategy of inserting fluorine to improve brain penetration [75]. Indeed, compound **36c** maintained the selectivity over the hERG channel and exhibited minimal changes in physicochemical properties and potency compared to **36b** and also exhibited significantly improved *in vivo* efficacy and K_p. Resolution of **36c** identified the (*S*)-enantiomer as the eutomer, and this compound was selected for progression into the clinic as **RG1678** (bitopertin).

pK_a and Ionization State

The majority of CNS drugs contain a basic center. Wager et al. found that a mean pK_a value for marketed CNS drugs and Pfizer CNS clinical candidates was 8.4 [43]. There are two schools of thought as to whether this fact reflects an inherent benefit of compound basicity on brain exposure or is merely due to bias in the dataset resulting from that fact that many of the currently prescribed CNS drugs target monoamine receptors or transporters, which have ligand pharmacophores that contain basic centers. Pajouhesh and Lenz proposed that the optimal pK_a range for CNS drugs is pK_a 7.5–10.5 [47], and although the lower limit espoused by this range would currently be considered overly conservative, the upper limit is consistent with more recent analyses of P-gp efflux and probably contributes to the upper pK_a limit because highly basic molecules are associated with P-gp substrate liability (*vide infra*) [43]. In a review of P-gp data for 1000 compounds, Petrauskas proposed a cutoff of pK_a < 8 for compounds designed to avoid possible P-gp interactions, although many CNS drugs, including the SSRIs and tricyclics, exceed this criterion [34]. The lower pK_a limit probably results from a combination of enhanced P-gp liability for acids and the inherently reduced average passive permeability of acids and zwitterions relative to neutral and basic molecules [45].

TABLE 18.13 A7 NAChR inhibitors: reduce PK_a to improve permeability [76]

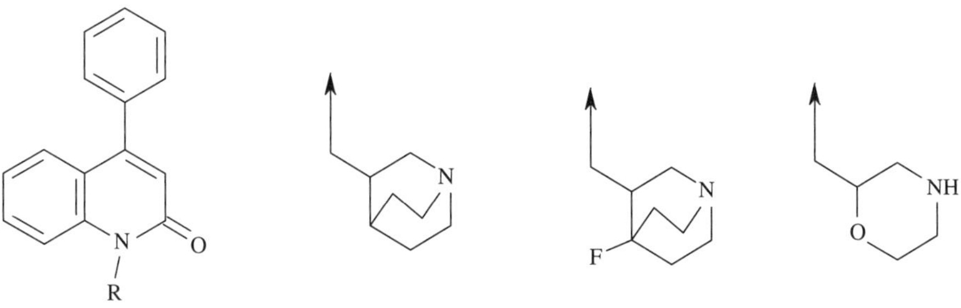

37 38 39

Compound	α7 nAChR EC_{50} nM	pK_a	ER[a]	P_{app}[b] (nm/s)	K_p
37	150	10.1	23	900	<0.02
38	5800	7.6	3.5	380	ND
39	260	8.1		550	0.9

[a]Caco-2 P_{app} (B to A)/P_{app} (A to B).
[b]PAMPA, pH 7.4.

A strategy of controlling pK_a was adopted by McDonald and colleagues at Bristol-Myers Squibb to improve brain penetration in a quinolone series of α7 nicotinic acetylcholine receptors (nAChR) agonists for the potential treatment of cognitive deficits in patients suffering from schizophrenia [76]. One of their early leads, quinuclidine **37**, was a potent α7 agonist that displayed good permeability in a PAMPA (P_{app} = 900 nm/s). However, despite this good permeability, **37** was found to have a very poor brain penetration in rats with a K_p value less than 0.002. Follow-up experiments revealed that all of the tested quinuclidines in the series had very high efflux ratios in a bidirectional Caco-2 assay, which suggested that they are possible substrates for transporter-mediated efflux (Table 18.13).

Based on the hypothesis that the observed ER was related to the high basicity of quinuclidine (pK_a ~10), the authors explored a series of quinolone analogues that contained less basic amines, including 4-fluoroquinuclidine **38** (pK_a = 7.6). Indeed, the high ER that was observed in the more basic analogues proved to be no issue for this compound (ER = 0.6); however, the drop in basicity was also accompanied by a substantial loss in α7 activity (EC_{50} = 5.8 μM). Interestingly, the morpholine analogue **39** retained α7 potency (EC_{50} = 260 nM) while displaying near unity ER in the Caco-2 assay and consequently achieving a K_p value in rats of 0.9 (a concomitant improvement in $C_{u,b}$ was not reported).

A similar pK_a-lowering strategy has been applied by a group at Novartis in their efforts to improve the brain penetration of a series of macrocyclic ethanol amine inhibitors of BACE-1 [77]. The initial compounds in this series (as represented by **40** (Table 18.14)) were potent and selective BACE-1 inhibitors; however, they also

TABLE 18.14 BACE-1 inhibitors: reduce PK_a to improve P_{app} and brain penetration [77].

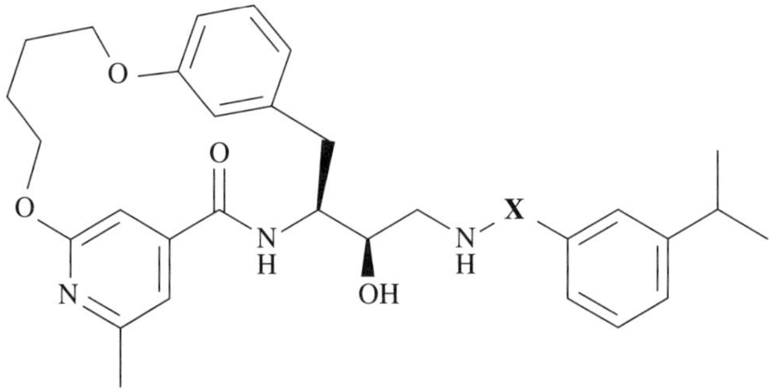

Compound	X	BACE-1 IC_{50} nM	pK_a	P_{app} [a] (10^{-6}cm/s)	ER[a]	C_b (μM)	Aβ40[b] (%)
40	CH_2	32	8.5	1.3	23	0.04	7
41	cPr	15	7.3	4.0	3.5	0.32	72

[a]MDR1-MDCK.
[b]APP51/16 tg mouse; dose: 60 μmol/kg.

displayed poor permeabilities and high efflux ratios in the MDR1-MDCK cell line (P_{app} = 1.4 10^{-6}cm/s; ER = 23). To address these issues, the authors focused on a systematic investigation of the environment around the ethanolamine group, which is the enzyme catalytic aspartate-binding motif. This approach included the insertion of small fluorinated alkyl groups, such as CHF_2 and CF_3, into the benzylic position next to the amine, which indeed improved the passive permeability and reduced the ER (data not shown). However, the strong electron-withdrawing groups rendered the compounds inactive, presumably because some level of basicity is required for the interaction with the enzyme catalytic aspartates. Interestingly, cyclopropane analogues, such as **41** (pK_a = 7.3), not only exhibited slightly improved BACE-1 potencies but also exhibited significantly improved passive permeabilities and reduced ERs (P_{app} = 4 10^{-6}cm/s; ER = 3.5).

When tested *in vivo*, **41** was present in significant concentrations in the brain (0.32 μM) and produced a profound reduction in brain levels of Ab40 (72%) in contrast to the unsubstituted benzylamine **40** (C_b = 0.04 μM; Aβ40 = 7%). Unfortunately, the introduction of the cyclopropyl group into the ethanolamine motif led to the loss of selectivity over the closely related aspartyl proteases cathepsin D and E, which prevented this series from progressing further.

In contrast to amines, carboxylic acids are generally associated with poor brain penetration [27] due to a combination of multiple factors that include high plasma protein bindings, poor passive permeabilities, and high efflux ratios. This effect was observed for the first time with terfenadine (**42**, Fig. 18.4), a first-generation

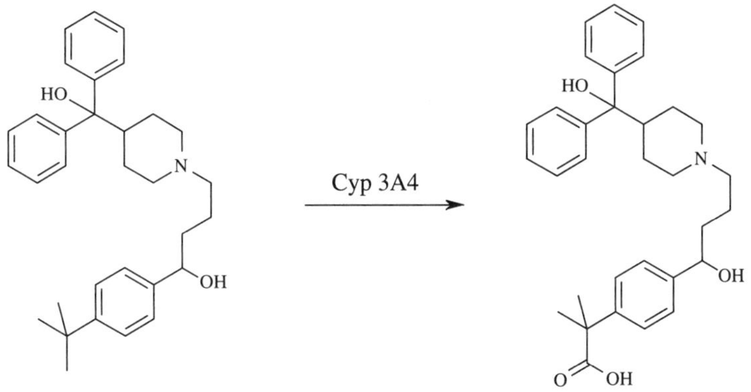

42, Terfenadine
H1 IC_{50} = 1 nM
hERG K_i = 56 nM

43, Fexofenadine
H1 IC_{50} = 15 nM
hERG K_i = 23,000 nM

FIGURE 18.4 Terfenadine and its main metabolite, fexofenadine [80].

antihistamine withdrawn from the market in 1997 due to instances of cardiac arrhythmia, which was subsequently connected to its potent hERG block. Interestingly, it was later discovered that its principle metabolite, carboxylate **43**, not only accounts for all of the therapeutic effects of terfenadine but also exhibits significantly reduced CNS side effects, such as sedation, and greatly improved hERG selectivity [78]. This metabolite was subsequently marketed as fexofenadine, the first of the second-generation antihistamines that are characterized by a lack of the central side effects that are intrinsic to the prior generations.

This improved side effect profile has been attributed to the low CNS exposure of the zwitterionic fexofenadine. It was subsequently shown that terfenadine has a high passive permeability (P_{app} = 28.5 × 10^{-6} cm/s), rapidly penetrates the CNS, and does not exhibit markedly different brain uptakes in wild-type and mdr1 KO mice. However, fexofenadine has low passive permeability (P_{app} = 6.6 × 10^{-6} cm/s), displays slow brain uptake, and has steady-state brain-to-plasma ratios of 0.005 in mdr1 $^{(+/+)}$ mice and 0.27 in mdr1 $^{(-/-)}$ mice that result in an *in vivo* efflux ratio of approximately 50. Therefore, the presence of the carboxylic acid moiety in fexofenadine simultaneously reduces passive permeability relative to terfenadine and acts as a recognition element for P-gp efflux [79, 80].

Despite the fact that zwitterions were initially introduced to limit the CNS exposure of first-generation antihistamines, examples of brain-penetrant zwitterionic molecules have been reported in the literature. One such example is a series of melanin-concentrating hormone receptor-1 (MCHR-1) antagonists reported by Mihalic and colleagues at Amgen [81]. The development of MCHR-1 antagonists for the treatment of obesity and mood disorders is an intensely pursued research area in which progress has been severely hampered by hERG channel block and

TABLE 18.15 MCHR-1 antagonists: zwitterions can be brain-penetrant [81]

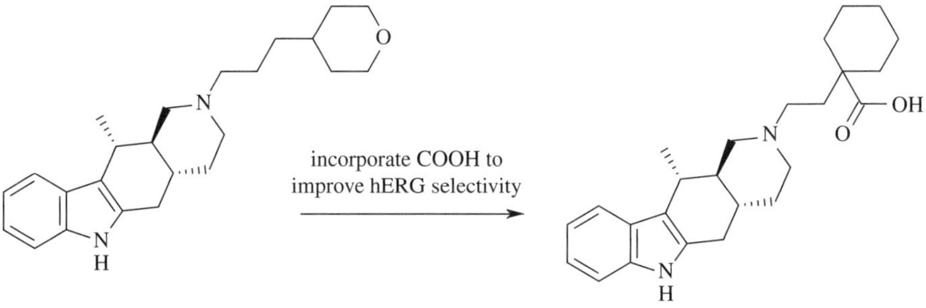

44 **45**

Compound	MCHR-1 IC_{50} (nM)	hERG[a] IC_{50} (μM)	CL[b] (l/h/kg)	K_p	C_{CSF}/C_p	C_b (μM)
44	1	0.03	0.2	—	—	0.04
45	0.6	>5	0.55	0.4	0.21	0.32

[a] Rb^+ efflux.
[b] APP51/16 tg mouse; dose: 60 μmol/kg.

related cardiovascular risks [82]. Achieving hERG selectivity in their MCHR-1 antagonist series also presented a challenge for the Amgen group. Their lead compound **44** was a potent MCHR-1 antagonist (IC_{50} = 1 nM) with good CNS exposure, a good overall ADME profile, and demonstrated efficacy in reducing food consumption in mice (Table 18.15). However, it was later discovered that **44** was also a potent hERG inhibitor, which prevented its further development. Consequently, the authors considered the incorporation of a carboxylic group into their scaffold.

As a potassium cation channel, hERG has evolved to stabilize positive charge within its central cavity, which may, at least in part, explain why many hERG blockers contain basic amine functionality that can be protonated under normal physiological conditions. This may also rationalize the fact that the presence of a functionality that is negatively charged at physiological pH, such as a carboxylic group, is almost universally detrimental to hERG binding [83]. Indeed, the replacement of the tetrahydropyran in **44** with a cyclohexyl carboxyl group in **45** resulted not only in the maintenance of MCHR-1 potency (IC_{50} = 0.6 nM) but also the near-complete elimination of hERG block (IC_{50} > 5 μM). Importantly, at an oral dose of 10 mg/kg, the compound achieved CNS exposure (C_{CSF}/C_p = 0.21) that was sufficient to produce robust *in vivo* efficacy in mouse models of obesity. The compound did not affect the body weights of MCHR-1 KO mice, which suggests that the observed reduction in food intake in the wild-type animals was a consequence of MCHR-1 antagonism. Based on these data, compound **45** (AMG 076) was advanced to the clinic. A similar zwitterion approach to improve selectivity was successfully employed in the design of dual H1/5-HT_{2A} antagonists that advanced into the clinical development for the treatment of sleep disorders [84].

Molecular Weight

The CNS drugs tend to be smaller than non-CNS therapeutics. The mean MW of CNS drugs launched between 1983 and 2002 is 310 [43] whereas, in contrast, the mean MW for oral therapeutics (including CNS drugs) is 377 [48]. Levin proposed, on the basis of brain capillary permeability measurements in anesthetized rats, that there is an MW cutoff for passive brain permeability somewhere between MW 400 and 657 [85]. Moreover, following an analysis of CNS versus non-CNS drugs, van de Waterbeemd proposed that the value should be MW less than 450 [50]. Of course, MW reduction often has coincidental beneficial effects on other parameters, such as PSA and cLogP, which are, as discussed earlier, frequently employed CNS optimization strategies in their own right. This approach is typically initiated by deletion studies to establish minimum pharmacophore, which is then optimized while closely monitoring the LE [86] and LLE [53].

An example of this strategy was reported by Verhoest et al. at Pfizer [87]. The goal was to identify a CNS-penetrant selective κ-opioid receptor (KOR) antagonist for treatment of depression and substance abuse. The starting point for their work was a large and lipophilic HTS hit **46** (Table 18.16). It also displayed low selectivity over the μ-opioid receptor (MOR) and had undesirable ADME properties, including high human liver microsomal clearance (>300 (ml/min)/kg) and suspected P-gp efflux (ER = 2.56). By any measure this was a very challenging starting point for a CNS program. However, the Pfizer group elected to leverage the fact that this chemotype was amenable to high-throughput synthesis, and rapidly evaluate the scope for optimization. Physicochemical property constraints for the library synthesis were set around preferred CNS drug properties: 0–1 HBDs, MW <425 and cLogD <3.

The library successfully identified that the peripheral aryl ring in the indolene in **46** could be replaced by smaller heterocyclic and aliphatic amines, exemplified by the *sec*-butylamine in **47**. This compound displayed significantly improved properties relative to the starting hit **46**, that is, the LE improved from 0.24 to 0.33, and the LLE improved from 3.82 to 6.20. Interestingly, a couple of analogues that fell outside of the proposed physicochemical constraints were suboptimal with respect to clearance (**48a**, >304 (ml/min/kg)) or MDR efflux ratio (**48b**, ER = 14.0). Compound **47** exhibited good brain penetration in rats ($AUC_{0\text{–}4h}$ $K_{p,uu}$ ~1, $AUC_{0\text{–}4h}$ CSF/free plasma 1.2, $AUC_{0\text{–}4h}$ CSF/free brain ~1.4) with no evidence of active efflux. Following further evaluations compound **47** was selected for progression into the clinical trials.

Molecular Flexibility and Rotational Bonds

Veber and colleagues at GSK have highlighted the importance of molecular flexibility, as measured by the number of rotational bonds (RB), for predicting rat oral bioavailability and permeation rates in an artificial membrane permeation assay [25]. They showed that increased molecular flexibility has a negative effect on the passive permeation. Consistent with the supposition that such effects would have a more

TABLE 18.16 KOR antagonists: reduce MW & HBD count to improve P_{app} [87]

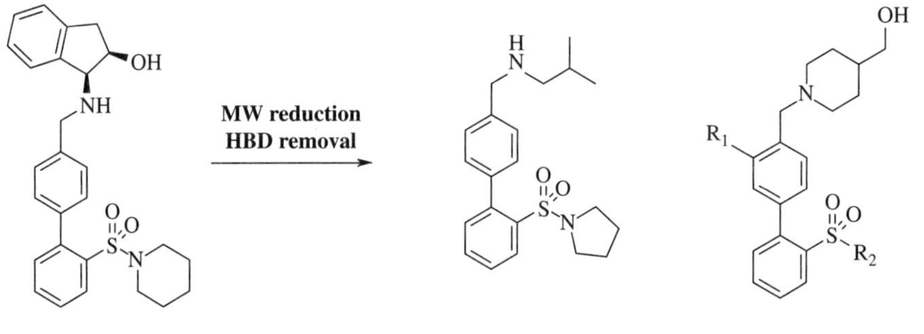

46	47	48a-b
LE 0.24 LLE 3.82	LE 0.33 LLE 6.20	

Compound	R1	R²	MW	cLogD	KOR K_i[a] (nM)	HLM[b] (ml/min)/kg	ER
46	—	—	462	4.1	9	>300	2.56
47	—	—	372.5	2.3	3.0	27.6	1.96
48a	F		460.6	3.6	8.8	>304	1.50
48b	H	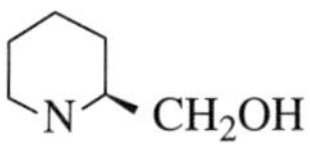	472.6	1.5	1.1	163	14.0

[a]Human KOR binding.
[b]Human liver microsome intrinsic clearance.

profound influence at the BBB, Leeson and Davis demonstrated that the average rotational bond count of oral CNS drugs (mean 4.7, median 4.5) within a dataset of oral drugs (1983–2002 NCE list) was reduced relative to the global average across all therapeutic areas (mean 6.4, median 6) [48]. This finding led to the proposed guideline of a rotatable bond count less than 8 as an attribute of a successful CNS drug candidate [47].

Rigidification often involves the introduction of new ring systems into the molecule, which results in significant changes to other physicochemical properties, including the H-bond count, MW, cLogP, and PSA; hence, the reports in which this approach has been used in isolation are rare, that is, see Table 18.9. However, it is a commonly used strategy wherein both covalent [68, 88–90] and noncovalent (intramolecular H-bonding) manifolds have been invoked [91, 30, 29]. For example, intramolecular H-bonding and the resulting reductions in polarity and flexibility have been postulated to contribute to the improved brain exposure of the NK_1 antagonist PD 174424 **49** [91] and the $S1P_1$ agonist **50** [29] (Fig. 18.5).

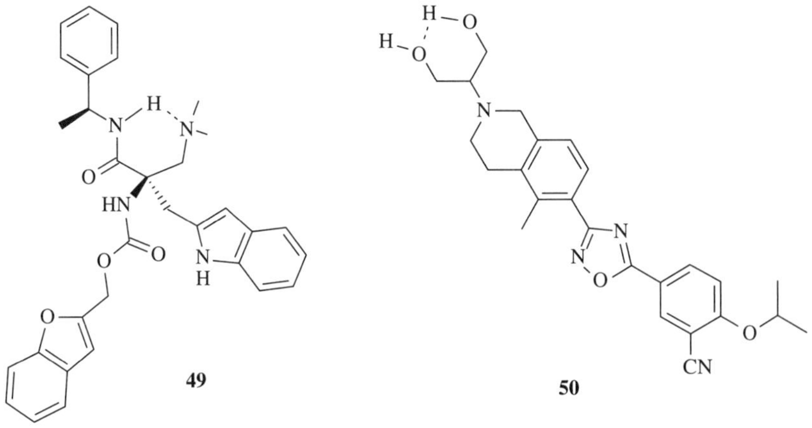

FIGURE 18.5 Reducing molecular flexibility can improve P_{app} [91, 30, 29].

SUMMARY AND OUTLOOK

Achieving optimal BBB penetration is a unique and major challenge in CNS drug discovery [4]. In order to address this challenge successfully and enable an effective optimization process, good understanding of structure–brain exposure relationship is essential. It is important to understand that total brain concentration, still commonly reported as a measure of CNS exposure, is more a measure of nonspecific binding to brain tissue rather than pharmacologically relevant concentrations. Reliance on total brain levels is frequently misleading and inevitably and unnecessarily leads to large and highly lipophilic molecules. Over recent years a large body of evidence has accumulated to show that unbound brain concentration is more reflective of the target site concentration and, ultimately, drug *in vivo* efficacy.

Due to relatively simple and straightforward sampling, CSF concentrations are commonly used as a surrogate marker for unbound drug concentrations in the brain. However, CSF can only serve as a good surrogate for drugs that are not P-gp substrates and have good membrane permeability. For highly effluxed drugs $K_{p,uu,CSF}$ was found to overpredict $K_{p,uu,b}$, which can be attributed to either a lower efflux capacity of the BCSFB compared to the BBB [92], or a possibility that the BCSFB has P-gp that moves substrates from blood into the CSF. It is, therefore, important to consider these three parameters, namely C_{CSF}, P-gp, and P_{app}, in addition to standard ADME parameters, to guide medicinal chemistry strategy and optimization efforts for CNS targets.

These three additional parameters significantly narrow the optimal CNS physicochemical properties space, that is, CNS drugs are generally smaller, have fewer HBDs, and smaller PSA when compared to non-CNS drugs [41]. Median values derived from analysis of marketed CNS drugs offer useful guidance when defining a desirable CNS candidate profile [43]:

cLogP = 2.8
cLogD = 1.7

MW = 305.3 Da
TPSA = 44.8 Å
pK_a = 8.4
HBD = 1
RB = 4.5

Molecular descriptors related to H-bonding are often dominant parameters affecting unbound brain exposure. This is due to the additive combination of detrimental effects that H-bonding is generally associated with, namely, poor passive permeability and increased risk of interactions with efflux transporters. For example, addition of two HBDs to the structure of a centrally acting drug can result in a twofold reduction of unbound brain exposure [92].

Contrary to early reports based on total brain concentrations, lipophilicity is not correlated with unbound brain exposure. Without an active transport mechanism in operation, a certain minimum lipophilicity is required to facilitate permeation through the hydrophobic BBB membrane. As the size of the drug molecule increases, the lipophilicity of the molecule needs to increase to maintain the same chance of success in crossing the BBB [44]. However, once that minimum lipophilicity is achieved, which is defined as sufficient to provide $P_{app} > 200$ nm/s, further increases in LogD/LogP are likely to be detrimental, increasing the extent of nonspecific binding and introducing a number of additional metabolic and toxicological liabilities [53].

Balancing multiple and often opposing physicochemical properties in SAR optimization to address deficits in one without affecting other parameters is one of the greatest challenges for medicinal chemists across all therapeutic areas. As a consequence of additional parameters that need to be considered for optimal brain exposure, the complexities presented to medicinal chemists in CNS drug discovery are even greater [4]. To be successful in this field access to relevant and high-quality data, together with the ability to translate those data effectively into the medicinal chemistry design strategy, is critical.

REFERENCES

[1] Abbott NJ, Rönnbäck L, Hansson E. Astrocyte–endothelial interactions at the blood–brain barrier. Nat. Rev. Neurosci. 2006; 7:41–53.

[2] Pardridge WM. The blood-brain barrier: bottleneck in brain drug development. NeuroRx J. Am. Soc. Exp. Neurother 2005; 2:3–14.

[3] Shen DD, Artru AA, Adkison KK. Principles and applicability of CSF sampling for the assessment of CNS drug delivery and pharmacodynamics. Adv. Drug Deliv Rev. 2004; 56:1825–1827.

[4] Rankovic Z, Bingham M. Medicinal Chemistry Challenges in CNS Drug Discovery. In: Rankovic Z, Bingham M, Nestler E, Hargreaves R, editors. Drug Discovery for Psychiatric Disorders. London: Royal Society of Chemistry; 2013. p 465–509.

[5] Pardridge WM. Blood–brain barrier delivery. Drug Discov Today 2007; 12: 54–61.

[6] Cecchelli R, Berezowski V, Lundquist S, Culot M, Renftel M, Dehouck M-P, Fenart L. Modelling of the blood–brain barrier in drug discovery and development. Nat. Rev. Drug Discov. 2007; 6:50–661.

[7] Pardridge WM. Log(BB), PS products and in silico models of drug brain penetration. Drug Discov. Today 2004; 9:392–393.

[8] Liu X, Chen C. Strategies to optimize brain penetration in drug discovery. Curr. Opin. Drug Discov. Dev. 2005; 8: 505–512.

[9] Liu X, Smith BJ, Chen C, Callegari E, Becker SL, Chen X, Cianfrogna J, Doran AC, Doran SD, Gibbs JP, Hosea N, Liu J, Nelson F, Szewc MA, Deusen JV. Use of a physiologically based pharmacokinetic model to study the time to reach brain equilibrium: an experimental analysis of the role of blood–brain barrier permeability, plasma protein binding, and brain tissue binding. J. Pharmacol. Exp. Therapeut. 2005; 313:1254–1262.

[10] Maurer TS, Debartolo DB, Tess DA, Scott DO. Relationship between exposure and non-specific binding of thirty-three central nervous system drugs in mice. Drug Metabol. Dispos. 2005; 33:175–181.

[11] Hilgert M, Noldner M, Chatterjee SS, Klein J. KA-672 inhibits rat brain acetylcholinesterase *in vitro* but not *in vivo*. Neurosci. Lett. 1999; 263:193–196.

[12] Gupta A, Hammarlund-Udenaes M, Chatelain P, Massingham R, Jonsson EN. Stereoselective pharmacokinetics of cetirizine in the guinea pig: role of protein binding. Biopharm. Drug Dispos. 2006; 27: 291–297.

[13] Johnson DJ, Forbes IT, Watson SP, Garzya V, Stevenson GI, Walker GR, Mudhar HS, Flynn SJ, Wyman PA, Smith PW, Murkitt GS, Lucas AJ, Mookherjee CR, Watson JM, Gartlon JE, Bradford AM, Brown F. The discovery of a series of N-substituted 3-(4-piperidinyl)-1,3-benzoxazolinones and oxindoles as highly brain penetrant, selective muscarinic M1 agonists. Bioorg. Med. Chem. Lett. 2010; 20:5434–5438.

[14] Polli JW, Baughman TM, Humphreys JE, Jordan KH, Mote AL, Salisbury JA, Tippin TK, Serabjit-Singh CJ. P-glycoprotein influences the brain concentrations of cetirizine (Zyrtec), a second-generation non-sedating antihistamine. J. Pharm. Sci. 2003; 92:2082–2089.

[15] Simon FER, Simons KJ. H1 antihistamines: current status and future directions. World Allergy Organ J. 2008; 1:145–155.

[16] Di L, Rong H, Feng B. Demystifying brain penetration in central nervous system drug discovery. J. Med. Chem. 2013; 56:2–12.

[17] Sadiq MW, Borgs A, Okura T, Shimomura K, Kato S, Deguchi Y, Jansson B, Bjoerkman S, Terasaki T, Hammarlundudenaes M. Diphenhydramine active uptake at the blood–brain barrier and its interaction with oxycodone *in vitro* and *in vivo*. J. Pharm. Sci. 2011; 100:3912–3923.

[18] Kageyama T, Nakamura M, Matsuo A, Yamasaki Y, Takakura Y, Hashida M, Kanai Y, Naito M, Tsuruo T, Minato N, Shimohama S. The 4F2hc/LAT1 complex transports L-DOPA across the blood-brain barrier. Brain Res. 2000; 879:115–121.

[19] Doran A, Obach RS, Smith BJ, Hosea NA, Becker S, Callegari E, Chen C, Chen X, Choo E, Cianfrogna J, Cox LM, Gibbs J P, Gibbs MA, Hatch H, Hop CE, Kasman IN, LaPerle J, Liu J, Liu X, Logman M, Maclin D, Nedza FM, Nelson F, Olson E, Rahematpura S, Raunig D, Rogers S, Schmidt K, Spracklin DK, Szewc M, Troutman M, Tseng E, Tu M, Van Deusen JW, Venkatakrishnan K, Walens G, Wang EQ, Wong D, Yasgar AS, Zhang C. The impact of P-glycoprotein on the disposition of drugs targeted for indications of the central nervous system: evaluation using the MDR1A/1B knockout mouse model. Drug Metab. Dispos. 2005; 33:165–174.

[20] Wang X, Ratnaraj N, Patsalos PN. The pharmacokinetic inter-relationship of tiagabine in blood, cerebrospinal fluid and brain extracellular fluid (frontal cortex and hippocampus). Seizure 2004; 13:574–581.

[21] Martin I. Prediction of blood–brain barrier penetration: are we missing the point? Drug Discov. Today 2004; 9:161–162.

[22] Debruyne D. Clinical pharmacokinetics of fluconazole in superficial and systemic mycoses. Clin. Pharmacokinet. 1997; 33:52–77.

[23] Smith DA, Di L, Kerns EH. The effect of plasma protein binding on *in vivo* efficacy: misconceptions in drug discovery. Nat. Drug Discov. Rev. 2010; 9:929–939.

[24] Reichel A. Addressing central nervous system (CNS) penetration in drug discovery: basics and implications of the evolving new concept. Chem. Biodivers. 2009; 9:2030–2049.

[25] Veber DF, Johnson SR, Chen H-Y, Smith BR, Ward KW, Kopple KD. Molecular properties that influence the oral bioavailability of drug candidates. J. Med. Chem. 2002; 45:2615–2623.

[26] Giacomini KM, Huang S-M, Tweedie DJ, Benet LZ, Brouwer KLR, Chu X, Dahlin A, Evers R, Fischer V, Hillgren KM, Hoffmaster KA, Ishikawa T, Keppler D, Kim RB, Lee CA, Niemi M, Polli JW, Sugiyama Y, Swaan PW, Ware JA, Wright SH, Yee SW, Zamek-Gliszczynski MJ, Zhang L. Membrane transporters in drug development. Nat. Rev. Drug Discov. 2010; 9:215–236.

[27] Hitchcock SA. Structural modifications that alter the P-glycoprotein efflux properties of compounds. J. Med. Chem. 2012; 55:4877–4895.

[28] Doan MKM, Humphreys JE, Webster LO, Wring SA, Shampine LJ, Serabjit-Singh CJ, Adkinson KK, Polli JW. Passive permeability and P-glycoprotein-mediated efflux differentiate central nervous system (CNS) and non-CNS marketed drugs. J. Pharmacol. Exp. Therapeut. 2002; 303:1029–1037.

[29] Demont EH, Arpino S, Bit RA, Campbell CA, Deeks N, Desai S, Dowell SJ, Gaskin P, Gray JRJ, Harrison LA, Haynes A, Heightman TD, Holmes DS, Humphreys PG, Kumar U, Morse MA, Osborne GJ, Panchal T, Philpott KL, Taylor S, Watson R, Willis R, Witherington J. J. Discovery of a Brain-Penetrant S1P3-Sparing Direct Agonist of the S1P1 and S1P5 Receptors Efficacious at Low Oral Dose. Med. Chem. 2011; 54:6724–6733.

[30] Seelig A. A general pattern for substrate recognition by P-glycoprotein. Eur. J. Biochem. 1998; 251:252–261.

[31] Gombar VK, Polli JW, Humphreys JE, Wring SA, Serabjit-Singh CS. Predicting P-glycoprotein substrates by a quantitative structure-activity relationship model. J. Pharm. Sci. 2004; 93:957–968.

[32] Zhang EY, Phelps MA, Cheng C, Elkins S, Swaan PW. Modeling of active transport systems. Adv. Drug Deliv. Rev. 2002; 54:329–354.

[33] Broccatelli F, Carosati E, Neri A, Frosini M, Goracci L, Oprea TI, Cruciani G. A novel approach for predicting P-glycoprotein (ABCB1) inhibition using molecular interaction fields. J. Med. Chem. 2011; 54:1740–1751.

[34] Didziapetris R, Japertas P, Avdeef A, Petrauskas A. Classification analysis of P-glycoprotein substrate specificity. J. Drug Target. 2003; 11:391–406.

[35] Cunningham CW, Mercer SL, Hassan HE, Traynor JR, Eddington ND, Coop A. Opioids and efflux transporters. Part 2: Pglycoprotein substrate activity of 3- and 6-substituted morphine analogs. J. Med. Chem. 2008; 51:2316–2320.

[36] Yuan Y, Li G, He H, Stevens DL, Kozak P, Scoggins KL, Mitra P, Gerk PM, Selley DE, Dewey WL, Zhang Y. Characterization of 6α- and 6β-N-heterocyclic substituted naltrexamine derivatives as novel leads to development of mu opioid receptor selective antagonists. ACS Chem. Neurosci. 2011; 2:346–351.

[37] Kuhn B, Mohr P, Stahl M. Intramolecular hydrogen bonding in medicinal chemistry. J. Med. Chem. 2010; 53:2601–2611.

[38] Linnet K, Ejsing TB. A review on the impact of P-glycoprotein on the penetration of drugs into the brain. Focus on psychotropic drugs. Eur. Neuropsychopharmacol. 2008; 18:157–169.

[39] Wakenhut F, Allan GA, Fish PV, Fray MJ, Harrison AC, McCoy R, Phillips SC, Ryckmans T, Stobie A, Westbrook D, Westbrook SL, Whitlock GA. N-[(3S)-Pyrrolidin-3-yl]benzamides as novel dual serotonin and noradrenaline reuptake inhibitors: impact of small structural modifications on P-gp recognition and CNS penetration. Bioorg. Med. Chem. Lett. 2009; 19:5078–5081.

[40] Lipinski CA, Lombardo F, Dominy BW, Feeney PJ. Experimental and computational approaches to estimate solubility and permeability in drug discovery and development settings. Adv. Drug. Deliv. Rev. 1997; 23:3–26.

[41] Ghose AK, Herbertz T, Hudkins RL, Dorsey BD, Mallamo JP. Knowledge-based, central nervous system (CNS) lead selection and lead optimization for CNS drug discovery. ACS Chem. Neurosci. 2012; 3:50–68.

[42] Wager TT, Hou X, Verhoest PR, Villalobos A. Moving beyond rules: the development of a central nervous system multiparameter optimization (CNS MPO) approach to enable alignment of druglike properties. ACS Chem. Neurosci. 2010; 1:435–439.

[43] Wager TT, Chandrasekaran RY, Hou X, Troutman MD, Verhoest PR, Villalobos A, Will Y. Defining desirable central nervous system drug space through the alignment of molecular properties, *in vitro* ADME, and safety attributes. ACS Chem. Neurosci. 2010; 1:420–434.

[44] Waring MJ. Defining optimum lipophilicity and molecular weight ranges for drug candidates-molecular weight dependent lower logD limits based on permeability. Bioorg. Med. Chem. Lett. 2009; 19:2844–2851.

[45] Gleeson MP. Generation of a set of simple, interpretable ADMET rules of thumb. J. Med. Chem. 2008; 51:817–834.

[46] Hitchcock SA, Pennington LD. Structure – brain exposure relationships. J. Med. Chem. 2006; 49:7559–7583.

[47] Pajouhesh H, Lenz GR. Medicinal chemical properties of successful central nervous system drugs. NeuroRx 2005; 2:541–553.

[48] Leeson PD, Davis AM. Time-related differences in the physical property profiles of oral drugs. J. Med. Chem. 2004; 47:6338–6348.

[49] Norinder U, Haeberlein M. Computational approaches to the prediction of the blood–brain distribution. Adv. Drug Deliv. Rev. 2002; 54:291–313.

[50] Van de Waterbeemd H, Camenisch G, Folkers G, Chretien JR, Raevsky OA. Estimation of blood-brain barrier crossing of drugs using molecular size and shape, and H-bonding descriptors. J. Drug Target. 1998; 6:151–165.

[51] Gratton JA, Abraham MH, Bradbury MW, Chadra HS. Molecular factors influencing drug transfer across the blood-brain barrier. J. Pharm. Pharmacol. 1997; 49:1211–1216.

[52] Van de Waterbeemd H, Kansy M. Hydrogen-bonding capacity and brain penetration. Chimia 1992; 46:299–303.

[53] Leeson PD, Springthorpe B. The influence of drug-like concepts on decision-making in medicinal chemistry. Nat. Rev. Drug Discov. 2007; 6:881–890.

[54] Leo A, Hansch C, Elkins D. Partition coefficients and their uses. Chem. Rev. 1971; 71:525–616.

[55] Bergström CAS, Strafford M, Lazorova L, Avdeef A, Luthman K, Artursson P. Absorption classification of oral drugs based on molecular surface properties. J. Med. Chem. 2003; 46:558–570.

[56] Tantishaiyakul V. Prediction of Caco-2 cell permeability using partial least square multivariate analysis. Pharmazie 2001; 56:407–411.

[57] Martin YC. A bioavailability score. J. Med. Chem. 2005; 48:3164–3170.

[58] Egan WJ, Merz Jr KM, Baldwin JJ. Prediction of drug absorption using multivariate statistics. J. Med. Chem. 2000; 43:3867–3877.

[59] Smith QR, Takasato Y. Kinetics of amino acid transport at the blood–brain barrier studied using an in situ brain perfusion technique. Ann. NY Acad. Sci. 1986; 481:186–201.

[60] Goodwin JT, Clark DE. In silico predictions of blood-brain barrier penetration: considerations to "keep in mind". J. Pharm. Exp. Therapeut. 2005; 315:477–483.

[61] Murakami H, Takanaga H, Matsuom H, Ohtani H, Sawada Y. Comparison of blood-brain barrier permeability in mice and rats using in situ brain perfusion technique. Am. J. Physiol. Heart Circ. Physiol. 2000; 279:H1022–1028.

[62] Liu X, Tu M, Kelly RS, Chen C, Smith BJ. Development of a computational approach to predict blood-brain barrier permeability. Drug Metabol. Dispos. 2004; 32:132–139.

[63] Middleton DS, Andrews M, Glossop P, Gymer G, Jessiman A, Johnson PS, MacKenny M, Pitcher MJ, Rooker T, Stobie A, Tang K, Morgan P. Designing rapid onset selective serotonin re-uptake inhibitors. Part 1: Structure–activity relationships of substituted (1S,4S)-4-(3,4 dichlorophenyl)-N-methyl-1,2,3,4-tetrahydro-1-naphthaleneamine. Bioorg. Med. Chem. Lett. 2006; 16:1434–1439.

[64] Smith PW, Wyman PA, Lovell P, Goodacre C, Serafinowska HT, Vong A, Harrington F, Flynn S, Bradley DM, Porter R, Coggon S, Murkitt G, Searle K, Thomas DR, Watson JM, Martin W, Wu Z, Dawson LA. New quinoline NK3 receptor antagonists with CNS activity. Bioorg. Med. Chem. Lett. 2009; 19:837–840.

[65] Bürli RW, Luckhurst CA, Aziz O, Matthews KL, Yates D, Kathy. Lyons A, Beconi M, McAllister G, Breccia P, Stott AJ, Penrose SD, Wall M, Lamers M, Leonard P, Müller I, Richardson CM, Jarvis R, Stones L, Hughes S, Wishart G, Haughan AF, O' Connell C, Mead T, McNeil H, Vann J, Mangette J, Maillard M, Beaumont V, Munoz-Sanjuan I, Dominguez C. Design, synthesis, and biological evaluation of potent and selective class IIa histone deacetylase (HDAC) inhibitors as a potential therapy for Huntington's disease. J. Med. Chem. 2013; 56:9934–9954.

[66] Österberg T, Norinder U. Prediction of polar surface area and drug transport processes using simple parameters and PLS statistics. J. Chem. Inform. Comput. Sci. 2000; 40:1408–1411.

[67] Abraham MH. The factors that influence permeation across the blood-brain barrier. Eur. J. Med. Chem. 2004; 39:235–240.

[68] Ahmed M, Briggs MA, Bromidge SM, Buck T, Campbell L, Deeks NJ, Garner A, Gordon L, Hamprecht DW, Holland V, Johnson CN, Medhurst AD, Mitchell DJ, Moss SF, Powles J, Seal JT, Stean TO, Stemp G, Thompson M, Trail B, Upton N, Winborn K, Witty DR. Bicyclic heteroaryl piperazines as selective brain penetrant 5-HT6 receptor antagonists. Bioorg. Med. Chem. Lett. 2005; 15:4867–4871.

[69] Hu E, Kunz RK, Chen N, Rumfelt S, Siegmund A, Andrews K, Chmait S, Zhao S, Davis C, Chen H, Lester-Zeiner D, Ma J, Biorn C, Shi J, Porter A, Treanor J, Allen JR. Design, optimization, and biological evaluation of novel keto-benzimidazoles as potent and selective inhibitors of phosphodiesterase 10A (PDE10A). J. Med. Chem. 2013; 56:8781–8792.

[70] Ghosh AK, Brindisi M, Tang J. Developing β-secretase inhibitors for treatment of Alzheimer's disease. J. Neurochem. 2012; 120:71–83.

[71] Brodney MA, Barreiro G, Ogilvie K, Hajos-Korcsok E, Murray J, Vajdos F, Ambroise C, Christoffersen C, Fisher K, Lanyon L, Liu JH, Nolan CE, Withka JM, Borzilleri KA, Efremov I, Oborski CE, Varghese A, O'Neill BT. Spirocyclic sulfamides as β-secretase 1 (BACE-1) inhibitors for the treatment of Alzheimer's disease: utilization of structure based drug design, watermap, and CNS penetration studies to identify centrally efficacious inhibitors. J. Med. Chem. 2012; 55:9224–9239.

[72] Kelder J, Grootenhuis PDJ, Bayada DM, Delbressine LPC, Ploemen J-P. Polar molecular surface as a dominating determinant for oral absorption and brain penetration of drugs. Pharmaceut. Res. 1999; 16:1514–1519.

[73] Clark DE. Rapid calculation of polar molecular surface area and its application to the prediction of transport phenomena. 1. Prediction of intestinal absorption. J. Pharmaceut. Sci. 1999; 88:807–814.

[74] Pinnard E, Alanine A, Alberati D, Bender M, Borroni E, Bourdeaux P, Brom V, Burner S, Fischer H, Hainzl D, Halm R, Hauser N, Jolidon S, Lengyel J, Marty H-P, Meyer T, Moreau J-L, Mory R, Narquizian R, Nettekoven M, Norcross RD, Puellmann B, Schmid P, Schmitt S, Stalder H, Wermuth R, Wettstein JG, Zimmerli D. Selective GlyT1inhibitors: discovery of[4-(3-fluoro-5-trifluoromethylpyridin-2-yl)piperazin-1-yl][5-methanesulfonyl-2-((S)-2,2,2-trifluoro-1-methylethoxy)phenyl]methanone (RG1678), a promising novel medicine to treat schizophrenia. J. Med. Chem. 2010; 53:4603–4614.

[75] Shengguo S, Adejare A. Fluorinated molecules as drugs and imaging agents in the CNS. Curr. Top. Med. Chem. 2006; 14:1457–1464.

[76] McDonald IM, Mate RA, Zusi FC, Huang H, Post-Munson DJ, Ferrante MA, Gallagher L, Bertekap Jr. RL, Knox RJ, Robertson BJ, Harden DG, Morgan DG, Lodge NJ, Dworetzky SI, Olson RE, Macor JE. Discovery of a novel series of quinolone α7 nicotinic acetylcholine receptor agonists. Bioorg. Med. Chem. Lett. 2013; 23:1684–1688.

[77] Lerchner A, Machauer R, Betschart C, Veenstra S, Rueeger H, McCarthy C, Tintelnot-Blomley M, Jaton A-L, Rabe S, Desrayaud S, Enz A, Staufenbiel M, Paganetti P, Rondeau J-M, Neumann U. Macrocyclic BACE-1 inhibitors acutely reduce Aβ in brain after po application. Bioorg. Med. Chem. Lett. 2010; 20:603–607.

[78] Rampe D, Wible B, Brown AM, Dage RC. Effects of terfenadine and its metabolites on a delayed rectifier K1 channel cloned from human heart. Mol. Pharmacol. 1993; 44:1240–1245.

[79] Doan MKM, Wring SA, Shampine LJ, Jordan KH, Bishop JP, Kratz J, Yang E, Serabjit-Singh CJ, Adkison KK, Polli JW. Steady-state brain concentrations of antihistamines in rats: interplay of membrane permeability, P-glycoprotein efflux and plasma protein binding. Pharmacology 2004; 72:92–98.

[80] Zhao R, Kalvass JC, Yanni SB, Bridges AS, Pollack GM. Fexofenadine brain exposure and the influence of blood-brain barrier P-glycoprotein after fexofenadine and terfenadine administration. Drug Metabol. Dispos. 2009; 37:529–535.

[81] Mihalic JT, Fan F, Chen X, Chen X, Fu Y, Motani A, Liang L, Lindstrom M, Tang L, Chen L-L, Jaen J, Dai K, Li L. Discovery of a novel melanin concentrating hormone receptor 1 (MCHR1) antagonist with reduced hERG inhibition. Bioorg. Med. Chem. Lett. 2012; 22:3781–3785.

[82] Högberg T, Frimurer TM, Sasmal PK. Melanin concentrating hormone receptor 1 (MCHR1) antagonists—still a viable approach for obesity treatment? Bioorg. Med. Chem. Lett. 2012; 22:6039–6047.

[83] Jamieson C, Moir EM, Rankovic Z, Wishart G. Medicinal chemistry of hERG optimizations: Highlights and hang-ups. J. Med. Chem. 2006; 49:5029–5050.

[84] Gianotti M, Botta M, Brough S, Carletti R, Castiglioni E, Corti C, Dal-Cin M, Fratte SD, Korajac D, Lovric M, Merlo G, Mesic M, Pavone F, Piccoli L, Rast S, Roscic M, Sava A, Smehil M, Stasi L, Togninelli A, Wigglesworth MJ. Novel spirotetracyclic zwitterionic dual H1/5-HT2A receptor antagonists for the treatment of sleep disorders. J. Med. Chem. 2010; 53:7778–7795.

[85] Levin VA. Relationship of octanol/water partition coefficients to rat brain capillary permeability. J. Med. Chem. 1980; 23: 682–684.

[86] Hopkins AL, Groom CR, Alex A. Ligand efficiency: a useful metric for lead selection. Drug Discov. Today 2004; 9:430–431.

[87] Verhoest PR, Sawant-Basak A, Parikh V, Hayward M, Kauffman GW, Paradis V, McHardy SF, McLean S, Grimwood S, Schmidt AW, Vanase-Frawley M, Freeman J, Van Deusen J, Cox L, Wong D, Liras S. Design and discovery of a selective small molecule kappa opioid antagonist (2-methyl-N-((2′-(pyrrolidin-1-ylsulfonyl)biphenyl-4-yl) methyl)propan-1-amine, PF-4455242). J. Med. Chem. 2011; 54:5868–5877.

[88] Bromidge SM, Brown AM, Clarke SE, Dodgson K, Gager T, Grassam HL, Jeffrey PM, Joiner GF, King FD, Middlemiss DN, Moss SF, Newman H, Riley G, Routledge C. Wyman P. 5-Chloro-N-(4-methoxy-3-piperazin-1-yl-phenyl)-3-methyl-2-benzothiophenesulfonamide (SB-271046): a potent, selective, and orally bioavailable 5-HT6 receptor antagonist. J. Med. Chem. 1999; 42:202–205.

[89] Trani G, Baddeley SM, Briggs MA, Chuang TT, Deeks NJ, Johnson CN, Khazragi AA, Mead TL, Medhurst AD, Milner PH, Quinn LP, Ray AM, Rivers DA, Stean TO, Stemp G, Trail BK, Witty DR. Tricyclic azepine derivatives as selective brain penetrant 5-HT6 receptor antagonists. Bioorg. Med. Chem. Lett. 2008; 18:5698–5700.

[90] Cowart M, Pratt JK, Stewart AO, Bennani YL, Esbenshade TE, Hancock AA. A new class of potent non-imidazole H(3) antagonists: 2-aminoethylbenzofurans. Bioorg. Med. Chem. Lett. 2004; 14:689–693.

[91] Ashwood VA, Field MJ, Horwell DC, Julien-Larose C, Lewthwaite RA, McCleary S, MC, Raphy J, Singh L. Utilization of an intramolecular hydrogen bond to increase the CNS penetration of an NK1 receptor antagonist. J. Med. Chem. 2001; 44:2276–2285.

[92] Fridén M, Winiwarter S, Jerndal G, Bengtsson O, Wan H, Bredberg U, Hammarlund-Udenaes M, Antonsson M. Structure-brain exposure relationships in rat and human using a novel data set of unbound drug concentrations in brain interstitial and cerebrospinal fluids. J. Med. Chem. 2009; 52:6233–6243.

19

CASE STUDIES OF CNS DRUG OPTIMIZATION—MEDICINAL CHEMISTRY AND CNS BIOLOGY PERSPECTIVES

KEVIN J. HODGETTS

Laboratory for Drug Discovery in Neurodegeneration, Harvard NeuroDiscovery Center, Cambridge, MA, USA; Brigham and Women's Hospital, Cambridge, MA, USA

Traditionally, increasing lipophilicity was used to optimize blood–brain barrier (BBB) permeability (rate), and brain-to-plasma (B/P) ratios were used to determine BBB penetration (extent) of drug molecules. Although both are important considerations, lipophilicity and B/P ratios alone are insufficient to predict or characterize pharmacologically relevant brain exposures of drugs. Over the years, other structural properties (e.g., MW, PSA, H-bonds, and pK_a) have become increasingly important considerations for permeability, and current approaches take a more balanced viewpoint [1]. Furthermore, and based on the free drug hypothesis, brain penetration is expressed in terms of unbound drug levels in brain, plasma, and cerebrospinal fluid (CSF), and free brain concentration/free plasma concentration is used to more accurately predict and rationalize CNS target engagement and *in vivo* effects than B/P ratio [2]. The most successful medicinal chemistry strategies for increasing brain penetration are (i) to design new molecules with physicochemical properties that are favorable for passive diffusion across the endothelial cells of the BBB; and (ii) to design new molecules to circumvent efflux by transporters, in particular, P-glycoprotein (P-gp). In this chapter, case studies taken from the literature will

Blood–Brain Barrier in Drug Discovery: Optimizing Brain Exposure of CNS Drugs and Minimizing Brain Side Effects for Peripheral Drugs, First Edition. Edited by Li Di and Edward H. Kerns.

illustrate the medicinal chemistry optimization of lead molecules to improve brain permeability and penetration. Consideration as to how the structural modifications affect physicochemical properties during the optimization of brain penetration will be made. The case studies will be divided into two categories: (1) optimization of passive permeability; and (2) strategies for the mitigation of P-gp efflux.

OPTIMIZATION OF PASSIVE PERMEABILITY

Passive diffusion through the cellular membrane is considered to be the major route of drug permeation through the BBB [3]. The requirements for passive diffusion by a molecule are conveniently defined by a number of structural properties, which include lipophilicity, molecular weight (MW), polar surface area (PSA), numbers of hydrogen-bond acceptors (HBA) and hydrogen-bond donors (HBD), and charge. An increasing number of rules and guidelines that predict BBB permeability based upon these structural properties have been put forward. The important structural properties that influence BBB permeation and that are commonly used in guidelines based upon these properties will be described briefly, followed by representative case studies that utilize these properties to optimize compounds for brain penetration (exposure). Ideally, structural modifications should be based on unbound brain-to-unbound plasma ratio (K_{puu}); here, B/P ratio is used assuming binding to plasma is similar to binding to brain tissues, as the field is going through a paradigm shift on using K_{puu} regularly during medicinal chemistry design [2].

Structural Properties

(i) Lipophilicity (e.g., Log*D* and *c*Log*P*). Historically, lipophilicity was the first of the structural properties to be recognized as important for CNS penetration. As early as 1897, Overton observed correlations between oil–water partition coefficients and narcotic potencies of a series of compounds in tadpoles [4]. By the 1960s, Hansch introduced the octanol/water system, which became the standard for measuring partition coefficients, and he demonstrated that BBB permeation is optimal when the Log*P* values are in the range of 1.5–2.7, with the mean value of 2.1 [5]. In line with Hansch's findings, the mean value for *c*Log*P* for 74 marketed CNS drugs has been reported by Leeson to be 2.5 [6]. A number of different guidelines for predicting BBB permeation based on structural properties have been proposed [3, 7–13], which suggests preferred ranges for Log*P* of between 1 and 5 (Table 19.1). In addition to permeability, lipophilicity plays an important role in other pharmacokinetic properties that may affect the efficacy of a CNS drug. For example, high lipophilicity will reduce aqueous solubility, and it can contribute to excessive volumes of distribution, and increased metabolic liability.

(ii) Molecular weight (MW). CNS drugs tend to have lower MW compared with other therapeutics; for marketed CNS drugs, the mean value of MW is 310, whereas the average MW of all marketed oral drugs is 377 [6]. Rules for

TABLE 19.1 Rules of thumb for CNS penetration [3, 5–13]

	Hansch [5]	Leeson [6]	Van de Waterbeemd [7]	Kelder [8]	Lipinski RO5 [9]	Lipinski BBB [10]	Mahar Doan [11]	Pajouhesh [12]	Hitchcock [3]	Wagner [13]	
										a	b
MW	—	310	<450	—	<500	<400	319	<450	<450	≤360	>500
Log$P$	1.5–2.7	2.5	—	—	≤5	≤5	3.43	<5	2–4	≤3	>5
Log$D$	—	—	—	—	—	—	—	—	2–4	≤2	>4
PSA	—	—	≤90	60–70	—	—	40	60–70	<70	40–90	<20, >120
HBD	—	1.5	—	—	≤5	≤3	0.85	<3	0–1	≤0.5	>3.5
HBA	—	3.74	—	—	≤10	≤7	3.56	<7	—	—	—
pK_a	—	—	—	—	—	—	—	7.5–10.5	—	≤8	>10

a, more desirable range; b, less desirable range.

predicting BBB tend to place the MW cutoff for BBB penetration in the 400–500 Da range (Table 19.1).

(iii) Polar surface area (PSA). The importance of PSA as a predictor for BBB penetration was first introduced by van de Waterbeemd and Kansy in the early 1990s [7]. The PSA for marketed CNS drugs (estimated at 50–60 Å^2) is significantly less than that for all marketed oral drugs (estimated at 100–110 Å^2). The suggested cutoff for BBB penetration is, generally, to keep PSA values below 70–90 Å^2 (Table 19.1). As the PSA of a molecule rises above this cutoff, the potential for poor passive permeability increases. However, a compound with a quaternary nitrogen atom may have a low PSA value, but it is highly unlikely to permeate a membrane via passive diffusion [14]. Conversely, some compounds with very high PSA values may still cross membranes, if they are substrates for active transport systems or if the polarity is masked (e.g., internal H-bonding) [15].

(iv) Hydrogen-bonding (H-bonding). Closely related to PSA and the potential to cross the BBB is the H-bonding capability of a molecule. On average, marketed CNS drugs contain 2.1 HBAs and 1.5 HBDs, considerably less than the 3.7 HBAs and 1.8 HBDs for all oral drugs [6].

(v) Charge. The majority of marketed CNS drugs contain a basic amine group. At physiological pH (e.g., pH 7.4), basic amines exist in equilibrium between the charged and neutral forms, with the greater fraction of neutrals favoring partition into membrane lipids and brain permeation. By contrast, a strong acid ($pK_a < 4$) or a strong base ($pK_a > 10$) will be fully ionized at physiological pH, and these compounds rarely penetrate the BBB unless by active transport [16]. Guidelines limiting the pK_a of a compound to between 4 and 10 for a CNS compound have been suggested [14].

Physicochemical Properties: Solubility and Passive Permeability

The interaction of a compound with its physical environment determines its physicochemical properties (e.g., solubility and passive permeability). This interaction is determined by the interplay of the structural properties outlined earlier. Taking pK_a as an example, the pK_a determines the degree of ionization, and it has a major effect on solubility and permeability. Ionized molecules are more polar and more soluble in aqueous solution than in their neutral form. However, ionized molecules are less permeable than in their neutral form. In the neutral form, the molecule is more lipophilic and permeates via passive diffusion. For an orally administered CNS drug to be efficacious, it needs to cross two biological barriers, the gastrointestinal (GI) tract and the BBB, before it can reach its intended target. (For a target that is inside the cell, the drug also needs to cross the cell membrane.) While formulation has been an effective tactic to affect dissolution and absorption across the GI tract, solubility remains an important requirement for a CNS drug, as demonstrated by Alelyunas, who determined the solubility of 98 marketed CNS drugs in pH 7.4 buffer [17]. Over 85% of the drugs tested had high aqueous solubility (>100 μM),but only seven drugs

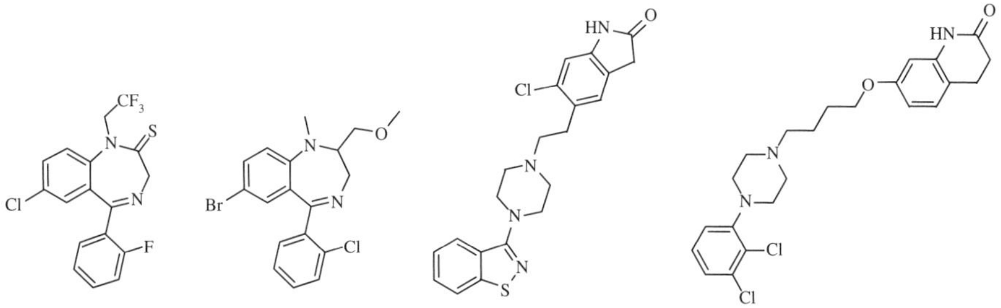

Quazepam	Metaclazepam	Ziprasidone	Aripiperazole
MW = 387	MW = 394	MW = 413	MW = 448
PSA = 16	PSA = 25	PSA = 48	PSA = 45
cLogP = 5.51	cLogP = 4.95	cLogP = 4.67	cLogP = 4.55
HBD = 0	HBD = 0	HBD = 1	HBD = 1
		pKa = 7.09	pKa = 7.46

FIGURE 19.1 Low-solubility CNS drugs are the exception.

had low solubility (<10 μM), two of which had very poor solubility less than 1 μM. Of the seven drugs with poor solubility, three are no longer on the market, while the remaining four (Fig. 19.1) are all administered at relatively low daily doses (e.g., <0.5 mg/kg: quazepam (15 mg); metaclazepam (15 mg); ziprasidone (20 mg); and aripiperazole (10 mg)). Although the structural properties of these compounds are generally within Lipinski rules for oral absorption, the combination of high MW, high lipophilicity, and low PSA predicts the potential for poor solubility. Presumably, the attributes of high potency and high permeability for these four drugs may compensate for their low aqueous solubility. It should be stressed, however, that these four drugs are very much the exceptions to the norm. Indeed, Lipinski described the estimation, which is used at Pfizer, to determine the minimum acceptable thermodynamic solubility required for an orally active drug with low, medium, and high permeability values at a particular clinical dose [18]. For example, to achieve oral absorption, a compound with medium intestinal permeability and a projected human potency of 1 mg/kg (e.g., 50–100 mg dose) needs a minimum aqueous solubility of 52 μg/ml. Thus, for a drug with an MW of 310 (the average MW for a CNS drug), it would require an aqueous solubility of approximately 165 μM.

Biochemical Properties: Metabolic Stability and Plasma Protein Binding

As the interaction of a compound with its physical environment determines its physicochemical properties, the interaction of a compound with proteins determines its biochemical properties (e.g., microsomal stability and plasma protein binding (PPB)). This interaction is also determined by the interplay of the structural properties outlined earlier. Although Phase I and II metabolisms have been observed in BBB endothelial cells [19], the rate of metabolism at the BBB is likely small. However, the first-pass metabolism of an orally administered drug significantly influences bioavailability, thus reducing plasma concentration; this is one of the primary reasons molecules do not reach adequate brain concentration. Improving metabolic stability and lowering intrinsic clearance are, therefore, important strategies, not only for increasing oral bioavailability, but also for increasing brain exposure. Protein binding in blood and binding in brain tissue are important considerations for CNS drugs, as, according to the free drug hypothesis, only the free drug can pass through cell membranes by passive diffusion and interact with the therapeutic target. CNS drugs are generally moderately basic molecules that bind to both human serum albumin and α-acid glycoprotein, and many CNS drugs have high protein binding (>99%). The use of PPB, therefore, can be misleading during the lead optimization phase, and it should only be used in conjunction with exposure data to convert total drug to free drug and to give insight into PK/PD relationships. The key parameter is the free drug concentration in brain in the biophase surrounding the therapeutic target. For drugs with good BBB permeability, and that are not transporter substrates, the free drug concentration in brain is approximately the same as in plasma [20].

In summary, the structural properties considered important for BBB permeation and brain penetration are intertwined, and changing a molecule to adjust one attribute will inevitably change other properties. These changes, in turn, will lead to changes in solubility, permeability, and metabolic stability. Clearly, the complex structure of the BBB makes brain penetration a complicated process which cannot always be rationalized on the basis of a single parameter. The following examples illustrate that rather than focusing on changing a single attribute (e.g., lipophilicity) and bringing its value in line with a particular structural guideline (e.g., *c*Log*P* 1–5), the overall balance of structural properties needs to be considered and optimized to design compounds for optimal brain penetration. The medicinal chemistry tactics of (i) conformational constraint; (ii) bioisosteric replacement; (iii) incorporation of fluorine; and (iv) increasing metabolic stability are used to optimize structural properties and passive permeability and to improve brain penetration.

Conformational Constraint Conformational constraint is the introduction of a more rigid, generally cyclic region into the molecule. Such a change is usually accompanied by a reduction in (i) number of H-bond donors; (ii) PSA; and (iii) number of conformations, leading to an improvement in passive permeability with minimal loss of the functional groups that are required for activity. In the following example, conformational constraint of a polar moiety lowers PSA and the number of HBDs, and increases lipophilicity.

Acylguanidine to a 2-Aminopyridine In the BACE1 inhibitor program for Alzheimer's disease at Wyeth [21], the discovery team found that the initial HTS hit (the acylguanidine BACE inhibitor (**1**)), which demonstrated promising activity, had a poor B/P ratio (0.04) (Fig. 19.2). The highly polar nature of the acylguanidine moiety was considered likely to limit the permeability of **1** (PSA = 85 Å^2). Modification of the acylguanidine moiety to reduce PSA and to improve permeability led to the identification of the 2-aminopyridine group as an excellent replacement for the acylguanidine. The resulting 2-aminopyridine **2** retained similar BACE1 activity as **1**, but it had a significantly lower PSA (–43 Å^2) and a dramatically increased B/P ratio (1.7). Further optimization through modulation of the two phenyl substituents led to the analog **3**, with improved BACE1 activity and a B/P ratio close to unity (1.1).

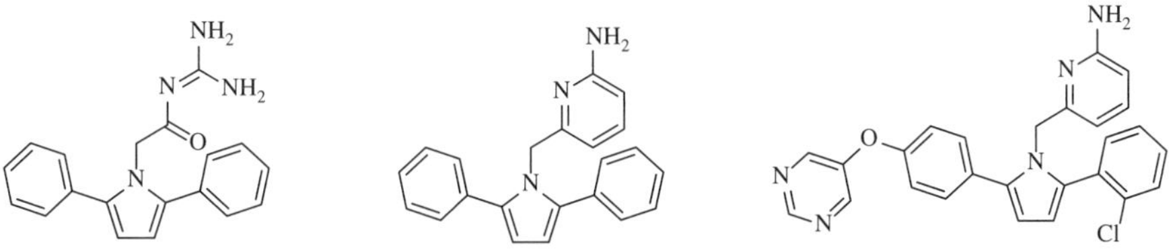

1 BACE1 IC_{50} = 3.7 uM
$AUC_{brain/plasma}$ = 0.04
MW = 318
cLogP = 2.7
PSA = 85 Å^2
pKa = 8.3
HBD = 4

2 BACE1 IC_{50} = 5.2 uM
$AUC_{brain/plasma}$ = 1.70
MW = 325
cLogP = 4.9
PSA = 42 Å^2
HBD = 2

3 BACE1 IC_{50} = 0.1 uM
$AUC_{brain/plasma}$ = 1.10
MW = 453
cLogP = 5.1
PSA = 76 Å^2
pKa = 6.9
HBD = 2

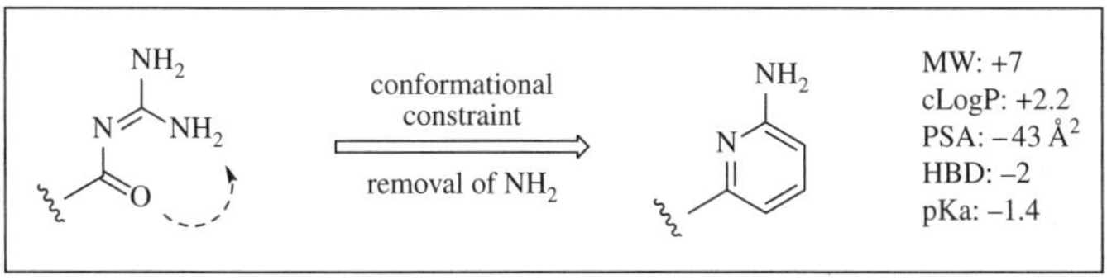

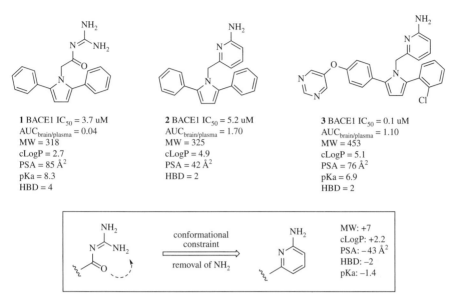

FIGURE 19.2 Conformational constraint of the acylguanidine group led to improved brain penetration.

Interestingly, although there was a dramatic improvement in activity and BBB permeation on going from the HTS hit **1** to the optimized analog **3**, the overall change in PSA was very small (−9 Å^2). Rather, the structural changes between the acylguanidine **1** and the 2-aminopyridine **3** led to a significant increase in lipophilicity (+2.2), a reduction in pK_a to 6.9, and, crucially, to a decrease in the number of HBDs from four to two. Combined, the effects of these changes resulted in a more balanced physicochemical profile and the observed improvement in brain exposure.

Sulfonamide NH to an Indole Brain-penetrant 5-HT_6 receptor antagonists have potentially beneficial effects in the treatment of learning and memory disorders, such as Alzheimer's disease. Researchers at GlaxoSmithKline identified benzazepine **4** as a potent and selective 5-HT_6 receptor antagonist lead (pK_i = 8.7) (Fig. 19.3) [22]. PK studies in rat indicated that compound **4** had a promising brain-to-blood ratio of 0.5. It was reasoned that conformational constraint and removal of the sulfonamide NH group would further improve permeability and lead to a higher brain-to-blood ratio. Incorporation of the NH group into an indole ring gave the tricyclic compound **5**, which had encouraging potency as a functional antagonist at the 5-HT_6 receptor (pK_i = 8.8). Unfortunately, compound **5** had poor *in vitro* metabolic stability in rat microsomes (CLi = 7.1 ml/min/g liver). Substitutions to the arylsulfone and indole positions of the tricyclic core were investigated, and the 3-chloro substituted phenylsulfone **6** was identified with optimal potency and improved metabolic stability (CLi = 2.5 ml/min/g liver). Encouragingly, oral dosing of compound **6** in rat gave a brain-to-blood ratio of 5:1, confirming that the use of conformational constraint and reduction in HBDs had improved brain penetration. However, plasma concentrations of **6** were low, suggesting

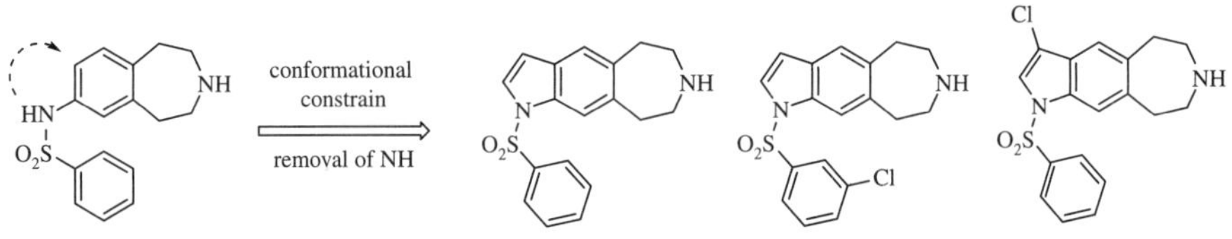

4 $5HT_6$ pKi= 8.7
brain/blood = 0.5:1
MW = 302
cLogP = 2.3
PSA = 58 Å^2
HBD = 2

5 $5HT_6$ pKi = 8.8
brain/blood = nd
MW = 326
cLogP = 2.8
PSA = 49 Å^2
HBD = 1

6 $5HT_6$ pKi = 8.8
brain/blood = 5:1
MW = 361
cLogP = 3.4
PSA = 49 Å^2
HBD = 1

7 $5HT_6$ pKi = 8.6
brain/blood = nd
MW = 361
cLogP = 3.4
PSA = 49 Å^2
HBD = 1

FIGURE 19.3 Conformational constraint: fewer H-bond donors.

that further improvement in metabolic stability was required. The 3-position of the indolyl core was identified as a potential metabolic soft spot. Introduction of a chlorine substituent at the 3-position gave compound **7**, which retained activity and had improved *in vitro* metabolic stability in rat microsomes (CLi = 1.7 ml/min/g liver) relative to both **5** and **6**. The brain-to-blood ratio for compound **7** was not reported, but **7** was progressed to a rat *ex vivo* binding assay and showed an ED_{50} = 4 mg/kg following oral dosing, suggesting brain penetration of compound **7**.

Bioisosteric Replacement To enhance the desired biological or physical properties of a compound without making significant changes in chemical structure, exchanging one functional group for another functional group is a commonly used tactic in medicinal chemistry [23]. Heterocyclic compounds bearing multiple HBDs are often associated with poor permeability. The bioisosteric replacement of a heterocyclic NH group by an alternative heterocycle, in which the NH is either capped or replaced, is commonly used to improve permeability.

Imidazole Replaced by Furan A brain-penetrant and selective B-Raf inhibitor may have utility as a neuroprotective agent for the treatment of stroke. The imidazole **8** was discovered by scientists at GlaxoSmithKline to be a potent and selective inhibitor of B-Raf kinase (Fig. 19.4) [24]. However, compound **8** had poor brain penetration (brain-to-blood ratio 0.14), which was far from optimal for this indication. To improve the CNS penetration of the series (target brain-to-blood ratio > 0.5), bioisosteres of the imidazole core that reduced PSA and HBDs were investigated, which led to the discovery of the furan analog **9**. The furan **9** retained similar B-Raf activity as **8**, but it had a lower PSA, one fewer HBD, and increased lipophilicity, which resulted in a significantly improved brain-to-blood ratio (1.33–1). Interestingly, the furan **10**, an isomer of **9** with identical structural properties (e.g., MW, PSA, H-bonds, pK_a), had a lower brain-to-blood ratio (0.56).

Incorporation of Fluorine Fluorine has a small atomic radius and is able to be both electron-withdrawing and lipophilic, while maintaining a unique ability to accept H-bonds and block metabolic sites. Strong bases (pK_a > 10) will likely be fully ionized at a physiological pH of 7.4, and are unlikely to cross membranes by passive

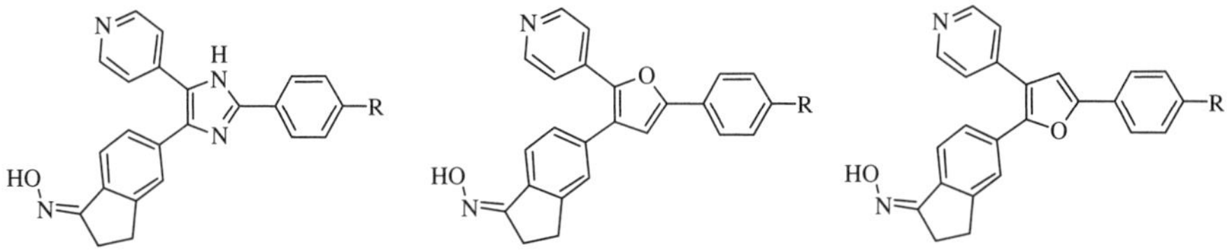

8 B-Raf K_d = 0.3 nM
brain/blood = 0.14
MW = 453
cLogP = 3.7
PSA = 82 Å^2
HBD = 2

9 B-Raf K_d = 0.5 nM
brain/blood = 1.33
MW = 453
cLogP = 4.2
PSA = 67 Å^2
HBD = 1

10 B-Raf K_d = 0.9 nM
brain/blood = 0.56
MW = 453
cLogP = 4.2
PSA = 67 Å^2
HBD = 1

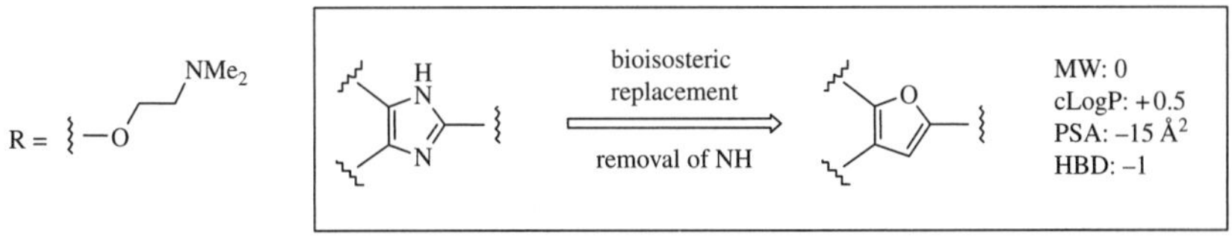

FIGURE 19.4 Bioisosteric replacement of the imidazole by furan led to significant changes in PSA, HDB, and improved brain penetration.

diffusion in this form. Strategies to improve brain penetration of strongly basic amines include reduction in the basicity of the nitrogen through the addition of an appropriately positioned electron-withdrawing group [25]. The strategic use of fluorine to improve brain penetration through modulation of pK_a is highlighted in the following example.

Modulation of pK_a Neuropeptide Y (NPY) is a 36 amino acid peptide widely distributed in both the CNS and in peripheral neurons. NPY has a range of biological actions that include stimulation of food intake and control of mood. As part of a project to discover potent selective NPY Y1 receptor antagonists, researchers at Glaxo [26] discovered a novel series of carbazole derivatives (Fig. 19.5). The piperidine-substituted analog **11** had high affinity for NPY Y1, but poor permeability and brain penetration in rat (B/P = 0.12). It was hypothesized that the strongly basic character of the piperidine group (pK_a = 11) was a major contributing factor for the poor pharmacokinetic profile and brain exposure of **11**. A focused exploration around the piperidine side-chain was made with modifications targeting reduction of the pK_a of the secondary amine. Studies indicated that affinity can be maintained or even improved, while reducing the basicity of the nitrogen. Methylation of the piperidine nitrogen (compound **12**) lowered the pK_a below 10, eliminated the HBD, and, as a result, compound **12** significantly improved brain penetration (B/P = 0.8). Introduction of an electron-withdrawing fluorine at the β-position of a piperidine will significantly lower the pK_a on the amine. For example, the pK_a of the *cis*-isomer **13** (pK_a = 7.9) and the *trans*-isomer **14** (pK_a = 6.5) were significantly lower than for the parent piperidine **11** (pK_a = 11). As expected, an improvement in the pharmacokinetic parameters was seen with compounds bearing less basic amine groups. Indeed, with other parameters effectively constant, a trend toward greater brain penetration was

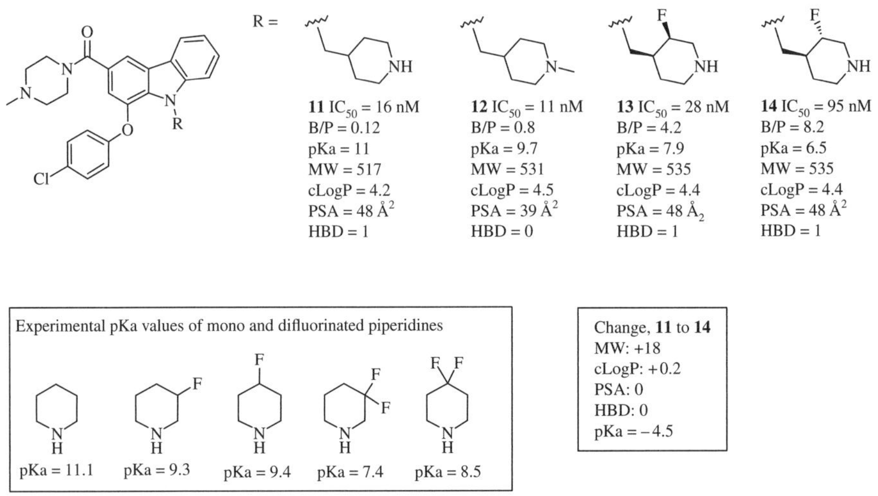

FIGURE 19.5 Fluoro-substitution to lower amine basicity.

observed with decreasing pK_a. Thus, the B/P ratios of compounds **11**, **12**, **13**, and **14** were measured as 0.12, 0.8, 4.2, and 8.2, respectively, as the pK_a decreased from 11 to 9.7, to 7.9, and to 6.5, respectively. An assessment of P-gp liability was not reported (e.g., MDR1-MDCK); however, the poor brain penetration observed for compound **11** could be due to P-gp efflux. P-gp efflux is often associated with highly basic amines [15], and brain penetration improved as the amine basicity was reduced (see Section "Tactics for the Mitigation of P-gp Efflux").

Modulation of pK_a and Increased Lipophilicity 5-HT_{5A} receptor antagonists have potential use in the treatment of mood disorders, including schizophrenia. Peters at Hoffmann-La Roche [27] identified the novel cyclic guanidine **15** as a promising lead 5-HT_{5A} antagonist (Fig. 19.6). Mouse PK studies on **15**, however, indicated resistance to brain penetration and a modest B/P ratio of 0.2. Since the lipophilicity of **15** (Log$D_{7.4}$=0) was in the low range for CNS drugs (Table 19.1), a variety of substituents were attached at the 2-NH_2 substituent in an effort to increase lipophilicity and brain penetration. Due to its straightforward synthesis, compound **17** was chosen as a model compound (see box in Figure 19.6). Attachment of lipophilic alkyl chains to the 2-amino group of **17**, however, did not lead to the higher Log*D* values expected from *c*Log*P* calculations, which predict a lipophilicity increase of approximately 0.5 log units per methylene unit (e.g., compare **18** and **19**). Instead, a strategy that lowered the guanidine basicity was investigated. It was found that the attachment of small, lipophilic, and electron-withdrawing fluoro-ethyl substituents (e.g., difluoro analog **20** and trifluoromethyl analog **21**) was a better way to improve Log$D_{7.4}$, by increasing not only the lipophilicity, *c*Log*P*, but also by lowering the high basicity (pK_a) of the guanidine core. The reduction of pK_a leads to a smaller pH-dependent term in Equation 19.1 and, thus, contributes to the increase in the measured lipophilicity (Log$D_{7.4}$). To further increase the lipophilicity, the lipophilic, electron-withdrawing 5-chloro substituent

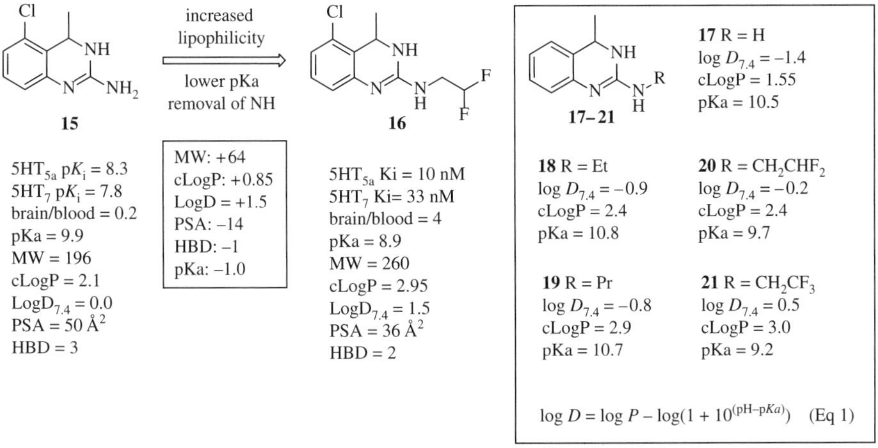

FIGURE 19.6 Fluoro-substitution to modulate pK_a and lipophilicity.

was reintroduced (e.g., **16**), which further lowered the basicity of the guanidine core. Compound **16** had similar potency as the lead **15**, but with a lower pK_a, lower measured lipophilicity, and one fewer HBD. Compound **16** was evaluated in a mouse pharmacokinetic experiment and showed an improved B/P ratio as compared to **15** (~20-fold increase), with otherwise similar pharmacokinetic properties.

Increasing Metabolic Stability The first-pass metabolism of an orally administered drug is one of the primary reasons that a molecule may not reach adequate brain levels. In such cases, improving metabolic stability and lowering clearance is an important strategy for increasing free brain drug concentration.

Scaffold Hopping to Improve Metabolic Stability Scaffold hopping has been widely applied by medicinal chemists to discover equipotent compounds having novel backbones with improved properties [28]. In this example, scaffold hopping and optimization of ring substituents were used to identify microsomally stable and brain-penetrant CRF-1 receptor antagonists as potential anxiolytic and antidepressant agents [29]. The lead molecule at Neurogen (e.g., isoquinoline **22**) was highly lipophilic ($c\text{Log}P=8.1$) and gave very poor oral bioavailability ($F=5\%$) in a rat PK experiment (Fig. 19.7). In an effort to increase hydrophilicity and to improve general pharmacokinetic properties, the bicyclic isoquinoline **22** was replaced with less lipophilic, monocyclic pyrimidone cores (e.g., **23**). However, despite lower lipophilicity and increased solubility relative to the isoquinoline **22**, very poor oral exposures in rat were observed. The low oral exposures were attributed to metabolic liabilities (rat microsomes $T_{1/2}=3.6$ min), which proved difficult to overcome within the pyrimidine series. To circumvent this issue, other monocyclic heterocycles, as replacements of the pyrimidine core, were examined, and a series of pyrazines were discovered with improved microsomal stability. Following optimization of the aryl substituents, pyrazine **24** was identified. Although lipophilicity had increased and aqueous solubility was poor, the

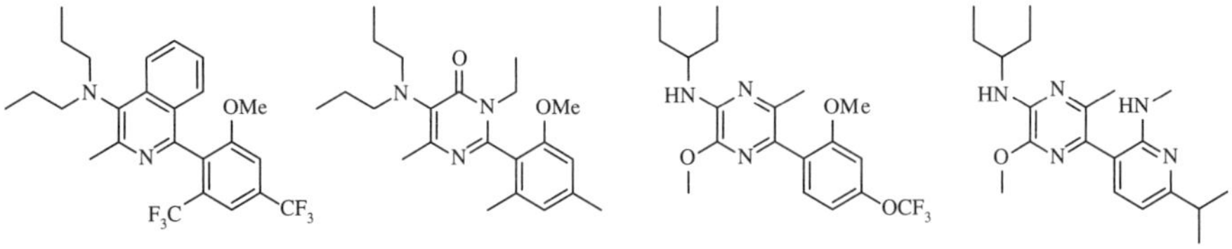

22 hCRF-1 *Ki* = 8 nM	**23** hCRF-1 *Ki* = 21 nM	**24** hCRF-1 *Ki* = 1 nM	**25** hCRF-1 *Ki* = 2 nM
cLogP = 8.1	cLogP = 3.55	cLogP = 6.0	cLogP = 5.1
MW = 484	MW = 371	MW = 399	MW= 357
PSA = 25 $Å^2$	PSA = 45 $Å^2$	PSA = 64 $Å^2$	PSA = 70 $Å^2$
pKa = not determined	pKa = not determined	pKa = 4.0	pKa = 7.4
HDB = 0	HDB = 0	HBD = 1	HBD = 2
Sol pH 7.4 < 0.5 μM	Sol pH 7.4 = 4 μM	Sol pH 7.4 < 1 μM	Sol pH 7.4 = 5 μM
Rat PK	**Rat PK**	**Rat PK**	**Rat PK**
%F = 5	%F = 2	%F = 45	%F = 51
$T_{1/2}$ = 5.4 h	$T_{1/2}$ = 3.3 h	$T_{1/2}$ = 41 h	$T_{1/2}$ = 1.8 h
B/P = nd	B/P = 0.9 to 1	B/P = 5.4 to1	B/P = 1.7 to1

FIGURE 19.7 Optimization of CRF-1 receptor antagonists.

stability (rat microsomes $T_{1/2}$>120 min) was considerably increased, and excellent oral exposure (%F=45) and high B/P ratio (5.4) were observed. In order to quantitatively evaluate the rat cortical CRF-1 receptor occupancy of **24** following oral dosing, *ex vivo* binding methods were utilized. Dose–response studies determined the dose of **24** required for 50% receptor occupancy (*ex vivo* ID_{50}) to be 2.0 mg/kg, and the plasma concentration that was associated with 50% occupancy of brain CRF-1 receptors (*ex vivo* IC_{50}) was determined to be 71 ng/ml. However, the development of **24** was halted following mass balance studies with [^{14}C]-**24** in rat, which indicated extensive accumulation in adipose tissue. This finding, in addition to a very long terminal half-life in rat ($T_{1/2}$=41 h), raised a potential safety issue for chronic dosing in humans. To lower the propensity to partition into adipose tissue, a strategy to reduce the lipophilicity and to increase aqueous solubility of molecules was followed. It was discovered that the aryl ring could be replaced by an appropriately substituted pyridine (e.g., **25**). The weakly basic (pK_a=7.4) pyridine **25** was more soluble and less lipophilic than **24**. In a rat PK study, **25** demonstrated good oral bioavailability (%F=51), a B/P ratio closer to unity (1.7), and a considerably shorter half-life. Mass balance studies in rat indicated that **25** did not accumulate in adipose tissue. Furthermore, *ex vivo* binding determined improved *in vivo* potency for **25**; an *ex vivo* ID_{50}=0.7 mg/kg with an *ex vivo* IC_{50}=15 ng/ml. In summary, the optimization of isoquinoline **22**, resulting in pyrazine **25**, significantly improved calculated structural properties (e.g., *c*Log*P*: −3.0, MW: −127, and PSA: +45), further resulting in enhanced solubility and metabolic stability. Although there are now two HBDs in **25**, the isopentyl amine NH potentially forms a "masking" H-bond with the pyrazine methoxy substituent.

TACTICS FOR THE MITIGATION OF P-gp EFFLUX

P-gp (or MDR1) is a 170 kDa transmembrane glycoprotein encoded by the ABCB1 gene in humans. P-gp is one of the most important members of the super family of 51 human genes, which encode ATP-binding cassette (ABC) transporters using the energy of ATP hydrolysis to actively extrude molecules across a lipid membrane.

P-gp was the first ABC transporter to be cloned, and it is widely expressed in particular at the BBB and BCSFB. Because of this localization, P-gp can play a significant role in the disposition of drugs targeted for the CNS, which are P-gp substrates. P-gp has a broad substrate specificity, accepting a wide range of structurally diverse substrates. A crystal structure of P-gp reveals that it has a large ligand-binding site, allowing simultaneous binding of various ligands at different positions within the site [30].

The importance of these factors on CNS exposure has propelled testing for P-gp substrates to earlier stages of the drug discovery processes. *In vitro* assays and *in vivo* models for assessing the potential for P-gp efflux have been discussed in earlier chapters. The most commonly used system is the Madin–Darby canine kidney (MDCK) epithelial cells that express the human ABCB1 gene by transfection (MDR1-MDCK). In this type of assay, permeability measurements are made in the apical-to-basal (AB) and basal-to-apical (BA) directions across a confluent cell monolayer, in the absence and presence of a specific P-gp inhibitor. Moderate AB permeability in parental MDCK and LLC-PK assays is considered to reside in the 10×10^{-6} cm/s range; rates above that value are considered to be high permeability. The BA/AB ratio, or efflux ratio (ER), of these permeability values can be used to provide an estimate of P-gp involvement. Generally, values above a cutoff determined experimentally are used to rank-order potential P-gp substrates. Given its lower cost, higher throughput, and absence of transport proteins, the parallel artificial membrane permeability assay (PAMPA) is used as a filter assay to triage compounds with sufficient passive permeability to test in MDCK assays. The impact of P-gp *in vivo* can be assessed by comparing the exposure of compounds in wild-type mdr1a (+/+) versus mdr1a (–/–) KO mice. Alternatively, a number of small-molecule P-gp inhibitors (e.g., cyclosporine, verapamil) have been used to create "chemical" P-gp knockouts, with the advantage that these inhibitors expand the scope of studies into species beyond mice [31].

Discovering that a compound is a P-gp substrate is not necessarily the end of the discovery road for the compound. A number of CNS drugs are now known to be P-gp substrates. However, in addition to potentially poor brain exposure, there are a number of risks to developing a P-gp substrate. For example, unknown unbound brain concentration in humans, low confidence in PK/PD translation across species, and interindividual variability in both oral and brain exposure could result in greater differences in efficacy, significant increases in peripheral drug exposure, and an increased risk of drug–drug interactions, particularly with another P-gp-interacting compound. It is, therefore, strongly recommended that, should P-gp efflux be identified for a series, SAR around P-gp must be developed, and that mitigation strategies also must be developed. The predominant approaches for mitigation of P-gp are the modulation of H-bonding characteristics and modulation of amine basicity. Examples of P-gp mitigation using these strategies are given in the following section.

Scaffold Change to Eliminate an NH Group

As part of the cannabinoid CB_2 agonist program for pain at Glaxo, the 6-azaindole **26** was identified as a potent (CB_2 $EC_{50} = 11$ nM) and selective (CB_1 $EC_{50} > 10\,\mu M$) lead (Fig. 19.8) [32]. However, **26** had very low CNS penetration (mouse brain-to-blood ratio of <0.05), and it was then demonstrated to be a P-gp substrate (ER = 74). The

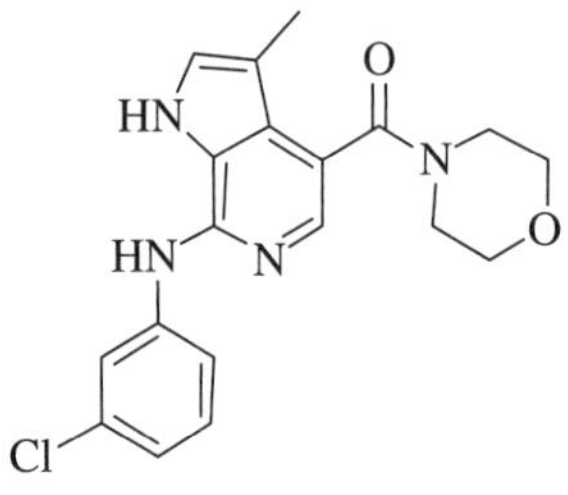

26 CB_2 EC_{50} = 11 nM
MW = 371
PSA = 66
cLogP = 2.44
HBD = 2
pKa = 5.8
Sol pH 5 = 160 μg/ml
ER = 74
brain/blood <0.05 : 1

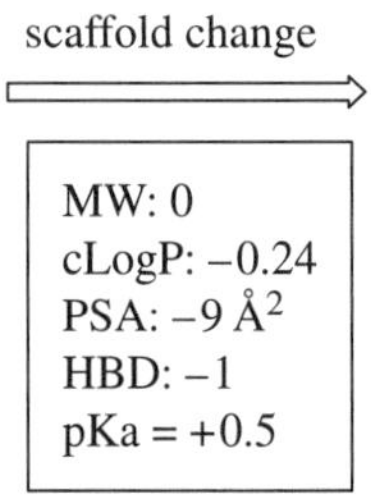

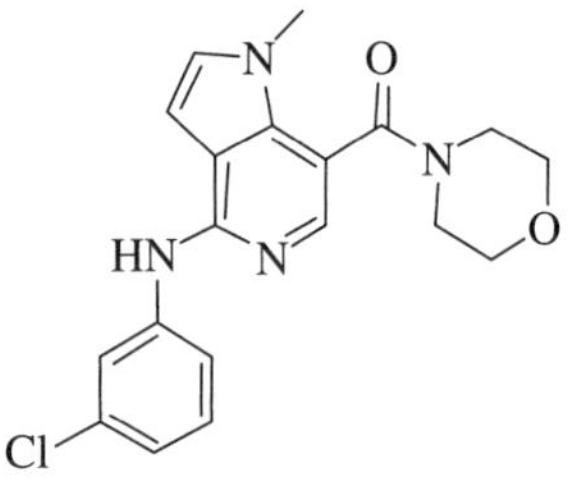

27 CB_2 EC_{50} = 5 nM
MW = 371
PSA = 57
cLogP = 2.2
HBD = 1
pKa = 6.1
Sol pH 5 = 0.59 mg/ml
ER = 2.9
brain/blood 1.04 : 1

FIGURE 19.8 Scaffold change eliminates NH group.

structural properties of **26** are in line with those required for good permeability, but the presence of two HBDs could be associated with the P-gp interaction. To remove one HBD, **26** was *N*-methylated, but this led to a significant loss in activity. A creative solution was to prepare the isomeric 5-aza indole analog **27**, which preserved the CB_2 potency (CB_2 EC_{50} = 5 nM) but eliminated the aza indole NH. Significantly, compound **27** had an improved solubility in simulated intestinal fluid (pH 5.0) and an improvement in ER (2.9), which indicated low affinity for P-gp compared to **26** (ER = 74). In rat PK studies, **27** had excellent oral bioavailability (F = 82%) and a brain-to-blood ratio close to unity (1.04). Compound **27** was efficacious in a chronic joint pain model, with efficacy equivalent to rofecoxib.

Replacement of HBDs

Inhibition of phosphodiesterase 10A (PDE10A) has emerged as a potential target to treat schizophrenia. Following an HTS screening campaign at Amgen [33], benzimidazole **28** was identified as a potent hit (IC_{50} = 9.7 nM) (Fig. 19.9). Compound **28** had acceptable passive permeability, but high efflux in human and rat transfected MDR1-MDCK cells (ER = 25 and 13, respectively). Substitution of the pyridine ring on **28** with morpholine produced **29**, which resulted in an improved potency (IC_{50} = 1.1 nM) and one fewer HBD. Interestingly, this brought about a decrease of the ER in humans (11) but an increase in rats (33). Because the co-crystal structure of **28** in PDE10A suggested that the N−H bond on the amine linker was not essential, the amine linker on **29** was replaced with a carbonyl group. The resulting keto-benzimidazole **30**(HBD = 1) had slightly reduced PDE10A potency and exhibited low ER in humans and rats (0.9 and 1.0, respectively). Compound **30** was advanced into a liquid chromatography–mass spectrometry (LC-MS)/MS receptor occupancy assay to assess its CNS target coverage; it gave 21% receptor occupancy at 10 mg/kg and 55% receptor occupancy at 30 mg/kg.

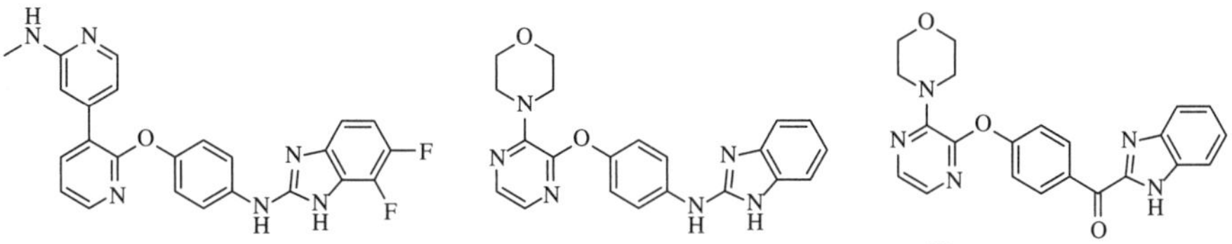

28 PDE10A IC_{50} = 9.7 nM
MW: 444
PSA: 82
cLogP: 5.0
HBD: 3
hER = 25
rER = 13

29 PDE10A IC_{50} = 1.1 nM
MW: 388
PSA: 83
cLogP: 3.4
HBD: 2
hER = 11
rER = 33

30 PDE10A IC_{50} = 4.5 nM
MW: 401
PSA: 88
cLogP: 2.85
HBD: 1
hER = 0.9
rER = 1.0

FIGURE 19.9 Reducing HBDs in a series of PDE10A inhibitors.

Scaffold Change to Eliminate an NH Group

In a secondPDE10A project example from Amgen [34], the NH-linked benzimidazole **31** had 3 HBDs and a very high ER (77). In this example, the hydrogen bond donor in the benzimidazole, itself, was capped or replaced by another heteroatom. The N-methyl analog **32** retained activity, which indicated that the benzimidazole NH was not essential for PDE10A activity, retained passive permeability, and decreased efflux (ER = 6). The benzoxazole **33** was fivefold less active than **31**, but it showed a further reduction in efflux (ER = 3.4). Finally, the benzothiazole **34** gave the most significant improvement in lowering efflux (ER = 1.1), although the passive permeability also was significantly reduced ($P_{app} < 10 \times 10^{-6}$ cm/s) (Table 19.2).

TABLE 19.2 The effect of changing the heterocycle on P-gpefflux ratio

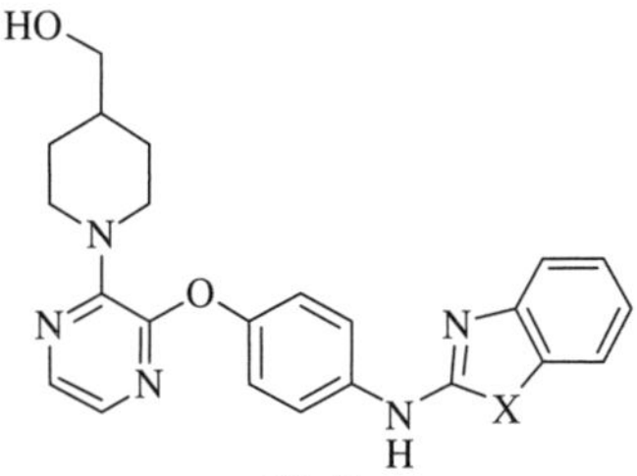

31-34

Compound	X	PDE10A IC_{50} (nM)	P_{app} [a]	ER [b]	MW	*c*Log*P*	PSA	HBD
31	NH	0.09	42	77	415	4.4	81	3
32	NMe	0.06	39	6	429	4.6	73	2
33	O	0.51	24	3.4	416	4.5	79	2
34	S	0.20	8	1.1	432	5.8	69	2

[a] Apparent permeability rates (10^{-6} cm/s) through porcine proximal tubule cells (LLC-PK1 cell line).
[b] Efflux ratio.

Intramolecular Hydrogen-Bonding

In an elegant example from a BACE1 inhibitor program for Alzheimer's disease at Amgen, the lead compound, azachromane **35**, had promising cell activity (IC_{50} = 12 nM) and good passive permeability (P_{app} = 21 × 10^{-6} cm/s) (Fig. 19.10) [35]. However, **35** was a strong substrate for P-gp efflux in both rats (ER = 43) and humans (ER = 18). Attempts to remove any of the three HBDs resulted in a significant loss in activity. However, an improvement in the P-gp-mediated efflux of analogs of **35** was obtained by introducing flanking groups that were capable of forming an internal H-bond with the amide NH group. For example, the methoxy ether compound **36** had good cell activity (IC_{50} = 28 nM) and a significant improvement in both rat (7) and human (5) ER.

Electron-Withdrawing Groups (e.g., Fluorine) to Reduce Hydrogen-Bond Capability

In the next example, the electron-withdrawing characteristics of fluorine are successfully exploited to effectively lower the P-gp liability within a series of bradykinin (BK) B1 receptor antagonists, leading to improved brain exposure and efficacy in animal models [36]. The biaryl compound **37** was identified as having moderate binding affinity for hBK B1 and promising pharmacokinetic properties (Table 19.3). However, **37** proved to be a substrate for P-gp-mediated efflux, and, as a result, would be predicted to have poor brain exposure in man (Table 19.3, ER = 16). Optimization strategies of the biaryl region incorporated fluorine in both aryl rings, resulting in compound **38**, which showed a significant boost in BK B1 potency and oral bioavailability in rat, yet very little change in P-gp efflux (ER = 18.4). In an attempt to reduce the P-gp liability of the series, replacement of the trifluoropropionamide group was investigated. While pentafluoropropionamide

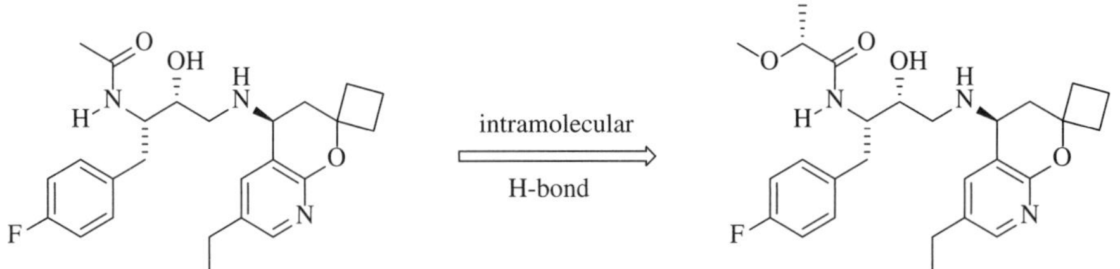

35 BACE IC_{50} = 12 nM
MW= 484
cLogP = 4.1
PSA = 83 Å^2
HBD = 3
hER = 18
rER= 43

36 BACE IC_{50} = 28 nM
MW = 528
cLogP = 4.2
PSA = 92 Å^2
HBD = 3
hER = 5
rER = 7

FIGURE 19.10 Modulation of P-gp efflux through intramolecular hydrogen-bonding.

TABLE 19.3 The effect of fluorine substitution on P-gp efflux ratio

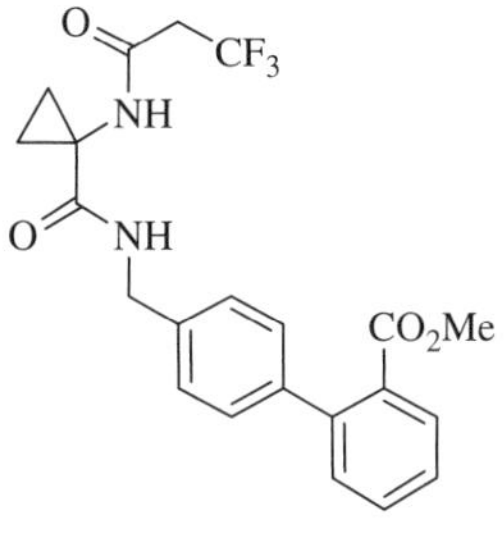

37

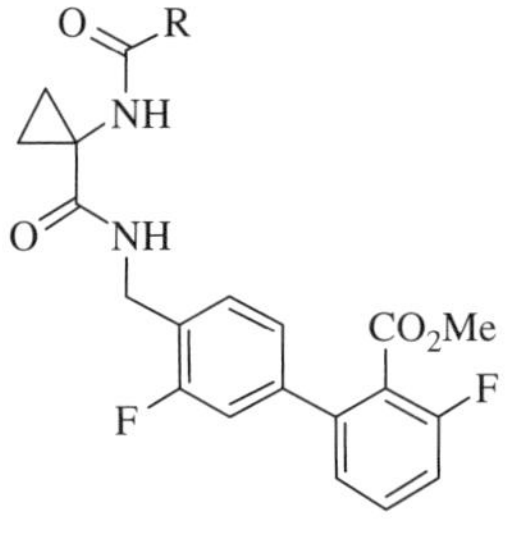

38–40

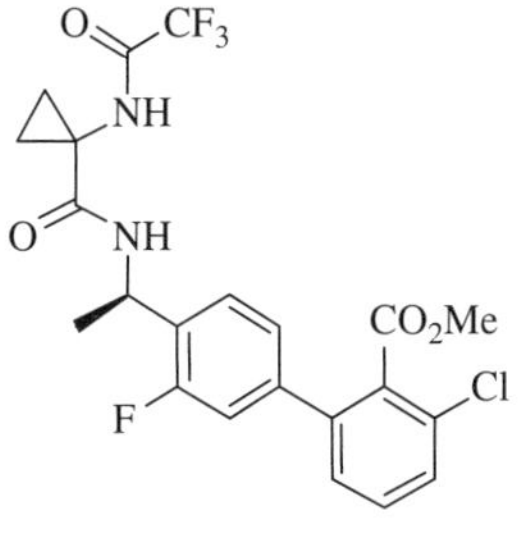

41

Compound	R	hBK B1	ER[a]	P_{app}[b]	MW	c Log*P*	PSA	HBD
37	—	63	16	18	434	2.97	84	2
38	CH_2CF_3	0.81	18.4	23	470	3.29	84	2
39	CF_2CF_3	2.95	2.2	31	506	3.94	84	2
40	CF_3	1.47	4.1	23	456	3.34	84	2
41	—	0.4	1.9	34	487	4.06	84	2

[a] MDR1 directional transport efflux ratio (B/A)/(A/B).
[b] Passive permeability (10^{-6} cm/s).

39 was threefold less active against hBK B1, it had dramatically reduced efflux potential (ER = 2.2). Replacement of the difluoromethylene for a trifluoroacetamide provided **40** with improved binding affinity, albeit with a slight gain in P-gp efflux potential (ER = 4.1). Overall, the trifluoromethyl analog **40** exhibited a good balance between hBK B1 potency and reduced P-gp liability. The electron-withdrawing trifluoromethyl group proved essential for lowering the P-gp liability. It was proposed that the more electron-deficient amide had less H-bonding capability, contributing to decreased recognition by P-gp [37, 38]. Further optimization led to the identification of the chloro analog **41**, which demonstrated minimal P-gp efflux (ER = 1.9) and oral efficacy in complete Freund's adjuvant (CFA)-induced hyperalgesia in a humanized mouse. In a transgenic rat model expressing human B1 receptors, **41** achieved 90% CNS B1 receptor occupancy at a brain concentration of 520 nM.

Modulation of Amine Basicity

In a previous example, fluorine was used to modulate H-bonding capability. In this example [39], fluorine is used to modulate the basicity of an amine to improve P-gp liability. The piperidine-substituted imidazole **42** had excellent potency as an α4β2 nAChR potentiator and possible utility in Parkinson's disease and schizophrenia. However, imidazole **42** was determined to be a substrate for P-gp (ER > 10). To reduce amine basicity and decrease the potential for P-gp efflux, the 4-fluoropiperidine **43** was prepared. The 4-fluoropiperidine **43** retained excellent potency and had significantly reduced P-gp efflux liability (ER ~1). CNS penetration of **43** was

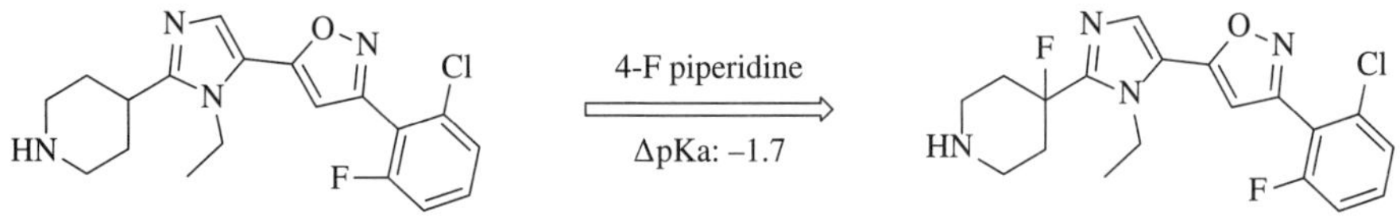

42 EC_{50} = 3 nM
MW = 375
cLogP = 3.3
PSA = 49
ER (MDR1-LLC-PK1) >10

43 EC_{50} = 3 nM
MW = 393
cLogP = 2.9
PSA = 49
ER (MDR1-LLC-PK1) ~ 1

FIGURE 19.11 Fluoro-substitution to modulate pK_a.

observed in rodents following intraperitoneal dosing at 5 mg/kg and showed a total brain concentration of 6.5 μM.

CONCLUSION

In summary, optimal brain exposure can be achieved through medicinal chemistry strategies that focus on the following aspects:

(i) The design of new molecules with structural and physicochemical properties that are favorable for passive diffusion across the BBB
(ii) The design of new molecules that circumvent efflux by transporters such as P-gp

The structural requirements for passive diffusion across the BBB are fairly well understood, and a number of interrelated structural attributes are used for the design of new molecules. Adjusting one attribute will invariably alter others, and a balancing act is required to make the best compromises between physicochemical properties and potency at the intended target. In addition, P-gp can play a significant role in limiting the CNS exposure of drugs which are P-gp substrates. The predominant approach for mitigation of P-gp efflux is to reduce the number or modulation of H-bonding characteristics.

REFERENCES

[1] (a) Wager, T., Villalobos, A., Verhoest, P.R., Hou, X., Shaffer, C.L. (2011). Strategies to optimize the brain availability of central nervous system drug candidates. *Expert Opin. Drug Discov. 6*, 371–381. (b) Fridén, M., Winiwarter, S., Jerndal, G., Bengtsson, O., Wan, H., Bredberg, U., Hammarlund-Udenaes, M., Antonsson, M. (2009). Structure-brain exposure relationships in rat and human using a novel data set of unbound drug concentrations in brain interstitial and cerebrospinal fluids. *J. Med. Chem. 52*, 6233–6243.

[2] (a) Hammarlund-Udenaes, M., Fridén, M., Syvänen, S., Gupta, A. (2008). On the rate and extent of drug delivery to the brain. *Pharm. Res. 25*, 1737–1750. (b) Liu, X., Smith, B.J.,

Chen, C., Callegari, E., Becker, S.L., Chen, X., Cianfrogna, J., Doran, A.C., Doran, S.D., Gibbs, J.P., Hosea, N., Liu, J., Nelson, F.R., Szewc, M.A., VanDeusen, J. (2006). Evaluation of cerebrospinal fluid concentration and plasma free concentration as a surrogate measurement for brain free concentration. *Drug Metab. Dispos. 34*, 1443–1447.(c) Lin, J.L. (2008). CSF as a surrogate for assessing CNS exposure: an industrial perspective. *Curr. Drug Metab. 9*, 46–59.(d)Smith, D.A., Di, L., Kerns, E.H. (2010). The effect of plasma protein binding on in vivo efficacy: misconceptions in drug discovery. *Nat. Rev. Drug Discov. 9*, 929–939.(e)Jeffrey, P., Summerfield, S. (2010). Assessment of the blood-brain barrier in CNS drug discovery. *Neurobiol. Dis. 37*, 33–37.

[3] Hitchcock, S., Pennington, L.D. (2006). Structure–brain exposure relationships. *J. Med. Chem. 49*, 1–25.

[4] Overton, E. (1897). Osmotic properties of cells in the bearing on toxicology and pharmacology. *Z. Phys. Chem. 22*, 189–209.

[5] (a) Leo, A., Hansch, C., Elkins, D. (1971). Partition coefficients and their uses. *Chem. Rev. 71*, 525–615. (b) Hansch, C., Leo, A.J. Substituent Constant for Correlation Analysis in Chemistry and Biology, John Wiley & Sons, Inc., New York, 1979.

[6] Leeson, P.D., Davis, A.M. (2004). Time-related differences in the physical property profiles of oral drugs. *J. Med. Chem. 47*, 6338–6348.

[7] (a) van de Waterbeemd, H., Kansy, M. (1992). Hydrogen bonding capacity and brain penetration. *Chimia 46*, 299–303. (b) van de Waterbeemd, H., Camenisch, G., Folkers, G., Chretien, J.R., Raevsky, O.A. (1998). Estimation of blood-brain barrier crossing of drugs using molecular size and shape and H-bonding descriptors. *J. Drug Target. 6*, 151–165.

[8] Kelder, J., Grootenhuis, P.D., Bayada, D.M., Delbressine, L.P., Ploemen, J.P. (1999). Polar molecular surface as a dominating determinant for oral absorption and brain penetration of drugs. *Pharm. Res. 16*, 1514–1519.

[9] Lipinski, C.A., Lombardo, F., Dominy, B.W., Feeney, P.J. (2001). Experimental and computational approaches to estimate solubility and permeability in drug discovery and development settings. *Adv. Drug Deliv. Rev. 46*, 3–26.

[10] Lipinski, C.A. (1999). Drew University Medical Chemistry Special Topics Course. July.

[11] Mahar Doan, K.M., Humphreys, J.E., Webster, L.O., Wring, S.A., Shampine, L.J., Serabjit-Singh, C.J., Adkison, K.K., Polli, J.W. (2002). Passive permeability and P-glycoprotein-mediated efflux differentiate central nervous system (CNS) and non-CNS marketed drugs. *J. Pharmacol. Exp. Ther. 303*, 1029–1037.

[12] Pajouhesh, H., Lenz, G.R. (2005). Medicinal chemical properties of successful central nervous system drugs. *NeuroRx 2*, 541–553.

[13] (a) Wager, T.T., Chandrasekaran, R.Y., Hou, X., Troutman, M.D., Verhoest, P.R., Villalobos, A., Will, Y. (2010). Defining desirable central nervous system drug space through the alignment of molecular properties, in vitro ADME, and safety attributes. *ACS Chem. Neurosci. 1*, 420–434. (b) Wager, T.T., Hou, X., Verhoest, P.R., Villalobos, A. (2010). Moving beyond rules: the development of a central nervous system multiparameter optimization (CNS MPO) approach to enable alignment of druglike properties. *ACS Chem. Neurosci. 1*, 435–449.

[14] Fischer, H., Gottschlich, R., Seelig, A. (1998). Blood-brain barrier permeation: molecular parameters governing passive diffusion. *J. Membrane Biol. 165*, 201–211.

[15] Hitchcock, S. (2012). Structural modifications that alter the P-glycoprotein efflux properties of compounds. *J. Med. Chem. 55*, 4877–4895.

[16] Austin, R.P., Davis, A.M., Manners, C.N. (1995). Partitioning of ionizing molecules between aqueous buffers and phospholipids vesicles. *J. Pharm. Sci. 84*, 1180–1183.

[17] Alelyunas, Y.W., Empfield, J.R., McCarthy, D., Spreen, R.C., Bui, K., Pelosi-Kilby, L., Shen, C. (2010). Experimental solubility profiling of marketed CNS drugs, exploring solubility limit of CNS discovery candidate. *Bioorg. Med. Chem. Lett. 20*, 7312–7316.

[18] Lipinski, C.A. (2000). Drug-like properties and the causes of poor solubility and poor permeability. *J. Pharmacol. Toxicol. Meth. 44*, 235–249.

[19] el-Bacha, R.S., Minn, A. (1999). Drug metabolizing enzymes in cerebrovascular endothelial cells afford a metabolic protection to the brain. *Cell. Mol. Biol. 45*, 15–23.

[20] Kalvass, J.C., Maurer, T.S., Pollack, G.M. (2007). Use of plasma and brain unbound fractions to assess the extent of brain distribution of 34 drugs: comparison of unbound concentration ratios to in vivo P-glycoprotein efflux ratios. *Drug Metabol. Dispos. 35*, 660–666.

[21] Malamas, M.S., Barnes, K., Hui, Y., Johnson, M., Lovering, F., Condon, J., Fobare, W., Solvibile, W., Turner, J., Hu, Y., Manas, E.S., Fan, K., Olland, A., Chopra, R., Bard, J., Pangalos, M.N., Reinhart, P., Robichaud, A.J. (2010). Novel pyrrolyl 2-aminopyridines as potent and selective human β-secretase (BACE1) inhibitors. *Bioorg. Med. Chem. Lett. 20*, 2068–2073.

[22] Trani, G., Baddeley, S.M., Briggs, M.A., Chuang, T.T., Deeks, N.J., Johnson, C.N., Khazragi, A.A., Mead, T.L., Medhurst, A.D., Milner, P.H., Quinn, L.P., Ray, A.M., Rivers, D.A., Stean, T.O., Stemp, G., Trail, B.K., Witty, D.R. (2008). Tricyclic azepine derivatives as selective brain penetrant 5-HT6 receptor antagonists. *Bioorg. Med. Chem. Lett. 18*, 5698–5700.

[23] Meanwell, N.A. (2011). Synopsis of some recent tactical application of bioisosteres in drug design. *J. Med. Chem. 54*, 2529–2591.

[24] Takle, A.K., Bamford, M.J., Davies, S., Davis, R.P., Dean, D.K., Gaiba, A., Irving, E.A., King, F.D., Naylor, A., Parr, C.A., Ray, A.M., Reith, A.D., Smith, B.B., Staton, P.C., Steadman, J.G., Stean, T.O., Wilson, D.M. (2008). The identification of potent, selective and CNS penetrant furan-based inhibitors of B-raf kinase. *Bioorg. Med. Chem. Lett. 18*, 4373–4376.

[25] Hodgetts, K.J., Combs, K.J., Elder, A.M., Harriman, G.C. (2010). The role of fluorine in the discovery and optimization of CNS agents: modulation of drug-like properties. *Ann. Rep. Med. Chem. 45*, 429–448.

[26] Leslie, C.P., Di Fabio, R., Bonetti, F., Borriello, M., Braggio, S., Dal Forno, G., Donati, D., Falchi, A., Ghirlanda, D., Giovannini, R., Pavone, F., Pecunioso, A., Pentassuglia, G., Pizzi, D.A., Rumboldt, G., Stasi, L. (2007). Novel carbazole derivatives as NPY Y1 antagonists. *Bioorg. Med. Chem. Lett. 15*, 1043–1046.

[27] Peters, J.U., Lübbers, T., Alanine, A., Kolczewski, S., Blasco, F., Steward, L. (2008). Cyclic guanidines as dual 5-HT5A/5-HT7 receptor ligands: optimising brain penetration. *Bioorg. Med. Chem. Lett. 18*, 262–266.

[28] Böhm, H.J., Flohr, A., Stahl, M. (2004). Scaffold hopping. *Drug Discov. Today Technol. 1*, 217–224.

[29] Hodgetts, K.J., Ge, P., Yoon, T., De Lombaert, S., Brodbeck, R., Gulianello, M., Kieltyka, A., Horvath, R.F., Kehne, J.H., Krause, J.E., Maynard, G.D., Hoffman, D., Lee, Y., Fung, L., Doller, D. (2011). Discovery of *N*-(1-ethylpropyl)-[3-methoxy-5-(2-methoxy-4-trifluoromethoxyphenyl)-6-methyl-pyrazin-2-yl]amine **59** (NGD 98-2): an orally active corticotropin releasing factor-1 (CRF-1) receptor antagonist. *J. Med. Chem. 54*, 4187–4206.

[30] Aller, S.G., Yu, J., Ward, A., Weng, Y., Chittaboina, S., Zhuo, R., Harrell, P.M., Trinh, Y.T., Zhang, Q., Urbatsch, I.L., Chang, G. (2009). Structure of P-glycoprotein reveals a molecular basis for poly-specific drug binding. *Science 323*, 1718–1722.

[31] Hendrikse, N.H., Schinkel, A.H., de Vries, E.G., Fluks, E., Van der Graaf, W.T., Willemsen, A.T., Vaalburg, W., Franssen, E.J. (1998). Complete in vivo reversal of P-glycoprotein pump function in the blood-brain barrier visualized with positron emission tomography. *Br. J. Pharmacol. 124*, 1413–1418.

[32] Giblin, G.M., Billinton, A., Briggs, M., Brown, A.J., Chessell, I.P., Clayton, N.M., Eatherton, A.J., Goldsmith, P., Haslam, C., Johnson, M.R., Mitchell, W.L., Naylor, A., Perboni, A., Slingsby, B.P., Wilson, A.W. (2009). Discovery of 1-[4-(3-chlorophenylamino)-1-methyl-1H-pyrrolo[3,2-c]pyridin-7-yl]-1-morpholin-4-ylmethanone (GSK554418A), a brain penetrant 5-azaindole CB2 agonist for the treatment of chronic pain. *J. Med. Chem. 52*, 5785–5788.

[33] Hu, E., Kunz, R.K., Chen, N., Rumfelt, S., Siegmund, A., Andrews, K., Chmait, S., Zhao, S., Davis, C., Chen, H., Lester-Zeiner, D., Ma, J., Biorn, C., Shi, J., Porter, A., Treanor, J., Allen, J.R. (2013). Design, optimization, and biological evaluation of novel keto-benzimidazoles as potent and selective inhibitors of phosphodiesterase 10A (PDE10A). *J. Med. Chem. 56*, 8781–8792.

[34] Rzasa, R.M., Hu, E., Rumfelt, S., Chen, N., Andrews, K.L., Chmait, S., Falsey, J.R., Zhong, W., Jones, A.D., Porter, A., Louie, S.W., Zhao, X., Treanor, J.J., Allen, J.R. (2012). Discovery of selective biaryl ethers as PDE10A inhibitors: improvement in potency and mitigation of Pgp-mediated efflux. *Bioorg. Med. Chem. Lett. 22*, 7371–7375.

[35] Weiss, M.M., Williamson, T., Babu-Khan, S., Bartberger, M.D., Brown, J., Chen, K., Cheng, Y., Citron, M., Croghan, M.D., Dineen, T.A., Esmay, J., Graceffa, R.F., Harried, S.S., Hickman, D., Hitchcock, S.A., Horne, D.B., Huang, H., Imbeah-Ampiah, R., Judd, T., Kaller, M.R., Kreiman, C.R., La, D.S., Li, V., Lopez, P., Louie, S., Monenschein, H., Nguyen, T.T., Pennington, L.D., Rattan, C., San Miguel, T., Sickmier, E.A., Wahl, R.C., Wen, P.H., Wood, S., Xue, Q., Yang, B.H., Patel, V.F., Zhong, W. (2012). Design and preparation of a potent series of hydroxyethylamine containing β-secretase inhibitors that demonstrate robust reduction of central β-amyloid. *J. Med. Chem. 55*, 9009–9024.

[36] Kuduk, S.D., Di Marco, C.N., Chang, R.K., Wood, M.R., Schirripa, K.M., Kim, J.J., Wai, J.M.C., DiPardo, R.M., Murphy, K.L., Ransom, R.W., Harrell, C.M., Reiss, D.R., Holahan, M.A., Cook, J., Hess, J.F., Sain, N., Urban, M.O., Tang, C., Prueksaritanont, T., Pettibone, D.J., Bock, M.G. (2007). Development of orally bioavailable and CNS penetrant biphenylaminocyclopropane carboxamide bradykinin B1 receptor antagonists. *J. Med. Chem. 50*, 272–282.

[37] Seelig, A. (1998). A general pattern for substrate recognition by P-glycoprotein. *Eur. J. Biochem. 251*, 252–261.

[38] Seelig, A., Landwojtowicz, E. (2000). Structure-activity relationship of P-glycoprotein substrates and modifiers. *Eur. J. Pharm. Sci. 12*, 31–40.

[39] Albrecht, B.K., Berry, V., Boezio, A.A., Cao, L., Clarkin, K., Guo, W., Harmange, J.C., Hierl, M., Huang, L., Janosky, B., Knop, J., Malmberg, A., McDermott, J.S., Nguyen, H.Q., Springer, S.K., Waldon, D., Woodin, K., McDonough, S.I. (2008). Discovery and optimization of substituted piperidines as potent, selective, CNS-penetrant alpha4beta2 nicotinic acetylcholine receptor potentiators. *Bioorg. Med. Chem. Lett. 18*, 5209–5212.

20

DESIGNING PERIPHERAL DRUGS FOR MINIMAL BRAIN EXPOSURE

PETER BUNGAY,[1] SHARAN BAGAL,[2] AND ANDY PIKE[1]

[1]Pfizer Neusentis, Great Abington, Cambridge, UK
[2]Worldwide Medicinal Chemistry, Pfizer Neusentis, Great Abington, Cambridge, UK

RATIONALE FOR MINIMIZING BRAIN EXPOSURE OF DRUGS AIMED AT PERIPHERAL TARGETS

Advantages of CNS Restriction

The blood–brain barrier (BBB), resulting from tight junctions between, and various biochemical processes within the endothelial cells of brain capillaries is likely to have evolved as a protective mechanism that limits the access of ingested xenobiotic compounds to tissues of the central nervous system (CNS). As a result, drug molecules are subject to varying degrees of exclusion from the brain (CNS restriction), presenting both challenges and opportunities to drug discovery projects. For CNS targets, considerable drug design effort is often required to overcome the passive and biochemical aspects of the BBB and achieve therapeutically relevant exposure. However, for peripheral drug targets, CNS exposure may be seen as detrimental either because

a) the drug target is also expressed in the CNS where its modulation may evoke undesirable side-effects;
b) drug concentrations in the CNS are sufficiently high to result in off-target pharmacological action, again leading to undesired side effects;

Blood–Brain Barrier in Drug Discovery: Optimizing Brain Exposure of CNS Drugs and Minimizing Brain Side Effects for Peripheral Drugs, First Edition. Edited by Li Di and Edward H. Kerns.

c) in the research phase, CNS side effects can confound the interpretation of pharmacological data in preclinical studies.

Several examples exist where side effects have been attributed to on-target actions of a drug in the CNS that lead to issues of clinical safety and tolerability and may limit dose or patient compliance [1, 2]. Whilst exquisite pharmacological selectivity can address off target activities, it is often difficult to achieve, and it is not always possible to identify all off-target liabilities in advance of preclinical or clinical *in vivo* studies. Therefore, when dealing with peripherally located drug targets, a more generic drug design strategy is to target restricted CNS access whilst maintaining appropriate exposure in peripheral tissues. This chapter will describe how CNS restriction of orally delivered drugs can be designed, evaluated, and achieved in compounds that act as substrates for active efflux transporters in the BBB and will consider the potential risks in employing this approach.

Evidence That CNS Restriction Can Reduce Incidence of Side-Effects

The advantage of restricting access to the CNS when targeting peripheral receptors is illustrated by several examples in the literature.

The first-generation histamine H1 antagonists, including drugs such as diphenhydramine and hydroxyzine, are effective in treatment of allergic disorders, but cause somnolence or sedation due to interaction with H1 receptors in the CNS. Second-generation histamine H1 antagonists, for example, cetirizine, display mild sedation [3], whilst the most recently developed compounds, for example, fexofenadine, have been found to produce relatively little somnolence at therapeutic doses. In Mdr1 knockout (KO) mice that lack expression of the efflux transporter P-glycoprotein (P-gp), the brain penetration of nonsedating antihistamines cetirizine, loratadine, and desloratadine was increased in comparison with wild-type animals, whereas in the cases of the sedating antihistamines diphenhydramine, hydroxyzine, and triprolidine, this difference was not observed [4]. Thus, the improved toleration profile in the nonsedating antihistamines is thought to result mainly from their CNS restriction rather than any pharmacological differentiation [5].

Several antimuscarinic agents used for the treatment of overactive bladder, for example, tolterodine, oxybutynin, darifenacin, and fesoterodine, act by binding to muscarinic receptors in the bladder detrusor muscle. Side effects, such as somnolence, and cognitive effects in older subjects, have been reported for oxybutynin, which is CNS penetrant and, therefore, binds to muscarinic receptors in the CNS. However others, such as darifenacin and 5-hydroxymethyltolterodine (active metabolite of fesoterodine), are not associated with CNS side effects and display CNS restriction [6].

The antidiarrhoeal δ-opioid agonist loperamide is generally free of typical opioid-related side effects (e.g., sedation and respiratory depression) at high clinical doses. Although poor intestinal absorption and relatively low systemic exposure may contribute to this side effect profile, it may nevertheless be at least partly attributable to its low degree of brain penetration [7].

In all of these examples, the ability of the CNS restricted agents to act as substrates for efflux transporters in the BBB underpins their differing degrees of brain penetration and toleration.

PROPERTIES OF THE BBB

Passive Permeability of the BBB

The microvessels that supply blood to the brain provide a barrier to the free exchange of blood-borne solutes due to efficient tight junctions [8] between adjacent brain microvascular endothelial cells (BMECs) that severely restrict the passage of solutes between adjacent cells (paracellular movement), and confer high electrical resistance and low ion conductance [9]. In order to traverse this endothelium, compounds have to cross through the lipophilic environment of the plasma membrane on both the apical (capillary lumen side) and basolateral (brain tissue side) membranes of the BMEC layer (transcellular movement). Therefore, the physicochemical properties of a compound including, lipophilicity, molecular weight, hydrogen bond potential, and polar surface area (PSA), influence rates of diffusion of compounds across the BBB. Some authors have described the passive permeability of the BBB as being lower than other cell layers, due to high membrane lipid packing density [10], whilst others suggest that the BBB has high permeability with similar absorptive capacity to hepatocytes [11]. Although some of the intrinsic properties of the BBB may not be unique, they nevertheless are instrumental in providing a highly effective barrier to diffusion via paracellular and transcellular pathways.

Role of Efflux Transporters in the BBB

A critical feature of BMECs is the presence of ATP-dependent transporter proteins, including P-glycoprotein (P-gp, ABCB1) and breast cancer resistance protein (BCRP, ABCG2), expressed on the apical membrane (blood capillary lumen side) (Fig. 20.1). Compounds acting as substrates for efflux transporter proteins in the BBB are subject to ejection from the BMEC apical membrane and/or cytoplasm [12, 13], resulting in efflux back into the capillary lumen. The roles of P-gp and BCRP as efflux transporters in the BBB has been demonstrated through the use of Mdr1 and Bcrp knockout (KO) studies in mouse [14–18], and rat [19]. As a result of the greater effect of P-gp knockout on the brain distribution of P-gp substrates than of BCRP knockout on BCRP substrates, it has been generally concluded that P-gp plays a quantitatively more important role than BCRP in rodents. More recently, quantitative mass spectrometry has shown that P-gp expression was approximately threefold higher than BCRP in mouse brain microvessels, whereas in human, the levels are comparable [20, 21], which perhaps suggests that the relative contribution of various transporters to efflux effects may be different between species. Roles for other efflux transporters, including multidrug resistance-associated protein 4 (MRP4, ABCC4) and the SLC organic anion transporter OAT3, in the BBB have been suggested on the basis of mouse gene knockout studies. In human, a role for OAT3 has been suggested in the CNS restriction of the antiviral drug Ro64–0802 [22], while MRP4 has been implicated as a transporter involved in limiting the chemotherapeutic action of topotecan in the brain [23]. However, quantified levels of

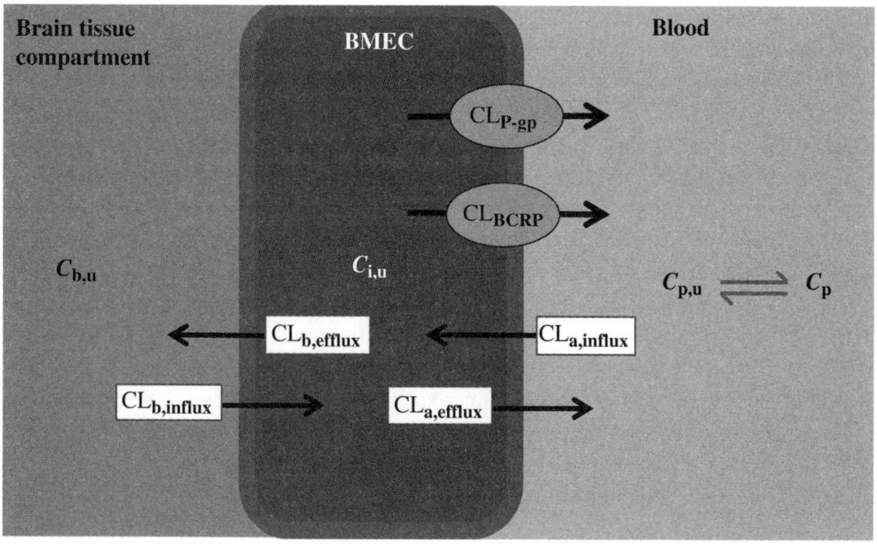

FIGURE 20.1 Illustration of drug movements across the BBB. This diagram considers the distribution of drug across three compartments of blood, the intracellular compartment of brain microvascular endothelial cells (BMEC), and brain tissue. The blood concentration (C_p) is in equilibrium with the unbound plasma concentration ($C_{p,u}$) that is available for distribution across the BMEC apical plasma membrane according to passive influx ($CL_{a,influx}$) and efflux ($CL_{a,efflux}$) clearances. For P-gp and/or BCRP substrates, active efflux clearances ($CL_{P\text{-}gp}$ and CL_{BCRP}) in excess of passive influx will result, at steady state, in unbound intracellular concentrations ($C_{i,u}$) lower than $C_{p,u}$. Passive equilibration from within the BMEC due to passive influx and efflux clearances across the basolateral membrane ($CL_{b,influx}$ and $CL_{b,efflux}$) will then result in unbound brain tissue concentrations ($C_{b,u}$) lower than $C_{p,u}$. According to Kodaira et al., (2010), the unbound brain tissue:plasma concentration ratio, $K_{pu,u}$, will be given by $K_{pu,u} = (CL_{a,influx} \times CL_{b,efflux})/(CL_{b,influx} \times (CL_{a,efflux} + CL_{P\text{-}gp} + CL_{BCRP}))$ (see Ref. 28).

MRP4 in human brain microvessels were found to be 2–3% of those of P-gp and BCRP, with OAT3 below the level of quantification [20]. Combined with a more limited range of substrates for MRP4, this suggests a limited, or at least highly specific role, for these transporters and that P-gp and BCRP are the dominant transporters maintaining CNS restriction.

DESIGN STRATEGY FOR CNS-RESTRICTION OF DRUGS

Contributions of Passive Permeability and Active Efflux

The CNS penetration of a compound depends on the balance of influx and efflux due to passive and active movements across the BBB, so strategies aimed at designing compounds that are excluded from the brain could address both aspects. According to the free drug hypothesis, at steady state, the free drug concentration is the same on both sides of any biomembrane, provided that the drug is not subject to active transport [24, 25]. This means that if efflux rate significantly exceeds rates permitted by passive permeability, asymmetric distribution across the BBB will occur at steady state. Whilst the rate at which steady state across the BBB can be achieved is influenced by passive processes, including permeability and tissue binding [26], the extent of CNS penetration depends upon the rates of active efflux. Therefore, a rational approach to take to achieve CNS restriction in orally administered drugs is to design compounds that are substrates of efflux transporters at the BBB. However, in cases where passive permeability is very low, the time required for steady state to be attained across the BBB may be much longer than the time taken for plasma concentrations to reach steady state due to drug clearance. This suggests that a design strategy based on maintaining low passive permeability may also succeed for an oral drug that is capable of absorption via the paracellular pathway in the intestine as junctions in the gastrointestinal tract are not as tight as in the BBB [27].

Targeting Active Efflux by P-gp and BCRP

The contributions of transport processes to net brain tissue distribution have been analyzed by consideration of the passive and active uptake and efflux clearances across the BBB [28, 29] (Fig. 20.1). This analysis has been applied to explain the greater magnitude of changes in brain distribution of dual P-gp/BCRP substrates observed in Mdr1a,b and Bcrp triple KO mice [17], when compared with results in separate single P-gp or BCRP KO animals. Although the effect of the triple KO has been described as an apparent synergy, it is not thought to require a functional interaction related to the co-localization or catalytic properties of P-gp and BCRP. Rather, by considering the fractional contribution of efflux transporters to steady-state distribution, the apparent synergy can be rationalized without the need to invoke a cooperative interaction [29, 30]. For example, if the fraction of drug excreted (f_e) via P-gp is 0.75 and f_e via BCRP = 0.15, the changes in brain exposure of a P-gp KO would be 1/(1–0.75) = 4-fold and of BCRP would be 1/(1–0.15) = 1.2-fold, whereas the change in the P-gp + BCRP KO would be 1/(1–0.75–0.15) = 10-fold. Whilst these analyses show that it is unlikely that P-gp and BCRP act synergistically in the BBB, the efflux contributions of each would nevertheless still reasonably be expected to be additive. In the absence of knowledge of individual substrate kinetic parameters, the relative contribution of P-gp and BCRP to efflux is difficult to estimate, but the comparable expression of P-gp and BCRP in human brain microvessels suggests the potential for significant contributions from both. Therefore, an ideal approach to maximizing CNS restriction is to target both P-gp and BCRP as it has the potential advantages of obtaining the highest possible efflux effect and of incorporating redundancy that might address the potential for population variation in transporter activity and expression. Polymorphisms of the MDR1 [31] and BCRP [32] genes have been demonstrated that may affect drug disposition. For example, a single nucleotide polymorphism (SNP) at position 3435 of exon 26 (C3435T) has been associated with reduced expression of P-gp in the duodenum [33]. SNPs at this gene locus in combination with others in MDR1 have been linked to increases in loperamide exposure and opioid side effects (pupil size decrease) caused by coadministration of the P-gp inhibitor quinidine [34], although it is not clear to what extent the side effects in this study resulted from increased systemic exposure as opposed to changes in CNS restriction. A general lack of consistency in the observed effects of MDR1 SNPs has been described and the multifactorial influences on drug disposition in relevant clinical studies has been emphasized, as these may confound the interpretation that changes in drug disposition are attributable to MDR1 gene polymorphisms [31, 35]. A BCRP SNP (421C>A) is associated with impaired BCRP expression and with increased exposure of BCRP substrates such as rosuvastatin, possibly as a result of an effect on intestinal absorption [36]. However, as is the case for P-gp, the functional relevance of BCRP polymorphisms at the BBB is not yet established.

Potential for Drug–Drug Interaction at the BBB

A drug–drug interaction (DDI), potentially leading to unwanted CNS penetration, could theoretically arise if a P-gp substrate is co-dosed with a P-gp inhibitor. However, a review of the risk of clinical modulation of efflux transport at the BBB by the International Transporter Consortium concluded that at clinical doses, unbound systemic concentrations of P-gp inhibitors are likely to be too low to inhibit P-gp *in vivo*, and, therefore, there is a low risk of DDI via inhibition of efflux at the BBB [30]. Thus, considering the free drug exposures expected at the BBB, only a very potent P-gp inhibitor could be expected to elicit a significant effect. As systemic unbound concentrations of orally administered drugs are generally well below K_m, K_i, or IC_{50} values for P-gp, both DDI's, or saturation of P-gp at the BBB, are unlikely to occur [30]. Although a clinical study with the potent designed P-gp inhibitor tariquidar (P-gp $K_i = 16$ nM) demonstrated an increase in [^{11}C]-verapamil distribution into human brain (maximal effect 2.7-fold increase), it was delivered by intravenous infusion with the intent of achieving inhibition of P-gp *in vivo* [37]. Increases in CSF:plasma concentration ratio of 1.6- to 5.8-fold and brain uptake ratio of up to 2-fold have been determined in clinical studies of interaction between P-gp substrates and inhibitors, but again, at relatively high doses of precipitant drugs. Estimating the degree of inhibition of P-gp in clinical investigations is often difficult, and it has been suggested that substrate and inhibitor probes other than those investigated so far could display interactions of higher magnitude [38]. Nevertheless, an examination of clinical relevance of P-gp inhibition for loperamide CNS effects at recommended doses [7] suggested a low likelihood of DDI at the BBB.

RISK OF LOW OR VARIABLE ABSORPTION IN THE INTESTINE WHEN TARGETING ACTIVE EFFLUX

Efflux Transport in the Intestinal Epithelium

Bioavailability of orally administered drugs depends on their absorption across the intestinal epithelium (IE). In common with the BBB, the IE contains cellular junctions (although overall allowing more paracellular permeation of small polar molecules than the BBB) and expresses the transporters P-gp and BCRP which can act as a barrier to absorption of xenobiotics, suggesting that building efflux transporter substrate potential into a drug risks reducing its oral bioavailability. There are also instances whereby increasing oral doses of a drug lead to progressive saturation of efflux transporters in the intestine and cause greater than proportional increases in exposure with dose [39, 40]. Nevertheless, at drug doses commonly prescribed for clinical use, the range of drug concentrations likely to exist in the gastrointestinal lumen following an oral dose are in the range over which saturation of P-gp will occur in the IE, assuming that the K_m for P-gp is likely to be in the range 1–100 μM, limiting the influence of P-gp on gastrointestinal absorption [41]. Other factors have been described [42] that would limit the influence of P-gp on absorption, including override of efflux by P-gp by high rates of passive diffusion across a steep concentration gradient. As the structure–activity relationship (SAR) for P-gp substrates displays strong overlap with cytochrome P450 (CYP3A) substrate SAR, it is often difficult to deconvolute the influences of efflux transport and metabolism on bioavailability and DDI between P-gp substrates in the gastrointestinal tract. However, a review of clinical interaction between digoxin, which is not appreciably metabolized by CYP enzymes, and drugs known to be P-gp substrates or inhibitors, showed that few cases (4 out of 123) led to greater than twofold increase in digoxin AUC [43]. Presently, our ability to accurately predict absorption of P-gp and BCRP substrates is limited until more quantitative information on intestinal transporter expression become available that can be combined with authentic inputs for modeling and simulation. So, whilst the risk of low or variable absorption due to efflux transport may be considered to be low in theory, it is recommended that evaluation of absorption is incorporated within the screen sequence, particularly in cases where low solubility and/or slow dissolution may limit effective concentrations in the gastrointestinal tract.

Balancing Physicochemical Properties for Oral Absorption and CNS Restriction

The physicochemical properties of orally administered compounds aimed at peripheral drug targets should be compatible with those required for absorption across the intestinal epithelium and be appropriately balanced to achieve CNS restriction. The familiar "rule of five" based on analysis of advanced clinical candidates identified molecular weight (MW) <500, lipophilicity, defined by logP or calculated logP (ClogP), of <5, number of hydrogen bond donors (HBD) <5 and number of hydrogen bond acceptors (HBA) <10 as favoring development of orally active drugs. Similarly, MW <500, polar surface area (PSA) <140 Å^2 and number of rotatable bonds of <10 have been associated with good oral absorption [44]. Separate retrospective analyses have associated good CNS penetration with median ClogP of 2.8, MW of 305.3, PSA of 44.8 Å^2, HBD of 1 and pK_a of 8.4 [45] and MW <450 and PSA <70 Å^2 [46]. Conversely, for non-CNS drugs, target ranges of PSA of 60–120 Å^2 and HBD or 2–3 have been suggested [2]. Hence, it is possible to identify an area of compatibility of MW of 450–500 and PSA of 70–140 Å^2 that favors restriction from the CNS whilst allowing good absorption in the gastrointestinal tract. The impact of physicochemical characteristics on passive permeability cannot easily be separated from their effects on the ability of compounds to act as efflux transporter substrates. An analysis of 45,000 P-gp efflux data points tested in the MDR-1 MDCK (P-gp) assay [1] suggested the probability of a compound being a P-gp substrate increases significantly above MW of 400 and PSA of 80 Å^2 which is consistent with P-gp possessing large binding site(s) that interact with HBA and HBD. The ability of compounds to act as substrates of P-gp can therefore be optimized using specific SAR for P-gp [47], which can often involve positional effects of HBA.

Evidence That Intestinal Absorption Can Be Maintained with CNS Restriction

There are several examples of drugs that are substrates of P-gp and/or BCRP, and are CNS restricted but that possess good oral bioavailability, including the antitumor agent imatinib [48] and the antiviral protease inhibitors ritonavir and indinavir [1]. Imatinib possesses a MW of 493, PSA of 86 Å^2, 6 HBA, and 2 HBD and is a substrate of P-gp and BCRP *in vitro* [17]. The high oral bioavailability of imatinib in human (97–98%) [49] is consistent with complete absorption whilst its brain penetration is restricted, as determined in mice by brain:plasma concentration ratio, in baboons by [^{11}C] positron emission tomography [50] and in nonhuman primates by cerebrospinal fluid (CSF):plasma concentration ratio [51]. Indinavir (MW 613, PSA 118 Å^2, 7 HBA, and 4 HBD) is a substrate of P-gp and was shown to be restricted in its brain penetration in mice as a result of P-gp mediated efflux [52]. Nevertheless, indinavir has estimated intestinal absorption following an oral dose of approximately 80% [53].

A strategy to deliver CNS restriction with good oral absorption is suggested whereby efflux, mediated by P-gp and BCRP, is balanced against passive influx at a level that allows rates of transporter-mediated efflux to effectively restrict penetration of the BBB but still allows good absorption across the IE. This approach has been used to design a series of CNS restricted histamine H3 antagonists with high oral bioavailability [1] in order to minimize clinical adverse events, such as insomnia, that would otherwise be observed. For example, in the H3 antagonist PF-3731237 (MW 395.5, PSA 85 Å^2, 5 HBA, and 2 HBD), optimization of passive permeability and introduction of P-gp and BCRP substrate activity enabled CNS restriction to be achieved in tissue partition experiments in rat (unbound brain:plasma ratio of 0.1) whilst maintaining good oral bioavailability (54%). Brain receptor occupancy data confirmed that CNS restriction was sustained over 7 days of dosing and electroencephalography data demonstrated the desired therapeutic index for efficacy over insomnia [1].

SCREEN SEQUENCES

In vitro Assays for Transporters

Transcellular permeability assays are widely deployed early in drug discovery programs as relatively rapid means of evaluating compounds as substrates of drug transporters, and to help develop SARs. Assays for P-gp and BCRP are conducted in polarized adherent cell monolayers cultured to confluence on semi-permeable membrane that separates apical (A) and basolateral (B) incubation chambers. The immortalized cell line MDCK (Madin Darby canine kidney) is commonly employed, having been transfected with the MDR1 or BCRP gene to express either P-gp or BCRP, respectively [15, 17]. The rate of test compound transport across the cell monolayer from donor chamber A or B into the corresponding acceptor chamber is measured, usually by LC–MS detection. Apparent permeability (P_{app}) is determined in each of the A–B and B–A directions and an efflux ratio ($ER = P_{app}$ B–A/P_{app} A–B) is calculated to quantify asymmetry in flux due to transporter activity. In the case of P-gp, results from the transfected cell line MDCK-MDR1 have been compared with brain penetration data in rodents, and an ER of >2.5 proposed to give an experimentally robust indication that compounds are P-gp substrates [15, 54]. However, in practice the numerical value and precision of ER judged to represent efflux substrates will need to be determined by each laboratory undertaking the transwell assay and the relationship between ER and *in vivo* CNS penetration investigated. A subpopulation of MDCKII cells with low expression of endogenous canine P-gp (MDCK-LE, formerly known as RRCK) has also been developed [55] that can be used to rank order passive permeability of compounds.

In Silico Models

A commonly used method of designing a P-gp substrate is to develop an understanding of the physicochemical space most likely to lead to P-gp efflux (MW>400, PSA>80) combined with leveraging the historical P-gp data generated in the assay of choice (e.g., MDCK-MDR1) to create an *in silico* mathematical model. *In silico* models, such as Cubist modeling methodology in which the entire P-gp dataset is used to create a collection of rules, where each rule has an associated multivariate linear model [56], can be applied prospectively to predict the likelihood of a compound being a P-gp substrate prior to synthesis of the compound. Compounds being evaluated for synthesis can be prioritized based on the likely P-gp activity predicted by the *in silico* model and, once synthesized and assessed in the P-gp assay, the new efflux data are fed back into the model to further enhance its predictability. Application of such *in silico* models allows a mathematical consideration of P-gp SAR when designing P-gp substrates.

Measurement of CNS Penetration *In Vivo*

Several reviews have described the application of methods for studying delivery of drugs to the brain [1, 57, 58]. The extent of CNS penetration can routinely be estimated in rodent species (rat and mouse), and efflux transporter gene knockout animals can be used to assess transporter contribution to exclusion. Ideally, test compound is administered such that concentrations in brain and blood or plasma are measured under conditions of steady-state distribution between blood and CNS tissue. Hence, intravenous infusion may be an optimum delivery method, although single dose oral, subcutaneous, or intraperitoneal administration are often combined with tissue sampling over a time course to provide tissue concentrations integrated over time (area under curve, AUC). Following sacrifice of animals, the brain is removed and, as it may not be possible to ensure complete removal of blood from brain tissue unless a perfusion of the brain tissue is undertaken, a correction for residual blood has been suggested [59]. In accord with the free drug hypothesis, comparison of unbound concentrations of compound in brain and plasma is needed to authentically estimate pharmacologically relevant tissue partition [25, 60, 61]. Unbound concentrations in brain ($C_{b,u}$) and plasma ($C_{p,u}$) are estimated by multiplying total measured plasma and brain concentrations by the unbound fraction (F_u) in each tissue measured separately using equilibrium dialysis [62]. The unbound brain:plasma concentration ratio, $Kp_{u,u}$ ($C_{b,u}/C_{p,u}$) is used as an index of CNS penetration, whereby values significantly lower than 1 indicate CNS restriction. CSF sampling has often featured as a surrogate $C_{b,u}$ estimation [63, 64]. Although CSF is a simpler matrix for bioanalysis than blood or plasma, sampling is prone to blood contamination that can result in a substantial overprediction of CNS penetration. Moreover, transporter expression and orientation at the blood–CSF barrier and the BBB differ [63] so that CSF: $C_{p,u}$ data may not accurately represent CNS penetration [30, 65] and has often been found to overpredict [60, 64, 66]. However, CSF analysis generally indicates impairment in brain exposure when P-gp is involved and remains one of the few means available to assess CNS penetration in human, bearing in mind that drug concentrations would usually be determined in lumbar spine CSF that may not directly reflect brain concentrations [63, 65]. *In vivo* microdialysis may offer a direct approach to measuring $C_{b,u}$ with continuous sampling to provide concentration/time profiles of a drug within a distinct region of the brain [64]. However, the technique can only be applied preclinically and can be very technically challenging, requiring specialist equipment and

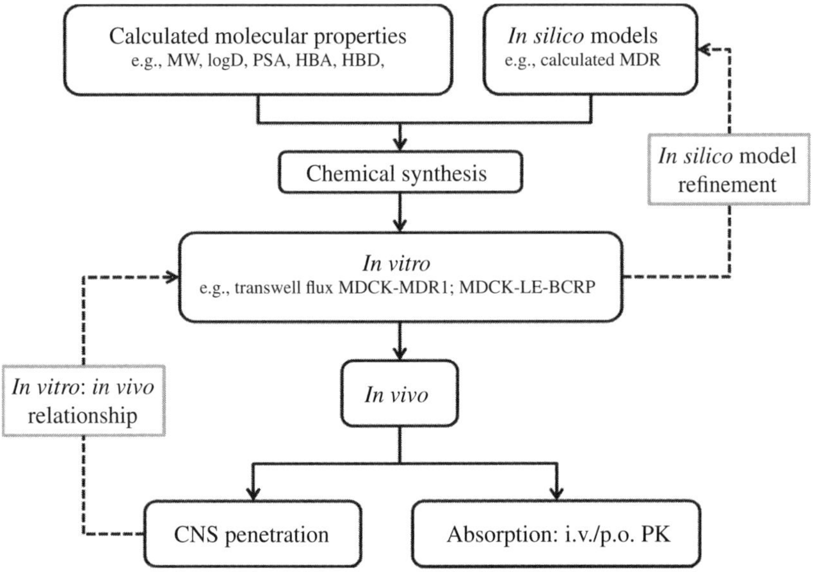

FIGURE 20.2 Proposed screen sequence to identify CNS-restricted compounds with high oral absorption. The translation of experimentally determined parameters between elements of the screening cascade is established under the conditions of the individual laboratory, and potentially of each chemical series investigated. For example, a relationship between efflux ratio in an *in vitro* P-gp assay and brain penetration *in vivo* may be established and subsequently used to build and refine an *in silico* model for P-gp interaction. As screening cycles are completed, an increase in the ability of the sequence to predict CNS restriction is expected.

expertise. From the repertoire of *in silico*, *in vitro* and *in vivo* techniques described, a screen sequence is suggested that addresses the design and evaluation of drugs with minimal brain exposure (Fig. 20.2).

CONCLUSION

The design of compounds with affinity for the BBB drug efflux transporters P-gp and BCRP is a pragmatic strategy to address drug safety by maximizing CNS restriction when the drug target is peripherally expressed. Recent assessment of the risk of clinical modulation of efflux transport at the BBB suggests a low likelihood of this occurring and the occurrence of genetic variation within P-gp and BCRP has so far not been clearly linked with functionally significant effects on CNS penetration in humans. Screen sequences can be designed that incorporate prescreening (*in silico*), *in vitro* and *in vivo* methods. This enables preselection of compounds for synthesis prior to evaluation as transporter substrates and demonstration of CNS restriction *in vivo*. Whilst the likelihood of poor or variable intestinal absorption due to efflux transport is theoretically low at drug doses commonly employed, this risk should be evaluated on a case-by-case basis.

REFERENCES

[1] Cole, S.; Bagal, S.; El-Kattan, A.; Fenner, K.; Hay, T.; Kempshall, S.; Lunn, G.; Varma, M.; Stupple, P.; Speed, W., Full efficacy with no CNS side-effects: unachievable panacea or reality?? DMPK considerations in design of drugs with limited brain penetration. *Xenobiotica* **2012,** 42, 11–27.

[2] Wager, T. T.; Liras, J. L.; Mente, S.; Trapa, P., Strategies to minimize CNS toxicity: in vitro high-throughput assays and computational modeling. *Expert Opin. Drug Metab. Toxicol.* **2012,** *8*, 531–542.

[3] Tashiro, M.; Kato, M.; Miyake, M.; Watanuki, S.; Funaki, Y.; Ishikawa, Y.; Iwata, R.; Yanai, K., Dose dependency of brain histamine H1 receptor occupancy following oral administration of cetirizine hydrochloride measured using PET with [11C]doxepin. *Hum. Psychopharmacol.* **2009,** *24*, 540–548.

[4] Chen, C.; Hanson, E.; Watson, J. W.; Lee, J. S., P-glycoprotein limits the brain penetration of nonsedating but not sedating H1-antagonists. *Drug Metab. Dispos.* **2003,** *31*, 312–318.

[5] Chishty, M.; Reichel, A.; Siva, J.; Abbott, N. J.; Begley, D. J., Affinity for the P-glycoprotein efflux pump at the blood-brain barrier may explain the lack of CNS side-effects of modern antihistamines. *J. Drug Target.* **2001,** *9*, 223–228.

[6] Callegari, E.; Malhotra, B.; Bungay, P. J.; Webster, R.; Fenner, K. S.; Kempshall, S.; LaPerle, J. L.; Michel, M. C.; Kay, G. G., A comprehensive non-clinical evaluation of the CNS penetration potential of antimuscarinic agents for the treatment of overactive bladder. *Br. J. Clin. Pharmacol.* **2011,** *72*, 235–246.

[7] Vandenbossche, J.; Huisman, M.; Xu, Y.; Sanderson-Bongiovanni, D.; Soons, P., Loperamide and P-glycoprotein inhibition: assessment of the clinical relevance. *J. Pharm. Pharmacol.* **2010,** *62*, 401–412.

[8] Kniesel, U.; Wolburg, H., Tight junctions of the blood-brain barrier. *Cell. Mol. Neurobiol.* **2000,** *20*, 57–76.

[9] Abbott, N. J.; Patabendige, A. A. K.; Dolman, D. E. M.; Yusof, S. R.; Begley, D. J., Structure and function of the blood-brain barrier. *Neurobiol. Dis.* **2010,** *37*, 13–25.

[10] Seelig, A., The role of size and charge for blood-brain barrier permeation of drugs and fatty acids. *J. Mol. Neurosci.* **2007,** *33*, 32–41.

[11] Fagerholm, U., The highly permeable blood-brain barrier: an evaluation of current opinions about brain uptake capacity. *Drug Discov. Today* **2007,** *12*, 1076–1082.

[12] Schinkel, A. H., P-Glycoprotein, a gatekeeper in the blood-brain barrier. *Adv. Drug Deliv. Rev.* **1999,** *36*, 179–194.

[13] Hammarlund-Udenaes, M.; Friden, M.; Syvaenen, S.; Gupta, A., On the rate and extent of drug delivery to the brain. *Pharm. Res.* **2008,** *25*, 1737–1750.

[14] Schinkel, A. H.; Wagenaar, E.; Mol, C. A. A. M.; van, D. L., P-glycoprotein in the blood-brain barrier of mice influences the brain penetration and pharmacological activity of many drugs. *J. Clin. Invest.* **1996,** *97*, 2517–2524.

[15] Feng, B.; Mills, J. B.; Davidson, R. E.; Mireles, R. J.; Janiszewski, J. S.; Troutman, M. D.; de, M. S. M., In vitro P-glycoprotein assays to predict the in vivo interactions of P-glycoprotein with drugs in the central nervous system. *Drug Metab. Dispos.* **2008,** *36*, 268–275.

[16] Doran, A.; Obach, R. S.; Smith, B. J.; Hosea, N. A.; Becker, S.; Callegari, E.; Chen, C.; Chen, X.; Choo, E.; Cianfrogna, J.; Cox, L. M.; Gibbs, J. P.; Gibbs, M. A.; Hatch, H.; Hop, C. E. C. A.; Kasman, I. N.; LaPerle, J.; Liu, J.; Liu, X.; Logman, M.; Maclin, D.; Nedza, F. M.; Nelson, F.; Olson, E.; Rahematpura, S.; Raunig, D.; Rogers, S.; Schmidt, K.; Spracklin, D. K.; Szewc, M.; Troutman, M.; Tseng, E.; Tu, M.; Van, D. J. W.; Venkatakrishnan, K.; Walens, G.; Wang, E. Q.; Wong, D.; Yasgar, A. S.; Zhang, C., The impact of P-glycoprotein on the disposition of drugs targeted for indications of the central nervous system: evaluation using the MDR1A/1B knockout mouse model. *Drug Metab. Dispos.* **2005,** *33*, 165–174.

[17] Zhou, L.; Schmidt, K.; Nelson, F. R.; Zelesky, V.; Troutman, M. D.; Feng, B., The effect of breast cancer resistance protein and P-glycoprotein on the brain penetration of flavopiridol, imatinib mesylate (Gleevec), prazosin, and 2-methoxy-3-(4-(2-(5-methyl-2-phenyloxazol-4-yl)ethoxy)phenyl)propanoic acid (PF-407288) in mice. *Drug Metab. Dispos.* **2009,** *37*, 946–955.

[18] Enokizono, J.; Kusuhara, H.; Ose, A.; Schinkel, A. H.; Sugiyama, Y., Quantitative investigation of the role of breast cancer resistance protein (Bcrp/Abcg2) in limiting brain and testis penetration of xenobiotic compounds. *Drug Metab. Dispos.* **2008,** *36*, 995–1002.

[19] Bundgaard, C.; Jensen, C. J. N.; Garmer, M., Species comparison of in vivo P-glycoprotein-mediated brain efflux using mdr1a-deficient rats and mice. *Drug Metab. Dispos.* **2012,** *40*, 461–466.

[20] Uchida, Y.; Ohtsuki, S.; Katsukura, Y.; Ikeda, C.; Suzuki, T.; Kamiie, J.; Terasaki, T., Quantitative targeted absolute proteomics of human blood-brain barrier transporters and receptors. *J. Neurochem.* **2011,** *117*, 333–345.

[21] Shawahna, R.; Uchida, Y.; Decleves, X.; Ohtsuki, S.; Yousif, S.; Dauchy, S.; Jacob, A.; Chassoux, F.; Daumas-Duport, C.; Couraud, P.-O.; Terasaki, T.; Scherrmann, J.-M., Transcriptomic and quantitative proteomic analysis of transporters and drug metabolizing enzymes in freshly isolated human brain microvessels. *Mol. Pharm.* **2011,** *8*, 1332–1341.

[22] Ose, A.; Ito, M.; Kusuhara, H.; Yamatsugu, K.; Kanai, M.; Shibasaki, M.; Hosokawa, M.; Schuetz, J. D.; Sugiyama, Y., Limited brain distribution of [3R,4R,5S]-4-acetamido-5-amino-3-(1-ethylpropoxy)-1-cyclohexene-1-carboxylate phosphate (Ro 64-0802), a pharmacologically active form of oseltamivir, by active efflux across the blood-brain barrier mediated by organic anion transporter 3 (Oat3/Slc22a8) and multidrug resistance-associated protein 4 (Mrp4/Abcc4). *Drug Metab. Dispos.* **2009,** *37*, 315–321.

[23] Leggas, M.; Adachi, M.; Scheffer, G. L.; Sun, D.; Wielinga, P.; Du, G.; Mercer, K. E.; Zhuang, Y.; Panetta, J. C.; Johnston, B.; Scheper, R. J.; Stewart, C. F.; Schuetz, J. D., Mrp4 confers resistance to topotecan and protects the brain from chemotherapy. *Mol. Cell. Biol.* **2004,** *24*, 7612–7621.

[24] Smith, D. A.; Di, L.; Kerns, E. H., The effect of plasma protein binding on in vivo efficacy: misconceptions in drug discovery. *Nat. Rev. Drug Discov.* **2010,** *9*, 929–939.

[25] Di, L.; Rong, H.; Feng, B., Demystifying brain penetration in central nervous system drug discovery. *J. Med. Chem.* **2013,** *56*, 2–12.

[26] Liu, X.; Smith, B. J.; Chen, C.; Callegari, E.; Becker, S. L.; Chen, X.; Cianfrogna, J.; Doran, A. C.; Doran, S. D.; Gibbs, J. P.; Hosea, N.; Liu, J.; Nelson, F. R.; Szewc, M. A.; Van, D. J., Use of a physiologically based pharmacokinetic model to study the time to reach brain equilibrium: an experimental analysis of the role of blood-brain barrier permeability, plasma protein binding, and brain tissue binding. *J. Pharmacol. Exp. Ther.* **2005,** *313*, 1254–1262.

[27] Anderson, J. M.; Van, I. C. M., Physiology and function of the tight junction. *Cold Spring Harb. Perspect. Biol.* **2009,** *1*, a002584.

[28] Kusuhara, H.; Sugiyama, Y., In vitro-in vivo extrapolation of transporter-mediated clearance in the liver and kidney. *Drug Metab. Pharmacokinet.* **2009,** *24*, 37–52.

[29] Kodaira, H.; Kusuhara, H.; Ushiki, J.; Fuse, E.; Sugiyama, Y., Kinetic analysis of the cooperation of P-glycoprotein (P-gp/Abcb1) and breast cancer resistance protein (Bcrp/Abcg2) in limiting the brain and testis penetration of erlotinib, flavopiridol, and mitoxantrone. *J. Pharmacol. Exp. Ther.* **2010,** *333*, 788–796.

[30] Kalvass, J. C.; Polli, J. W.; Bourdet, D. L.; Feng, B.; Huang, S. M.; Liu, X.; Smith, Q. R.; Zhang, L. K.; Zamek-Gliszczynski, M. J., Why clinical modulation of efflux transport at the human blood-brain barrier Is unlikely: the ITC evidence-based position. *Clin. Pharmacol. Ther. (N. Y., NY, U. S.)* **2013,** *94*, 80–94.

[31] Marzolini, C.; Paus, E.; Buclin, T.; Kim, R. B., Polymorphisms in human MDR1 (P-glycoprotein): recent advances and clinical relevance. *Clin. Pharmacol. Ther. (St. Louis, MO, U. S.)* **2004,** *75*, 13–33.

[32] Giacomini, K. M.; Balimane, P. V.; Cho, S. K.; Eadon, M.; Edeki, T.; Hillgren, K. M.; Huang, S. M.; Sugiyama, Y.; Weitz, D.; Wen, Y.; Xia, C. Q.; Yee, S. W.; Zimdahl, H.; Niemi, M., International transporter consortium commentary on clinically important transporter polymorphisms. *Clin. Pharmacol. Ther. (N. Y., NY, U. S.)* **2013,** *94*, 23–26.

[33] Hoffmeyer, S.; Burk, O.; Von, R. O.; Arnold, H. P.; Brockmoller, J.; Johne, A.; Cascorbi, I.; Gerloff, T.; Roots, I.; Eichelbaum, M.; Brinkmann, U., Functional polymorphisms of the human multidrug-resistance gene: multiple sequence variations and correlation of one allele with P-glycoprotein expression and activity in vivo. *Proc. Natl. Acad. Sci. U. S. A.* **2000,** *97*, 3473–3478.

[34] Skarke, C.; Jarrar, M.; Schmidt, H.; Kauert, G.; Langer, M.; Geisslinger, G.; Loetsch, J., Effects of ABCB1 (multidrug resistance transporter) gene mutations on disposition and central nervous effects of loperamide in healthy volunteers. *Pharmacogenetics* **2003,** *13*, 651–660.

[35] Cascorbi, I., Role of pharmacogenetics of ATP-binding cassette transporters in the pharmacokinetics of drugs. *Pharmacol. Ther.* **2006,** *112*, 457–473.

[36] Keskitalo, J. E.; Zolk, O.; Fromm, M. F.; Kurkinen, K. J.; Neuvonen, P. J.; Niemi, M., ABCG2 polymorphism markedly affects the pharmacokinetics of atorvastatin and rosuvastatin. *Clin. Pharmacol. Ther. (N. Y., NY, U. S.)* **2009,** *86*, 197–203.

[37] Bauer, M.; Zeitlinger, M.; Karch, R.; Matzneller, P.; Stanek, J.; Jaeger, W.; Boehmdorfer, M.; Wadsak, W.; Mitterhauser, M.; Bankstahl, J. P.; Loescher, W.; Koepp, M.; Kuntner, C.; Mueller, M.; Langer, O., Pgp-mediated interaction between (R)-[11C]verapamil and tariquidar at the human blood-brain barrier: a comparison with rat data. *Clin. Pharmacol. Ther. (N. Y., NY, U. S.)* **2012,** *91*, 227–233.

[38] Eyal, S.; Hsiao, P.; Unadkat, J. D., Drug interactions at the blood-brain barrier: fact or fantasy? *Pharmacol. Ther.* **2009,** *123*, 80–104.

[39] Tubic, M.; Wagner, D.; Spahn-Langguth, H.; Bolger, M. B.; Langguth, P., In Silico Modeling of Non-linear drug absorption for the P-gp substrate talinolol and of consequences for the resulting pharmacodynamic effect. *Pharm. Res.* **2006,** *23*, 1712–1720.

[40] Chiou, W. L.; Chung, S. M.; Wu, T. C.; Ma, C., A comprehensive account on the role of efflux transporters in the gastrointestinal absorption of 13 commonly used substrate drugs in humans. *Int. J. Clin. Pharmacol. Ther.* **2001,** *39*, 93–101.

[41] Tachibana, T.; Kato, M.; Sugiyama, Y., Prediction of nonlinear intestinal absorption of CYP3A4 and P-glycoprotein substrates from their in vitro km values. *Pharm. Res.* **2012,** *29*, 651–668.

[42] Lin, J. H., How significant is the role of P-glycoprotein in drug absorption and brain uptake? *Drugs Today* **2004,** *40*, 5–22.

[43] Fenner, K. S.; Troutman, M. D.; Kempshall, S.; Cook, J. A.; Ware, J. A.; Smith, D. A.; Lee, C. A., Drug-drug interactions mediated through P-glycoprotein: clinical relevance and in vitro-in vivo correlation using digoxin as a probe drug. *Clin. Pharmacol. Ther. (N. Y., NY, U. S.)* **2009,** *85*, 173–181.

[44] Veber, D. F.; Johnson, S. R.; Cheng, H.-Y.; Smith, B. R.; Ward, K. W.; Kopple, K. D., Molecular properties that influence the oral bioavailability of drug candidates.*J. Med. Chem.* **2002,** *45*, 2615–2623.

[45] Wager, T. T.; Chandrasekaran, R. Y.; Hou, X.; Troutman, M. D.; Verhoest, P. R.; Villalobos, A.; Will, Y., Defining desirable central nervous system drug space through the alignment of molecular properties, in vitro ADME, and safety attributes. *ACS Chem. Neurosci.* **2010,** *1*, 420–434.

[46] Testa, B.; van de Waterbeemd, H.; Folkers, G.; Guy, R.; Editors, Pharmacokinetic Optimization in Drug Research: Biological, Physicochemical, and Computational Strategies. (Proceedings of the Second Symposium held March 2000, LogP2000 at the University of Lausanne). Verlag Helvetica Chimica Acta, Zürich: *2001*, 51–64.

[47] Hitchcock, S. A., Structural modifications that alter the P-glycoprotein efflux properties of compounds. *J. Med. Chem.* **2012,** *55*, 4877–4895.

[48] Peng, B.; Lloyd, P.; Schran, H., Clinical pharmacokinetics of imatinib. *Clin. Pharmacokinet.* **2005,** *44*, 879–894.

[49] Peng, B.; Dutreix, C.; Mehring, G.; Hayes, M. J.; Ben-Am, M.; Seiberling, M.; Pokorny, R.; Capdeville, R.; Lloyd, P., Absolute bioavailability of imatinib (Glivec) orally versus intravenous infusion. *J. Clin. Pharmacol.* **2004,** *44*, 158–162.

[50] Kil, K.-E.; Ding, Y.-S.; Lin, K.-S.; Alexoff, D.; Kim, S. W.; Shea, C.; Xu, Y.; Muench, L.; Fowler, J. S., Synthesis and positron emission tomography studies of carbon-11-labeled imatinib (Gleevec). *Nucl. Med. Biol.* **2007,** *34*, 153–163.

[51] Neville, K.; Parise, R. A.; Thompson, P.; Aleksic, A.; Egorin, M. J.; Balis, F. M.; McGuffey, L.; McCully, C.; Berg, S. L.; Blaney, S. M., Plasma and cerebrospinal fluid pharmacokinetics of imatinib after administration to nonhuman primates. *Clin. Cancer Res.* **2004,** *10* (Copyright (C) 2014 American Chemical Society (ACS). All Rights Reserved.), 2525–2529.

[52] Kim, R. B.; Fromm, M. F.; Wandel, C.; Leake, B.; Wood, A. J. J.; Roden, D. M.; Wilkinson, G. R., The drug transporter P-glycoprotein limits oral absorption and brain entry of HIV-1 protease inhibitors. *J. Clin. Invest.* **1998,** *101*, 289–294.

[53] Williams, G. C.; Sinko, P. J., Oral absorption of the HIV protease inhibitors: a current update. *Adv. Drug Deliv. Rev.* **1999,** *39*, 211–238.

[54] Doan, K. M. M.; Humphreys, J. E.; Webster, L. O.; Wring, S. A.; Shampine, L. J.; Serabjit-Singh, C. J.; Adkison, K. K.; Polli, J. W., Passive permeability and P-glycoprotein-mediated efflux differentiate central nervous system (CNS) and non-CNS marketed drugs. *J. Pharmacol. Exp. Ther.* **2002,** *303*, 1029–1037.

[55] Di, L.; Whitney-Pickett, C.; Umland, J. P.; Zhang, H.; Zhang, X.; Gebhard, D. F.; Lai, Y.; Federico, J. J.; Davidson, R. E.; Smith, R.; Reyner, E. L.; Lee, C.; Feng, B.; Rotter, C.; Varma, M. V.; Kempshall, S.; Fenner, K.; El-kattan, A. F.; Liston, T. E.; Troutman, M. D., Development of a new permeability assay using low-efflux MDCKII cells. *J. Pharm. Sci.* **2011,** *100*, 4974–4985.

[56] Keefer, C. E.; Kauffman, G. W.; Gupta, R. R., Interpretable, probability-based confidence metric for continuous quantitative structure-activity relationship models. *J. Chem. Inf. Model.* **2013,** *53*, 368–383.

[57] Di, L.; Kerns, E. H.; Carter, G. T., Strategies to assess blood-brain barrier penetration. *Expert Opin. Drug Discov.* **2008,** *3*, 677–687.

[58] Hammarlund-Udenaes, M.; Bredberg, U.; Friden, M., Methodologies to assess brain drug delivery in lead optimization. *Curr. Top. Med. Chem. (Sharjah, United Arab Emirates)* **2009,** *9*, 148–162.

[59] Friden, M.; Ljungqvist, H.; Middleton, B.; Bredberg, U.; Hammarlund-Udenaes, M., Improved measurement of drug exposure in the brain using drug-specific correction for residual blood. *J. Cereb. Blood Flow Metab.* **2010,** *30*, 150–161.

[60] Kalvass, J. C.; Maurer, T. S., Influence of nonspecific brain and plasma binding on CNS exposure: implications for rational drug discovery. *Biopharm. Drug Dispos.* **2002,** *23*, 327–338.

[61] Read, K. D.; Braggio, S., Assessing brain free fraction in early drug discovery. *Expert Opin. Drug Metab. Toxicol.* **2010,** *6*, 337–344.

[62] Waters, N. J.; Jones, R.; Williams, G.; Sohal, B., Validation of a rapid equilibrium dialysis approach for the measurement of plasma protein binding. *J. Pharm. Sci.* **2008,** *97*, 4586–4595.

[63] Lin, J. H., CSF as a surrogate for assessing CNS exposure: an industrial perspective. *Curr. Drug Metab.* **2008,** *9*, 46–59.

[64] Liu, X.; Van, N. K.; Yeo, H.; Vilenski, O.; Weller, P. E.; Worboys, P. D.; Monshouwer, M., Unbound drug concentration in brain homogenate and cerebral spinal fluid at steady state as a surrogate for unbound concentration in brain interstitial fluid. *Drug Metab. Dispos.* **2009,** *37*, 787–793.

[65] de, L. E. C. M.; Danhof, M., Considerations in the use of cerebrospinal fluid pharmacokinetics to predict brain target concentrations in the clinical setting: implications of the barriers between blood and brain. *Clin. Pharmacokinet.* **2002,** *41*, 691–703.

[66] Friden, M.; Winiwarter, S.; Jerndal, G.; Bengtsson, O.; Wan, H.; Bredberg, U.; Hammarlund-Udenaes, M.; Antonsson, M., Structure-brain exposure relationships in rat and human using a novel data set of unbound drug concentrations in brain interstitial and cerebrospinal fluids. *J. Med. Chem.* **2009,** *52*, 6233–6243.

21

CASE STUDIES OF NON-CNS DRUGS TO MINIMIZE BRAIN PENETRATION—NONSEDATIVE ANTIHISTAMINES

Andrew Crowe

School of Pharmacy and CHIRI-Biosciences, Curtin University, Perth, WA, Australia

INTRODUCTION

Antihistamines are a popular class of medicines in the modern era of ever-increasing allergies. Although food allergies are on the rise in many countries [1], the annual spring onslaught of allergic rhinitis, urticaria, hives, or other peripheral inflammatory conditions brought on by huge histamine releases in the body brings with it increasing visits to the pharmacy for antihistamine therapies. One of the reasons that antihistamines have become so popular in the past 15 years is that one of the largest debilitating adverse effects has been largely overcome with recent iterations of these drugs [2, 3]. Sedation along with cognitive impairment were two of the key limiting adverse effects preventing widespread use by sufferers who need to function for work and leisure activities after seeking remedies for allergic conditions.

Antihistamines, or more accurately H_1 receptor reverse agonists, but still commonly referred to as H_1 receptor blockers, have been with us for over 60 years. There are three other receptor subclasses that histamine can bind in the body, but only histamine activation of the H_1 receptors causes the majority of allergic conditions we see clinically. Until the 1980s, the only antihistamines available, of which a number are still available today, were ones that crossed the blood–brain barrier (BBB) and caused CNS effects due to H_1 receptors being widely dispersed throughout parenchymal brain tissue as well as in the peripheral organs and blood vessels [4]. Only the

Blood–Brain Barrier in Drug Discovery: Optimizing Brain Exposure of CNS Drugs and Minimizing Brain Side Effects for Peripheral Drugs, First Edition. Edited by Li Di and Edward H. Kerns.

cerebellum and midbrain lack the receptor in any significant quantity [5]. The role of the H_1 receptor in the brain is not to induce inflammation, as in the periphery, but instead to trigger pathways that change aspects of cognition, including suppressing the waking state, as well as some thermal regulation, learning behaviors, and some contribution to emotional state [2, 6].

Although the sedating antihistamines have been with us for a long time, clinical trials done when these drugs were released lacked the scientific vigor, such as appropriate controls, that would be expected today. It was not until clinical studies were done for the later second-generation antihistamines, with first-generation ones used as positive controls for CNS effects, that we could statistically interpret the extent of cognition based effects of both first- and second-generation antihistamines after targeting CNS receptors.

BACKGROUND TO ANTIHISTAMINE DEVELOPMENT

The first-generation antihistamines can be characterized within five classes. There are the ethylenediamines, such as pyrilamine; ethanolamine esters, such as diphenhydramine; and doxylamine. There are also the propylamines such as pheniramine and triprolidine and the cyclizines, which include hydroxyzine as well as some tricyclics such as cyproheptadine. The second-generation antihistamines are even more varied in their groupings (Fig. 21.1). Some of the more recent second-generation antihistamines are modifications of older antihistamines, as illustrated using the arrows shown in Figure 21.1, which designate the precursor and active metabolite formed. For example, levocetirizine is the R isoform of the stock racemic drug that is cetirizine, which itself is a metabolite of a sedating antihistamine, hydroxyzine. Desloratadine is a metabolite of loratadine and fexofenadine is a metabolite of terfenadine, although we no longer use terfenadine due to its unacceptable risk of cardiac events.

With the universal use of less-sedating antihistamines since the early 1990s, at first glance it would appear as an anomaly that the first-generation antihistamines are still available for use today. However, what may be an adverse effect when treating peripheral inflammation can be useful in its own right when sedation is required. Patients with insomnia benefit from administration of diphenhydramine or doxylamine [7]. H_1 receptors are also present at the vestibular apparatus [8], and activation of these receptors has a role in emetic behavior, especially relating to motion sickness; so, in addition to those mentioned earlier, dimenhydrinate and promethazine are useful for antiemetic action [9]. Second-generation antihistamines have no effect on the vestibular apparatus [10]. Some of the early antihistamines were not very specific for the H_1 receptor and would also bind to muscarinic receptors and occasionally to serotonin receptors, adding to the adverse effect profile of these drugs.

Psychomotor and Sedation Analysis

Reports attempting to piece together the CNS effects of second-generation antihistamines relative to the first-generation ones in the early years of second-generation drugs availability, using drugs like loratadine, cetirizine, terfenadine, and astemizole,

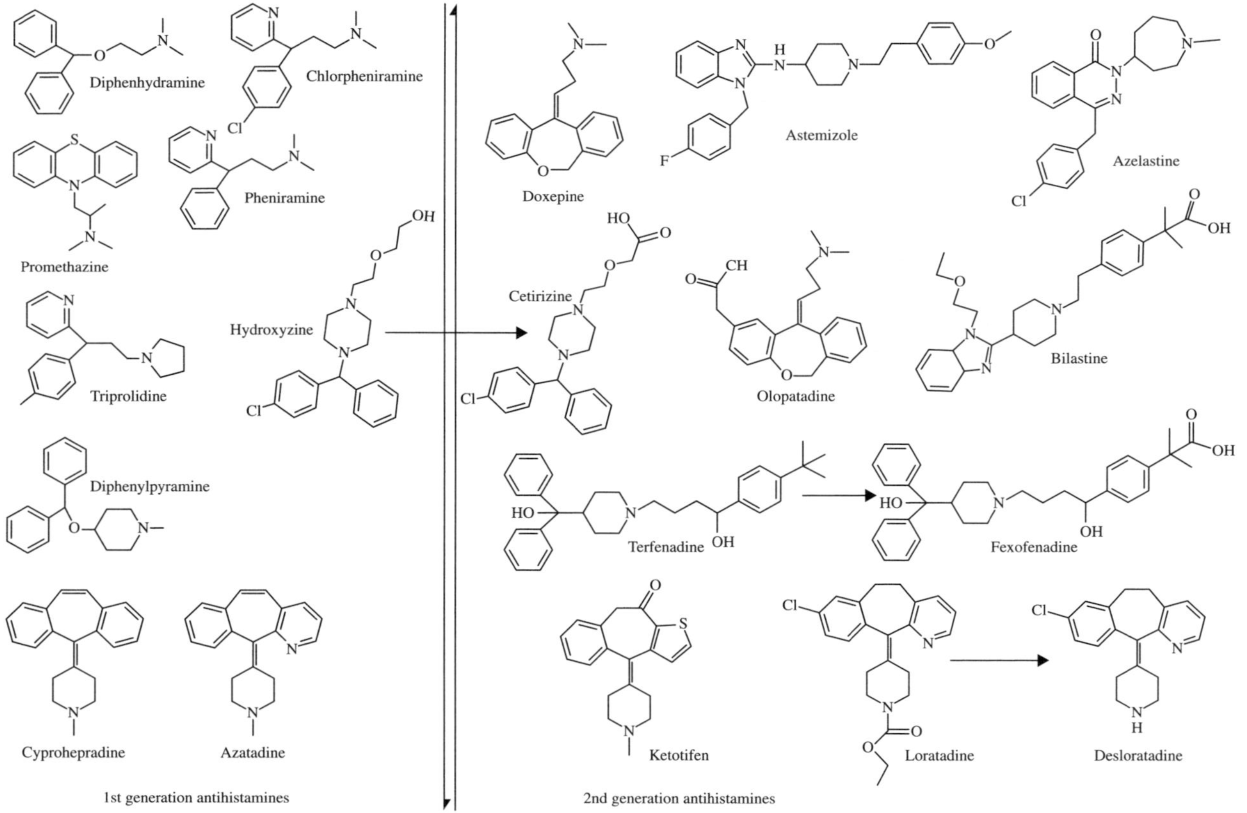

FIGURE 21.1 Structures of both first- and second-generation antihistamines. The arrows leading from one antihistamine to another highlight the metabolism of a precursor antihistamine into a new antihistamine. For loratadine and terfenadine, this metabolism can occur within the human body, while hydroxyzine requires fungal metabolism to convert it to cetirizine.

were all using different doses, administration times, time periods, and conditions of analysis. One mini review in 1995 attempted to compress the information of the day into a relative risk and did show less evident sedation for the second-generation drugs [11]. In the intervening years though, some of the studies have become more uniform in their approach and used cross over studies to allow intrasubject validation as part of the analysis.

Clinical studies examining mental capacity and ability to concentrate on tasks such as driving [12–14] or flying are common studies used to compare one set of antihistamines to another. A common theme of all of these studies showed that drugs classed as less sedating do just that and that the ability to conduct tasks is diminished upon switching these drugs for older first-generation antihistamines that are known to cross the BBB [2, 13, 15–17]. To illustrate the benefit of second-generation antihistamines having little BBB inflow, especially relating to cognition effects, one study showed loratadine, in conjunction with montelukast caused no additional sleepiness at normal 10 mg doses, as well as visual tracking being unaffected and, thus, concentration not altered in their pilots and flight crew volunteers [18]. Conversely, diphenhydramine caused many hours of sedation, tracking errors, and at least a 7% worsening in vigilance to undergo procedural tests relevant to flight during the 6 h study [18]. Additional studies done at normal atmospheric pressure showed the same outcome for diphenhydramine [13, 19]. Such studies illustrate the significance to pilots, paramedics, and others in charge of rapidly moving vehicles of not taking first-generation antihistamines in the 6 h preceding their jobs and that taking such over the counter medications is at least as problematic for alertness and complex function as having significant blood alcohol levels.

Long-term treatment of another second-generation antihistamine (olopatadine) showed that after 20 days of treatment with 10 mg/day, in over 80 patients, only one patient showed signs of sleepiness, based on the study's internal ratings scale, and one gained weight [20]. Further placebo-controlled PET scanning results examining the competitive removal of antihistamines to dislodge radiolabeled doxepin, have proven a popular noninvasive way to examine brain H_1 receptor occupancy after oral administration [5, 20]. Here olopatadine was shown to only have a 15% occupancy of CNS H_1 receptors [21].

With up to one-third of all randomized clinical trials not being presented in a public forum, with no final publications [22], this does introduce caveats regarding interpretations on the BBB penetration in patients or volunteers, as not all information regarding sedation, cognition changes, mood effects, or other CNS-related adverse actions of new drugs would be reported prior to release of medications to market. We are unaware, naturally, of what is done but not published in the antihistamine area; but if this area of drug development is consistent with the total body of drug trial data [22], then there are likely to be unpublished trials adding to the evidence pool of cognition changes, or lack thereof, already completed that would assist further development into the future.

Nevertheless, many tests have been constructed to assess the subtle differences in cognition that may be caused by antihistamines. Most of the tests involve a computer screen interaction, such as being presented with a number of words and then needing to recall them, timing the exposure to the words prior to a bulk recollection, or doing simple associations of numbers to each other when simultaneously present on the screen, to a sequence of numbers that need some type of pattern recognition [16]. A common tool has been the digit symbol substitution where subjects are given a pen and paper and need to convert numbers on a screen into symbols as subjects follow a conversion grid provided to them [2, 16, 23]. Some of the cognitive decline discovered in these tests is not consciously perceived by the patients as a concern. They believe they are not sedated, and, thus, there is no concept that more subtle degradation of normal habits could be occurring. Wilken's 2003 article [24] on desloratadine provided a clear summary of the rationale behind these tests, where vigilance, divided attention, working memory, and psychomotor speed are important determinants. Vigilance is a critical factor to determine and measures a subject's ability to focus on a task for extended periods of time, especially when that task is monotonous, such as driving for long stretches. In addition, divided attention is a measure of how quickly and accurately two simultaneous actions can be achieved. Working memory requires the absorbing of information, further cognitive process applied to this information, and an answer produced. Psychomotor speed works with hand–eye coordination and simultaneous thinking skills and reasoning, which tests higher cognitive thought [24].

First-Generation Antihistamines

Diphenhydramine In human trials to examine the lack of CNS effects of many second-generation antihistamines, diphenhydramine appears as a popular positive control, because sedation and cognitive decline occur rapidly with this drug [25]. Interestingly, one study claimed that up until year 2000, many studies only showed sedation, but a meta-analysis showed no other cognitive decline in memory and reaction time [17]. However, in the intervening decade, the data from the 50 mg diphenhydramine literature would indeed suggest that cognitive decline occurs with this drug [18, 26–29]. Diphenhydramine has been shown to cause greater cognition impairment than alcohol intoxication [2], so combinations of alcohol plus a first-generation antihistamine were very likely to cause significant disruption to cognitive brain function in patients [13]. Interestingly, diphenhydramine in one study did not affect memory tracking [19], suggesting some psychomotor testing is more relevant for histamine action than others.

Chlorpheniramine One study on chlorpheniramine examined occupancy of CNS H_1 receptors using PET scanning and C^{11}-doxepin removal as signs of binding. Psychomotor tests were conducted and associated with the PET scanning results of receptor occupancy. It was shown that the binding increased dramatically when the injected dose was increased from 2 to 5 mg of chlorpheniramine [30]. What was most significant about this work though was the evidence that cognitive impairment occurred at doses that did not elicit sleepiness. Not only could the two types of effects on the body be distinguished, but patients on such medication may believe they are unaffected, allowing them to conduct complex tasks with only the signs of sedation worthy of consideration. Thus, increased error generation and attention, which may be considered careless, may eventuate without the conscious awareness of the subject.

Promethazine Promethazine has not been used as regularly as diphenhydramine as a positive CNS affecting drug when establishing second-generation antihistamine effects, but it is equally potent at causing sedation and other CNS issues. In one study, where it was used as a positive control alongside desloratadine, promethazine was shown to impair object tracking and reaction time when registering the location on the screen of symbols. There were also more missed responses when trying to maintain sustained attention and subjects on 25 mg promethazine also substituted more incorrect numbers when trying to recode numbers quickly [16].

Second-Generation Antihistamines One of the earliest second-generation antihistamines with reduced lipophilicity was terfenadine. Back in 1982, it was shown to be below detection limits in the brain of rats after oral gavage. Peak levels came about quickly though, with peak tissue readings within an hour of a radiolabeled dose of terfenadine, mostly in the lungs [31]. Cetirizine, like fexofenadine, exists as zwitterions at physiological pH values and, as such, have relatively low Log Ds of around 1.5 [32] and 2.48 [33], respectively. They also have very high albumin binding in the circulation, which assists reduction of brain uptake relative to their parent compounds, hydroxyzine [32] and terfenadine [32, 33]. These physicochemical characteristics form the bulk of the reasons why these drugs do not get access through the BBB as much as their precursors and the first-generation antihistamines. In addition, being substrates for P-glycoprotein, which is discussed in detail in Section "P-Glycoprotein-Mediated Efflux", simply means that this efflux protein can mop up the low amounts that would get into the brain, providing the final avenue for removal of drug from the brain. However, if any drug does make it into the brain proper, then differences in affinity, or receptor occupancy, for the H_1 receptor is integral to the functionality of these H_1 receptor blockers. As an example of the complexities of such binding, one group calculated that the peripheral receptor occupancy of desloratadine in the circulation was 71% at 4 h; yet, there was only a 34% maximum wheal inhibition at that time, while fexofenadine and levocetirizine, with over 90% receptor occupancy, gave 100% reduction in wheal inhibition at 4 h. However, at 24 h, desloratadine still has 43% receptor occupancy, and has almost the same wheal inhibition compared to 4 h (32%), while fexofenadine dropped to 12% occupancy and 15% inhibition at 24 h. Levocetirizine, with its 57% occupancy, still had 60% wheal inhibition after a day [34]. These numbers suggest potency results from a combination of factors, including the generation of active metabolites. Another study by Gillard and others showed that the binding affinity for the second-generation antihistamines was lowest for loratadine, with a K_i of 16 nM, fexofenadine, cetirizine, and levocetirizine had K_i values between 2 and 10 nM and desloratadine was 0.4 nM. Binding affinity is inversely proportional to K_i, so low K_i is equivalent to high affinity. Thus, desloratadine has the highest affinity for the H_1 receptor of all second-generation drugs [35].

Molimard has suggested that a volume of distribution of an antihistamine between 0.1 and 0.6 l/kg would be an optimal zone for peripheral action [36].

Levocetirizine is one drug that falls within this range [36]. It has been shown that lipophilicity changes alone are not critical to changing the brain uptake and receptor occupancy of H_1 receptor substrates, or any drug for that matter. Instead, the additional factor of time to reach brain equilibrium must be taken into consideration. When this second measure is considered, we have two classes of brain uptake compounds: ones that spike early in brain parenchyma and also reach a plateau quickly, and others that slowly increase over many hours [37]. The key question to ask is, are some of the second-generation antihistamines moving from one pattern of absorption to another?

Cetirizine/Levocetirizine Not surprisingly, there was very little effect on adults after 10 mg cetirizine loading. Normal psychomotor test results were evident, including eye tracking of varying difficulties, as well as divided attention, memory tracking tests [19], and changes to the perceived rate at which flashing lights appear to stop flickering (critical flicker fusion test) [25]. There doesn't appear to be any significant brain penetration in children either, based on observable tiredness evident in a pediatric study using 6- to11-year-olds [38]. Tashiro found H_1 receptor binding of 26% in the cortex of Japanese volunteers (using a higher than normal dose of 20 mg) [39]. This level of binding caused some perceived sedation in their study. Thus, if all it takes is a doubling of a standard dose to induce significant CNS H_1 binding, and therefore increase sedation, then overcoming mechanisms responsible for keeping cetirizine out of the brain, like coadministering drugs that block P-gp function, may result in a normal 10 mg dose inducing effects similar to 20 mg doses as seen in this Japanese study [39].

Loratadine A small study on six women in 1987 showed that loratadine caused some cognitive impairment, with memory of 12-digit sequences being poorer than normal and reaction times being slightly delayed, but only with four times normal doses (40 mg), showing no obvious CNS changes at 10 and 20 mg [40]. With normal doses being 10 mg, circulating levels from either increased dose or drug transporter changes would need to be substantially higher than the low baseline to start causing CNS issues similar to sedating antihistamines. A decade later, additional cognitive studies using standard doses of loratadine also showed no cognitive changes after acute (single dose) or 5 days of daily dosing using a standard array of memory, vigilance, and fatigue studies [41, 42]. However, a recent study showed a decline in reaction time in a left- and right-sided decision test, where patients were given either loratadine or promethazine in a three way crossover study. They matched each other in their diminished response times, which suggested a central autonomic nervous system effect was associated with loratadine [43]. In addition, with PET scanning studies showing a 11% occupancy of central H_1 receptors for loratadine [44], one could see that loradatine does enter the brain, which may not follow the assumption that a second-generation antihistamine is not sedating because of the BBB keeping the drug from the CNS. This is at odds with Bradley and Nicholson [40], who in 1987 stated that only 40 mg of loratadine caused any CNS cognitive impairment and that nothing was noted at 10 mg.

Desloratadine Desloratadine has been shown to have a greater affinity for the H_1 receptor than most of the other less-sedating antihistamines [35], which would suggest that even low amounts of brain uptake may increase sedation. However, this is not evident from the literature. Berger [23] mentioned that no evidence of CNS action was forthcoming back in 2005. This is likely due to P-gp-mediated efflux of desloratadine. Other studies showed that desloratadine had no impact on driving or flying behavior when used at normal 5 mg doses [14, 45]. However, the same could not be said for the first-generation antihistamine, diphenhydramine, used in both studies. When used at 50 mg doses, it consistently produced worse driving skills, including following distance, lane change tracking, and the qualitative self-assessment on the Stanford sleepiness scale [14], as well as vigilance on complex tasks and error-making [45]. These studies were done on subjects that were not suffering with sessional allergic rhinitis. Wilken and colleagues [24] performed a hay fever-induced study in 2003, adding placebo, diphenhydramine or desloratadine, to patients exposed to allergic pollen prior to conducting a battery of memory, psychomotor, vigilance, and computational studies. Essentially, whether a patient is normal and takes a second-generation antihistamine, especially one like desloratadine, or whether they are suffering with the effects of allergic rhinitis prior to drug therapy, no significant differences from placebo have been shown. As with the previous healthy subjects, addition of diphenhydramine at 50 mg to these hay fever sufferers led to worsening of all cognitive studies tested, in addition to the self-reporting of increased sleepiness [24].

Studies in children have also confirmed that desloratadine is not able to cross the young brain either, as no obvious effects of sedation were noted. No other adverse effects at doses that were up to 50% of an adult dose for children above 6 years, and a quarter of the adult dose in children between 2 and 5 years either [46]. Meanwhile, desloradatine did result in a decrease in memory recall of simple numbers within a psychomotor analysis study in adult volunteers. This was the only test of many in this study to show any desloratadine related effect [16]. Interestingly, there appeared to be some learned behavior as the study progressed, as subjects improved in their ability to conduct some of the psychomotor tests, so desloratadine appeared to put subjects back to an unlearnt state, while the first-generation positive control, promethazine, actually hindered the ability to remember what numbers were flashed on the screen [16].

Fexofenadine Fexofenadine has poor absorption, in comparison to the other antihistamines, and needs much greater doses to be administered to get peak plasma levels of over 200 ng/ml [47]. Thus, it appears the approach taken to alleviate brain penetration was to be to create a drug that has poor absorption from the onset. A small amount crosses the gut when a high enough dose is given, although this has more to do with the physicochemical properties of the drug than any P-gp-mediated efflux. The small amount getting into the circulation has very little prospect of diffusing into any tissue, let alone through the BBB, by virtue of its zwitterion nature at physiological pH and albumin binding [32]. A clinical study using 6–11 year old children showed how poor fexofenadine was at suppressing wheals and flares after oral administration in comparison to cetirizine [48], and this likely related to the poor absorption properties of this antihistamine on a mass basis compared to the other drugs. Nevertheless, some studies have shown 45% increases in the absorption rate of fexofenadine when drugs known to inhibit P-glycoprotein, such as itraconazole [49] and grapefruit juice [50], are used. It should be noted that grapefruit juice, and more specifically the hyperforin contained within it, is used as an inducer of P-gp in sustained doses, yet blocks P-gp in acute doses. The act of blocking P-gp encourages gastrointestinal cells to upregulate RNA transcription and subsequent P-gp translation in the cells over many days [51–54]. It is also possible that there could be some differences of P-gp affinity between the two enantiomers of fexofenadine [55].

One study, published in 1999, used multiple doses of fexofenadine up to 180 mg. Although it did appear, 180 mg doses were suggesting some cognitive decline was occurring, especially with the critical flicker fusion thresholds where pulsing light appears to become a continuously projected light source, the statistical evaluation stated this was not significantly different from controls. Only the first-generation antihistamine, promethazine, used in parallel to fexofenadine caused significant decline in this flicker test [42]. More often though, the clinical studies using fexofenadine have used quite low doses of fexofenadine, like a driving study in Iowa [13]. Only 60 mg of fexofenadine was used to show a similar lack of sedation compared to alcohol. Combination studies were not undertaken in this study, and it certainly would be useful to know whether alcohol's effects were elevated in such a setting with combination effects, as this could be a realistic setting. Nevertheless, the fact that 50 mg of diphenhydramine gave significantly worse results for lane keeping, braking speeds, and braking distance upon sudden obstacles in the road, compared to subjects with a blood alcohol level of 0.1% [13], which in some countries would result in an automatic suspension of driving rights, highlights the severity of first-generation antihistamines in cognition dysfunction due to crossing the BBB. Interestingly, even though subjects were given a very low dose of fexofenadine, they still did significantly worse in the collision avoidance test than under placebo controlled circumstances [13]. Some dual alcohol-antihistamine studies have been done with other drugs as well. Bilastine, at a 20 mg dose, did not add to the alcohol mediated sedation effects, but 80 mg bilastine caused similar levels of sedation to 10 mg cetirizine [56].

P-Glycoprotein-Mediated Efflux P-gp is known to prevent a number of drugs from traversing the BBB [57]. P-gp expressed at the BBB is structurally identical to P-gp expressed at other barrier sites, such as the kidney, liver, testes, placenta, and gastrointestinal tract. Recent evidence quantitating P-gp expression in human microvessels has shown at least a twofold difference between individuals [58], which would contribute to patient variability in CNS effects of some antihistamines. In the gut, a higher drug concentration will saturate the transporter and allow the drug to pass into the systemic circulation, so being a P-gp substrate does not necessarily predict oral availability. However, once in the circulation, the concentrations of drugs diminish rapidly to mid-to-low micromolar concentrations, or even high nanomolar concentrations. In this range of concentrations, P-gp and other active efflux proteins are able to function to keep many drugs from passing the BBB.

Loratadine Many studies have examined the difference between first- and second-generation antihistamines, both *in vitro* and using brain penetration in P-gp (mdr1a) knockout mice [59]. Doran et al. [60] showed no difference in area under the curve (AUC) brain penetration (measured AUC_{brain}/AUC_{plasma}) of diphenhydramine, hydroxyzine, or triprolidine between the control and knockout mice with ratios all over 3.6, irrespective of P-gp status of the animals. However, loratadine's brain AUC in mdr1a P-gp knockout mice was double that of wild-type mice (3.30 vs 1.6) [61]. Double knockout mice studies (both mdr1a and mdr1b removed) also showed a twofold elevation in brain loratadine [60]. *In vitro* studies, conducted in our laboratory, found no evidence of P-gp mediated efflux. Transport rates were very high in both directions within bidirectional studies, having a passive permeability above 26×10^{-6} cm/s. This suggested that loratadine was either a very weak substrate for P-gp, or not a substrate at all [62] with rapid absorption and good volume of distribution. However, one report has suggested that loratadine is a weak inhibitor of P-gp [63].

Other *in vitro* studies that show loratadine to have some affinity to P-gp include work done by Uchida and co-workers in Japan from 2011. Efflux was again very poor, with their efflux ratio being 1.75-fold greater in basolateral (B) to apical (A) transport. In addition, the raw apparent permeability data in the A to B direction was over 30×10^{-6} cm/s, which is very high and suggested reasonable transport through the BBB. In their study, it had the second highest A to B transport in the mdr1a expressing cells, just behind diazepam [64], which suggests loratadine would get through the BBB. Note though that these are murine cell models, with mouse mdr1a transfected, not human P-gp. In fact, based on the rationale that in vitro results are more likely to over-exaggerate P-gp-mediated efflux, due to the cells having more P-gp expression than normal BBB cells *in vivo* [65], low levels of *in vitro* affinity are likely to reflect no substantial effects in a patient at all.

If loratadine has such poor affinity for P-gp, what is keeping it out of the brain? Perhaps, it is effluxed by BCRP instead, especially when it has recently been shown that BCRP is at least as prevalent, if not the dominant efflux protein, in the human BBB [58]. However, this is speculation, as there are no reports in the literature examining loratadine's affinity for BCRP. Loratadine's physicochemical properties are more closely aligned with the sedating antihistamines and, without having a clear picture of any efflux system providing adequate defence against this drug, protein binding and affinity of the parent loratadine to the H_1 receptor might explain the clearly limited CNS effects evident from patients taking this therapy. It has been shown that loratadine has one of the weakest H_1 receptor binding affinities of the second-generation antihistamines [35, 66]. Both loratadine and fexofenadine needed about 180 nM concentrations to inhibit both desloratadine and pyrilamine binding to the H_1 receptor. In comparison, azelastine only needed 1 nM and diphenhydramine 2.5 nM [66]. Thus, could this contribute to less sedation in the brain and less CNS receptor occupancy, or is it kept out of the brain through a different efflux transporter, given that P-gp affinity seems quite low? Even now, some of these questions have yet to be answered, as our understanding of loratadine still has some way to go with regard to receptor affinity and brain transport rates.

Desloratadine Desloratadine is an active metabolite of loratadine, and it appears that P-gp has much greater affinity toward it than the parent drug. Desloratadine's brain-to-plasma AUC was less than 1 in control mice and over 14 in knockout mice [61], which was a higher brain-to-plasma AUC than any of the sedating antihistamines, implying good brain transport when P-gp is blocked. It also has an efflux ratio of 7 using transport through Caco-2 cell monolayers [62], suggesting desloratadine is kept out of the brain to a large degree by P-gp, and that its inhibition would result in high brain penetration.

Cetirizine Cetirizine has been shown to be a P-gp substrate *in vitro* [28, 61, 62] and in animal studies [28]. *In vitro* studies showed cetirizine to have between a four- to sixfold greater efflux ratio at concentrations expected in the circulation [62, 67]. Pharmacokinetic studies in volunteers showed that when cetirizine is orally administered alone and then with a potent P-gp inhibitor (e.g., ritonavir), T_{max} increased from 0.75 to 2.0 h and half-life increased from 7.85 h to almost 12 h [68]. Unfortunately, as this study only examined pharmacokinetic properties of the drug interaction, there was no data on the likely increased CNS effects that may have ensued from such P-gp inhibition over the 19 days of the study [68].

In horses, which also have high P-glycoprotein expression in the gut and other tissues [69], it was shown that use of the P-gp inhibitor ivermectin increased brain uptake of cetirizine [70]. Interestingly, from a kinetic viewpoint, a large increase in plasma levels after a 90 minute pre-dosing with ivermectin was not observed, which suggests that drug levels reaching the intestinal lumen may be enough to saturate available P-gp efflux transporters for oral absorption into the blood circulation [70]. This indicates that overcoming P-gp action on a drug at the gut level is relatively easy, simply by substantially increasing the dose. Drug concentrations in the circulation will be substantially less though, and this may allow P-gp to efflux drugs at non-saturating concentrations that would limit brain exposure, even if they were not excluded from the gastrointestinal tract initially. In mice though, brain-to-plasma AUC ratios never exceeded 0.08, even with P-gp inhibition, which does question the clinical relevance of cetirizine being a P-gp substrate if other properties of the drug are effectively resulting in only low-level uptake in the brain [61]. Polli and colleagues also showed a brain to plasma ratio of 0.08 for cetirizine, although the drug's brain to freely available plasma concentration ratio was closer to 0.5. In addition, the knockout mice had over a sixfold higher total brain to total plasma concentration ratio, and an eightfold higher free brain to free plasma concentration drug ratio [67]. These studies illustrate the difficulty in using animal data in its raw form to interpret the absorption in humans. Although the effect of P-gp on cetirizine was consistent, the condition of the BBB and the locations of microdialysis within the brain may affect the amount of drug detected. In addition, a first-generation antihistamine, diphenhydramine, had circulating unbound drug concentration six times higher in rats than in humans [29], suggesting that it is difficult to directly relate transport properties of unbound drug in plasma to the organs as comparable to humans if such large differences in unbound percentages of drug persist between species.

Cetirizine's precursor drug, the sedating antihistamine hydroxyzine, has also been tested *in vitro* for P-gp-mediated efflux and has shown no such affinity. Instead, it had very high passive permeability of over 26×10^{-6} cm/s, which was indicative of total absorption [28]. In addition, the same study examined brain penetration in Sprague–Dawley rats and using microdialysis probe collections of brain fluid, with calculating the unidirectional transfer constant (K_{in}), hydroxyzine showed over 2.0 ml/min/g of brain weight. This is a very high value, signaling rapid brain uptake. This was similar for diphenhydramine, at over 2.2 ml/min/g, and chlorpheniramine, at 1.2 ml/min/g [28]. Hydroxyzine transport in a double knockout mice study also showed absolutely no evidence of any P-gp action [60].

Fexofenadine A study that examined fexofenadine levels in tissues of wild-type mice compared to mdr1a knockout mice showed that end point brain levels increased from 0.71 to 6.3 ng/g (8.8-fold increase), with a commensurate increase in plasma levels from 4.1 to 19 (4.6-fold increase) [71]. Schinkel's group examined the brain to plasma ratio and in their wild-mice type compared to mdr1a knockout mice, values increased from 0.17 to 0.33 for fexofenadine, which was only a twofold increase. These raw values suggested that even after P-gp inhibition, brain uptake values were still very low compared to plasma, which suggests P-gp would not be a significant player in keeping this antihistamine out of the brain [72]. Parallel studies done by Tahara and co-workers in mdr1a/1b double knockouts also showed that when fexofenadine was infused, whether mice were wild type or knockout, organ uptake was low and brain uptake was elevated threefold in the P-gp knockout mice, but from a very low baseline [73]. The constant infusion model used in these studies, resulted in plasma fexofenadine at 400–500 nM, which was higher than at any time during oral administration, even to the mdr1a/1b knockout mice, yet brain to plasma ratios were still less than 0.15 [73]. This indicated the inherently poor ability of this drug to cross the BBB even when P-gp is absent.

Bioavailability for fexofenadine was reported as only 2%, and our own in vitro data in Caco-2 cells supports this very poor oral absorption for fexofenadine [62], which, although increased if P-gp is inhibited in the gut, would not increase significantly above 5%. Biliary and urinary secretion of fexofenadine does not appear to be affected by P-gp activity [73], which means removal from the body is similar, irrespective of P-gp inhibitory status.

We showed (unpublished work in our lab) that fexofenadine had poor induction properties for P-gp, unlike its precursor, terfenadine, which increased P-gp expression by about 50% *in vitro*. This confirmed functional studies by Perloff et al. [74], who showed Rh123 efflux increased in human gastrointestinal cells after exposure to terfenadine, but not to fexofenadine. Thus, fexofenadine has no inhibitory ability on P-gp and only behaves as a substrate.

Bilastine/Olopatadine Other antihistamines developed to date that have not been discussed extensively in this chapter include bilastine and olopatadine. These have also both been shown to be P-gp substrates [75, 76]. Thus, P-gp affinity does appear to play a role in limiting most second-generation antihistamine brain uptake, except for loratadine. P-gp is not the only transporter involved for some of these newer less sedating antihistamines however, and other transporters such as BCRP are present at high levels at the BBB alongside P-gp [58].

SUMMARY

It has been expected by sufferers of acute rhinitis, chronic urticarial allergies, and general allergies that recent antihistamines, promoted as safe enough to use without requiring a prescription in many countries, have adverse effects so minor as to not require a second thought. Thus, there can be a nonchalant attitude to their possible sedation effects or their more subtle cognitive changes, as discussed here. In actual fact, taking the medication before using heavy machinery, such as motor vehicles, farm equipment, or even workshop equipment, could result in a drug-induced inattention, leading to harm to self or others, especially when in conjunction with other potentially sedating medication, or P-gp inhibitor drugs. Thus, transport through the BBB, or lack thereof, is a critical factor in harm minimization for this drug group that is so widely utilized around the world. This chapter provides detailed evidence that cognitive changes, brought on by exposure to antihistamines, can be complex, with sedation only being an obvious end point to cognitive change. Alleviating the core adverse effect by limiting brain uptake, as in the modern drugs, or changing the affinity for the receptor, while still maintaining peripheral action, may still lead allergy sufferers to exhibit cognitive decline, especially if they dose themselves at above recommended levels. Thus, it is granted that this class of drugs is a great boon for the quality of life of many patient groups, but caution for possible adverse effects still needs to be considered when taking these drugs.

ACKNOWLEDGMENTS

Many thanks go to Paul Murray from the School of Pharmacy, Curtin University, for his assistance with the drawing of the structures of the antihistamines.

REFERENCES

[1] Ben-Shoshan M., Turnbull E., Clarke A. (2012) Food allergy: temporal trends and determinants. *Current Allergy and Asthma Reports*, 12(4):346–372.

[2] Simons F. E. (2004) Advances in H1-antihistamines. *The New England Journal of Medicine*, 351(21):2203–2217.

[3] Timmerman H. (1999) Why are non-sedating antihistamines non-sedating? *Clinical and Experimental Allergy*, 29(Suppl 3):13–18.

[4] Simons F. E. (2002) Comparative pharmacology of H1 antihistamines: clinical relevance. *The American Journal of Medicine*, 113(Suppl 9A):38S–46S.

[5] Mochizuki H., Kimura Y., Ishii K., Oda K., Sasaki T., Tashiro M., Yanai K., Ishiwata K. (2004) Quantitative measurement of histamine H(1) receptors in human brains by PET and [11C]doxepin. *Nucleur Medicine and Biology*, 31(2):165–171.

[6] Welch M. J., Meltzer E. O., Simons F. E. (2002) H1-antihistamines and the central nervous system. *Clinical Allergy and Immunology*, 17:337–388.

[7] Sproule B. A., Busto U. E., Buckle C., Herrmann N., Bowles S. (1999) The use of non-prescription sleep products in the elderly. *International Journal of Geriatric Psychiatry*, 14(10):851–857.

[8] Housley G. D., Norris C. H., Guth P. S. (1988) Histamine and related substances influence neurotransmission in the semicircular canal. *Hearing Research*, 35(1):87–97.

[9] Wood C. D. (1979) Antimotion sickness and antiemetic drugs. *Drugs*, 17(6):471–479.

[10] Cheung B. S., Heskin R., Hofer K. D. (2003) Failure of cetirizine and fexofenadine to prevent motion sickness. *Annales of Pharmacotherapy*, 37(2):173–177.

[11] Hindmarch I. (1995) Psychometric aspects of antihistamines. *Allergy*, 50(24 Suppl): 48–54.

[12] O'Hanlon J. F., Ramaekers J. G. (1995) Antihistamine effects on actual driving performance in a standard test: a summary of Dutch experience, 1989–94. *Allergy*, 50(3): 234–242.

[13] Weiler J. M., Bloomfield J. R., Woodworth G. G., Grant A. R., Layton T. A., Brown T. L., McKenzie D. R., Baker T. W., Watson G. S. (2000) Effects of fexofenadine, diphenhydramine, and alcohol on driving performance. A randomized, placebo-controlled trial in the Iowa driving simulator. *Annales of Internal Medicine*, 132(5):354–363.

[14] Vuurman E. F., Rikken G. H., Muntjewerff N. D., de Halleux F., Ramaekers J. G. (2004) Effects of desloratadine, diphenhydramine, and placebo on driving performance and psychomotor performance measurements. *European Journal of Clinical Pharmacology*, 60(5):307–313.

[15] Hindmarch I., Johnson S., Meadows R., Kirkpatrick T., Shamsi Z. (2001) The acute and sub-chronic effects of levocetirizine, cetirizine, loratadine, promethazine and placebo on cognitive function, psychomotor performance, and weal and flare. *Current Medical Research and Opinion*, 17(4):241–255.

[16] Nicholson A. N., Handford A. D., Turner C., Stone B. M. (2003) Studies on performance and sleepiness with the H1-antihistamine, desloratadine. *Aviation, Space, and Environmental Medicine*, 74(8):809–815.

[17] Bender B. G., Berning S., Dudden R., Milgrom H., Tran Z. V. (2003) Sedation and performance impairment of diphenhydramine and second-generation antihistamines: a meta-analysis. *The Journal of Allergy and Clinical Immunology*, 111(4):770–776.

[18] Valk P. J., Simons M. (2009) Effects of loratadine/montelukast on vigilance and alertness task performance in a simulated cabin environment. *Advances in Therapy*, 26(1):89–98.

[19] Verster J. C., Volkerts E. R., van Oosterwijck A. W., Aarab M., Bijtjes S. I., De Weert A. M., Eijken E. J., Verbaten M. N. (2003) Acute and subchronic effects of levocetirizine and diphenhydramine on memory functioning, psychomotor performance, and mood. *The Journal of Allergy and Clinical Immunology*, 111(3):623–627.

[20] Ohmori K., Hayashi K., Kaise T., Ohshima E., Kobayashi S., Yamazaki T., Mukouyama A. (2002) Pharmacological, pharmacokinetic and clinical properties of olopatadine hydrochloride, a new antiallergic drug. *Japanese Journal of Pharmacology*, 88(4):379–397.

[21] Tashiro M., Mochizuki H., Sakurada Y., Ishii K., Oda K., Kimura Y., Sasaki T., Ishiwata K, Yanai K. (2006) Brain histamine H receptor occupancy of orally administered antihistamines measured by positron emission tomography with (11)C-doxepin in a placebo-controlled crossover study design in healthy subjects: a comparison of olopatadine and ketotifen. *British Journal of Clinical Pharmacology*, 61(1):16–26.

[22] Jones C. W., Handler L., Crowell K. E., Keil L. G., Weaver M. A., Platts-Mills T. F. (2013) Non-publication of large randomized clinical trials: cross sectional analysis. *British Medical Journal*, 347:f6104.

[23] Berger W. E. (2005) The safety and efficacy of desloratadine for the management of allergic disease. *Drug Safety: An International Journal of Medical Toxicology and Drug Experience*, 28(12):1101–1118.

[24] Wilken J. A., Kane R. L., Ellis A. K., Rafeiro E., Briscoe M. P., Sullivan C. L., Day J. H. (2003) A comparison of the effect of diphenhydramine and desloratadine on vigilance and cognitive function during treatment of ragweed-induced allergic rhinitis. *Annals of Allergy, Asthma & Immunology*, 91(4):375–385.

[25] Gandon J. M., Allain H. (2002) Lack of effect of single and repeated doses of levocetirizine, a new antihistamine drug, on cognitive and psychomotor functions in healthy volunteers. *British Journal of Clinical Pharmacology*, 54(1):51–58.

[26] Chen X., Zhang Y., Zhong D. (2004) Simultaneous determination of chlorpheniramine and pseudoephedrine in human plasma by liquid chromatography-tandem mass spectrometry. *Biomedical Chromatography*, 18(4):248–253.

[27] Neuhaus W., Mandikova J., Pawlowitsch R., Linz B., Bennani-Baiti B., Lauer R., Lachmann B., Noe C.R. (2012) Blood-brain barrier in vitro models as tools in drug discovery: assessment of the transport ranking of antihistaminic drugs. *Die Pharmazie*, 67(5):432–439.

[28] Obradovic T., Dobson G. G., Shingaki T., Kungu T., Hidalgo I. J. (2007) Assessment of the first and second generation antihistamines brain penetration and role of P-glycoprotein. *Pharmaceutical Research*, 24(2):318–327.

[29] Sadiq M. W., Borgs A., Okura T., Shimomura K., Kato S., Deguchi Y., Jansson B., Björkman S., Terasaki T., Hammarlund-Udenaes M. (2011) Diphenhydramine active uptake at the blood-brain barrier and its interaction with oxycodone in vitro and in vivo. *Journal of Pharmaceutical Sciences*, 100(9):3912–3923.

[30] Okamura N., Yanai K., Higuchi M., Sakai J., Iwata R., Ido T., Sasaki H., Watanabe T., Itoh M. (2000) Functional neuroimaging of cognition impaired by a classical antihistamine, d-chlorpheniramine. *British Journal of Pharmacology*, 129(1):115–123.

[31] Leeson G. A., Chan K. Y., Knapp W. C., Biedenbach S. A., Wright G. J., Okerholm R. A. (1982) Metabolic disposition of terfenadine in laboratory animals. *Arzneimittel-Forschung*, 32(9a):1173–1178.

[32] Pagliara A., Testa B., Carrupt P. A., Jolliet P., Morin C., Morin D., Urien S., Tillement J. P., Rihoux J. P. (1998) Molecular properties and pharmacokinetic behavior of cetirizine, a zwitterionic H1-receptor antagonist. *Journal of Medicinal Chemistry*, 41(6): 853–863.

[33] Mahar Doan K. M., Wring S. A., Shampine L. J., Jordan K. H., Bishop J. P., Kratz J., Yang E., Serabjit-Singh C. J., Adkison K. K., Polli J. W. (2004) Steady-state brain concentrations of antihistamines in rats: interplay of membrane permeability, P-glycoprotein efflux and plasma protein binding. *Pharmacology*, 72(2):92–98.

[34] Gillard M., Benedetti M. S., Chatelain P., Baltes E. (2005) Histamine H1 receptor occupancy and pharmacodynamics of second generation H1-antihistamines. *Inflammation Research*, 54(9):367–369.

[35] Gillard M., Christophe B., Wels B., Peck M., Massingham R., Chatelain P. (2003) H1 antagonists: receptor affinity versus selectivity. *Inflammation Research*, 52 Suppl 1: S49–S50.

[36] Molimard M., Diquet B., Benedetti M. S. (2004) Comparison of pharmacokinetics and metabolism of desloratadine, fexofenadine, levocetirizine and mizolastine in humans. *Fundamental and Clinical Pharmacology*, 18(4):399–411.

[37] Liu X., Smith B. J., Chen C., Callegari E., Becker S. L., Chen X., Cianfrogna J., Doran A. C., Doran S. D., Gibbs J. P., Hosea N., Liu J., Nelson F. R., Szewc M. A., Van Deusen J.. (2005) Use of a physiologically based pharmacokinetic model to study the time to reach brain equilibrium: an experimental analysis of the role of blood-brain barrier permeability, plasma protein binding, and brain tissue binding. *The Journal of Pharmacology and Experimental Therapeutics*, 313(3):1254–1262.

[38] Simons F. E., Simons K. J. (2005) Levocetirizine: pharmacokinetics and pharmacodynamics in children age 6 to 11 years. *The Journal of Allergy and Clinical Immunology*, 116(2):355–361.

[39] Tashiro M., Sakurada Y., Iwabuchi K., Mochizuki H., Kato M., Aoki M., Funaki Y., Itoh M., Iwata R., Wong D. F., Yanai K. (2004) Central effects of fexofenadine and cetirizine: measurement of psychomotor performance, subjective sleepiness, and brain histamine H1-receptor occupancy using 11C-doxepin positron emission tomography. *Journal of Clinical Pharmacology*, 44(8):890–900.

[40] Bradley C. M., Nicholson A. N. (1987) Studies on the central effects of the H1-antagonist, loratadine. *European Journal of Clinical Pharmacology*, 32(4):419–421.

[41] Kay G. G., Berman B., Mockoviak S. H., Morris C. E., Reeves D., Starbuck V., Sukenik E., Harris A. G. (1997) Initial and steady-state effects of diphenhydramine and loratadine on sedation, cognition, mood, and psychomotor performance. *Archives of Internal Medicine*, 157(20):2350–2356.

[42] Hindmarch I., Shamsi Z., Stanley N., Fairweather D. B. (1999) A double-blind, placebo-controlled investigation of the effects of fexofenadine, loratadine and promethazine on cognitive and psychomotor function. *British Journal of Clinical Pharmacology*, 48(2):200–206.

[43] Kavanagh J. J., Grant G. D., Anoopkumar-Dukie S. (2012) Low dosage promethazine and loratadine negatively affect neuromotor function. *Clinical Neurophysiology*, 123(4):780–786.

[44] Kubo N., Senda M., Ohsumi Y., Sakamoto S., Matsumoto K., Tashiro M., Okamura N., Yanai K. (2011) Brain histamine H1 receptor occupancy of loratadine measured by positron emission topography: comparison of H1 receptor occupancy and proportional impairment ratio. *Human Psychopharmacology*, 26(2):133–139.

[45] Valk P. J., Van Roon D. B., Simons R. M., Rikken G. (2004) Desloratadine shows no effect on performance during 6 h at 8,000 ft simulated cabin altitude. *Aviation, Space, and Environmental Medicine*, 75(5):433–438.

[46] Gupta S., Khalilieh S., Kantesaria B., Banfield C. (2007) Pharmacokinetics of desloratadine in children between 2 and 11 years of age. *British Journal of Clinical Pharmacology*, 63(5):534–540.

[47] Drescher S., Schaeffeler E., Hitzl M., Hofmann U., Schwab M., Brinkmann U., Eichelbaum M., Fromm M. F. (2002) MDR1 gene polymorphisms and disposition of the P-glycoprotein substrate fexofenadine. *British Journal of Clinical Pharmacology*, 53(5):526–534.

[48] Simons F. E., Semus M. J., Goritz S. S., Simons K. J. (2003) H1-antihistaminic activity of cetirizine and fexofenadine in allergic children. *Pediatric Allergy and Immunology*, 14(3):207–211.

[49] Shimizu M., Uno T., Sugawara K., Tateishi T. (2006) Effects of itraconazole and diltiazem on the pharmacokinetics of fexofenadine, a substrate of P-glycoprotein. *British Journal of Clinical Pharmacology*, 61(5):538–544.

[50] Wang Z., Hamman M. A., Huang S. M., Lesko L. J., Hall S. D. (2002) Effect of St John's wort on the pharmacokinetics of fexofenadine. *Clinical Pharmacology and Therapeutics*, 71(6):414–420.

[51] Bailey D., Malcolm J., Arnold O., Spence J. (1998) Grapefruit juice-drug interactions. *Journal of Clinical Pharmacology*, 46:101–110.

[52] Banfield C., Gupta S., Marino M., Lim J., Affrime M. (2002) Grapefruit juice reduces the oral bioavailability of fexofenadine but not desloratadine. *Clinical Pharmacokinetics*, 41(4):311–318.

[53] Kupferschmidt H., Fattinger K., Ha H., Follath F., Krähenbühl S. (1998) Grapefruit juice enhances the bioavailability of the HIV protease inhibitor saquinavir in man. *British Journal of Clinical Pharmacology*, 45:355–359.

[54] Murdoch D., Goa K. L., Keam S. J. (2003) Desloratadine: an update of its efficacy in the management of allergic disorders. *Drugs*, 63(19):2051–2077.

[55] Miura M., Uno T., Tateishi T., Suzuki T. (2007) Pharmacokinetics of fexofenadine enantiomers in healthy subjects. *Chirality*, 19(3):223–227.

[56] Montoro J., Mullol J., Davila I., Ferrer M., Sastre J., Bartra J., Jáuregui I., del Cuvillo A., Valero A. (2011) Bilastine and the central nervous system. *Journal of Investigational Allergology & Clinical Immunology*, 21 Suppl 3:9–15.

[57] Broccatelli F., Carosati E., Cruciani G., Oprea T. I. (2010) Transporter-mediated efflux influences CNS side effects: ABCB1, from antitarget to target. *Molecular Informatics*, 29(1–2):16–26.

[58] Uchida Y., Ohtsuki S., Katsukura Y., Ikeda C., Suzuki T., Kamiie J., Terasaki T. (2011) Quantitative targeted absolute proteomics of human blood-brain barrier transporters and receptors. *Journal of Neurochemistry*, 117(2):333–345.

[59] Schinkel A. H., Smit J. J., van Tellingen O., Beijnen J. H., Wagenaar E., van Deemter L., Mol C. A., van der Valk M. A., Robanus-Maandag E. C., te Riele H. P., Berns A. J. M., Borst P. (1994) Disruption of the mouse mdr1a P-glycoprotein gene leads to a deficiency in the blood-brain barrier and to increased sensitivity to drugs. *Cell*, 77(4):491–502.

[60] Doran A., Obach R. S., Smith B. J., Hosea N. A., Becker S., Callegari E., Chen C., Chen X., Choo E., Cianfrogna J., Cox L. M., Gibbs J. P., Gibbs M. A., Hatch H., Hop C. E., Kasman I. N., Laperle J., Liu J., Liu X., Logman M., Maclin D., Nedza F. M., Nelson F., Olson E., Rahematpura S., Raunig D., Rogers S., Schmidt K., Spracklin D. K., Szewc M., Troutman M., Tseng E., Tu M., Van Deusen J. W., Venkatakrishnan K., Walens G., Wang E. Q., Wong D., Yasgar A. S., Zhang C. (2005) The impact of P-glycoprotein on the disposition of drugs targeted for indications of the central nervous system: evaluation using the MDR1A/1B knockout mouse model. *Drug Metabolism and Disposition*, 33(1):165–174.

[61] Chen C., Hanson E., Watson J., Lee J. (2003) P-glycoprotein limits the brain penetration of nonsedating but not sedating H1-antagonists. *Drug Metabolism and Disposition*, 31(3):312–318.

[62] Crowe A., Wright C. (2012) The impact of P-glycoprotein mediated efflux on absorption of 11 sedating and less-sedating antihistamines using Caco-2 monolayers. *Xenobiotica; the Fate of Foreign Compounds in Biological Systems*, 42(6):538–549.

[63] Wang E. J., Casciano C. N., Clement R. P., Johnson W. W. (2001) Evaluation of the interaction of loratadine and desloratadine with P-glycoprotein. *Drug Metabolism and Disposition*, 29(8):1080–1083.

[64] Uchida Y., Ohtsuki S., Kamiie J., Terasaki T. (2011) Blood-brain barrier (BBB) pharmacoproteomics: reconstruction of in vivo brain distribution of 11 P-glycoprotein substrates based on the BBB transporter protein concentration, in vitro intrinsic transport activity, and unbound fraction in plasma and brain in mice. *Journal of Pharmacology and Experimental Therapeutics*, 339(2):579–588.

[65] Linnet K., Ejsing T. B. (2008) A review on the impact of P-glycoprotein on the penetration of drugs into the brain. Focus on psychotropic drugs. *European Neuropsychopharmacology*, 18(3):157–169.

[66] Anthes J. C., Gilchrest H., Richard C., Eckel S., Hesk D., West R. E., Jr., Williams S. M., Greenfeder S., Billah M., Kreutner W., Egan R. E. (2002) Biochemical characterization of desloratadine, a potent antagonist of the human histamine H(1) receptor. *European Journal of Pharmacology*, 449(3):229–237.

[67] Polli J. W., Baughman T. M., Humphreys J. E., Jordan K. H., Mote A. L., Salisbury J. A., Tippin T. K., Serabjit-Singh C. J. (2003) P-glycoprotein influences the brain concentrations of cetirizine (Zyrtec), a second-generation non-sedating antihistamine. *Journal of Pharmaceutical Science*, 92(10):2082–2089.

[68] Peytavin G., Gautran C., Otoul C., Cremieux A. C., Moulaert B., Delatour F., Melac M., Strolin-Benedetti M., Farinotti R. (2005) Evaluation of pharmacokinetic interaction between cetirizine and ritonavir, an HIV-1 protease inhibitor, in healthy male volunteers. *European Journal of Clinical Pharmacology*, 61(4):267–273.

[69] Linardi R. L., Stokes A. M., Andrews F. M. (2013) The effect of P-glycoprotein on methadone hydrochloride flux in equine intestinal mucosa. *Journal of Veterinary Pharmacology and Therapeutics*, 36(1):43–50.

[70] Olsen L., Ingvast-Larsson C., Bondesson U., Brostrom H., Tjalve H., Larsson P. (2007) Cetirizine in horses: pharmacokinetics and effect of ivermectin pretreatment. *Journal of Veterinary Pharmacology and Therapeutics*, 30(3):194–200.

[71] Cvetkovic M., Leake B., Fromm M., Wilkinson G., Kim R. (1999) OATP and P-glycoprotein transporters mediate the cellular uptake and excretion of fexofenadine. *Drug Metabolism and Disposition*, 27(8):866–871.

[72] Schinkel A. H., Wagenaar E., Mol C. A., van Deemter L. (1996) P-glycoprotein in the blood-brain barrier of mice influences the brain penetration and pharmacological activity of many drugs. *Journal of Clinical Investigation*, 97(11):2517–2524.

[73] Tahara H., Kusuhara H., Fuse E., Sugiyama Y. (2005) P-glycoprotein plays a major role in the efflux of fexofenadine in the small intestine and blood-brain barrier, but only a limited role in its biliary excretion. *Drug Metabolism and Disposition*, 33(7): 963–968.

[74] Perloff M. D., Stormer E., von Moltke L. L., Greenblatt D. J. (2003) Rapid assessment of P-glycoprotein inhibition and induction in vitro. *Pharmaceutical Research*, 20(8): 1177–1183.

[75] Lucero M. L., Gonzalo A., Ganza A., Leal N., Soengas I., Ioja E., Gedey S., Jahic M., Bednarczyk D. (2012) Interactions of bilastine, a new oral H(1) antihistamine, with human transporter systems. *Drug and Chemical Toxicology*, 35 Suppl 1:8–17.

[76] Mimura N., Nagata Y., Kuwabara T., Kubo N., Fuse E. (2008) P-glycoprotein limits the brain penetration of olopatadine hydrochloride, H1-receptor antagonist. *Drug Metabolism and Pharmacokinetics*, 23(2):106–114.

PART 5

CASE STUDIES IN CNS DRUG DISCOVERY

22

CASE STUDY 1: THE DISCOVERY AND DEVELOPMENT OF PERAMPANEL

ANTONIO LAURENZA,[1] JIM FERRY,[1] HAICHEN YANG,[1] SHIGEKI HIBI,[2] TAKAHISA HANADA,[2,3] AND ANDREW SATLIN[1]

[1] *Eisai Neuroscience Product Creation Unit, Woodcliff Lake, NJ, USA*
[2] *Global Biopharmacology, Neuroscience & General Medicine Product Creation System, Eisai Co Ltd, Tsukuba, Ibaraki, Japan*
[3] *Center for Tsukuba Advanced Research Alliance, Graduate School of Life and Environmental Sciences, University of Tsukuba, Tsukuba, Ibaraki, Japan*

INTRODUCTION

Since the introduction of phenytoin in 1938, there has been extensive research into the design and development of antiepileptic drugs (AEDs), with particularly notable advances over the past 20 years [1, 2]. Of the licensed AEDs, many target similar pathways, including voltage-gated sodium, calcium, and potassium channels, and γ-aminobutyric acid (GABA) receptors, transporters, and enzymes, with some drugs having multiple mechanisms of action [3, 4]. However, despite the availability of more than 20 AEDs, approximately 20–40% of patients with newly diagnosed epilepsy will be refractory to treatment or become refractory over time [1, 5]. One possible reason may be that AEDs are not delivered effectively to the brain due to the presence of efflux transporters such as P-glycoprotein (P-gp) in the blood–brain barrier [6]. A need, therefore, exists for therapies that are different from currently available AEDs, in particular for agents that can circumvent some of the issues associated with penetration of the central nervous system (CNS) and that are active at new molecular targets.

Blood–Brain Barrier in Drug Discovery: Optimizing Brain Exposure of CNS Drugs and Minimizing Brain Side Effects for Peripheral Drugs, First Edition. Edited by Li Di and Edward H. Kerns.

In recent years, neuronal α-amino-3-hydroxy-5-methyl-4-isoxazole propionic acid (AMPA)-type glutamate receptors have been identified as potential targets for new AED therapies. AMPA receptors mediate fast excitatory glutamate neurotransmission and have a key role in seizure initiation and propagation of seizure activity [7]. Although several AMPA receptor antagonists have been investigated as potential agents for the treatment of epilepsy, they have not succeeded in clinical development for several reasons, including crossing the blood–brain barrier [8].

Here, we provide an overview of the discovery and development of perampanel (Fycompa®, Eisai), which, in 2012, became the first noncompetitive AMPA receptor antagonist to receive approval from the U.S. Food and Drug Administration and the European Medicines Agency for the adjunctive treatment of partial-onset seizures, with or without secondary generalization, in patients aged 12 years and older [9, 10]. As of November 2013, perampanel is approved for use in over 30 countries (please consult the Prescribing Information in your country).

AMPA AS A TARGET FOR EPILEPSY TREATMENT

Glutamate is the most important excitatory neurotransmitter in the brain, and increased glutamatergic transmission has been associated with the etiology of epilepsy [11, 12]. Two distinct types of glutamate receptors exist: metabotropic glutamate receptors and ionotropic glutamate receptors. AMPA receptors, together with kainate and *N*-methyl-D-aspartate (NMDA) receptors, belong to the ionotropic glutamate receptor family and mediate extremely rapid synaptic excitatory responses to glutamate [13, 14]. Of the ionotropic glutamate receptors, AMPA receptors are the most widespread in the mammalian brain [7], and AMPA and NMDA are the main types found at excitatory synapses [8].

In the late 1980s, inhibition of non-NMDA ionotropic glutamate receptors was found to diminish or eliminate epileptiform activity in the rat hippocampus, cortex, and neocortex [15–19]. By contrast, selective blockade of NMDA receptors alone had no or minimal effect on epileptiform discharges *in vitro* [19]. These early observations supported a key role for AMPA receptors in the initiation of epileptiform activity and provided the basis for further research on AMPA receptor antagonists as agents for the treatment of epilepsy.

The quinoxalinediones were the first-reported selective AMPA receptor antagonists to achieve potent AMPA receptor blockade through a competitive interaction at the glutamate recognition site [8, 20]. However, early compounds were associated with several serious shortcomings. For example, 6-cyano-7-nitroquinoxaline-2, 3-dione was associated with low *in vivo* activity due to poor penetration of the blood–brain barrier [8]. By contrast, although the lipophilic analog 2,3-dioxo-6-nitro-1,2,3, 4-tetrahydrobenzo[*f*]quinoxaline-7-sulfonamide (NBQX) was systemically active [21, 22], it was poorly water soluble in the neutral pH range, resulting in precipitation in the kidneys and, thus, nephrotoxicity. Although structural modifications were made to NBQX to try to increase solubility, the inclusion of a polar moiety led to a further decrease in permeability across the blood–brain barrier [23, 24]. *In vivo* anticonvulsant activity was subsequently achieved with later quinoxalinediones; for example, YM90K (6-[1*H*-imidazol-1-yl]-7-nitro-2,3[1*H*,4*H*]-quinoxalinedione), YM872 ([2,3-dioxo-7-(1*H*-imidazol-1-yl)-6-nitro-1,2,3,4-tetrahydro-1-quinoxalinyl]-acetic acid monohydrate), and ZK200775 (1,2,3,4-tetrahydro-7-morpholinyl-2, 3-dioxo-6-[trifluoromethyl)quinoxalin-1-yl]methylphosphonate), as well as with structurally dissimilar competitive AMPA receptor antagonists (e.g., RPR117824 [9-(carboxymethyl)-4-oxo-5,10-dihydro-4*H*-imidazo[1,2-*a*]indeno[1,2-*e*]pyrazine-2-carboxylic acid]) [25–31]. However, none of these succeeded in clinical development.

The late 1980s and early 1990s saw the characterization of a novel class of selective noncompetitive AMPA receptor antagonists, the 2,3-benzodiazepines [32–34]. Mechanistically distinct from competitive AMPA receptor antagonists, non-competitive receptor antagonists interact with an allosteric binding site distinct from the AMPA recognition site [32]. It was anticipated that if noncompetitive receptor antagonists could be targeted to the CNS and have the ability to cross the blood–brain barrier, they could potentially have greater efficacy than competitive receptor antagonists. This is because the therapeutic effect of noncompetitive AMPA receptor antagonists is mediated independently of glutamate levels or polarization state of the synaptic membrane, and they have only minimal effects on normal glutamatergic activity [13].

GYKI 52466 (4-(8-methyl-9*H*-1,3-dioxolo[4,5-*h*][2, 3]benzodiazepin-5-yl)-benzenamine hydrochloride) was the prototype noncompetitive AMPA antagonist and demonstrated broad-spectrum action in animal models of epilepsy [35, 36]. The GYKI 52466 analog, talampanel, progressed to clinical evaluation for the treatment of epilepsy and other conditions with glutamate-related mechanisms as their underlying pathogenesis (e.g., recurrent malignant glioma and amyotrophic lateral sclerosis) [37–39]. In a Phase II, double-blind, crossover study in patients with refractory partial seizures ($n=49$), oral talampanel (25, 60, or 75 mg three times daily) as add-on therapy significantly reduced seizure frequency ($P=0.001$) with a median reduction in average weekly seizure rate (all types) of 21%. In addition, almost 80% (30/38) of patients who completed the study experienced fewer seizures on talampanel compared with placebo [37]. However, the relatively short half-life of talampanel (mean 3 h (range 1.6–5.8 h)) [40] necessitated multiple daily dosing in clinical trials [37–39].

Safety issues have limited the development of other AMPA receptor antagonists. For example, a Phase II dose-escalation trial of the selective, competitive AMPA receptor antagonist ZK 200775 in patients with acute ischemic stroke ($n=61$) was stopped prematurely due to reductions in consciousness (stupor and coma) in 8 of 13 patients who received a total ZK 200775 dose of 525 mg in 48 h by continuous infusion [41].

THE CONCEPTION OF PERAMPANEL

Historically, the discovery and development of new AEDs has involved either the screening of large numbers of compounds for antiseizure activity or the further optimization of existing AEDs. In a move away from this traditional approach, Eisai

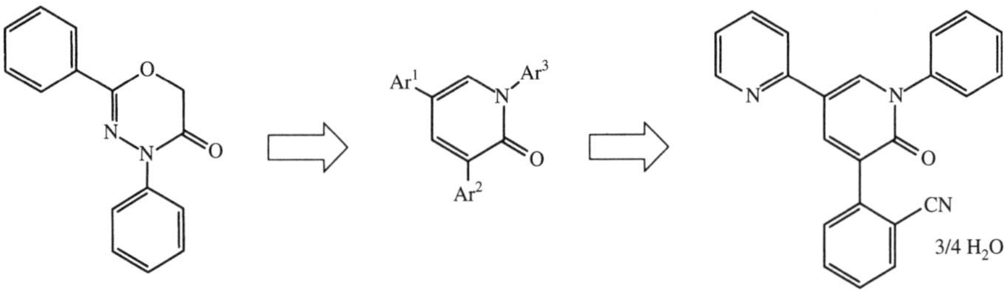

1 2 3

FIGURE 22.1 Chemical structures of (**1**) 2,4-diphenyl-4*H*-[1,3,4]oxadiazin-5-one, (**2**) 1,3,5-triaryl-1*H*-pyridin-2-one, and (**3**) perampanel.

Research Laboratories conducted a focused discovery program [42] to identify an orally active, noncompetitive, potent AMPA receptor antagonist with a favorable pharmacokinetic and safety profile that could permeate across the blood-brain barrier.

To explore potential new compounds with different structures, two high-throughput screening assays were used [43, 44]. The rat cortical neuron AMPA-induced cell death assay served as a functional assay of AMPA receptor blockade, whilst a [^{3}H]AMPA-binding assay was used for the detection of compounds that acted as competitive AMPA receptor antagonists [44]. To provide a more direct assay of AMPA receptor antagonist activity, and to allow elimination of false positives, a membrane-permeant calcium-sensitive dye was used to test potential compounds for their ability to inhibit AMPA-induced Ca^{2+} influx in rat primary neuron cultures [43].

Although the [^{3}H]AMPA-binding assay failed to identify any potential compounds, the functional assay identified 2,4-diphenyl-4*H*-[1,3,4]oxadiazin-5-one, a noncompetitive AMPA receptor antagonist with a half maximal inhibitory concentration (IC_{50}) of approximately 5 μM for AMPA receptor blockade [44]. Using 2,4-diphenyl-4*H*-[1,3,4]oxadiazin-5-one as a starting point, medicinal chemistry efforts were subsequently focused on developing an orally absorbed drug that had improved potency and was able to penetrate the blood–brain barrier. This strategy resulted in the identification of the pyridine core structure, 1,3,5-triaryl-1*H*-pyridin-2-one, as the preferred scaffold for further optimization [43]. Several potential clinical compounds were built around this pyridine core, but it was modification to optimize oral efficacy that led to the identification of perampanel (2-[2-oxo-1-phenyl-5-pyridin-2-yl-1,2-dihydropyridin-3-yl]benzonitrile) (Fig. 22.1). With a pK_a ($-\log_{10}$ of the acid dissociation constant, K_a) of 3.24, perampanel (weak base) dissolves readily at acidic pHs, such that it is completely absorbed from the gastrointestinal tract, maximizing oral absorption [44]. The molecular weight of perampanel is 349.4 g/mol, calculated partition coefficient (Clog*P*, a measure of lipophilicity) is 4.2, topological polar surface area is 58.7 square angstroms, the number of hydrogen bond donors is 0, and the number of hydrogen bond acceptors is 4 (data on file). These values are broadly consistent with the general properties of druggable compounds [45].

THE PRECLINICAL EVALUATION OF PERAMPANEL—*IN VITRO* PHARMACOLOGY

In vitro functional and binding studies have provided confirmatory evidence that perampanel is an orally active, selective, noncompetitive antagonist of the AMPA receptor. In cultured rat cerebral cortical neurons, perampanel inhibited AMPA-induced increases in intracellular Ca^{2+} in a concentration-dependent manner, with an IC_{50} of 93 nM (32.5 ng/ml) [24]. This compared favorably with an IC_{50} of 12.5 μM for the prototypical noncompetitive AMPA antagonist GYKI 52466 [24]. In contrast, perampanel showed only weak inhibition of Ca^{2+} increases induced by NMDA (100 μM) in rat cortical neurons, and only at the highest concentration tested; the maximal inhibitory response was 18% with perampanel 30 μM compared with 85% with the noncompetitive NMDA receptor antagonist MK801 ([5*S*,10*R*]-[+]-5-methyl-10,11-dihydro-5*H*-dibenzo(*a*,*d*)cyclohepten-5,10-imine maleate; 1 μM) [24].

The selectivity of perampanel was further investigated in the rat hippocampus by determining its inhibitory activity on the synaptic responses known to be selectively mediated by AMPA, NMDA, or kainate receptors. Perampanel inhibited AMPA receptor-mediated field excitatory postsynaptic potentials (f-EPSPs) in rat hippocampal slices with an IC_{50} of 0.23 μM and was approximately 34-fold more potent than GYKI 52466 (IC_{50} 7.8 μM). Complete inhibition of AMPA receptor-mediated f-EPSPs was reported with perampanel 3 μM [46]. However, at doses at least 100-fold higher, perampanel (10 μM) failed to reduce NMDA or kainate receptor-mediated f-EPSPs, which were completely blocked by the competitive NMDA antagonists D-APS (30 μM) and NBQX (10 μM) [46]. These findings were consistent with the theory that selective AMPA receptor antagonism was the primary mode of action of perampanel.

Analysis of whole cell voltage-clamp recordings in cultured rat hippocampal neurons expressing native AMPA receptors indicated that perampanel binds with similar affinity to the open and closed states of AMPA receptors, and does not affect AMPA receptor desensitization, having no effect on the decay time course of AMPA-evoked currents [47]. In addition, high (30 μM) perampanel concentrations had no effect on NMDA-evoked patch clamp currents in this cell culture [47].

Assessment of binding affinity using radiolabeled perampanel ([^{3}H]perampanel) in rat forebrain membranes provided additional evidence of a noncompetitive interaction between perampanel and AMPA receptors. Binding of [^{3}H]perampanel to rat forebrain membranes was not significantly reduced by glutamate (1 mM), AMPA (0.1 mM), or NBQX (0.1 mM); however, [^{3}H]perampanel was displaced by the noncompetitive AMPA receptor antagonists CP465022 (IC_{50} 21.1 nM) and GYKI 52466 (IC_{50} 23.3 nM), suggesting a common site of action on the AMPA receptor for all three noncompetitive antagonists [24].

Preclinical data suggested that perampanel had good oral absorption and the potential to be a novel therapeutic target in the CNS. A critical additional requirement for success was the ability of perampanel to cross the blood–brain barrier. This barrier controls the distribution of drugs into the brain [48] and, in particular, efflux transporters such as P-gp and the breast cancer resistance protein (BCRP) stop many drugs from entering and accumulating in the brain [49, 50]. The permeability of perampanel remained consistent between native human cell lines and P-gp- and BCRP-overexpressing cell lines, indicating that perampanel is not a substrate for these blood–brain barrier efflux transporter proteins. In fact, perampanel was found to be a weak inhibitor of P-gp and BCRP (IC_{50} 18.1 and 12.8 µmol/l, respectively) [44, 51]. As such, these efflux transporters were not considered to limit the access of perampanel to the brain.

THE PRECLINICAL EVALUATION OF PERAMPANEL—*IN VIVO* PHARMACOLOGY

In vivo studies showed that perampanel demonstrated broad-spectrum antiseizure activity when tested in a range of animal models of epilepsy, including those seizures induced by electrical, chemical, and sensory stimuli [24]. Oral perampanel was a more potent anticonvulsant than carbamazepine and sodium valproate in protecting mice from tonic–clonic generalized seizures in the maximal electric shock test (50% effective dose [ED_{50}]: 1.6 vs 21 and 460 mg/kg, respectively) and audiogenic-induced seizure test (ED_{50}: 0.47 vs 6.1 and 160 mg/kg, respectively), as well as from absence or myoclonic seizures in subcutaneous pentylenetetrazole-induced seizure tests (0.94 vs >100 and 350 mg/kg, respectively) [24]. In the rat amygdala-kindling model of temporal lobe epilepsy, perampanel (10 mg/kg orally) significantly increased the after-discharge threshold (ADT; $P<0.05$ vs vehicle) and significantly attenuated motor seizure duration, after-discharge duration, and seizure severity recorded at 50% higher intensity than the ADT current ($P<0.05$ for all measures vs vehicle) [24].

Perampanel has also been studied in animal models of refractory epilepsy. Using a rat amygdala-kindling model with a strong stimulus intensity (3-fold higher than the ADT) to elicit drug-resistant partial-onset seizures, Wu et al. [52] reported synergistic reductions in electroencephalogram seizure duration, motor seizure duration, and seizure score when low-dose perampanel (0.75 mg/kg) was administered in combination with the AEDs levetiracetam (50 mg/kg), lamotrigine (20 mg/kg), carbamazepine (20 mg/kg), and sodium valproate (200 mg/kg) [52]. Perampanel also terminated seizures in an animal model of benzodiazepine-resistant status epilepticus. Using the lithium-pilocarpine rat model of status epilepticus, perampanel terminated seizures in all 6 of 6 rats tested when administered intravenously 60 min after pilocarpine administration. By contrast, GYKI 52466 (50 mg/kg intravenously) only terminated seizures in 2 out of 4 rats when administered at 30 min [53].

Using the 6 Hz electroshock-induced seizure test, which is a refractory psychomotor seizure model, sodium channel-blocking AEDs are at best only weakly effective. Some drugs, for example, phenytoin and lamotrigine, demonstrate a reduced inhibitory response with increasing stimulation intensity [54]. Using the 6 Hz electroshock seizure test, perampanel (1–8 mg/kg orally) conferred dose-dependent protection against seizures in mice. Notably, the ED_{50} for perampanel in this setting was similar at both 32 mA (2.1 mg/kg) and 44 mA (2.8 mg/kg) [24]. Furthermore, there was evidence that perampanel acted synergistically with other AEDs, with greater reductions in the incidence of seizures when perampanel was administered in combination with phenytoin (10 mg/kg), carbamazepine (20 mg/kg), and sodium valproate (100 mg/kg) [24].

Perampanel was reported to cause dose-dependent motor impairment in mice and rats using the rotarod performance test, which assesses motor coordination (median toxic dose of 1.8 and 9.1 mg/kg in mice and rats, respectively). These doses were very close to the levels of perampanel needed to reduce seizure activity, indicating that perampanel has a narrow therapeutic window. Such CNS-depressant effects were not unexpected with perampanel, given the key role played by AMPA receptors in mediating glutamatergic excitatory neurotransmission in the CNS, although preclinical studies are not reliable indicators of the clinical therapeutic index of AEDs [24].

THE PRECLINICAL EVALUATION OF PERAMPANEL—PHARMACOKINETICS

Preliminary pharmacokinetic evaluation in a small number of mice and rats was conducted as part of the early development of perampanel. In these studies, brain, plasma, and cerebrospinal fluid concentrations of perampanel indicated that perampanel freely penetrates the brain [43] and supports previous evidence that perampanel is not a substrate for blood–brain barrier transporter proteins [9, 44]. Brain-to-plasma concentration ratios following oral administration of perampanel to mice (3 mg/kg) and rats (10 mg/kg) were 1.06 and 1.14, respectively (Table 22.1). The cerebrospinal fluid-to-unbound plasma concentration ratio for perampanel (0.5 mg/kg intraperitoneally) was 1.14 in mice [24, 43], again supporting previous evidence that perampanel freely crosses the blood–brain barrier.

An overview of the pharmacokinetic parameters for perampanel in rats is shown in Table 22.2. Although the terminal elimination half-life of perampanel was short in animals (1.4–2.4 h in rats, 5.3–7.6 h in dogs and monkeys) [51], liver microsome studies suggested a very low metabolic turnover rate in human liver microsomes (clearance 0.009 μl/min/mg protein) [43]. Consistent with this, perampanel was reported to have a long half-life in Phase I pharmacokinetic studies in healthy male volunteers: the half-life of perampanel ranged from 52 to 129 h, following single-dose administration (0.2–8.0 mg) and from 66 to 90 h after multiple-dose administration

TABLE 22.1 Brain/plasma or cerebrospinal fluid/unbound plasma concentration ratios for perampanel in male mice or rats[a]

Species	Dose	Sampling time (min)	Brain/plasma concentration ratio	Cerebrospinal fluid/unbound plasma concentration ratio
Mice	3 mg/kg p.o.	60	1.06	NT
Mice	0.5 mg/kg i.p.	20	NT	1.14
Rats	10 mg/kg p.o.	30	1.14	NT

Each value represents the mean of two or three animals. i.p., intraperitoneal injection; NT, not tested; and p.o., oral administration.

[a] Reprinted [adapted] with permission from Hibi, S., et al. Discovery of 2-(2-oxo-1-phenyl-5-pyridin-2-yl-1, 2-dihydropyridin-3-yl)benzonitrile (perampanel): a novel, noncompetitive a-amino-3-hydroxy-5-methyl-4-isoxazolepropanoic acid (AMPA) receptor antagonist. *Journal of Medicinal Chemistry* [2012], *55*, 10584-10600. Copyright 2013 American Chemical Society.

TABLE 22.2 Pharmacokinetic parameters for perampanel (i.v. and p.o.) in male rats[a]

Perampanel 1 mg/kg (i.v.)		Perampanel 1 mg/kg (p.o.)	
CL (l/h/kg)	1.82	C_{max} (μg/ml)	0.17
V_{dss} (l/kg)	4.56	T_{max} (h)	0.50
AUC (μg/ml·h)	0.56	AUC (μg/ml·h)	0.36
$T_{1/2}$ (h)	2.37	F (%)	64.3

Each parameter is calculated from the mean of plasma concentrations of three animals. AUC, area under the concentration–time curve; CL, clearance; F, ratio of $AUC_{(0\text{-inf})}$ values after oral (p.o.) and intravenous (i.v.) administrations; $T_{1/2}$, half-life; T_{max}, time to meet peak plasma concentration; V_{dss}, volume of distribution at steady state.

[a]Reprinted [adapted] with permission from Hibi, S., et al. Discovery of 2-(2-oxo-1-phenyl-5-pyridin-2-yl-1,2-dihydropyridin-3-yl)benzonitrile (perampanel): a novel, noncompetitive a-amino-3-hydroxy-5-methyl-4-isoxazolepropanoic acid (AMPA) receptor antagonist. *Journal of Medicinal Chemistry* [2012], *55*, 10584-10600 [43]. Copyright 2013 American Chemical Society.

(1–6 mg/day for 14 days) [55]. The combination of a low metabolic rate in human liver microsomes and a long half-life suggested that perampanel was potentially suitable for once-daily administration in humans.

Following oral administration, perampanel is rapidly and almost completely absorbed in humans, with low systemic clearance and bioavailability approaching 100% [9, 44, 55]. In healthy volunteers, peak plasma concentrations of perampanel were observed within approximately 1.0 h and steady-state plasma concentrations were reached within 14 days of multiple-dose administration [55]. Perampanel undergoes negligible first-pass metabolism but is extensively metabolized via oxidation and sequential glucuronidation, principally by the cytochrome (CYP) enzyme CYP3A4. The resulting metabolites are thought to have negligible contribution to the therapeutic activity of perampanel, given their very low plasma concentrations relative to unchanged perampanel and their inability to inhibit AMPA receptors [44]; they are excreted via the fecal and urinary routes.

Perampanel does not act as an enzyme inducer or inhibitor; however, the pharmacokinetic profile of perampanel is affected by enzyme-inducing drugs, specifically inducers of CYP3A4. The CYP3A4-inducer carbamazepine (300 mg twice daily) increased the clearance of perampanel 3-fold, and decreased the maximum plasma concentration and bioavailability area under the concentration–time curve from 0 to time infinity [$AUC_{(0\text{-inf})}$] by 26 and 67%, respectively. This resulted in an average half-life of 25 h. Ketoconazole, a CYP3A4 inhibitor, increased the exposure $AUC_{(0\text{-inf})}$ of perampanel by 20% [51]. Plasma protein binding of perampanel is approximately 95% [9, 44].

THE CLINICAL EVALUATION OF PERAMPANEL

As a result of promising preclinical and pharmacokinetic data, perampanel was further evaluated in an extensive step-wise clinical development program across a large, multinational population of adolescent and adult patients with refractory

Phase II[a]	Phase III[b]	Long-term extension
Study 206 [56] *n*=153 (18–70 years) Uncontrolled POS despite ≥3 AEDs in past 2 years PER 1–4 mg/day (QD or BID) vs PL 8-week dose titration[c] + 4-week maintenance	**Study 304** [57] *n*=388 (≥12 years) Uncontrolled POS ± SG on 1–3 AEDs Adjunctive PER 8 or 12 mg/day vs PL 6-week dose titration (2-mg increment per week) + 13-week maintenance 1° endpoint: % change in seizure frequency (50% responder rate for EU registration)	**Study 207** [61] Patients (*n*=138 enrolled) who completed a Phase II study PER dose titration to ≤12 mg/day 12-week dose titration (2-mg increment every 2 weeks) + up to 424-week maintenance (≈8 years)
Study 208 [56] *n*=48 (18–70 years) Uncontrolled POS despite ≥3 AEDs in past 2 years PER ≤12 mg/day (QD) vs PL 12-week dose titration[c] + 4-week maintenance	**Study 305** [58] *n*=386 (≥12 years) Uncontrolled POS ± SG on 1–3 AEDs Adjunctive PER 8 or 12 mg/day vs PL 6-week dose titration (2-mg increment per week) + 13-week maintenance 1° endpoint: % change in seizure frequency (50% responder rate for EU registration)	**Study 307** [60] Patients (*n*=1218 enrolled) who completed a Phase III study PER dose titration (≤12 mg/day, QD) 16-week blinded conversion period (2-mg increment every 2 weeks) + 256-week maintenance + 4-week follow-up
	Study 306 [59] n=706 (≥12 years) Uncontrolled POS ± SG on 1–3 AEDs Adjunctive PER 2, 4, or 8 mg/day vs PL 6-week dose titration (2-mg increment per week) + 13-week maintenance 1° endpoint: % change in seizure frequency (50% responder rate for EU registration)	

AED, antiepileptic drug; BID, twice daily; PER perampanel; PL, placebo; POS, partial-onset seizures; QD, once daily; SG, secondary generalized.
[a]Multicenter, randomized, double-blind, placebo-controlled, dose-escalation trials.
[b]Multicenter, randomized, double-blind, placebo-controlled trials (a pooled analysis of the three Phase III studies has been published [64]).
[c]Patients who did not tolerate treatment during titration were permitted to revert to the previous dose level.

FIGURE 22.2 Overview of perampanel clinical studies.

partial-onset seizures who were receiving 1–3 concomitant AEDs. The clinical program included two Phase II dose-finding trials, in which the perampanel dose was titrated up to 4 or 12 mg once daily [56], three randomized, double-blind, placebo-controlled Phase III registration trials, in which perampanel was titrated up to 8 or 12 mg once daily [57–59], and two long-term extension studies (Fig. 22.2) [60, 61]. Here, we provide an overview of the efficacy and safety data from these studies of perampanel, with the key findings summarized in Tables 22.3 and 22.4. For a more comprehensive review, please refer to the recent articles by Kerling and Kasper [62] and Serratosa et al. [63].

In the two Phase II dose-escalation studies, which were designed to guide the minimum and maximum doses of perampanel to be tested in Phase III clinical trials, perampanel was well tolerated at doses up to 12 mg/day [56]. In Study 206, 82.4% of patients tolerated perampanel 4 mg/day, whilst in Study 208, it was estimated that 97, 55, and 44% of patients tolerated perampanel doses of 4, 8, and 12 mg/day, respectively. Although not powered to detect a statistically significant difference in terms of efficacy, responder rate (defined as the proportion of patients experiencing a ≥50% reduction in seizure frequency) was higher and there was a greater improvement in seizure frequency with perampanel versus placebo in both studies (Table 22.3) [56]. Within the therapeutic dose range (perampanel 4–12 mg/day) and assuming 95% protein binding with a free fraction of 5%, free plasma concentrations were 0.03–0.07 μM meaning that concentrations of perampanel at the AMPA receptor would be much lower than the IC_{50} [44]. This suggests that a low level of inhibition of AMPA receptors is sufficient for protection against seizures [8, 46].

Based on the results of the Phase II trials, the efficacy and tolerability of perampanel (2–12 mg/day) were investigated in the three pivotal Phase III, double-blind, placebo-controlled clinical trials. The intent-to-treat population across the three Phase III studies (Fig. 22.2) included 1478 patients [57–59]. In Study 306 in which patients were enrolled from Europe, Asia, and Australia, and received perampanel 2, 4, or 8 mg/day, a clear dose-dependent treatment effect was observed for both median percent change in seizure frequency and 50% responder rate; the difference versus placebo reached statistical significance for perampanel 4 and 8 mg/day for both endpoints (Table 22.3) [59]. In Studies 304 and 305, patients received higher once-daily perampanel doses of 8 or 12 mg/day [57, 58]; patients were enrolled from North and Latin America in Study 304, and from Europe, North America, and Australia in Study 305. In Study 304, median percent change in seizure frequency was significantly higher with perampanel 8 and 12 mg/day versus placebo; however, the higher responder rates with perampanel were not statistically significant versus placebo (Table 22.3). From pooled Phase III pivotal trial data, a subgroup analysis by geographic region found a higher placebo response rate in patients from Latin America [65], causing a treatment-by-region interaction that was statistically significant. However, the reason for this higher placebo response rate is unknown since there were no differences in demographic or baseline characteristics unique to Latin America. Geographic region was also not found to influence the relationship between efficacy responses and plasma concentrations of perampanel, providing further evidence that across all geographic regions there is a consistent degree of improvement in seizure control across the effective dose range (4–12 mg) [66]. By contrast, for perampanel doses of 8 and 12 mg/day in Study 305, both median percent change in seizure frequency from baseline per 28 days and responder rates were significantly greater compared with placebo (Table 22.1).

A subsequent analysis of pooled Phase III pivotal trial data further supported these findings (Table 22.3) [64]. In addition, and from an integrated actual (last) dose analysis of data from the three pivotal Phase III trials and a long-term extension study (Study 307), in patients who were able to tolerate perampanel 12 mg/day, the higher dose was associated with additional benefits in seizure control compared with 8 mg/day [67]. Although only interim data have been published to date, efficacy data from the two long-term extension studies indicate that the degree of seizure control observed in the Phase II and III studies is maintained during long-term therapy (Table 22.3) [60, 61]. Available pharmacokinetic/pharmacodynamic data, based on

TABLE 22.3 Overview of Efficacy Data

	Phase II		Phase III			Long-term extension		Pooled analysis
	Study 206 (n = 153) [56]	Study 208 (n = 48) [56]	Study 304 (n = 388) [57]	Study 305 (n = 386) [58]	Study 306 (n = 706) [59]	Study 207 Phase II[a] (n = 138) [61]	Study 307 Phase III[b] (n = 1186) [60]	Phase III[c] (n = 1480) [64]
Baseline seizure frequency/month (median)	PER QD: 16.6[d] PER BID: 26.4[d] Placebo: 19.6[d]	PER QD: 17.6[d] Placebo: 17.3[d]	PER 8 mg/day: 14.3 12 mg/day: 12.0 Placebo: 13.7	PER 8 mg/day: 13.0 12 mg/day: 13.7 Placebo: 11.8	PER 2 mg/day: 10.1 4 mg/day: 10.0 8 mg/day: 10.9 Placebo: 9.3	9.0	11.2	PER 2 mg/day: 10.1 4 mg/day: 10.0 8 mg/day: 12.2 12 mg/day: 13.0 Placebo: 11.1
Median change in seizure frequency per 28 days vs baseline (%)	PER: –25.7 Placebo: –19.5	PER: –39.6 Placebo: +2.1	PER 8 mg/day: –26.3* 12 mg/day: –34.5* Placebo: –21.0	PER 8 mg/day: –30.5* 12mg/day: –17.6* Placebo: –9.7	PER 2 mg/day: –13.6 4 mg/day: –23.3* 8 mg/day: –30.8* Placebo: –10.7	Overall (n = 138): –31.5 1 year (n = 89): –43.7 2 year (n = 66): –52.0 3 year (n = 52): –49.7 4 year (n = 18): –48.4	1 year (n = 588): –47.2[e] 2 year (n = 19): –56.0[e]	PER 2 mg/day: –13.6 4 mg/day: –23.3* 8 mg/day: –28.8* 12 mg/day: –27.2* Placebo: –12.8
Responder rates (%)[f]	PER: 30.7 (n = 31/101) Placebo: 21.6 (n = 11/51)	PER: 39.5 (n = 15/38) Placebo: 22.2 (n = 2/9)	PER 8 mg/day: 37.6 12 mg/day: 36.1 Placebo: 26.4	PER 8 mg/day: 33.3* 12 mg/day: 33.9* Placebo: 14.7	PER 2 mg/day: 20.6 4 mg/day: 28.5* 8 mg/day: 34.9* Placebo: 17.9	1 year (n = 89): 43.8 2 year (n = 66): 51.5 3 year (n = 52): 49.0 4 year (n = 18): 50.0	1 year (n = 588): 47.6 2 year (n = 19): 63.2	PER 2 mg/day: 20.6 4 mg/day: 28.5* 8 mg/day: 35.3* 12 mg/day: 35.0* Placebo: 19.3

(Continued)

TABLE 22.3 (*Continued*)

	Phase II		Phase III			Long-term extension		Pooled analysis
	Study 206 ($n=153$) [56]	Study 208 ($n=48$) [56]	Study 304 ($n=388$) [57]	Study 305 ($n=386$) [58]	Study 306 ($n=706$) [59]	Study 207 Phase II[a] ($n=138$) [61]	Study 307 Phase III[b] ($n=1186$) [60]	Phase III[c] ($n=1480$) [64]
Median change in frequency of CP+SG per 28 days vs baseline (%)	—	—	PER 8 mg/day: –33.0* 12 mg/day: –33.1* Placebo: –17.9	PER 8 mg/day: –32.7* 12 mg/day: –21.9* Placebo: –8.1	PER 2 mg/day: –20.5 4 mg/day: –31.2* 8 mg/day: –38.7* Placebo: –17.6	–44.7[g]	—	PER 2 mg/day: –20.5 4 mg/day: –31.2* 8 mg/day: –35.6* 12 mg/day: –28.6* Placebo: –13.9
Patients free of seizures (%)	—	—	PER 8 mg/day: 2.2 12 mg/day: 1.5 Placebo: 0.0	PER 8 mg/day: 2.3 12 mg/day: 5.0 Placebo: 1.5	PER 2 mg/day: 1.9 4 mg/day: 4.4 8 mg/day: 4.8 Placebo: 1.2	2.9[h]	7.1 (12-month data only)	PER 2 mg/day: 1.9 4 mg/day: 4.4* 8 mg/day: 3.5* 12 mg/day: 4.1* Placebo: 1.0

BID, twice daily; CP, complex partial; PER, perampanel; QD, once daily; SG, secondary generalized seizures; yr, year(s).

[a] Patients were recruited from the Phase II Studies 206 and 208.

[b] Patients were recruited from the Phase III Studies 304, 305, and 306, and data shown are from the interim cutoff date.

[c] Pooled analysis of the Phase III Studies 304, 305, and 306.

[d] Mean.

[e] In the last 13-week interval.

[f] Percentage of patients with ≥50% reduction in seizure frequency.

[g] Of the 131 patients with CP±SG across the entire open-label extension treatment phase (titration period 12 weeks; maintenance period up to 424 weeks). Responder rate for group was 45.8%.

[h] By the interim cutoff date.

*$P<0.05$ versus placebo.

pooled data from the three pivotal Phase III trials, suggest that within the perampanel dose range (2–12 mg), a linear relationship exists between increasing plasma concentration and clinical response. Increasing average perampanel plasma concentration at steady state was associated with a reduction in seizure frequency and an increased probability of a patient being a responder (≥50% reduction in seizure frequency) [66].

In Phase II and III studies, perampanel 2–12 mg/day was associated with an acceptable safety profile, with most adverse events being of mild-to-moderate intensity [56–59]. The most common adverse events reported in Phase III clinical trials of perampanel and during long-term follow-up are summarized in Table 22.4. No clinically significant changes were reported in terms of vital signs, laboratory values, or electrocardiogram parameters. In accordance with the CNS-depressant effects observed in preclinical studies, the most commonly reported adverse events in the three perampanel Phase III pivotal studies were dizziness, somnolence, and headache, occurring with an incidence of 10.0–47.9, 9.3–18.2, and 8.5–15.0%, respectively, with perampanel 2–12 mg/day [57–59]. Available pharmacokinetic/pharmacodynamic data pooled from these Phase III pivotal studies found a significantly higher predicted probability of dizziness, somnolence, fatigue, irritability, gait disturbance, weight increase, dysarthria and euphoric mood for higher doses of perampanel [65]. Based on the dose-dependent trend that was generally evident for dizziness and somnolence, both these adverse events are listed in the "warnings and precautions" section of the European Summary of Product Characteristics and the U.S. Prescribing Information [9, 10]. Such CNS-related symptoms are not unique to perampanel, and they are commonly reported with a range of AEDs, including lamotrigine, carbamazepine, and gabapentin [68–70].

Other adverse events occurring in ≥10% of perampanel-treated patients during Phase III evaluation included fatigue, fall, ataxia, and irritability. As the risk of falls may be a result of increased dizziness and somnolence, the "warnings and precautions" section of the current prescribing guidelines highlights that patients should be assessed for falls and injuries, and gait disturbance [9, 10]. Psychiatric and behavioral adverse events (classified as either hostility or aggression) were reported in a higher number of patients receiving perampanel in the three pivotal Phase III studies compared with placebo; the rate increased with a higher dose and was related to irritability. A total of 12 patients receiving perampanel experienced serious psychiatric treatment-emergent adverse events (the most common was aggression; $n = 3$) compared with four patients receiving placebo (no aggression) [64]. Current prescribing guidelines warn of the risk of serious psychiatric and behavioral reactions, and recommend monitoring patients for psychiatric events. In cases of persistent severe or worsening psychiatric symptoms or behaviors, dose reduction of perampanel or permanent discontinuation may be necessary [9, 10].

A pooled analysis of safety data from the three pivotal Phase III studies generally reflected the data from the individual studies [63]. Furthermore, no unexpected adverse events or new safety concerns have been identified to date in the two open-label extension trials [60, 61] (Table 22.4).

TABLE 22.4 Overview of the Most Common Adverse Events in Perampanel Phase III Studies.

	Study 304 [57]			Study 305 [58]		
	PER: 8 mg (n=133)	**PER: 12 mg** (n=134)	**Placebo** (n=121)	**PER: 8 mg** (n=129)	**PER: 12 mg** (n=121)	**Placebo** (n=136)
Any adverse event	117 (88.0)	123 (91.8)	100 (82.6)	112 (86.8)	104 (86.0)	93 (68.4)
Any treatment-related adverse event	99 (74.4)	108 (80.6)	58 (47.9)	89 (69.0)	94 (77.7)	65 (47.8)
Any adverse event leading to study/drug discontinuation	9 (6.8)	26 (19.4)	8 (6.6)	12 (9.3)	23 (19.0)	6 (4.4)
Any adverse event leading to dose reduction/interruption	30 (22.6)	45 (33.6)	6 (5.0)	27 (20.9)	34 (28.1)	5 (3.7)
Any serious adverse event	8 (6.0)	9 (6.7)	6 (5.0)	10 (7.8)	12 (9.9)	7 (5.1)
Adverse events in ≥10% (any treatment group)[c]						
Dizziness	50 (37.6)	51 (38.1)	12 (9.9)	42 (32.6)	58 (47.9)	10 (7.4)
Somnolence	24 (18.0)	23 (17.2)	16 (13.2)	16 (12.4)	22 (18.2)	4 (2.9)
Headache	20 (15.0)	18 (13.4)	16 (13.2)	11 (8.5)	16 (13.2)	18 (13.2)
Fatigue	—	—	—	17 (13.2)	20 (16.5)	11 (8.1)
Fall	13 (9.8)	17 (12.7)	8 (6.6)	—	—	—
Irritability	10 (7.5)	19 (14.2)	6 (5.0)	—	—	—
Ataxia	8 (6.0)	16 (11.9)	0	—	—	—
Nausea	—	—	—	—	—	—
Vomiting	—	—	—	—	—	—

All data shown are number (%) of patients. PER, perampanel.

[a] At the interim cutoff date.

[b] Maximum daily dose of perampanel patients were exposed to; one patient received perampanel <4 mg/day and has not been included in table.

[c] Adverse event incidence rate for each study is only reported if adverse events occurred in ≥10% of patients in at least one treatment group.

Reprinted with permission from French, J.A., et al. Adjunctive perampanel for refractory partial-onset seizures: randomized phase III study 304. *Neurology* [2012], *79*, 589–596. Copyright 2013 American Academy of Neurology; reprinted with permission from French, J.A., et al. Evaluation of adjunctive perampanel in patients with refractory partial-onset seizures: results of randomized global phase III study 305. *Epilepsia* [2013], *54*, 117–125. Copyright 2013 International League Against Epilepsy; reprinted with permission from Krauss, G.L., et al. Randomized phase III study 306: adjunctive perampanel for refractory partial-onset seizures. Neurology [2012], *78*, 1408–1415. Copyright 2013 American Academy of Neurology; reprinted with permission from Krauss, G.L., et al. Perampanel, a selective, noncompetitive a-amino-3-hydroxy-5-methyl-4-isoxazolepropionic acid receptor antagonist, as adjunctive therapy for refractory partial-onset seizures: interim results from phase III, extension study 307. *Epilepsia* [2013], *54*, 126–134. Copyright 2013 International League Against Epilepsy) [57–60].

CONCLUSIONS

The development of perampanel is a valuable example of how a focused drug discovery program was used to identify an AMPA receptor antagonist that was able to overcome the pharmacokinetic challenges observed with prior AMPA receptor antagonists, including a lack of blood–brain barrier permeability and a short half-life.

Study 306 [59]				Study 307 (long-term extension)[a] [60]		
PER: 2 mg ($n = 180$)	**PER: 4 mg** ($n = 172$)	**PER: 8 mg** ($n = 169$)	**Placebo** ($n = 185$)	**PER: 4 mg** ($n = 15$)[b]	**PER: >4–8 mg** ($n = 86$)[b]	**PER: >8–12 mg** ($n = 1084$)[b]
111 (61.7)	111 (64.5)	121 (71.6)	101 (54.6)	13 (86.7)	83 (96.5)	940 (86.7)
67 (37.2)	77 (44.8)	96 (56.8)	59 (31.9)	13 (86.7)	82 (95.3)	832 (76.8)
12 (6.7)	5 (2.9)	12 (7.1)	7 (3.8)	6 (40.0)	23 (26.7)	127 (11.7)
3 (1.7)	12 (7.0)	29 (17.2)	6 (3.2)	10 (66.7)	71 (82.6)	386 (35.6)
6 (3.3)	6 (3.5)	6 (3.6)	9 (4.9)	2 (13.3)	11 (12.8)	144 (13.3)
18 (10.0)	28 (16.3)	45 (26.6)	18 (9.7)	9 (60.0)	51 (59.3)	461 (42.5)
22 (12.2)	16 (9.3)	27 (16.0)	12 (6.5)	3 (20.0)	23 (26.7)	214 (19.7)
16 (8.9)	19 (11.0)	18 (10.7)	16 (8.6)	2 (13.3)	24 (27.9)	172 (15.9)
—	—	—	—	2 (13.3)	13 (15.1)	128 (11.8)
—	—	—	—	—	—	—
—	—	—	—	—	—	—
—	—	—	—	0	10 (11.6)	63 (5.8)
—	—	—	—	2 (13.3)	7 (8.1)	72 (6.6)
—	—	—	—	2 (13.3)	0	61 (5.6)

Perampanel was characterized by high oral absorption, a long half-life supporting once-daily dosing, and was found to freely cross the blood–brain barrier. Perampanel is the first AMPA receptor antagonist to be approved in an epilepsy indication, offering a potential treatment for patients refractory to other AEDs and paving the way for the further development and application of this class of compounds. In the future, effective AMPA receptor antagonists may be useful for the management and treatment of other epileptic conditions. Perampanel has shown broad-spectrum antiepileptic activity, including efficacy against absence or myoclonic seizures and primary generalized tonic–clonic seizures in animal models. Phase III evaluation (NCT01393743) of perampanel for the treatment of patients aged 12 years or older with primary generalized tonic–clonic seizures is ongoing.

DISCLOSURES

A Laurenza, J Ferry, H Yang, S Hibi, T Hanada, and A Satlin are employees of Eisai, Inc.

ACKNOWLEDGMENTS

Editorial support was provided by Hannah FitzGibbon and Julie Adkins of Complete Medical Communications, and was funded by Eisai Inc.

REFERENCES

[1] Bialer, M. and White, H.S. (2010). Key factors in the discovery and development of new antiepileptic drugs. *Nature Reviews Drug Discovery, 9*, 68–82.

[2] Brodie, M.J. (2010). Antiepileptic drug therapy the story so far. *Seizure, 19*, 650–655.

[3] Rogawski, M.A. and Löscher, W. (2004). The neurobiology of antiepileptic drugs. *Nature Reviews Neuroscience, 5*, 553–564.

[4] Landmark, C.J. (2007). Targets for antiepileptic drugs in the synapse. *Medical Science Monitor, 13*, RA1–RA7.

[5] French, J.A. (2007). Refractory epilepsy: clinical overview. *Epilepsia, 48(Suppl* 1*)*, 3–7.

[6] Banks, W.A. (2008). Developing drugs that can cross the blood-brain barrier: applications to Alzheimer's disease. *BMC Neuroscience, 9(Suppl* 3*)*, S2.

[7] Rogawski, M.A. (2011). Revisiting AMPA receptors as an antiepileptic drug target. *Epilepsy Currents, 11*, 56–63.

[8] Rogawski, M.A. (2013). AMPA receptors as a molecular target in epilepsy therapy. *Acta Neurologica Scandinavica, 127(Suppl s*197*)*, 9–18.

[9] European Medicines Agency. (2012). Fycompa summary of product characteristics. Available at http://www.ema.europa.eu/docs/en_GB/document_library/EPAR_-_Product_Information/human/002434/WC500130815.pdf (accessed on August 11, 2014).

[10] Food and Drug Administration. (2012). Fycompa prescribing information. Available at http://www.accessdata.fda.gov/drugsatfda_docs/label/2012/202834lbl.pdf (accessed on August 11, 2014).

[11] Meldrum, B.S. (2000). Glutamate as a neurotransmitter in the brain: review of physiology and pathology. *Journal of Nutrition, 130*, 1007S–1015S.

[12] Watkins, J.C. and Jane, D.E. (2006). The glutamate story. *British Journal of Pharmacology, 147(Suppl* 1*)*, S100–S108.

[13] De Sarro, G., Gitto, R., Russo, E., Ibbadu, G.F., Barreca, M.L., De Luca, L., Chimirri, A. (2005). AMPA receptor antagonists as potential anticonvulsant drugs. *Current Topics in Medicinal Chemistry, 5*, 31–42.

[14] Mayer, M.L. and Armstrong, N. (2004). Structure and function of glutamate receptor ion channels. *Annual Review of Physiology, 66*, 161–181.

[15] Hwa, G.G. and Avoli, M. (1991). The involvement of excitatory amino acids in neocortical epileptogenesis: NMDA and non-NMDA receptors. *Experimental Brain Research, 86*, 248–256.

[16] Jones, R.S. and Lambert, J.D. (1990). Synchronous discharges in the rat entorhinal cortex in vitro: site of initiation and the role of excitatory amino acid receptors. *Neuroscience, 34*, 657–670.

[17] Lee, W.L. and Hablitz, J.J. (1989). Involvement of non-NMDA receptors in picrotoxin-induced epileptiform activity in the hippocampus. *Neuroscience Letters, 107*, 129–134.

[18] McBain, C.J., Boden, P., Hill, R.G. (1988). The kainate/quisqualate receptor antagonist, CNQX, blocks the fast component of spontaneous epileptiform activity in organotypic cultures of rat hippocampus. *Neuroscience Letters, 93*, 341–345.

[19] Neuman, R.S., Ben-Ari, Y., Cherubini, E. (1988). Antagonism of spontaneous and evoked bursts by 6-cyano-7-nitroquinoxaline-2,3-dione (CNQX) in the CA3 region of the in vitro hippocampus. *Brain Research, 474*, 201–203.

[20] Drejer, J. and Honoré, T. (1988). New quinoxalinediones show potent antagonism of quisqualate responses in cultured mouse cortical neurons. *Neuroscience Letters, 87*, 104–108.

[21] Honoré, T., Davies, S.N., Drejer, J., Fletcher, E.J., Jacobsen, P., Lodge, D., Nielsen, F.E. (1988). Quinoxalinediones: potent competitive non-NMDA glutamate receptor antagonists. *Science, 241*, 701–703.

[22] Sheardown, M.J., Nielsen, E.O., Hansen, A.J., Jacobsen, P., Honoré, T. (1990). 2,3-Dihydroxy-6-nitro-7-sulfamoyl-benzo(F)quinoxaline: a neuroprotectant for cerebral ischemia. *Science, 247*, 571–574.

[23] Weiser, T. (2005). AMPA receptor antagonists for the treatment of stroke. *Current Drug Targets. CNS and Neurological Disorders, 4*, 153–159.

[24] Hanada, T., Hashizume, Y., Tokuhara, N., Takenaka, O., Kohmura, N., Ogasawara, A., Hatakeyama, S., Ohgoh, M., Ueno, M., Nishizawa, Y. (2011). Perampanel: a novel, orally active, noncompetitive AMPA-receptor antagonist that reduces seizure activity in rodent models of epilepsy. *Epilepsia, 52*, 1331–1340.

[25] Hara, H., Yamada, N., Kodama, M., Matsumoto, Y., Wake, Y., Kuroda, S. (2006). Effect of YM872, a selective and highly water-soluble AMPA receptor antagonist, in the rat kindling and rekindling model of epilepsy. *European Journal of Pharmacology, 531*, 59–65.

[26] Kodama, M., Yamada, N., Sato, K., Kitamura, Y., Koyama, F., Sato, T., Morimoto, K., Kuroda, S. (1999). Effects of YM90K, a selective AMPA receptor antagonist, on amygdala-kindling and long-term hippocampal potentiation in the rat. *European Journal of Pharmacology, 374*, 11–19.

[27] Krampfl, K., Schlesinger, F., Cordes, A.L., Bufler, J. (2006). Molecular analysis of the interaction of the pyrazine derivatives RPR119990 and RPR117824 with human AMPA-type glutamate receptor channels. *Neuropharmacology, 50*, 479–490.

[28] Mattes, H., Carcache, D., Kalkman, H.O., Koller, M. (2010). Alpha-amino-3-hydroxy-5-methyl-4-isoxazolepropionic acid (AMPA) antagonists: from bench to bedside. *Journal of Medicinal Chemistry, 53*, 5367–5382.

[29] Mignani, S., Bohme, G.A., Birraux, G., Boireau, A., Jimonet, P., Damour, D., Genevois-Borella, A., Debono, M.W., Pratt, J., Vuilhorgne, M., Wahl, F., Stutzmann, J.M. (2002). 9-Carboxymethyl-5*H*,10*H*-imidazo[1,2-*a*]indeno[1,2-*e*]pyrazin-4-one-2-carbocylic acid (RPR117824): selective anticonvulsive and neuroprotective AMPA antagonist. *Bioorganic & Medicinal Chemistry, 10*, 1627–1637.

[30] Turski, L., Huth, A., Sheardown, M., McDonald, F., Neuhaus, R., Schneider, H.H., Dirnagl, U., Wiegand, F., Jacobsen, P., Ottow, E. (1998). ZK200775: a phosphonate quinoxalinedione AMPA antagonist for neuroprotection in stroke and trauma. *Proceedings of the National Academy of Sciences of the United States of America, 95*, 10960–10965.

[31] Takahashi, M., Kohara, A., Shishikura, J., Kawasaki-Yatsugi, S., Ni, J.W., Yatsugi, S., Sakamoto, S., Okada, M., Shimizu-Sasamata, M., Yamaguchi, T. (2002). YM872: a selective, potent and highly water-soluble alpha-amino-3-hydroxy-5-methylisoxazole-4-propionic acid receptor antagonist. *CNS Drug Reviews, 8*, 337–352.

[32] Donevan, S.D. and Rogawski, M.A. (1993). GYKI 52466, a 2,3-benzodiazepine, is a highly selective, noncompetitive antagonist of AMPA/kainate receptor responses. *Neuron, 10*, 51–59.

[33] Donevan, S.D., Yamaguchi, S., Rogawski, M.A. (1994). Non-*N*-methyl-D-aspartate receptor antagonism by 3-*N*-substituted 2,3-benzodiazepines: relationship to anticonvulsant activity. *Journal of Pharmacology and Experimental Therapeutics, 271*, 25–29.

[34] Szabados, T., Gigler, G., Gacsályi, I., Gyertyán, I., Lévay, G. (2001). Comparison of anticonvulsive and acute neuroprotective activity of three 2,3-benzodiazepine compounds, GYKI 52466, GYKI 53405, and GYKI 53655. *Brain Research Bulletin, 55*, 387–391.

[35] Chapman, A.G., Smith, S.E., Meldrum, B.S. (1991). The anticonvulsant effect of the non-NMDA antagonists, NBQX and GYKI 52466, in mice. *Epilepsy Research, 9*, 92–96.

[36] Yamaguchi, S., Donevan, S.D., Rogawski, M.A. (1993). Anticonvulsant activity of AMPA/kainate antagonists: comparison of GYKI 52466 and NBQX in maximal electroshock and chemoconvulsant seizure models. *Epilepsy Research, 15*, 179–184.

[37] Chappell, A.S., Sander, J.W., Brodie, M.J., Chadwick, D., Lledo, A., Zhang, D., Bjerke, J., Kiesler, G.M., Arroyo, S. (2002). A crossover, add-on trial of talampanel in patients with refractory partial seizures. *Neurology, 58*, 1680–1682.

[38] Iwamoto, F.M., Kreisl, T.N., Kim, L., Duic, J.P., Butman, J.A., Albert, P.S., Fine, H.A. (2010). Phase 2 trial of talampanel, a glutamate receptor inhibitor, for adults with recurrent malignant gliomas. *Cancer, 116*, 1776–1782.

[39] Pascuzzi, R.M., Shefner, J., Chappell, A.S., Bjerke, J.S., Tamura, R., Chaudhry, V., Clawson, L., Haas, L., Rothstein, J.D. (2010). A phase II trial of talampanel in subjects with amyotrophic lateral sclerosis. *Amyotrophic Lateral Sclerosis, 11*, 266–271.

[40] Langan, Y.M., Lucas, R., Jewell, H., Toublanc, N., Schaefer, H., Sander, J.W., Patsalos, P.N. (2003). Talampanel, a new antiepileptic drug: single- and multiple-dose pharmacokinetics and initial 1-week experience in patients with chronic intractable epilepsy. *Epilepsia, 44*, 46–53.

[41] Walters, M.R., Kaste, M., Lees, K.R., Diener, H.C., Hommel, M., De Keyser, J., Steiner, H., Versavel, M. (2005). The AMPA antagonist ZK 200775 in patients with acute ischaemic stroke: a double-blind, multicentre, placebo-controlled safety and tolerability study. *Cerebrovascular Diseases, 20*, 304–309.

[42] Satlin, A., Kramer, L.D., Laurenza, A. (2013). Development of perampanel in epilepsy. *Acta Neurologica Scandinavica, 127(Suppl s197)*, 3–8.

[43] Hibi, S., Ueno, K., Nagato, S., Kawano, K., Ito, K., Norimine, Y., Takenaka, O., Hanada, T., Yonaga, M. (2012). Discovery of 2-(2-oxo-1-phenyl-5-pyridin-2-yl-1,2-dihydropyridin-3-yl)benzonitrile (perampanel): a novel, noncompetitive α-amino-3-hydroxy-5-methyl-4-isoxazolepropanoic acid (AMPA) receptor antagonist. *Journal of Medicinal Chemistry, 55*, 10584–10600.

[44] Rogawski, M.A. and Hanada, T. (2013). Preclinical pharmacology of perampanel, a selective non-competitive AMPA receptor antagonist. *Acta Neurologica Scandinavica, 127(Suppl s*197*)*, 19–24.

[45] Pajouhesh, H. and Lenz, G.R. (2005). Medicinal chemical properties of successful central nervous system drugs. *NeuroRx, 2*, 541–553.

[46] Ceolin, L., Bortolotto, Z.A., Bannister, N., Collingridge, G.L., Lodge, D., Volianskis, A. (2012). A novel anti-epileptic agent, perampanel, selectively inhibits AMPA receptor-mediated synaptic transmission in the hippocampus. *Neurochemistry International, 61*, 517–522.

[47] Rogawski M.A., Chen, C.-Y., Matt, L., Hell, J.W. (2012). Blocking mechanism of the AMPA receptor antagonist perampanel. *Epilepsy Currents, 13(Suppl* 1*)*, 6, abs 1.013.

[48] Tamai, I. and Tsuji, A. (2000). Transporter-mediated permeation of drugs across the blood-brain barrier. *Journal of Pharmaceutical Sciences, 89*, 1371–1388.

[49] Schinkel, A.H. (1999). P-Glycoprotein, a gatekeeper in the blood-brain barrier. *Advanced Drug Delivery Reviews, 36*, 179–194.

[50] Löscher, W. and Potschka, H. (2005). Blood-brain barrier active efflux transporters: ATP-binding cassette gene family. *NeuroRx, 2*, 86–98.

[51] European Medicines Agency. (2012). Fycompa assessment report. Available at http://www.ema.europa.eu/docs/en_GB/document_library/EPAR_-_Public_assessment_report/human/002434/WC500130839.pdf (accessed on August 11, 2014).

[52] Wu, T. and Hanada, T. (2013). Anti-seizure effects of perampanel in combination with other antiepileptic drugs (AEDs) in a rat amygdala kindling model. *Epilepsia, 54(Suppl* 3*)*, 69, abs P202.

[53] Ido, K. and Hanada, T. (2013). Therapeutic time window of perampanel for the termination of diazepam-resistant status epilepticus (SE) in a lithium-pilocarpine rat model. *Epilepsia, 54(Suppl* 3*)*, 69, abs P203.

[54] Barton, M.E., Klein, B.D., Wolf, H.H., White, H.S. (2001). Pharmacological characterization of the 6 Hz psychomotor seizure model of partial epilepsy. *Epilepsy Research, 47*, 217–227.

[55] Templeton, D. (2009). Pharmacokinetics of perampanel, a highly selective AMPA-type glutamate receptor antagonist. *Epilepsia, 50(Suppl* 11*)*, 98, abs 1.199.

[56] Krauss, G.L., Bar, M., Biton, V., Klapper, J.A., Rektor, I., Vaiciene-Magistris, N., Squillacote, D., Kumar, D. (2012). Tolerability and safety of perampanel: two randomized dose-escalation studies. *Acta Neurologica Scandinavica, 125*, 8–15.

[57] French, J.A., Krauss, G.L., Biton, V., Squillacote, D., Yang, H., Laurenza, A., Kumar, D., Rogawski, M.A. (2012). Adjunctive perampanel for refractory partial-onset seizures: randomized phase III study 304. *Neurology, 79*, 589–596.

[58] French, J.A., Krauss, G.L., Steinhoff, B.J., Squillacote, D., Yang, H., Kumar, D., Laurenza, A. (2013). Evaluation of adjunctive perampanel in patients with refractory partial-onset seizures: results of randomized global phase III study 305. *Epilepsia, 54*, 117–125.

[59] Krauss, G.L., Serratosa, J.M., Villanueva, V., Endziniene, M., Hong, Z., French, J., Yang, H., Squillacote, D., Edwards, H.B., Zhu, J., Laurenza, A. (2012). Randomized phase III study 306: adjunctive perampanel for refractory partial-onset seizures. *Neurology, 78*, 1408–1415.

[60] Krauss, G.L., Perucca, E., Ben-Menachem, E., Kwan, P., Shih, J.J., Squillacote, D., Yang, H., Gee, M., Zhu, J., Laurenza, A. (2013). Perampanel, a selective, noncompetitive α-amino-3-hydroxy-5-methyl-4-isoxazolepropionic acid receptor antagonist, as adjunctive therapy for refractory partial-onset seizures: interim results from phase III, extension study 307. *Epilepsia, 54*, 126–134.

[61] Rektor, I., Krauss, G.L., Bar, M., Biton, V., Klapper, J.A., Vaiciene-Magistris, N., Kuba, R., Squillacote, D., Gee, M., Kumar, D. (2012). Perampanel Study 207: long-term open-label evaluation in patients with epilepsy. *Acta Neurologica Scandinavica, 126*, 263–269.

[62] Kerling, F. and Kasper, B.S. (2013). Efficacy of perampanel: a review of clinical trial data. *Acta Neurologica Scandinavica, 127(Suppl s197)*, 25–29.

[63] Serratosa, J.M., Villanueva, V., Kerling, F., Kasper, B.S. (2013). Safety and tolerability of perampanel: a review of clinical trial data. *Acta Neurologica Scandinavica, 127(Suppl s197)*, 30–35.

[64] Steinhoff, B.J., Ben-Menachem, E., Ryvlin, P., Shorvon, S., Kramer, L., Satlin, A., Squillacote, D., Yang, H., Zhu, J., Laurenza, A. (2013). Efficacy and safety of adjunctive perampanel for the treatment of refractory partial seizures: a pooled analysis of three phase III studies. *Epilepsia, 54*, 1481–1489.

[65] Kramer, L., Perucca, E., Ben-Menachem, E., Kwan, P., Shih, J., Squillacote, D., Yang, H., Zhu, J., Laurenza, A. (2012). Perampanel, a selective, non-competitive AMPA receptor antagonist, as adjunctive therapy in patients with refractory partial-onset seizures: a dose–response analysis from phase III studies. *Neurology, 78 (Meeting Abstracts 1)*, abs P06.117.

[66] Gidal, B.E., Ferry, J., Majid, O., Hussein, Z. (2013). Concentration-effect relationships with perampanel in patients with pharmacoresistant partial-onset seizures. *Epilepsia, 54*, 1490–1497.

[67] Kramer, L.D., Satlin, A., Krauss G.L., French, J., Perucca, E., Ben-Menachem, E., Kwan, P., Shih, J.J., Laurenza, A., Yang, H., Zhu, J., Squillacote, D. (2014). Perampanel for adjunctive treatment of partial-onset seizures: a pooled dose–response analysis of Phase III studies. *Epilepsia, 55*, 423–431.

[68] Kennedy, G.M. and Lhatoo, S.D. (2008). CNS adverse events associated with antiepileptic drugs. *CNS Drugs, 22*, 739–760.

[69] Perucca, P., Carter, J., Vahle, V., Gilliam, F.G. (2009). Adverse antiepileptic drug effects: toward a clinically and neurobiologically relevant taxonomy. *Neurology, 72*, 1223–1229.

[70] Rektor, I. (2012). Perampanel, a novel, non-competitive, selective AMPA receptor antagonist as adjunctive therapy for treatment-resistant partial-onset seizures. *Expert Opinion on Pharmacotherapy, 14*, 225–235.

23

CASE STUDY 2: THE DISCOVERY AND DEVELOPMENT OF THE MULTIMODAL ACTING ANTIDEPRESSANT VORTIOXETINE

CHRISTOFFER BUNDGAARD,[1] ALAN L. PEHRSON,[2] CONNIE SÁNCHEZ,[2] AND BENNY BANG-ANDERSEN[3]

[1]*H. Lundbeck A/S, Department of Discovery DMPK, Valby, Denmark*
[2]*Lundbeck Research USA, Department of Neuroscience, Paramus, NJ, USA*
[3]*H. Lundbeck A/S, Department of Medicinal Chemistry, Valby, Denmark*

Abbreviations:

5-HT	serotonin
BBB	blood–brain barrier
CNS	central nervous system
DMPK	drug metabolism and pharmacokinetics
FDA	Food and Drug Administration
GABA	gamma-aminobutyric acid
IA	intrinsic activity
iv	intravenous
MoA	mechanism of action
MAOI	monoamine oxidase inhibitor
MDCK	Madin–Darby canine kidney
MDD	major depressive disorder
MPO	multiparameter optimization
NET	norepinephrine transporter

Blood–Brain Barrier in Drug Discovery: Optimizing Brain Exposure of CNS Drugs and Minimizing Brain Side Effects for Peripheral Drugs, First Edition. Edited by Li Di and Edward H. Kerns.

PD pharmacodynamic
PET positron emission tomography
PK pharmacokinetic
SERT serotonin transporter
SNRI serotonin and norepinephrine reuptake inhibitor
SSRI selective serotonin reuptake inhibitor
TCA tricyclic antidepressant
WHO World Health Organization

INTRODUCTION

According to the World Health Organization (WHO), depression is the leading cause of disability worldwide—and with more than 350 million people suffering from depression, it is a major contributor to the global disease burden [1]. Depression has major implications for everyday function at work or school and in the family and—in the worst case, lead to suicide. Substantial challenges remain to be surmounted by the psychiatric research community in order to meet the needs of depressed patients. In addition to poor access to trained health care personnel and antidepressant medication in some parts of the world, current antidepressants have a slow therapeutic onset and limited efficacy. Today's antidepressants largely build on knowledge gained from careful studies of the mechanism of action (MoA) of the serendipitously discovered tricyclic and monoamine oxidase inhibitor antidepressants (TCA and MAOI, respectively). Thus, removal of antagonism of postsynaptic receptors from the TCAs led to the development of the selective serotonin (5-HT) reuptake inhibitors (SSRIs) and the serotonin norepinephrine reuptake inhibitors (SNRIs), which mainly offered improved tolerability. Studies of the impact of neuroadaptive feedback mechanisms on SSRI effects during the mid-1990s led to the hypothesis that concomitant blockade of somatodendritic 5-HT_{1A} autoreceptors and the serotonin transporter (SERT) would lead to a faster and/or enhanced antidepressant effect, compared to an SSRI, due to greater and more rapid increases in central extracellular 5-HT than seen with an SSRI [2]. Thus, in the mid-1990s, pharmaceutical companies initiated drug discovery projects aiming for antidepressants with this dual MoA. However, it turned out to be very difficult to design compounds with this pharmacological profile. Later, it was realized that increasing brain 5-HT more than an SSRI could be achieved by targeting other 5-HT receptor subtypes or desensitizing 5-HT_{1A} autoreceptors in combination with SERT inhibition, as discussed later in this chapter.

Vortioxetine, which was approved by the Food and Drug Administration (FDA) on September 30, 2013, for treatment of major depressive disorder (MDD) originates from the line of research aiming at an enhanced therapeutic effect compared to SSRIs through modulation of neuroadaptive feedback mechanisms. This research led to a new class of multimodal acting antidepressants, which act via SERT inhibition and receptor modulation [3]. So far, two such agents, vilazodone and vortioxetine, have made it to the market. The subsequent sections tell the story of vortioxetine's discovery and development, and the impact that careful attention to engaging specific biological targets in the brain has played in its successful development.

THE ROUTE TOWARD SELECTING VORTIOXETINE AS A DRUG CANDIDATE FOR TREATMENT OF MDD

The drug discovery project leading to vortioxetine was initiated in January 2001, with the aim of finding dual-acting compounds, targeting the SERT and the 5-HT$_{2C}$ receptor in order to increase the extracellular 5-HT level beyond that seen with an SSRI. Researchers at Lundbeck and Groningen University collectively demonstrated that blockade of 5-HT$_{2C}$ receptors could potentiate the neurochemical and antidepressant-like behavioral action of SSRIs in rodent models [4]. For example, intracerebral microdialysis studies in rats using acute administration of the 5-HT$_{2C}$ receptor selective antagonists SB-242084 and RS-102221 substantially augmented the serotonergic response to citalopram in the ventral hippocampus, that is, 900% of baseline for the combination versus 400–500% for citalopram alone. The selective 5-HT$_{2C}$ receptor antagonists did not alter the 5-HT level when given alone [4].

At the start of the project, assessment of drug metabolism and pharmacokinetic (DMPK) properties during lead optimization was evolving as a new scientific discipline integrated in the drug discovery process at Lundbeck. The overall DMPK criteria that need to be fulfilled before advancing a compound into the clinic include satisfactory human predicted pharmacokinetics (PK) and a pharmacokinetic/pharmacodynamic (PK/PD) profile that allows for oral administration at reasonable dosing intervals with minimal intra- and interindividual variability and low risk of drug–drug interactions. Several DMPK properties determine this profile, including metabolic stability, solubility, membrane permeability, and CYP450- or transporter-mediated interaction liability. Optimizing some of these basic drug-like properties came to play a central role in the lead optimization toward vortioxetine.

An iterative lead-finding process was initiated early on in the project to find molecules with combined 5-HT$_{2C}$ receptor antagonism and SERT inhibition, and a number of distinct chemical series were identified. We chose to advance only one chemical series, and compound 1 is an early lead from this aryl piperazine series (Table 23.1). Compound 1 displayed the desired *in vitro* profile on 5-HT$_{2C}$ receptors and SERT (Table 23.1); however, the DMPK profile was unsatisfactory due to its poor metabolic stability in human liver microsomes and its potent inhibition of CYP2D6.

The observation that only minor structural changes within this series resulted in significant changes of both *in vitro* target affinities and especially of *in vitro* DMPK parameters (Table 23.1) was an important reason for choosing the aryl piperazine series. For example, removing the methyl group in the central benzene ring of compound 1 led to compound 2, with substantially improved *in vitro* stability in human liver microsomes and reduced CYP2D6 inhibition potential (Table 23.1). The compounds of the series were generally of low molecular weight, and their lipophilicity could be kept within an acceptable range. A retrospective analysis of lead compound 1 shows that it, with a log$D_{7.4}$ of 2.7 and molecular weight of 314 Da, is situated within the boundaries of the Golden Triangle proposed by Johnson et al. [5], supporting the idea that the series was a good starting point for lead optimization.

TABLE 23.1 Structure–activity relationships in the aryl piperazine series showing that minor structural changes resulted in significant changes in target affinities and DMPK parameters.

Parameter	Unit	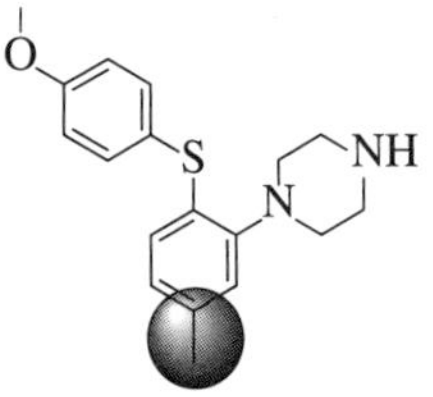 **1**	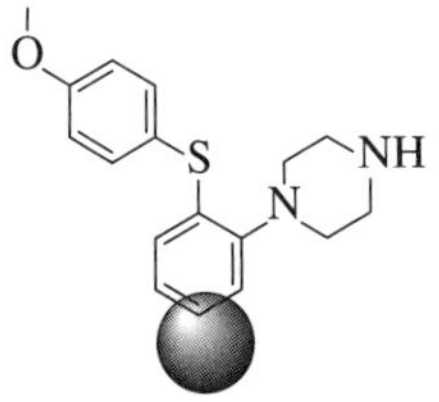 **2**	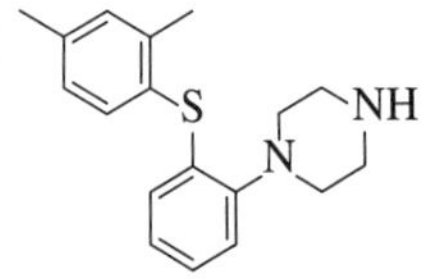 Vortioxetine
SERT (IC_{50})	nM	7.9[a]	8.0[a]	5.3/5.4[a,b]
5-HT_{2C} (K_i)	nM	13[c]	90[c]	180[a]
5-HT_{3A} (K_i)	nM	190[a]	36[a]	23/3.7[a,b]
5-HT_{1A} (K_i)	nM	4000[a]	130[a]	39/15[a,b]
hCL_{int}	l/h/kg	2.4[c]	0.4[c]	0.4[a]
rCL_{int}	l/h/kg	18[c]	15[c]	13[a]
CYP2D6 inhibition (IC_{50})	μM	0.10[c]	1.9[c]	9.8[a]

[a] Ref. 6.

[b] Two values are given. The first value was obtained in early research from the same assays as the other data in the table and is given for comparison. The second value is validated by more determinations and are the final values.

[c] Unpublished data.

CL_{int}: intrinsic metabolic clearance in liver microsomes in human (hCL_{int}) and rat (rCL_{int})

During the lead optimization program aiming at finding a combined 5-HT_{2C} receptor antagonist and SERT inhibitor with good drug-like DMPK properties, vortioxetine showed up in the screening cascade, displaying the desired *in vitro* target activity at both SERT and 5-HT_{2C} receptors. As important, it also had a favorable predicted human DMPK profile, as discussed in further detail in the Section "Vortioxetine's Properties as a CNS Drug". During further testing it became clear that the preliminary 5-HT_{2C} receptor activity was less pronounced than originally anticipated. However, in parallel with this understanding, vortioxetine was tested in our main mechanistic *in vivo* screening assay, a rat microdialysis assay. This assay measured extracellular 5-HT in ventral hippocampus after 3 days of continuous dosing via subcutaneous osmotic minipumps, followed by determination of brain SERT occupancies by *in vivo* binding. In this assay, vortioxetine increased the extracellular 5-HT level beyond that of an SSRI at SERT brain occupancy levels as low as 40–50% [6, 7]. At least 80% SERT occupancy is needed for SSRIs to produce a significant increase of 5-HT in this assay (unpublished data). Having realized that vortioxetine had weaker effects on 5-HT_{2C} receptors than first anticipated, we started to search for other mechanisms that could explain this significant *in vivo* response. Thus, we tested vortioxetine against a panel of ion channels, G protein-coupled receptors, enzymes, and transporters. The high affinity antagonist properties of vortioxetine for 5-HT_3 receptors (discussed in more detail later in this chapter) were found to at least partly explain the enhanced release of 5-HT it engendered. Furthermore, it was realized that vortioxetine is an agonist at the 5-HT_{1A} receptor, which would lead to faster desensitization of inhibitory somatodendritic 5-HT_{1A} autoreceptors than an SSRI, and thereby result in further increases of extracellular brain 5-HT. Consequently, one part of the project was redefined based on these findings and targeted toward finding molecules with combined 5-HT_3 receptor antagonism, 5-HT_{1A} receptor agonism and SERT inhibition. A detailed structure–activity relationship study around vortioxetine was recently published [6]. One minor structural change, removing the methyl group in compound 1 to give compound 2, led to substantial effect on the affinity for both 5-HT_3 and 5-HT_{1A} receptors (Table 23.1). In conclusion, vortioxetine is to a large extent the result of serendipity and persistence in understanding the unexpected findings in the project; we were looking for compounds with a specific target profile, but ended up testing vortioxetine in our *in vivo* microdialysis assay before realizing that the compound did not fulfill the predefined target profile. However, we followed up on this discovery and have since continuously built on vortioxetine's emerging pharmacological profile in order to reveal and understand its full potential.

VORTIOXETINE'S PROPERTIES AS A CNS DRUG

In drug discovery, *in vitro* and *in vivo* DMPK studies play a key role in ensuring that the drug-like properties and the developability potential are in place before the compound is advanced to the clinic. Besides basic DMPK properties such as metabolic stability and CYP450-mediated interaction liability, an additional important DMPK parameter to be dealt with is the rate and extent of blood–brain barrier (BBB) penetration. The BBB is a tightly controlled membrane with endothelial cells expressing numerous uptake and efflux transporters, in addition to extensive tight junctions [8], and therefore presents a potential obstacle for compounds designed to exert effects in the central nervous system (CNS).

With a molecular weight of 298 Da, a log P of 4.7, and a log $D_{7.4}$ of 3.1, vortioxetine is situated at the inside edge of the Golden Triangle proposed by Johnson et al. [5]. Such balanced physical–chemical profile has been shown to provide optimal chances of achieving ideal outcomes in terms of permeability and metabolic stability. Table 23.2 provides a summary of the basic *in vitro* and *in vivo* DMPK parameters for vortioxetine. This compound exhibits medium intrinsic *in vitro* permeability across Madin–Darby canine kidney (MDCK) cell monolayers (6×10^{-6} cm/s) and low intrinsic clearance in human liver microsomes and hepatocytes (0.4 and 0.8 l/h/kg, respectively, relative to a human liver blood flow of 1.3 l/h/kg). In rodents, however, the metabolic stability across the chemical series is poor, including that of vortioxetine, which has values of 13 l/h/kg in rat liver microsomes and hepatocytes relative

TABLE 23.2 Summary of *in vitro* and *in vivo* DMPK properties of vortioxetine in rats and humans

Parameter	Unit	Rat	Human
CL_{int}	l/h/kg	13[a]	0.4[a]
CL_{iv}	l/h/kg	4[b]	0.4[c]
V_{ss}	l/kg	13[b]	34[c]
F	%	7[b]	75[c]
$T_{½}$	h	2.5	60[c]
Plasma free fraction	%	0.8[a]	1.3[a]
Brain free fraction	%	0.2[d]	n/a
MDCK P_{app}	cm/s × 10^{-6}	n/a	6.2[e]
MDCK efflux ratio	–	n/a	1.9[e]

[a] Ref. 6.
[b] Ref. 7.
[c] Ref. 10.
[d] Unpublished data.
[e] Data on file (NDA).

CL_{int}, intrinsic metabolic clearance in liver microsomes; CL_{iv}, *in vivo* systemic clearance; V_{ss}, volume of distribution at steady state; F, oral bioavailability; $T_{½}$, effective elimination half-life; MDCK P_{app}, apparent permeability in the apical to basal direction; efflux ratio, ratio between P_{app} in the basal-apical and apical-basal direction; n/a, not applicable.

to a rat liver blood flow of 4.8 l/h/kg [6]. However, as the systemic clearance also was high in rats following intravenous (iv) administration, an intraspecies *in vitro–in vivo* clearance correlation was indicated when applying the well-stirred model to predict rat hepatic clearance [9]. Along these lines, the low human intrinsic clearance observed *in vitro* was predicted to result in low-medium clearance *in vivo* in humans following *in vitro–in vivo* interspecies scaling, which was confirmed in clinical studies in healthy volunteers, that is, 0.4 l/h/kg relative to a human liver blood flow of 1.3 l/h/kg after iv administration [10]. From the same clinical iv data, the volume of distribution was found to be high (34 l/kg). This extensive distribution was also in line with what could be predicted from rat pharmacokinetic studies [7]. Systemic clearance and distribution volume are the sole determinants of the systemic elimination half-life which is around 60 h for vortioxetine in humans. Prediction of human elimination half-life is often one of the key components during candidate selection. Since systemic clearance and volume of distribution can be directly deduced only from iv administration, it is important to have these pharmacokinetic data available in the clinic for appropriate retrospective assessment of the predicted PK parameters from rodents. Moreover, the favorable predicted and observed clearance of vortioxetine in humans highlights the fact that poor rodent PK in a given chemical series should not necessarily disqualify compounds from moving forward, as exemplified here with vortioxetine.

During lead optimization, target affinities were measured by *in vitro* screens using cloned cells expressing the target of interest. To assess how *in vitro* affinities related to *in vivo* target occupancy, *in vivo* displacement studies in rodents were conducted early in the lead optimization program. With the use of suitable selective radioligands such as 3-amino-4-(2-dimethylaminomethyl dimethylaminomethylphenylsulfanyl)-benzonitrile) ([^{11}C]-DASB) targeting the SERT, this technique is a powerful tool to quantitate target engagement in the CNS and obtain an understanding of the PK/PD relationship [11–14]. Moreover, these occupancy studies also served as a direct tool to assess the extent of brain penetration. Acutely, vortioxetine dose-dependently occupied the SERT in rats, thus verifying its brain penetration [7, 15]. Following three days of continuous treatment via minipumps in rats, a significant increase in brain extracellular 5-HT was observed using intracerebral microdialysis with only 41% SERT occupancy [7]. This indicated that relevant functional effects at the neurotransmitter level could be obtained at low SERT occupancy compared to classical SSRIs and SNRIs, which typically require 80% SERT occupancy for a functional and therapeutic effect [16, 17]. In a positron emission tomography (PET) study of vortioxetine in healthy men performed at different oral doses, the relationship between plasma concentrations and SERT occupancy showed that doses of 5–10 mg/day resulted in occupancies around 40–55% [18]. Later, a wide range of efficacy studies in depressed patients confirmed that these doses were therapeutically effective and well tolerated [19–21].

In vivo target occupancy studies were used as a powerful mechanistic and predictive tool in understanding vortioxetine's pharmacology and translating its effects from animals to humans. Predicting and/or comparing *in vivo* potency between species based on plasma exposure is a key element of translational PK/PD assessments, with the basic assumption that effective drug exposure at the site of action is the same between species [22]. However, this assumption implies that important species-dependent parameters need to be accounted for when translating plasma exposure between species. First, target affinity may differ between species, requiring information on *in vitro* K_i in the involved species. Second, it is widely recognized that the unbound concentrations are responsible for engaging the target and are therefore the relevant exposure metric to consider [23, 24]. Thus, species differences in plasma protein binding also need to be accounted for. Third, species differences in drug uptake and/or efflux transporters at the BBB level may lead to species differences in CNS penetration. For vortioxetine, retrospective PK/PD evaluation showed that approximately 10-fold higher total plasma concentrations were required in rats compared to humans to achieve the same SERT occupancy [7, 18]. However, vortioxetine displays approximately a sixfold weaker SERT K_i in rats (8.9 ± 1.2 nM) compared to humans (1.6 ± 0.4 nM), thus increasing the predicted required equivalent exposure in rats correspondingly. As to plasma protein binding, vortioxetine is highly bound to plasma proteins in both rats (0.76 ± 0.52% free) and humans (1.25 ± 0.48% free), making it difficult to quantitatively appraise whether this parameter would have implications for translating effective drug exposure between these species. Regarding brain penetration, low involvement of active efflux at the BBB from P-glycoprotein (P-gp) is suggested from *in vitro* permeability assessment (Table 23.2). Overall, it seems likely that the species difference in SERT affinity is the main contributor to the observed difference between rats and humans in the SERT occupancy PK/PD relationship.

TABLE 23.3 *In vitro* binding affinities and functional activities of vortioxetine at human and rat targets expressed in recombinant cell lines

		Human		Rat	
Target	Function	Binding affinity K_i (nM)	Functional potency IC_{50}/EC_{50} (nM), IA (%)	Binding affinity K_i (nM)	Functional potency IC_{50}/EC_{50} (nM), IA (%)
$5\text{-}HT_3$	Ant	3.7[a]	12[a]	1.1[b]	0.18[a]
$5\text{-}HT_7$	Ant	19[a]	450[c]	200[c]	2080[c]
$5\text{-}HT_{1D}$	Ant	54[d]	370[e]	3.7[c]	260[e]
$5\text{-}HT_{1B}$	Part ago	33[a]	120(55)[c]	16[c]	340(40)[f]
$5\text{-}HT_{1A}$	Ago	15[a]	200[b]	230[b]	nd
SERT	Inhib	1.6[a]	5.4[a]	8.9[d]	5.3[a]

[a]Ref. 6.
[b]Ref. 26.
[c]Ref. 7.
[d]Ref. 27.
[e]Data on file (NDA).
[f]Unpublished data.
nd: Not determined.
Ant, antagonist function; Part ago, partial agonist; Inhib, inhibition; IA, intrinsic activity shown in parentheses.

During the characterization of vortioxetine as an active CNS compound in this drug discovery program, *in vivo* target occupancy was applied as the absolute measure of brain penetration. Characterization of the relationship between occupancy and plasma exposure allowed for estimations or predictions of target occupancy during *in vivo* pharmacological and behavioral experiments, which greatly supports the mechanistic interpretation of such data. In addition, vortioxetine's low brain free fraction (below 0.2% in rats and mice, unpublished data) highlights the importance of putting such binding data into context and emphasizes the fact that low free fraction in the target tissue does not preclude efficiency, in line with modern BBB paradigms concerning free fraction concepts and CNS drug discovery [23, 24].

From an *in silico* perspective, another evolving concept is the application of the multiparameter optimization (MPO) algorithm to predict CNS drug-likeness. The MPO function is based on specific physical–chemical parameters calculated from the chemical structure (lipophilicity, molecular weight, topological polar surface area, number of hydrogen bond donors, and pK_a) resulting in a score between 0 and 6 [25]. A retrospective analysis has shown that around 75% of marketed CNS drugs score ≥4. With a calculated CNS MPO score for vortioxetine of only 2.7, such holistic assessment would predict a suboptimal CNS drug likeness. Hence, despite the lack of what is sometimes viewed as apparent desirable drug-like attributes such as low MPO score, poor rodent PK, and very low free fraction in tissues, an aggregate view with assimilated data interpretations and predictions increases the likelihood of success.

UNVEILING THE FULL POTENTIAL OF VORTIOXETINE

Linkage of *In Vitro* and *In Vivo* Target Profiles, Species Comparisons and Relation between Clinically Used Doses and Preclinical Equivalents

To explore the full clinical potential of vortioxetine, a comprehensive preclinical program is being undertaken to understand vortioxetine's MoA. As alluded to in the Section "The Route toward Selecting Vortioxetine as a Drug Candidate for Treatment of MDD", the target profile defined in the drug discovery program that led to vortioxetine was SERT inhibition combined with 5-HT$_{1A}$ receptor agonism and 5-HT$_3$ receptor antagonism [6]. However, additional *in vitro* studies revealed that vortioxetine also is a 5-HT$_{1D}$ receptor antagonist, a partial 5-HT$_{1B}$ receptor agonist and a 5-HT$_7$ receptor antagonist (Table 23.3). Furthermore, comparison of rat and human properties revealed important species differences for some targets. Specifically, vortioxetine's potency at 5-HT$_{1A}$ and 5-HT$_7$ receptors is 10–15 times weaker for rat than human, which indicates that the contributions from these targets might be underestimated in rats compared to human.

The dose–SERT occupancy relationship had been established in human and rat by the use of PET imaging and *in vivo* binding methods, respectively. An attempt was also made to investigate 5-HT$_{1A}$ receptor occupancy in humans using the 5-HT$_{1A}$ receptor antagonist radiotracer [^{11}C]WAY100-635. These data suggested that vortioxetine displayed little or no 5-HT$_{1A}$ receptor occupancy in humans [28]. However, using a 5-HT$_{1A}$ receptor antagonist radiotracer such as [^{11}C]WAY100-635 may be an inappropriate method of estimating receptor occupancies for vortioxetine, a drug that acts as a full agonist. The 5-HT$_{1A}$ receptor complex can exist in a coupled or decoupled state with its G protein signaling complex, and the decoupled state of the 5-HT$_{1A}$ receptor is associated with low agonist affinity [29, 30] and a low probability of modulating second messenger signaling when stimulated by an agonist. For antagonists such as WAY100-635, however, the association state of the receptor with its G protein complex has no bearing on its 5-HT$_{1A}$ receptor affinity [31]. Therefore, antagonist radiotracers may lead to artificially low occupancy estimations at mechanistically relevant 5-HT$_{1A}$ receptor populations. Since no validated agonist PET radiotracers existed for 5-HT$_{1A}$ receptors or for vortioxetine's other receptor targets, we developed assays to establish dose–occupancy relationships in rodents. This would enable the understanding of the relationship between the *in vitro* and *in vivo* binding potencies. Additionally, since the SERT occupancies were determined both in human and rat, this strategy would allow us to create a bridge between clinical and preclinical measures. Furthermore, establishing the dose–occupancy relationships allowed clinically equivalent doses to be used in all preclinical studies, and allowed the possibility to mimic the human level of 5-HT$_{1A}$ and 5-HT$_7$ receptor occupancy in rats by combining vortioxetine with an appropriate dose of a 5-HT$_{1A}$ receptor agonist and/or a 5-HT$_7$ receptor antagonist. As some of the available radioligands were not appropriate for *in vivo* binding, we developed *ex vivo* binding assays for vortioxetine targets and used autoradiography with rat brain slices. The dose–occupancy relationship in rat brain slices for all targets is shown in Figure 23.1, except for 5-HT$_{1D}$ receptors, for which there was no appropriate radioligand. The

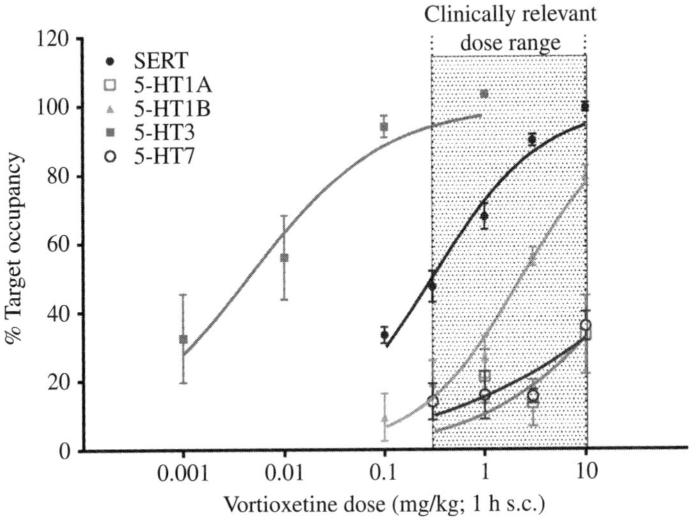

FIGURE 23.1 Relationship between vortioxetine dose and corresponding target occupancy measured by ex vivo binding in rats dosed subcutaneously. Data are presented as mean ± SEM (n = 4–8). Dotted area reflects clinically relevant dose range in rats predicted from human SERT occupancy PET studies.

ex vivo potency ranking is similar to the *in vitro* potencies of vortioxetine, with 5-HT_3 receptor affinity being the most potent and 5-HT_{1A} and 5-HT_7 receptors the least potent. *Ex vivo* SERT occupancy was determined in rodents over a range of vortioxetine doses, using [^{3}H]DASB or [^{3}H]escitalopram, and was in agreement with the previous results using *in vivo* binding methodology [7, 15]. Given that SERT occupancies estimated using PET imaging were in the range of 50–90% at clinically relevant vortioxetine doses (ranging from 5 to 30 mg/day [18, 28]), rodent doses corresponding to these SERT occupancies (from approximately 0.3 to 10 mg/kg) were taken to be clinically equivalent (Fig. 23.1). For each receptor target, the chosen radioligand had a similar intrinsic efficacy to vortioxetine at the target in question, with the exception of the 5-HT_{1B} receptor, where a 5-HT_{1B} receptor antagonist was used to estimate receptor occupancy in the rodent CNS. This was because no partial agonist radioligands were available for this target at the time. Thus, vortioxetine dose–receptor occupancy curves for the 5-HT_{1A}, 5-HT_{1B}, 5-HT_3, and 5-HT_7 receptors were developed using [^{3}H]8-OH-DPAT [7, 15], [^{3}H]GR125743 [15, 32], [^{3}H]LY278584 [15, 32], and [^{3}H]SB269970 (unpublished data), respectively. It was found that clinically equivalent vortioxetine doses produced high occupancy of the 5-HT_3 and 5-HT_{1B} receptors over the clinically equivalent dose range defined above (Fig. 23.1), suggesting that each of these receptor mechanisms are most likely engaged at clinically relevant doses. For the 5-HT_{1A} and 5-HT_7 receptors, vortioxetine achieved occupancies of approximately 35% at the highest clinically equivalent dose. Given the fact that vortioxetine's affinity for the rodent 5-HT_{1A} and 5-HT_7 receptors are approximately 10- to 15-fold lower than at human versions of these receptors, it is expected that these receptor targets are also at meaningful occupancy levels when vortioxetine is used in humans.

Differentiation of Vortioxetine from Currently Used Antidepressants

Since most of the currently used antidepressants exert their pharmacological activity through inhibition of the SERT and/or norepinephrine (NET) transporters, we have focused on exploring the impact of vortioxetine's receptor activities on the net pharmacological activity in preclinical assays relevant for therapeutic activities and tolerability. Thus, the ability to assess the level of target occupancies has been crucial for the interpretation of the biological activities measured. Since published studies using selective tool compounds have indicated that each of vortioxetine's receptor activities individually had positive effects on different aspects of cognitive functions [33], our research is focused on exploration of the antidepressant as well as cognition-enhancing properties of vortioxetine.

Vortioxetine shows antidepressant- and anxiolytic-like activity in standard animal models that are sensitive to SSRIs and/or SNRIs. However, unlike fluoxetine and duloxetine, vortioxetine also showed antidepressant-like activity in a progesterone withdrawal model of depression [34]. Progesterone is metabolized into the neuroactive steroid, allopregnanolone, which modulates the expression pattern [35] and function of gamma-aminobutyric acid (GABA) receptors [36] and results in hyperexcitable GABA neurons during progesterone withdrawal [35, 37]. Both a 5-HT$_3$ receptor antagonist and a 5-HT$_{1A}$ receptor agonist had antidepressant-like effects in the progesterone withdrawal model. Since central 5-HT$_3$ receptors are expressed almost exclusively on GABAergic interneurons, where they act as a serotonin-mediated stimulatory ion channel [38], vortioxetine's 5-HT$_3$ receptor antagonism would be expected to reduce GABAergic inhibition. Additionally, a subpopulation of 5-HT$_{1A}$ heteroreceptors are also expressed on GABAergic interneurons, where they exert an inhibitory influence on GABA neurotransmission [39], implying that vortioxetine's 5-HT$_{1A}$ receptor agonism might also reduce GABAergic inhibition.

To investigate vortioxetine in a nonclinical model of depression-related cognitive dysfunction, we used among others a 5-HT depletion model, since clinical research has demonstrated that tryptophan depletion impairs memory function [40, 41]. In rodents, we found that 5-HT depletion using the irreversible tryptophan hydroxylase inhibitor 4-chloro-DL-phenylalanine methyl ester lead to impairments in novel object recognition and spontaneous alternation that could be reversed using vortioxetine, but not escitalopram or duloxetine [32, 42]. A 5-HT$_3$ receptor antagonist (ondansetron) reversed the 5-HT depletion-induced deficits in novel object recognition, while a 5-HT$_{1A}$ receptor agonist (flesinoxan) reversed deficits in both novel object recognition and spontaneous alternation [32]. Thus, the progesterone withdrawal and 5-HT depletion models highlighted the importance of 5-HT$_3$ and 5-HT$_{1A}$ receptors, raising the possibility that modulating GABAergic neurotransmission and increased downstream activation of neural targets such as glutamatergic pyramidal cells contribute to vortioxetine's effects on mood and cognitive function. In both these animal models, measurement of *ex vivo* target occupancy levels enabled us to conclude that these receptor mechanisms were relevant at the doses used in these studies. Subsequent mechanistic studies support the hypothesis that vortioxetine reduces GABAergic neurotransmission and increases glutamatergic activation, for example, vortioxetine blocked 5-HT-induced increases in inhibitory postsynaptic currents of hippocampus pyramidal neurons [43] and stimulated the firing rate at cortical pyramidal neurons [44]. In both studies, 5-HT_3 receptor antagonism was found to play an important role. Based on these data, vortioxetine's MoA may be mediated in part by downstream modulation of GABA and glutamate neurotransmission.

Clinical Efficacy

Vortioxetine has repeatedly shown antidepressant efficacy in placebo-controlled trials [19–21], although it should be noted that vortioxetine has not shown statistically significant separation from placebo in every clinical trial in MDD [45–47]. Vortioxetine is well tolerated, with the most common adverse event across clinical trials being nausea [19–21]. The clinically approved dose range of 5–20 mg corresponds to SERT occupancies of about 50–80%, which is in line with the originally predicted relevant range in the first microdialysis studies in rats. A recent clinical study in elderly depressed patients [21] found in its predefined exploratory endpoints that vortioxetine significantly improved cognitive performance compared to placebo. These clinical data are in line with the preclinical studies, and thus supports the idea that vortioxetine may also have a favorable effect on cognitive function, although more data are needed to draw a firm conclusion.

CONCLUSION

The extensive use of target occupancy assessments has been instrumental throughout the discovery program that led to the discovery of vortioxetine and its further characterization. By including a combined *in vivo* SERT occupancy assessment and *in vivo* microdialysis as the main mechanistic *in vivo* assay in the screening cascade, the unique profile of vortioxetine was captured at an early stage. Traditionally, screening cascades are constructed as a filter, with the least resource-intensive tests first. However, it may be a better strategy for drug discovery projects aiming at multitarget molecules to go directly to the use of resource-intensive assays, since the number of molecules with the targeted profile will be low and complex models will be more informative. The implication of this approach was that even though it was realized that vortioxetine did not fulfill the requirements of the predefined *in vitro* target profile during retesting, the compound was not discarded. On the contrary, an alternative branch was added to the discovery project exploring the chemical space around vortioxetine. This would never have happened in a conventional screening cascade.

The addition of *in vivo* SERT occupancy determinations to the DMPK assessment of drug-like properties also ensured selection of drugs with optimal drug properties and facilitated clinical dose finding and the transition from the preclinical to clinical development in spite of important species differences with respect to target affinity, metabolic stability and other DMPK measures. Finally, broadening the occupancy measurements to include almost all of vortioxetine's targets played an important role for the interpretation of preclinical data and their translation to a clinical setting, especially for a compound like vortioxetine with target-dependent interspecies affinity differences.

REFERENCES

[1] WHO Fact sheet No 369. WHO Fact sheet No 369. 2012.

[2] Artigas, F., Romero, L., de Montigny, C., & Blier, P. (1996) Acceleration of the effect of selected antidepressant drugs in major depression by 5-HT1A antagonists. Trends Neurosci 19, 378–383.

[3] Nutt, D.J. (2009) Beyond psychoanaleptics—can we improve antidepressant drug nomenclature? J Psychopharmacol 23, 343–345.

[4] Cremers, T.I., Giorgetti, M., Bosker, F.J., Hogg,S., Arnt,J., Mork,A., Honig,G., Bogeso,K.P., Westerink,B.H., den Boer,H., Wikstrom,H.V., & Tecott,L.H. (2004) Inactivation of 5-HT(2C) receptors potentiates consequences of serotonin reuptake blockade. Neuropsychopharmacology 29, 1782–1789.

[5] Johnson, T.W., Dress, K.R., & Edwards, M. (2009) Using the golden triangle to optimize clearance and oral absorption. Bioorg Med Chem Lett 19, 5560–5564.

[6] Bang-Andersen, B., Ruhland, T., Jorgensen, M., Smith, G., Frederiksen, K., Jensen, K.G., Zhong, H., Nielsen, S.M., Hogg, S., Mork, A., & Stensbol, T.B. (2011) Discovery of 1-[2-(2,4-dimethylphenylsulfanyl)phenyl]piperazine (Lu AA21004): A novel multimodal compound for the treatment of major depressive disorder. J Med Chem 54, 3206–3221.

[7] Mork, A., Pehrson, A., Brennum, L.T., Nielsen, S.M., Zhong, H., Lassen, A.B., Miller, S., Westrich, L., Boyle, N.J., Sánchez, C., Fischer, C.W., Liebenberg, N., Wegener, G., Bundgaard, C., Hogg, S., Bang-Andersen, B., & Stensbol, T.B. (2012) Pharmacological effects of Lu AA21004: A novel multimodal compound for the treatment of major depressive disorder. J Pharmacol Exp Ther 340, 666–675.

[8] Abbott, N.J., Patabendige, A.A., Dolman, D.E., Yusof, S.R., & Begley, D.J. (2010) Structure and function of the blood-brain barrier. Neurobiol Dis 37, 13–25.

[9] Rowland, M., & Tozer, T.N. (1995) Clinical Pharmacokinetics: Concepts and Applications, 3rd ed. Lippincott Williams & Wilkins, Media, PA, pp. 165–168.

[10] Areberg, J., Sogaard, B., & Hojer, A.M. (2012) The clinical pharmacokinetics of Lu AA21004 and its major metabolite in healthy young volunteers. Basic Clin Pharmacol Toxicol 111, 198–205.

[11] Alavijeh, M.S., & Palmer, A.M. (2010) Measurement of the pharmacokinetics and pharmacodynamics of neuroactive compounds. Neurobiol Dis 37, 38–47.

[12] Grimwood, S., & Hartig, P.R. (2009) Target site occupancy: Emerging generalizations from clinical and preclinical studies. Pharmacol Ther 122, 281–301.

[13] Bourdet, D.L., Tsuruda, P.R., Obedencio, G.P., & Smith, J.A. (2012) Prediction of human serotonin and norepinephrine transporter occupancy of duloxetine by pharmacokinetic/pharmacodynamic modeling in the rat. J Pharmacol Exp Ther 341, 137–145.

[14] Melhem, M. (2013) Translation of central nervous system occupancy from animal models: Application of pharmacokinetic/pharmacodynamic modeling. J Pharmacol Exp Ther 347, 2–6.

[15] Pehrson, A.L., Cremers, T., Betry, C., van der Hart, M.G., Jorgensen, L., Madsen, M., Haddjeri, N., Ebert, B., & Sánchez, C. (2013) Lu AA21004, a novel multimodal antidepressant, produces regionally selective increases of multiple neurotransmitters—a rat microdialysis and electrophysiology study. Eur Neuropsychopharmacol 23, 133–145.

[16] Meyer, J.H. (2007) Imaging the serotonin transporter during major depressive disorder and antidepressant treatment. J Psychiatry Neurosci 32, 86–102.

[17] Kreilgaard, M., Smith, D.G., Brennum, L.T., & Sanchez, C. (2008) Prediction of clinical response based on pharmacokinetic/pharmacodynamic models of 5-hydroxytryptamine reuptake inhibitors in mice. Br J Pharmacol 155, 276–284.

[18] Areberg, J., Luntang-Jensen, M., Sogaard, B., & Nilausen, D.O. (2012) Occupancy of the serotonin transporter after administration of Lu AA21004 and its relation to plasma concentration in healthy subjects. Basic Clin Pharmacol Toxicol 110, 401–404.

[19] Alvarez, E., Perez, V., Dragheim, M., Loft, H., & Artigas, F. (2012) A double-blind, randomized, placebo-controlled, active reference study of Lu AA21004 in patients with major depressive disorder. Int J Neuropsychopharmacol 15, 589–600.

[20] Henigsberg, N., Mahableshwarkar, A.R., Jacobsen, P., Chen, Y., & Thase, M.E. (2012) A randomized, double-blind, placebo-controlled 8-week trial of the efficacy and tolerability of multiple doses of Lu AA21004 in adults with major depressive disorder. J Clin Psychiatry 73, 953–959.

[21] Katona, C., Hansen, T., & Olsen, C.K. (2012) A randomized, double-blind, placebo-controlled, duloxetine-referenced, fixed-dose study comparing the efficacy and safety of Lu AA21004 in elderly patients with major depressive disorder. Int Clin Psychopharmacol 27, 215–223.

[22] Gabrielsson, J., Dolgos, H., Gillberg, P.G., Bredberg, U., Benthem, B., & Duker, G. (2009) Early integration of pharmacokinetic and dynamic reasoning is essential for optimal development of lead compounds: Strategic considerations. Drug Discov Today 14, 358–372.

[23] Read, K.D., & Braggio, S. (2010) Assessing brain free fraction in early drug discovery. Expert Opin Drug Metab Toxicol 6, 337–344.

[24] Smith, D.A., Di, L., & Kerns, E.H. (2010) The effect of plasma protein binding on *in vivo* efficacy: Misconceptions in drug discovery. Nat Rev Drug Discov 9, 929–939.

[25] Wager, T.T., Hou, X., Verhoest, P.R., & Villalobos, A. (2010) Moving beyond rules: The development of a central nervous system multiparameter optimization (CNS MPO) approach to enable alignment of druglike properties. ACS Chem Neurosci 1, 435–449.

[26] Sánchez, C., Westrich, L., Zhong, H., Nielsen, S.M., Boyle, N.J., Hentzer, M., Frederiksen, K., & Stensbol, T.B. (2012) *In vitro* effects of the multimodal antidepressant Lu AA21004 at human and rat 5-HT1A, 5-HT1B, and 5-HT7 receptors, and 5-HT transporters. Eur Neuropsychopharmacol 22[Suppl. 2], S245–S246.

[27] Westrich, L., Pehrson, A., Zhong, H., Nielsen, S.M., Frederiksen, K., Stensbol, T.B., Boyle, N., Hentzer, M., & Sanchez, C. (2012) *In vitro* and *in vivo* effects of the multimodal antidepressant vortioxetine (Lu AA21004) at human and rat targets. Int J Psych Clin Pract 5, 47.

[28] Stenkrona, P., Halldin, C., & Lundberg, J. (2013) 5-HTT and 5-HT1A receptor occupancy of the novel substance vortioxetine (Lu AA21004). A PET study in control subjects. Eur Neuropsychopharmacol 23, 1190–1198.

[29] Assie, M.B., Cosi, C., & Koek, W. (1999) Correlation between low/high affinity ratios for 5-HT(1A) receptors and intrinsic activity. Eur J Pharmacol 386, 97–103.

[30] Emerit, M.B., el Mestikawy, S., Gozlan, H., Rouot, B., & Hamon, M. (1990) Physical evidence of the coupling of solubilized 5-HT1A binding sites with G regulatory proteins. Biochem Pharmacol 39, 7–18.

[31] Kobilka, B. (1992) Adrenergic receptors as models for G protein-coupled receptors. Annu Rev Neurosci 15, 87–114.

[32] du Jardin, K.G., Jensen, J.B., Sánchez, C., & Pehrson, A.L. (2014) Vortioxetine dose-dependently reverses 5-HT depletion-induced deficits in spatial working and object recognition memory: A potential role for 5-HT receptor agonism and 5-HT receptor antagonism. Eur Neuropsychopharmacol 24, 160–171.

[33] Mork, A., Montezinho, L.P., Miller, S., Trippodi-Murphy, C., Plath, N., Li, Y., Gulinello, M., & Sánchez, C. (2013) Vortioxetine (Lu AA21004), a novel multimodal antidepressant, enhances memory in rats. Pharmacol Biochem Behav 105, 41–50.

[34] Li, Y., Raaby, K.F., Sanchez, C., & Gulinello, M. (2013) Serotonergic receptor mechanisms underlying antidepressant-like action in the progesterone withdrawal model of hormonally induced depression in rats. Behav Brain Res 256, 520–528.

[35] Gulinello, M., Gong, Q.H., & Smith, S.S. (2002) Progesterone withdrawal increases the alpha4 subunit of the GABA(A) receptor in male rats in association with anxiety and altered pharmacology—a comparison with female rats. Neuropharmacology 43, 701–714.

[36] Vanover, K.E., Rosenzweig-Lipson, S., Hawkinson, J.E., Lan, N.C., Belluzzi, J.D., Stein, L., Barrett, J.E., Wood, P.L., & Carter, R.B. (2000) Characterization of the anxiolytic properties of a novel neuroactive steroid, Co 2-6749 (GMA-839; WAY-141839; 3alpha, 21-dihydroxy-3beta-trifluoromethyl-19-nor-5beta-pregnan-20-one), a selective modulator of gamma-aminobutyric acid(A) receptors. J Pharmacol Exp Ther 295, 337–345.

[37] Sundstrom, I., Nyberg, S., & Backstrom, T. (1997) Patients with premenstrual syndrome have reduced sensitivity to midazolam compared to control subjects. Neuropsychopharmacology 17, 370–381.

[38] Puig, M.V., Santana, N., Celada, P., Mengod, G., & Artigas, F. (2004) In vivo excitation of GABA interneurons in the medial prefrontal cortex through 5-HT3 receptors. Cereb Cortex 14, 1365–1375.

[39] Llado-Pelfort, L., Santana, N., Ghisi, V., Artigas, F., & Celada, P. (2012) 5-HT1A receptor agonists enhance pyramidal cell firing in prefrontal cortex through a preferential action on GABA interneurons. Cereb Cortex 22, 1487–1497.

[40] Riedel, W.J., Klaassen, T., Deutz, N.E., van Someren, A., & van Praag, H.M. (1999) Tryptophan depletion in normal volunteers produces selective impairment in memory consolidation. Psychopharmacology (Berl) 141, 362–369.

[41] Sobczak, S., Riedel, W.J., Booij, I., Aan Het, R.M., Deutz, N.E., & Honig, A. (2002) Cognition following acute tryptophan depletion: Difference between first-degree relatives of bipolar disorder patients and matched healthy control volunteers. Psychol Med 32, 503–515.

[42] Jensen, J.B., du Jardin, K.G., Song, D., Budac, D., Smagin, G., Sanchez, C., & Pehrson, A.L. (2014) Vortioxetine, but not escitalopram or duloxetine, reverses memory impairment induced by central 5-HT depletion in rats: Evidence for direct 5-HT receptor modulation. Eur Neuropsychopharmacol 24, 148–159.

[43] Dale, E., Zhang, H., Leiser, S.C., Xiao, Y., Lu, D., Yang, C.R., Plath, N., Sánchez, C. (2014) Vortioxetine disinhibits pyramidal cell function and enhances synaptic plasticity in the rat hippocampus. J Psychopharmacol 28, 891–902.

[44] Riga, M.S., Celada, P., Sánchez, C., & Artigas, F. (2013) Role of 5-HT3 receptors in the mechanism of action of the investigational antidepressant vortioxetine. Eur Neuropsychopharmacol 23[Suppl. 2], S393–S394.

[45] Jain, R., Mahableshwarkar, A.R., Jacobsen, P.L., Chen, Y., & Thase, M.E. (2013) A randomized, double-blind, placebo-controlled 6-wk trial of the efficacy and tolerability of 5 mg vortioxetine in adults with major depressive disorder. Int J Neuropsychopharmacol 16, 313–321.

[46] Mahableshwarkar, A.R., Jacobsen, P.L., & Chen, Y. (2013) A randomized, double-blind trial of 2.5 mg and 5 mg vortioxetine (Lu AA21004) versus placebo for 8 weeks in adults with major depressive disorder. Curr Med Res Opin 29, 217–226.

[47] Baldwin, D.S., Loft, H., & Dragheim, M. (2012) A randomised, double-blind, placebo controlled, duloxetine-referenced, fixed-dose study of three dosages of Lu AA21004 in acute treatment of major depressive disorder (MDD). Eur Neuropsychopharmacol 22, 482–491.

PART 6

DRUG DELIVERY TECHNIQUES TO CNS

24

BRAIN DELIVERY USING NANOTECHNOLOGY

HUILE GAO[1] AND XINGUO JIANG[2]

[1] Key Laboratory of Drug Targeting and Drug Delivery Systems, West China School of Pharmacy, Sichuan University, Chengdu, China
[2] Key Laboratory of Smart Drug Delivery (Fudan University), Ministry of Education; Department of Pharmaceutics Sciences, School of Pharmacy, Fudan University, Shanghai, China

INTRODUCTION

Currently, central nervous system (CNS) disorders including brain tumors, neurodegenerative diseases, and cerebrovascular diseases are serious and increasing threats to human health. The blood–brain barrier (BBB) is a major obstacle for delivering therapeutics to the brain and/or diseased cells in the brain [1]. Normally, nutrients, such as hexoses, amino acids, and neuropeptides are transported through specific transporters on the BBB. Beside nutrients and other molecules having transporter- or receptor-mediated uptake, only small lipophilic molecules (<500 Da) sufficiently cross the BBB to reach an efficacious brain concentration [2]. Almost 100% of the macromolecular drug and over 98% of the small molecular drug candidates have low penetration at this barrier [3]. Intracranial drug delivery is a useful but inconvenient method, with poor compliance and risk of infection and edema.

The development of nanotechnology provides various nanoparticulated systems with particle size between 1 and 1000 nm, named nanoparticles (NPs) herein, liposomes, organic NPs, polymersomes, micelles, dendrimers, inorganic NPs, carbon nanotubes, fullerenes, and so on [4], which can encapsulate the therapeutics and act

Blood–Brain Barrier in Drug Discovery: Optimizing Brain Exposure of CNS Drugs and Minimizing Brain Side Effects for Peripheral Drugs, First Edition. Edited by Li Di and Edward H. Kerns.

as carriers for these therapeutics. NP-based brain targeted delivery can be achieved through receptor-mediated, transporter-mediated, and adsorptive-mediated BBB transportation and pharmacological disruption of the BBB [5]. Targeted drug delivery systems are promising modes of treating CNS diseases due to several distinguishing characteristics [6]. First, systemically administered targeted delivery systems that envelope drugs can convey them into the brain. Second, producing targeted delivery systems with excellent traits is possible due to the rapid developments in materials science and nanotechnology. And last but not least, progress in biology and etiology has provided strategies that allow targeted delivery systems to enter the brain and reach the sites of disease.

In this chapter, recent advancements in NP-based brain targeted delivery systems are discussed along with the limitations and the future directions of brain-targeted delivery systems. As the intravenous injection is the main route for access of NP-based brain targeting delivery systems to the body, the description and discussion in this chapter is all acquiescently based on intravenously injected NP.

BBB TARGETING DELIVERY

As mentioned in Section “Introduction”, for most CNS disorders, the BBB is the main obstacle that restricts the brain access of therapeutics. Consequently, there are several strategies applied in BBB targeting delivery, which are based on the biological and physical characters of the BBB.

Decorating Delivery Systems with BBB Receptor Ligands

Many receptors are highly expressed on the BBB, including the transferrin receptor (TfR), the low-density lipoprotein receptor, the insulin receptor, the insulin-like growth factor receptor, the diphtheria toxin receptor, the nicotinic acetylcholine receptor (nAChR), and scavenger receptor class B type [5]. Thus, delivery systems can be decorated with ligands of these receptors to mediate their penetration of the BBB, which has been the most common and successful strategy in BBB targeting delivery.

NP functionalized with transferrin (Tf), the specific ligand for TfR, was used for targeted delivery [7]. In this study, the Tf-modified NP (Tf-NP) delivered significantly more cargo to brain tumors than that of unmodified NP. Treatment with doxorubicin-loaded Tf-NP significantly increased the median survival time of brain tumor-bearing rats, which was 70% longer than that of rats treated with a doxorubicin solution.

Coupling Antibodies with BBB Receptor Ligand Delivery Systems

Endogenous Tf may inhibit the internalization of Tf-NP mediated by the TfR. Coupled antibodies directed against these receptors can avoid this problem because the interaction between antibodies and receptors cannot be competitively inhibited by the endogenous ligands of the receptors. Thus, the antibodies for the corresponding receptors can be anchored onto the surface of NP for BBB targeting delivery. OX26, an antibody for TfR, has been used for brain-targeted delivery of NC1900, a peptide that can be used for the treatment of various neurological disorders [8]. The average ratio of the amount of drug in brain tissue for OX26-NP was 0.136%, which was 2.62-fold higher than with unmodified NP. Consequently, NC1900-loaded OX26-NP displayed best treatment outcome for Alzheimer's disease (AD) bearing rats as determined by the scopolamine-induced learning and memory impairments in a water maze task.

Coupling Peptides and Aptamers with BBB Receptor Ligand Delivery Systems

The application of proteins and antibodies are restricted by their instability and immunogenicity. Small molecules, such as peptides and aptamers, are better choices. CDX is a peptide redesigned from the loop II region of candoxin, a ligand for nAChR [9]. CDX was anchored onto the end of the materials to form targeting NP. The CDX-functionalized NP accumulated in brain tumors at significantly higher levels than that of unmodified NP. Paclitaxel-loaded targeting NP prolonged the median survival time of brain tumor-bearing mice to 27 days, which was significantly longer than that obtained with unmodified NP (a median survival time of 20 days). Another peptide, TGN, is a BBB-specific ligand discovered through phage display by our groups, which was decorated onto NP to deliver NAP (an activity-dependent neuroprotective protein) into the brain of the AD rat model created by intracerebroventricular injection of $A\beta_{1\text{–}40}$. The TGN-modified NP proved to deliver approximately fourfold more cargo to the brain than that of unmodified NP [10].

Transporter Ligand Coupled Micelle Delivery

In addition to receptors, transporters on the BBB can also be used for brain-targeted delivery, including amino acid transporters, hexose transporters, and monocarboxylate transporters [5]. The transporter for glutathione is highly expressed at the BBB. G-Technology is a glutathione conjugated liposomes based delivery system developed by the company ToBBB (the Netherlands) [11]. In a microdialysis study on rats, intravenous injection of glutathione-poly ethylene glycol (PEG) liposomes could deliver a fourfold higher fluorescent tracer to brain compared with PEG control liposomes [12]. Several drugs were successfully delivered into brain by G-Technology. One of them, glutathione PEGylated liposomal doxorubicin was under evaluation of clinical Phase I/II, which may become the first targeted nanomedicine approved in the world, because of the well-defined system, favorable pharmacokinetics, safety, and humanly applicable ligand.

Electrostatic Micelle Delivery

Due to electrostatic forces, positively charged delivery systems interact with the negatively charged BBB through adsorption-mediated endocytosis [5]. Cationized albumin is an example of this. Our group conjugated cationized albumin to NP

(CBSA-NP) to deliver the tumor necrosis factor-related apoptosis-inducing ligand (TRAIL) gene and aclarubicin to the brain [13, 14]. The aclarubicin concentrations in the tumors of CBSA-NP-treated mice were 2.6- and 3.3-fold higher than those of mice treated with unmodified NPs and a solution of aclarubicin 1 h post-injection and were 2.7- and 6.6-fold higher after 24 h, respectively. Four cycles of treatment with aclarubicin-loaded CBSA-NPs significantly increased the median survival time of brain tumor-bearing mice. Repeated injections of TRAIL gene-loaded CBSA-NPs also provided a better antibrain tumor effect than did the unmodified NPs. Additionally, cationized immunoglobulins, cationized monoantibodies, and histone have brain targeting properties via the same mechanism [5]. However, poor tissue selectivity and BBB specificity are the predominant problems of adsorptive-mediated targeting.

Chemical Enhancement of BBB Permeation

Temporarily open BBB by physical and pharmacological methods can also be used for brain targeting delivery. Borneol is a bicyclic monoterpene that is widely used in Traditional Chinese Medicine, and can enhance drug permeation through biological membranes including the BBB [15]. Utilizing this function, Zhang et al. intravenously injected NP into rats, followed by an oral dose of borneol. The coadministration resulted in 1.86-fold higher accumulation of NP in brain compared to administrating NP solely [15]. Other physical methods, such as ultrasound, can also increase the brain access of NP. However, the open BBB can elevate the delivery of both desired NP and therapeutics and undesired toxins in blood to the brain, which may lead to unpredictable side effects. Additionally, the disruption on the BBB may result in persistent harm to the function of the BBB.

Intranasal delivery can bypass the BBB, leading to enhanced brain access of NP and therapeutics, which is discussed in Chapter 25.

BBB Targeted Delivery Overview

Using brain-targeted delivery, the therapeutic outcome indeed improved, accompanied by the increased access of therapeutics to the normal and diseased brain cells. However, the distribution in the brain, particularly the diseased cells/normal brain cells ratio, is very important because the therapeutics generally do not exhibit cell-type selection. Distribution in the normal brain tissue may cause serious side effects.

TUMOR CELL TARGETING DELIVERY

In certain conditions, such as high-grade brain tumors, the BBB is unintegrated due to the rapid amplification of brain tumor cells and the formation of neovasculature [16]. Consequently, nanotherapeutics can reach the brain tumor bed directly via the enhanced permeability and retention (EPR) effect and display a brain-tumor treatment effect [17]. Brain tumor cells overexpress several receptors, including epidermal growth factor receptor, matrix metalloproteinase-2 (MMP-2), integrins, interleukin 13 receptor, nucleolin, TfR, and low-density lipoprotein receptor [18]. Studies of brain tumor-targeted delivery were generally based on these receptors.

Recently, several technologies were developed to screen for peptides or aptamers that possess high binding efficiency and high specificity. Phage display and SELEX screens are examples, and redesigning the existing ligands is another useful approach [9, 10, 19]. The AS1411 aptamer that binds nucleolin was discovered using SELEX. AS1411 modified NP (AS1411-NP) displayed approximately two-fold higher uptake by and localization in brain tumor cells compared with unmodified NP. *In vivo*, paclitaxel-loaded AS1411-NP effectively slowed tumor growth (81.68% slower than controls) and prolonged the median survival time of brain tumor-bearing mice (72% longer than controls), which was significantly better than the results obtained using unmodified NP [20]. The GMT-8 aptamer that selectively binds U87 cells was also used for brain tumor therapy and it exhibited an elevated antibrain tumor effect [21].

To further decrease localization in the non-target sites, ligands can be coated while circulating in the blood, then uncoated in specific microenvironments. Low molecular weight protamine (LMWP), a cell-penetrating peptide, was coated with a short cationic peptide through a linker, PLGLAG, which can be cleaved by MMP-2 [22]. This molecule is called active cell penetrating peptide (ACP). ACP-modified NP displayed significantly lower distribution in the liver, spleen, heart, and lungs; but much higher distribution in brain tumors. Consequently, paclitaxel-loaded ACP-modified NP prolonged the median survival time of brain tumor-bearing mice 39% longer than did LMWP-modified NP. Alternatively, activatable ligands can be coated with PEG via pH-sensitive, esterase-sensitive, or reduction-sensitive chemical bonds [6].

Codelivering chemotherapeutics with genes or proteins can synergistically enhance the antitumor effect. Our lab encapsulated TRAIL protein and doxorubicin in liposomes for brain tumor therapy [23]. Low dose doxorubicin sensitizes brain tumors cells to TRAIL protein. Incubating U87 cells for 12 h with either 37 ng/ml of TRAIL protein or 1.0 μg/ml of doxorubicin did not significantly inhibit cell growth (inhibition effect <25%); however, the combination of these two drugs effectively inhibited cell growth (inhibition effect >50%). The median survival time of mice treated with the codelivery liposomes was 50 and 23% longer than that of mice treated with TRAIL protein-loaded or doxorubicin-loaded liposomes, respectively.

Utilizing targeting ligands may further improve the antitumor effect. Zhan et al. encapsulated the TRAIL gene in CDX-modified micelles and encapsulated paclitaxel in RGD-modified micelles, which bind nAChR and integrin $\alpha_v\beta_3$, respectively [9, 24]. The median survival time of the brain tumor-bearing mice treated with these codelivered therapeutics (33.5 days) was significantly longer than those of mice treated solely with paclitaxel micelles (25.5 days) or TRAIL gene micelles (24.5 days). Codelivery of the TRAIL gene and paclitaxel in liposomes anchored to angiopep-2 also showed promising antibrain tumor effects [25].

However, the efficacy of brain tumor targeting delivery systems is restricted by poor access to infiltrated tumor cells where the BBB is intact. Consequently, the survival time is indeed prolonged, but recurrence cannot be prevented. Moreover, brain tumor-targeted delivery is useless for low grade brain tumors, in which the EPR effect is absent. Additionally, due to the integrity of BBB, the diseased cells targeting delivery cannot be applied into the treatment of other CNS disorders.

DUAL BBB AND DISEASE CELL TARGETING DELIVERY

Dual targeting delivery systems can target both the BBB and diseased cells, fully conquering the treatment barriers. Ideally, the first ligand should be dissociated after penetration of the BBB to minimize the unfavorable effects of this ligand on the targeted diseased cells.

The TGN peptide is a BBB-specific ligand discovered through phage display and the AS1411 aptamer is a specific ligand for nucleolin that is highly expressed on tumor cells [10, 26]. We recently functionalized NP with these two ligands [27] and they showed high brain tumor selectivity and accumulation. Docetaxel-loaded dual modified NP provided the best treatment effect in brain tumor-bearing mice. The median survival time of dual targeted NP-treated mice was 28% longer than that of mice treated with solely targeted NP. QSH is a peptide that has good affinity with $A\beta_{1-42}$, which is the main component of amyloid plaque. Taking this into consideration, Zhang et al. dual functionalized TGN and QSH onto NP to deliver therapeutics to the brains of AD mice [28]. Compared with TGN modified NP, the dual modified NP could deliver more cargo to the brain hippocampus, where the $A\beta_{1-42}$ was preimplanted, suggesting the modification with QSH could improve the cell selectivity of NP in brain. However, the TGN peptide could not be detached after entering brain, which may lead to its distribution in normal brain.

Fusion proteins that combine the active domains of two ligands can also be used for this purpose. Alternatively, if receptors are highly expressed on both the BBB and diseased brain cells, such as the low-density lipoprotein receptor-related protein (LRP) [29–31], the corresponding ligands could serve as dual targeting ligands to conquer both barriers with only one ligand. Angiopep-2 is a peptide derived from the Kunitz domain that possesses high binding affinity for LRP [32]. Angiopep-2 modification enhanced the BBB penetration of NP and led to higher gene expression in the brain [33]. Angiopep-2 modified NP encapsulating paclitaxel effectively prolonged the median survival time of brain tumor-bearing mice by 20% compared with unmodified NP [34].

Compared with BBB-targeting delivery systems and diseased cell-targeting delivery systems, the dual targeting delivery systems improved the access of drugs to brain diseases cells and the selectivity between diseased brain cells and normal brain cells. However, these systems are still not intelligent enough, which is not due only to the uncleaved ligands. Also, therapeutics release is not controlled and circumvent-response in these systems, leading to the undesired release of the drugs in normal tissues and the consequential side effects to the normal tissues.

THE PITFALLS AND REALITIES OF BRAIN TARGET DELIVERY

Most brain target delivery systems were claimed to be "designed" to "target" the BBB and/or diseased cells that were mediated by the surfaced modified ligands (Section "BBB Targeting Delivery", "Tumor Cell Targeting Delivery", and "Dual BBB and Disease Cell Targeting Delivery"). However, these NP are far from well-designed and there are several concerns needing to be addressed.

First, NP could not actually target specific sites and cells. In other words, the modified ligand cannot act as motor to "drive" or take the NP into a specific site. Actually, the interaction between ligands and receptors/transporters occurs only in a distance as short as 0.5 nm [35]. That means the *in vivo* behavior of NP mostly depends on the characteristics of tissues, the flow of blood and the intrinsic properties of NP, such as particle size, shape, and surface charge. The modification with ligands can statistically increase the possibility of interaction between NP and targeted cells, when the NP occasionally distribute nearby the cells. This description apparently doesn't deny the value of ligand modification. In fact, although the benefit of increased targeted site localization is still controversial, the ligand modification indeed can improve cell internalization [36, 37]. However, to emphasize the importance of related research, there is an aura of hyperbole that surrounds the field of targeted delivery, which we should avoid and honestly evaluate the contribution of ligand modification.

Second, the strength of the receptor–ligand interaction, density, length, and rigidity of ligands can affect the interaction [38]. Through dissipative particle dynamics, the interaction between ligand-modified NP and the lipid layer has been evaluated. The different force strength of the receptor–ligand interaction (i.e., adhesion energy) is defined by the difference ($\Delta\alpha$) between the interaction parameter of the receptor and ligand head and the interaction parameter of the ligand head and other particles (which is a given value in a certain situation). Interaction strength is positively correlated with $\Delta\alpha$. In other words, high receptor–ligand affinity is useful for increasing particle-membrane adhesion and particle engulfment by the membrane. The length of ligands also affects the interaction, even at an equal interaction strength. Increasing the length of ligands is helpful for attachment, due to the greater adhesion area that results from increased particle deformation and decreased bending energy caused by the larger effective volume of the ligand-modified particle. However, ligands tend to "form a bond" with receptors due to their high affinity, while long ligands easily become stretched and move away from their original orientation, resulting in the absence of ligands on the top surface of the particle. This absence of ligands can suppress further engulfment. In other words, compared to short ligands, particles modified with long ligands more easily attach to the membrane but are harder to engulf. This phenomenon is similar to the interactions of different particle shapes with cells [39]. The uniform distribution of ligands on the particle is useful for total engulfment of the particle, which can be achieved by increasing the ligand density and rigidity. Hydrophilic ligand-modified NPs tend to complicate total engulfment, because the exposed hydrophilic surface of the particle (due to the lack of ligands) barely interacts with the lipid membrane. Therefore, optimizing the hydrophobic/lipophobic properties of ligands is useful for total particle engulfment. The linker between NP and ligands should also be optimized because the linker can also affect the interaction between ligands and receptors. For example, Tf decorated onto NP through a given length of PEG could increase the cellular uptake of NP in a TfR-dependent manner, however, directly conjugating Tf onto NP could not [40].

Third, protein corona may hinder the specific interaction between ligands and receptors. The interaction between NP and plasma protein occurs as soon as the NPs are introduced into the blood, leading to formation of protein corona and the consequential modification on *in vivo* distribution [41]. The protein corona consists by over 50 kinds of proteins from blood, which can be divided into opsonin and dysopsonin according to their influence on blood circulation time of NP [42]. The formation of protein corona influences not only the behavior and distribution of NP, but it also covers the ligands that were decorated onto the NP. As discussed, Tf can specifically bind with TfR. The Tf modification could, obviously, increase uptake of NP by Tf-overexpressed cells through a TfR-dependent pathway, when the incubation system is free of serum [40]. However, in a Tf-depleted serum-containing incubation system, the specific interaction between Tf on the NP and TfR on the cells was decreased and, ultimately, became negligible. Thus, it could be concluded that the presence of serum caused specific target loss. A consistent result was observed in the interaction between a strained cycloalkyne and bicyclononyne on the NP and an azide on a silicon substrate, which was used as model interaction between ligand and receptor [43]. In serum-free system, the reaction could result in a large amount of accumulation of NP on the silicon substrate. The presence of serum could significantly decrease the accumulation of the NP. According to these studies, the targeting ability of ligands, whether for large proteins or small molecules, could be hindered by the plasma protein adsorption. Thus, when constructing targeting delivery systems, this must be taken into consideration, and the influence of plasma protein adsorption on the interaction between ligands and receptors should be systemically evaluated.

Last, the parameters of NP are could not specific control. Particle size is a critical parameter that affects the *in vivo* behavior and tissue distribution of NP. For example, the pore cutoff size of the U87 brain tumor model is 7–100 nm, while the size is even smaller in the RG2 brain tumor model [44], suggesting only particles in that size range may diffuse from blood into tumor matrix. However, for most biodegradable NP, such as liposomes, "mean particle size is 100 nm" is only a statistical description, which may actually refer to a mixture of particles that range from 50 to 150 nm. Thus, a large percentage of the liposomes may be directly prevented, by physical barriers, from accessing by the brain tumor. Consequently, to make the NP more controllable, the preparation method needs much attention. Particle replication in non-wetting templates® (PRINT®) is a newly emerged method that can produce uniform NP [45], which may address the pitfalls of brain target delivery. However, no targeted drug delivery system has been reported using this technology. Other parameters, such as drug loading capacity and the release profile, are also poorly controlled, which may influence the delivery efficiency and treatment outcome.

PERSPECTIVE

Targeted delivery is a promising branch of nanotechnology. The potential benefits include noninvasive administration, optimized drug distribution, elevated treatment outcome, reduced systemic side effects, and improved compliance. To further improve the treatment outcome, researchers should understand the irregular conditions of CNS disorders, which are the basis of targeted delivery.

Although elevated treatment outcomes were generally observed in recent studies, the drug distribution is far from ideal. Most drugs (over 95%) are distributed to non-targeted sites. Modification with targeting ligands may enhance the distribution to the targeted site. Because the targeting efficiency mainly depended on the properties of NP, optimizing the pharmacokinetic behavior is a critical aspect of targeted delivery systems.

Translation from the laboratory into the clinic is the foremost challenge in the application of targeted delivery systems. Currently, only a few nanotherapeutics are commercially available, including doxorubicin-loaded liposomes, paclitaxel micelles, and NP with albumin-bound paclitaxel. In addition to the complex production procedure for most nanotherapeutics, the safety of nanomaterials is a major concern. Currently, only several materials are approved for injection, including polylactide, poly lactide-polyglycolide, and poly ε-caprolactone. Most materials used for gene delivery and/or with good photoelectric properties cannot be used in human, because of the systemic toxicity. Natural materials are superior in this regard. More research should focus on functionalizing natural materials with the required properties or reducing the toxicity of existing materials that have excellent characteristics.

The application of targeting ligands is also a big concern. Although many ligands were used in brain targeted delivery, the biocompatibility of these ligands was not assured for human use as the ligands must be safe, effective, and economic.

Overall, the application of targeted delivery systems for CNS disorders depends on developments in CNS biology, nanotechnology, and materials science. By combining the improvements achieved in these fields, targeted delivery systems can play an important role in the management of CNS disorders.

ACKNOWLEDGMENTS

The work was granted by the Sichuan University Starting Foundation for Young Teachers (2014SCU11044) and the National Basic Research Program of China (973 Program, 2013CB932504).

REFERENCES

[1] Wohlfart, S., Gelperina, S., Kreuter, J. (2012). Transport of drugs across the blood-brain barrier by nanoparticles. *Journal of Controlled Release*, 161, 264–273.

[2] Pardridge, W. M. (1998). CNS drug design based on principles of blood-brain barrier transport. *Journal of Neurochemistry*, 70, 1781–1792.

[3] Pardridge, W. M. (2007). Drug targeting to the brain. *Pharmaceutical Research*, 24, 1733–1744.

[4] Srikanth, M., Kessler, J. A. (2012). Nanotechnology-novel therapeutics for CNS disorders. *Nature Review Neurology*, 8, 307–318.

[5] Guo, L., Ren, J., Jiang, X. (2012). Perspectives on brain-targeting drug delivery systems. *Current Pharmaceutical Biotechnology*, 13, 2310–2318.

[6] Gao, H., Pang, Z., Jiang, X. (2013). Targeted delivery of nano-therapeutics for major disorders of the central nervous system. *Pharmaceutical Research*, 30, 2485–2498.

[7] Pang, Z., Gao, H., Yu, Y., Guo, L., Chen, J., Pan, S., Ren, J., Wen, Z., Jiang, X. (2011). Enhanced intracellular delivery and chemotherapy for glioma rats by transferrin-conjugated biodegradable polymersomes loaded with Doxorubicin. *Bioconjugate Chemistry*, 22, 1171–1180.

[8] Pang, Z., Lu, W., Gao, H., Hu, K., Chen, J., Zhang, C., Gao, X., Jiang, X., Zhu, C. (2008). Preparation and brain delivery property of biodegradable polymersomes conjugated with OX26. *Journal of Controlled Release*, 128, 120–127.

[9] Zhan, C., Li, B., Hu, L., Wei, X., Feng, L., Fu, W., Lu, W. (2011). Micelle-based brain-targeted drug delivery enabled by a nicotine acetylcholine receptor ligand. *Angewandte Chemie International Edition in England*, 50, 5482–5485.

[10] Li, J., Feng, L., Fan, L., Zha, Y., Guo, L., Zhang, Q., Chen, J., Pang, Z., Wang, Y., Jiang, X., Yang, V. C., Wen, L. (2011). Targeting the brain with PEG-PLGA nanoparticles modified with phage-displayed peptides. *Biomaterials*, 32, 4943–4950.

[11] Gaillard, P. J., Visser, C. C., Appeldoorn, C. C. M., Rip, J. (2012). Enhanced brain drug delivery: Safely crossing the blood–brain barrier. *Drug Discovery Today: Technologies*, 9, e155–e160.

[12] Rip, J., Chen, L., Hartman, R., van den Heuvel, A., Reijerkerk, A., van Kregten, J., van der Boom, B., Appeldoorn, C., de Boer, M., Maussang, D., de Lange, E. C., Gaillard, P. J. (2014). Glutathione PEGylated liposomes: Pharmacokinetics and delivery of cargo across the blood-brain barrier in rats. *Journal of Drug Targeting*, 22(5): 460–467.

[13] Lu, W., Wan, J., Zhang, Q., She, Z., Jiang, X. (2007). Aclarubicin-loaded cationic albumin-conjugated PEGylated nanoparticle for glioma chemotherapy in rats. *International Journal of Cancer*, 120, 420–431.

[14] Lu, W., Sun, Q., Wan, J., She, Z., Jiang, X. G. (2006). Cationic albumin-conjugated PEGylated nanoparticles allow gene delivery into brain tumors via intravenous administration. *Cancer Research*, 66, 11878–11887.

[15] Zhang, L., Han, L., Qin, J., Lu, W., Wang, J. (2013). The use of borneol as an enhancer for targeting aprotinin-conjugated PEG-PLGA nanoparticles to the brain. *Pharmaceutical Research*, 30, 2560–2572.

[16] Agarwal, S., Sane, R., Oberoi, R., Ohlfest, J. R., Elmquist, W. F. (2011). Delivery of molecularly targeted therapy to malignant glioma, a disease of the whole brain. *Expert Reviews in Molecular Medicine*, 13, e17.

[17] Gao, H., Pang, Z., Pan, S., Cao, S., Yang, Z., Chen, C., Jiang, X. (2012). Antiglioma effect and safety of docetaxel-loaded nanoemulsion. *Archives of Pharmacal Research*, 35, 333–341.

[18] Liu, Y., Lu, W. (2012). Recent advances in brain tumor-targeted nano-drug delivery systems. *Expert Opinion on Drug Delivery*, 9, 671–686.

[19] Bayrac, A. T., Sefah, K., Parekh, P., Bayrac, C., Gulbakan, B., Oktem, H. A., Tan, W. (2011). *In vitro* selection of DNA aptamers to glioblastoma multiforme. *ACS Chemical Neuroscience*, 2, 175–181.

[20] Guo, J., Gao, X., Su, L., Xia, H., Gu, G., Pang, Z., Jiang, X., Yao, L., Chen, J., Chen, H. (2011). Aptamer-functionalized PEG-PLGA nanoparticles for enhanced antiglioma drug delivery. *Biomaterials*, 32, 8010–8020.

[21] Gao, H., Qian, J., Yang, Z., Pang, Z., Xi, Z., Cao, S., Wang, Y., Pan, S., Zhang, S., Wang, W., Jiang, X., Zhang, Q. (2012). Whole-cell SELEX aptamer-functionalised poly(ethyleneglycol)-poly(epsilon-caprolactone) nanoparticles for enhanced targeted glioblastoma therapy. *Biomaterials*, 33, 6264–6272.

[22] Gu, G., Xia, H., Hu, Q., Liu, Z., Jiang, M., Kang, T., Miao, D., Tu, Y., Pang, Z., Song, Q., Yao, L., Chen, H., Gao, X., Chen, J. (2013). PEG-co-PCL nanoparticles modified with MMP-2/9 activatable low molecular weight protamine for enhanced targeted glioblastoma therapy. *Biomaterials*, 34, 196–208.

[23] Guo, L., Fan, L., Pang, Z., Ren, J., Ren, Y., Li, J., Chen, J., Wen, Z., Jiang, X. (2011). TRAIL and doxorubicin combination enhances antiglioblastoma effect based on passive tumor targeting of liposomes. *Journal of Controlled Release*, 154, 93–102.

[24] Zhan, C., Wei, X., Qian, J., Feng, L., Zhu, J., Lu, W. (2012). Co-delivery of TRAIL gene enhances the antiglioblastoma effect of paclitaxel *in vitro* and *in vivo*. *Journal of Controlled Release*, 160, 630–636.

[25] Sun, X., Pang, Z., Ye, H., Qiu, B., Guo, L., Li, J., Ren, J., Qian, Y., Zhang, Q., Chen, J., Jiang, X. (2012). Co-delivery of pEGFP-hTRAIL and paclitaxel to brain glioma mediated by an angiopep-conjugated liposome. *Biomaterials*, 33, 916–924.

[26] Ireson, C. R., Kelland, L. R. (2006). Discovery and development of anticancer aptamers. *Molecular Cancer Therapeutics*, 5, 2957–2962.

[27] Gao, H., Qian, J., Cao, S., Yang, Z., Pang, Z., Pan, S., Fan, L., Xi, Z., Jiang, X., Zhang, Q. (2012). Precise glioma targeting of and penetration by aptamer and peptide dual-functioned nanoparticles. *Biomaterials*, 33, 5115–5123.

[28] Zhang, C., Wan, X., Zheng, X., Shao, X., Liu, Q., Zhang, Q., Qian, Y. (2014). Dual-functional nanoparticles targeting amyloid plaques in the brains of Alzheimer's disease mice. *Biomaterials*, 35, 456–465.

[29] Maletinska, L., Blakely, E. A., Bjornstad, K. A., Deen, D. F., Knoff, L. J., Forte, T. M. (2000). Human glioblastoma cell lines: Levels of low-density lipoprotein receptor and low-density lipoprotein receptor-related protein. *Cancer Research*, 60, 2300–2303.

[30] Ito, S., Ohtsuki, S., Terasaki, T. (2006). Functional characterization of the brain-to-blood efflux clearance of human amyloid-beta peptide (1–40) across the rat blood-brain barrier. *Neuroscience Research*, 56, 246–252.

[31] Li, H., Qian, Z. M. (2002). Transferrin/transferrin receptor-mediated drug delivery. *Medicinal Research Reviews*, 22, 225–250.

[32] Demeule, M., Regina, A., Che, C., Poirier, J., Nguyen, T., Gabathuler, R., Castaigne, J. P., Béliveau, R. (2008). Identification and design of peptides as a new drug delivery system for the brain. *Journal of Pharmacology and Experimental Therapeutics*, 324, 1064–1072.

[33] Ke, W., Shao, K., Huang, R., Han, L., Liu, Y., Li, J., Kuang, Y., Ye, L., Lou, J., Jiang, C. (2009). Gene delivery targeted to the brain using an angiopep-conjugated polyethylene-glycol-modified polyamidoamine dendrimer. *Biomaterials*, 30, 6976–6985.

[34] Xin, H., Sha, X., Jiang, X., Zhang, W., Chen, L., Fang, X. (2012). Antiglioblastoma efficacy and safety of paclitaxel-loading angiopep-conjugated dual targeting PEG-PCL nanoparticles. *Biomaterials*.

[35] Lammers, T., Kiessling, F., Hennink, W. E., Storm, G. (2012). Drug targeting to tumors: Principles, pitfalls and (pre-) clinical progress. *Journal of Controlled Release*, 161, 175–187.

[36] Gao, H., Yang, Z., Zhang, S., Cao, S., Shen, S., Pang, Z., Jiang, X. (2013). Ligand modified nanoparticles increases cell uptake, alters endocytosis and elevates glioma distribution and internalization. *Science Reports*, 3, 2534.

[37] Kirpotin, D. B., Drummond, D. C., Shao, Y., Shalaby, M. R., Hong, K., Nielsen, U. B., Marks, J. D., Benz, C. C., Park, J. W. (2006). Antibody targeting of long-circulating lipidic nanoparticles does not increase tumor localization but does increase internalization in animal models. *Cancer Research*, 66, 6732–6740.

[38] Gao, H., He, Q. (2014). The interaction of nanoparticles with plasma proteins and the consequent influence on nanoparticles behavior. *Expert Opinion on Drug Delivery*, 11(3), 409–420.

[39] Sharma, G., Valenta, D. T., Altman, Y., Harvey, S., Xie, H., Mitragotri, S., Smith, J. W. (2010). Polymer particle shape independently influences binding and internalization by macrophages. *Journal of Controlled Release*, 147, 408–412.

[40] Salvati, A., Pitek, A. S., Monopoli, M. P., Prapainop, K., Bombelli, F. B., Hristov, D. R., Kelly, P. M., Åberg, C., Mahon, E., Dawson, K. A. (2013). Transferrin-functionalized nanoparticles lose their targeting capabilities when a biomolecule corona adsorbs on the surface. *Nature Nanotechnology*, 8, 137–143.

[41] Nel, A. E., Madler, L., Velegol, D., Xia, T., Hoek, E. M., Somasundaran, P., Klaessig, F., Castranova, V., Thompson, M. (2009). Understanding biophysicochemical interactions at the nano-bio interface. *Nature Materials*, 8, 543–557.

[42] Aggarwal, P., Hall, J. B., McLeland, C. B., Dobrovolskaia, M. A., McNeil, S. E. (2009). Nanoparticle interaction with plasma proteins as it relates to particle biodistribution, biocompatibility and therapeutic efficacy. *Advanced Drug Delivery Reviews*, 61, 428–437.

[43] Mirshafiee, V., Mahmoudi, M., Lou, K., Cheng, J., Kraft, M. L. (2013). Protein corona significantly reduces active targeting yield. *Chemical Communications (Cambridge)*, 49, 2557–2559.

[44] Zhan, C., Lu, W. (2012). The blood-brain/tumor barriers: Challenges and chances for malignant gliomas targeted drug delivery. *Current Pharmaceutical Biotechnology*, 13, 2380–2387.

[45] Rolland, J. P., Maynor, B. W., Euliss, L. E., Exner, A. E., Denison, G. M., DeSimone, J. M. (2005). Direct fabrication and harvesting of monodisperse, shape-specific nanobiomaterials. *Journal of the American Chemical Society*, 127, 10096–10100.

25

INTRANASAL DELIVERY TO THE CENTRAL NERVOUS SYSTEM

LISBETH ILLUM
IDentity, The Park, Nottingham, UK

INTRODUCTION

Delivery of drugs to the CNS via the nasal route is an area of increasing interest due to the possibility of circumventing the blood–brain barrier (BBB) by exploiting direct transport pathways between nose and brain and, thereby, allowing drugs that do not easily cross the BBB, such as hydrophilic small molecules, peptides, and proteins, access to the various brain regions. Such improved access provides the possibility of optimizing the efficacy of treatment of chronic neurodegenerative conditions of the CNS, such as Parkinson's disease and Alzheimer's disease, as well as acute conditions, such as meningitis and stroke.

When lipophilic drugs are administered nasally, in general, such drugs are absorbed fast and efficiently across the nasal mucosa and reach the systemic circulation often with a bioavailability up to 100% and a plasma profile resembling that after an intravenous administration. Once in the bloodstream, these molecules can diffuse through the BBB and reach the brain. The degree of this diffusion is dependent on the log P and molecular size of the drug [1], as well as dependent on the site of deposition in the nasal cavity; only a minor portion of the dose of these drugs will reach the brain through a direct nose-to-brain pathway.

Polar molecules in general only cross the nasal membrane in limited amounts; and hence, only a minor part reaches the systemic circulation, with resultant bioavailabilities of less than 10% for small polar drugs and less than 1% for macromolecules.

Blood–Brain Barrier in Drug Discovery: Optimizing Brain Exposure of CNS Drugs and Minimizing Brain Side Effects for Peripheral Drugs, First Edition. Edited by Li Di and Edward H. Kerns.

Furthermore, such polar drugs do not readily cross the BBB unless by some form of naturally occurring receptor or carrier-mediated transport mechanism, as is for example the case for insulin [2].

In my opinion, it has been shown in numerous studies in animal models, such as rats, sheep, and monkeys, that it is possible to achieve direct nose-to-brain delivery from the nasal cavity via the olfactory region or via the trigeminal nerves or nerve endings in the respiratory tissue for a range of drugs that only to a limited degree can cross the BBB [3–12]. This fact was disputed in a paper by Merkus and van den Berg [13] who evaluated about 100 published papers on delivery of drug to the brain via the nasal route in animal models and concluded that, based on their criteria, only 12 papers presented sound experimental designs and, of these, only 2 studies in rats were found to provide results that indicated direct transport of the drug from the nose to the CNS. I believe the criteria applied were too stringent; and furthermore, a wealth of papers has since been published showing strong evidence of direct nose-to-brain delivery in animal models.

Due to ethical problems associated with such studies, only few papers have shown evidence of direct nose-to-brain transport in man by evaluating pharmacokinetic data after nasal administration [5, 14–16], whereas many papers have been published that evaluate the pharmacodynamic effects of nasally administered drugs compared with drugs given via a parenteral route, in terms of brain functions and therapeutic efficiency. On the basis of results from such studies in man, it is, in general, accepted that these results indicate at least some degree of direct nose-to-brain targeting, although Merkus and van den Berg [13] also dispute these results [13, 17–28].

From studies performed in animal models, it has been found that, in general, the amount of drug that reaches the CNS after nasal application (certainly when applied in a simple formulation) is well below 1% of administered dose. Considering that the olfactory area in most animal models, such as rats, is significantly larger (50% of nasal surface area) than in man (about 2.5% of nasal surface area), the transport that takes place in animal models, at least using the olfactory epithelium pathway, can be considered as an overestimation. For a drug administered nasally to man, it will be difficult to reach the olfactory region that is placed high up in the nasal cavity at and above the superior conchae, since it has been demonstrated that the nasal anatomical configuration affects the airflow and only about 15–20% of the airflow reaches the olfactory region in man [29].

Hence, it is important to exploit special delivery devices, such as the Bi-directional™ technology (OptiNose), Controlled Particle Dispersion® technology (Kurve Technology, Inc.), or the Precision Olfactory Delivery (POD) technology (Impel NeuroPharma), that are able to target the olfactory area and develop optimal drug delivery systems designed to enable the transport across the nasal mucosa (e.g., the olfactory epithelium), involving bioadhesives, absorption enhancers, and/or micro- or nanoparticulate delivery systems.

The development of novel delivery systems for optimizing the delivery of drugs via the nasal cavity to the CNS has been described extensively in the literature in the past 10 years. Such published papers mostly describe the use of systems, such as hydrogels [e.g., 30], microemulsions [e.g., 31, 32], and especially nanoparticles [e.g., 12, 33–37]. Further, several reviews have been published in this area, for example, Mathison et al. [38], Frey [39], Illum [6, 40], Lochhead and Thorne [41], Zhu et al. [42], Djupesland [28], and Tayebati et al. [43].

This chapter sets out to discuss the scientific developments of nose-to-brain delivery of drugs in the past 10 years and the use of novel delivery systems and devices for optimization of the amount of drug reaching the brain. The chapter also gives some background information in terms of the anatomy and physiology of the nasal cavity and the transport pathways from the nasal cavity to the CNS.

NASAL ANATOMY AND PHYSIOLOGY

The nasal anatomy and physiology has been comprehensively described in textbooks and scientific reviews, including Mygind and Dahl [44], Moran et al. [45], Hilger [46], Illum [4, 5] and Mistry et al. [47]. Hence, only details that are necessary for understanding more fully the potential of an efficient transport of drugs and the factors affecting nose-to-brain delivery from the olfactory nerve region and through the trigeminal nerve system will be described here.

Nasal Cavity

The nose consists of the outer visible nasal vestibule that starts at the nostrils and ends at the nasal valve and the inner non-visible nasal cavity. The nose is divided, along the long axis, into two non-connected halves by the nasal septum. Each side comprises the nasal vestibule and the nasal cavity, with the respiratory region and the olfactory region. Anteriorly, the nasal cavity opens out to the facial site through the nasal valve, a narrow triangular-shaped slit situated at the top of the nostril, with a surface area of about 0.3 cm^2. Posteriorly, the nasal cavity opens into the nasopharynx by way of the posterior nasal apertures. The nasopharynx is considered the airway region posterior to the end of the nasal septum and proximal to the termination of the soft palate. Each nasal passage of the nasal cavity is defined by a septal wall, a lateral wall, a floor, and a roof. The surface of the nasal cavity is lined by a mucous membrane that is well vascularized and covered by a mucous layer. The superior, middle, and inferior conchae project from the lateral wall into the lumen of the nasal cavity, thereby increasing the inner surface area of the nasal cavity. This is important for humidification, filtering, and heating of the inhaled air (Fig. 25.1).

The nasal cavity (the main chamber) in man is about 5–8 cm long, and the total surface area has been reported by various researchers to be between 150 and 200 cm^2 [48–50]. The olfactory epithelium is limited to about 2.5% of the total surface area (~4.5–6.0 cm^2). In rats, which is the animal model mostly used for nose-to-brain delivery studies, the total surface area of a 16-week-old (288 g) rat has been reported to be 13.3 cm^2 and, of this, the olfactory epithelium occupies about 50% (~6.7 cm^2). For a 7 kg monkey, the dimensions are 61.6 cm^2 total surface area, and the olfactory epithelium constitutes about 13% of the area (~8.0 cm^2). It is also important to note that the ratio between nasal surface area and body weight varies significantly between the three species: rat, 53.6; monkey, 8.8; and man, 2.2. It should also be noted that

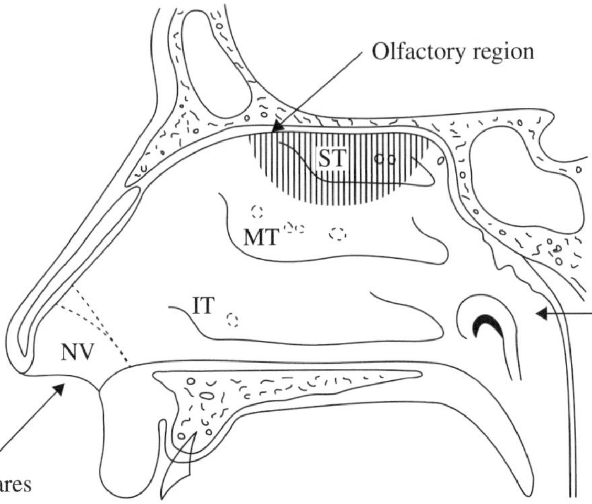

FIGURE 25.1 Lateral wall of the human nasal cavity. NV, nasal vestibule; IT, inferior turbinate; MT, middle turbinate; ST, superior turbinate.

there are significant differences in the nasal morphology between species, since the human nose has a relatively simple structure, with breathing as the primary function (microsmatic), whereas, the noses of most animal models comprise more complex structures, where olfaction is the primary function (macrosmatic). Hence, these differences in morphology between the animal species and humans vary considerably and should be taken into account when evaluating data for nose-to-brain delivery.

The rate and direction of airflow during respiration is modified by the nasal valve. Airflow during normal tidal breathing creates velocities of 18 m/s, with predominantly laminar airflow through the nasal passages. When a subject sniffs, airflow can reach 32 m/s, resulting in a local turbulent airflow downstream from the nasal valve [51, 52]. This is caused by the progressive narrowing of the nasal valve with increasing inspiratory flow rate. During normal breathing about 5–10% of the inspired air reaches the olfactory region; and when sniffing, this fraction can increase to 20% [53]. Hence, most of the inspired air and hence formulations given nasally will not reach the olfactory region in man.

The nasal vestibule is covered with stratified squamous epithelium that resembles skin and from the nasal valve region in the anterior part of the nose gradually transitions into pseudostratified columnar epithelium, i.e., the respiratory epithelium. The olfactory epithelium is situated in the roof of the nasal cavity, partly on the nasal septum and partly on the superior and middle conchae.

Nasal Respiratory Epithelium

The respiratory epithelium, a pseudostratified columnar epithelium, is covered by microvilli; and due to the large surface area created by the microvilli, the respiratory epithelium is considered the major site for systemic absorption of nasally

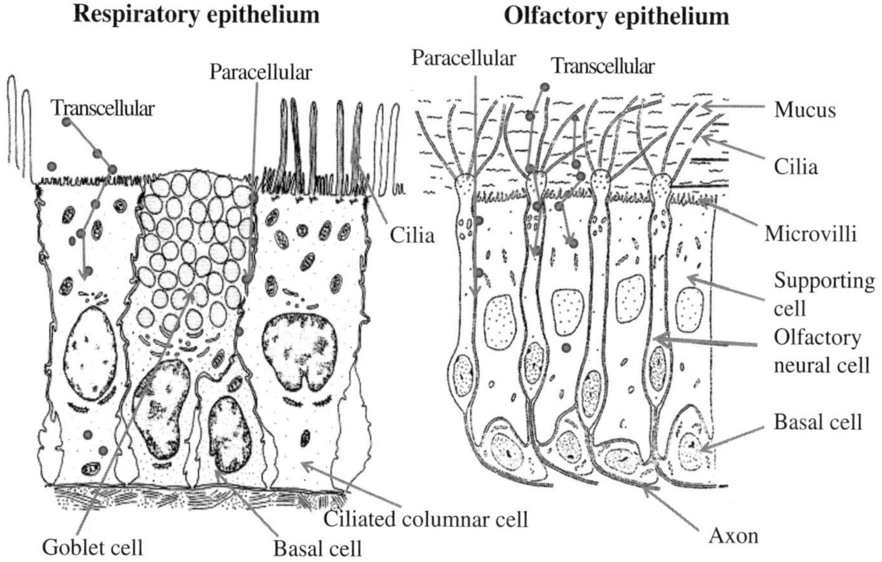

FIGURE 25.2 The respiratory and olfactory epithelia. The respiratory epithelium comprises ciliated or non-ciliated pseudostratified columner epithelial cells, goblet cells providing mucus, and basal cells. The olfactory epithelium comprises olfactory neural cells with long immovable cilia, supporting cells, and basal cells that are able to replace the neural cells. The schematic drawing also shows the paracellular (through the tight junctions) and the transcellular routes of transepithelial passage of nanoparticles. (Adapted from Mygind N and Dahl R, 1998. Anatomy, physiology and function of the nasal cavities in health and disease. *Adv. Drug Del. Rev.* 29:3–12. With permission from Elsevier.)

administered drugs. The respiratory epithelium comprises four different cell types: ciliated pseudostratified columnar cells, non-ciliated pseudostratified columnar cells, goblet cells, and basal cells (Fig. 25.2). Under the luminal surface epithelium lies the lamina propria with numerous blood and lymphatic vessels, nerves, glands, and mesenchymal cells. The respiratory epithelium is covered by a mucus layer, consisting of an upper viscous gel layer (2–4 µm thick), floating on a lower serous fluid layer (or sol layer) (3–5 µm thick) surrounding the cilia.

Cilia are long, thin projections of 2–4 µm length originating on the surface of the columnar cells. They are mobile and through a synchronized beating (1000 strokes/min), they propel the viscous part of the mucus layer, covering the cells, posteriorly toward the nasopharynx with a speed of about 5 mm/min, while the lower sol layer of the mucus remains stationary. This constitutes the so-called mucociliary clearance mechanism.

Nasal Olfactory Epithelium

The nasal olfactory epithelium is situated at the top of the nasal cavity just under the cribriform plate of the ethmoid bone, which separates the nasal cavity from the brain. The olfactory nerves emerge from the synapses in the olfactory bulbs, through holes in the cribriform plate, and together with the sustentacular (supporting) cells and the basal cells form the olfactory epithelial tissue placed on the upper part of the septum and the superior conchae. It has also been reported that less dense olfactory filaments and olfactory tissue can be found on the anterior and posterior parts of the middle conchae [54].

The olfactory epithelium is a modified (pseudostratified) columnar epithelium and comprises, apart from the olfactory sensory neurons, also the sustentacular cells, which, by ensheathing the neural cells, provide mechanical support to the cells, and two types of basal cells that can differentiate into neural cells (globose basal cells) or sustentacular cells and cells of the Bowman's glands (horizontal basal cells), and replace these as needed (Fig. 25.2). The number of olfactory neurons is estimated to be 6×10^6 with a density of 30,000 neurons per mm^2 [45]. Below the epithelium lies the lamina propria, containing the Bowman's glands that are interspersed between the olfactory nerve bundles. Bowman's glands contain a high amount of a mucosubstance that contributes to the mucous layer covering the luminal surface of the olfactory epithelium. Furthermore, the olfactory epithelium also comprises a small amount of cells with numerous microvilli (microvillus cells), although the function of these cells are not known [50, 55–57].

The olfactory sensory neurons terminate at the apical surface of the epithelium as small bulbous olfactory knobs from which extend numerous, up to 200 μm long, immobile cilia intertwined with each other and with the microvilli in the surface fluid. The nerve axon originates from the base of the nerve cell and passes through the lamina propria, where it bundles together with other axons in a nerve bundle or nerve fascicle, which is ensheathed by glial cells (Schwann cells). The nerve bundles cross the cribriform plate through small holes and synapse in the olfactory bulbs situated in the cranial cavity. From here, the nerve projections reach the amygdale, the prepyriform cortex, the anterior olfactory nucleus, and the entorhinal cortex, as well as the hippocampus, hypothalamus, and the thalamus.

It should be noted that since the cilia on the surface of the epithelium are non-mobile, the mucous will not be cleared by a mucociliary clearance mechanism; however, it is envisaged that the mucous will clear due to overproduction and also due to the normal upright position of a human.

Nasal Trigeminal Nerve System

The trigeminal nerve is the largest of the cranial nerves and is responsible for the sensation in the face and certain motor functions, such as biting and chewing. Its name derives from the fact that each trigeminal nerve, one on each side of the pons, has three major branches: the ophthalmic nerve, the maxillary nerve, and the mandibular nerve (Fig. 25.3). The pons is part of the brainstem that links the medulla oblongata and the thalamus in the brain. The ophthalmic (V_1), the maxillary (V_2), and the mandibular (V_3) nerves converge on the trigeminal ganglion (or Gasserian ganglion), analogous to the dorsal root ganglia of the spinal cord, which is located within Meckel's cave, and contains the cell bodies of incoming sensory nerve fibers. From the trigeminal ganglion, a single large sensory root enters the brainstem at the level

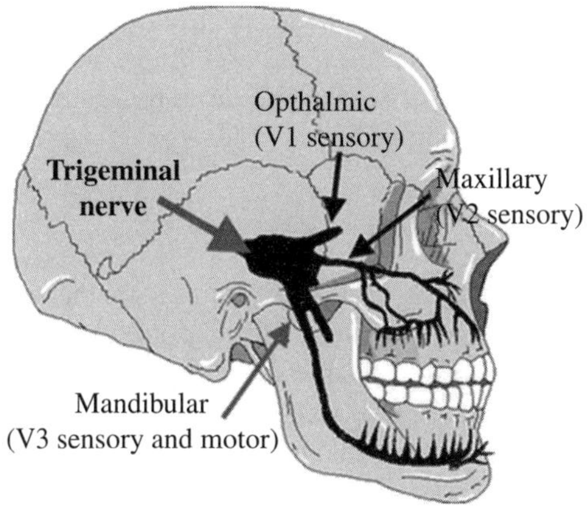

FIGURE 25.3 Lateral view of skull illustrating the areas innervated by the trigeminal nerve and its branches.

of the pons. The ophthalmic, maxillary, and mandibular branches leave the skull through three separate openings at the base of the skull: the superior orbital fissure, the foramen rotundum, and the foramen ovale.

The areas of cutaneous distribution of the three branches of the trigeminal nerve have sharp borders with relatively little overlap. Hence, an injection of local anesthetics in the mandibular nerve results in the numbing of the teeth on one side of the jaw, well known from dentistry.

The ophthalmic and the maxillary nerves only carry sensory information, whereas the mandibular nerve also participates in motor functions. The ophthalmic nerve (V_1), carries sensory information from the scalp and forehead; the upper eyelid; the conjunctiva and cornea of the eye; the nose (including its tip); the nasal mucosa; the frontal sinuses; parts of the meninges and the maxillary nerve (V_2) from the lower eyelid and cheek; the nares and upper lip; the upper teeth and gums; the nasal mucosa; the palate and roof of the pharynx; the maxillary, ethmoid, and sphenoid sinuses; and parts of the meninges. The mandibular nerve (V_3) carries sensory information from the lower lip, the lower teeth and gums, the chin and jaw, parts of the external ear, and parts of the meninges.

Hence, in mammals, trigeminal nerve endings from the ophthalmic and the maxillary nerves are distributed throughout the nasal respiratory epithelium, although these nerve endings (as opposed to the olfactory nerves) are not directly exposed to the lumen of the nasal cavity, but lie in the epithelium near the epithelial surface stopping at the level of the tight junctions [58]. In humans, non-myelinated nerves approach the respiratory mucosa in bundles of up to 200 axons, devoid of perineural sheaths [59]; and in dogs, corpuscular endings (300–500 μm long and 100–250 μm wide) displaying bulbous, laminar, and varicose expansions are found near the lumen in the respiratory epithelium on the septum and the nasal conchae [60]. Drug molecules deposited in the nasal cavity may reach the nerve endings, dependent on the extent to which their physicochemical characteristics enable crossing of the epithelium using extracellular and transcellular routes. The neurons have a direct transsynaptic connection with the CNS, through the pons, and the rest of the hindbrain, and a part of the trigeminal nerve also passes through the cribriform plate and is, hence, connected to the forebrain [61].

Epithelial Cell Barrier

The respiratory and olfactory epithelia are similar in that they are both tightly connected by intercellular junctions that surround the cells of the epithelia. These junctional complexes are narrow belt-like structures comprising, from the epithelial surface in direction of the basal surface, the zona occludens (what is normally called the tight junction), the zona adherence and the macula adherence (Fig. 25.4). Tight junctions comprise a branching network of sealing strands, each strand acting independently from the others. Each strand is formed from a row of transmembrane proteins embedded in both plasma membranes, with extracellular domains joining one another directly. The major types of proteins are the claudins and the occludins. These associate with different peripheral membrane proteins such as ZO-1 located on the intracellular side of the plasma membrane, which anchor the strands to the actin component of the cytoskeleton. Thus, tight junctions join together the cytoskeletons of adjacent cells and form a regulatable semipermeable diffusion barrier between the epithelial cells [62]. In the nasal cavity, the diameter of the tight junctions (the pore size) has been measured to be 3.9–8.4 Å. It was shown by Van Itallie et al. [63] in studies in Caco-2, MDCK II, MDCK C7 cells and in pig ileum that the number of pores per area of monolayer (pore density) is different among cell types, with the density being highest for the Caco-2 cells compared to the other cell lines tested. Interestingly, they also found that the paracellular flux can be modeled on

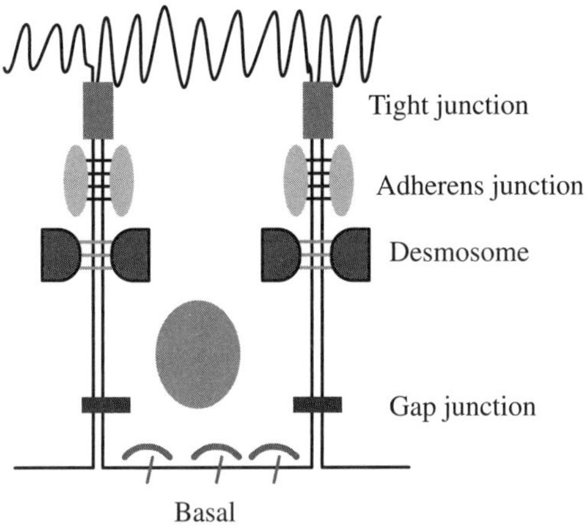

FIGURE 25.4 Epithelial cell barrier—intercellular junctions.

at least two components: (i) high-capacity, size-restrictive pores, with clearly defined aperture size, for substances less than 4Å in radius, and (ii) a second, relative size-independent pathway, where diffusion occurs through intercellular spaces, at least for substances up to a radius of 7Å. These observations are in line with those of Stevenson et al. [64].

When considering transport of drugs across epithelial cell barriers, it is important also to consider the size of the drug molecules and of any particulate carrier such as nanoparticles and whether size wise, a paracellular transport through an open or closed tight junction is physically possible or whether only the trancellular/intracellular route is available for large molecules and/or carrier systems. The size and, potentially, also secondary and tertiary structures of biological molecules, are determined by their physicochemical environment; and hence, it is difficult to relate size to an exact molecular weight. As an example, the size of an insulin monomer (MW 5808Da) has been described as 26.8Å in diameter and peroxidase (MW 44,000Da) as 40Å, which is disproportional to the size of the insulin molecule [65]. Cytochrome C (MW 13,000Da) has a diameter of 30Å [66], Lysozyme (MW 14,300Da) from chicken egg white is an ellipsoid with diameters 35 and 45Å, whereas the hydrodynamic radius of hen lysozyme (MW 16,000Da) is 20.5Å in the native folded state and 34.6Å under denaturing conditions [67]. On the other hand, the branched polysaccharide dextran labeled with fluorescein isothiocyanate shows a linear increase in hydrodynamic radius of 23Å for the 10kDa dextran, 45Å for the 40kDa dextran, and 60Å for the 70kDa (data provided by Sigma-Aldrich). As reported by Costantino et al. [68], absorption enhancers can open tight junctions both in the respiratory and the olfactory epithelium, with an increase in diameter of about 10–15 times; however, the maximal diameter reached was shown to be of the order of 15nm.

TRANSPORT ROUTES AND MECHANISMS

As have been described by a range of authors in various detailed reviews and papers, it is generally accepted that drugs administered nasally can reach the brain using three main pathways: (i) absorption across the nasal respiratory epithelium into the systemic circulation and, from there, across the BBB into the brain (systemic pathway), (ii) direct paracellular or transcellular transport via the olfactory neurons (olfactory neural pathway) or the olfactory epithelial cells (olfactory epithelial pathway), or (iii) transport via the trigeminal nerves (trigeminal pathway) [4, 7, 41, 50, 69, 70].

Only drugs that are sufficiently lipophilic to readily cross the nasal membrane and that can pass the BBB in sufficient quantities are able to exploit the "systemic pathway" and rapidly be transported from the nasal cavity to the brain tissue. Hence, for such drugs there would be little difference in brain levels between parenterally and nasally administered drugs.

Drug formulations can be internalized into the knob-like swellings of the primary neurons located on the surface of the olfactory epithelium, by a mechanism of pinocytosis or endocytosis, depending on the type of formulation and drug, and be transported by intracellular axonal transport along the neuron in the anterograde direction. The nerves form bundles with other nerves and transverse the cribriform plate through holes and reach the olfactory bulbs, located in the forebrain directly above. It has been shown that this transport pathway is slow, taking up to 24 h for the drug to reach the CNS [71]. Such neuronal transport has, for example, been shown for insulin-like growth factor 1 (IGF-1) [7], wheat germ agglutinin–horseradish peroxidase [72, 73], hypocretin-1 [74], interferon-β [75], wheat germ agglutinin conjugated polyethylene glycol–polylactic acid (PEG–PLA) nanoparticles [37] and aluminum lactate [76], apart from that shown for vira and microorganisms [77]. Several dendrites are emitted from the first-order synapse, in the olfactory bulbs, further into the various areas of the brain (mainly the anterior parts), which can, hence, be reached by the engulfed formulations.

The "extracellular olfactory pathway" transports material by paracellular diffusion through the tight junctions, between the olfactory cells and the supporting cells, to the underlying lamina propria and enters the perineural space surrounding the nerves, either through a loosely adherent epithelium surrounding the nerve, or through the tight junctions of the adherent epithelium if the epithelium is closely adherent. This transport pathway relies on the presence of a direct connection between the submucosa and the subarachnoid extensions (the perineural space surrounding the olfactory nerve). From here, the formulations can access the cerebrospinal fluid (CSF) and also the olfactory bulb, as described earlier. From the CSF, the formulations can be distributed, via the bulk-flow mechanism, and reach the brain interstitial fluid in the brain tissue.

As described earlier, the trigeminal nerves innervate, for the major part in the respiratory epithelium, but a small portion also branches into the olfactory epithelium, and enters the CNS in the pons and, hence, creates entry points to both the caudal and the rostral brain [61]. As opposed to the olfactory nerves, the trigeminal nerve endings are situated in the epithelium below the surface and are not exposed in the nasal cavity. However, it is anticipated that drugs or drug formulations diffusing through the epithelia, or transcytosing across other epithelial cells, would be able to reach the nerve endings and/or the perineural space (most likely) surrounding the nerves dispersed in the epithelium. A range of papers have been published showing transport of drugs to the brain via the trigeminal nerves, such as interferon-β1b [8, 75], hypocretin-1 [74, 78], and IGF-1 [7, 8].

Nanoparticles can pass epithelial membranes either transcellularly or paracellularly, dependent on their size. As discussed earlier, in order to pass through tight junctions the nanoparticles will have to be less than 20 nm in diameter. In general, nanoparticles used for drug delivery are larger (50–500 nm) and would be expected to be endocytosed and transported intracellularly by trigeminal or olfactory neural cells.

It is not always possible to deduce, from the results of a nose-to-brain study, whether the drug (or drug formulation) is transported by the "neural pathway" or the "olfactory epithelial pathway" or by both, since most animal studies of this kind are terminated after, at most, 4 h. However, it is not likely that intracellular neural transport would be the main mode of transport from the nose to the brain, as supported by a large number of publications showing drug appearing rapidly in the CNS and dynamic effects of the drug seen within 1 h after administration [7, 74, 79, 80]. Lochhead and Thorne [41] have attempted to calculate the likely transport time for intracellularly transported drugs (within the olfactory neurons) from the olfactory epithelium to the olfactory bulb and found this to be up to 2.7 h in pikes, whereas for intracellular transport, within the trigeminal nerves to the brainstem, this transport time was calculated to be up to 13 h. These estimates were in agreement with data published for transport of WGA-HRP in mice. On the other hand, using albumin as a model, it was calculated that it would take, at most, 0.33 h for an extracellular bulk flow to the olfactory bulb within perivascular spaces and 1.7 h for extracellular bulk flow to the brainstem within perivascular spaces associated with the peripheral trigeminal system.

TRANSPORT OF DRUGS FROM NASAL CAVITY TO CNS IN ANIMAL MODELS

The transport of drugs from the nasal cavity to the CNS in animal models has been thoroughly discussed in many reviews, among others Illum [4], Mistry et al. [47], Dhuria et al. [70], Misra and Kher [81], and Patel et al. [82].

For most of the published studies, there is no indication of the degree to which the nasally administered drug reaches the brain directly from the nasal cavity. However, some papers have attempted to calculate the percentage of drug that reaches the brain; and in most circumstances, this is less than 0.1% of the amount of drug administered. Therefore, in recent years most of the papers published have been dedicated to novel delivery systems that were able to increase the amount of drug transported directly to the CNS. Hence, this section will be limited to providing illustrative examples of what progress has been achieved, in terms of improving targeting to the brain, using systems such as bioadhesive emulsions, absorption enhancers, nanoparticles, or functionalized nanoparticles, especially selected to interact with the olfactory epithelium and is not a full review of all papers published in this area.

Nano- and Microemulsions

One of the delivery systems that has been widely evaluated for enhancement of nose-to-brain delivery of different drugs is micro- and nanoemulsions. Hence, Zhang et al. [83] evaluated the effect of using an oil-in-water (o/w) microemulsion (comprising Labrafil, Cremophor RH 40, ethanol and water) on enhancement of the brain uptake of nimodipine in rats. The olfactory bulb concentration was threefold that after intravenous administration, and the drug targeting efficiency (DTE) was significantly higher. In a similar way, Vyas et al. [84] studied the nose-to-brain targeting of microemulsions of zolmitriptan (Z) in a rat model, the study comprising the nasal administration of a zolmitriptan mucoadhesive microemulsion (ZMME), a microemulsion (ZME), a zolmitriptan solution, and intravenously administered ZME. The microemulsion comprised medium chain triglyceride (oil), caprylocaproyl macrogol glyceride (surfactant), a mixture of purified diethylene glycol monoethyl ether, and fatty acid ester of polyglycerol (co-surfactant) and water. Polycarbophil was added as the mucoadhesive agent. Brain/blood uptake ratios were found to be 0.70, 0.56, 0.27, and 0.13, respectively, showing the superior performance of the ZMME. The DTEs were 1.89, 2.55, and 5.33 for the ZS, ZME, and the ZMME, respectively, showing increased brain targeting with the mucoadhesive emulsion compared with the non-mucoadhesive emulsion and the drug solution. It was also shown, by the calculated drug targeting percentage (DTP) of 81%, that most of the drug that reached the brain after nasal administration in the ZMME formulation was due to a direct transport from the nasal cavity. Gamma scintigraphy studies also showed that the brain uptake was much more pronounced after the ZMME formulation compared to the others. A similar study was done in rabbits for clonazepam comparing the brain uptake after intranasal administration of the drug in a simple solution, in a mucoadhesive microemulsion (CMME), a microemulsion (CME) and an intravenous injection of CME [85]. The microemulsion comprised medium chain triglyceride (oil), polyoxyethylene-35-ricinoleate (surfactant), polysorbate 80 (cosurfactant), and propylene glycol (anhydrous continuous phase). The mucoadhesive agent was polycarbophil. The results were similar to those from the zolmitriptan study, with highly improved targeting to the brain for the CMME formulation. Again, gamma scintigraphy studies confirmed the superior targeting of the mucoadhesive formulation to the brain. The same group later carried out similar experiments in rats for risperidone administered nasally in a mucoadhesive and non-mucoadhesive nanoemulsion and as a rispiridone solution [86]. The nanoemulsion consisted of capmul (oil), tween 80 (surfactant), a mixture of transcutol, and propylene glycol (co-surfactant) and water. The mucoadhesive agent was chitosan. The results obtained were similar to the results for the zolmitriptan and clonazepam studies, showing improved direct nose-to-brain transport with the mucoadhesive formulation and the results were, again, confirmed by gamma scintigraphy. Kumar et al. [87] also investigated olanzapine and obtained similar positive results for a mucoadhesive nanoemulsion. A different group [88] investigated the possibility of delivering an anti-HIV drug (with low permeability across the BBB) directly to the brain for treatment of neuro-AIDS, by delivering the drug nasally in a nanoemulsion. They found that compared to the plain drug suspension and the IV injection, the drug administered in the nanoemulsion gave a higher brain concentration, a higher DTE, and a higher DTP. Further, gamma scintigraphy demonstrated the higher transport to the brain for the nanoemulsion formulation.

Absorption Enhancers

Kanazawa et al. [89] produced block copolymer micelles (MPEG-PLC) and micelles modified with cell-penetrating peptide (MPEG-PLC-Tat) with encapsulated coumarin and evaluated the brain uptake after nasal administration compared to an intravenous injection. The amount of coumarin in the brain was found to be significantly higher for the (MPEG-PLC-Tat) than for the (MPEG-PLC) formulation or for the intravenous administration of coumarin, showing that the cell-penetrating peptide promoted the direct transport from the nasal cavity to the brain. In a similar study, the authors investigated the nasal delivery of a model siRNA (Alexa-dextran) complexed with the MPEG-PLC-Tat delivery systems [90]. It was found that even for this large siRNA (MW 10,000 Da), the concentration in the brain was significantly higher when delivered nasally complexed with the PEG-PLC-Tat (2%/g brain tissue, ~ total 3.4%) than that resulting from an intravenous injection (0.2%/g brain tissue) or after a nasal administration of the Alexa-dextran in a simple solution (0.45%/g brain tissue). Vaka et al. [91] used chitosan as a membrane modulating agent and increased the uptake of neurotropic factor 13-fold in the brain in rats after nasal administration in a chitosan solution as compared to a solution without chitosan. It was further found that the chitosan formulation significantly decreased immobility time in rats subjected to immobilization stress. The transport of Zidovudine to the brain from the nasal cavity was investigated by Ved and Kim [92], who used the gelling agent Poloxamer 407 in combination with the absorption enhancing agent *n*-tridecyl-β-D-maltoside (TDM/TR-1) for nasal delivery of the drug. The uptake into the CSF, six different regions of the brain, and the blood was compared in rabbits with that after an intravenous administration and a nasal control solution of the drug. The TDM/TR-1 formulation greatly enhanced the brain and the CSF concentrations of the drug, both compared to the IN control (~45-fold) and the IV injection (~56-fold) at the same time as the plasma concentrations for the two nasal formulations remained low (Fig. 25.5). On the basis of the IV AUC and the brain AUC for TDM/TR-1 and, taking into account the weight of the rabbit and the rabbit brain weight and blood volume, it is possible to estimate a brain "bioavailability" of about 8% for the TDM/TR-1 formulation, compared to 0.1% for the simple intranasal formulation. The authors suggested that the drug had been transported via a direct olfactory pathway from the nasal cavity to the brain.

Micellar Nanocarriers

Jain et al. [93] developed micellar nanocarriers attempting to improve the direct transport of zolmitriptan (^{99m}Tc labeled) into the brains of rats after nasal administration. The micellar nanocarriers (24.2 ± 0.73 nm) were produced by dissolving zolmitriptan in Transcutol P and benzyl alcohol and mixing this solution with a mixture of the surfactants, Pluronic F127, PEG 400, and Vitamin E-TPGS. An IV injection and a simple nasal solution of the drug served as controls. The micellar nanocarriers enabled a significantly higher uptake of the labeled drug into the brain (five- to sixfold) compared with the two controls. Brain localization studies showed that the drug was predominantly located in the olfactory bulb (mainly olfactory pathway) with lesser amounts reaching the midbrain, cerebellum, and pons (mainly trigeminal pathway). Micellar nanocarriers comprising a mixture of Pluronics L121 and P123 were developed by Abdelbary and Tadros [32] and evaluated for the improved transport of olanzapine to the brain of rats after nasal administration compared to a simple solution formulation and an IV injection of the drug. The authors found a significantly higher brain concentration for the micellar formulation than for any of the two controls, with DTE being 5.20 for the micellar nasal formulation and 3.39 for the simple nasal formulation.

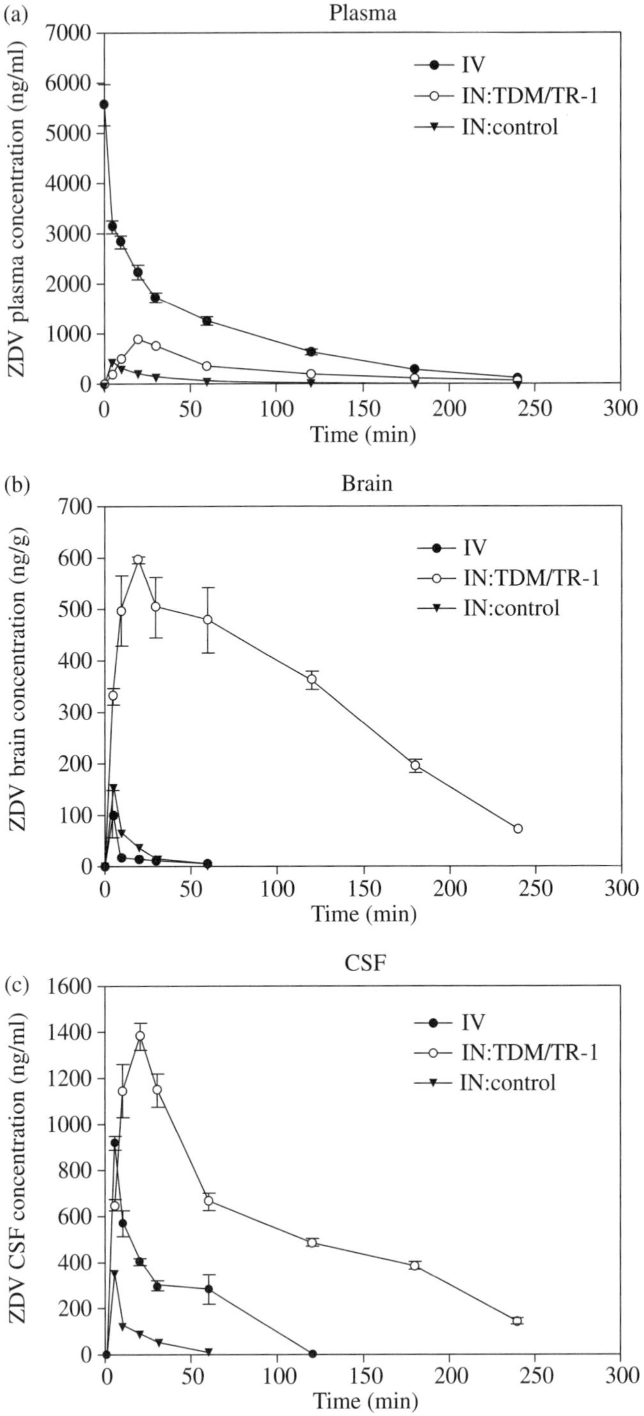

FIGURE 25.5 Mean ZDV concentration-time profile in (a) plasma, (b) brain, and (c) CSF after IV injection and IN administration at 1 mg dose in rabbits. Data represented as mean ± SEM. (Reprinted from Ved P M and Kim K, 2011. Poly(ethylene oxide/propylene oxide) copolymer thermo-reversible gelling system for the enhancement of intranasal zidovudine delivery to the brain. *Int. J. Pharm.* 411:1–9. With permission from Elsevier.)

Solid Lipid Nanoparticles

Solid lipid nanoparticles (148 nm) containing risperidone were prepared using a solvent emulsification-solvent evaporation method for the evaluation of targeting to the CNS after nasal administration [94]. Pharmacokinetic and biodistribution studies in mice showed that the brain/blood ratio 1 h after administration was 1.36 for the solid lipid nanoparticle formulation compared to 0.17 for IV administration of a simple drug solution and 0.78 for the IV administration of the nanoparticles. There was no simple drug solution given nasally as a control. Gelatine nanostructured lipid carriers, prepared using a w/w emulsion/freeze-drying method, were evaluated for their ability to carry fibroblast GF (bFGF) from the nasal cavity to the brain in hemiparkinsonian rats compared with gelatin nanoparticles and other controls [95]. A higher level of bFGF was found in the olfactory bulb and striatum of the rats after nasal administration of the bFGF-GNLs compared to the gelatin nanoparticles (bFGF-GNs). The bFGF-GNLs also better stimulated dopaminergic function in surviving synapsis and may also play a neuroprotective role.

Nanoparticles

The use of nanoparticles for nose-to-brain delivery of drugs has been described in detail in a previous paper by Illum [40]. A vast number of studies have been carried out in the past 5–10 years, using different types of nanoparticles and different types of drugs for improved delivery of the drugs from the nasal cavity to the brain. In general, it is found that nanoparticles are able to provide a higher drug concentration in the brain compared with simple solutions of the drug and, if studied, also better pharmacodynamics effects. Hence, only a few illustrative examples are given here.

Seju et al. [96] evaluated olanzapine-loaded PLGA nanoparticles (~90 nm) in the rat model and found a 6.35- and 10.86-fold higher uptake in the brain after nasal administration of olanzapine-loaded PLGA nanoparticles compared to an olanzapine solution given by the IV route or intranasally. In a similar way, Md et al. [12] found a higher blood/brain ratio (0.69) after nasal administration of bromocriptine-loaded chitosan nanoparticles (161 nm) compared with a drug solution administered nasally (0.47) and by the intravenous route (0.05). The DTP for the nanoparticles was calculated to be 84% showing direct nose-to-brain delivery. The study also found a reversal in catalepsy and akinesia behavior, especially in haloperidol induced rats given the nasal nanoparticle formulation. An evaluation of leucine-enkephalin (Leu-Enk)-loaded trimethyl chitosan (TMC) nanoparticles (443 nm) for nose-to-brain delivery found a higher accumulation of fluorescent marker NBD-F labeled Leu-Enk, using fluorescent microscopy of mouse brain sections, when administered intranasally by TMC nanoparticles than when a simple solution was given nasally [97]. Furthermore, the enhancement in brain uptake resulted in a significant improvement in the antinociceptive effect of Leu-Enk, as measured by hot plate and acid induced writhing assays. Hague et al. [98] evaluated venlafaxine loaded alginate nanoparticles (173 nm) (VLF AG-NP) in rats for treatment of depression and found that a nasal administration of VLF AG-NP significantly improved the behavioral analysis parameters, such as swimming, climbing and immobility, compared with the nasal administration of a VLF solution and oral tablets. The pharmacokinetic results supported this finding in that the brain concentration of VLF was significantly higher when administering the VLF AG-NP nasally compared to the VLF solution given by the IN and the IV routes.

Functionalized Nanoparticles

A range of papers (especially from Fudan University, Shanghai) has been published in the past 7 years that evaluates the effect of functionalizing the surface of nanoparticles with lectins or similar molecules, which attach themselves to specific receptors in the nasal cavity or, specifically, on the olfactory epithelium, for improvement in transport of the associated drug to the brain. It has generally been found that this approach enables a further improvement in transport of drugs to the brain over non-functionalized nanoparticles.

Hence, Gao et al. [99] presented early data on wheat germ agglutinin (WGA) modified PEG-PLA nanoparticles containing the model drug coumarin that showed a twofold increase in brain concentration after nasal administration compared to administration of the drug in combination with the non-functionalized nanoparticles. In a later paper, the neuroprotective peptide, vasoactive intestinal peptide (VIP) was incorporated into the functionalized nanoparticles and nasally administered to mice [100]. Measuring the AUC_{1-12h} of VIP in the brain, it was found that, compared to a nasal solution, the non-funtionalized and the WGA-functionalized nanoparticles enhanced the brain AUC by 3.5–4.7- and 5.6–7.7-fold, respectively, showing the superior performance of the functionalized nanoparticles. The results were supported by pharmacodynamics studies in terms of improvement in spatial memory in ethylcholine aziridium treated rats. It was further found that the WGA-functionalized nanoparticles accumulated to a higher degree in the olfactory epithelium than in the respiratory epithelium. The authors attempted to quantify the uptake into the brain, calculated by dividing the radioactivity count per gram brain tissue by the administered dose per gram body weight of the animal and suggested a 28% "bioavailability." However, this calculation is flawed since the body weight of the rat vastly overwhelms the weight of the rat brain. It was later shown by the same group that ^{125}I –labeled WGA-PEG-PLA nanoparticles were transported into the brain via both the olfactory and the trigeminal nerve pathways [37]. Similar increases in brain targeting were found for coumarin-loaded PEG-PLA nanoparticles modified with the lectin, Ulex europeus agglutinin I, that binds specifically to receptors on the olfactory epithelium [101] and for odorranalectin-modified PEG-PLA (OL-PEG-PLA) nanoparticles loaded with urocortin peptide (UCN) [102]. The higher uptake of OL-PEG-PLA nanoparticles in the brain region, as monitored by an in vivo imaging system, was supported by improved pharmacodynamic effects in hemiparkinsonian rats. Similar results have also been found by the same group for protamine-functionalized PEG-PLA nanoparticles [103], Solanum tuberosum lectin-conjugated PLGA nanoparticles [104], and lactoferrin modified PEG-co-PCL nanoparticles [105]. In general, sufficient information is not given in these papers to estimate the improved brain "bioavailability."

Stem Cell Delivery

Maybe one of the most exciting new developments in this area is studies showing that stem cells administered nasally can reach different parts of the brain and treat local diseases. Hence, for example, Danielyan et al. [106] found in a 6-OHDA lesion model of Parkinson's disease that intranasally administered mesenchymal stem cells (MSCs) reached the olfactory bulb, cortex, hippocampus, striatum, cerebellum, brain stem, and the spinal cord and that the MSC targeted preferentially the lesioned sites and damaged areas of the brain. It was also found that 24% of the cells survived for at least 4.5 months and of these 3% were still proliferative in the brain of the lesioned rats. It was also shown that the MSC treatment was able to increase tyrosine hydroxylase levels and prevented dopamine loss in the lesioned striatum and substantia nigra and was also able to eliminate the 6-OHDA induced increase in tunnel staining. Furthermore, a general improvement in motor function was also observed in the animals. Previously, the group had also identified that the MSC reached the CNS via two migration routes after crossing the cribriform plate above the olfactory epithelium: (i) migration into the olfactory bulb and to other parts of the brain and (ii) entry into the CSF with movement along the surface of the cortex followed by entrance into the brain parenchyma. The delivery of cells was enhanced by hyaluronidase treatment applied intranasally 30 min prior to the application of cells [107].

TRANSPORT OF DRUGS FROM NASAL CAVITY TO CNS IN MAN

Studies in the literature evaluating the potential for direct nose-to-brain delivery of drugs are overwhelmingly carried out in animal models. As described earlier, there are many differences both morphologically and physiologically between the nasal passages of animal models and that of man, and hence one cannot, without caution, directly extrapolate results obtained in animal to man. However, for obvious reasons, most of the published studies carried out in man do not directly measure the rate and degree of transport into the CNS from the nasal cavity. Instead, most studies measure the pharmacodynamic effects of the drugs on the CNS, such as effect on related brain potentials or measurable pharmacological effects. Early studies in man have previously been reviewed by, for example, Illum [5], Lochhead and Thorne [41] and for insulin Freiherr et al. [108], and this chapter will focus on more recently reported studies.

Pharmacodynamic Evidence of Direct Nose to Brain Transport

In early studies, a group in Lűbeck, Germany headed by Kern, Fehm, and Born, evaluated the physiological effects (as measured by increase in certain brain potentials) of nasal administration of different peptides. Hence, for arginine-vasopressin [17] and cholecyctokinin-8 [18, 19], significantly enhanced event-related brain potentials for nasally administered drugs compared to intravenously administered drugs were found, although the plasma concentration of the drugs were lower when administered nasally. Kern et al. [20] found that corticotropin-releasing hormone, after nasal administration, was able to decrease gastric acid secretion, and hence increase gastric pH, similar to the effect seen after intracerebral administration; while after intravenous administration, the pH decreased in human volunteers. In volunteers, similar plasma levels were seen after nasal and intravenous administration of angiotensin II, and resulted in similar rises in blood pressure, whereas subsequent blood pressure profiles were different for the two routes of administration. These studies indicated that the drug reached the brain by direct transport from the nasal cavity.

The effect of acute and chronic nasal administration of melanocortin melanocyte-stimulating hormone, adrenocorticotropin 4–10 (ACTH 4–10) on brain potentials and attention was evaluated by Smolnik et al. [109, 110] in healthy volunteers performing a selective attention task and showed diminished focusing of attention for acute administration, indicating a direct effect on the brain function. Furthermore, the administration of the same drug nasally to healthy subjects gave rise to reduced body fat and body weight, although no change was seen in ACTH 4–10 plasma levels [111]. Concurrently, a significant increase in ACTH 4–10 CSF levels were found, indicating direct transport of the ACTH 4–10 from the nasal cavity to the CNS, albeit injection controls were not included in the studies.

A study in volunteers by Kern et al. [21] evaluated whether the nasal administration of insulin (20 IU every 15 min for 60 min) affected the auditory-evoked brain potential (AEP) measured, while the subjects performed an odd ball task. No IV insulin control was included in the study. The nasal insulin was found to reduce the amplitude of components of the AEP, although neither the serum insulin levels nor the glucose levels in the blood were changed, indicating a direct transport of the insulin form the nasal cavity to the brain. Similar indications were also obtained in a later study [22].

It has been shown that neuropeptides, such as growth hormone-releasing hormone, do not readily pass the BBB [112] and, hence, effects on the CNS (effect on sleep pattern) are not evident after intravenous administration of this hormone [113]. Perras et al. [114] administered the hormone nasally to both young and old subjects before bedtime and showed a reduction in the sleep-induced elevation of GH during early sleep compared to nasal placebo control. In a similar study, the total sleep time in the second half of the night, after nasal administration of vasopressin was increased compared to a control [115]. It was suggested for both studies that the drugs acted directly on the brain rather than via a peripheral mediated action, although the vasopressin drug was found in the systemic circulation and was also found to be actively transported across the BBB [112]. However, vasopressin given nasally had also been found to be more potent in changing stimulus-evoked brain potential compared to vasopressin administered intravenously [17].

In order to evaluate the cognitive effect of nasally administered insulin, subjects of normal body weight were treated with nasal insulin (160 IU/day) or placebo in studies by Benedict et al. [116] and Hallschmid et al. [117]. The delayed recall of words significantly improved (placebo ~2.92 ± 1.00, insulin 6.20 ± 1.03), whereas immediate recall, non-declarative memory, and selective attention were not affected. The authors suggested that since their studies have repeatedly shown that insulin does not alter blood glucose and serum insulin levels after nasal administration, this indicated that the insulin does not enter the systemic circulation in substantial quantities [15, 116, 117].

The effect of intranasal insulin on cognition was evaluated by Reger et al. [118] who administered either a nasal saline placebo or nasal insulin (20 or 40 IU) to 26 memory-impaired subjects and 35 control subjects and tested cognition 15 min after administration. Intranasal insulin did not change plasma insulin or glucose levels, indicating no systemic transfer from the nasal cavity. The study found that the insulin treatment facilitated recall on two measures of verbal memory in memory impaired ε4-adults. The authors suggested that the insulin had reached the brain via an extraneural pathway.

Craft et al. [119] examined the effects of intranasal administration of insulin (20 or 40 IU a day for 4 months), using a ViaNase™ device, on cognition, function, cerebral glucose metabolism, and cerebral fluid biomarkers in 104 adults with amnestic mild cognitive impairment or Alzheimer's disease as compared to control. Treatment with 20 IU of insulin improved delayed memory ($P<0.05$), and both doses of insulin (20 and 40 IU) preserved caregiver-rated functional ability ($P<0.01$). Both insulin doses also preserved general cognition. CSF biomarkers did not change for insulin-treated participants as a group. In exploratory analyses, changes in CSF Aβ42 levels and tau protein-to-Aβ42 ratios were associated with cognitive and functional changes for insulin-treated participants. Placebo-assigned participants showed decreased fludeoxyglucose F 18 uptake in the parietotemporal, frontal, precuneus, and cuneus regions and insulin-minimized progression. Since insulin is not readily absorbed in the nasal cavity into the systemic circulation this study indicates that a proportion of the drug is transported directly from nose-to-brain, using the olfactory and/or the trigeminal pathways. The ViaNase device is designed to deliver drugs to the olfactory region, as described later.

Jauch-Chara et al. [120] similarly evaluated the effect of nasally administered insulin (40 IU) on cerebral energy levels by measuring ATP and PCr levels by ^{31}P magnetic resonance spectroscopy (^{31}P-MRS), and also assessed the food intake in normal-weight men. It was demonstrated that the cerebral high-energy phosphate content increased after nasal insulin administration compared with placebo, and this, in turn, suppressed the food intake. The fact that the peripheral glucose metabolism remained unaffected by the insulin administration throughout the experiment indicated that only a negligible amount of insulin could have reached the systemic circulation and potentially crossed the BBB and the blood-CSF barriers by receptor mediated transcytosis [121]. This suggests that the insulin reached the brain from the nasal cavity via a direct transport mechanism.

In a study in 12 healthy volunteers, the same dose of midazolam (3.4 mg) was administered intravenously or nasally using either a conventional nasal spray or a novel breath actuated bidirectional (Optinose) nasal delivery device designed to deliver drugs to the olfactory region, in order to evaluate whether this new device would improve the nasal and brain uptake of the drug [122]. Pharmacokinetic calculations were done using non-compartmental modeling and sedation was assessed by a subjective 0–10 NRS-scale. The study indicated (although not definitively proven) that when using the novel device, there was a higher degree of direct nose-to-brain uptake than for the traditional spray, which again resulted in a better degree of sedation.

In a similar study from the same group, Luthringer et al. [25] used the bidirectional nasal delivery device to study the pharmacodynamics and the effect of sumatriptan on quantitative electroencephalography (EEG) in patients with migraine following a glyceryl trinitrate (GTN) challenge. Blood level C_{max} and AUCs were lower after nasal administration compared to subcutaneous administration. Using the GTN challenge, the nasal dosing had similar effects on EEG and headache pain to those after subcutaneous sumatriptan, despite much lower systemic exposure. These results indicate that a direct nose-to-brain transport of sumatriptan had taken place.

Direct Evidence of Nose to Brain Transport

As discussed in a previous review [5], although the authors in earlier studies for ethical reasons did not measure drug concentrations in the brain nor the CSF, these studies in my opinion provide compelling evidence that the nasal administration of the drugs resulted in effects on the CNS or the functions of the CNS that were not seen after intravenous administration. Also, it was shown in some studies that the levels of drug in the blood after nasal administration were either unaffected (peptides) or lower than after intravenous administration. It is, hence, very unlikely that the effect on the CNS originate from drug passing into the systemic circulation and then across the BBB or the blood–CSF barriers. However, some studies have been reported in the literature that attempt to directly measure the transport of drugs from the nasal cavity to the CNS in man. These will be briefly reported here.

Okuyama [123], using a simple nasal spray device, sprayed a solution ^{99m}TC-DTPA (diethylenetriamine pentaacetate) mixed with hyaluronidase deep into the nasal cavity of subjects, lying on their backs for easy access to the olfactory region, and recorded by gamma scintigraphy the amount of activity appearing in the juxta-cribriform laminal intracranial space. The study was only performed in two subjects: a 60-year-old healthy male and a 67-year-old female with dysfunction of air-borne scenting. A significant rise in cerebral radioactivity was only found in the female subject, possibly due to a perceived increased olfactory permeability. These studies could not be considered a significant proof of direct transport from nose-to-brain in humans.

Later studies have evaluated not only the pharmacodynamic effects on the CNS but also compared the CSF concentrations of a drug after nasal administration of drugs and placebo formulations. Following on from the studies described earlier, the Lűbeck group [15] evaluated the serum and CSF concentrations of three different peptides (melanocortin 4–10 (MSH/ACTH 4–10), vasopressin and insulin) in male and female volunteers after nasal administration. The study was placebo controlled but was not compared to intravenous injections of the drugs. Significant increases in CSF concentrations compared with placebo were found for each of the nasally administered peptides. In terms of the serum concentrations of the drugs for MSH/ACTH 4–10 and for insulin, no significant increases were found; whereas, for vasopressin, the serum levels increased significantly, showing transport from the nasal cavity into the systemic circulation. This data strongly indicate that at least MSH/ACTH 4–10 and insulin reached the CSF via a direct pathway (olfactory and/or trigeminal) from the nasal cavity. It should be considered though that endogenous insulin may have masked any small amount of insulin reaching the circulation from the nasal cavity and that physiological increases in insulin serum levels can produce significant increases in CSF insulin levels [2]. For vasopressin, it is not clear whether any of the drug reached the CSF directly from the nasal cavity or whether all the drug originated from the systemic circulation. But support for (at least some) direct transport can be found in the studies described earlier [17–19] where it was shown that vasopressin resulted in event-related effects on the brain potential after nasal, but not after intravenous administration.

Merkus et al. [16] performed a study in postoperative neurology patients, where the patients, as part of the operation, were fitted with a CSF drain from the site of the operation. The patients were administered either hydroxycobalamin (five patients) or melatonin (three patients) using either a nasal spray device or an intravenous injection. CSF and arterial blood samples were collected for up to 180 min. The data were expressed as individual AUC values and as a CSF ratio: ((AUC CSF IN/AUC plasma IN)/(AUC CSF IV/AUC plasma IV)). It was found that for the most lipophilic drug, melatonin (log $P \sim 1.65$), there was no difference in the AUC values for plasma or CSF, whether the drug was given nasally or intravenously. For the CSF ratio, based on the individual ratios for each patient, the mean CSF ratio was calculated to be 0.71, indicating that less drug reached the CSF after nasal application than by intravenous injection. For the more hydrophilic drug, hydroxocobalamin (log P not available) when calculating the CSF ratio, the authors found a value of 1.00, and concluded that there was no evidence of a direct nose to CSF pathway. However, as discussed in detail by Illum [5], in her opinion, the calculations were done incorrectly; and when calculated on an individual patient basis, a CSF ratio of 1.61 indicated direct nose-to-brain transport for this drug.

Wall et al. [124] evaluated the distribution of zolmitriptan after nasal administration in healthy volunteers using positron emission tomography (PET) and MRI scan to obtain anatomical information. They found that the zolmitriptan rapidly reached all regions of the brain studied, reaching a concentration of 2 nM (0.5 μg/l) after 30 min, 3.5 nM (1.0 μg/l), (equivalent to one-fifth of plasma concentration after intravenous infusion) after 1 h and 1.5 μg/l after 2 h. At 5 min after nasal administration, the concentration in the CNS had reached 0.5 nM. The authors concluded that zolmatriptan enters the brain parenchyma in humans, achieving an uptake rate and concentration compatible with a central mode of action. Since, zolmitriptan is hydrophilic, has a low nasal bioavailabilty (~14%) and does not readily cross the BBB [125], it is unlikely that the drug would rapidly have reached the brain via the "systemic transport" route. Furthermore, Bergstrom et al. [126] modeled the data from this study and concluded that the data supported a rapid brain availability after nasal administration.

As far as I am aware, no other studies have been published evaluating directly the nose-to-brain transport. However, from the earlier studies, there is a convincing indication that a direct transport route is present and working in man.

SPECIFIC NOSE-TO-BRAIN DELIVERY DEVICES

As discussed in the Section "Introduction", it is difficult for applied nasal formulations to reach the olfactory region in the nasal cavity of man, since this is placed high up in the nasal cavity, above the superior conchae. It has been shown that only about 15–20% of the normal nasal airflow reaches this region in man [29]. The low exposure to nasal airflow would likely limit the fraction of nasally applied drugs reaching the brain. A range of novel nasal devices based on different concepts have been developed that are designed to deposit the nasally applied formulation specifically to the olfactory region, for example, Bi-Directional Technology™ (Optinose), Controlled Particle Dispersion Technology, or POD technology (Impel NeuroPharma). Some examples of clinical evaluation will be given here.

The breath-activated Bi-Directional device consists of a sealing nosepiece that the user slides into one nostril until it forms a seal with the flexible soft tissue of the nostril opening, which triggers an expansion of the nasal valve, and a mouthpiece that is placed in the mouth and through which the user exhales. When exhaling against the resistance of the device the soft palate is automatically elevated to close the access from the nose to the rest of the respiratory system. The airstream that enters one nostril containing the drug formulation from the device, passes around the nasal septum and exits through the other nostril. By optimizing design parameters, it has been possible to target specific sites within the nasal cavity, such as the olfactory region.

A range of studies with the Bi-Directional device has been published evaluating the deposition of drug in the nasal cavity, to include the targeting efficiency to the olfactory region, and also the pharmacokinetics and pharmacodynamics of drug delivery when compared to other devices [25, 26, 122, 127]. For example, it was shown in two studies in volunteers by scintigraphy that the Bi-Directional device compared to a standard pump spray device provided less deposition in the non-ciliated nasal vestibule and a significant greater deposition to the upper posterior regions in the main nasal cavity [26, 127]. It was further shown that sumatriptan, after nasal dosing using the Bi-Directional device, had similar effects on EEG and headache pain to those after subcutaneous sumatriptan, despite much lower systemic exposure [25]. It was also found that midazolam sedation after use of the bidirectional spray followed intravenous sedation closely, while sedation after using the traditional spray was less pronounced, although the bioavailabilities using both devices were the same [122].

Impel NeuroPharma's POD device delivers aerosolized drugs to the nasal cavity by mixing a drug solution in the device body with pressurized gas (e.g., nitrogen). The device body ends in a custom-fit aerosol nozzle with a 0.8-mm outside diameter, which is fitted to a small length (2.0 mm) of metal cylinder, with two spiral fluid passages in which the fluid/gas mixture travels. This enables the mixture of the nitrogen and liquid drug to create an aerosol output to enhance penetration into the nasal cavity towards the cribriform plate area while minimizing pressure on the nasal epithelia. A liner is placed over the outside of the metal tip in order to protect the nasal epithelia from being damaged by the nozzle during use [128].

In a human proof-of-concept study using SPECT imaging, it was shown that the POD device successfully deposited a therapeutic amount of radiolabeled tripeptide into the deep nasal cavity (>50% in the olfactory region) and then enabled rapid and significant delivery to the CNS (www.impelneuropharma.com). Recently, it was reported that a study in seven volunteers, by means of SPECT, showed the successful delivery of a small off the shelf peptide attached to a radioactive tracer deep into the upper nasal cavity and into the brain using the POD device. The press release did not mention a comparative IV injection of the peptide (www.xconomy.com).

The Controlled Particle Dispersion (CPD®) Technology Platform (Kurve Technology) takes the form of a nasal nebulizer and a spray bottle and, by using an electric atomizer, produces controlled fine droplet dispersions. The device releases a metered dose into the nebulizer chamber, covering the subject's nose, which is then inhaled by breathing regularly for 2 min. The device contains six critical-to-function design parameters that can be adjusted to create the optimal delivery environment for a given formulation. Using the principal of vortical flow, the CPD ViaNase device effectively disrupts inherent nasal cavity airflows to deliver formulations to the entire nasal cavity. By varying control parameters, CPD can target specific nasal regions, including the olfactory region and the paranasal sinuses, while minimizing peripheral deposition to the lungs and stomach. Gamma scintigraphy studies in human volunteers have confirmed the ability of the ViaNase device to saturate the entire nasal cavity, including the olfactory regions and the paranasal sinuses. Studies showed the average area of intranasal distribution was as much as 300% greater for Kurve's ViaNase device compared to traditional spray bottles. In all cases, ViaNase demonstrated a greater propensity for delivery of droplets to the paranasal sinuses than current methods (www.kurvetech.com).

As is evident from the discussion above, the different special devices have not all as yet been fully evaluated in terms of their ability to target the olfactory region in comparison with 'off the shelf' simple devices, nor have they sufficiently been tested in terms of their ability to deliver sufficient drug to the brain to obtain a relevant therapeutic outcome. It can be expected that such data will be available in the coming years.

CONCLUSION

This chapter sets out to discuss the scientific evidence and the potential for enhanced delivery of drugs directly from the nasal cavity to the CNS in animal and in man and how it is possible to exploit novel delivery approaches in order for the amount of drug actually reaching the brain after nasal administration to be enhanced from the normally low uptake (<0.1%). Furthermore, the chapter also gives some necessary background information in terms of anatomy, physiology of the nose, and also up-to-date information on transport routes for nose-to-brain delivery. It is evident from the sections "Nano- and Microemulsions," "Absorption Enhancers," "Micellar Nanocarriers," "Solid Lipid Nanoparticles," "Nanoparticles," "Functionalized Nanoparticles," and "Stem Cell Delivery" discussing new delivery approaches, that a lot of progress has been made in the past 10 years with systems exploiting micro-emulsions, micellar nanocarriers, nanoparticles, and nanoparticles functionalized with lectin molecules, which attach themselves to specific receptors on the olfactory epithelium or the respiratory epithelium and, hence, have a better chance of delivering the drug for transport across the membrane. For all of these novel approaches, great enhancement in brain uptake has been shown compared to simple solution formulations and to a parenteral administration of the drug. Of special interest is the possibility of delivering stem cells via the nose to the brain to achieve long survival times in the brain and improvement in neurodegenerative diseases, such as Parkinson's disease. It is also apparent that the discussion of the evidence for nose-to-brain delivery in man provides strong substantiation for this route of delivery. It is also noteworthy that a range of novel delivery devices are being developed and has already shown great promise of delivery of the nasal formulations to the appropriate place in the nasal cavity in higher quantities than is the case for simple nasal spray systems. What we hopefully will see in the near future is an evaluation of the novel delivery systems, combined with the novel nasal devices, and testing for efficacy in man.

REFERENCES

[1] Temsamani J, 2002. Delivering drugs to the brain—beating the blood brain barrier. *Eur. Biopharm. Rev.* Autumn:72–75.

[2] Schwartz M W, Sipols A, Kahn S E, Lattemann D F, Taborsky D J, Bergman R N, Woods S C, and Porte D, 1990. Kinetics and specificity of insulin uptake from plasma into cerebrospinal fluid. *Am. J. Physiol.* 259:E378–E383.

[3] Chou K J and Donovan M D, 1998. Distribution of antihistamine into the CSF following intranasal delivery. *Biopharm. Drug Dispos.* 18:335–346.

[4] Illum L, 2000. Transport of drugs from the nasal cavity to the central nervous system. *Eur. J. Pharm. Sci.* 11:1–18.

[5] Illum L, 2004. Is nose-to-brain transport of drugs in man a reality? *J. Pharm. Pharmacol.* 56:3–17.

[6] Illum L, 2007. Nanoparticulate systems for nasal delivery of drugs: a real improvement over simple systems? *J. Pharm. Sci.* 96:473–483.

[7] Thorne R G, Pronk G J, Padmanabhan V, and Frey W H, 2004. Delivery of insulin-like growth factor-I to the rat brain and spinal cord along olfactory and trigeminal pathways following intranasal administration. *Neuroscience* 127:481–496.

[8] Thorne R G, Hanson L R, Ross T M, Tung D, and Frey W H II, 2008. Delivery of interferon-beta to the monkey nervous system following intranasal administration. *Neuroscience* 152:785–797.

[9] Johnson N J, Hanson L R, and Frey W H II, 2010. Trigeminal pathways deliver a low molecular weight drug from the nose to the brain and orofacial structures. *Mol. Pharm.* 7:884–893.

[10] Schiöth H B, Craft S, Brooks S J, Frey W H II, and Benedict C, 2012. Brain insulin signalling and Alzheimer's disease: current evidence and future directions. *Mol. Neurobiol.* 46:4–10.

[11] Veening J G and Olivier B, 2013. Intranasal administration of oxytocin: behavioral and clinical effects, a review. *Neurosci. Biobehavior. Rev.* 37:1445–1465.

[12] Md S, Khan R A, Mustafa G, Chuttani K, Baboota S, Sahni J K, and Ali J, 2013. Bromocriptine loaded chitosan nanoparticles intended for direct nose to brain delivery: pharmacodynamic, pharmacokinetic and scintigraphy study in mice model. *Eur. J. Pharm. Sci.* 48:393–405.

[13] Merkus F W and van den Berg M P, 2007. Can nasal drug delivery bypass the blood-brain barrier?: questioning the direct transport theory. *Drugs R D* 8:133–144.

[14] Fehm H L, Perras B, Smolnik R, Kern W, and Born J, 2000. Manipulating neuropeptidergic pathways in humans: a novel approach to neuropharmacology. *Eur. J. Pharmacol.* 405:43–54.

[15] Born J, Lange T, Kern W, McGregor G P, Bickel U, and Fehn H L, 2002. Sniffing neuropeptides: a transnasal approach to the human brain. *Nat. Neurosci.* 5:514–516.

[16] Merkus P, Guchelaar H J, Bosch A, and Merkus F W H M, 2003. Direct access of drugs to the human brain after intranasal drug administration? *Neurology* 60:1669–1671.

[17] Pietrowsky R, Claassen L, Frercks H, Fehm H L, and Born J, 1996. A nose-to-brain pathway for psychotropic peptides: evidence from a brain evoked potential study with cholecystokinin. *Psychoneuroendocrinology* 21:559–572.

[18] Pietrowsky R, Struben C, Molle M, Fehm H L, and Born J, 1996. Brain potential changes after intranasal vs. intravenous administration of vasopressin: evidence for a direct nose-to-brain pathway for peptide effects in humans. *Biol. Psychiatry* 39:332–340.

[19] Pietrowsky R, Thiemann A, Kern W, Fehm H L, and Born J, 2001. Time course of intranasally administered cholecystokinin-8 on central nervous effects. *Neuropsychobiology* 43:254–259.

[20] Kern W, Schiefer B, Schwarzenburg J, Strange E F, Born J, and Fehm H L, 1997. Evidence for central nervous effects of corticotropin-releasing hormone on gastric acid secretion in humans. *Neuroendocrinology* 65:291–298.

[21] Kern W, Born J, Schreiber H, and Fehm H L, 1999. Central nervous system effects of intranasally administered insulin during euglycemia in men. *Diabetes* 48:557–563.

[22] Kern W, Peters A, Fruehwald-Schultes B, Deininger E, Born J, and Fehm H L, 2001. Improving influence of insulin on cognitive functions in humans. *Neuroendocrinology* 74:270–280.

[23] Derad I, Willeke K, Pietrowsky R, Born J, and Fehm H L, 1998. Intranasal angiotensin II directly influences central nervous regulation of blood pressure. *Am. J. Hypertens.* 11:971–977.

[24] Lindhardt K, Gizurarson S, Stefansson S B, Olafsson D R, and Beckgaard E, 2001. Electroencephalographic effects and serum concentrations after intranasal and intravenous administration of diazepam to healthy volunteers. *Br. J. Clin. Pharmacol.* 52:521–527.

[25] Luthringer R, Djupesland P G, Sheldrake C D, Flint A, Boeijinga P, Danjou P, Demazieres A, and Hewson G, 2009. Rapid absorption of sumatriptan powder and effects on glyceryl trinitrate model of headache following intranasal delivery using a novel bi-directional device. *J. Pharm. Pharmacol.* 61:1219–1228.

[26] Djupesland P G, Skretting A, Winderen M, and Holand T, 2006. Breath actuated device improves delivery to target sites beyond the nasal valve. *Laryngoscope* 116:466–472.

[27] Djupesland P G, Mahmoud R, Messina J, and Miller P, 2011. Rate of systemic absorption of sumatriptan may not explain differences in headache response suggestion the potential for an additional route to the site of action. *Cephalalgia* 31(Suppl 1):PS1–PS55.

[28] Djupesland P G, 2013. Nasal drug delivery devices: characteristics and performance in a clinical perspective—a review. *Drug Deliv. Transl. Res.* 3:42–62.

[29] Keyhani K, Scherer P W, and Mozell M M, 1997. A numerical model of nasal odorant transport for the analysis of human olfaction. *J. Theor. Biol.* 186:279–301.

[30] Chen X, Zhi F, Jia X, Zhang X, Ambardekar R, Meng Z, Paradkar A R, Hu Y, and Yang Y, 2013. Enhanced brain targeting of curcumin by intranasal administration of a thermosensitive poloxamer hydrogel. *J. Pharm. Pharmacol.* 65:807–816.

[31] Sintov A C, Levy H V, and Botner S, 2010. Systemic delivery of insulin via the nasal route using a new microemulsion system: in vitro and in vivo studies. *J. Control. Rel.* 148:168–176.

[32] Abdelbary G A and Tadros M I, 2013. Brain targeting of olanzapine via intranasal delivery of core-shell di-functional block copolymer mixed nanomicellar carriers: in vitro characterization, ex vivo estimation of nasal toxicity and in vivo biodistribution studies. *Int. J. Pharm.* 452:300–310.

[33] Elder A, Gelein R, Silva V, Feikert T, Opanashuk L, Carter J, Potter R, Maynard A, Ito Y, Finkelstein J, and Oberdörster G, 2006. Translocation of inhaled ultrafine manganese oxide particles to the central nervous system. *Environ. Health Perspect.* 114:1172–1178.

[34] Sahni J K, Doggui S, Ali J, Baboota S, Dao L, and Ramassamy C, 2011. Neurotherapeutic applications of nanoparticles in Alzheimer's disease. *J. Control. Rel.* 152:208–238.

[35] Fazil M, Md S, Haque S, Kumar M, Baboota S, Sahni J K, and Ali J, 2012. Development and evaluation of rivastigmine loaded chitosan nanoparticles for brain targeting. *Eur. J. Pharm. Sci.* 47:6–15.

[36] Lucchini R G, Dorman D C, Elder A, and Veronesi B, 2012. Neurological impacts from inhalation of pollutants and the nose-brain connection. *Neurotoxicology.* 33:838–841.

[37] Liu Q, Shen Y, Chen J, Gao X, Feng C, Wang L, Zhang Q, and Jiang X, 2012. Nose-to-brain transport pathways of wheat germ agglutinin conjugated PEG-PLA nanoparticles. *Pharm. Res.* 29:546–558.

[38] Mathison S, Nagilla R, and Kompella U B, 1998. Nasal route for direct delivery of solutes to the central nervous system: fact or fiction? *J. Drug Target.* 5:415–441.

[39] Frey W H II, 2002. Bypassing the blood-brain barrier to deliver therapeutic agents to the brain and spinal cord. *Drug Deliv. Tech.* 2:46–49.

[40] Illum L, 2014. The potential for nose-to-brain delivery of drugs. In: Gehr P and Tsuda A, eds. *Nanoparticles in the lungs: Environmental exposure and drug delivery.* CRC Press/ Taylor & Francis Group, Boca Raton, pp. 289–315.

[41] Lochhead J J and Thorne R G, 2012. Intranasal delivery of biologics to the central nervous system. *Adv. Drug Del. Rev.* 64:614–628.

[42] Zhu J, Jiang Y, Xu G, and Liu X, 2012. Intranasal administration: a potential solution for cross-BBB delivering neurotrophic factors. *Histol. Histopathol.* 27:537–548.

[43] Tayebati S K, Nwankwo I E, and Amenta F, 2013. Intranasal drug delivery to the central nervous system: present status and future outlook. *Curr. Pharm. Des.* 19:510–526.

[44] Mygind N and Dahl R, 1998. Anatomy, physiology and function of the nasal cavities in health and disease. *Adv. Drug Del. Rev.* 29:3–12.

[45] Moran D T, Rowley J C, Jafek B W, and Lowell M A, 1982. The fine structure of the olfactory mucosa in man. *J. Neurocytol.* 11:721–746.

[46] Hilger P A, 1989. *Applied anatomy and physiology of the nose*. In: *Otolaryngology. A textbook of ear, nose and throat diseases*. W. B. Saunders, Philadelphia, pp. 177–195.

[47] Mistry A, Stolnik S, and Illum L, 2009. Nanoparticles for direct nose-to-brain delivery of drugs. *Int. J. Pharm.* 379:146–157.

[48] Guilmette R A, Wicks J D, and Wolff R K, 1989. Morphometry of human nasal airways in vivo using magnetic resonance imaging. *J. Aerosol. Med.* 2:2273–2287.

[49] Andersen I, Lundqvist G R, and Proctor D F, 1971. Human nasal mucosal function in a controlled climate. *Arch. Environ. Health* 23:408–420.

[50] Harkema J R, Carey S A, and Wagner J G, 2006. The nose revisited: a brief review of the comparative structure, function and toxicologic pathology of the nasal epithelium. *Toxicol. Pathol.* 34:252–269.

[51] Cole P, 1992. Nasal and oral airflow resistors. Site, function and assessment. *Arch. Otolaryngol. Head Neck Surg.* 118:790–793.

[52] Swift D L and Proctor D F, 1977. Access of air to the respiratory tract. In: Brain D J, Proctor D F, Reid L M, eds. *Respiratory defence mechanisms*. Marcel Dekker, New York, pp. 63–93.

[53] Douek E, 1974. *The sense of smell and its abnormalities*. Livingstone, Edinburgh.

[54] Leopold D A, Hummel T, and Schwob J E, 2000. Anterior distribution of human olfactory epithelium. *Laryngoscope* 110:417–421.

[55] Iwai N, Zhou Z, Roop D R, and Behringer R R, 2008. Horizontal basal cells are multipotent progenitors in normal and injured adult olfactory epithelium. *Stem Cells* 26:1298–1306.

[56] Caggiano M, Kauer J S, and Hunter D D, 1994. Globose basal cells are neuronal progenitors in the olfactory epithelium: a lineage analysis using a replication-incompetent retrovirus. *Neuron* 13;339–352.

[57] Menco B P and Jackson J E, 1997. A banded topography in the developing rat's olfactory epithelial surface. *J. Comp. Neurol.* 388:293–306.

[58] Finger T E, St Jeor V L, Kinnamon J C, and Silver W L, 1990. Ultrastructure of substance P- and CGRP-immunereactive nerve fibers in the nasal epithelium of rodents. *J. Comp. Neurol.* 294:293–305.

[59] Cauna N, Hinderer K H, and Wentges R T, 1969. Sensory receptor organs of the human nasal respiratory mucosa. *Am. J. Anat.* 124:187–209.

[60] Yamamoto Y, Kondo A, Atoji Y, Tsubone H, and Suzuki Y, 1998. Morphology of intraepithelial corpuscular nerve endings in the nasal respiratory mucosa of the dog. *J. Anat.* 193:581–586.

[61] Schaefer M L, Bottger B, Silver W L, and Finger T E, 2002. Trigeminal collaterals in the nasal epithelium and olfactory bulb: a potential route for direct modulation of olfactory information by trigeminal stimuli. *J. Comp. Neurol.* 444:221–226.

[62] Madara J L, 2000. Modulation of tight junctional permeability. *Adv. Drug Del. Rev.* 41:251–253.

[63] Van Itallie C M, Holmes J, Bridges A, Gookin J L, Coccaro M R, Proctor W, Colegio O R, and Anderson J M, 2008. The density of small tight junction pores varies among cell types and is increased by expression of claudin-2. *J. Cell Sci.* 121:298–305.

[64] Stevenson B R, Anderson J M, and Bullivant S, 1988. The epithelial tight junctions: structure, function and preliminary biochemical characterisation. *Mol. Cell Biochem.* 83:129–145.

[65] Shorten P R, McMahon C D, and Soboleva T K, 2007. Insulin transport within skeleton muscle transverse tubule network. *Biophys. J.* 93:3001–3007.

[66] Kabanov V A, Skobeleva V B, Rogacheva V B, and Zezin A B, 2004. Sorption of proteins by slightly cross-linked polyelectrolyte hydrogels: kinetics and mechanism. *J. Phys. Chem. B.* 108:1485–1490.

[67] Wilkins K D, Grimshaw B S, Receveur V, Dobson C M, Jones J A, and Smith L J, 1999. Hydrodynamic radii of native and denatured proteins measured by pulse field gradient NMR techniques. *Biochemistry* 38:16424–16431.

[68] Costantino H R, Illum L, Brandt G, Johnson P, and Quay S C, 2007. Intranasal delivery: physiochemical and therapeutic aspects. *Int. J. Pharm.* 337:1–24.

[69] Thorne R G and Frey W H, 2001. Delivery of neurotropic factors to the central nervous system. *Clin. Pharmacokinet.* 40:907–946.

[70] Dhuria S V, Hanson L R, and Frey W H II, 2010. Intranasal delivery to the central nervous system: mechanisms and experimental considerations. *J. Pharm. Sci.* 99:1654–1673.

[71] Kristensson K and Olsson Y, 1971. Uptake of exogenous proteins in mouse olfactory cells. *Acta Neuropathol.* 19:145–154.

[72] Itaya S K, 1987. Anterograde transynaptic transport of WGA-HRP in rat olfactory pathways. *Brain Res.* 409:205–214.

[73] Thorne R G, Emory C R, Ala T A, and Frey W H II, 1995. Quantitative analysis of the olfactory pathway for drug delivery to the brain. *Brain Res.* 692:278–282.

[74] Hanson L R, Martinez P M, Taheri S, Kamsheh L, Mignot E, and Frey W H II, 2004. Intranasal administration of hypocretin 1 (orexin A) bypasses the blood-brain barrier & targets the brain: a new strategy for the treatment of narcolepsy. *Drug Del. Tech.* 4:66–70.

[75] Ross T M, Martinez P M, Renner J C, Thorne R G, Hanson L R, and Frey W H II, 2004. Internasal administration of interferon beta bypasses the blood brain barrier to target the central nervous system and cervical lymph nodes: a non-invasive treatment strategy for multiple sclerosis. *J Neuroimmunol* 151:66–77.

[76] Perl D P and Good P F, 1987. The association of aluminium Alzheimer's disease, and neurofibrillary tangles. *J. Neural. Transm. Suppl.* 24:205–211.

[77] Kristensson K, 2011. Microbes roadmap to neurons. *Nat. Rev. Neurosci.* 12:345–357.

[78] Dhuria S V, Hanson L R, and Frey W H II, 2009. Intranasal drug targeting of hypocretin-1 (orexin-A) to the central nervous system. *J. Pharm. Sci.* 98:2501–2515.

[79] Banks W A, During M J, and Niehoff M L, 2004. Brain uptake of the glucagon-like peptide-1 antagonist extendin (9–39) after intranasal administration. *J. Pharmacol. Exp. Ther.* 309:469–475.

[80] Charlton S T, Whetstone J, Fayinka S T, Read K D, Illum L, and Davis S S, 2008. Evaluation of direct transport pathways of glycine receptor antagonists from the nasal cavity to the central nervous system in the rat model. *Pharm. Res.* 25:1531–1543.

[81] Misra A and Kher G, 2012. Drug delivery systems from nose to brain. *Curr. Pharm. Biotechnol.* 13:2355–2379.

[82] Patel Z, Patel B, Patel S, Pardeshi C, 2012. Nose to brain targeted drug delivery bypassing the blood-brain barrier: an overview. *Drug Invent. Today* 4:610–615.

[83] Zhang Q, Jiang X, Jiang W, Lu W, Su L, and Shi Z, 2004. Preparation of nimodipine-loaded microemulsion for intranasal delivery and evaluation on the targeting efficiency to the brain. *Int. J. Pharm.* 275:85–96.

[84] Vyas T, Babbar A K, Sharma R K, and Misra A, 2005. Intranasal mucoadhesive microemulsions of zolmitriptan: preliminary studies on brain targeting. *J. Drug Target.* 13:317–324.

[85] Vyas T, Babbar A K, Sharma R K, Singh S, and Misra A, 2006. Intranasal mucoadhesive microemulsions of clonazepam: preliminary studies on brain targeting. *J. Pharm. Sci.* 95:570–580.

[86] Kumar M, Misra A, Babbar A K, Mishra A K, Mishra P, and Pathak K, 2008. Intranasal nanoemulsion based brain targeting drug delivery system of risperidone. *Int. J. Pharm.* 358:285–291.

[87] Kumar M, Misra A, Babbar A K, Mishra A K, Mishra P, and Pathak K, 2008. Mucoadhesive nanoemulsion-based intranasal drug delivery system of olanzapine for brain targeting. *J. Drug Target.* 16:806–814.

[88] Mahajan H S, Mahajan M S, Nerkar P P, and Agrawal A, 2014. Nanoemulsion-based intranasal drug delivery system of saquinavir mesylate for brain targeting. *Drug Deliv.* 21:148–154.

[89] Kanazawa T, Taki H, Tanaka K, Takashima Y, and Okada H, 2011. Cell penetrating peptide-modified block copolymer micelles promote direct delivery via intranasal administration. *Pharm. Res.* 28:2130–2139.

[90] Kanazawa T, Akiyama F, Kakizaki S, Takashima Y, and Seta Y, 2013. Delivery of siRNA to the brain using a combination of nose-to-brain delivery and cell-penetrating peptide-modified nano-micelles. *Biomaterials* 34:9220–9226.

[91] Vaka S R, Murthy S N, Balaji A, and Repka M A, 2012. Delivery of brain-derived neurotrophic factor via nose-to-brain pathway. *Pharm. Res.* 29:441–447.

[92] Ved P M and Kim K, 2011. Poly(ethylene oxide/propylene oxide) copolymer thermo-reversible gelling system for the enhancement of intranasal zidovudine delivery to the brain. *Int. J. Pharm.* 411:1–9.

[93] Jain R, Nabar S, Dandekar P, and Patravale V, 2010. Micellar nanocarriers: potential nose-to-brain delivery of zolmitriptan as novel migraine therapy. *Pharm. Res.* 27:655–664.

[94] Patel S, Chavhan S, Soni H, Babbar A K, Mathur R, Mishra A K, and Sawant K, 2010. Brain targeting of risperidone-loaded solid lipid nanoparticles by intranasal route. *J. Drug Target.* 19:468–474.

[95] Zhao Y-Z, Li X, Lu C-T, Lin M, Chen L-J, Xiang Q, Zhang M, Jin R-R, Jiang X, Shen X-T, Li X-K, and Cai J, 2014. Gelatin nanostructured lipid carriers-mediated intranasal delivery of basic fibroblast growth factor enhances functional recovery in hemiparkinsonian rats. *Nanomedicine* 10:755–764.

[96] Seju U, Kumar A, and Sawant K K, 2011. Development and evaluation of olanzapine-loaded PLGA nanoparticles for nose-to-brain delivery: in vitro and in vivo studies. *Acta Biomater.* 7:4169–4176.

[97] Kumar M, Pandey R S, Patra K C, Jain S K, Soni M L, Dangi J S, and Madan J, 2013. Evaluation of neuropeptide loaded trimethyl chitosan nanoparticles for nose to brain delivery. *Int. J. Biol. Macromol.* 61:189–195.

[98] Hague S, Md S, Sahni J K, Ali J, and Baboota S, 2014. Development and evaluation of brain targeted intranasal alginate nanoparticles for treatment of depression. *J. Psychiatr. Res.* 48:1–12.

[99] Gao X, Tao W, Lu W, Zhang Q, Zhang Y, Jiang X, and Fu S, 2006. Lectin-conjugated PEG-PLA nanoparticles: preparation and brain delivery after intranasal administration. *Biomaterials* 27:3482–3490.

[100] Gao X, Chen J, Tao W, Zhu J, Zhang Q, Chen H, and Jiang X, 2007. UEA I-bearing nanoparticles for brain delivery following intranasal administration. *Int. J. Pharm.* 340:207–215.

[101] Gao X, Wu B, Zhang Q, Chen J, Zhu J, Zhang W, Rong Z, Chen H, and Jiang X, 2007. Brain delivery of vasoactive intestinal peptide enhanced with the nanoparticle conjugated with wheat germ agglutinin following intranasal administration. *J. Control. Rel.* 121:156–167.

[102] Wen Z, Yan Z, Hu K, Pang Z, Cheng X, Guo L, Zhang Q, Jiang X, Fang L, and Lai R, 2011. Odorranalectin-conjugated nanoparticles: preparation, brain delivery and pharmacodynamics study on Parkinson's disease following intranasal administration. *J. Control. Rel.* 151:131–138.

[103] Xia H, Gao X, Gu G, Liu Z, Zeng N, Hu Q, Song Q, Yao L, Pang Z, Jiang X, Chen J, and Chen H, 2011. Low molecular weight protamine-functionalised nanoparticles for drug delivery to the brain after intranasal administration. *Biomaterials* 32:9888–9898.

[104] Chen J, Zhang C, Liu Q, Shao X, Feng C, Shen Y, Zhang Q, and Jiang X, 2012. Solanum tuberrosum lectin-conjugated PLGA nanoparticles for nose-to-brain delivery: in vivo and in vitro evaluations. *J. Drug Target* 20:174–184.

[105] Liu Z, Jiang M, Kang T, Miao D, Gu G, Song Q, Yao L, Hu Q, Tu Y, Pang Z, Chen H, Jiang X, Gao X, and Chen J, 2013. Lactoferin-modified PEG-co-PCL nanoparticles for enhanced brain delivery of NAP peptide following intranasal administration. *Biomaterials* 34:3870–3881.

[106] Danielyan L, Schafer R, von Ameln-Mayerhof A, Bernhard F, Verleysdonk S, Buadze M Lourhmati A, Klopfer T, Schaumann F, Schmid B, Koehle C, Proksch B, Weissert R, Reichardt H M, van den Brandt J, Buniatian G H, Schwab M, Gleiter C H, and Frey W H 2nd, 2011. Therapeutic efficacy of intranasally delivered mesenchymal stem cells in a rat model of Parkinson's disease. *Rejuvenation Res.* 14:3–16.

[107] Danielyan L, Schäfer R, von Ameln-Mayerhofer A, Buadze M, Geisler J, Klopfer T, Burkhardt U, Proksch B, Verleysdonk S, Ayturan M, Buniatian G H, Gleiter C H, and Frey W H 2nd, 2009. Intranasal delivery of cells to the brain. *Eur. J. Cell Biol.* 88:315–324.

[108] Freiherr J, Hallschmid M, Frey W H II, Brunner Y F, Chapman CD, Holscher C, Craft S, De Felice F G, and Benedict C, 2013. Intranasal insulin as a treatment for Alzheimer's disease: a review of basic research and clinical evidence. *CNS Drugs* 27:505–514.

[109] Smolnik R, Molle M, Fehm H L, and Born J, 1999. Brain potential and attention after acute and subchronic intranasal administration of ACTH 4–10 and desacetyl-alpha-MSH in humans. *Neuroendocrinology* 70:63–72.

[110] Smolnik R, Perras B, Molle M, Fehm H L, and Born J, 2000. Event-related brain potentials and working memory function in healthy humans after single-dose and prolonged intranasal administration of adrenocorticotropin 4–10 and desacetyl-alpha-melanocyte stimulation hormone. *J. Clin. Psychopharmacol.* 20:445–454.

[111] Fehm H L, Smolnik R, Kern W, McGregor G P, Bickel U, and Born J, 2001. The melanocortin melanocyte-stimulating hormone/adrenocorticotropin 4–10 decreases body fat in humans. *J. Clin. Endocrinol. Metab.* 86:1144–1148.

[112] Zlokovic B V, Segal M B, Davson H, Lipovac M N, Hyman S, and McComb J G, 1990. Circulating neuroactive peptides and the blood-brain and blood-cerebrospinal fluid barriers. *Endocrinol. Exp.* 24:9–17.

[113] Kupfer D J, Jarrett D B, and Ehlers C L, 1991. The effect of GRF on the EEG sleep of normal males. *Sleep* 14:87–88.

[114] Perras B, Marshall L, Kohler G, Born J, and Fehm H L, 1999. Sleep and endocrine changes after intranasal administration of growth hormone-releasing hormone in young and aged humans. *Psychoneuroendocrinology* 24:743–757.

[115] Perras B, Pannenborg H, Marshall L, Pietrowsky R, Born J, and Fehm H L, 1999. Beneficial treatment of age-related sleep disturbance with prolonged intranasal vasopressin. *J. Clin. Psychopharmacol.* 19:28–36.

[116] Benedict C, Hallschmid M, Hatke A, Schultes B, Fehm H L, Born J, and Kern W, 2004. Intranasal insulin improves memory in humans. *Psychoneuroendocrinology* 29:1326–1334.

[117] Hallschmid M, Benedict C, Schultes B, Fehm H L, Born J, and Kern W, 2004. Intranasal insulin reduces body fat in men but not in women. *Diabetes* 53:3024–3029.

[118] Reger M A, Watson G S, Frey W H II, Baker L D, Cholerton B, Keeling M L, Belongia M A, Fishel M A, Plymate S R, Schellenberg G D, Cherrier M M, and Craft S, 2006. Effects of intranasal insulin on cognition in memory-impaired older adults: modulation by APOE genotype. *Neurobiol. Aging* 27:451–458.

[119] Craft S, Baker L D, Montine T J, Minoshima S, Watson G S, Claxton A, Arbuckle M, Callaghan M, Tsai E, Plymate S R, Green P S, Leverenz J, Cross D, and Gerton B, 2012. Intranasal insulin therapy for Alzheimer disease and amnestic mild cognitive impairment: a pilot clinical trial. *Arch. Neurol.* 69:29–38.

[120] Jauch-Chara K, Friedrich A, Rezmer M, Melchert U H, Scholand-Engler H G, Hallschmid M, and Oltmanns K M, 2012. Intranasal insulin suppresses food intake via enhancement of brain energy levels in humans. *Diabetes* 61:2261–2268.

[121] Baura G D, Foster D M, Porte D Jr, Kahn S E, Bergman R N, Cobelli C, and Schwartz M W, 1993. Saturable transport of insulin from plasma into the central nervous system of dogs in vivo. *J. Clin. Invest.* 92:1824–1830.

[122] Dale O, Nilsen T, Loftsson T, Hjorth Tønnesen H, Klepstad P, Kaasa S, Holand T, and Djupesland P G, 2006. Intranasal midazolam: a comparison of two delivery devices in human volunteers. *J. Pharm. Pharmacol.* 58:1311–1318.

[123] Okuyama S, 1997. The first attempt at radioisotopic evaluation of the integrity of the nose-brain barrier. *Life Sci.* 60:1881–1884.

[124] Wall A, Kagedal M, Bergstrom M, Jacobsson E, Nilsson D, Antoni G, Frandberg P, Gustavsson S A, langstrom B, and Yates R, 2005. Distribution of zolmitriptan into the CNS in healthy volunteers: a positron emission tomography study. *Drugs R D* 6:139–147.

[125] Tfelt-Hansen P C, 2010. Does sumatriptan cross the blood-brain barrier in animals and man? *J. Headache Pain* 11:5–12.

[126] Bergstrom M, Yates R, Wall A, Kagedal M, Syvanen S, and Langstrom B, 2006. Blood-brain barrier penetration of zolmitriptan-modelling of positron emission tomography data. *J. Pharmacokinet. Pharmacodyn.* 33:75–91.

[127] Djupesland P G and Skretting A, 2009. Nasal deposition and clearance in man: comparison of a bidirectional powder device and a traditional liquid spray pump. *J. Aerosol. Med. Pulm. Drug Del.* 25:280–289.

[128] Hoekman J D and Ho R J Y, 2011. Enhanced analgesic responses after preferential delivery of morphine and fentanyl to the olfactory epithelium in rats. *Anesth. Analg.* 113:641–651.

PART 7

FUTURE PROSPECTS IN BLOOD-BRAIN BARRIER UNDERSTANDING AND DRUG DISCOVERY

26

FUTURE PERSPECTIVES

N. Joan Abbott
Institute of Pharmaceutical Science, King's College London, London, UK

INTRODUCTION

This volume has provided a very comprehensive overview on the challenges associated with designing and delivering drugs to the central nervous system (CNS), as well as the converse demands of treatments for peripheral disorders where CNS side-effects need to be minimized. It has mapped the history of the field and documented the refinement of ideas, understanding, and methods for measurement. It has described the advanced chemistry producing effective medicines and presented examples from successful case histories that met the challenges with practical solutions. The detail presented can enable both novices and experts in the field to gain valuable insights to guide future research.

This book brings us up to date on approaches applied so far and documented progress. However, it acknowledges that progress has been slow in discovering treatments for CNS diseases; for example in the case of Alzheimer's disease, no new treatments have been produced since 2004 (Chapter 4). The volume nevertheless provides a very helpful framework, enabling us to return to some basic questions, and to rethink the field to map out possible directions for the future, with the aim of incorporating better understanding of the blood–brain barrier (BBB) and making the next 20 years of CNS drug discovery and delivery more productive than the past 20 years.

Blood–Brain Barrier in Drug Discovery: Optimizing Brain Exposure of CNS Drugs and Minimizing Brain Side Effects for Peripheral Drugs, First Edition. Edited by Li Di and Edward H. Kerns.

CURRENT STATUS OF BBB KNOWLEDGE AND PHARMACOKINETIC IMPLICATIONS

The Blood–Brain and Blood–Cerebrospinal Fluid Barriers, and Dynamics of Brain Fluids

Recent work has contributed to understanding of the physiology of the brain endothelial cells forming the BBB, and their interaction with the associated cells of the neurovascular unit (NVU) including pericytes, astrocytes, microglia, and neurons [1]. Some of the interactions involve the extracellular matrix components of the endothelial and parenchymal basement membranes, and include important contributions to induction, maintenance, regulation, and repair of the barrier. The choroid plexus (CP) responsible for secretion of cerebrospinal fluid (CSF) is now known to be made up of a heterogeneous population of cells with different transport functions [2, 3]. Improved methods for monitoring the flow of CSF [4] and interstitial fluid (ISF) [5], and their interactions show that brain fluids work as an integrated homeostatic system, mediating and regulating the distribution of nutrients and other agents to the cells of the brain, and contributing to their clearance [6].

Transport across the BBB and CP

There has been progress in molecular understanding of the mechanisms responsible for BBB integrity, especially the role and regulation of junctional proteins: tight junctions [7–9] and *adherens* junctions [10–12], and the mechanisms underlying endothelial apical–basal polarization [13] affecting membrane properties, transporters, and transcytosis. The list of identified transporters continues to grow, many of them with identified genes, but it is clear that further transporters remain to be identified (Chapter 7) [14]. 3D mapping of the pharmacophore of the large neutral amino acid carrier LAT1 provides new opportunities for drug design and prodrug delivery [15], but the detailed mechanisms for operation of the transporters capable of binding a wide range of diverse substrates, such as the ABC efflux transporters (Chapter 6) and the transporters of organic cations and anions, are still unclear. There has been some success in developing and testing protein vectors (Chapter 8) [16] and nanocarriers (Chapter 24) [17–19] capable of delivering drugs to the CNS across the BBB. However, there is concern about the degree to which the larger constructs can move through the brain parenchyma [20], and they have yet to translate into effective therapeutics. It is clear that development in several fields is required, including better understanding of mechanisms of receptor-mediated transcytosis (RMT) at the brain endothelium [21] and of the significance of the different pattern of transport in the brain endothelium and CP [3]. Many of these differences have implications for both CNS function and drug delivery.

OBSTACLES TO PROGRESS IN CNS DRUG DISCOVERY

The targets in CNS disorders are often unclear. For peripheral disorders, it has often proved possible to identify a drug target, and to follow the "one disease, one target, one drug" model for success (Chapter 2)—for example, insulin for diabetes or effective treatments for hypertension. However, for many CNS disorders, such as most of the affective disorders, the target is unclear and existing treatments are much less evidence-based [22–24]. Much better understanding of disease etiology and time-course is required (Chapter 4).

Treatments directed against neurodegenerative disease are given too late. For neurodegenerative disorders such as Alzheimer's and Parkinson's diseases, by the time symptoms appear, too many neurons have died, so that rescue and repair is impossible. Current treatments directed at "end-stage disease" can at best only ameliorate symptoms, and possibly slow progression, with the aim of preserving some function [25, 26].

The obvious target may not be the best one to treat. In Alzheimer's disease, most recent drug discovery and delivery programs have focused on efforts to tackle beta-amyloid (Aβ) accumulation and tauopathy, aiming to modify the tau–amyloid cascade seen as critical to development of the pathology [27–29]. This has involved complex strategies including passive and active immunization either to create peripheral sinks for Aβ, or to enter the brain and reduce amyloid plaques [30–32]. However, with improved understanding, both of the nature of plaque deposits and the vulnerability of the brain vasculature treated to reverse long-term Aβ accumulation [33–35], alternative treatments are emerging, focusing on inhibition of beta-secretase 1 (BACE1) [36] or reducing neuronal death [37–39]. These also have the advantage that some specific small-molecule drug treatments may prove effective [40, 41].

Many CNS disorders are broad categories based on symptoms, but can include very heterogeneous pathologies. This is well recognized, for example for dementia, brain tumors, and epilepsy, but may apply much more widely, including for multiple sclerosis [42] and stroke [43]. Early differentiation and stratification are needed to provide more homogeneous patient subpopulations, with a better chance they will show similar targets and benefit from similar drug treatments [44–46].

Classical CNS treatments have been "neurocentric"; for many, including dementias, the brain microvasculature and the BBB may be better or additional targets. There is increasing recognition that microhemorrhages and BBB dysfunction are present in many CNS pathologies, providing alternative treatment sites [47–49]. Given modern understanding of the importance of interactions between the cell types within the NVU also provides additional targets for combination or sequential therapies [50–52].

Traditional pharmacokinetic and pharmacodynamic (PKPD) modeling of the CNS has used simplistic models. Classical [53] and more recent studies with modern imaging techniques show the importance of bulk flow of both brain ISF [6, 54] and CSF [4, 55], and diurnal patterns of flow capable of influencing CNS drug PK [56, 57], with relevance for dosing schedules.

There has been little recognition or recruitment of endogenous protection and repair mechanisms. In the past, the focus was on designing drugs to arrest or reverse pathology. However, there is growing awareness that the endogenous protective and repair processes of the body and the CNS, including those of the BBB, are involved in day-to-day maintenance and can be recruited, both to maintain health and for a more natural recovery from pathology [58–62].

IMPROVEMENTS FOR THE FUTURE

From this catalogue of obstacles, certain future directions and activities are clear.

Need to Identify Better Biomarkers of Disease

This would permit better stratification of patients and allow more accurate monitoring of response to different drug regimens. Biomarkers already used include identified molecules in peripheral blood [63, 64] and in CSF [65], imaging methods applied to the living CNS (magnetic resonance imaging (MRI) and positron emission tomography (PET) [66, 67]); and, where available, biopsy material for example, from brain tumors and resectioned epileptic tissue. Of these, peripheral blood is particularly useful as its collection is relatively noninvasive, it reflects changes over time, and it is increasingly seen as a "high-content" resource. Thus, in addition to circulating plasma proteins and cytokines, and populations of precursor cells and leukocytes, peripheral blood contains microparticles or vesicles including exosomes shed from cells, both normal and pathological [68]. The detailed mapping of the composition and content of these structures is providing whole databases of markers related to different physiological and pathological functions of the cells in contact with the blood (especially endothelium, including brain endothelium) [69, 70], and also the cells in tissues beyond the vasculature, including the brain.

Identifying Potential Biomarkers from Postmortem Material

Postmortem material from human brain has helped in identification of proteins involved in and, in some cases, responsible for CNS pathologies and neuroprotection [71]. If such central biomarkers prove to be reflected in markers in blood (see section "Need to Identify Better Biomarkers of Disease") and peripheral tissues, then many of the insights into the disease process, currently only available postmortem, may become much more accessible and at earlier timepoints, opening up the possibility for targeted treatment to particular cohorts of at-risk individuals.

Toward Personalized Medicine—Making More Use of "Omics"

With determination of the full genome for individuals now much more affordable, and with better methods for deriving information from proteomics, generation of personal profiles to inform drug treatment is becoming more realistic, helping both in the generation of knowledge (e.g., patient stratification, see section "Need to Identify Better Biomarkers of Disease") and in devising individual treatments [72–74]. Other individual "omics" are also proving relevant, including the metabolome (metabolic profile) and microbiome (intestinal microflora, influencing a large number of physiological and pathological processes, including those at the BBB [75, 76]. It is already envisaged that personal information of this kind will soon be contained in a credit card or even implantable chip format, able to be read and integrated in medical diagnosis and treatment.

Making Use of a Broader Chemical Space for Small-Molecule Drugs, Including Transporter Substrates

The pharmaceutical industry has tended to focus on compounds that cross the BBB by passive diffusion, and that are not significantly affected by efflux transporters, especially P-glycoprotein, on the basis that this should make predicting CNS exposure simpler (e.g., Chapter 2). Apart from a few compounds tailored to interact with LAT1 (including cytotoxic agents for oncology), uptake transporters have been largely ignored as routes into the CNS (Chapter 7). However, it is clear that uptake transporters can, in some cases, counteract the effects of efflux mechanisms, allowing effective CNS exposure [77] and, indeed, many compounds and drugs with CNS effects are now known to enter the brain via transporters, including some only recently identified (e.g., oxycodone via pyrilamine transporter [14]). Moreover, knowledge of the SAR of the many BBB transport systems for organic cations will help in the design of drugs either to enter the brain effectively via such mechanisms, or to be excluded to reduce potential CNS side effects (e.g., to reduce sedative effects of H1 histamine antagonists). Taking into account transporter properties and their drug specificities helps in the development of drugs occupying a broader chemical space, with potentially higher specificity for the CNS than for peripheral tissues.

Incorporating More Accurate Understanding of Brain Fluid Dynamics to Predict CNS PK

Earlier work using tracers injected into the brain or CNS fluid compartments led to estimates of the rates of CSF and ISF turnover and mapping of their routes for clearance. More recent work has added to the evidence that brain ISF is flowing, like CSF, and at a 5–10× higher rate than earlier estimated [78]. This is leading to revision of compartmental models used to predict drug distribution and PK in brain [79, 80].

Drug Treatments to Complement Endogenous Mechanisms for Protection and Repair

It is now recognized that the brain has endogenous mechanisms for protection and "running repairs," and that intelligent drug design strategies should, where possible, recruit these mechanisms to amplify the natural protection against pathologies in at risk individuals, and/or reduce the severity of CNS pathologies when they occur [81]. This can include the concept of "hormesis," where a low dose of a drug or toxic agent, or low level stressor, such as oxidative stress, can upregulate mechanisms capable of protecting the system from subsequent stress, as demonstrated for stroke [82, 83].

Integrating *In Silico*, *In Vitro*, and *In Vivo* Studies to Predict CNS Drug Exposure

The earliest *in silico* (computer-based) modeling of BBB permeability gave predictions for passive permeability of compounds across the BBB, using physchem (physical chemistry) parameters. When *in vitro* and *in vivo* data are included, outliers show the influence of uptake and efflux transporters. With access to a range of *in vitro* cell-based models (including human), more sophisticated modeling will be possible, generating physiologically-based pharmacokinetic (PBPK) models with extrapolation to humans [84]. As more data from human PET imaging and microdialysis becomes available, these models will become more reliable and capable of dealing with disturbances such as pathological state, and progressive changes (e.g., with aging).

FUTURE "TWO-PRONGED" APPROACH TO CNS DRUG DISCOVERY AND DELIVERY

It is clear from this discussion that while current CNS pharmaceutical effort will continue to be directed along traditional lines, using well-honed tools and methods, newer and complementary approaches will be developed to tackle disease states much earlier, ideally before significant neuronal cell death has occurred. Better understanding of the physiology and pathology of CNS barrier layers will contribute to this effort. Reducing the morbidity and mortality due to CNS disorders especially those associated with old age is now a realistic prospect.

REFERENCES

[1] Abbott NJ (2014). Anatomy and physiology of the blood-brain barriers. In: Drug delivery to the brain, physiological concepts, methodologies, and approaches. Hammarlund-Udenaes M, de Lange ECM, Thorne RG (Eds.). AAPS Advances in the Pharmaceutical Sciences Series 10. New York: Springer Press, pp. 3–21.

[2] Saunders NR, Daneman R, Dziegielewska KM, Liddelow SA (2013). Transporters of the blood-brain and blood-CSF interfaces in development and in the adult. Mol Aspects Med 34(2–3):742–752.

[3] Strazielle N, Ghersi-Egea JF (2013). Physiology of blood-brain interfaces in relation to brain disposition of small compounds and macromolecules. Mol Pharm 10(5):1473–1491.

[4] Penn RD, Linninger A (2009). The physics of hydrocephalus. Pediatr Neurosurg 45(3):161–174.

[5] Iliff JJ, Wang M, Zeppenfeld DM, Venkataraman A, Plog BA, Liao Y, et al., (2013). Cerebral arterial pulsation drives paravascular CSF-interstitial fluid exchange in the murine brain. J Neurosci 33(46):18190–18199.

[6] Iliff JJ, Wang M, Liao Y, Plogg BA, Peng W, Gundersen GA, Benveniste H, Vates GE, Deane R, Goldman SA, Nagelhus EA, Nedergaard M (2012). A paravascular pathway facilitates CSF flow through the brain parenchyma and the clearance of interstitial solutes, including amyloid *β*. Sci Transl Med 4(147):147ra111.

[7] Obermeier B, Daneman R, Ransohoff RM (2013). Development, maintenance and disruption of the blood-brain barrier. Nat Med 19(12):1584–1596.

[8] Gonçalves A, Ambrósio AF, Fernandes R (2013). Regulation of claudins in blood-tissue barriers under physiological and pathological states. Tissue Barriers 1(3):e24782.

[9] Dyrna F, Hanske S, Krueger M, Bechmann I (2013). The blood-brain barrier. J Neuroimmune Pharmacol 8(4):763–773.

[10] Dejana E, Giampietro C (2012). Vascular endothelial-cadherin and vascular stability. Curr Opin Hematol 19(3):218–223.

[11] Bravi L, Dejana E, Lampugnani MG (2014). VE-cadherin at a glance. Cell Tissue Res 355(3):515–522.

[12] Giannotta M, Trani M, Dejana E (2013). VE-cadherin and endothelial adherens junctions: Active guardians of vascular integrity. Dev Cell 26(5):441–454.

[13] Lampugnani MG, Orsenigo F, Rudini N, Maddaluno L, Boulday G, Chapon F, Dejana E (2010). CCM1 regulates vascular-lumen organization by inducing endothelial polarity. J Cell Sci 123(Pt. 7):1073–1080.

[14] Tachikawa M, Uchida Y, Ohtsuki S, Terasaki T (2014). Recent progress in blood-brain barrier and blood-CSF barrier transport research: Pharmaceutical relevance for drug delivery to the brain. In: Drug delivery to the brain, physiological concepts, methodologies and approaches. Hammarlund-Udenaes M, de Lange ECM, Thorne RG (Eds.). AAPS Advances in the Pharmaceutical Sciences Series 10, New York: Springer Press, pp. 23–62.

[15] Ylikangas H, Peura L, Malmioja K, Leppänen J, Laine K, Poso A, Lahtela-Kakkonen M, Rautio J (2012). Structure–activity relationship study of compounds binding to large amino acid transporter 1 (LAT1) based on pharmacophore modeling and *in situ* rat brain perfusion. Eur J Pharm Sci 48(3):523–531.

[16] Gabathuler R (2014). Development of new protein vectors for the physiological delivery of large therapeutic compounds to the CNS. In: Drug delivery to the brain, physiological concepts, methodologies and approaches. Hammarlund-Udenaes M, de Lange ECM, Thorne RG (Eds.). AAPS Advances in the Pharmaceutical Sciences Series 10, New York: Springer Press, pp.455–484.

[17] Pehlivan SB (2013). Nanotechnology-based drug delivery systems for targeting, imaging and diagnosis of neurodegenerative diseases. Pharm Res 30(10):2499–2511.

[18] Pinzón-Daza ML, Campia I, Kopecka J, Garzón R, Ghigo D, Riganti C (2013). Nanoparticle- and liposome-carried drugs: New strategies for active targeting and drug delivery across blood-brain barrier. Curr Drug Metab 14(6):625–640.

[19] Gaillard PJ, Visser CC, de Boer M, Appeldoorn CCM, Rip J (2014). Blood-to-brain drug delivery using nanocarriers. In: Drug delivery to the brain, physiological concepts, methodologies and approaches. Hammarlund-Udenaes M, de Lange ECM, Thorne RG (Eds.). AAPS Advances in the Pharmaceutical Sciences Series 10, New York: Springer Press, pp. 433–454.

[20] Wolak DJ, Thorne RG (2013). Diffusion of macromolecules in the brain: Implications for drug delivery. Mol Pharm 10(5):1492–1504.

[21] de Lange EC (2012). The physiological characteristics and transcytosis mechanisms of the blood-brain barrier (BBB) Curr Pharm Biotechnol 13(12):2319–2327.

[22] Grady MM, Stahl SM (2013). Novel agents in development for the treatment of depression. CNS Spectr 18(Suppl.1):37–40.

[23] D'Souza P, Jago C (2014). Spotlight on depression: A pharma matters report. Drugs Today (Barc) 50:251–267.

[24] Finberg JP (2014). Update on the pharmacology of selective inhibitors of MAO-A and MAO-B: Focus on modulation of CNS monoamine neurotransmitter release. Pharmacol Ther 143:133–152.

[25] Devos D, Moreau C, Dujardin K, Cabantchik I, Defebvre L, Bordet R (2013). New pharmacological options for treating advanced Parkinson's disease. Clin Ther 35(10):1640–1652.

[26] Schneider LS, Mangialasche F, Andreasen N, Feldman H, Giacobini E, Jones R, Mantua V, Mecocci P, Pani L, Winblad B, Kivipelto M (2014). Clinical trials and late-stage drug development for Alzheimer's disease: An appraisal from 1984 to 2014. J Intern Med 275(3):251–283.

[27] Skaper SD (2012). Alzheimer's disease and amyloid: Culprit or coincidence? Int Rev Neurobiol 102:277–316.

[28] Singh S, Kushwah AS, Singh R, Farswan M, Kaur R (2012). Current therapeutic strategy in Alzheimer's disease. Eur Rev Med Pharmacol Sci 16(12):1651–1664.

[29] Tayeb HO, Murray ED, Price BH, Tarazi FI (2013). Bapineuzumab and solanezumab for Alzheimer's disease: Is the 'amyloid cascade hypothesis' still alive? Expert Opin Biol Ther 13(7):1075–1084.

[30] Panza F, Frisardi V, Solfrizzi V, Imbimbo BP, Logroscino G, Santamato A, Greco A, Seripa D, Pilotto A (2012). Immunotherapy for Alzheimer's disease: From anti-β-amyloid to tau-based immunization strategies. Immunotherapy 4(2):213–238.

[31] Robert R, Wark KL (2012). Engineered antibody approaches for Alzheimer's disease immunotherapy. Arch Biochem Biophys 526(2):132–138.

[32] Lannfelt L, Relkin NR, Siemers ER (2014). Amyloid-β-directed immunotherapy for Alzheimer's disease. J Intern Med 275(3):284–295.

[33] Yamada M (2012). Predicting cerebral amyloid angiopathy-related intracerebral hemorrhages and other cerebrovascular disorders in Alzheimer's disease. Front Neurol 3:64. DOI: 10.3389/fneur.2012.00064.

[34] Carare RO, Hawkes CA, Jeffrey M, Kalaria RN, Weller RO (2013). Review: Cerebral amyloid angiopathy, prion angiopathy, CADASIL, and the spectrum of protein elimination failure angiopathies (PEFA) in neurodegenerative disease with a focus on therapy. Neuropathol Appl Neurobiol 39(6):593–611.

[35] Carare RO, Teeling JL, Hawkes CA, Püntener U, Weller RO, Nicoll JA, Perry VH (2013). Immune complex formation impairs the elimination of solutes from the brain: Implications for immunotherapy in Alzheimer's disease. Acta Neuropathol Commun 1(1):48. DOI: 10.1186/2051-5960-1-48.

[36] Butini S, Brogi S, Novellino E, Campiani G, Ghosh AK, Brindisi M, Gemma S (2013). The structural evolution of β-secretase inhibitors: A focus on the development of small-molecule inhibitors. Curr Top Med Chem 13(15):1787–1807.

[37] Iqbal K, Liu F, Gong CX (2014). Alzheimer disease therapeutics: Focus on the disease and not just plaques and tangles. Biochem Pharmacol 88(4):631–639.

[38] Hardy J, Bogdanovic N, Winblad B, Portelius E, Andreasen N, Cedazo-Minguez A, Zetterberg H (2014). Pathways to Alzheimer's disease. J Intern Med 275(3):296–303.

[39] Dias KS, Viegas C Jr. (2014). Multi-target directed drugs: A modern approach for design of new drugs for the treatment of Alzheimer's disease. Curr Neuropharmacol 12(3):239–255.

[40] Moreno JA, Halliday M, Molloy C, Radford H, Verity N, Axten JM, Ortori CA, Willis AE, Fischer PM, Barrett DA, Mallucci GR (2013). Oral treatment targeting the unfolded protein response prevents neurodegeneration and clinical disease in prion-infected mice. Sci Transl Med 5(206):206ra138.

[41] Knowles JK, Simmons DA, Nguyen TV, Vander Griend L, Xie Y, Zhang H, Yang T, Pollak J, Chang T, Arancio O, Buckwalter MS, Wyss-Coray T, Massa SM, Longo FM (2013). Small molecule p75NTR ligand prevents cognitive deficits and neurite degeneration in an Alzheimer's mouse model. Neurobiol Aging 34(8):2052–2063.

[42] Metz I, Weigand SD, Popescu BF, Frischer JM, Parisi JE, Guo Y, Lassmann H, Brück W, Lucchinetti CF (2014). Pathologic heterogeneity persists in early active multiple sclerosis lesions. Ann Neurol 75:728–738.

[43] Kimmelman J, Mogil JS, Dirnagl U (2014). Distinguishing between exploratory and confirmatory preclinical research will improve translation. PLoS Biol 12(5):e1001863.

[44] Wiens AL, Martin SE, Bertsch EC, Vance GH, Stohler RA, Cheng L, Badve S, Hattab EM (2014). Luminal subtypes predict improved survival following central nervous system metastasis in patients with surgically managed metastatic breast carcinoma. Arch Pathol Lab Med 138(2):175–181.

[45] Grefkes C, Fink GR (2014). Connectivity-based approaches in stroke and recovery of function. Lancet Neurol 13(2):206–216.

[46] Ye Z, Altena E, Nombela C, Housden CR, Maxwell H, Rittman T, Huddleston C, Rae CL, Regenthal R, Sahakian BJ, Barker RA, Robbins TW, Rowe JB (2014). Selective serotonin reuptake inhibition modulates response inhibition in Parkinson's disease. Brain 137(Pt. 4):1145–1155.

[47] Zlokovic BV (2011). Neurovascular pathways to neurodegeneration in Alzheimer's disease and other disorders. Nat Rev Neurosci 12(12):723–738.

[48] Ronaldson PT, Davis TP (2012). Blood-brain barrier integrity and glial support: mechanisms that can be targeted for novel therapeutic approaches in stroke. Curr Pharm Des 18(25):3624–3644.

[49] Yates PA, Desmond PM, Phal PM, Steward C, Szoeke C, Salvado O, Ellis KA, Martins RN, Masters CL, Ames D, Villemagne VL, Rowe CC AIBL Research Group (2014). Incidence of cerebral microbleeds in preclinical Alzheimer disease. Neurology 82(14):1266–1273.

[50] Enzmann G, Mysiorek C, Gorina R, Cheng YJ, Ghavampour S, Hannocks MJ, Prinz V, Dirnagl U, Endres M, Prinz M, Beschorner R, Harter PN, Mittelbronn M, Engelhardt B, Sorokin L (2013). The neurovascular unit as a selective barrier to polymorphonuclear granulocyte (PMN) infiltration into the brain after ischemic injury. Acta Neuropathol 125(3):395–412.

[51] Jullienne A, Badaut J (2013). Molecular contributions to neurovascular unit dysfunctions after brain injuries: Lessons for target-specific drug development. Future Neurol 8(6):677–689.

[52] Sagare AP, Bell RD, Zhao Z, Ma Q, Winkler EA, Ramanathan A, Zlokovic BV (2013). Pericyte loss influences Alzheimer-like neurodegeneration in mice. Nat Commun 4:2932. DOI: 10.1038/ncomms3932.

[53] Abbott NJ (2004). Evidence for bulk flow of brain interstitial fluid: Significance for physiology and pathology. Neurochem Int 45(4):545–552.

[54] Rangroo Thrane V, Thrane AS, Plog BA, Thiyagarajan M, Iliff JJ, Deane R, Nagelhus EA, Nedergaard M (2013). Paravascular microcirculation facilitates rapid lipid transport and astrocyte signaling in the brain. Sci Rep 3:2582.

[55] Battal B, Kocaoglu M, Bulakbasi N, Husmen G, Tuba Sanal H, Tayfun C (2011). Cerebrospinal fluid flow imaging by using phase-contrast MR technique. Br J Radiol 84(1004):758–765.

[56] Nilsson C, Ståhlberg F, Thomsen C, Henriksen O, Herning M, Owman C (1992). Circadian variation in human cerebrospinal fluid production measured by magnetic resonance imaging. Am J Physiol 262(1 Pt. 2):R20–24.

[57] Xie L, Kang H, Xu Q, Chen MJ, Liao Y, Thiyagarajan M, O'Donnell J, Christensen DJ, Nicholson C, Iliff JJ, Takano T, Deane R, Nedergaard M (2013). Sleep drives metabolite clearance from the adult brain. Science 342(6156):373–377.

[58] Suzuki H, Hasegawa Y, Kanamaru K, Zhang JH (2010). Mechanisms of osteopontin-induced stabilization of blood-brain barrier disruption after subarachnoid hemorrhage in rats. Stroke 41(8):1783–1790.

[59] Ji RR, Xu ZZ, Strichartz G, Serhan CN (2011). Emerging roles of resolvins in the resolution of inflammation and pain. Trends Neurosci 34(11):599–609.

[60] Terrando N, Eriksson LI, Ryu JK, Yang T, Monaco C, Feldmann M, Jonsson Fagerlund M, Charo IF, Akassoglou K, Maze M (2011). Resolving postoperative neuroinflammation and cognitive decline. Ann Neurol 70(6):986–995.

[61] Xing C, Hayakawa K, Lok J, Arai K, Lo EH (2012). Injury and repair in the neurovascular unit. Neurol Res 34(4):325–330.

[62] Nguyen AQ, Cherry BH, Scott GF, Ryou MG, Mallet RT (2014). Erythropoietin: Powerful protection of ischemic and post-ischemic brain. Exp Biol Med (Maywood). Mar 4. DOI: 10.1177/1535370214523703.

[63] Holdhoff M, Yovino SG, Boadu O, Grossman SA (2013). Blood-based biomarkers for malignant gliomas. J Neurooncol 113(3):345–352.

[64] Mapstone M, Cheema AK, Fiandaca MS, Zhong X, Mhyre TR, MacArthur LH, Hall WJ, Fisher SG, Peterson DR, Haley JM, Nazar MD, Rich SA, Berlau DJ, Peltz CB, Tan MT, Kawas CH, Federoff HJ (2014). Plasma phospholipids identify antecedent memory impairment in older adults. Nat Med 20(4):415–418.

[65] Stefani A, Olivola E, Stampanoni Bassi M, Pisani V, Imbriani P, Pisani A, Pierantozzi M (2013). Strength and weaknesses of cerebrospinal fluid biomarkers in Alzheimer's disease and possible detection of overlaps with frailty process. CNS Neurol Disord Drug Targets 12(4):538–546.

[66] Jack CR Jr, Holtzman DM (2013). Biomarker modeling of Alzheimer's disease. Neuron 80(6):1347–1358.

[67] Chen Z, Zhong C (2013). Decoding Alzheimer's disease from perturbed cerebral glucose metabolism: implications for diagnostic and therapeutic strategies. Prog Neurobiol 108:21–43.

[68] Sáenz-Cuesta M, Osorio-Querejeta I, Otaegui D (2014). Extracellular vesicles in multiple sclerosis: What are they telling us? Front Cell Neurosci 8:100. DOI: 10.3389/fncel.2014.00100.

[69] Huang SH, Wang L, Chi F, Wu CH, Cao H, Zhang A, Jong A (2013). Circulating brain microvascular endothelial cells (cBMECs) as potential biomarkers of the blood-brain barrier disorders caused by microbial and non-microbial factors. PLoS One 8(4):e62164.

[70] van Ierssel SH, Jorens PG, Van Craenenbroeck EM, Conraads VM (2014). The endothelium, a protagonist in the pathophysiology of critical illness: Focus on cellular markers. Biomed Res Int 2014:985813. DOI: 10.1155/2014/985813.

[71] Lu T, Aron L, Zullo J, Pan Y, Kim H, Chen Y, Yang TH, Kim HM, Drake D, Liu XS, Bennett DA, Colaiácovo MP, Yankner BA (2014). REST and stress resistance in ageing and Alzheimer's disease. Nature 507(7493):448–454.

[72] Ostrom Q, Cohen ML, Ondracek A, Sloan A, Barnholtz-Sloan J (2013). Gene markers in brain tumors: What the epileptologist should know. Epilepsia 54(Suppl. 9):25–29.

[73] Lionetto L, Gentile G, Bellei E, Capi M, Sabato D, Marsibilio F, Simmaco M, Pini LA, Martelletti P (2013). The omics in migraine. J Headache Pain 14:55.

[74] Zameel Cader M. The molecular pathogenesis of migraine: New developments and opportunities. Hum Mol Genet 22(R1):R39–44.

[75] Chen X, D'Souza R, Hong ST (2013). The role of gut microbiota in the gut-brain axis: Current challenges and perspectives. Protein Cell 4(6):403–414.

[76] Hornig M (2013). The role of microbes and autoimmunity in the pathogenesis of neuropsychiatric illness. Curr Opin Rheumatol 25(4):488–795.

[77] Kalvass JC, Maurer TS, Pollack GM (2007). Use of plasma and brain unbound fractions to assess the extent of brain distribution of 34 drugs: Comparison of unbound concentration ratios to *in vivo* p-glycoprotein efflux ratios. Drug Metab Dispos 35(4):660–666.

[78] Groothuis DR, Vavra MW, Schlageter KE, Kang EW, Itskovich AC, Hertzler S, Allen CV, Lipton HL (2007). Efflux of drugs and solutes from brain: The interactive roles of diffusional transcapillary transport, bulk flow and capillary transporters. J Cereb Blood Flow Metab 27(1):43–56.

[79] Westerhout J, Smeets J, Danhof M, de Lange EC (2013). The impact of P-gp functionality on non-steady state relationships between CSF and brain extracellular fluid. J Pharmacokinet Pharmacodyn 40(3):327–342.

[80] Westerhout J, van den Berg DJ, Hartman R, Danhof M, de Lange EC (2014). Prediction of methotrexate CNS distribution in different species—Influence of disease conditions. Eur J Pharm Sci 57:11–24.

[81] Zhao J, Redell JB, Moore AN, Dash PK (2011). A novel strategy to activate cytoprotective genes in the injured brain. Biochem Biophys Res Commun 407(3):501–506.

[82] Alfieri A, Srivastava S, Siow RC, Modo M, Fraser PA, Mann GE (2011). Targeting the Nrf2-Keap1 antioxidant defence pathway for neurovascular protection in stroke. J Physiol 589(Pt. 17):4125–4136.

[83] Stowe AM, Wacker BK, Cravens PD, Perfater JL, Li MK, Hu R, Freie AB, Stüve O, Gidday JM (2012). CCL2 upregulation triggers hypoxic preconditioning-induced protection from stroke. J Neuroinflammation 9:33.

[84] Abbott NJ, Dolman DEM, Yusof SR, Reichel A (2014). *In vitro* models of CNS barriers. In: Drug delivery to the brain, physiological concepts, methodologies and approaches. Hammarlund-Udenaes M, de Lange ECM, Thorne RG (Eds.). AAPS Advances in the Pharmaceutical Sciences Series 10, New York: Springer Press, pp. 163–197.

INDEX

Blood–Brain Barrier in Drug Discovery: Optimizing Brain Exposure of CNS Drugs and Minimizing Brain Side Effects for Peripheral Drugs, First Edition. Edited by Li Di and Edward H. Kerns.